Petra + Angus King 1997

TRANSLATIONAL CONTROL

COLD SPRING HARBOR MONOGRAPH SERIES

The Lactose Operon
The Bacteriophage Lambda
The Molecular Biology of Tumour Viruses
Ribosomes
RNA Phages
RNA Polymerase
The Operon
The Single-Stranded DNA Phages
Transfer RNA:
  Structure, Properties, and Recognition
  Biological Aspects
Molecular Biology of Tumor Viruses, Second Edition:
  DNA Tumor Viruses
  RNA Tumor Viruses
The Molecular Biology of the Yeast Saccharomyces:
  Life Cycle and Inheritance
  Metabolism and Gene Expression
Mitochondrial Genes
Lambda II
Nucleases
Gene Function in Prokaryotes
Microbial Development
The Nematode *Caenorhabditis elegans*
Oncogenes and the Molecular Origins of Cancer
Stress Proteins in Biology and Medicine
DNA Topology and Its Biological Implications
The Molecular and Cellular Biology of the yeast *Saccharomyces:*
  Genome Dynamics, Protein Synthesis, and Energetics
  Gene Expression
Transcriptional Regulation
Reverse Transcriptase
The RNA World
Nucleases, Second Edition
The Biology of Heat Shock Proteins and Molecular Chaperones
Arabidopsis
Cellular Receptors for Animal Viruses
Telomeres
Translational Control

TRANSLATIONAL CONTROL

Edited by

John W.B. Hershey
University of California, Davis

Michael B. Mathews
Cold Spring Harbor Laboratory

Nahum Sonenberg
McGill University, Montreal

COLD SPRING HARBOR LABORATORY PRESS
1996

TRANSLATIONAL CONTROL
Monograph 30

Printed in the United States of America
Book design by Emily Harste

ISBN 0-87969-458-0
ISSN 0270-1847
LC 95-74767

All Cold Spring Harbor Laboratory Press publications may be ordered directly from Cold Spring Harbor Laboratory, 10 Skyline Drive, Plainview, New York 11803. Phone: 1-800-843-4388 in Continental U.S. and Canada. All other locations: (516) 349-1930. FAX: (516) 349-1946.

Contents

Preface, vii

1 **Origins and Targets of Translational Control, 1**
M.B. Mathews, N. Sonenberg, and J.W.B. Hershey

2 **The Pathway and Mechanism of Eukaryotic Protein Synthesis, 31**
W.C. Merrick and J.W.B. Hershey

3 **A Comparative View of Initiation Site Selection Mechanisms, 71**
R.J. Jackson

4 **Binding of Initiator Methionyl-tRNA to Ribosomes, 113**
H. Trachsel

5 **Protein Kinases That Phosphorylate eIF2 and eIF2B, and Their Role in Eukaryotic Cell Translational Control, 139**
M.J. Clemens

6 **Translational Control Mediated by Upstream AUG Codons, 173**
A.P. Geballe

7 **Translational Control of *GCN4:* Gene-specific Regulation by Phosphorylation of eIF2, 199**
A.G. Hinnebusch

8 **mRNA 5′ Cap-binding Protein eIF4E and Control of Cell Growth, 245**
N. Sonenberg

9 **Translational Control during Heat Shock, 271**
R.F. Duncan

10 **Regulation of Protein Synthesis by Calcium, 295**
A.C. Nairn and H.C. Palfrey

11 **Masked and Translatable Messenger Ribonucleoproteins in Higher Eukaryotes, 319**
A.S. Spirin

12 **Translational Control of Ferritin, 335**
T.A. Rouault, R.D. Klausner, and J.B. Harford

13 **Translational Control of Ribosomal Protein mRNAs in Eukaryotes, 363**
O. Meyuhas, D. Avni, and S. Shama

14 **Ribosomal Protein S6 Phosphorylation and Signal Transduction, 389**
H.B.J. Jefferies and G. Thomas

15 **Translational Control of Developmental Decisions, 411**
M. Wickens, J. Kimble, and S. Strickland

16 **Poly(A) Metabolism and Translation: The Closed-loop Model, 451**
A. Jacobson

17 **Dynamics of Poly(A) Addition and Removal during Development, 481**
J.D. Richter

18 **Interactions between Viruses and the Cellular Machinery for Protein Synthesis, 505**
M.B. Mathews

19 **Initiation of Translation by Picornavirus RNAs, 549**
E. Ehrenfeld

20 **Adenovirus and Vaccinia Virus Translational Control, 575**
R.J. Schneider

21 **Translational Control in Cells Infected with Influenza Virus and Reovirus, 607**
M.G. Katze

22 **Translationally Coupled Degradation of mRNA in Eukaryotes, 631**
N.G. Theodorakis and D.W. Cleveland

23 **Regulatory Recoding, 653**
J.F. Atkins and R.F. Gesteland

24 **Mammalian Ribosomes: The Structure and the Evolution of the Proteins, 685**
I.G. Wool, Y.-L. Chan, and A. Glück

25 **Genetics of Mitochondrial Translation, 733**
T.D. Fox

26 **Control of Translation Initiation in Prokaryotes, 759**
H.O. Voorma

Index, 779

Preface

Regulation of gene expression occurs at multiple levels: transcription, messenger RNA processing and stability, translation, and protein modification and turnover. Translational control is defined narrowly as modulation of the efficiency of translation of mRNAs and more broadly to include translation-coupled regulation of mRNA stability. Control at the level of protein synthesis allows cells to respond rapidly to changes in physiological conditions, since activation or repression of mRNAs can occur essentially instantaneously, whereas regulation at the level of transcription entails a considerable time lag before a precursor RNA is processed and mRNA can accumulate or decay in the cytoplasm. Numerous examples of regulation of gene expression at the translational level are now known and serve as the focus of this monograph.

Our understanding of the pathways of transcription and translation is based on experiments that began in the 1950s and identified the major elements involved, such as DNA-dependent RNA polymerase and ribosomes. Whereas early studies of transcription employed largely genetic approaches and were concerned with defining *cis*-acting DNA sequence elements, characterization of the translational machinery and pathway relied almost exclusively on biochemical methods. Through fractionation of the translational apparatus and reconstitution of active protein synthesis systems in vitro, the major macromolecular components such as tRNAs and soluble factors were identified and characterized. During the 1960s and early 1970s, a detailed pathway of protein synthesis was defined in prokaryotes and that for eukaryotes lagged only a few years behind. Thus, by 1980, the molecular interactions of the major components of protein synthesis were understood at a remarkably sophisticated biochemical level and a few examples of translational control were elucidated. In contrast, the biochemical complexity of the transcriptional apparatus was only just beginning to unfold.

During the 1980s, activity continued on generating more refined prokaryotic ribosome structures and on characterizing the soluble factors. Advances were discussed at numerous international meetings concerned primarily with ribosome structure and function. At the same time, an appreciation developed of the great extent that translational control contributes to regulating gene expression. As cloning technologies enabled determination of mRNA levels in cells, many examples were found where a change in the synthesis rate of a protein is not accompanied by a corresponding change in its mRNA level, implying translation control. Since there was no major international meeting dealing with these phenomena, a small workshop on translational control was was held in 1985 at the Banbury Center, Cold Spring Harbor. Brief reports by the participants were published in 1986 by Cold Spring Harbor Laboratory in a slim volume entitled *Translational Control.* The response to the workshop more than justified a large-scale meeting, and Michael Mathews, together with Brian Safer and John Hershey, organized in 1987 what was to be the first in a series of meetings at Cold Spring Harbor Laboratory on the topic of translational control. This monograph is inspired in large part by these and subsequent meetings held in 1989, 1992, and 1994. Following the 1992 meeting, John Inglis of the Cold Spring Harbor Laboratory Press appoached John Hershey and suggested developing a book on translational control. John requested that Michael Mathews and Nahum Sonenberg join him as co-editors in this endeavor, and planning for the book began in late 1993.

With the cooperation of the chapter authors, we have assembled a monograph that aims to identify the major currents of research in translational control and to provide sufficient background information needed for understanding the molecular mechanisms involved. The book focuses almost entirely on translational control in eukaryotes, as regulation in prokaryotes is planned to be covered in another monograph. Chapters are concerned with basic mechanisms, the role of phosphorylation, regulation by *trans*-acting proteins, effects of viral infection, regulation during early development, mRNA stability, *inter alia*. The book is intended to provide sufficient background detail and examples to guide researchers in their studies of other examples of translational control. We are convinced that controls at the level of protein synthesis will assume greater importance for regulating gene expression in cell proliferation, development, and cell differentiation and for integrating the various metabolic pathways in the cell. It is our hope that this monograph will assist and enhance the future development of these fields.

We thank the contributors for their extensive efforts and diligence in responding promptly and their patience and goodwill in dealing with our numerous editorial comments. They deserve full credit for the success of this monograph. We also thank the staff of the Cold Spring Harbor Laboratory Press, John Inglis, Nancy Ford, Dorothy Brown, and especially Joan Ebert for her help in guiding us through the preparation and completion of the monograph.

J.W.B. Hershey
M.B. Mathews
N. Sonenberg

October, 1995

TRANSLATIONAL CONTROL

1

Origins and Targets of Translational Control

Michael B. Mathews
Cold Spring Harbor Laboratory
Cold Spring Harbor, New York 11724

Nahum Sonenberg
Department of Biochemistry
McGill University, Montreal
Quebec H3G 1Y6, Canada

John W.B. Hershey
Department of Biological Chemistry
School of Medicine
University of California
Davis, California 95616

ORIGINS OF TRANSLATIONAL CONTROL

The central idea of translational control proclaims that gene expression can be regulated by the efficiency of utilization of messenger RNA in specifying protein synthesis. This notion emerged only a few years after the articulation of the central dogma of molecular biology (Crick 1958) and very soon after the formulation of the messenger hypothesis. In 1961, Jacob and Monod perceived that "the synthesis of individual proteins may be provoked or suppressed within a cell, under the influence of specific external agents, and. . . the relative rates at which different proteins are synthesized may be profoundly altered, depending on external conditions." They pointed out that such regulation "is absolutely essential to the survival of the cell," and went on to advance the concept of an unstable RNA intermediary between gene and protein as a key feature of their elegant model for transcriptional control (Jacob and Monod 1961). The idea that this mRNA could be subject to differential utilization depending on the circumstances was accorded scant attention in the bacterial culture of the time, but it was taken up enthusiastically by workers in other fields, to the extent that 10 years later, one writer could allude to the "now classical conclusion" that eggs contain translationally silent mRNA that is activated upon fertilization (Humphreys 1971).

The term Translational Control was certainly in use as early as 1968, by which date at least four clearly distinct exemplars had been recognized and were already coming under mechanistic scrutiny. The groundwork for these four paradigms—developing embryos, reticulocytes, virus- and phage-infected cells, and higher cells responding to stimuli ranging from heat to hormones and starvation to mitosis—had all been laid by the middle of the 1960s. They founded a thriving and expanding field of study which has advanced from its largely eukaryotic origins to embrace prokaryotes (although not yet the archaebacteria, as far as we are aware).

A Short History of Translation

The genesis of the translational control field took place at a time when translation itself was in its infancy; many (although not all) of the reactions had been observed, but most of the components were not yet characterized and mechanistic details were essentially unknown. To place the origins of translational control in context, we briefly outline the development of protein synthesis.

Biochemical investigations of the process began in the latter half of the 1950s, at the same time as the view of proteins as unique, nonrandom linear arrays of just 20 amino acid residues was solidifying. Enabled by the availability of radioactive amino acids as tracers, biochemistry ran ahead of genetics, as it has continued to do in this field until the relatively recent advent of cloning and the systematic exploitation of the yeast system. Siekevitz and Zamecnik (1951) produced a cell-free preparation from rat liver that incorporated amino acids into protein, showing that energy was required in the form of ATP and GTP. The system was refined by stages and resolved into subfractions including a microsomal fraction that included ribosomes attached to fragments of intracellular membrane (for review, see Zamecnik 1960). It is salutary to recall that this was accomplished in advance of an understanding of the central role of RNA in the flow of genetic information to protein and in an era when theories of protein synthesis via enzyme assembly and peptide intermediates were entertained along with template theories (Campbell and Work 1953). Further biochemical work demonstrated that the ribonucleoprotein particles later called ribosomes comprise the site of protein synthesis, but it was not until the early 1960s that polysomes were observed and their function appreciated in light of the messenger hypothesis (Marks et al. 1962; Warner et al. 1963).

At much the same time, the role of aminoacyl-tRNA was being established. The existence of an intermediate, activated amino acid state was detected (Hultin and Beskow 1956) and characterized (Hoagland et al. 1959), then understood as the physical manifestation of the adaptor RNA predicted on theoretical grounds (Crick 1958). Once its function had been realized, the name transfer RNA rapidly displaced the original descriptive term, "soluble" RNA. Later, chemical modification of the amino acid moiety of a charged tRNA confirmed that it is the RNA component which decodes the template (Chapeville et al. 1962). Thus, responsibility for the fidelity of information transfer from nucleic acid to protein rests in part on the aminoacyl-tRNA synthetases, which became the first macromolecular components of the protein synthetic apparatus to be purified (Berg and Ofengand 1958). These, together with the other enzymes, or protein "factors" as they became known, were steadily characterized and purified such that nearly all of the protein components have been known for more than 20 years. Yet, new ones continue to be reported (see, e.g., Wiedmann et al. 1994), and, even today, there is no certainty that the full complement of protein factors involved in translation has been identified.

It was genetics rather than biochemistry that supplied the missing cornerstone of the protein synthetic system, mRNA. According to the messenger hypothesis, the ribosomes and other components of the protein synthesis machinery constitute a relatively stable decoding and synthetic apparatus that is programmed by an unstable template (Jacob and Monod 1961). This insight received rapid confirmation in bacteria (Brenner et al. 1961; Gros et al. 1961) and in bacterial cell-free systems. The discovery that poly(U) can direct the synthesis of polyphenylalanine (Nirenberg and Matthaei 1961) was particularly fruitful, greatly speeding the elucidation of the genetic code by the mid 1960s. Because of the greater stability of most eukaryotic mRNAs, the applicability of the messenger hypothesis to higher cells was less readily apparent. Nonetheless, the existence of a class of rapidly labeled RNA, heterogeneous in size and with distinct chromatographic properties, was soon recognized. Its essential features as informational intermediary were confirmed and it was universally accepted several years before the discovery (in the early 1970s) of 5′ caps and 3′ poly(A), the modern hallmarks of most eukaryotic mRNAs. The mRNA concept immediately revolutionized thinking about gene expression in all cells.

To appreciate the pace at which protein synthesis advanced during the decade of the 1960s, it is instructive to compare the Cold Spring Harbor Symposium volume of 1962 (on Cellular Regulatory Mechanisms) with that of 1970, a much thicker book devoted to a narrower topic (the Mechanism of Protein Synthesis). By the end of the decade, much of the translational apparatus had been characterized (although much also remained to be done), many problems of regulation had been laid out, and translational control came to receive increasing attention.

General Features of Translational Control

In a multistep, multifactorial pathway like that of protein synthesis, regulation can be exerted at many levels. Examples of translational control are indeed found at different levels, but the overwhelming preponderance of known instances—including all of the earliest cases recognized—is at the level of initiation. This empirical observation conforms to the biological (and logical) principle that it is more efficient to govern a pathway at its outset than to interrupt it in midstream and have to deal with the resultant logjam of recyclable components and the accumulation of intermediates as by-products. Nevertheless, well-characterized cases do occur at later steps in the translational pathway, especially at the elongation level where it seems that a translational block may be imposed as a safety measure to halt further peptide bond formation.

One of the chief virtues of translation as a site of regulation is that it offers the possibility of rapid response to external stimuli without invoking nuclear pathways for mRNA synthesis and transport. Predictably, the first cases to be recognized were those in which it was simplest to establish, if it was not self-evident, that transcription and other nuclear events were not responsible. By the same token, the relative scarcity of prokaryotic examples and their generally later recognition can be largely attributed to the lack of a nuclear barrier between the sites of mRNA synthesis and translation. The greater speed of macromolecular synthesis in bacteria and their lesser dependence on mRNA processing are other factors. These circumstances allow for a coupling of transcription and translation that all but obviates the need for translational control. That it occurs at all in prokaryotes is due to the exigencies of particular circumstances and to the potency of translational control mechanisms.

The earliest cases of translational control to be explored in depth, in fertilized invertebrate eggs and mammalian reticulocytes, were those in which the departure from the transcription-based regulatory model was the most obvious and extreme. Protein synthesis is abruptly turned on (in eggs) and off (in reticulocytes) in the absence of ongoing transcription. A further distinction that made it easier to define and study these two particular cases is that the regulation is apparently indiscriminate in that it affects protein synthesis generically, rather than the synthesis of specific proteins. Not all translational control is of this type, however, and it is useful to distinguish here between global and selective controls (sometimes referred to, rather misleadingly, as quantitative and qualitative controls). Global controls, such as those operating in eggs and reticulocytes, impact the entire complement of mRNAs within a cell, switching their translation on or off or modulating it by degrees in unison. This kind of regulation is usually implemented by adjustments in the activity of general components of the protein synthesis machinery that act in a nonspecific manner. Selective controls, on the other hand, affect a subset of the mRNAs within a cell, in the extreme case a single species only. This can be accomplished through mechanisms that target ligands to individual mRNAs or classes of mRNA, but it is achieved more commonly by exploiting the differential sensitivity of mRNAs to changes in the activity of general components of the translation system. Although examples of all these exist and are discussed at length in this monograph, in the context of the historical origins of translational control, it should come as no surprise that the earliest examples were mainly of the global variety and that (with notable exceptions) definitive evidence in favor of selective translational control accumulated more slowly.

PARADIGMS OF TRANSLATIONAL CONTROL

In large part, the origins of translational control can be traced to the confluence of four early streams of investigation, which still continue to flow. Their early courses are described below, followed by an example of elongation control.

Sea Urchin Eggs

The eggs of sea urchins and other invertebrates provide a striking example of regulated gene expression which, it was quickly realized, did not harmonize with the emerging theme of transcriptional control. These cells are essentially quiescent until they are galvanized into action by fertilization. Egg ribosomes synthesize protein at a very low rate but are triggered to incorporate amino acids within a few minutes of fertilization (Hultin 1961). Although the rate of protein synthesis accelerates rapidly after fertilization, there is little or no concomitant RNA synthesis (Hultin 1961; Nemer 1962; Gross and Cousineau 1963). Translation in enucleated eggs can be activated parthenogenetically (Denny and Tyler 1964). Moreover, actinomycin D fails to block the first wave of increased translation, which lasts for several hours, and both cell division and many morphogenetic events proceed unimpeded by the transcriptional inhibition. A second wave of increased protein synthesis is prevented by actinomycin D, however, presumably because this wave does depend on new mRNA synthesis (Gross et al. 1964). Such observations are explained by the fact that the eggs contain preexisting mRNA in a form that is not translated until some stimulus dependent on fertilization is received. In principle, the limitation could be due to a deficiency in the translational machinery, but unequivocal evidence in this direction has been more difficult to obtain. For example, a comparison of polysome sizes and translation rates in eggs and embryos did not disclose any defect in the apparatus itself (Humphreys 1969). On the other hand, a good deal of evidence points to a defect in the availability of mRNA. Consistent with the conclusion that mRNA is largely sequestered in eggs, deproteinized egg RNA can be translated in a cell-free system (Maggio et al. 1964). The ribosomes from eggs—unlike those from embryos—display little intrinsic protein synthetic activity, although they are able to translate added poly(U) (Nemer 1962; Wilt and Hultin 1962), suggesting that they possess latent translational capacity. Egg mRNA exists in a masked form: Cytoplasmic messenger ribonucleoprotein (mRNP) particles have been observed (Spirin and Nemer 1965), and some studies even indicated that the template could be activated by trypsin treatment, presumably by removing masking proteins (Monroy et al. 1965). Since the assembly of masked mRNP complexes must take place during oogenesis, the sea urchin system exemplifies a reversible process of mRNA repression and activation. Developments in this arena are discussed later in this volume (see Richter; Spirin; Wickens et al.).

Reticulocytes

These immature red cells have endowed researchers with a unique and especially dynamic system for studying the mechanism and control of translation. Because mammalian reticulocytes are enucleate, unlike those of most vertebrates, it was taken for granted that the regulation of protein production would be exercised at the translational level, an assumption that has been borne out in numerous studies. More than 90% of the protein made in the reticulocyte is hemoglobin, which consists of two α-globin and two β-globin chains together with four molecules of heme, an iron-containing porphyrin. In the intact rabbit reticulocyte, the synthesis of heme parallels that of globin (Kruh and Borsook 1956), and subsequent work showed that globin synthesis is controlled by the availability of heme or of ferrous ions (Bruns and London 1965). The phenomenon was made experimentally accessible by the development of the highly active unfractionated reticulocyte lysate translation system

(Lamfrom and Knopf 1964), which became the forerunner of the widely used messenger-dependent system of Pelham and Jackson (1976). Regulation by heme is reversible in intact cells, and, to a limited extent, the repression of protein synthesis that ensues in the reticulocyte lysate soon after heme deprivation can also be rescued by restoring the heme level. When globin synthesis is inhibited in cells or extracts, the polysomes dissociate to monosomes (Hardesty et al. 1963; Waxman and Rabinowitz 1966), arguing that heme is involved in regulating translation initiation. Contrary to intuitive expectation, there is no necessary linkage between the role of heme as the prosthetic group of globin and its role as translational regulator. The effects of heme deprivation on protein synthesis in the reticulocyte or its lysate are mimicked by unrelated stimuli such as double-stranded RNA (dsRNA) and oxidized glutathione (Ehrenfeld and Hunt 1971; Kosower et al. 1971) and extend to all mRNAs in the reticulocyte lysate (Mathews et al. 1973). Such observations imply that a general mechanism of translational control is being invoked: In each of the conditions under which protein synthesis is down-regulated, inhibitors—now known to be the eIF2 kinases HCR (or HRI) and PKR (or DAI)—are activated (see Clemens; Trachsel; both this volume). By 1977, a unifying scheme could be advanced (Farrell et al. 1977), centering on the phosphorylation of the α-subunit of initiation factor eIF2 and the loading of the 40S ribosomal subunit with Met-$tRNA_i$. This mechanism has been found to have wide applicability in cells and tissues responding to a broad range of stimuli.

Virus-infected Cells

During the 1960s, it came to be appreciated that cellular protein synthesis is suppressed during infection with many viruses (see Ehrenfeld; Katze; Mathews; Schneider; all this volume). This inhibition may begin before the onset of viral protein synthesis and without any apparent interference with cellular mRNA production or stability. In poliovirus infection, an early example, the shutoff of host-cell translation can be complete within 2 hours after infection and is followed by a wave of viral protein synthesis (Summers et al. 1965). The first phase is accompanied by the reduction of polysomes to monosomes without any effect on the elongation or termination phases of protein synthesis (Penman and Summers 1965; Summers and Maizel 1967). In the second phase, virus-specific polysomes form (Penman et al. 1963), evidence that initiation has become selective for a class of mRNA—in this case viral, rather than cellular. Later studies showed that cellular mRNA remains intact in the infected cell (Leibowitz and Penman 1971) and is translatable in a cell-free system, although it is not translated in the infected cell. Furthermore, the inhibition extends to the mRNAs of several other viruses introduced together with poliovirus in a double infection (Ehrenfeld and Lund 1977), indicative of a general effect. Although circumstantial evidence aroused suspicions that viral dsRNA and PKR might be responsible for the phenomenon, later work incriminated a modification of the cap-binding complex, eIF4F. Cleavage of the eIF4G subunit of this complex prevents cap-dependent initiation on cellular mRNAs but does not interfere with initiation on the viral mRNA which occurs by internal ribosome entry (see Ehrenfeld; Mathews; both this volume).

Bacteriophage f2 provided the first evidence for translational control in a prokaryotic system, as well as the first clear cases of mechanisms specific for the synthesis of individual protein species. The phage RNA genome encodes four polypeptides, the maturation protein, coat protein, lysis protein, and replicase, that are initiated individually but produced at dissimilar rates. Several regulatory interactions among them are now known. One was revealed by the observation that a nonsense mutation early in the cistron coding for the viral coat protein down-regulates replicase synthesis (Lodish and Zinder 1966). Apparently, passage of ribosomes through a critical region of the coat protein cistron is required to melt RNA structure and allow replicase translation. In contrast, a second nonsense mutation leads to overproduction of the replicase, suggesting that the coat protein acts as a repressor of replicase translation. This inference has been amply confirmed, and the binding of the coat protein to the hairpin structure containing the replicase AUG has become one of the best-characterized cases of RNA-protein interaction (Witherell et al. 1991). Subsequent studies have disclosed translational control mechanisms in the DNA phages as well as in bacterial genes themselves (see Mathews; Voorma; both this volume), but it was eukaryotic systems that made most of the early running.

Physiological Stimuli

The cells and tissues of higher organisms have been reported to regulate the expression of individual genes or of whole classes of genes at the translational level in response to a wide variety of stimuli or conditions. These include cell state changes, such as mitosis (Steward et al. 1968; Hodge et al. 1969; Fan and Penman 1970) and differentiation (Heywood 1970; Mathews 1970); stress resulting from heat shock (McCormick and Penman 1969), treatment with noxious substances or the incorporation of amino acid analogs (Thomas and Mathews 1984); and normal cellular responses to ions (Drysdale and Munro 1965) and hormones (Eboué-Bonis et al. 1963; Garren et al. 1964; Martin and Young 1965; Tomkins et al. 1965). Not in every case was the evidence for regulation at the translational level complete, and in a few instances, the trail has gone cold or been erased upon more detailed examination, but the accumulated volume of information added conviction to the view that translational control is both widespread and important. One of the chief stumbling blocks in this arena lay in determining that the level at which control was exerted was indeed translational. This can be a difficult task in nucleated cells, let alone in a tissue or whole animal (or plant), and it was addressed in various ways. A popular approach exploited selective inhibitors of transcription or translation, such as actinomycin D and cycloheximide, but the results were liable to be complicated (if not confounded) by side effects of the drugs or their indirect sequelae in complex systems. Another argument that could be made for an effect at the translational level, although not without some reservations, came from its rapidity (see below). Timing alone cannot provide a definite assignment, however, and the most convincing proofs often came from subsequent investigations of the underlying biochemical processes—for example, by demonstrating changes in polysome profiles or initiation factor phosphorylation states as discussed below and in a number of chapters in this volume. The ultimate goal is to achieve an understanding of the regulatory mechanisms set in train by the stimuli applied, and within this wide array of phenomena lie many of the challenges for the future.

Secretory Proteins

No overview of the principal themes of translational control would be complete if it dwelt exclusively on the initiation phase. One of the best-studied examples of regulation during the elongation phase is found in the synthesis of proteins that are destined for secretion. These are made on polysomes that are attached to the endoplasmic reticulum, isolated from cellular homogenates in the form of microsomes. In the early 1970s, it began to seem likely that ribosomes become associated with cell membranes only after protein synthesis has been initiated (Lisowska-Bernstein et al. 1970; Rosbash 1972). Contemporaneously, the existence of what came to be called a signal peptide was reported on an immunoglobulin light chain (Milstein et al. 1972) and other secreted proteins (Devillers-Thiery et al. 1975). These findings lent substance to the signal hypothesis (Blobel and Sabatini 1971), which suggested that an amino-terminal sequence might ensure secretion, and prompted the development of cell-free systems that enabled the biochemical dissection of the secretory pathway (Blobel and Dobberstein 1975). One of the surprising discoveries to emerge was the involvement in secretion of a ribonucleoprotein particle, the signal recognition particle (SRP), which interacts with the signal peptide, the ribosome, and the endoplasmic reticulum. Remarkably, binding of the SRP to a nascent signal peptide protruding from the ribosome causes translational arrest in the absence of cell membranes (Walter and Blobel 1981). This elongation block is relieved when the ribosome docks with the endoplasmic reticulum, allowing the protein chain to be completed and simultaneously translocated across the lipid bilayer. It has been speculated that this mechanism serves to ensure cotranslational protein export and to prevent the accumulation of secretory proteins in an improper subcellular compartment (the cytosol). Interestingly, a similar rationale has been offered to account for control at the elongation level during heat shock (for review, see Duncan, this volume). In this situation, it has been proposed that a translational arrest is imposed to prevent the synthesis of proteins that might be abnormally folded. Thus, elongation blocks might be used under exceptional circumstances, to preserve cellular integrity when it is threatened by the production of protein at the wrong time or in the wrong place, or perhaps in the event of a sudden shortage of energy or an essential metabolite.

WHAT LIMITS PROTEIN SYNTHESIS IN PRINCIPLE?

Given that translational controls are so widespread in eukaryotic cells, it is appropriate to examine the fundamental principles on which these controls are based. Translational control is defined as a change in the rate (efficiency) of translation of one or more mRNAs, i.e., the number of completed protein products changes per mRNA per unit time. It is generally believed that during protein synthesis, the number of protein chains initiated is about the same as the number of proteins completed; in other words, few nascent polypeptides abort and fall off the ribosome (Tsung et al. 1989). Therefore, under steady-state conditions, the number of initiation events per unit time approximates the number of protein products produced during the same time interval. It follows logically that the rate of protein synthesis is determined by the number of initiation events, i.e., the rate of initiation. What determines the number of initiation events per unit time? Four major parameters may influence or define the rate of protein synthesis. Each is considered briefly below.

The Activity of the Protein Synthesis Machinery

Numerous examples exist of cells that possess ribosomes and mRNAs in excess of those actively engaged in protein synthesis. This may occur if a single translational component (e.g., a soluble factor) is limiting in amount or if one or more components have reduced specific activities. Such regulation frequently involves the phosphorylation status of translational components, as detailed in numerous chapters in this monograph. Regulation of the overall activity of the translational apparatus is expected to affect the translation of essentially all mRNAs. As argued earlier by Lodish (1976), down-regulation of the initiation steps that occur prior to the binding of mRNAs is expected to lead to greater inhibition of those mRNAs whose initiation rate constants are relatively low ("weak" mRNAs), as compared to "strong" mRNAs. Reciprocally, activation of such steps may stimulate more greatly the translation of weak mRNAs. Alteration of the activities of components that interact with mRNAs and affect their binding to ribosomes also would be expected to generate differential effects on the translation of the mRNA population (Godefroy-Colburn and Thach 1981).

The Rate of Elongation

The initiation rate on an mRNA can be inhibited if a ribosome, having already initiated, vacates the initiation region too slowly. A ribosome bound at the AUG initiator codon occupies about 12–15 nucleotides (4–5 codons) downstream from the AUG and about 20 nucleotides upstream. Another ribosome can occupy the initiation site only after the first ribosome has moved about seven codons down the mRNA. When the time needed to vacate the initiation region approaches or exceeds the time required for initiation, the elongation rate becomes limiting. In general, it is believed that the elongation rate is about the same for all mRNAs (3–8 amino acids per second per ribosome), because measurements of a few specific examples gave similar results in this range (Lodish and Jacobsen 1972; Palmiter 1974). Nevertheless, the rate of elongation is not uniform throughout the coding region of an mRNA, as pausing may occur at specific locations, possibly due to the occurrence of rare codons or RNA secondary structure (Wolin and Walter 1988). If ribosome pausing occurs such that it impedes initiation, mRNA efficiency is decreased. The question of which translation phase is rate-limiting, initiation or elongation/termination, is addressed in greater detail below.

The Amount or Efficiency of mRNAs

The level of mRNA in the cytoplasm is determined by the rate of transcription, the proportion of primary transcripts that are processed and transported into the cytoplasm, and the degradation rate of cytoplasmic mRNAs. In actively translating mammalian cells, mRNAs often are found entirely in polysomes, as shown for actin (Endo and Nadal-Ginard 1987); thus, the rates of synthesis of such specific proteins are mRNA-limited. However, total mRNA in the cytoplasm frequently appears to be present in excess, with about 30% of the mRNA in cultured cells present as free mRNP particles (Geoghegan et al. 1979; Kinniburgh et al. 1979; Ouellette et al. 1982). Therefore, the level of mRNA appears not to limit the overall number of translational initiation events in these cells. In cells exhibiting low translational activity, many mRNAs are repressed and apparently unavailable to the translational apparatus (masked), as seen most dramatically in oocytes and unfertilized eggs as described above, but also in somatic cells in culture (Lee and Engelhardt 1979). Such repression sometimes appears to be all or none, as some mRNAs aie distributed bimodally in polysome profiles; a fraction of the specific mRNA is completely repressed (nontranslating mRNP particles), whereas a portion is actively translated as large polysomes (Yenofsky et al. 1982; Agrawal and Bowman 1987). In instances of specific regulation of protein synthesis, mRNA repression and availability to the translational apparatus likely have a dominant role, for example, in the translation of ferritin mRNA (see Rouault et al., this volume) and ribosomal protein mRNAs (see Meyuhas et al., this volume). Furthermore, individual activated mRNAs differ greatly in their efficiencies of translation as deduced from polysome sizes, thereby contributing to regulation of gene expression. These innate efficiencies are determined in large part by the primary and higher-order structures of the mRNAs (for review, see Jackson; Mathews; Merrick and Hershey; Voorma; all this volume).

The Abundance of Ribosomes

The cellular levels of ribosomes may be rate-limiting under some circumstances. Cells active in protein synthesis, for example, liver cells from fed rats, engage 90–95% of their ribosomes in protein synthesis (Henshaw et al. 1971), suggesting that still higher rates of protein synthesis might have been possible were there a greater number of ribosomes. On the other hand, in translationally repressed cells, such as liver cells from fasted rats (Henshaw et al. 1971) or in quiescent cells in culture (Duncan and McConkey 1982; Meyuhas et al. 1987), fewer than half of the ribosomes may be actively translating mRNAs. The level of ribosomes surely is not limiting in these cells, since a rapid increase in the rate of protein synthesis can be induced within 20 minutes, before the assembly of more ribosomes is possible (Duncan and McConkey 1982). Translation also may be limited by the levels (as opposed to specific activities) of other components of the translational apparatus, e.g., eIF2 and eIF4F, the latter likely through its eIF4E subunit (see Sonenberg; this volume). When amino acids become limiting, global protein synthesis is rapidly repressed by inhibiting the activity of initiation factors (Clemens et al. 1987).

WHICH PHASE OF PROTEIN SYNTHESIS IS RATE-LIMITING AND REGULATED?

The analysis above identifies three ways in which the rate of protein synthesis may be limited and thus regulated over a relatively short time period (on the order of minutes): the rate of initiation, the rate of elongation/termination, and the repression/activation of mRNAs/mRNPs. How is the rate of protein synthesis measured and how is the rate-limiting step identified? The overall rate of protein synthesis can be measured by assaying the time course of incorporation into protein of radioactively labeled amino acids added to the culture medium. The method is complicated only by the uncertainty of the specific radioactivities of the precursors within cells, as intracellular de novo synthesis and degradation of proteins may influence these values. A second method measures the absolute number of active ribosomes and the elongation rate, from which the number of amino acids incorporated per unit time can be calculated. The elongation rate is obtained by dividing the number of amino acids in the average protein by the ribosome transit time (Fan and Penman 1970), the time it takes to translate an average-sized mRNA. The fraction of total ribosomes that is active is assessed by high-salt sucrose gradient centrifugation of cell lysates to generate a polysome profile: Active ribosomes in polysomes are separated from nonactive, free ribosomal subunits. The second method, although more laborious than the measurement of labeled amino acid incorporation, is not complicated by uncertainties of amino acid specific radioactivities. Both methods serve to analyze global rates of protein synthesis. The relative synthesis rates of specific proteins can be measured by radioactively labeling proteins, followed by immunoprecipitation or fractionation of proteins by high-resolution two-dimensional gel electrophoresis.

Which phase of protein synthesis is rate-limiting, initiation or elongation/termination? Although most mRNAs are thought to be limited by their initiation rate, others are limited by the rate of elongation/termination. Therefore, the question is best addressed to specific mRNAs rather than to the whole population. Insight into which phase is rate-limiting is gained by an examination of polysome profiles, where the specific mRNA is located in the sucrose gradient fractions by hybridization techniques. The rate of initiation, i.e., the number of initiation events per minute, can be calculated from the number of ribosomes translating an mRNA (polysome size) and the ribosome transit time (the time required for the ribosome to traverse the mRNA). As elegantly determined for ovalbumin mRNA in chick oviducts (Palmiter 1975), ovalbumin polysomes average 12 ribosomes and the ribosome transit time is 1.3 minutes, giving a rate of initiation of 9.2 events per minute (or one initiation every 6.5 seconds). Since the elongating ribosome requires only about 2 seconds to vacate the initiation site, it is clear that the initiation rate is slower than potentially possible and thus is rate-limiting. Parenthetically, if the number of mRNA molecules in the polysomes is known, an absolute rate of specific protein synthesis can be calculated.

A second way to determine whether initiation or elongation/termination is rate-limiting for an mRNA is to treat cells with low concentrations of an elongation (e.g., cycloheximide and sparsomycin) or initiation (e.g., pactamycin) inhibitor. If translation of the specific mRNA is limited by the elongation rate, its synthesis will be sensitive to the inhibitors of elongation. Conversely, if initiation is rate-limiting, such mRNAs will be insensitive to elongation inhibitors but sensitive to initiation inhibitors. For example, when mRNAs encoding α- and β-globin (Lodish and Jacobsen 1972) or reovirus proteins (Walden et al. 1981) were analyzed, initiation was the sensitive step. Because the majority of mRNAs in cells are resistant to low concentrations of cycloheximide, it is thought that the translation of most mRNAs is limited at the initiation phase.

Further evidence that the rate of initiation limits the translation of most mRNAs is obtained by examining polysome sizes from sucrose gradients. On the average, ribosomes in polysomes occur once every 80–100 nucleotides. For example, the average polysome size for globins is about five ribosomes per mRNA, or one ribosome per 90 nucleotides. When protein synthesis is inhibited by cycloheximide such that elongation becomes rate-limiting, polysomes increase in size (to more than 12 ribosomes per globin mRNA, for example). Therefore, polysome densities of one ribosome per 30–40 nucleotides are possible. This approaches the limit for close packing, since a ribosome occupies about 30 nucleotides of mRNA. That average polysome densities are much less is due to the relatively low rate of initiation.

Changes in the size (number of ribosomes per mRNA) or amounts (amplitude) of polysomes may be diagnostic of the phase of global protein synthesis that is being modulated. If the size of polysomes decreases, either initiation is inhibited or elongation/termination is stimulated, or a combination of both occurs. Conversely, an increase in polysome size can be caused by an increased rate of initiation and/or a decreased rate of elongation/termination. To interpret polysome profiles unambiguously, it is advisable to measure the elongation rate by determining the ribosome transit time and average length of mRNAs being translated. In cases where the overall rate of protein synthesis is repressed and polysomes are smaller, initiation has clearly been inhibited. Regulation of a specific mRNA is readily evaluated by these methods, since the average size of its polysomes is readily determined by hybridization techniques with cloned probes. Repression or activation of protein synthesis need not always affect polysome size. Instead, the number of translating mRNAs may be affected by masking mRNAs or mobilizing them into polysomes. In this case, there is a change in the *amount* (i.e., amplitude) of polysomes, but the average *size* of the polysomes may remain the same.

Are there cases where the elongation rate is regulated? Examination of a number of specific mRNAs shows that rather modest changes in the rate of elongation are found following treatment of cells with hormones and other agents. A dramatic example is the fivefold stimulation of the rate of elongation of tyrosine aminotransferase seen when rat hepatoma cells are treated with dibutyryl-cAMP (Roper and Wicks 1978). Similarly, the elongation rate on vitellogenin mRNA drops about fourfold when cockerel liver explants are treated with 17β-estradiol (Gehrke and Ilan 1987). Even small changes in the elongation rate will affect the efficiencies of those mRNAs that are elongation-limited; whether or not moderate inhibition of elongation affects initiation-limited mRNA expression depends on the degree that initiation is limiting.

TARGETS AND MECHANISMS OF TRANSLATIONAL CONTROL

Having defined the rate-limiting steps in the protein synthesis pathway, we now turn to the means by which its regulation is accomplished in the cell. Translational control is realized through multiple mechanisms that target structural features of the mRNA and *trans*-acting components; the latter may be either protein or (less commonly) RNA in nature. The survey that follows takes stock of the principal targets of translational control and the mechanisms which they coordinate, giving reference to chapters in this monograph where these topics are considered in greater detail.

mRNA

The intrinsic translational efficiency of an mRNA is dependent on several *cis*-acting elements, which also have critical roles in the regulation of mRNA utilization (see Jackson; Merrick and Hershey; Voorma; all this volume). It is convenient to divide the *cis*-acting elements into two categories: those that act alone or with general translation factors and those whose actions are mediated by specific *trans*-acting factors.

In prokaryotes, the first category is of overriding importance. Translational efficiency is heavily influenced by mRNA primary structure, especially the Shine-Dalgarno sequence, as well as by the degree of secondary structure that can be modulated by various mechanisms (Voorma, this volume). In eukaryotes, *cis*-acting elements distributed along the length of the mRNA modulate translational efficiency. Primary structure, notably the 5′ cap, the sequence flanking the initiator AUG (its "context"), and the presence of upstream AUG triplets all determine translational efficiency (Jackson; Merrick and Hershey; both this volume). Secondary structure, particularly in the 5′-untranslated region (5′UTR), can also have a determinative role. Upstream open reading frames (uORFs) participate in translational control in yeast and higher eukaryotes. Regulation of the translation of uORF-containing mRNAs is dependent on many factors, including the amino acid sequence encoded by the uORF, the length of intercistronic regions, and the sequence context of the termination codon of the uORFs (Geballe; Hinnebusch; both this volume).

Within the coding sequence of some mRNAs are elements that signal ribosome frameshifting, hopping, termination codon read-through, and the incorporation of selenocysteine. Some of these processes are known to be regulated. For example, ribosomal frameshifting is regulated in both eukaryotes (in antizyme) and prokaryotes (in RF2 and tryptophanase) (Atkins and Gesteland, this volume). It is also possible that in some cases, ribosome hopping is modulated by the efficiency of translation initiation (Engelberg-Kulka and Schoulaker-Schwartz 1994).

Cis-acting elements belonging to the second category also occur throughout the mRNA. The iron-responsive element (IRE) is a sequence- and structure-specific negative regulatory element, found in the 5′UTR of ferritin mRNA (and subsequently in other mRNAs), that modulates its translation in accordance with the level of cellular iron. This regulation is mediated by a *trans*-acting iron repressor protein (IRP) that binds to the IRE and inhibits translation (Rouault et al., this volume). It is reasonable to expect that other such *negative* mRNA-specific *trans*-acting regulators of translation are awaiting discovery. *Positive* mRNA-specific regulators of translation have been described in bacteriophages. For example, the Com protein of bacteriophage Mu activates translation of *mom* mRNA by binding near its initiation site and altering its secondary structure (Mathews; Voorma; both this volume). Although no factor with similar activity has yet been reported in eukaryotes, several proteins interact with the internal ribosome entry site (IRES) of picornavirus RNAs and stimulate their translation (Ehrenfeld; Jackson; both this volume).

The past decade has seen the surprising discovery that the 3′UTR is a rich repository of *cis*-acting elements that determine mRNA stability and localization in the cytoplasm and also serve to regulate translation initiation. These controls are most likely mediated by *trans*-acting factors (Jacobson; Richter; Wickens et al.; all this volume). Most such examples of translational control occur during early development, but some cases have been described in somatic cells. An unusual case is seen in the developmentally regulated *Caenorhabditis elegans* gene *lin-14*. Translation of this mRNA is inhibited by a short (22-nucleotide) RNA transcribed from the *lin-4* gene, which can base pair with sequences in the 3′UTR of the *lin-4* mRNA. At the 3′end of eukaryotic mRNAs, the poly(A) tail also has an important role as an enhancer of translation. Intriguingly, the poly(A) tail acts in synergy with the mRNA 5′cap structure, and the translational activity of the poly(A) tail may be mediated by the poly(A)-binding protein (Jacobson, this volume).

mRNA stability is an important determinant of cytoplasmic mRNA levels and therefore of protein synthesis. In many instances, translation has a direct role in determining mRNA stability, as mRNA degradation may be coupled to translation (Jacobson; Theodorakis and Cleveland; both this volume). Most but not all of the *cis*-acting elements that trigger mRNA degradation are localized to the 3′UTR; the poly(A) tail influences the degradation of mRNAs via the poly(A)-binding protein, and short-lived mRNAs possess sequence-specific elements that mediate mRNA degradation. A separate pathway exists to degrade mRNAs that contain premature termination codons (nonsense-mediated decay). This pathway has most probably evolved to prevent the synthesis of truncated proteins that might function in a dominant negative manner. It is puzzling that this degradative pathway operates in the cytoplasm in yeast, whereas it is nuclear in mammals. The nuclear mode of nonsense-mediated decay poses intriguing questions concerning the mechanism whereby nonsense codons are recognized in the nucleus, and the possible coupling between translation and nuclear-cytoplasmic mRNA transport.

Initiation Factors

The effects of the various *cis*-acting elements in the mRNA 5′UTR are modulated through the activity of initiation factors and other *trans*-acting factors. Phosphorylation of initiation factors provides the chief means to control the rate of mRNA binding. Several factors that promote mRNA binding to ribosomes (eIF3, eIF4B, eIF4E, and eIF4G) are phosphorylated, and the phosphorylation status of these proteins correlates positively with both translational and growth rates of the cell (Sonenberg, this volume). The phosphorylation state of these initiation factors is modulated in a wide variety of circumstances and affects translation during infection with viruses, after heat shock, or in response to growth factors and hormones (Duncan; Mathrews; Schneider; Sonenberg; all this volume). Although there is some biochemical evidence that the phosphorylation of eIF4E potentiates its cap-binding activity, for eIF4B and eIF4G, the consequences of phosphorylation are not yet established. Phosphorylation of eIF4A in plants might also have a role in regulating translation initiation (Sonenberg, this volume).

Phosphorylation of eIF2 also has a central role in regulating translation by affecting the binding of Met-tRNA$_i$. In contrast to the eIF4 group, phosphorylation of eIF2 inactivates its ability to recycle, as the exchange of GDP for GTP on the factor is blocked, leading to inhibition of translation (Trachsel, this volume). Phosphorylation of eIF2, like that of the eIF4 proteins, has a role in differentiation and occurs under conditions of stress, including heat shock (Duncan, this volume), viral infection (Clemens; Katze; Mathews; Schneider; all this volume), and serum deprivation (Clemens, this volume). Extensive analyses of the mechanisms of eIF2 phosphorylation led to the identification and characterization of two mammalian protein kinases, PKR and HCR, the former having a key role in the antiviral host defense mechanism that is mediated by interferons (Clemens, this volume). Furthermore, an eIF2 kinase in yeast, GCN2, regulates translation reinitiation on the 5′UTR of GCN4 and mediates the response to amino acid deprivation (Hinnebusch, this volume). Thus, phosphorylation of eIF2 controls the rate of reinitiation of translation on mRNAs that contain uORFs. Phosphorylation also controls the activity of eIF2B, the eIF2 guanine nucleotide exchange factor (Clemens; Hinnebusch; Trachsel; all this volume).

Apart from phosphorylation, translation initiation factor activity can be modulated in principle by other reversible or irreversible modifications. One important example that occurs as a result of infection with certain picornaviruses is the cleavage of eIF4G. This cleavage is responsible in part for the shutoff of host-protein synthesis after viral infection (Ehrenfeld; Mathews; both this volume).

An important recent development is the discovery that initiation factor activity can be modulated by proteins that interact with initiation factors. For example, polypeptides that bind eIF4E and inhibit cap-dependent translation initiation have been identified; their activity is modulated by phosphorylation under the control of growth factors and hormones (Sonenberg, this volume). Similarly, eIF2 activity may be modulated by an accessory protein, p67, that binds to eIF2 and prevents its phosphorylation by eIF2 kinases (Clemens, this volume).

Elongation Factors

Elongation rates are also modulated by phosphorylation, particularly through the activity of the translation elongation factor eEF2. This factor undergoes phosphorylation in response to growth-promoting stimuli, calcium ion fluxes, and other agents, to affect translation (Nairn and Palfrey, this volume). eEF2 and the other elongation factors are also altered posttranslationally by other modifications. For example, eEF2 is a substrate for ADP-ribosylation by diphtheria toxin on the unique diphthamide residue (derived from histidine). There is evidence that diphthamide has a role in polypeptide chain elongation (Merrick and Hershey, this volume). Both bacterial EF1A and eukaryotic eEF1A also contain modifications, but their functions are not yet clear.

Ribosomes

Phosphorylation of ribosomal proteins may also affect translational initiation. Of these, ribosomal protein S6 provides the best-studied example: Its phosphorylation promotes the initiation of translation on mRNAs encoding ribosomal proteins and elongation factors. Recent studies have revealed that the mechanism underlying this selectivity involves an oligopyrimidine tract in the 5′UTR of the target mRNAs and have shed light on the signal transduction pathways that link growth-promoting stimuli to S6 phosphorylation (Jefferies and Thomas; Meyuhas et al.; both this volume).

WHY CONTROL TRANSLATION?

Thus far, we have considered the basis and principles of translational control. As mentioned above, there is a clear-cut rationale for regulating a biochemical pathway at its first step; this principle holds true, by and large, for protein synthesis, in that regulation is most often exercised at the initiation phase. From a broader perspective, however, matters become less clear-cut. Viewing gene expression in totality, translation occupies a position somewhere in the middle of a complex pathway that begins with transcription, continues with RNA processing and transport, and ends with protein translocation, modification, folding, and assembly. Each of these steps is known to be regulated in one or another biological system. Yet, two of the steps in this grand scheme, transcription and translation, are especially critical for the cell. Both are biosynthetic steps in which the cell makes large investments of energy. Consequently, both are steps at which the cell's expenditure of resources is checked. Indeed, transcription is subject to a multitude of controls. So, why control translation, too? And where and when is this option exercised?

To these frequently asked questions there is no single answer. Rather, there are several compelling reasons for cells to deploy translational control in their arsenal of regulatory mechanisms. Some of the advantages offered by translational control are considered briefly below. Evidently, the benefits more than compensate for the energetic and other penalties paid for the privilege of exerting regulation over a downstream reaction in a long pathway.

Directness and Rapidity

Immediacy is the most conspicuous advantage of translational control over transcriptional and other nuclear control mechanisms. Whereas transcriptional control affects the first step in the flow of genetic information, translational control affects the last step. When control is applied at a step prior to translation, the cell has to confront subsequent biochemical reactions (splicing, nuclear transport, etc.) that might be rate-limiting and inevitably entail a delay in implementing changes in protein synthesis. No such time lag applies in the case of translational control.

Reversibility

Most translational controls are effected by reversible modifications of translation factors, chiefly through phosphorylation. The readily reversible nature of translational control mechanisms is economical in energetic terms, a feature that is of particular biological significance in energy-deprived cells.

Fine Control

There are numerous examples of genes that are under both transcriptional and translational control (e.g., ornithine decarboxylase, cyclin D1, and p53). In most instances, the changes in transcription rates are considerably greater in magnitude than the changes in translation rates. Thus, regulation of gene expression at the translational level provides a means for fine control.

Regulation of Large Genes

Some genes are extremely long (e.g., dystrophin, >2000 kb), and their transcription is estimated to take an extended period of time (>24 hours for dystrophin). It is reasonable to assume that if their expression needs to be regulated in a relatively short period, it is likely to be accomplished at the level of translation.

Systems That Lack Transcriptional Control

In some systems (e.g., reticulocytes, oocytes, and RNA viruses), there is little or no opportunity for transcriptional control, and gene expression is modulated mostly at the translational level. The widespread use of translational controls to regulate gene expression during development suggests that this mode of control preceded transcriptional control in evolution. Such a hypothesis is consistent with the notion of the existence of an RNA world prior to the emergence of DNA. Is it therefore possible that translational control was more prevalent early in evolution and that we are now witnessing only the relics of such control mechanisms?

Spatial Control

Regulation of the site of protein synthesis within the cell can generate concentration gradients of proteins. Such gradients are known to affect the translational efficiency of other specific mRNAs that determine patterning in early development (Wickens et al., this volume).

Flexibility

Because of the wide variety of mechanisms for translational control, it can be focused by specific effector mechanisms on a single or a few gene(s) or cistrons (such as the coat protein and replicase of RNA phages, antizyme, and ferritin); alternatively, by influencing general factors, it can encompass whole classes of mRNAs (as in heat shock and virus-induced host-cell shutoff). Such flexibility affords the cell a powerful and adaptable means to regulate gene expression.

FUTURE TRENDS

One prediction that can be made with confidence is that translational control will prove to be more prevalent than hitherto thought. Because the recognition of translational control normally requires the measurement of two parameters—the rate of protein synthesis and the concentration of the corresponding mRNA—it is likely that many cases still await discovery. Another safe prediction is that the mechanisms exploited to control gene expression at this level will prove to be even more diverse than currently contemplated.

As knowledge accumulates on the structure and functioning of the basal translational machinery, additional regulatory opportunities will be recognized. Some of these are already beginning to materialize. For example, many mRNAs bear long 5′UTRs and 3′UTRs which contain elements that control translation rates. A challenge for the future is to identify and characterize these elements and—most importantly—to identify the *trans*-acting factors that mediate their function. In all probability, the ferritin IRP will come to serve as a prototype for a large group of translational repressor proteins, and it is likely that mRNA-specific translational enhancer proteins will also be discovered. Furthermore, much still remains to be learned about both the scanning and internal ribosome entry mechanisms of polypeptide chain initiation and about the factors that mediate the two processes.

In other arenas, there are issues relevant to translational control whose outlines have barely been discerned. Some of the relevant studies are in their infancy; some have not been pursued sufficiently to allow a mechanistic understanding of the phenomena in question; and some will require improvements in technology before they will yield. The following are examples of these emerging research areas.

Role of the Cytoskeleton

Several lines of evidence indicate that the association of mRNAs with structural elements in the cytoplasm correlates with their translation. Certain viral mRNAs associate with the cytoskeleton only when translated (Cervera et al. 1981). Furthermore, cellular mRNAs dissociate from the cytoskeleton upon infection with poliovirus, concomitantly with the shutoff of host-protein synthesis (Lenk and Penman 1979). The importance of the cytoskeleton for translation is also evident from the findings that translational components including ribosomes and initiation factors are physically associated with the cytoskeleton (Howe and Hershey 1984). It is noteworthy that one translation factor (eEF1A) has been shown to have an important role in maintaining the integrity of the cytoskeleton and thus may be regarded as a critical structural element of the cytoskeleton (Owen et al. 1992). Despite the many studies that link the cytoskeleton and translation, it is presently unclear whether mRNA association is a *requirement* for translation or a *consequence* of translation.

mRNA Localization

The cytoskeleton also serves to localize mRNAs to specific regions of the cytoplasm (Singer 1993). For example, Lawrence and Singer (1986) showed that β-actin mRNA is localized to peripheral sites in the cell that are involved with motility. Furthermore, maternal mRNA localization in the embryo is mediated by cytoskeletal elements that act through the 3′UTR (Yisraeli et al. 1990). It is clear that a future field of research will focus on the interaction of the translation machinery with the structural elements of the cytoplasm, and the implications of these associations for translational control.

Molecular Chaperones

Another aspect of translational control that is just beginning to emerge is the involvement of chaperone proteins, including heat shock proteins. For example, heat shock proteins appear to bind to nascent polypeptide chains and facilitate elongation (Frydman et al. 1994; Hansen et al. 1994), and a heat shock protein, SIS1, has been implicated in translation initiation in yeast (Zhong and Arndt 1993).

mRNPs

A time-honored but still poorly understood aspect of translational control is the function of RNA-binding proteins that are found in mRNP particles (Spirin, this volume). These proteins possess nonspecific RNA-binding activity and are associated with both translating polysomes and masked mRNP particles that are translationally inactive. Some of these proteins (e.g., FRGY 2) inhibit translation in oocytes and thus have a role in translational control during development. The mechanisms by which these proteins exert their effects are unknown, but models involving protein-induced changes in mRNA structure and condensation have been advanced. Tertiary structures for mRNAs and mRNPs will provide an urgently needed basis for addressing these and other fundamental questions in RNA metabolism.

Initiation and the 3′ End

The involvement of the 3′UTR in regulating the translation of certain mRNAs is well established (Jacobson; Richter; Wickens et al.; all this volume), yet knowledge of the underlying molecular mechanisms is rudimentary. One attractive explanation holds that the 5′and 3′ends of mRNAs communicate with each other, but direct biochemical evidence for such communication is lacking. It is also not known whether ribosomes dissociate from the mRNA after terminating translation, and whether they are then shunted from the 3′to the 5′end of the mRNA to reinitiate translation. Understanding how termination is controlled (Hinnebusch, this volume) may shed light on the relationship between the termination and initiation phases of translation.

Mindful of the surprises that have sprung from the study of translational control during its long history, we venture to hazard as a final prediction that unlooked-for insights will continue to illuminate fields of research as diverse as memory and maize and as remote as aging is from zygosis.

ACKNOWLEDGMENTS

The authors' work has been supported by grants from the National Institutes of Health (M.B.M. and J.W.B.H.) and from the Medical Research Council and National Cancer Institute of Canada (N.S.).

REFERENCES

Agrawal, M.G. and L.H. Bowman. 1987. Transcriptional and translational regulation of ribosome protein formation during mouse myoblast differentiation. *J. Biol. Chem.* **262:** 4868–4875.

Berg, P. and E.J. Ofengand. 1958. An enzymatic mechanism for linking amino acids to RNA. *Proc. Natl. Acad. Sci.* **44:** 78–86.

Blobel, G. and B. Dobberstein. 1975. Transfer of proteins across membranes. II. Reconstitution of functional rough microsomes from heterologous components. *J. Cell Biol.* **67:** 852–862.

Blobel, G. and D.D. Sabatini. 1971. Ribosome-membrane interaction in eukaryotic cells. In *Biomembranes* (ed. L.A. Mason), vol. 2, pp. 193–195. Plenum Press, New York.

Brenner, S., F. Jacob, and M. Meselson. 1961. An unstable intermediate carrying information from genes to ribosomes for protein synthesis. *Nature* **190:** 576–581.

Bruns, G.P. and I.M. London. 1965. The effect of hemin on the synthesis of globin. *Biochem. Biophys. Res. Commun.* **18:** 236–242.

Campbell, P.N. and T.S. Work. 1953. Biosynthesis of proteins. *Nature* **171:** 997–1001.

Cervera, M., G. Dreyfuss, and S. Penman. 1981. Messenger RNA is translated when associated with the cytoskeletal framework in normal and VSV-infected HeLa cells. *Cell* **23:** 113–120.

Chapeville, F., F. Lipmann, G. von Ehrenstein, B. Weisblum, W.J. Ray, and S. Benzer. 1962. On the role of soluble ribonucleic acid in coding for amino acids. *Proc. Natl. Acad. Sci.* **48:** 1086–1092.

Clemens, M.J., A. Galpine, S.A. Austin, R. Panniers, E.C. Henshaw, R. Duncan, J.W.B. Hershey, and J.W. Pollard. 1987. Regulation of polypeptide chain initiation in Chinese hamster ovary cells with a temperature-sensitive leucyl-tRNA synthetase: Changes in phosphorylation of initiation factor eIF-2 and in the activity of the guanine nucleotide exchange factor GEF. *J. Biol. Chem.* **262:** 767–771.

Crick, F.H.C. 1958. On protein synthesis. *Symp. Soc. Exp. Biol.* **12:** 138–163.

Denny, P.C. and A. Tyler. 1964. Activation of protein biosynthesis in non-nucleate fragments of sea urchin eggs. *Biochem. Biophys. Res. Commun.* **14:** 245–249.

Devillers-Thiery, A., T. Kindt, G. Scheele, and G. Blobel. 1975. Homology in aminoterminal sequence of precursors to pancreatic secretory proteins. *Proc. Natl. Acad. Sci.* **72:** 5016–5020.

Drysdale, J.W. and H.N. Munro. 1965. Failure of actinomycin D to prevent induction of liver apoferritin after iron administration. *Biochim. Biophys. Acta* **103:** 185–188.

Duncan, R. and E.H. McConkey. 1982. Rapid alterations in initiation rate and recruitment of inactive RNA are temporally correlated with S6 phosphorylation. *Eur. J. Biochem.* **123:** 539–544.

Eboué-Bonis, D., A.M. Chambaut, P. Volfin, and H. Clauser. 1963. Action of insulin on the isolated rat diaphragm in the presence of actinomycin D and puromycin. *Nature* **199:** 1183–1184.

Ehrenfeld, E. and T. Hunt. 1971. Double-stranded poliovirus RNA inhibits initiation of protein synthesis by reticulocyte lysates. *Proc. Natl. Acad. Sci.* **68:** 1075–1078.

Ehrenfeld, E. and H. Lund. 1977. Untranslated vesicular stomatitis virus messenger RNA after poliovirus infection. *Virology* **80:** 297–308.

Endo, T. and B. Nadal-Ginard. 1987. Three types of muscle-specific gene expression in fusion-blocked rat skeletal muscle cells: Translational control in EGTA-treated cells. *Cell* **49:** 515–526.

Engelberg-Kulka, H. and R. Schoulaker-Schwartz. 1994. Regulatory implications of translational frameshifting in cellular gene expression. *Mol. Microbiol.* **11:** 3–8.

Fan, H. and S. Penman. 1970. Regulation of protein synthesis in mammalian cells. II. Inhibition of protein synthesis at the level of initiation during mitosis. *J. Mol. Biol.* **50:** 655–670.

Farrell, P.J., K. Balkow, T. Hunt, R.J. Jackson, and H. Trachsel. 1977. Phosphorylation of initiation factor eIF-2 and the control of reticulocyte protein synthesis. *Cell* **11:** 187–200.

Frydman, J., E. Nimmesgern, K. Ohtsuka, and F.U. Hartl. 1994. Folding of nascent polypeptide chains in a high molecular mass assembly with molecular chaperones. *Nature* **370:** 111–117.

Garren, L.D., R.R. Howell, G.M. Tomkins, and R.M. Crocco. 1964. A paradoxical effect of actinomycin D: The mechanism of regulation of enzyme synthesis by hydrocortisone. *Proc. Natl. Acad. Sci.* **52:** 1121–1129.

Gehrke, L. and J. Ilan. 1987. Regulation of messenger RNA translation at the elongation step during estradiol-induced vitellogenin synthesis in avian liver. In *Translational regulation of gene expression* (ed. J. Ilan), pp. 165–186. Plenum Press, New York.

Geoghegan, T., S. Cereghini, and G. Brawerman. 1979. Inactive mRNA-protein complexes from mouse sarcoma-180 ascites cells. *Proc. Natl. Acad. Sci.* **76:** 5587–5591.

Godefroy-Colburn, T. and R.E. Thach. 1981. The role of mRNA competition in regulating translation. IV. Kinetic model. *J. Biol. Chem.* **256:** 11762–11773.

Gros, F., H. Hiatt, W. Gilbert, G.G. Kurland, R.W. Risebrough, and J.D. Watson. 1961. Unstable ribonucleic acid revealed by pulse labelling of *Escherichia coli. Nature* **190:** 581–585.

Gross, P.R. and G.H. Cousineau. 1963. Effects of actinomycin D on macromolecule synthesis and early development in sea urchin eggs. *Biochem. Biophys. Res. Commun.* **10:** 321–326.

Gross, P.R., L.I. Malkin, and W.A. Moyer. 1964. Templates for the first proteins of embryonic development. *Proc. Natl. Acad. Sci.* **51:** 407–414.

Hansen, W.J., V.R. Lingappa, and W.J. Welch. 1994. Complex environment of nacent polypeptide chains. *J. Biol. Chem.* **269:** 26610–26613.

Hardesty, B., R. Miller, and R. Schweet. 1963. Polyribosome breakdown and hemoglobin synthesis. *Proc. Natl. Acad. Sci.* **50:** 924–931.

Henshaw, E.C., C.A. Hirsch, B.E. Morton, and H.H. Hiatt. 1971. Control of protein synthesis in mammalian tissues through changes in ribosome activity. *J. Biol. Chem.* **246:** 436–446.

Heywood, S.M. 1970. Specificity of mRNA binding factor in eukaryotes. *Proc. Natl. Acad. Sci.* **67:** 1782–1788.

Hoagland, M.B., P.C. Zamecnik, and M.L. Stephenson. 1959. A hypothesis concerning the roles of particulate and soluble ribonucleic acids in protein synthesis. In *A symposium on molecular biology* (ed. R.E. Zirkle), pp. 105–114. University of Chicago Press, Chicago, Illinois.

Hodge, L.D., E. Robbins, and M.D. Scharff. 1969. Persistence of messenger RNA through mitosis in HeLa cells. *J. Cell Biol.* **40:** 497–507.

Howe, J.G. and J.W.B. Hershey. 1984. Translational initiation factor and ribosome association with the cytoskeletal framework fraction from HeLa cells. *Cell* **37:** 85–93.

Hultin, T. 1961. Activation of ribosomes in sea urchin eggs in response to fertilization. *Exp. Cell Res.* **25:** 405–417.

Hultin, T. and G. Beskow. 1956. The incorporation of ^{14}C-L-leucine into rat liver proteins in vitro visualized as a two-step reaction. *Exp. Cell Res.* **11:** 664–666.

Humphreys, T. 1969. Efficiency of translation of messenger-RNA before and after fertilization in sea urchins. *Dev. Biol.* **20:** 435–458.

———. 1971. Measurements of messenger RNA entering polysomes upon fertilization of sea urchin eggs. *Dev. Biol.* **26:** 201–208.

Jacob, F.C. and J. Monod. 1961. Genetic regulatory mechanisms in the synthesis of proteins. *J. Mol. Biol.* **3:** 318–356.

Kinniburgh, A.J., M.D. McMullen, and T.E. Martin. 1979. Distribution of cytoplasmic Poly(A^+)RNA sequences in free messenger ribonucleoprotein and polysomes of mouse ascites cells. *J. Mol. Biol.* **132:** 695–708.

Kosower, N.S., G.A. Vanderhoff, B. Benerofe, T. Hunt, and E.M. Kosower. 1971. Inhibition of protein synthesis by glutathione disulfide in the presence of glutathione. *Biochem. Biophys. Res. Commun.* **45:** 816–821.

Kruh, J. and H. Borsook. 1956. Hemoglobin synthesis in rabbit reticulocytes *in vitro*. *J. Biol. Chem.* **220:** 905–915.

Lamfrom, H. and P.M. Knopf. 1964. Initiation of haemoglobin synthesis in cell-free systems. *J. Mol. Biol.* **9:** 558–575.

Lawrence, J.B. and R.H. Singer. 1986. Intracellular localization of messenger RNAs for cytoskeletal proteins. *Cell* **45:** 407–415.

Lee, G.T.-Y. and D.L. Engelhardt. 1979. Peptide coding capacity of polysomal and nonpolysomal messenger RNA during growth of animal cells. *J. Mol. Biol.* **129:** 221–233.

Leibowitz, R. and S. Penman. 1971. Regulation of protein synthesis in HeLa cells. III. Inhibition during poliovirus infection. *J. Virol.* **8:** 661–668.

Lisowska-Bernstein, B., M.E. Lamm, and P. Vassalli. 1970. Synthesis of immunoglobulin heavy and light chains by the free ribosomes of a mouse plasma cell tumor. *Proc. Natl. Acad. Sci.* **66:** 425–432.

Lodish, H.F. 1976. Translational control of protein synthesis. *Annu. Rev. Biochem.* **45:** 39–72.

Lodish, H.F. and M. Jacobsen. 1972. Regulation of hemoglobin synthesis. Equal rates of translation and termination of α- and β-globin chains. *J. Biol. Chem.* **247:** 3622–3629.

Lodish, H.F. and N.D. Zinder. 1966. Mutants of the bacteriophage f2 VIII, control mechanisms for phage-specific syntheses. *J. Mol. Biol.* **19:** 333–348.

Maggio, R., M.L. Vittorelli, A.M. Rinaldi, and A. Monroy. 1964. *In vitro* incorporation of amino acids into proteins stimulated by RNA from unfertilized sea urchin eggs. *Biochem. Biophys. Res. Commun.* **15:** 436–441.

Marks, P.A., E.R. Burka, and D. Schlessinger. 1962. Protein synthesis in erythroid cells. I. Reticulocyte ribosomes active in stimulating amino acid incorporation. *Proc. Natl. Acad. Sci.* **48:** 2163–2171.

Martin, T.E. and F.G. Young. 1965. An *in vitro* action of human growth hormone in the presence of actinomycin D. *Nature* **208:** 684–685.

Mathews, M.B. 1970. Tissue-specific factor required for the translation of a mammalian viral RNA. *Nature* **228:** 661–663.

Mathews, M.B., T. Hunt, and A. Brayley. 1973. Specificity of the control of protein synthesis by Haemin. *Nat. New Biol.* **243:** 230–233.

McCormick, W. and S. Penman. 1969. Regulation of protein synthesis in HeLa cells; translation at elevated temperatures. *J. Mol. Biol.* **39:** 315–333.

Meyuhas, O., E.A. Thompson, Jr., and R.P. Perry. 1987. Glucocorticoids selectively inhibit translation of ribosomal protein mRNAs in P1798 lymphosarcoma cells. *Mol. Cell. Biol.* **7:** 2691–2699.

Milstein, C., G.G. Brownlee, T.M. Harrison, and M.B. Mathews. 1972. A possible precursor of immunoglobulin light chains. *Nat. New Biol.* **239:** 117–120.

Monroy, A., R. Maggio, and A.M. Rinaldi. 1965. Experimentally induced activation of the ribosomes of the unfertilized sea urchin egg. *Proc. Natl. Acad. Sci.* **54:** 107–111.

Nemer, M. 1962. Interrelation of messenger polyribonucleotides and ribosomes in the sea urchin egg during embryonic development. *Biochem. Biophys. Res. Commun.* **8:** 511–515.

Nirenberg, M.W. and J.H. Matthaei. 1961. The dependence of cell-free protein synthesis in *E. coli* upon naturally occurring or synthetic polyribonucleotides. *Proc. Natl. Acad. Sci.* **47:** 1588–1602.

Ouellette, A.J., C.P. Ordahl, J. Van Ness, and R.A. Malt. 1982. Mouse kidney non-polysomal messenger ribonucleic acid: Metabolism, coding function, and translational activity. *Biochemistry* **21:** 1169–1177.

Owen, C.H., D.J. DeRosier, and J. Condeelis. 1992. Actin crosslinking protein EF-1a of *Dictyostelium discoideum* has a unique bonding rule that allows square-packed bundles. *J. Struct. Biol.* **109:** 248–254.

Palmiter, R.D. 1974. Differential rates of initiation on conalbumin and ovalbumin messenger ribonucleic acid in reticulocyte lysates. *J. Biol. Chem.* **249:** 6779–6787.

———. 1975. Quantitation of parameters that determine the rate of ovalbumin synthesis. *Cell* **4:** 189–197.

Pelham, H.R.B. and R.J. Jackson. 1976. An efficient mRNA-dependent translation system from reticulocyte lysates. *Eur. J. Biochem.* **67:** 247–256.

Penman, S. and D. Summers. 1965. Effects on host cell metabolism following synchronous infection with poliovirus. *Virology* **27:** 614–620.

Penman, S., K. Scherrer, Y. Becker, and J.E. Darnell. 1963. Polyribosomes in normal and poliovirus-infected HeLa cells and their relationship to messenger RNA. *Proc. Natl. Acad. Sci.* **49:** 654–662.

Roper, M.D. and W.D. Wicks. 1978. Evidence for acceleration of the rate of elongation of tyrosine aminotransferase nascent chains by dibutyryl cyclic AMP. *Proc. Natl. Acad. Sci.* **75:** 140–144.

Rosbash, M. 1972. Formation of membrane-bound polysomes. *J. Mol. Biol.* **65:** 413–422.

Siekevitz, P. and P.C. Zamecnik. 1951. In vitro incorporation of 1-^{14}C-DL-alanine into proteins of rat-liver granular fractions. *Fed. Proc.* **10:** 246–247.

Singer, R.H. 1993. RNA zipcodes for cytoplasmic addresses. *Curr. Biol.* **3:** 719–721.

Spirin, A.S. and M. Nemer. 1965. Messenger RNA in early sea-urchin embryos: Cytoplasmic particles. *Science* **150:** 214–217.

Steward, D.L., J.R. Schaeffer, and R.M. Humphrey. 1968. Breakdown and assembly of polyribosomes in synchronized chinese hamster cells. *Science* **161:** 791–793.

Summers, D.F. and J.V. Maizel. 1967. Disaggregation of HeLa cell polysomes after infection with poliovirus. *Virology* **31:** 550–552.

Summers, D.F., J.V. Maizel, and J.E. Darnell. 1965. Evidence for virus-specific noncapsid proteins in poliovirus-infected HeLa cells. *Proc. Natl. Acad. Sci.* **54:** 505–513.

Thomas, G.P. and M.B. Mathews. 1984. Alterations of transcription and translation in HeLa cells exposed to amino acid analogues. *Mol. Cell. Biol.* **4:** 1063–1072.

Tomkins, G.M., L.D. Garren, R.R. Howell, and B. Peterkofsky. 1965. The regulation of enzyme synthesis by steroid hormones: The role of translation. *J. Cell. Comp. Physiol.*

66: 137–151.

Tsung, K., S. Inouye, and M. Inouye. 1989. Factors affecting the efficiency of protein synthesis in *Escherichia coli:* Production of a polypeptide of more than 6000 amino acid residues. *J. Biol. Chem.* **264:** 4428–4433.

Walden, W.E., T. Godefroy-Colburn, and R.E. Thach. 1981. The role of mRNA competition in regulating translation. I. Demonstration of competition in vivo. *J. Biol. Chem.* **256:** 11739–11746.

Walter, P. and G. Blobel. 1981. Translocation of proteins across the endoplasmic reticulum. III. Signal recognition protein (SRP) causes signal sequence-dependent and site-specific arrest of chain elongation that is released by microsomal membranes. *J. Cell Biol.* **9:** 557–561.

Warner, J.R., P.M. Knopf, and A. Rich. 1963. A multiple ribosomal structure in protein synthesis. *Proc. Natl. Acad. Sci.* **49:** 122–129.

Waxman, H.S. and M. Rabinowitz. 1966. Control of reticulocyte polyribosome content and hemoglobin synthesis by heme. *Biochim. Biophys. Acta* **129:** 369–379.

Wiedmann, B., H. Sakai, T.A. Davis, and M. Wiedmann. 1994. A protein complex required for signal-sequence-specific sorting and translocation. *Nature* **370:** 434–440.

Wilt, F.H. and T. Hultin. 1962. Stimulation of phenylalanine incorporation by polyuridylic acid in homogenates of sea urchin eggs. *Biochem. Biophys. Res. Commun.* **9:** 313–317.

Witherell, W.G., J.M. Gott, and O.C. Uhlenbeck. 1991. Specific interaction between RNA phage coat proteins and RNA. *Prog. Nucleic Acid Res. Mol. Biol.* **40:** 185–220.

Wolin, S.L. and P. Walter. 1988. Ribosome pausing and stacking during translation of a eukaryotic mRNA. *EMBO J.* **7:** 3559–3569.

Yenofsky, R., I. Bergmann, and G. Brawerman. 1982. Messenger RNA species partially in a repressed state in mouse sarcoma ascites cells. *Proc. Natl. Acad. Sci.* **79:** 5876–5880.

Yisraeli, J.K., S. Sokol, and D.A. Melton. 1990. A two-step model for the localization of maternal mRNA in *Xenopus* oocytes: Involvement of microtubules and microfilaments in the translocation and anchoring of Vg1 mRNA. *Development* **108:** 289–298.

Zamecnik, P.C. 1960. Historical and current aspects of the problem of protein synthesis. *Harvey Lect.* **54:** 256–281.

Zhong, T. and K.T. Arndt. 1993. The yeast SIS1 protein, a DnaJ homolog, is required for the initiation of translation. *Cell* **73:** 1175–1186.

2

The Pathway and Mechanism of Eukaryotic Protein Synthesis

William C. Merrick
Department of Biochemistry
School of Medicine
Case Western Reserve University
Cleveland, Ohio 44106

John W.B. Hershey
Department of Biological Chemistry
School of Medicine, University of California
Davis, California 95616

Knowledge of the detailed mechanism of protein synthesis is essential for understanding translational controls. We are primarily concerned with how the various macromolecules of the translational apparatus interact to promote the process of protein synthesis in eukaryotic cells. The entire process is divided conveniently into three phases: initiation, elongation, and termination. This chapter focuses on how the soluble factors, namely, initiation factors, elongation factors, and release factors, catalyze the sequential binding and reaction of aminoacyl-tRNAs to ribosomes as dictated by the template messenger RNA. Details of transfer RNA structure and aminoacylation and ribosome structure lie outside the scope of this chapter. For reviews on the formation and functions of aminoacyl-tRNAs, see Carter (1993) and Söll (1993); see also the chapter by Wool (this volume) on ribosomal proteins and a book on ribosome structure/function (Nierhaus et al. 1993).

Insight into the process of protein synthesis emerged primarily from biochemical studies that utilized radioactively labeled amino acids and fractionated lysates derived from either bacterial or mammalian cells. The major macromolecular components were identified by purifying proteins and nucleic acids required to reconstitute translation in the test tube. Surprisingly, bacterial genetic approaches contributed only modestly to the identification of the greater than 200 macromolecular components that comprise the translational apparatus. Because the biochemical approach was so fruitful, subsequent in vitro studies on how these molecules interact proceeded rapidly. It is only recently that genetic studies with the yeast *Saccharomyces cerevisiae* or experiments using recombinant DNA techniques have enabled researchers to examine the mechanism of protein synthesis in vivo. Most of the information provided in the sections that follow is based on in vitro biochemical studies, and only in a few instances have the views been confirmed by in vivo experiments. Therefore, the reader is cautioned that the pathways proposed are working models and that corrections and fine tuning of these pathways are anticipated in the future. While attempting to describe the complex pathways clearly and simply, we shall point out those areas that are not yet firmly established. Because translational controls operate most frequently during the initiation phase, a description of this pathway is provided in greater detail.

THE INITIATION PHASE

During the process of initiation, the translational apparatus selects an mRNA and forms a ribosome initiation complex in which the anticodon of the initiator Met-tRNA$_i$ interacts with the initiator codon. This interaction precisely establishes the reading frame of the mRNA. The binding of Met-tRNA$_i$ and mRNA to ribosomes is promoted by at least 11 initiation factors (eIFs), and energy in the form of ATP and GTP hydrolysis is consumed. Initiation occurs either on mRNA in polysomes or with an as yet untranslated mRNA in the form of a messenger ribonucleoprotein particle (mRNP). We shall first consider the major structural features of mRNAs that are recognized during initiation and then discuss the five major steps of the pathway: dissociation of ribosomes into subunits, binding of Met-tRNA$_i$ to the 40S ribosomal subunit, binding of mRNA, recognition of the initiator codon, and junction with the 60S subunit. In the descriptions of each step below, we shall first delineate the major interactions in the pathway and then consider the structure and function of the initiation factors involved. A working model of the initiation pathway is given in Figure 1. A list of mammalian translation factors and their characteristics is provided in Table 1; accession numbers of cloned cDNAs and a comparison with yeast factor sequences are found in Table 2.

Features of mRNA Structure Recognized during Initiation

Different mRNAs are translated with different efficiencies, and the efficiency is nearly always determined by the rate of initiation. Structural elements that determine the innate efficiency of an mRNA include the m^{7}G cap and its accessibility; the initiator codon, which usually is AUG; the context surrounding the AUG, where ACCAUGG is especially strong in metazoans (but not yeasts); the length of the 5′-untranslated region

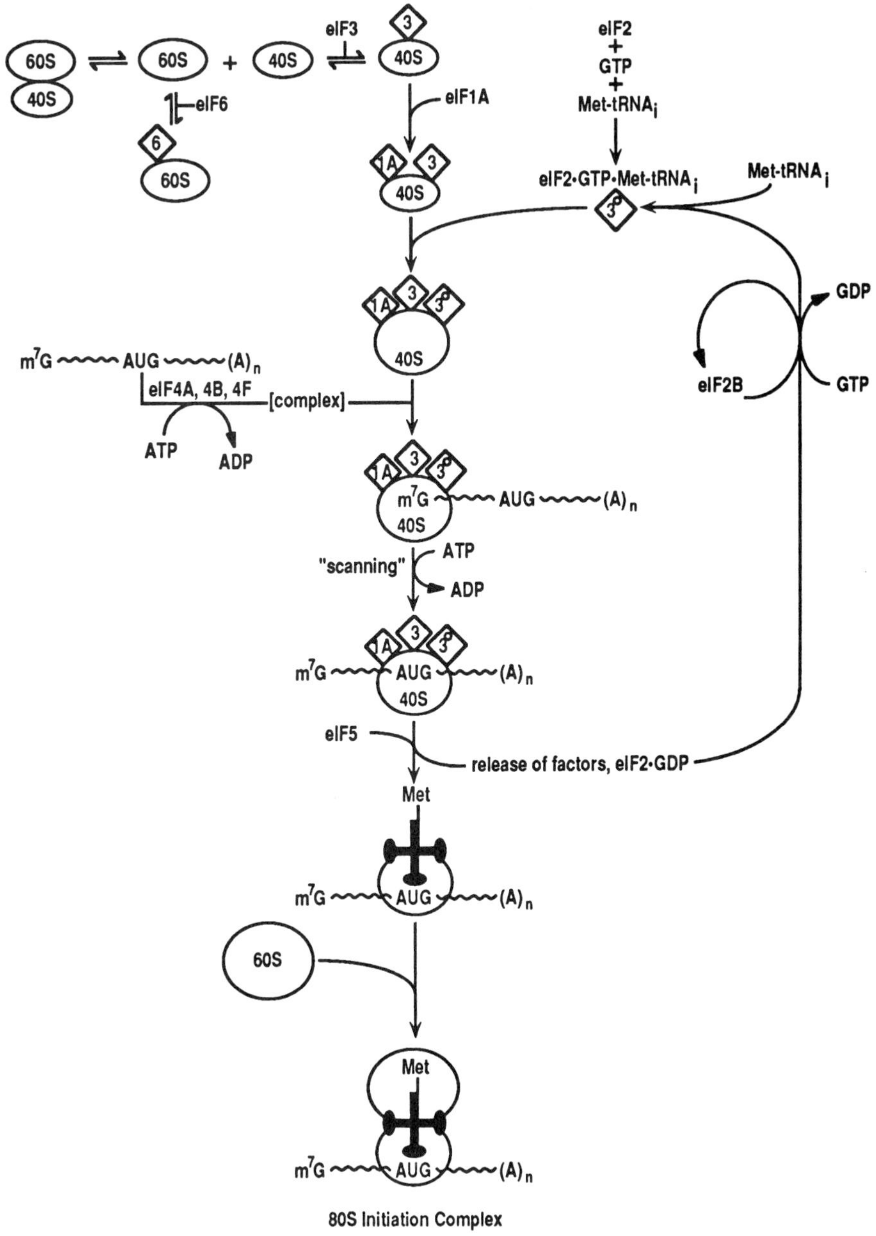

Figure 1 Model of pathway of initiation in eukaryotes. (Adapted from Merrick 1992.)

Table 1 Characteristics of Mammalian Translation Factors

		Mass (kD)			Previous
Name[a]	Subunits	SDS	sequence	Function	names
Initiation factors					
eIF1		15	12.6	enhances initiation complex formation; SUI1 homolog	eIF-1
eIF1A		17	16.5	pleiotropic: ribosomal dissociation; promotes Met-tRNA$_i$ binding	eIF-4C
eIF2		125		binds Met-tRNA$_i$ and GTP	eIF-2
	α	36	36.1	site of phosphorylation on Ser-51	
	β	37	38.4	binds Met-tRNA$_i$	
	γ	52	51.8	binds GTP, Met-tRNA$_i$	
eIF2B		270		guanine nucleotide exchange factor for eIF2	GEF
	α	30	33.7		
	β	39	39.0	binds GTP	
	γ	58		binds ATP	
	δ	66	57.1	binds ATP	
	ε	84	80.2	phosphorylated by CKII and GSK-3	
eIF2C		94		stabilizes ternary complex in presence of RNA	Co-eIF-2A
eIF3		650		dissociates ribosomes; promotes Met-tRNA$_i$ and mRNA binding	eIF-3
	p35	35			α
	p36	36	36.5		β
	p40	40	39.9		γ
	p44	44	46.4		δ
	p47	47			ε
	p66	66		binds RNA	ζ
	p115	115	105.3	major phosphorylated subunit	η
	p170	170			θ

eIF4A		46	44.4	ATPase, helicase, binds RNA	eIF-4A, 4β
eIF4B		80	69.8	binds RNA, promotes helicase activity	eIF-4B
eIF4F		250		binds m^7G caps, helicase	eIF-4, CBPII
	eIF4E	25	25.1	m^7G cap-binding subunit	eIF-4α, CBPI
	eIF4A	46	44.4	ATPase, helicase	eIF-4β
	eIF4G	220	153.4	binds eIF4A, eIF4E, and eIF3	eIF-4γ, p220
eIF5		58	48.9	promotes GTPase with eIF2 and ejection of eIFs	eIF-5
eIF6		25		binds to 60S ribosomes, promotes dissociation	eIF-3A
Elongation factors					
eEF1A		51	50.1	GTP-dependent binding of aminoacyl-tRNAs; GTPase	eEF-1α
eEF1B		114		guanine nucleotide exchange on eEF1A	eEF-1βγδ
	α	30	24.8	possesses GTP exchange activity	δ
	β	36	31.2	possesses GTP exchange activity	γ
	γ	48	50.1		β
eEF2		95	95.3	promotes translocation; GTPase; contains diphthamide	eEF-2
Termination factors					
eRF1		55	48.0	recognizes UAA, UAG, and UGA; promotes peptide hydrolysis	eRF
eRF3			55.8	GTPase; stimulates eRF1 activity	

[a]Translation factors are named here and throughout the volume according to a revised nomenclature suggested by a panel of researchers appointed by the Nomenclature Committee of the International Union of Biochemistry and Molecular Biology and chaired by Dr. Marianne Grunberg-Manago. The new nomenclature has not yet been approved by the IUBMB.

(5′UTR); the amount and stability of secondary structure in the 5′UTR; and the absence of AUGs or open reading frames (ORFs) upstream of the initiator AUG. These elements have been elucidated by analyses of the sequences of known mRNAs and by measuring the efficiencies of translation of mRNAs whose structures have been methodically altered.

m^7G Cap Structure and Accessibility

The m^7G cap structure enhances the translation of mRNAs in cell lysates, in microinjected *Xenopus* oocytes, and in cells transfected with mRNAs. Dependence on the m^7G cap structure varies with the translational system and with the specific mRNA being tested. Stimulation of translation by the m^7G cap structure and poly(A) tail is synergistic (Gallie 1991), although an explanation for how these two structures communicate is not available (see Jacobson, this volume). However, mammalian or yeast cell mRNAs may not be totally dependent on the m^7G cap, as mRNAs expressed by RNA polymerase I or III are translated in vivo (Palmer et al. 1993; Gunnery and Mathews 1995). Inhibition of in vitro protein synthesis by cap analogs also implies that the cap is important. The accessibility of the m^7G cap structure to eIF4F appears to be especially important for efficient protein synthesis (Lawson et al. 1988).

Initiator Codon

Most mRNAs begin translation at the 5′-proximal AUG, although other codons such as ACG and UUG are used infrequently. Initiation at non-AUG codons usually is less efficient in mammalian cells (Kozak 1989a) and does not occur at all in yeast (Cigan and Donahue 1987). The nucleotides surrounding the AUG contribute to initiator codon recognition; the consensus sequence GCCACCAUGG, where the initiator codon is underlined, is considered to be the strongest signal in mammalian cells. Most important is the purine at –3 (the A of the AUG is +1). If the AUG lacks a good consensus sequence, the ribosome may pass over it in a process called "leaky scanning" and initiate further downstream. In yeast, the nucleotide context around the initiator AUG appears to have a far lesser role.

Length of 5′UTR

The length of the 5′UTR of mammalian mRNAs ranges from 3 to more than 1000 nucleotides, although most lengths are about 50–70 nucleotides. There is no strict correlation of efficiency and length; however, leaders that are very short are less efficiently translated (Kozak 1991b). A systematic lengthening of the mRNA leader preceding the CAT reporter gene showed increasing efficiency of translation in vitro, especially at relatively high ionic strength (Kozak 1991a). In yeast, the average length of the 5′UTR is somewhat less, and cases are known where initiation occurs on the AUG directly adjacent to the 5′-cap structure (Ellis et al. 1987).

Secondary Structure

mRNAs lacking strong secondary structure in their 5′UTRs tend to be more efficiently translated than those possessing such structures. When a hairpin structure with a predicted stability of –50 kcal/mole is inserted into the 5′UTR of an mRNA, protein synthesis is inhibited (Pelletier and Sonenberg 1985). The inhibition is more severe if the secondary structure is close to the 5′terminus, as noted above. As most mRNA-coding regions contain considerable secondary structure, it is possible that such secondary structure facilitates initiation at the correct site, presumably by promoting pausing of the scanning 40S ribosome at the AUG. Although secondary structure is clearly important in determining the efficiency of translation, precise structures are difficult to define, in part because of possible long-range RNA-RNA interactions and because bound proteins influence the structures. A rigorous evaluation of the tertiary structure of the whole mRNA is needed but has never been attempted.

Upstream AUGs or ORFs

The presence of an upstream AUG decreases the frequency of initiation at the downstream AUG. This has been shown by introducing such AUGs into the 5′UTRs of mRNAs. Similarly, small upstream ORFs reduce translation of the downstream major ORF. Upstream ORFs found in native mRNAs often have a regulatory role, as described in detail by Geballe and by Hinnebusch (both this volume).

Dissociation of 80S Ribosomes

Because initiation of protein synthesis begins with the small ribosomal subunit, 80S ribosomes first must dissociate into subunits. At physiological Mg^{++} concentrations (~1–2 mM), 80S ribosomes are the predominant species, in dynamic equilibrium with their subunits. The equilibrium position is often measured by sucrose gradient centrifugation, a method that may itself influence the equilibrium position. Alternate methods, such as light scattering, are complicated by the tendency of eukaryotic ribosomal subunits to associate into larger complexes. A consequence of these difficulties is that molecular details of the dissociation of ribosomes are poorly understood.

Three initiation factors contribute to shifting the equilibrium position toward dissociation. eIF1A and eIF3 bind to the 40S ribosomal subunit and prevent its association with the 60S ribosomal subunit (Goumans et al. 1980). eIF6 binds to the 60S ribosomal subunit and inhibits its association with the smaller subunit. eIF1A also reduces the formation of 40S dimers (Goumans et al. 1980). The molar levels of eIF1A (Wei et al. 1995b) and eIF3 (Duncan and Hershey 1983) in HeLa cells are 20–50% of that of total ribosomes and therefore are sufficient to interact stoichiometrically with all of the native 40S and 60S subunits under most conditions.

eIF1A

eIF1A (17–22 kD) has been isolated as a single polypeptide chain from mammalian, plant, and yeast cells and also as a complex with eIF5 (Kemper et al. 1976; Trachsel et al. 1977; Benne et al. 1978). From the cloned cDNA sequences, the protein contains 143 amino acids in both mammals and plants (Dever et al. 1994) and 151 residues in yeast (Wei et al. 1995a) and is one of the most conserved of the initiation factors (Table 2). The factor activity is pleiotropic, stabilizing Met-tRNA$_i$ binding to 40S subunits and promoting mRNA binding (see sections below), as well as functioning in ribosome dissociation. Although no sequence motif of note is discernible in the protein, eIF1A is very polar, with 10 of the first 22 amino-terminal residues being basic and 13 of the last 20 carboxy-terminal residues being acidic. Although eIF1A overall is acidic, the basic amino terminus may explain the observations that eIF1A binds very tightly to RNA (Wei et al. 1995b) and to phosphocellulose (Kemper et al. 1976). eIF1A in plants is the most stable of the initiation factors as greater than 85% of its activity is maintained after heating to 90°C for 5 minutes (Seal et al. 1982).

eIF3

The largest of the eukaryotic initiation factors, mammalian eIF3 contains at least eight different polypeptide chains ranging from 35 kD to 170 kD and has an aggregate molecular mass of about 600,000 daltons (Benne and Hershey 1976; Safer et al. 1976; Schreier et al. 1977). As detailed in

Table 2 Comparison of translation factor sequences from mammals and yeast

	Mammalian			Yeast			
Name	mass (kD)	access. no.	species	gene	mass (kD)	access. no.	% ID
Initiation factors							
eIF1	12.7	Z19334 Z21478	human	*SUI1*	12.3	M77514	58
eIF1A	16.5	L18960	human	*TIF11*	17.4	U11585	65
eIF2α	36.2	J02646	human	*SUI2*	34.7	M25552	58
eIF2β	38.4	M29536	human	*SUI3*	31.6	M21813	42
eIF2γ	51.8	L19161	human	*GCD11*	57.9	L04268	71
eIF2Bα	33.7	U05821	rat	*GCN3*	34.0	M23356	42
eIF2Bβ	39.0	U31880	rat	*GCD7*	42.6	L07116	36
eIF2Bγ				*GCD1*	65.7	X07846	
eIF2Bδ	57.8	Z48225	rat	*GCD2*	70.9	X15658	36
eIF2Bε	80.2	U19511	rat	*GCD6*	81.2	L07115	30
eIF4AI	44.4	S00985	mouse	*TIF1*	45.0	X12813	65
eIF4AII	46.3	S00986	mouse	*TIF2*	44.6	X12814	
eIF4B	69.2	S12566	human	*TIF3*	48.5	X71996	26
eIF4E	25.1	M15353	human	*TIF45*	24.3	M21620	33
eIF4G	153.4	D12686	human	*TIF4631*	107.1	L16923	22
				TIF4632	103.9	L16924	21
eIF5	48.9	L11651	rat	*TIF5*	45.2	L10840	39
eIF5A	16.7	M23419	human	*TIF51*	17.1	M63541	63
Elongation factors							
eEF1A	50.1	X16869	human	*TEF1*	50.1	M15666	81
				TEF2		M15667	81
eEF1Bα	24.8	X60489	human				
eEF1Bβ	31.2	P29692	human	*TEF5*	22.6	D14080	50
eEF1Bγ	50.1	Z11531	human	*TEF3*	47.1	X67917	36
				TEF4	46.7	L01880	36
eEF2	95.3	Z11692	human	*EFT1*	93.3	M59369	67
				EFT2	93.3	M59370	67
Termination factors							
eRF1	48.0	M75715	human	*SUP45*	49.0	X04082	60
eRF3	55.8	X17644	human	*SUP35*	76.6	Y00829	53

the sections below, it is an RNA-binding protein complex that binds to 40S ribosomal subunits in the absence of other translational components, interacts with ternary complexes of eIF2, GTP, and Met-tRNA$_i$, stabilizes Met-tRNA$_i$ binding to 40S subunits, and is required for mRNA binding. Yeast eIF3 also contains eight subunits, but their masses are smaller, ranging from 16 kD to 135 kD (Naranda et al. 1994a). Yeast eIF3 can replace mammalian eIF3 in a mammalian in vitro assay for initiation, indicating strong conservation of function (Naranda et al. 1994a).

Biophysical and electron microscopic analyses of mammalian eIF3 show that it has the shape of a flat triangular prism with side dimensions of 17, 17, and 14 nm with a thickness of 7 nm (Lutsch et al. 1985). However, localization of the subunits in a three-dimensional structure of the factor has not yet been determined. The p66 subunit of mammalian eIF3 and the corresponding p62 subunit of yeast (GCD10) are the proteins primarily responsible for binding to RNA (Garcia-Barrio et al. 1995). The functions of the other subunits are not yet known, except for yeast p16 which is encoded by *SUI1* (T. Naranda et al., in prep.). Mutant forms of *SUI1* allow initiation from a UUG codon in yeast (Yoon and Donahue 1992), thereby implicating eIF3 in initiator codon recognition. A mammalian homolog of SUI1 corresponds to a protein called eIF1 (Kasperaitis et al. 1995), previously implicated in the initiation pathway (Trachsel et al. 1977; Benne et al. 1978). However, there is no evidence indicating that eIF1 is a component of mammalian eIF3.

eIF6

Like eIF3, eIF6 helps provide a pool of 40S subunits for the initiation pathway, although eIF6 binds exclusively to the 60S subunit (Russell and Spremulli 1979; Raychaudhuri et al. 1984). In total, it accounts for about 75% of the "anti-association activity" and by size and function would appear to be the best candidate to be the eukaryotic homolog to bacterial IF3 which binds to the small subunit (30S) and thereby prevents binding of the large subunit (50S). Unfortunately, there has been little work on eIF6 since its discovery and thus there has been no determination of its amino acid sequence that might allow for a more complete comparison of eIF6 and IF3.

Met-tRNA$_i$ Binding to the 40S Subunit

Prior to binding to the 40S ribosomal subunit, Met-tRNA$_i$ forms a ternary complex with eIF2 and GTP. This obligatory intermediate can be prepared in high yield with purified components at physiological concentrations (~1 μM), and its formation is readily monitored by nitrocellulose filtration with radioactively labeled Met-tRNA$_i$ or GTP. Two initiation factors, eIF2C and eIF3, act stoichiometrically to stabilize the ternary complex at low concentrations (<0.1 μM) (Gupta et al. 1990). eIF2C also prevents ternary complex disruption by RNA (Roy et al.

1988). It remains unclear whether these two initiation factors interact with the ternary complex off the ribosome or whether the stabilizations observed derive from interactions normally occurring on the 40S subunit.

GDP inhibits ternary complex formation, as the binary eIF2•GDP complex cannot bind Met-$tRNA_i$. Furthermore, when eIF2 completes a round of initiation, it is released from the ribosome as a binary complex with GDP (see below). For eIF2 to function in another round of initiation, it must exchange the GDP for GTP prior to ternary complex formation. This exchange reaction occurs slowly and requires catalysis by eIF2B. eIF2B binds to eIF2•GDP and promotes the guanine nucleotide exchange reaction (see Trachsel, this volume). Since GDP binds 400-fold more tightly than GTP to eIF2 (Rowlands et al. 1988), ternary complex formation in theory may be partly influenced by the GTP/GDP ratio and thus by the energy charge of the cell.

The ternary complex binds to 40S ribosomal subunits to form a 43S preinitiation complex. The Met-$tRNA_i$•40S complex is moderately stable and its formation can be measured by sucrose gradient centrifugation. Both eIF1A and eIF3 stabilize ternary complex binding and are present in the 43S preinitiation complex (Trachsel et al. 1977; Benne and Hershey 1978; Goumans et al. 1980). GTP carried on eIF2 is not hydrolyzed at the binding step, as nonhydrolyzable analogs of GTP promote Met-$tRNA_i$ binding as well. Met-$tRNA_i$ bound to 40S ribosomes lacking mRNA is detected in cell lysates (Smith and Henshaw 1975), indicating that the 43S preinitiation complex may be an intermediate in the initiation pathway (see Fig. 1). The formation of ternary complexes, their binding to 40S ribosomal subunits, and the regulation of these reactions are described in greater detail by Trachsel (this volume). The formation of ternary complexes, and thus the binding of Met-$tRNA_i$ to 40S ribosomes, is one of the most important sites of translational control.

eIF2

eIF2 has been purified in numerous laboratories from plants, animals, insects, and eukaryotic microorganisms. In essentially all cases, the purified initiation factor comprises three nonidentical subunits with molecular masses of approximately 52 kD (γ), 38 kD (β), and 36 kD (α). In those instances where only two eIF2 subunits (α and γ) are detected, either proteolysis of the β-subunit (Das et al. 1982) or failure to resolve the β and γ subunits by the SDS-gel electrophoresis system (Lloyd et al. 1980) is likely responsible. eIF2 appears to function as a stable trisubunit protein, and its concentration in HeLa cells is about 1 μM (0.5 eIF2 molecules per ribosome) (Duncan and Hershey 1983). It is essential for both Met-tRNA$_i$ and mRNA binding to ribosomes in vitro. In yeast, mutant forms of each of the subunits affect selection of the initiator codon, thereby implicating eIF2 in the process of initiation in intact cells (Donahue et al. 1988; Cigan et al. 1989; T.F. Donahue, pers. comm.).

The sequences of the three subunits are known from mammals and yeast through the cloning of their respective cDNAs or genes. Each subunit is essential for cell viability, as shown in yeast by gene deletion/disruption experiments (Donahue et al. 1988; Cigan et al. 1989; Hannig et al. 1993). The three protein sequences are highly conserved between yeast and humans, ranging from 42% to 71% identity (see Table 2). Furthermore, the human α and β subunits can replace their corresponding yeast counterparts in vivo (M. Kainuma and J. Hershey, unpubl.). Examination of the amino acid sequence motifs suggests specific functions for the subunits. Both human and yeast eIF2γ possess the three consensus elements found in GTP-binding proteins, suggesting that this subunit is responsible for GTP binding. eIF2γ is strongly homologous to bacterial EF1A (formerly called EF-Tu), especially in the G-domain, but also in domain 2 of EF1A. However, GTP analogs cross-link to both the γ and β subunits, suggesting that the β-subunit may be very close to the GTP-binding site. The β-subunit from either humans or yeast possesses a zinc finger motif (but no zinc is detected in the protein) which is involved in eIF2 function since mutations in the motif affect initiator codon selection in yeast (Donahue et al. 1988). eIF2β also contains three tracts of six to eight lysine residues in the amino-terminal half of the protein; their function is not known, but they may be involved in binding to RNA. There are reports that eIF2β also binds to ATP (Dholakia et al. 1989; Anthony et al. 1990; Gonsky et al. 1990), but no ATP-binding sequence elements are discernible in the sequence (see also Flynn et al. 1994). Met-tRNA$_i$ cross-links to both the β and γ subunits (Gaspar et al. 1994), suggesting that both proteins may contribute to its recognition and binding. eIF2α has no distinguishing sequence elements, but Ser-51 is the site of phosphorylation by eIF2α kinases (see Trachsel; Clemens; both this volume). eIF2 phosphorylation is one of the most important mechanisms for repressing global rates of protein synthesis.

eIF2B

Mammalian eIF2B has been characterized in numerous laboratories as a high-molecular-mass complex of five different polypeptide chains: α (26 kD), β (39 kD), γ (58 kD), δ (67 kD), and ε (82 kD). The cDNAs encoding the subunits have been cloned and sequenced (Table 2). eIF2B levels in mammalian cells vary from tissue to tissue or species to species (Oldfield et al. 1994) but are always less than that for eIF2. Furthermore, some differences in subunit mass between species have been noted (Oldfield et al. 1994). The mammalian subunits correspond well to the five subunits in yeast eIF2B identified through their cloned genes: GCD6 (82 kD), GCD2 (71 kD), GCD1 (58 kD), GCD7 (42 kD), and GCN3 (34 kD) (Bushman et al. 1993). The four largest subunits are essential for cell viability, but the α-subunit (GCN3) is not, although disruption of this gene causes a slow-growth phenotype (Hannig and Hinnebusch 1988).

Biochemical characterization of the mammalian initiation factor indicates that it binds ATP on the γ and δ subunits, GTP on the β-subunit, and NADPH (Dholakia et al. 1986, 1989; Oldfield and Proud 1992). Whereas the bound GTP may be the nucleotide that is exchanged for the GDP tightly bound to eIF2, the roles of ATP and NADPH are unclear, although it is possible that these molecules might regulate the activity of eIF2B allosterically. eIF2B activity also is regulated by phosphorylation of the ε-subunit (see Trachsel, this volume) and is an important site of translational control.

Insight into the interaction of eIF2B with eIF2 comes from genetic studies of the regulation of *GCN4* expression in yeast (see Hinnebusch, this volume). The inhibition of eIF2B activity by phosphorylated eIF2 requires the presence of GCN3, making this a regulatory subunit of eIF2B. Reversion analysis of strains carrying a mutation in *GCN2* that leads to constitutive phosphorylation of eIF2 identified mutations in the genes encoding GCN3, GCD2, and GCD7. The results suggest that stable eIF2B binding to phosphorylated eIF2 occurs through direct contacts of these subunits with eIF2α (Vazquez de Aldana and Hinnebusch 1994).

eIF2C

eIF2C stabilizes ternary complexes and 40S preinitiation complexes when RNA is present (Roy et al. 1988). The purified protein exhibits an apparent mass of 94 kD on SDS gels (Gupta et al. 1990) but is very susceptible to proteolysis. Western blot analyses of cell lysates suggest that even higher-molecular-weight forms of the protein may exist in vivo. Since the protein has been shown to function only under in vitro conditions with purified components, its physiological role in the cell remains to be elucidated.

mRNA Binding to Ribosomes and Initiator Codon Recognition

Two kinds of basic mechanisms of mRNA binding to ribosomes exist in eukaryotes: preinitiation complex binding to the 5′ terminus, followed by "scanning" down the mRNA until the initiator codon is recognized, and internal binding of the ribosome at or upstream of the initiator codon. We shall focus exclusively on the more prevalent scanning mechanism here; the internal initiation mechanism is described in detail by Jackson (this volume). For either mechanism, the binding of the 40S ribosomal complex occurs either on an mRNA already being translated (polysomes) or on an mRNA not yet being translated (mRNP particle). The structural features of mRNAs important in the scanning mechanism have been described above. The reader is reminded that mRNAs, either in mRNP particles or in polysomes, are associated with proteins not normally ascribed to the translational apparatus and that these proteins may affect the structure and reactivity of the mRNA in currently unknown ways (see Spirin, this volume).

The first step in the scanning pathway is recognition of the m^7G cap structure by eIF4E (see Sonenberg, this volume). eIF4E, together with eIF4A and eIF4G, is a subunit of the cap-binding protein complex, eIF4F. The affinity of eIF4F for capped mRNA is about 15-fold greater than that of eIF4E alone (Lawson et al. 1988), suggesting that eIF4F also may interact with other regions of the mRNA. Bound eIF4F together with eIF4B possesses ATP-dependent RNA helicase activity (described in detail under the *eIF4A* section below) which melts secondary structure near the 5′terminus of the mRNA. Next, the 40S preinitiation complex binds to the unfolded mRNA through an interaction of the eIF4G subunit with eIF3, the latter already bound to the 40S subunit. It remains to be determined if the RNA helicase activity occurs while the eIF4F-eIF4B complex is bound to the 40S subunit or if the initiation factors alone move down the mRNA, with the 40S complex following passively. Finally, there is no solid evidence that eIF4F and eIF2 bind to eIF3 on the 40S subunit at the same time. Clearly, the molecular details of mRNA binding to ribosomes remain poorly elucidated, despite the fact that this step is an important site of translational control.

There are variations or alternatives to the simple scanning mechanism described above. In one model, eIF4E alone binds the cap and then complexes with eIF4G which already is bound to the 40S ribosome through its interaction with eIF3 (Joshi et al. 1994). It is not certain that scanning involves a linear movement of the 40S complex along the mRNA until the AUG is recognized. Alternatively, a looping model, where the 5′ terminus is bound by eIF4F on the ribosome and the AUG simultaneously interacts with ribosome-bound Met-$tRNA_i$, is compatible with much of the evidence. Hopping or skipping over sequences in the 5′UTR may occur, as is indicated from studies of cauliflower mosaic virus mRNA (Fütterer et al. 1993) and the tripartite leader of late adenovirus mRNAs (see Schneider, this volume). For a detailed discussion of scanning mechanisms, see Jackson (this volume).

The scanning model predicts that the 5′-proximal AUG will serve as the initiator codon, and this is true for more than 90% of mRNAs (Kozak 1989b). Exceptions to the rule are found if the AUG is less than about ten nucleotides from the 5′ terminus (Kozak 1991b; Slusher et al. 1991), which may be due to a requirement for sufficient mRNA length to allow both cap and AUG interactions on the ribosomal surface. Another exception, called leaky scanning (Kozak 1989a), involves the passing over of AUG codons in a context that weakly matches the consensus sequence, ACCAUGG discussed above. The purine at –3 appears to have the most impact of any of the AUG context nucleotides. However, elements of the translational apparatus responsible for recognition of the consensus sequence have not been identified. From genetic studies in the yeast, *S. cerevisiae*, all three subunits of eIF2 and the smallest subunit of eIF3 (see above) are implicated in the AUG recognition process, as mutant forms (called *sui*) of these proteins enable yeast to initiate at a UUG codon (Donahue et al. 1988; Cigan et al. 1989); (T.F. Donahue, pers. comm.). The mutant initiation factors likely either stabilize the imperfect interaction between the Met-$tRNA_i$ anticodon and the UUG or fail to destabilize the interaction. In either case, a reduction in fidelity of initiation occurs.

eIF4A

eIF4A (46 kD) has been purified from a number of mammalian cells as well as from lower eukaryotes including yeast. It is one of the most abundant initiation factors, present in about three copies per ribosome in HeLa cells (Duncan and Hershey 1983). eIF4A is unusual in that it appears to function both as a single polypeptide (Schreier et al. 1977; Benne and Hershey 1978) and as a subunit of the eIF4F complex (Grifo et al. 1983; Conroy et al. 1990), since translation in a highly fractionated system requires both forms. Two mammalian cDNAs have been cloned that encode different forms of the factor which share 91% sequence identity (Nielsen and Trachsel 1988), but a functional difference such as exclusive incorporation into eIF4F could not be demonstrated (Conroy et al. 1990). eIF4A is the first initiation factor for which two separate, functional genes were identified. Two yeast genes encoding eIF4A have been cloned; the yeast factors are identical in sequence and are very similar (65% identity) to mammalian eIF4A. Yeast eIF4A is essential for cell growth, and depletion studies demonstrate that the protein functions in the initiation pathway (Blum et al. 1989). Given the high similarity, it is curious that mammalian eIF4A does not substitute for yeast eIF4A in vivo or in vitro (Prat et al. 1990).

eIF4A is an RNA-dependent ATPase (Grifo et al. 1984; Ray et al. 1985) that participates with eIF4B in bidirectional RNA helicase activity (Rozen et al. 1990). The factor is the prototypical member of the so-called DEAD box proteins that share the conserved DEAD sequence motif involved in ATP binding and that possess helicase activity. There are nine conserved sequence elements in RNA helicase proteins (Linder et al. 1989). Site-directed mutagenic studies of eIF4A indicate that most of these elements are critical for eIF4A function (Pause and Sonenberg 1992; Pause et al. 1993). Sonenberg and co-workers propose that when ATP binds to eIF4A, a conformational change is induced that allows RNA binding dependent on the HRIGRXXR region, and RNA binding in turn induces ATP hydrolysis followed by more stable RNA binding. One of the defective mutant forms of eIF4A is a dominant inhibitor of both cap-dependent and cap-independent (internal initiation) protein synthesis when added to a reticulocyte lysate (Pause et al. 1994a). Addition of exogenous eIF4F, and to a lesser extent wild-type eIF4A, relieves the inhibition. Because mutant eIF4A inhibits the eIF4F-dependent helicase assay, but not the eIF4A-dependent reaction, it is proposed that eIF4A functions primarily as a subunit of eIF4F and that it cycles through the eIF4F complex during translation (Pause et al. 1994a).

eIF4B

Mammalian eIF4B was purified as a homogeneous protein of about 80 kD as determined by SDS-PAGE, but the active form of the factor appears to be a homodimer (Abramson 1987). A mass of 70 kD is calculated from its cloned cDNA (Milburn et al. 1990), whereas plant eIF4B appears to be smaller (59 kD) (Seal et al. 1986; Browning et al. 1989), and the cloned yeast gene encodes a protein of 48.5 kD (Altmann et al. 1993; Coppolecchia et al. 1993). eIF4B is an RNA-binding protein that functions to promote the ATPase and helicase activities of eIF4F and eIF4A (Abramson et al. 1987; Rozen et al. 1990). It also stimulates mRNA binding to ribosomes in highly fractionated assay systems for initiation (Benne and Hershey 1978). Surprisingly, disruption of the yeast eIF4B gene is not lethal (Altmann et al. 1993; Coppolecchia et al. 1993), indicating either that the initiation factor is not essential or that another protein possesses eIF4B-like activity. The initiation factor exhibits multiple isoelectric variants when examined by two-dimensional isoelectric focusing–SDS-PAGE (Duncan and Hershey 1984). All but the most basic variant are phosphoproteins, but phosphorylation may not be the only posttranslational modification responsible for generation of the variant forms.

Structural elements important for RNA binding and helicase activity have been identified by deletion and point mutagenesis experiments. The eIF4B sequence contains an RNA recognition motif that may be involved in RNA binding, but the motif appears not to be the primary determinant, as an arginine-rich region (residues 385–423) has been identified that is esssential for RNA binding and helicase activity (Méthot et al. 1994; Naranda et al. 1994b). It is proposed that eIF4B recognizes the junction region between single- and double-stranded RNAs (Méthot et al. 1994). However, a detailed understanding of the mechanism of RNA helicase activity and the contribution of eIF4B is lacking.

eIF4F

eIF4F comprises three subunits: eIF4E (25 kD), eIF4A (46 kD), and eIF4G (154 kD). The trimeric complex is purified from mammalian cells by classical fractionation methods or by affinity chromatography with m^7G affinity columns. When fractionated on cation exchange columns, an active dimeric form containing only eIF4E and eIF4G may be obtained. Apparently, eIF4A can cycle in and out of the eIF4F complex (Yoder-Hill et al. 1993). Besides a two-subunit form of eIF4F, plants uniquely contain an isozyme form that contains 28-kD and 86-kD subunits (Browning et al. 1987). The subunits appear to be related to eIF4E and eIF4G, respectively, as evidenced by homologous sequences (Allen et al. 1992) and similar properties in biochemical assays of function (Abramson et al. 1988). Frequently, eIF3 and eIF4B copurify with eIF4F, suggesting that a higher-molecular-weight complex of all of these initiation factors may exist in cells.

eIF4E, the subunit responsible for recognition of the m^7G cap structure, can be purified as a single protein that binds to cap analogs with an association constant of 1×10^5 M^{-1} (Carberry et al. 1992). Its protein level in mammalian cells has been reported as either 0.2 copies (Duncan et al. 1987) or 0.02 copies (Hiremath et al. 1985) per ribosome, and eIF4E appears to be the limiting initiation factor. Furthermore, eIF4E is prevented from forming eIF4F complexes by an interaction with two proteins called eIF4E-binding proteins (4E-BPI and II), in effect further reducing the concentration of active protein (Pause et al. 1994b). The view that eIF4E is limiting is supported by the observation that overexpression of eIF4E cDNA or microinjection of the protein causes malignant transformation (Lazaris-Karatzas et al. 1990) or deregulation of cell growth (De Benedetti and Rhoads 1990). eIF4E is phosphorylated on Ser-209 in vivo (Joshi et al. 1995), and at the same site in vitro by protein kinase C (PKC) (Morley et al. 1991). Phosphorylation increases its affinity for caps by a factor of about 3 (Minich et al. 1994). For a detailed discussion of eIF4E phosphorylation, function, and regulation by 4E-BPs, see Sonenberg (this volume). In yeast, eIF4E has been purified, and its gene (*CDC33*) has been cloned and shown to be essential for growth (Altmann et al. 1987). Although the mammalian and yeast proteins share only 33% sequence identity, the mouse eIF4E cDNA can substitute for the yeast gene (Altmann et al. 1989), indicating that the functions of the two proteins are highly conserved.

The 46-kD subunit of eIF4F is identical to the noncomplexed form of eIF4A described above. Both isoforms of eIF4A are present in the eIF4F complex and these proteins exchange readily with the free forms. There is no evidence that the two isoforms function differently in translation.

eIF4G (formerly called p220) shows a complex pattern when eIF4F is examined by SDS-PAGE: Three forms with masses in the range of 210–220 kD are seen. However, the mammalian eIF4G cDNA codes for a protein of 154 kD. The cause for the slow migration of eIF4G in SDS gels is not known, but the region responsible lies near the amino terminus. In yeast, two genes encoding proteins corresponding to mammalian eIF4G have been identified that are only 53% identical to each other (Goyer et al. 1993). Disruption of both yeast genes results in a failure of cells to grow. eIF4F was originally characterized by its ability to reverse the inhibition of translation of capped mRNAs in extracts from poliovirus-infected cells (Grifo et al. 1983) and the eIF4G subunit was shown to be cleaved at Arg-486 (Lamphear et al. 1995). (For a full discussion of the effects of eIF4G cleavage by picornavirus proteases, see Ehrenfeld [this volume].) Both eIF4E and eIF4G in the eIF4F complex are phosphorylated by either PKC or the multipotential S6 kinase, leading to activation of the factor (Morley et al. 1991), but neither the sites of phosphorylation on eIF4G nor the specific effects on the activity of this subunit are known. A major function of eIF4G during mRNA binding appears to be to bring the eIF4E subunit close to eIF3. The amino-terminal half of eIF4G binds to eIF4E (Mader et al. 1995), whereas the carboxy-terminal half binds to eIF4A and eIF3 (Lamphear et al. 1995). Cleavage by poliovirus protease 2A and other picornavirus proteases in effect separates the eIF4E- and eIF3-binding regions, resulting in a failure to bring these two initiation factors into close proximity. It is also suggested that eIF4G has an RNA-binding domain that enables nonspecific binding to RNA.

When the synthesis of eIF4E is reduced in mammalian cells by an antisense strategy, protein synthesis is inhibited and polysomes become smaller (Joshi-Barve et al. 1992), all consistent with a role for this protein in the initiation phase. Interestingly, the accumulation of eIF4G also is reduced in these cells. eIF4G synthesis may represent one of the few instances in eukaryotes where the expression of the polypeptide chain is translationally autoregulated (Goyer et al. 1993). The 5′UTR of the yeast and human mRNAs contains upstream AUGs and a polypyrimidine tract, elements also found in the 5′UTR of poliovirus mRNA. Although not proven, this suggests that the mRNA might be translated via an internal initiation rather than a cap-dependent scanning mechanism. With a possibly reduced dependence on eIF4F, the mRNA would be more efficiently translated under conditions where the level of eIF4F activity is low.

Junction with the 60S Subunit

Following formation of the 40S preinitiation complex, where Met-tRNA$_i$ interacts with the AUG initiator codon of the mRNA, eIF5 recognizes the complex, presumably binds to the 40S complex, and promotes the hydrolysis of the GTP carried by eIF2 (Trachsel et al. 1977; Benne and Hershey 1978). eIF5 can be thought of as a fidelity factor, assessing the correctness of the Met-tRNA$_i$–initiator codon interaction. When GTP hydrolysis occurs, bound initiation factors exhibit reduced affinity for the 40S ribosome, as shown in in vitro sucrose gradient experiments with purified 40S subunits and initiation factors. eIF2 is released as a binary complex with GDP, whose recycling has been described above. After the initiation factors have dissociated, the 60S ribosomal subunit joins the 40S initiation complex to form the 80S initiation complex, which is then competent to enter the elongation phase of translation. Whether or not eIF6 has a role in 60S junction is unclear. However, evidence from *S. cerevisiae* indicates that the poly(A)-binding protein (PABP) may be involved in the 60S junction step (Sachs and Davis 1989). Loss of PABP activity results in reduced amounts of polysomes and to an increase in "half-mers" (polysomes with a 40S ribosome, possibly located at the initiator codon waiting for the junction step). Furthermore, some extragenic suppressors of a PABP mutant code for 60S ribosomal proteins (Sachs and Davis 1990).

It is possible that eIF1A also has a role during the junction step, since it has been isolated as a complex with the 150-kD form of eIF5 (Schreier et al. 1977). This possibility has not been pursued experimentally, however. eIF5A also is implicated late in the initiation pathway, since it stimulates the synthesis of methionyl-puromycin, dependent on the formation of 80S initiation complexes. eIF5A is uniquely modified on Lys-50 by the polyamine, spermidine, to form a novel amino acid residue called hypusine. Both the protein and its hypusine modification are required for cell viability in yeast (Schnier et al. 1991). However, depletion of the protein in yeast does not affect the rate of protein synthesis significantly (Kang and Hershey 1994), casting doubt that eIF5A has an essential role in the translation of most, if not all, mRNAs.

eIF5

eIF5 was first purified as a 125–150-kD protein that strongly stimulates assays for initiation (Schreier et al. 1977; Benne et al. 1978; Merrick 1979). Later, a smaller form (58 kD by SDS gels) with a similar activity was purified (Raychaudhuri et al. 1985). Antibodies to the 58-kD form do not recognize the larger form, indicating that the smaller form is not due to partial proteolysis of the larger (Ghosh et al. 1989). Convincing evidence for the existence of the smaller form was generated by the cloning of a mammalian cDNA encoding a 45-kD protein identical to the smaller form (Chakravarti et al. 1993). Later, a yeast gene was cloned which encodes a similar protein that is essential for cell viability (Chakravarti and Maitra 1993). The mammalian and yeast protein sequences are 39% identical, indicating moderate conservation of structure. Since preparations of the 150-kD form do not contain sufficient amounts of the 45-kD form to account for the factor activity, it is unclear whether or not the two size forms of eIF5 are isozymes or whether the larger form is an artifact. Although the eIF5 sequence possesses weak homology with GTP-binding proteins, it is likely that the factor induces a GTPase activity in eIF2 rather than directly catalyzing GTP hydrolysis. Both the 45-kD and 150-kD forms of eIF5 are phosphoproteins, but a possible role for phosphorylation in regulating the factor's activity has not been demonstrated.

Additional Features of the Initiation Pathway

Although the pathway of initiation depicted in Figure 1 is supported by much experimental evidence, the scheme should be considered as a working model rather than as dogma. Much of the evidence was generated by studying intermediates that are stable to analysis by sucrose gradient centrifugation and that were constructed in vitro with purified components. It is possible that some of these intermediates are not on the major pathway of initiation or that unidentified intermediates are kinetically favored but are not sufficiently stable to be detected. In particular, it is not certain that Met-$tRNA_i$ binds to the 40S subunit prior to mRNA, that eIF2 recycling occurs off the ribosome rather than while bound, and that the initiation factors function as separate entities rather than as a larger macromolecular complex. Nevertheless, analysis of cell lysates shows the presence of 40S complexes containing Met-$tRNA_i$ but not mRNA (Smith and Henshaw 1975), indicating that at least for some mRNAs, Met-$tRNA_i$ binding precedes mRNA binding. In contrast, studies of the yeast *GCN4* system indicate that when reinitiation occurs on an mRNA, Met-$tRNA_i$ binding can occur on the 40S subunit that is already scanning along the mRNA (see Hinnebusch, this volume). Thus, the reader is encouraged to consider all possible pathways when studying the translation of specific mRNAs.

What step of initiation is rate-limiting? Because 40S complexes that contain mRNA are not readily detected in reticulocyte lysates, it is thought that mRNA binding is rate-limiting and that the junction step is rapid once the scanning 40S subunit recognizes the initiator codon. On the other hand, analyses of ribosomes paused on mRNAs show that ribosomes remain at the initiator codon for a relatively long time, rather than entering quickly into the elongation phase of translation (Wolin and Walter 1988). There is no solid information for how long the 40S subunit needs to scan from the 5′cap to the initiator codon or for how rapidly RNA helicase action occurs.

Finally, a few comments are in order about how eukaryotic and prokaryotic initiation phases differ. Identification of the initiator codon in bacteria is based on two major features of the mRNA: a lack of secondary structure in the region to which the ribosome binds and RNA-RNA interactions between the ribosomal RNA and the mRNA, as well as between the ribosome-bound initiator tRNA and the initiator codon. In contrast, the scanning mechanism for initiation in eukaryotes relies on recognition of the m^7G-capped mRNA and the proximity of the initiator codon to the 5′terminus. The RNA-RNA interaction between the initiator tRNA and the initiator codon is preserved, but there appears to be no rRNA-mRNA interaction. Secondary structure also is less important because of the presence of an RNA helicase activity in cell cytoplasms. These differences may have arisen in part because transcription and translation are coupled spatially and termporally in prokaryotes but are uncoupled in eukaryotes. For more detailed comparisons of the prokaryotic and eukaryotic pathways, see Jackson (this volume) and Voorma (this volume).

THE ELONGATION AND TERMINATION PHASES

The elongation phase involves the sequential addition of amino acid residues to the carboxy-terminal end of the nascent peptide. It involves four major steps: binding of the aminoacyl-tRNA in the ribosomal A site, GTP hydrolysis and guanine nucleotide exchange on eEF1A (previously called eEF-1α), peptide bond formation, and translocation of the mRNA and peptidyl-tRNA on the ribosomal surface. The reaction cycle is quite rapid; in mammalian cells, up to eight amino acids are incorporated per second per elongating ribosome. The ribosome has three tRNA-binding sites: the A-site, where the aminoacyl-tRNA first binds; the P-site, where peptidyl-tRNA binds after the translocation reaction; and the E-site, where the stripped tRNA binds before it is ejected from the ribosome. The mechanism of this cyclic process has been studied in great detail in prokaryotes (for review, see Moldave 1985; Slobin 1990). It is thought, but hardly proven, that the process is very similar in eukaryotes. Since there are few studies of the eukaryotic pathway, we shall describe a pathway and mechanism based on studies of the bacterial system but using the eukaryotic elongation factors that promote the cycle (Fig. 2).

Aminoacyl-tRNA Binding

Aminoacyl-tRNAs form a ternary complex with eEF1A and GTP prior to interacting with the ribosome. Ternary complexes bind to the A-site, with the cognate ternary complex (those whose tRNA anticodon matches the mRNA's codon exposed in the A-site) binding at a rate comparable or slightly faster than that of noncognate complexes. Correct selection of the ternary complex is achieved in part because the rate of release of noncognate complexes is much greater than that of cognate complexes.

GTP Hydrolysis and Guanine Nucleotide Exchange on eIF1A

The GTPase reaction is catalyzed by eEF1A when bound to the ribosome. The reaction is rapid, with cognate complexes undergoing GTP hydrolysis much faster than noncognate complexes. However, some noncognate ternary complexes may undergo GTP hydrolysis before they dissociate from the ribosome. Following GTP hydrolysis, the resulting eEF1A•GDP complex leaves the ribosome in what is likely the rate-

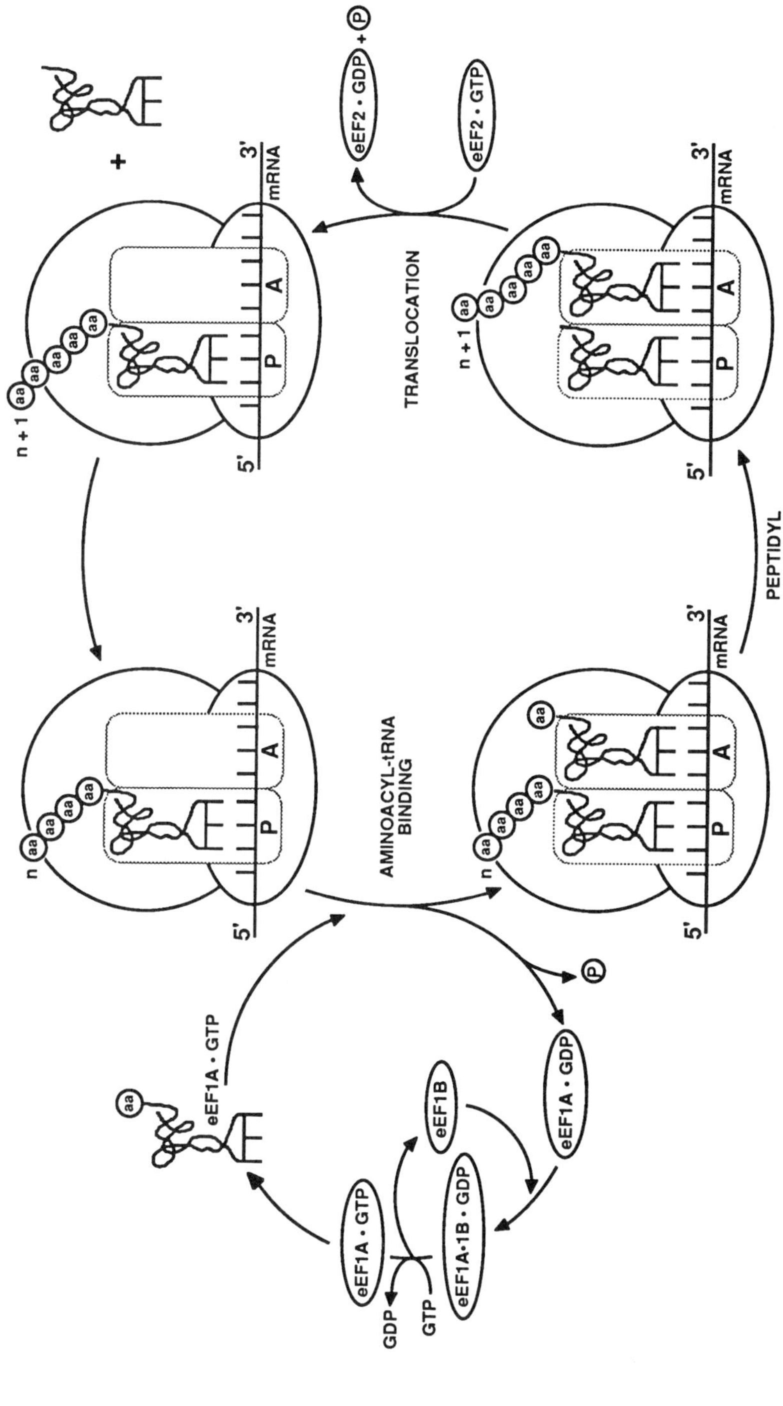

Figure 2 Model of pathway of elongation in eukaryotes. (Adapted from Merrick 1992.)

limiting step of the elongation phase. Prior to ejection of the factor, the whole ternary complex of aminoacyl-tRNA•eEF1A•GDP has the opportunity to dissociate, thus providing the possibility of a proofreading mechanism for the ejection of noncognate ternary complexes. For eEF1A to promote another round of aminoacyl-tRNA binding, the GDP must be exchanged for GTP. The exchange reaction is slow and is facilitated by eEF1B (see below). Interestingly, overexpression of eEF1A in yeast obviates the requirement for eEF1B (Kinzy and Woolford 1995).

Peptide Bond Formation

Following release of eEF1A•GDP, the aminoacyl-tRNA in the A-site is now competent to react with the peptidyl-tRNA in the P-site. The reaction is an O to N shift of the peptidyl moiety, which uses the high-energy content of the ester bond in the peptidyl-tRNA to promote the reaction. The products are a new peptidyl-tRNA that is one amino acid residue longer and a stripped tRNA. The peptide-forming reaction is catalyzed by the peptidyl transferase center of the 60S ribosome and may utilize both rRNA and ribosomal protein catalytic elements. The peptidyl- and stripped tRNAs, either during peptidyl transfer or immediately thereafter, shift their positions at their aminoacyl ends from the A and P sites to the E and P sites on the 60S subunit while keeping their anticodon ends complexed with the mRNA in the P and A sites of the 40S subunit, respectively. This partial translocation occurs without the expenditure of energy and likely is driven by a stronger binding site for the peptidyl portion of the tRNA in the 60S subunit P-site.

Translocation

Translocation of the mRNA and the anticodon ends of the two tRNAs is promoted by the binding of eEF2•GTP. Following GTP hydrolysis on the ribosome, eEF2•GDP is ejected. GTP hydrolysis presumably is not required for the actual shift since replacement of GTP with the nonhydrolyzable analog GMP-PNP results in the peptidyl-tRNA becoming reactive with puromycin, a drug that mimics the aminoacyl end of aminoacyl-tRNA in the A-site and therefore is considered to be diagnostic of peptidyl-tRNA in the P-site. Following translocation and GTP hydrolysis, peptidyl-tRNA lies in the P-site, stripped tRNA is in the E-site, and the A-site is vacant, awaiting the binding of the next ternary complex. The presence of stripped tRNA in the E-site appears to decrease the binding affinity of a ternary complex in the A-site. In yeast, ejection of stripped tRNA is promoted by eEF3, thereby enhancing ternary complex binding (Triana et al. 1993).

Termination

When a termination codon is exposed in the A-site, there is no ternary complex available to recognize the codon. Instead, a release factor binds and promotes hydrolysis of the peptidyl-tRNA in the P-site, presumably by the peptidyl transferase center. Only a single release factor, eRF1 (55 kD), is found in mammalian cells and this factor must therefore recognize all three termination codons (UAA, UAG, UGA). In fact, eRF1 recognizes four nucleotides in the A-site (Brown et al. 1990), the termination codon triplet plus an adjacent downstream nucleotide. In bacteria, RF1 and RF2 each functions as a heterodimer with RF3, a protein with GTPase activity that stimulates the rate of peptide release. In yeast, two proteins that were originally recognized as omnisuppressors, called SUP45 and SUP35, promote peptide chain termination. Finally, in bacteria, another factor called RF4 (formerly RRF) catalyzes the release of the ribosome from the mRNA (Ryoji et al. 1981), but no corresponding eukaryotic factor has yet been identified.

eEF1A

eEF1A (53 kD) promotes the GTP-dependent binding of aminoacyl-tRNAs to ribosomes. It is the most studied of the eukaryotic translation factors, in part because it is one of the most abundant cell proteins (1–3% of the soluble protein). More than 100 eEF1A sequences have been determined, thereby enabling the construction of phylogenetic trees. Whereas little structural information is available on eEF1A, a structure for the homologous bacterial factor, EF1A (formerly called EF-Tu), has been determined by X-ray crystallography for both the GTP and GDP conformations (Kjeldgaard and Nyborg 1992; Berchtold et al. 1993; Kjeldgaard et al. 1993). Furthermore, a model of the ternary complex with Phe-tRNA and GTP and the structure of archaebacterial EF1A are expected to be solved soon (J. Nyborg and M. Kjeldgaard; A. Zagari; both pers. comm.). As with the bacterial factor, the mammalian protein appears to consist of three domains. The first domain, the GTP-binding domain, is very similar to Ras and is highly conserved among the translation factors that bind GTP (Cavallius and Merrick 1993). The second and third domains have not yet been well characterized for function, but they are thought to be involved in the binding of the aminoacyl-tRNA. All three domains are sites for posttranslational modifications with methylated lysines appearing in all three domains, in contrast to the single methylated lysine in bacterial EF1A in the first domain (Cavallius and Merrick 1993). The modifications are neither highly conserved (Cavallius et al. 1993) nor necessary for function, as site-directed mutagenesis of the modified lysines (to arginines) yields a functional protein with similar properties in vivo and in vitro (J. Cavallius and W. Merrick, unpubl.). Higher eukaryotic eEF1A species also contain a posttranslational modification that has not yet been found on any other protein, namely, glycerylphosphorylethanolamine, which is attached to glutamic acid residues 301 and 374 in the mammalian protein (Dever et al. 1989; Whitehart et al. 1989).

eEF1A is encoded by multiple functional genes and there are at least 40 pseudogenes in the human diploid genome (Uetsuki et al. 1989). Analyses from *S. cerevisiae* (Nagata et al. 1976; Schirmaier and Philippsen 1984) and a number of higher eukaryotes including humans (Knudsen et al. 1993; Lee et al. 1993) suggest that there are three functional eEF1A species: one associated with embryogenesis, one specific for muscle, and one present in nonmuscle adult tissues. In addition to tissue-specific expression, eEF1A also is coordinately expressed with ribosomal proteins due to a polypyrimidine tract at the 5′end of its mRNA (see Meyuhas et al., this volume).

Besides simply promoting aminoacyl-tRNA binding to the A-site of ribosomes, eEF1A is implicated in translational fidelity (Song et al. 1989) and cell transformation (Tatsuka et al. 1992). Identifying physiologically relevant associations of eEF1A with other cellular components is problematic as eEF1A is a very abundant, basic protein. eEF1A has been reported to be part of mRNP particles (Greenberg and Slobin 1987), a part of the valyl-tRNA synthetase complex (Motorin et al. 1988; Bec et al. 1989), an actin-binding protein (Yang et al. 1990), a microtubule-severing protein (Shiina et al. 1994), a calmodulin-binding protein (Durso and Cyr 1994), a didemnin-binding protein (Crews et al. 1994; P.L. Toogood, pers. comm.), an endoplasmic-reticulum-associated protein (Hayashi et al. 1989), a mitotic-apparatus-associated protein (Ohta et al. 1990), an activator of phosphatidylinositol 4-kinase in carrots (Yang et al. 1993), and a component of the ubiquitin-dependent pathway for the degradation of amino-terminally blocked proteins (Gonen et al. 1994). The validity of these activities and how they might be regulated to provide the appropriate cellular response remain to be determined.

eEF1B

The factor (formerly eEF-1βγδ) catalyzes the exchange of GDP for GTP on eEF1A in a reaction thought to be similar to that of eIF2B activity on eIF2 (see above). The complex has been purified from a variety of eukaryotic cells and contains one copy each of the α-subunit (30 kD), β-subunit (36 kD), and γ-subunit (48 kD) (note: some older nomenclatures have reversed the order of subunits in relation to their masses). The cDNAs encoding all three subunits have been cloned and sequenced (Table 2). eEF1B is usually present in smaller amounts than eEF1A, although the exact ratio varies from tissue to tissue and from species to species (Nagata et al. 1976). The β-subunit is most responsible for facilitating the guanine nucleotide exchange reaction (Riis et al. 1990).

The α and β subunits display a high degree of sequence identity in their carboxyl portions, with the larger β-subunit also containing a leucine zipper motif in its amino-terminal region (Cormier et al. 1993; Sanders et al. 1993). However, none of the subunits shows an apparent homology with bacterial EF1B (formerly called EF-Ts) which facilitates guanine nucleotide exchange on bacterial EF1A. Although bacterial EF1A is homologous throughout its structure to the two eukaryotic factors that bind GTP and aminoacyl-tRNA, namely, eIF2γ and eEF1A, there appears to be no equivalent homolog of EF1B.

eEF2

eEF2 functions as a single polypeptide chain with a mass of about 95 kD. It is responsible for promoting the GTP-dependent translocation step and contains a GTP consensus element in its sequence similar to that of the other guanine-nucleotide-binding translation factors (Cavallius and Merrick 1993). eEF2 also contains a unique posttranslational modification; His-715 is modified to diphthamide (2-[3-carboxyamido-3-(trimethylamino)propyl]histidine) (van Ness et al. 1988; Rapp et al. 1989). On the basis of genetic studies in yeast, the modification requires at least five different steps (Chen and Bodley 1988) but is not essential for yeast growth (Omura et al. 1989). The His-715 residue must be completely modified to diphthamide in order to serve as a substrate for ADP-ribosylation by diphtheria toxin, which results in inhibition of eEF2 activity and protein synthesis. As eEF2 appears to be the only cytosolic protein that is ADP-ribosylated by diphtheria toxin, modification with [^{14}C]NAD has been used frequently as a method to quantitate levels of eEF2. Since ADP-ribosylation inactivates eEF2, it has been suggested that such regulation of eEF2 activity may occur in normal (uninfected) cells (Iglewski and Fendrick 1990). In yeast, a systematic substitution of diphthamide by the other 19 amino acids yielded three categories: mutants that are inhibitory to growth even in the presence of wild-type eEF2, mutants that are nonfunctional and noninhibitory, and mutants that can replace wild-type eEF2 but with a slower growth rate and reduced resistance to heat (Kimata and Kohno 1994). The results suggest that there may be an important function served by diphthamide in special circumstances. On the basis of structural comparisons of the EF1A•GMP-PNP•Phe-tRNA ternary complex and EF2, it appears that the diphthamide residue in eEF2 is in a position corresponding to the anticodon in the ternary complex (J. Nyborg and M. Kjeldgaard, pers. comm.). If so, the diphthamide residue may be an important contact site between eEF2 and the ribosome.

eEF3

eEF3 (125 kD) appears to be a unique translation factor that is found only in yeast and fungi (Belfield and Tuite 1993). Together with yeast or mammalian eEF1A and eEF2, it is required for protein synthesis with yeast ribosomes but not with mammalian ribosomes (Skogerson and Wakatama 1976; Kamath and Chakraburtty 1986). This has prompted the suggestion that a corresponding activity in higher eukaryotes may be part of the ribosome. Alternatively, eEF3 may be unique to yeast and fungi and may therefore be an excellent target for antifungal drugs (Harford 1995). eEF3 exhibits ribosome-dependent ATPase and GTPase activities and promotes polyphenylalanine synthesis in the yeast system in the presence of ATP (100 μM) and GTP (10–100 μM), which are normal physiological concentrations of the nucleotides (Skogerson 1979; Uritani and Miyazaki 1988). Experiments implicate eEF3 in promoting the ejection of stripped tRNA from the exit site of the ribosome following translation, thereby enhancing the binding of an aminoacyl-tRNA to the A-site (Triana et al. 1993).

Analysis of the amino acid sequence identifies three regions that are homologous to known proteins. The amino-terminal domain is homologous to *E. coli* ribosomal protein S5, the central domain contains the bipartite nucleotide-binding sequences found in the "ATP-binding cassette," and the carboxy-terminal domain contains a region homologous to aminoacyl-tRNA synthetases (Belfield and Tuite 1993). These sequence elements are compatible with eEF3 functioning in aminoacyl-tRNA binding. Further work will be necessary to determine whether or not there is indeed a ribosomal counterpart in higher eukaryotes.

eRF1

Mammalian eRF1 was purified from rabbits as a 110-kD complex with subunits of approximately 55 kD (Tate and Caskey 1974; Konecki et al. 1977). In vitro assays indicated a requirement for GTP and a nucleotide 3′ of the stop codon. An early report on the cloning of eRF1 found that it is highly similar to Trp-tRNA synthetase, but this has been refuted (Timchenko and Caskey 1994). On the basis of amino acid sequencing and in vitro assays of the expressed protein, a positive human cDNA has been identified recently (Frolova et al. 1994). As the expressed protein recognizes stop codons in a GTP-independent manner, it is called eRF1. A second component, analogous to bacterial RF3 and therefore termed eRF3, has been identified that causes eRF1 to be more efficient in assays that have become GTP-dependent. In yeast, the corresponding proteins appear to be SUP45 (49 kD; eRF1) and SUP35 (77 kD; eRF3), which on genetic grounds are suggested to interact. SUP35 is an eEF1A-like protein that contains the three GTP-binding consensus elements.

Reinitiation following Termination

Following termination, the 80S ribosome is thought to be released from the polysome and then to dissociate so that it can participate in another round of protein synthesis. Because most eukaryotic mRNAs are functionally monocistronic, even when possessing another ORF downstream from the first, reinitiation by the terminating ribosome usually does not occur. However, reinitiation after translation of short upstream ORFs is known to occur in the case of the yeast *GCN4* mRNA (see Hinnebusch, this volume) and in a number of other eukaryotic mRNAs (see Geballe, this volume). Cases where reinitiation occurs following translation of a long ORF are rarer, but a low frequency of expression of downstream cistrons has been reported, for example, in artificial constructs expressing xanthine-guanine phosphoribosyl transferase (Peabody and Berg 1986) and dihydrofolate reductase (Kaufman et al. 1987). Characterization of the *GCN4* mRNA indicates that two types of terminations occur, one resulting in a low frequency of reinitiation and the other allowing quite high reinitiation frequencies. How these two types of terminations differ mechanistically and how subsequent reinitiation occurs remain to be explained.

PERSPECTIVES

It is apparent that the pathway of eukaryotic protein synthesis is known in rich detail at a biochemical level. Although ester bonds are formed and converted into peptide bonds and GTP and ATP are hydrolyzed, a striking characteristic of the pathway is that very few covalent chemical bonds are made or broken. Instead, most of the reactions in the pathway involve noncovalent interactions of proteins and RNA. Since the rules governing protein-protein and protein-RNA interactions are known only imprecisely, it is not yet possible to understand all of the reactions at the molecular level. Detailed structural knowledge of the ribosome and its complexes remains to be elucidated, and there is a need for sophisticated kinetic studies of the various steps of the initiation, elongation, and termination pathways. We also can anticipate that our current understanding of the mechanism of protein synthesis will be confirmed or modified by in vivo studies employing yeast genetics. As our comprehension of the pathway increases, we shall be better able to elucidate the regulatory mechanisms that control gene expression at the translational level.

REFERENCES

Abramson, R.D. 1987. "mRNA-Specific eukaryotic initiaton factors." Ph.D. thesis. Case Western Reserve University, Cleveland, Ohio.

Abramson, R.D., T.E. Dever, T.G. Lawson, B.K. Ray, R.E. Thach, and W.C. Merrick. 1987. The ATP-dependent interaction of eukaryotic initiation factors with mRNA. *J. Biol. Chem.* **262:** 3826–3832.

Abramson, R.D., K.S. Browning, T.E. Dever, T.G. Lawson, R.E. Thach, J.M. Ravel, and W.C. Merrick. 1988. Initiation factors that bind mRNA: A comparison of mammalian factors with wheat germ factors. *J. Biol. Chem.* **263:** 5462–5467.

Allen, M.L., A.M. Metz, R.T. Timmer, R.E. Rhoads, and K.S. Browning. 1992. Isolation and sequence of the cDNAs encoding the subunits of the isozyme form of wheat protein synthesis initiation factor 4F. *J. Biol. Chem.* **267:** 23232–23236.

Altmann, M., C. Handschin, and H. Trachsel. 1987. mRNA cap-binding protein: Cloning of the gene encoding protein synthesis initiation factor eIF-4E from *Saccharomyces cerevisiae. Mol. Cell. Biol.* **77:** 998–1003.

Altmann, M., P.P. Mueller, J. Pelletier, N. Sonenberg, and H. Trachsel. 1989. The mammalian translation initiation factor 4E substitutes for its yeast homologue *in vivo. J. Biol. Chem.* **264:** 12145–12147.

Altmann, M., P.P. Mueller, B. Wimmer, F. Ruchti, S. Lanker, and H. Trachsel. 1993. A *Saccharomyces cerevisiae* homologue of mammalian translation initiation factor 4B contributes to RNA helicase activity. *EMBO J.* **12:** 3997–4003.

Anthony, D.D., T.G. Kinzy, and W.C. Merrick. 1990. Affinity labeling of eukaryotic initiation factor 2 and elongation factor 1αβγ with GTP analogs. *Arch. Biochem. Biophys.* **281:** 157–162.

Bec, G., P. Kerjan, X.D. Zha, and J.P. Waller. 1989. Valyl-tRNA synthetase from rabbit liver. I. Purification as a heterotypic complex in association with elongation factor 1. *J. Biol. Chem.* **264:** 21131–21137.

Belfield, G.P. and M.F. Tuite. 1993. Translation elongation factor 3: A fungus-specific translation factor? *Mol. Microbiol.* **9:** 411–418.

Benne, R. and J.W.B. Hershey. 1976. Purification and characterization of initiation factor IF-E3 from rabbit reticulocytes. *Proc. Natl. Acad. Sci.* **73:** 3005–3009.

———. 1978. The mechanism of action of protein synthesis initiation factors from rabbit reticulocytes. *J. Biol. Chem.* **253:** 3078–3087.

Benne, R., M.L. Brown-Luedi, and J.W.B. Hershey. 1978. The purification and characterization of protein synthesis initiation factors eIF-1, eIF-4C, eIF-4D and eIF-5 from rabbit reticulocytes. *J. Biol. Chem.* **253:** 3070–3077.

Berchtold, H., L. Reshetnikova, C.O.A. Reiser, N.K. Schirmer, M. Sprinzl, and R. Hilgenfeld. 1993. Crystal structure of active elongation factor Tu reveals major domain rearrangements. *Nature* **365:** 126–132.

Blum, S., M. Mueller, S.R. Schmid, P. Linder, and H. Trachsel. 1989. Translation in *Saccharomyces cerevisiae*: Initiation factor 4A-dependent cell-free system. *Proc. Natl. Acad. Sci.* **86:** 6043–6046.

Brown, C.M., P.A. Stockwell, C.N.A. Trotman, and W.P. Tate. 1990. Sequence analysis suggests that tetra-nucleotides signal the termination of protein synthesis in eukaryotes. *Nucleic Acids Res.* **18:** 6339–6345.

Browning, K.S., S.R. Lax, and J.M. Ravel. 1987. Identification of two messenger RNA cap binding proteins in wheat germ. *J. Biol. Chem.* **262:** 11228–11232.

Browning, K.S., L. Fletcher, S.R. Lax, and J.M. Ravel. 1989. Evidence that the 59 kDa protein synthesis initiation factor from wheat germ is functionally similar to the 80 kDa initiation factor 4B from mammalian cells. *J. Biol. Chem.* **264:** 8491–8494.

Bushman, J.L., A.I. Asuru, R.L. Matts, and A.G. Hinnebusch. 1993. Evidence that GCD6 and GCD7, translational regulators of *GCN4*, are subunits of the guanine nucleotide exchange factor for eIF-2 in *Saccharomyces cerevisiae*. *Mol. Cell. Biol.* **8:** 808–820.

Carberry, S.E., D.E. Friedland, R.E. Rhoads, and D.J. Goss. 1992. Binding of protein synthesis initiation factor 4E to oligoribonucleotides: Effects of cap accessibility and secondary structure. *Biochemistry* **31:** 1427–1432.

Carter, C.W.J. 1993. Cognition, mechanism, and evolutionary relationships in aminoacyl-tRNA synthetases. *Annu. Rev. Biochem.* **62:** 715–748.

Cavallius, J. and W.C. Merrick. 1993. Eukaryotic translation factors which bind GTP. In *GTPases in biology* (ed. B. Dickey and L. Birnbaumer), pp. 115–130. Springer-Verlag, Berlin.

Cavallius, J., W. Zoll, K. Chakraburtty, and W.C. Merrick. 1993. Characterization of yeast EF-1α: Non-conservation of post-translational modifications. *Biochim. Biophys. Acta* **1163:** 75–80.

Chakravarti, D. and U. Maitra. 1993. Eukaryotic translation initiation factor 5 from *Saccharomyces cerevisiae*: Cloning, characterization and expression of the gene encoding the 45,346 dalton protein. *J. Biol. Chem.* **268:** 10524–10533.

Chakravarti, D., T. Maiti, and U. Maitra. 1993. Isolation and immunochemical characterization of eukaryotic translation initiation factor 5 from *Saccharomyces cerevisiae*. *J. Biol. Chem.* **268:** 5754–5762.

Chen, J.C. and J.W. Bodley. 1988. Biosynthesis of diphthamide in *Saccharomyces cerevisiae*. *J. Biol. Chem.* **263:** 11692–11696.

Cigan, A.M. and T.F. Donahue. 1987. Sequence and structural features associated with translational initiator regions in yeast—A review. *Gene* **59:** 1–18.

Cigan, A.M., E.K. Pabrish, L. Feng, and T.F. Donahue. 1989. Yeast translation initiation suppressor *sui2* encodes the alpha subunit of eukaryotic initiation factor 2 and shares sequence identity with the human alpha subunit. *Proc. Natl. Acad. Sci.* **86:** 2784–2788.

Conroy, S.C., T.E. Dever, C.L. Owens, and W.C. Merrick. 1990. Characterization of the 46,000 dalton subunit of eIF-4F. *Arch. Biochem. Biophys.* **282:** 363–371.

Coppolecchia, R., P. Buser, A. Stotz, and P. Linder. 1993. A new yeast translation initiation factor suppresses a mutation in the eIF-4A RNA helicase. *EMBO J.* **12:** 4005–4011.

Cormier, P., H.B. Osborne, J. Morales, T. Bassez, O. Minella, R. Poulhe, R. Bellé, and O. Mulner-Lorillon. 1993. Elongation factor 1 contains two homologous guanine-nucleotide exchange proteins as shown from the molecular cloning of beta and delta subunits. *Nucleic Acids Res.* **21:** 743.

Crews, M.C., J.L. Collins, W.S. Lane, M.L. Snapper, and S.L. Schreiber. 1994. GTP-dependent binding of the antiproliferative agent didemnin to elongation factor 1α. *J. Biol. Chem.* **269:** 15411–15414.

Das, A., M.K. Bagchi, P. Ghosh-Dastidar, and N.K. Gupta. 1982. Protein synthesis in rabbit reticulocytes: A study of peptide chain initiation using native and β-subunit-depleted eukaryotic initiation factor 2. *J. Biol. Chem.* **257:** 1282–1288.

De Benedetti, A. and R.E. Rhoads. 1990. Overexpression of eukaryotic protein synthesis initiation factor 4E in HeLa cells results in aberrant growth and morphology. *Proc. Natl. Acad. Sci.* **87:** 8212–8216.

Dever, T.E., C.E. Costello, C.L. Owens, and W.C. Merrick. 1989. Location of seven post-translational modifications in rabbit EF-1α including dimethyllysine, trimethyllysine and glycerylphosphorylethanolamine. *J. Biol. Chem.* **264:** 20518–20525.

Dever, T.E., C.-L. Wei, L.A. Benkowski, K. Browning, W.C. Merrick, and J.W.B. Hershey. 1994. Determination of the amino acid sequence of rabbit, human and wheat germ protein synthesis factor eIF-4C by cloning and chemical sequencing. *J. Biol. Chem.* **269:** 3212–3218.

Dholakia, J.N., B.R. Francis, B.E. Haley, and A.J. Wahba. 1989. Photoaffinity labeling of the rabbit reticulocyte guanine nucleotide exchange factor and eukaryotic initiation factor 2 with 8-azidopurine nucleotides: Identification of GTP- and ATP-binding domains. *J. Biol. Chem.* **264:** 20638–20642.

Dholakia, J.N., T.C. Mueser, C.L. Woodley, L.J. Parkhust, and A.J. Wahba. 1986. The association of NADPH with the guanine nucleotide exchange factor from rabbit reticulocytes: A role of pyridine dinucleotides in eukaryotic polypeptide chain initiation. *Proc. Natl. Acad. Sci.* **83:** 6746–6750.

Donahue, T.F., A.M. Cigan, E.K. Pabrich, and B.C. Valavicius. 1988. Mutations at a Zn(II) finger motif in the yeast eIF-2 beta gene alter ribosomal start-site selection during the scanning process. *Cell* **54:** 621–632.

Duncan, R. and J.W.B. Hershey. 1983. Identification and quantitation of levels of protein synthesis initiation factors in crude HeLa cell lysates by two-dimensional polyacrylamide gel electrophoresis. *J. Biol. Chem.* **258:** 7228–7235.

———. 1984. Heat shock-induced translational alterations in HeLa cells. *J. Biol. Chem.* **259:** 11882–11889.

Duncan, R., S.C. Milburn, and J.W.B. Hershey. 1987. Regulated phosphorylation and low abundance of HeLa cell initiation factor eIF-4F suggest a role in translational control: Heat shock effects on eIF-4F. *J. Biol. Chem.* **262:** 380–388.

Durso, N.A. and R.J. Cyr. 1994. A calmodulin-sensitive interaction between microtubules and a higher plant homolog of elongation factor 1α. *Plant Cell* **6:** 893–905.

Ellis, S.R., A.K. Hopper, and N.C. Martin. 1987. Amino-terminal extension generated from an upstream AUG codon is not required for mitochondrial import of yeast N^2,N^2-dimethylguanosine-specific tRNA methyl transferase. *Proc. Natl. Acad. Sci.* **84:** 5172–5176.

Flynn, A., I.N.P. Shatsky, C.G. Proud, and A. Kaminsky. 1994. The RNA-binding properties of protein synthesis initiation factor eIF-2. *Biochim. Biophys. Acta* **1219:** 292–301.

Frolova, L., X. Le Goff, H.H. Rasmussen, S. Cheperegin, G. Drugeon, M. Kress, I. Arman, A.-L. Haenni, J.E. Celis, M. Philippe, J. Justesen, and L. Kisselev. 1994. A highly conserved eukaryotic protein family possessing properties of polypeptide chain release factor. *Nature* **372:** 701–703.

Fütterer, J., Z. Kiss-Laszlo, and T. Hohn. 1993. Nonlinear ribosome migration on cauliflower mosaic virus 35S RNA. *Cell* **73:** 789–802.

Gallie, D.R. 1991. The cap and poly(A) tail function synergistically to regulate mRNA translational efficiency. *Genes Dev.* **5:** 2108–2116.

Garcia-Barrio, M.T., T. Naranda, C.R. Vazquez de Aldana, R. Cuesta, A.G. Hinnebusch, J.W.B. Hershey, and M. Tamame. 1995. GCD10, a translational repressor of *GCN4*, is the RNA-binding subunit of eukaryotic translation initiation factor-3. *Genes Dev.* **9:** 1781–1796.

Gaspar, N.J., T.G. Kinzy, B.J. Scherer, M. Hümbelin, J.W.B. Hershey, and W.C. Merrick. 1994. Translation initiation factor eIF2: Cloning and expression of the human cDNA encoding the γ-subunit. *J. Biol. Chem.* **269:** 3415–3422.

Ghosh, S., J. Chevesich, and U. Maitra. 1989. Further characterization of eukaryotic initiation factor 5 from rabbit reticulocytes. *J. Biol. Chem.* **264:** 5134–5140.

Gonen, H., C. Smith, N.R. Siegel, W.C. Merrick, K. Chakraburtty, A.L. Schwartz, and A. Ciechanover. 1994. Protein synthesis elongation factor EF-1α is essential for ubiquitin-dependent degradation of N-α-acetylated proteins and may be substituted for by the bacterial elongation factor EF-Tu. *Proc. Natl. Acad. Sci.* **91:** 7648–7654.

Gonsky, R., M.A. Lebendiker, R. Harary, Y. Banai, and R. Kaempfer. 1990. Binding of ATP to eukaryotic initiation factor 2: Differential modulation of mRNA binding activity and GTP-dependent binding of methionyl-$tRNA_f$. *J. Biol. Chem.* **265:** 9083–9089.

Goumans, H., A. Thomas, A. Verhoeren, H.O. Voorma, and R. Benne. 1980. The role of eIF-4C in protein synthesis initiation complex formation. *Biochim. Biophys. Acta* **608:** 39–46.

Goyer, C., M. Altmann, H.S. Lee, A. Blanc, H. Trachsel, and H. Sonenberg. 1993. *TIF4631* and *TIF4632*: Two yeast genes encoding the high molecular weight subunits of the cap binding protein complex (eIF-4F) contain an RRM-like sequence and carry out an essential function. *Mol. Cell. Biol.* **13:** 4860–4874.

Greenberg, J.R. and L.I. Slobin. 1987. Eukaryotic elongation factor Tu is present in mRNA-protein complexes. *FEBS Lett.* **224:** 54–58.

Grifo, J.A., R.D. Abramson, C.A. Satler, and W.C. Merrick. 1984. RNA-stimulated ATPase activity of eukaryotic initiation factors. *J. Biol. Chem.* **259:** 8648–8654.

Grifo, J.A., S.M. Tahara, M.A. Morgan, A.J. Shatkin, and W.C. Merrick. 1983. New initiation factor activity required for globin mRNA translation. *J. Biol. Chem.* **258:** 5804–5810.

Gunnery, S. and M.B. Mathews. 1995. Functional mRNA can be generated by RNA polymerase III. *Mol. Cell. Biol.* **15:** 3597–3607.

Gupta, N.K., A.L. Roy, M.K. Nag, T.G. Kinzy, S. MacMillan, R.E. Hileman, T.E. Dever,

W. Wu, W.C. Merrick, and J.W.B. Hershey. 1990. New insights into an old problem: Ternary complex (Met-tRNA$_f$.eIF-2.GTP) formation in animal cells. In *Post-transcriptional control of gene expression* (ed. J.E.G. McCarthy and M.F. Tuite), pp. 521–526. Springer-Verlag, Berlin.

Hannig, E.M. and A.G. Hinnebusch. 1988. Molecular analysis of *GCN3*, a translational activator of *GCN4*: Evidence for posttranslational control of *GCN3* regulatory function. *Mol. Cell. Biol.* **8:** 4808–4820.

Hannig, E.M., A.M. Cigan, B.A. Freeman, and T.G. Kinzy. 1993. *GCD11*, a negative regulator of *GCN4* expression, encodes the γ subunit of eIF-2 in *Saccharomyces cerevisiae*. *Mol. Cell. Biol.* **13:** 506–520.

Harford, J.B. 1995. Translation-targeted therapeutics for viral diseases. *Gene Expr.* **4:** 357–367.

Hayashi, Y., R. Urade, S. Utsumi, and M. Kito. 1989. Anchoring of peptide elongation factor EF-1α by phosphatidylinositol at the endoplasmic reticulum membrane. *J. Biochem.* **106:** 560–563.

Hiremath, L.S., N.R. Webb, and R.E. Rhoads. 1985. Immunological detection of the messenger RNA cap-binding protein. *J. Biol. Chem.* **260:** 7843–7849.

Iglewski, W.J. and J.L. Fendrick. 1990. ADP ribosylation of elongation factor 2 in animal cells. In *ADP-ribosylating toxins and G proteins: Insights into signal transduction* (ed. J. Moss and M. Vaughn), pp. 511–524. American Society for Microbiology, Washington, D.C.

Joshi, B., R. Yan, and R.E. Rhoads. 1994. *In vitro* synthesis of human protein synthesis initiation factor 4γ and its localization on 43 and 48 S initiation complexes. *J. Biol. Chem.* **269:** 2048–2055.

Joshi, B., A.-L. Cai, B.D. Keiper, W.B. Minich, R. Mendez, C.W. Beach, J. Stepinski, R. Stolarski, E. Darzynkiewicz, and R.E. Rhoads. 1995. Phosphorylation of eukaryotic protein synthesis initiation factor 4E at Ser-209. *J. Biol. Chem.* **270:** 14597–14603.

Joshi-Barve, S., A. De Benedetti, and R.E. Rhoads. 1992. Preferential translation of heat shock mRNAs in HeLa cells deficient in protein synthesis initiation factors eIF-4E and eIF-4γ. *J. Biol. Chem.* **267:** 21038–21043.

Kamath, A. and K. Chakraburtty. 1986. Protein synthesis in yeast: Purification of elongation factor 3 from temperature-sensitive mutant 13-06 of the yeast *Saccharomyces cerevisiae*. *J. Biol. Chem.* **261:** 12596–12598.

Kang, H.A. and J.W.B. Hershey. 1994. Effect of initiation factor eIF-5A depletion on protein synthesis and proliferation of *Saccharomyces cerevisiae*. *J. Biol. Chem.* **269:** 3934–3940.

Kasperaitis, M.A.M., H.O. Voorma, and A.A.M. Thomas. 1995. The amino acid sequence of eukaryotic translation initiation factor 1 and its similarity to yeast initiation factor SUI1. *FEBS Lett.* **365:** 47–50.

Kaufman, R.J., P. Murtha, and M.V. Davies. 1987. Translational efficiency of polycistronic mRNAs and their utilization to express heterologous genes in mammalian cells. *EMBO J.* **6:** 187–193.

Kemper, W.M., K.W. Berry, and W.C. Merrick. 1976. Purification and properties of rabbit reticulocyte protein synthesis initiation factors M2Bα and M2Bβ. *J. Biol. Chem.* **251:** 5551–5557.

Kimata, Y. and K. Kohno. 1994. Elongation factor 2 mutants deficient in diphthamide formation show temperature-sensitive cell growth. *J. Biol. Chem.* **269:** 13497–13501.

Kinzy, T.G. and J.L.J. Woolford. 1995. Increased expression of *Saccharomocyes cerevisiae* translation elongation factor 1α bypasses the lethality of a *TEF5* null allele encoding EF-1β. *Genetics* **140:** 481–489.

Kjeldgaard, M. and J. Nyborg. 1992. Refined structure of elongation factor Tu from *Escherichia coli*. *J. Mol. Biol.* **223:** 721–742.

Kjeldgaard, M., P. Nissen, S. Thirup, and J. Nyborg. 1993. The crystal structure of elongation factor EF-Tu from *Thermus aquaticus* in the GTP conformation. *Structure* **1:** 35–50.

Knudsen, S.M., J. Frydenberg, B.F.C. Clark, and H. Leffers. 1993. Tissue-dependent variation in the expression of elongation factor 1α isoforms: Isolation and characterization of a cDNA encoding a novel variant of human elongation factor 1α. *Eur. J. Biochem.* **215:** 549–554.

Konecki, D.S., K.C. Aune, W.P. Tate, and C.T. Caskey. 1977. Characterization of reticulocyte release factor. *J. Biol. Chem.* **252:** 4514–4520.

Kozak, M. 1989a. Context effects and inefficient initiation at non-AUG codon in eukaryotic cell-free translation systems. *Mol. Cell. Biol.* **9:** 5073–5080.

———. 1989b. The scanning model for translation: An update. *J. Cell. Biol.* **108:** 229–241.

———. 1991a. Effects of long 5′ leader sequences on initiation by eukaryotic ribosomes *in vitro*. *Gene Expr.* **1:** 117–125.

———. 1991b. A short leader sequence impairs the fidelity of initiation by eukaryotic ribosomes *in vitro*. *Gene Expr.* **1:** 111–115.

Lamphear, B.J., R. Kirchweger, T. Skern, and R.E. Rhoads. 1995. Mapping of functional domains in eIF4G with picornaviral proteases. Implications for cap-dependent and cap-independent translational initiation. *J. Biol. Chem.* **270:** 21975–21983..

Lawson, T.G., M.H. Cladaras, B.K. Ray, K.A. Lee, R.D. Abramson, W.C. Merrick, and R.E. Thach. 1988. Discriminatory interaction of purified eukaryotic initiation factors 4F plus 4A with the 5′ ends of reovirus messenger RNAs. *J. Biol. Chem.* **263:** 7266–7276.

Lazaris-Karatzas, A., K.S. Montine, and N. Sonenberg. 1990. Malignant transformation by a eukaryotic initiation factor subunit that binds to mRNA 5′ cap. *Nature* **345:** 544–547.

Lee, S., A. Wolfraim, and E. Wang. 1993. Differential expression of S1 and elongation factor 1α during rat development. *J. Biol. Chem.* **268:** 24453–24459.

Linder, P., P.F. Lasko, M. Ashburner, P. Leroy, P.J. Nielsen, K. Nishi, J. Schnier, and P.P. Slonimski. 1989. Birth of the D-E-A-D box. *Nature* **337:** 121–122.

Lloyd, M.A., J.C. Osborne, B. Safer, G.M. Powell, and W.C. Merrick. 1980. Characterization of eukaryotic initiation factor 2 and its subunits. *J. Biol. Chem.* **255:** 1189–1193.

Lutsch, G., R. Benndorf, P. Westermann, J. Behlke, U.-A. Bommer, and H. Bielka. 1985. On the structure of native small ribosomal subunits and initiation factor eIF-3 isolated from rat liver. *Biomed. Biochim. Acta* **44:** K1–K7.

Mader, S., H., Lee, A. Pause, and N. Sonenberg. 1995. The translation initiation factor eIF-4E binds to a common motif shared by the translation factor eIF-4γ and the translational repressors, 4E-binding proteins. *Mol. Cell. Biol.* **15:** 4990–4997.

Merrick, W.C. 1979. Assays for eukaryotic protein synthesis. *Methods Enzymol.* **60:** 108–123.

———. 1992. Mechanism and regulation of eukaryotic protein synthesis. *Microbiol. Rev.* **56:** 291–315.

Méthot, N., A. Pause, J.W.B. Hershey, and N. Sonenberg. 1994. Translation initiation factor eIF-4B contains an RNA binding region that is distinct and independent from its ribonucleoprotein-consensus sequence. *Mol. Cell. Biol.* **14:** 3207–3216.

Milburn, S.C., J.W.B. Hershey, M.V. Davies, K. Kelleher, and R.J. Kaufman. 1990. Cloning and expression of eukaryotic initiation factor 4B cDNA: Sequence determination identifies a common RNA recognition motif. *EMBO J.* **9:** 2783–2790.

Minich, W.B., M. Luisa Balesta, D.J. Goss, and R.E. Rhoads. 1994. Chromatographic resolution of *in vivo* phosphorylated and nonphosphorylated eukaryotic translation initiation factor eIF-4E: Increased cap affinity of the phosphoryl form. *Proc. Natl. Acad. Sci.* **91:** 7668–7672.

Moldave, K. 1985. Eukaryotic protein synthesis. *Annu. Rev. Biochem.* **54:** 1109–1149.

Morley, S.J., T.E. Dever, D. Etchison, and J.A. Traugh. 1991. Phosphorylation of eIF-4F by protein kinase C or multipotential S6 kinase stimulates protein synthesis at initiation. *J. Biol. Chem.* **266:** 4669–4672.

Motorin, Y.A., A.D. Wolfson, A.F. Orlovsky, and K.L. Gladilin. 1988. Mammalian valyl-tRNA synthetase forms a complex with the first elongation factor. *FEBS Lett.* **238:** 262–264.

Nagata, S., K. Iwasaki, and Y. Kaziro. 1976. Distribution of the low molecular weight form of eukaryotic elongation factor 1 in various tissues. *J. Biochem.* **80:** 73–77.

Naranda, T., S.E. MacMillan, and J.W.B. Hershey. 1994a. Purified yeast translationl initiation factor eIF-3 is an RNA-binding protein complex that contains the PRT1 protein. *J. Biol. Chem.* **269:** 32286–32292.

Naranda, N., W.B. Strong, J. Menaya, B.J. Fabbri, and J.W.B. Hershey. 1994b. Two structural domains of initiation factor eIF-4B are involved in binding to RNA. *J. Biol. Chem.* **269:** 14465–14472.

Nielsen, P.J. and H. Trachsel. 1988. The mouse protein synthesis initiation factor 4A gene family includes two related functional genes which are differentially expressed. *EMBO J.* **7:** 2097–2105.

Nierhaus, K.H., F. Franceschi, A.R. Subramanian, V.A. Erdmann, and B. Wittmann-Liebold, eds. 1993. *The translational apparatus: Structure, function, regulation, evolution.* Plenum Press, New York.

Ohta, K., M. Toriyama, M. Miyazaki, H. Murofushi, S. Hosoda, S. Endo, and H. Sakai. 1990. The mitotic apparatus-associated 51 kDa protein from sea urchin eggs is a GTP-binding protein and is immunologically related to yeast polypeptide elongation factor 1α. *J. Biol. Chem.* **265:** 3240–3247.

Oldfield, S. and C.G. Proud. 1992. Purification, phosphorylation and control of the guanine-nucleotide-exchange factor from rabbit reticulocyte lysates. *Eur. J. Biochem.* **208:** 73–81.

Oldfield, S., B.L. Jones, D. Tanton, and C.G. Proud. 1994. Use of monoclonal antibodies to study the structure of eukaryotic protein synthesis initiation factor eIF-2B. *Eur. J. Biochem.* **221:** 399–410.

Omura, F., K. Kohno, and T. Uchida. 1989. The histidine residue of codon 715 is essential for function of elongation factor 2. *Eur. J. Biochem.* **180:** 1–8.

Palmer, T.D., A.D. Miller, R.H. Reeder, and B. McStay. 1993. Efficient expression of a protein coding gene under control of an RNA polymerase I promoter. *Nucleic Acids Res.* **15:** 3451–3457.

Pause, A. and N. Sonenberg. 1992. Mutational analysis of a DEAD box RNA helicase: The mammalian translation initiation factor eIF-4A. *EMBO J.* **11:** 2643–2654.

Pause, A., N. Méthot, and N. Sonenberg. 1993. The HRIGRR region of the DEAD box RNA helicase eukaryotic translation initiation factor 4A is required for RNA binding and ATP hydrolysis. *Mol. Cell. Biol.* **13:** 6789–6798.

Pause, A., N. Méthot, Y. Svitkin, W.C. Merrick, and N. Sonenberg. 1994a. Dominant negative mutants of mammalian translation initiation factor eIF-4A define a critical role for eIF-4F in cap-dependent and cap-independent initiation of translation. *EMBO J.* **13:** 1205–1215.

Pause, A., G.J. Belsham, A.-C. Gingras, O. Donzé, T.-A. Lin, J.C.J. Lawrence., and N. Sonenberg. 1994b. Insulin-dependent stimulation of protein synthesis *via* phosphorylation of a novel regulator of cap function. *Nature* **371:** 762–767.

Peabody, D.S. and P. Berg. 1986. Termination-reinitiation occurs in the translation of mammalian cell mRNAs. *Mol. Cell. Biol.* **6:** 2695–2703.

Pelletier, J. and N. Sonenberg. 1985. Insertion mutagenesis to increase secondary structure within the 5′ noncoding region of a eukaryotic mRNA reduces translational efficiency. *Cell* **40:** 515–526.

Prat, A., S.R. Schmid, P. Buser, S. Blum, H. Trachsel, P.J. Nielsen, and P. Linder. 1990. Expression of translation initiation factor 4A from yeast and mouse in *Saccharomyces cerevisiae. Biochim. Biophys. Acta* **1050:** 140–145.

Rapp, G., J. Klaudiny, G. Hagendorff, M.R. Luck, and K.H. Scheit. 1989. Complete sequence of the coding region of human elongation factor 2 (EF-2) by enzymatic amplification of cDNA from human ovarian granulosa cells. *Biol. Chem. Hoppe-Seyler* **370:** 1071–1075.

Ray, B.K., T.G. Lawson, J.C. Kramer, M.H. Claderos, J.A. Grifo, R.D. Abramson, W.C. Merrick, and R.E. Thach. 1985. ATP-dependent unwinding of messenger RNA structure by eukaryotic initiation factors. *J. Biol. Chem.* **260:** 7651–7658.

Raychaudhuri, P., A. Chaudhuri, and U. Maitra. 1985. Eukaryotic initiation factor 5 from calf liver is a single polypeptide chain protein of M_r=62,000. *J. Biol. Chem.* **260:** 2132–2139.

Raychaudhuri, P., E.A. Stringer, D.M. Valenzuela, and U. Maitra. 1984. Ribosomal subunit anti-association activity in rabbit reticulocytes. *J. Biol. Chem.* **259:** 11930–11935.

Riis, B., S.I.S. Rattan, B.F.C. Clark, and W.C. Merrick. 1990. Eukaryotic protein elongation factors. *Trends Biochem. Sci.* **15:** 420–424.

Rowlands, A.G., R. Panniers, and E.C. Henshaw. 1988. The catalytic mechanism of guanine-nucleotide-exchange factor action and competitive inhibition by phosphorylated eukaryotic initiation factor 2. *J. Biol. Chem.* **263:** 5526–5533.

Roy, A.L., D. Chakrabarti, B. Datta, R.E. Hileman, and N.K. Gupta. 1988. Natural mRNA is required for directing Met-tRNA$_f$ binding to 40S ribosomal subunits in animal cells: Involvement of Co-eIF-2A in natural mRNA-directed initiation complex formation. *Biochemistry* **27:** 8203–8209.

Rozen, F., I. Edery, K. Meerovitch, T.E. Dever, W.C. Merrick, and N. Sonenberg. 1990. Bidirectional RNA helicase activity of eucaryotic translation initiation factors 4A and 4F. *Mol. Cell. Biol.* **10:** 1134–1144.

Russell, D.W. and L.L. Spremulli. 1979. Purification and characterization of a ribosome dissociation factor (eukaryotic initiation factor) from wheat germ. *J. Biol. Chem.* **254:** 8796–8800.

Ryoji, M., J.W. Karpen, and A. Kaji. 1981. Further characterization of ribosome releasing factor and evidence that it prevents ribosomes from reading through a termination codon. *J. Biol. Chem.* **256:** 5798–5801.

Sachs, A.B. and R.W. Davis. 1989. The poly(A) binding protein is required for poly(A) shortening and 60S ribosomal subunit-dependent translation initiation. *Cell* **58:** 857–867.

———. 1990. Translation initiation and ribosomal biogenesis: Involvement of a putative helicase and RPL46. *Science* **247:** 1077–1079.

Safer, B., S.L. Adams, W.M. Kemper, K.W. Berry, M. Floyd, and W.C. Merrick. 1976. Purification and characterization of two initiation factors required for maximal activity of a highly fractionated globin mRNA translation system. *Proc. Natl. Acad. Sci.* **73:** 2584–2588.

Sanders, J., R. Raggiaschi, J. Morales, and W. Moeller. 1993. The human leucine zipper-containing guanine-nucleotide exchange protein elongation factor-1δ. *Biochim. Biophys. Acta* **1174:** 87–90.

Schirmaier, F. and P. Philippsen. 1984. Identification of two genes coding for the translational elongation factor EF-1α of *S. cerevisiae*. *EMBO J.* **3:** 3311–3315.

Schnier, J., H.G. Schwelberger, Z. Smit-McBride, H.A. Kang, and J.W.B. Hershey. 1991. Translation initiation factor 5A and its hypusine modification are essential for cell viability in the yeast *Saccharomyces cerevisiae*. *Mol. Cell. Biol.* **11:** 3105–3114.

Schreier, M.H., B. Erni, and T. Staehelin. 1977. Initiation of mammalian protein synthesis. I. Purification and characterization of seven initiation factors. *J. Mol. Biol.* **116:** 727–753.

Seal, S.N., A. Schmidt, and A. Marcus. 1982. A heat-stable protein synthesis initiation factor from wheat germ. *J. Biol. Chem.* **257:** 8634–8637.

Seal, S.N., A. Schmidt, A. Marcus, I. Edery, and N. Sonenberg. 1986. A wheat germ cap-site factor functional in protein chain initiation. *Arch. Biochem. Biophys.* **246:** 710–715.

Shiina, N., Y. Gotoh, N. Kubomura, A. Iwamatsu, and E. Nishida. 1994. Microtubule severing by elongation factor 1α. *Science* **266:** 282–285.

Skogerson, L. 1979. Separation and characterization of yeast elongation factors. *Methods Enzymol.* **60:** 676–685.

Skogerson, L. and E. Wakatama. 1976. A ribosome-dependent GTPase from yeast distinct from elongation factor 2. *Proc. Natl. Acad. Sci.* **73:** 73–76.

Slobin, L.I. 1990. Eucaryotic polypeptide chain elongation. In *Translation in Eukaryotes* (ed. H. Trachsel), pp. 149–175, The Telford Press, Caldwell, New Jersey.

Slusher, L.B., E.C. Gillman, N.C. Martin, and A.K. Hopper. 1991. mRNA leader length and initiation codon context determine alternative AUG selection for the yeast gene *MOD5*. *Proc. Natl. Acad. Sci.* **88:** 9789–9793.

Smith, K.E. and E.C. Henshaw. 1975. Binding of Met-tRNA$_f$ to native ribosomal subunits in Ehrlich ascites tumor cells. *J. Biol. Chem.* **250:** 6880–6884.

Söll, D. 1993. Transfer RNA: An RNA for all seasons. In *The RNA world* (ed. R.F. Gestland and J.F. Atkins), pp. 157–184, Cold Spring Harbor Laboratory Press, Cold Spring Harbor, New York.

Song, J.M., S. Picologlou, C.M. Grant, M. Firoozan, M.F. Tuite, and S. Liebman. 1989. Elongation factor EF-1α gene dosage alters translational fidelity in *Saccharomyces cerevisiae*. *Mol. Cell. Biol.* **9:** 4571–4575.

Tate, W.P. and C.T. Caskey. 1974. The mechanism of peptide chain termination. *Mol. Cell. Biochem.* **5:** 115–126.

Tatsuka, M., H. Mitsui, M. Wada, A. Nagata, H. Nojima, and H. Okayama. 1992. Elongation factor 1α gene determines susceptability to transformation. *Nature* **359:**

333–336.

Timchenko, L. and C.T. Caskey. 1994. The "eRF" clone corresponds to tryptophanyl-tRNA synthetase, not mammalian release factor. *Proc. Natl. Acad. Sci.* **91:** 2777–2780.

Trachsel, H., B. Erni, M.H. Schreier, and T. Staehelin. 1977. Initiation of mammalian protein synthesis. II. The assembly of the initiation complex with purified initiation factors. *J. Mol. Biol.* **116:** 755–767.

Triana, F.J., K.H. Nierhaus, J. Ziehler, and K. Chakraburtty. 1993. Defining the function of EF-3, a unique elongation factor in low fungi. In *The translational apparatus: Structure, function, regulation, evolution* (ed. K.H. Nierhaus et al.), pp. 327–338, Plenum Press, New York.

Uetsuki, T., A. Naito, S. Nagata, and Y. Kaziro. 1989. Isolation and characterization of the human chromosomal gene for polypeptide elongation factor 1α. *J. Biol. Chem.* **264:** 5791–5798.

Uritani, M. and M. Miyazaki. 1988. Characterization of the ATPase and GTPase activities of elongation factor 3 (EF-3) from yeasts. *J. Biochem.* **103:** 522–530.

van Ness, B.G., J.B. Howard, and J.W. Bodley. 1988. Isolation and properties of the trypsin-derived ADP-ribosyl peptide from diphtheria toxin-modified yeast elongation factor 2. *J. Biol. Chem.* **253:** 8687–8690.

Vazquez de Aldana, C.R. and A.G. Hinnebusch. 1994. Mutations in the GCD7 subunit of yeast guanine nucleotide exchange factor eIF-2B overcome the inhibitory effects of phosphorylated eIF2 on translation initiation. *Mol. Cell. Biol.* **14:** 3208–3222.

Wei, C.-L., M. Kainuma, and J.W.B. Hershey. 1995a. Characterization of yeast translation initiation factor 1A and cloning of its essential gene. *J. Biol. Chem.* **270:** 22788–22794.

Wei, C.-L., S.E. MacMillan, and J.W.B. Hershey. 1995b. Protein synthesis initiation factor eIF-1A is a moderately abundant RNA-binding protein. *J. Biol. Chem.* **270:** 5764–5771.

Whitehart, S.W., P. Shenbagamurthi, L. Chen, R.J. Cotter, and G.W. Hart. 1989. Murine EF-1α is posttranslationally modified by novel amide-linked ethanolamine-phosphoglycerol moieties. *J. Biol. Chem.* **264:** 14334–14341.

Wolin, S.L. and P. Walter. 1988. Ribosomal pausing and stacking during translation of a eukaryotic mRNA. *EMBO J.* **7:** 3559–3569.

Yang, F., M. Demma, V. Warren, S. Dharmawardhane, and J. Condeelis. 1990. Identification of an actin-binding protein from *Dictyostelium* as elongation factor 1α. *Nature* **347:** 494–496.

Yang, W., W. Burkhart, J. Cavallius, W.C. Merrick, and W.F. Boss. 1993. Purification and characterization of a phosphatidylinositol 4-kinase activator in carrot cells. *J. Biol. Chem.* **268:** 392–398.

Yoder-Hill, J., A. Pause, N. Sonenberg, and W.B. Merrick. 1993. The p46 subunit of eIF-4F exchanges with eIF-4A. *J. Biol. Chem.* **268:** 5566–5573.

Yoon, H. and T.F. Donahue. 1992. The *sui1* suppressor locus in *Saccharomyces cerevisiae* encodes a novel translation factor that functions during $tRNA^{met}_i$ recognition of the start codon. *Mol. Cell. Biol.* **12:** 248–260.

3

A Comparative View of Initiation Site Selection Mechanisms

Richard J. Jackson
Department of Biochemistry
University of Cambridge
Cambridge CB2 1QW, United Kingdom

Before plunging into details of the various models for translation initiation and the roles of the initiation factor proteins, it is worth first asking the question: What does initiation really involve or how do we define the stage at which the initiation process has been completed and elongation takes over? Viewed in this way, initiation can be defined quite simply as the process in which a special initiator tRNA, formyl-Met-$tRNA_f$ or Met-$tRNA_i$, is positioned in the P-site of a ribosome, which is itself located at the correct AUG codon (or in some cases, a non-AUG initiation codon) for translation of the downstream open reading frame (ORF). Here, P-site occupancy is defined in the usual way as reactivity toward puromycin. Quite apart from the distinctive formyl group in the case of prokaryotes, the initiator tRNA itself has several features distinguishing it from all other tRNAs, including Met-$tRNA_m$ used for elongation (Seong and Rajbhandary 1987; Wakao et al. 1989), and as insertion of a charged tRNA into the P-site is an event unique to the initiation process, it is hardly surprising that this needs special protein factors, IF2 in prokaryotic systems and eIF2 in eukaryotes.

Apart from this feature of initiation common to both prokaryotes and eukaryotes, another universal characteristic is that the initiation process starts with separated ribosomal subunits (Guthrie and Nomura 1968; Howard et al. 1970; Blumberg et al. 1979). It is not entirely clear why this has to be the case. Perhaps efficient access of the initiator tRNA into the P-site under physiological conditions is only possible by this route. In any case, the necessity to start with separate subunits yet under conditions when subunit association is likely to be favored provides an explanation for the requirement for further initiation factors, to aid subunit dissociation or to prevent association, and to regulate subunit joining to the appropriate stage in the initiation pathway.

THE SEQUENCE OF EVENTS IN THE INITIATION PATHWAY

Comparisons between Prokaryotes and Eukaryotes

Continuing with this theme of the sequence of events, some apparent differences between prokaryotes and eukaryotes emerge, in that for prokaryotic initiation, it is often assumed that the small ribosomal subunit binds to the messenger RNA before initiator tRNA is bound, whereas the converse is generally assumed for eukaryotes. However, this difference may be less absolute than is often believed. Jay and Kaempfer (1974, 1975) produced evidence that a 30S/formyl-Met-tRNA$_f$ complex, albeit a complex that was unstable to sucrose gradient centrifugation in unfixed samples, was a preferred kinetic intermediate in the formation of the 30S/mRNA/formyl-Met-tRNA$_f$ complex. However, evidence for a preferred 30S/mRNA intermediate was adduced by van Duin and his colleagues (Backendorf et al. 1980; van Duin et al. 1980). For an informative discussion of this controversy, see Gold et al. (1981). The thorough kinetic analyses of Gualerzi and his colleagues have in fact shown that with *Escherichia coli* ribosomes translating poly(A,U,G) and other synthetic mRNAs, the formation of the 30S/formyl-Met-tRNA$_f$/mRNA complex is random order and either route is possible (Gualerzi et al. 1977; Calogero et al. 1988; Gualerzi and Pon 1990). However, one would imagine that with initiation sites which have strong Shine-Dalgarno sequences, the pathway in which the 30S/mRNA complex forms before formyl-Met-tRNA$_f$ binds must be strongly favored.

In the case of the eukaryotic initiation pathway not only is a stable 40S/Met-tRNA$_i$ complex readily observable and can be shown to be an intermediate on the direct pathway (Darnbrough et al. 1973), but results of early experiments also implied that no 40S/mRNA complexes could form in the absence of eIF2 and Met-tRNA$_i$ (Hunter at al. 1977; Trachsel et al. 1977; Benne and Hershey 1978). However, these experiments could only detect complexes that are stable to sucrose density gradient centrifugation and could not rule out the existence of an unstable 40S/mRNA complex lacking initiator tRNA, which, some might argue, could scan completely through the mRNA and detach during the time required for the analysis. In contrast, the results of more recent work examining the relative utilization of two tandem initiation sites as a function of varying mRNA concentrations are most readily interpreted on the assumption that scanning 40S/mRNA complexes lacking Met-tRNA$_i$ can exist (Dasso et al. 1990). Moreover, the current model of scanning-reinitiation, as in the case of yeast GCN4 mRNA translation, carries the implication that after a ribosome has translated a short upstream ORF, the 40S ribosomal subunit resumes scanning along the mRNA without Met-tRNA$_i$ bound to it, at least in the early stages of this resumed migration (Hinnebusch 1994 and this volume). However, the putative 40S/mRNA complex lacking bound Met-tRNA$_i$ has never been proven either by sucrose gradients or by the sort of kinetic analysis carried out by the Gualerzi group.

Still focusing on the pathway and the sequence of events, one difference between prokaryotes and eukaryotes that has received little comment is that only in the eukaryotic pathway is there a need for a special factor, eIF5, to promote subunit joining (Merrick and Hershey, this volume). It is not immediately obvious why there should be this difference, but it is correlated with a difference in the stage at which GTP hydrolysis actually occurs. In the eukaryotic system, the use of non-hydrolyzable GTP analogs results in the accumulation of 40S/mRNA/Met-tRNA$_i$ complexes, apparently positioned at the initiation codon and blocked before the subunit joining step (Trachsel et al. 1977; Benne and Hershey 1978). In contrast, in prokaryotes, these analogs do not inhibit subunit joining, but they allow the formation of a 70S/mRNA/formyl-Met-tRNA complex, which, however, is not reactive to puromycin and fails to release IF2 for recycling (Kolakofsky et al. 1968; Benne et al. 1973).

Another difference, possibly related to those discussed above, is that subunit joining during prokaryotic initiation seems to be a very rapid step in comparison with the preceding steps, whereas in eukaryotes, it appears to be slower, proceeding at a rate of the same order as the previous stages. This can be inferred from the fact that low levels of nonintegral polysomes, consisting of an integral number of 80S ribosomes and a single additional 40S subunit, are frequently seen on sucrose gradient analysis of mammalian cell extracts even in the absence of inhibitors that block subunit joining (Hoerz and McCarty 1969). To my knowledge, such nonintegral polyribosomes have not been seen in similar experiments with prokaryotic systems. We do not know with absolute certainty whether the 40S subunits of eukaryotic nonintegral polysomes are actually positioned at the initiation codon waiting for the subunit joining step to take place or whether they are in the process of scanning through the 5′-untranslated region (5′UTR). However, given the fact that nonintegral polysomes can be seen with mRNAs that have short 5′UTRs (e.g., the endogenous α- and β-globin mRNAs of rabbit reticulocyte lysates) and that it has proven very difficult to trap 40S subunits that are scanning through the 5′UTR (Kozak 1989b), the most probable explanation is that they are positioned at the initiation codon, which implies that subunit joining is a relatively slow step in the pathway.

The Conundrum of Reinitiation after Termination

In principle, every cistron of a prokaryotic polycistronic mRNA could be accessed independently via its own Shine-Dalgarno (SD) sequence, and so the ribosomes that translate a downstream cistron need not necessarily have translated the upstream cistrons. Thus, in the case of the L10 ribosomal protein operon, the initiation frequency at the downstream L7/L12 cistron is thought to be at least fourfold higher than that for the 5′-proximal L10 cistron (Yates et al. 1981). Moreover, mutation of the initiation codon of an upstream cistron of a polycistronic mRNA should, in principle, have no influence on the expression of a downstream cistron, although in practice, polarity effects often interfere with such simple predictions.

There are also numerous examples of translational coupling, however, where the downstream cistron expression is dependent on proper translation of upstream cistrons. One explanation is that the SD motif or initiation codon of the downstream cistron is buried in secondary structure unless or until translation of the upstream cistron unwinds such secondary structure. This is believed to be the case in at least some ribosomal protein operons (Yates et al. 1981; Nomura et al. 1984) and is also the likely basis whereby translation of the RNA polymerase cistron of the RNA bacteriophages is coupled to translation of the upstream coat protein cistron (Berkhout and van Duin 1985). In other cases, an alternative explanation seems to hold, as exemplified by the coupling of bacteriophage fd gene VII synthesis to translation of the upstream gene V cistron. In this case, the gene VII initiation site has only a vestigial and very weak SD sequence, which seems to be recognized at very low efficiency unless ribosomes are "delivered" in close proximity as a result of translating the upstream gene V cistron (Ivey-Hoyle and Steege 1992). Similarly, potential initiation sites within an open reading frame can be activated by the presence of an in-frame stop codon located a short distance either upstream or downstream from the initiation codon (Adhin and van Duin 1990).

In these cases of translational coupling where the termination codon of the upstream cistron lies very close to the initiation codon of the downstream cistron, it is pertinent to ask whether the same individual ribosome translates both cistrons. The results with the fd gene V/gene VII mRNA are most readily explained if there was such "readthrough." Although subunit dissociation is thought to be the ultimate fate of ribosomes following termination, it is not clear whether this occurs instantaneously, nor what is the exact function in this event of the ribosome release factor protein described by Kaji's group (Hirashima and Kaji 1972; Ichikawa et al. 1989). Perhaps the 50S subunit is released immediately at termination, but the 30S subunit has the potential to remain bound transiently to the mRNA (Martin and Webster 1975) and to (re)initiate at the gene VII initiation site. The fact that a stop codon allows initiation at an otherwise silent AUG or UUG codon located up to 40 residues either upstream or downstream from the stop codon argues that not only do 30S subunits remain associated with the mRNA for a limited time following termination, but they are even capable of limited bidirectional scanning of the mRNA from the stop codon (Adhin and van Duin 1990). This scanning process resembles that of the eukaryotic initiation model insofar as it selects the nearest initiation codon from the termination codon; it differs in the fact that the scanning is limited to approximately 40 nucleotides in either direction. As for the bidirectionality, although eukaryotic ribosome scanning is usually considered to be unidirectional, there are in fact some reports of reinitiation at AUG codons located upstream of a translation termination site (see Peabody and Berg 1986; Peabody et al. 1986; Thomas and Capecchi 1986), exactly paralleling the prokaryotic example of Adhin and van Duin (1990).

With eukaryotes, there is also a conundrum, albeit somewhat different, concerning events following termination. With those mRNAs with short upstream ORFs, of which the best studied is yeast GCN4 mRNA, there is good evidence that ribosomes which translate the upstream ORF may retain competence to (re)initiate translation of the main ORF further downstream (Hinnebusch 1994 and this volume). The prevailing wisdom is that this will only occur if the upstream ORF is relatively short, and although this has not been tested systematically, it is certainly true that virtually all the best characterized examples involve ORFs of less than 20 codons. In addition, detailed studies of the GCN4 system have shown that the nature of codons just before the termination codon of the short ORF, and the sequences just downstream from the termination codon, can strongly influence whether (re)initiation will occur at the main ORF further downstream (Grant and Hinnebusch 1994). The reasons for this influence of both the length of the upstream short ORF and the sequence around its termination codon are completely unknown. Nor is it known what actually happens in circumstances that are unfavorable to reinitiation. Do the 40S ribosomal subunits resume scanning the mRNA but remain incompetent to reinitiate or do they detach from the mRNA at the termination codon of the short ORF?

SELECTION OF THE CORRECT INITIATION SITE

Initiation Site Selection in Prokaryotes

Thus far, the question of how the correct initiation site is recognized has been an issue peripheral to the sequence of events in the initiation pathway. The time has now come to correct this imbalance. In prokaryotes, despite the existence of some mRNAs lacking an SD motif, there is overwhelming evidence that for the vast majority of messages, the SD sequence is the essential but not necessarily the sole "identifier" element. The experiments of Steitz and Jakes (1975) demonstrating base pairing between the mRNA initiation site and the 3′end of 16S rRNA, and the dedicated ribosome experiments of Hui and de Boer (1987), are the foundation stones of this model, but there is a mass of supporting evidence, including results of the classy but somewhat baffling toeprinting assays developed relatively recently by Gold's group. The first toeprinting experiments failed to detect the expected binary complex between 30S ribosomal subunits and the mRNA SD motif (Hartz et al. 1989), but when the concentration of reverse transcriptase used for the primer extension was reduced, or the extension reaction was carried out at low temperature, then such complexes were observed (Hartz et al. 1991). So, just as sucrose gradient centrifugation assays carry the caveat that they only detect complexes that are stable in the gradient, toeprinting carries a caveat that complexes can only be detected if they are sufficiently stable or sufficiently long-lived to survive the onslaught of the reverse transcriptase during the primer extension step. Taken as a whole, the results of toeprinting assays show that 30S subunits can bind to the mRNA at the SD motif in the absence of other ligands. With such binary complexes, the reverse transcriptase penetrates to a point two to five nucleotides downstream from the 3′ G of the GGAGG SD sequence, whereas in the presence of initiator tRNA, the primer extension stops at a point 17–18 nucleotides further downstream, equivalent to 14–15 nucleotides downstream from the A of the AUG initiation codon (Hartz et al. 1991). Thus, in the 30S/mRNA binary complex, close contacts with the 30S subunit are limited to the SD motif and a few residues on either side, but in the ternary complex, the close contacts extend to 15 residues downstream from the initiation codon. This conclusion is consistent with the fact that site-specific cross-linking by UV-irradiation between the 16S rRNA and sites downstream from the initiation codon occurred only in the presence of initiator tRNA, whereas cross-linking to upstream sites was independent of tRNA (McCarthy and Brimacombe 1994). Under physiological conditions, the conversion of the binary complex to a stable ternary complex is accomplished by binding of initiator tRNA and initiation factors. This can be achieved in vitro by IF2 on its own, provided charged formyl-Met-$tRNA_f$ is present, or by IF3 on its own, in which case the $tRNA_f^{Met}$ need not be charged, and indeed, just the anticodon stem-loop alone suffices to fix the ribosomal subunit at the initiation codon (Hartz et al. 1989). Thus, in addition to its other roles, IF3 appears to monitor and stabilize the initiation codon/$tRNA_f$ anticodon interaction. IF1, on the other hand, seems to have no direct role in stabilizing 30S/mRNA interactions at the correct initiation site, but it does increase the yield of initiation complexes when 70S ribosomes are used (Hartz et al. 1989), which is possibly due to its dissociation factor activity.

That initiation efficiency is related to the number of complementary base pairs that can form between the SD motif and the 16S rRNA, and that there is a finite window of allowable spacing between the SD motif and the initiation codon, has been demonstrated in numerous studies of which the most definitive and informative is that of Ringquist et al. (1992), not least because all the mutations were made in a neighboring sequence background designed to eliminate secondary structure. The elegant experiments of de Smit and van Duin (1990a,b) on the effect of mutations around the initiation site of the RNA bacteriophage coat protein cistron demonstrate that the SD sequence and the initiation codon must be in unstructured regions to allow initiation; the frequency of initiation was proportional to the fraction of time during which these elements would be present in an open conformation. However, although the SD motif and the initiation codon must both be in unstructured regions, hairpin loops between these two elements seem to be allowed. In bacteriophage T4 gene *38* mRNA, the linear spacing between the SD sequence and the AUG initiation codon is so large that efficient initiation would not be expected. However, this intervening region can fold into a hairpin loop with eight base pairs (Gold 1988), which seems likely to exist in reality given its high GC content and the fact that it is closed by a UUCG tetraloop. This would have the effect of making the spacing between the SD sequence and the AUG codon close to the optimal. If this is indeed the case, it carries the implication that the mRNA lies in a slot or channel in the 70S ribosome, rather than being threaded through anything resembling a tube or hole.

Although some components of the prokaryotic translation machinery, notably IF3 and ribosomal protein S1 (Thomas and Szer 1982; Subramanian 1983), have been proposed to have RNA melting properties, they are not ATP-dependent helicases like eukaryotic eIF4A. It is interesting to note that although the original purifications of prokaryotic translation initiation factors were mainly done using assays for translation of the RNA bacteriophage MS2 (or Qβ, R17, f2, etc.) coat protein cistron, in which the initiation site is somewhat occluded in a hairpin structure (de Smit and van Duin 1990a,b), no ATP-dependent helicases were isolated as factors which stimulate translation initiation.

Although the SD motif is the critical sequence defining an initiation site, it is unlikely to be the only feature recognized by the initiating ribosome. A statistical analysis has shown that the sequence of ribosome-binding sites in *E. coli* is not random between positions –20 and +14 (relative to the A of the AUG codon). In support of this, it is found that when random fragments of *E. coli* genes are tested for their ability to function as initiation sites in a type of shotgun experiment, all of the active fragments (with one exception) were from authentic initiation sites. Moreover, all of the active fragments included the –20 to +15 region, but the lengths of sequences flanking this core were variable (Dreyfus 1988). It is thought that this nonrandomness of the approximately 35-nucleotide segment is not merely due to evolutionary pressure to eliminate secondary structure, but reflects the fact that there is contact between the ribosome and the mRNA throughout the element. Significantly, the length of mRNA protected by ribosome binding at the initiation site is about 35 nucleotides long and extends to about 15 nucleotides downstream from the initiation codon (Steitz and Jakes 1975; Hartz et al. 1989). However, despite the nonrandom sequence of this region, there is no obvious conserved consensus (apart from the SD motif), but rather a general preference for A residues, especially downstream from the SD motif, and to a lesser extent for U, with counterselection against C (Dreyfus 1988; Gold 1988).

Translational Enhancers in Prokaryotic mRNAs

There are numerous reports of other sequences that can influence the efficiency of initiation in prokaryotic systems. Some of these are found upstream of the SD motif, some downstream. Some lie within the –20 to +15 window of direct contact with the initiating ribosome and thus may contribute to the nonrandomness of the sequence of this region, but others lie outside this window. In every case, it has been suggested that these elements enhance initiation through base pairing with various motifs in 16S rRNA. However, the postulated complementarities exist only on paper, and there has been no attempt to prove their reality by constructing 16S rRNA genes bearing compensating mutations, as was done by Hui and de Boer (1987) to prove the pivotal importance of the SD sequence. Moreover, many of the translational enhancing motifs could only base pair completely at the proposed sites if phylogenetically conserved intramolecular base pairing in the 16S rRNA were disrupted. In addition, according to current models for the three-dimensional structure of 16S rRNA within the 30S subunit (McCarthy and Brimacombe 1994), any pairing between these "enhancers" and the 16S rRNA would probably have to occur preliminary to, rather than simultaneously with, pairing to the SD motif. A possible exception is a UGAUCC motif revealed by a data base search to be a common feature of highly expressed genes (Thanaraj and Pandit 1989) and postulated to pair with the 16S rRNA just upstream of the CCUCC at the 3′end (Fig. 1). However, simultaneous pairing at this site and at the SD motif would seem to be the exception rather than the rule, since in the majority of cases, the mRNA UGAUCC sequence is upstream of position –20, rather than adjacent to the SD sequence.

Among the experimentally proven upstream enhancer elements, there is the "Olins Box" (also known as the Epsilon sequence), a region immediately upstream of the SD motif of bacteriophage T7 gene *10* mRNA and found in other highly expressed late T7 genes, with a consensus of UUAACUUU. Provided the initiation site included an SD element, this motif conferred a strong enhancement of translation efficiency on heterologous mRNAs, especially those coded by nonbacterial genes which would presumably lack the nonrandom features of true bacterial initiation sites (Olins et al. 1988; Olins and Rangwala 1989). Surprisingly, this consensus sequence was shown to enhance translation efficiency not only when in its usual position upstream of the SD sequence, but also when placed a short distance downstream from the initiation codon. A translational enhancer with properties transferable to other genes has also been found upstream of the SD motif of the *atpE* gene of *E. coli*, and although the actual enhancer has not been exactly delineated, it is interesting to note that this region includes the Epsilon-like sequences UUAACU and UUAAUUUAC, with the first of these lying just outside the –20 to +15 contact window and the other yet further upstream (McCarthy et al. 1985, 1986). The Epsilon motif is complementary to a segment of 16S rRNA around nucleotide 460, although part of this segment is itself base-paired in the current 16S rRNA secondary structure models (Olins and Rangwala 1989).

There is also the well-documented, almost to the point of being over-advertised, effect of the Ω sequence, which is essentially the 5′UTR of

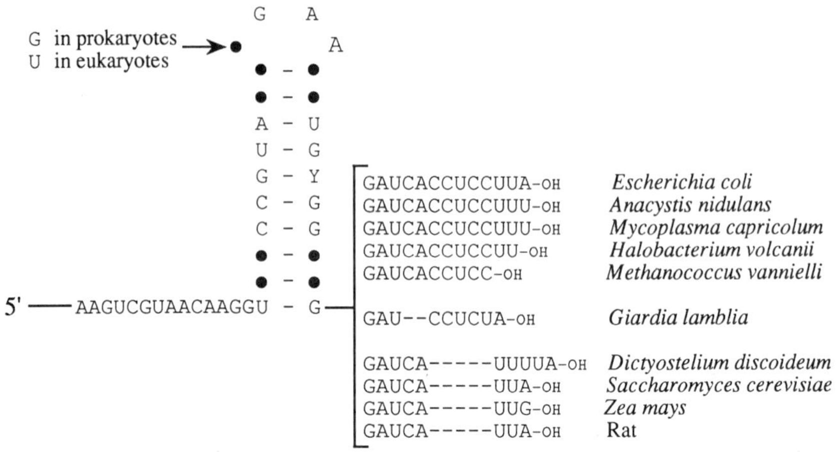

Figure 1 Conservation of the sequence at the 3′ end of the small ribosomal subunit rRNA. The selected sequences shown are from Guttell et al. (1985), except that of *Giardia lamblia* (Sogin et al. 1989). Among the species listed, there is complete conservation of (1) the sequence shown upstream of the stem-loop, (2) the length of the stem-loop, and (3) the sequence in three of the positions of the tetraloop. Individual sequences downstream from the stem-loop are given, with gaps (denoted by –) introduced to optimize alignment.

tobacco mosaic virus RNA. This 69-nucleotide element, and a 43-nucleotide truncated derivative of it, strongly enhances mRNA translation in *E. coli* if positioned upstream of the initiation codon of mRNAs lacking an SD motif, but it has a much smaller effect if there is an SD sequence in the usual position (Gallie and Kado 1989). The enhancing effect of Ω is augmented by duplication of the segment, can be transferred to several different mRNAs, and seems to be independent of the position of the motif with respect to the initiation codon. The Ω sequence lacks any G residues, and it has been suggested but not absolutely proven that the critical element is an ACAAUUAC motif, which is repeated three times in the Ω segment (Gallie and Kado 1989).

A downstream enhancer is found in the first part of the *glnS*-coding sequences of *E. coli*, as shown by the influence of mutations in this region on the translation of mRNAs encoding *glnS* fusion proteins (Faxen et al. 1991). The optimal sequence following the initiation codon was AGUGAAUCACAA. A similar sequence in bacteriophage T7 gene *0.3* protein was also found to have translation enhancing properties, but in this case, the motif was further downstream starting at the +15 position (Sprengart et al. 1990). A data base search showed the existence of this motif in many mRNAs but in a variable position, from immediately after the initiation codon to more than 30 nucleotides downstream. It is proposed that this might function by base pairing to the 1470–1480 region of 16S rRNA, although the whole of this segment is involved in conserved intramolecular base pairing. Perhaps the most striking case of a downstream determinant is that of the prophage form of bacteriophage λ*c*I mRNA in which the initiation codon is at the extreme 5′end, and so there is no SD motif nor any upstream sequences. Efficient translation of *c*I/*lacZ* fusions required 5′-proximal *c*I-coding sequences, mainly the first four codons that have the sequence 5′-AUGAGCACAAA (Shean and Gottesman 1992), a truncated variant of the downstream consensus noted above.

Amidst all this enthusiasm for finding complementarity between putative translational enhancer elements and 16S rRNA, one should not overlook a possible role for interactions between the mRNA and ribosomal proteins. Boni has noted that ribosomal protein S1 binds strongly to U-rich sequences found upstream of the SD element in many efficiently translated mRNAs (Boni et al. 1991). Current models envisage an exceedingly elongated structure for S1 (Subramanian 1983; Wallaczek et al. 1990), which therefore might act as a sort of "fishing-line" to capture and tether the mRNA to the 30S ribosomal subunit, before the initiation site is "landed" by the well-established SD motif/16S rRNA interaction. Clearly, prokaryotic initiation site selection is more complex than the standard textbooks would have us believe.

Does Initiation Site Selection in Eukaryotes Involve 18S rRNA/mRNA Pairing?

It is axiomatic that eukaryotic cellular and viral mRNAs lack a highly conserved motif in the position of the SD sequences of prokaryotic mRNAs. This correlates with the fact that although the sequence and structure at the 3′end of the eukaryotic 18S rRNA are very similar to that of the prokaryotic 16S rRNA, the ..CCUCC.. sequence in 16S rRNA that pairs with the SD element is precisely deleted in all eukaryotic cytoplasmic small ribosomal RNAs (Fig. 1), with the exception of one protozoan, *Giardia lamblia*, which retains it in a slightly corrupted form (Sogin et al. 1989).

What is not yet completely ruled out, however, is the possibility that eukaryotic mRNAs might have features such as the Olins Box of prokaryotic mRNAs which might be recognized by base pairing with rRNA. Given that the Olins Box (or Epsilon sequence) can apparently act as an enhancer even if present in different locations with respect to the initiation codon, it might be hard to spot such a conserved motif in eukaryotic mRNAs by the usual form of linear alignment of mRNA sequences that has become fashionable since the work of Kozak (1986a) highlighted the importance of the context of the initiation codon for efficiency of initiation site recognition. Nevertheless, there is such a mass of sequences in the data base that one would hope and expect that any common enhancer element should have been spotted. A further contraindication against the idea that initiation site recognition for the majority of mRNAs occurs by base pairing through Epsilon-like motifs is that such a mechanism would most likely lead to the ability to translate polycistronic mRNAs. That this is clearly not the case is shown by countless dicistronic mRNA constructs generated as controls for internal initiation, which give very low yields indeed of downstream cistron translation product. Any doubts on this issue are dispelled by the fact that viral RNAs that are structurally polycistronic (e.g., the genomic RNAs of alphaviruses and of tobacco mosaic virus) are nevertheless functionally monocistronic (Glanville et al. 1976; Hunter et al. 1976; Goelet et al. 1982; Kääriäinen et al. 1987).

For all cellular mRNAs and the vast majority of viral mRNAs, it thus seems unlikely that initiation site recognition is by base pairing between 18S rRNA elements and one or more mRNA motifs conserved in sequence but not necessarily in location in the mRNA. However, a possible exception is clearly those RNAs translated by internal initiation, where a *cis*-acting RNA element can promote direct ribosome entry to the initiation site of the downstream cistron in di- or even tricistronic mRNAs. Is the *cis*-acting RNA element, which is generally known as the IRES (for internal ribosome entry site), fulfilling the same function as the Epsilon and other initiation-enhancing motifs of prokaryotic mRNAs and probably base pairing with the 18S rRNA? This is an explanation for internal initiation that is far from implausible. However, the best studied IRESs, those of the picornaviruses, span a length of approximately 450 nucleotides, whereas all the various elements that have been proposed to enhance initiation in prokaryotes are confined within a much shorter length, from about 50 residues upstream of the initiation codon to a maximum of 50 residues downstream. Moreover, a phylogenetic comparison of the picornavirus IRESs shows that the secondary (and probably tertiary) structure is highly conserved, yet the conserved primary sequence motifs are quite short and mostly confined to unpaired regions (Agol 1991; Jackson et al. 1994; Kaminski et al. 1994b). Thus, if the picornavirus IRESs direct internal initiation through viral RNA/18S rRNA interactions, there must be numerous such RNA/RNA contacts, each over a very short length: a three-dimensional array of many contact points, each very weak, but perhaps collectively strong. In contrast, even the most fanciful view of prokaryotic initiation site recognition invokes a limited number of mRNA/16S rRNA interactions, each over at least five base pairs, and at least conceptually arranged in two-dimensional space, since there is no evidence that a conserved secondary/tertiary structure around the initiation codon is actually required for these enhancers to function.

THE SCANNING RIBOSOME MECHANISM

What Exactly Is Ribosome Scanning?

We are therefore left with the scanning ribosome model as the most plausible model for recognition of the correct initiation site of the vast majority of eukaryotic mRNAs (Kozak 1989a). In essence, this proposes that the primed 40S subunit with associated initiation factors and Met-$tRNA_i$ first binds to the mRNA close to the 5′-cap structure and then migrates in a 5′-3′direction, selecting usually the first AUG codon as the initiation site. But what exactly does scanning mean in mechanistic terms? Is it equivalent to a threading of the mRNA through the 40S ribosomal subunit, much as a bead is threaded on a string? A threading model, where the thread passes through the 40S subunit itself, implies that after termination of translation, the 40S subunit continues to migrate through the 3′UTR and poly(A) tail (albeit in a state where it is not competent to reinitiate) until it reaches the extreme physical 3′end of the mRNA. This seems inherently improbable, and there is in fact no evidence for 40S subunits moving through the 3′UTR. However, this is yet another area where absence of evidence is not evidence of absence, as all attempts to test the idea have to make assumptions about the stability of 40S/3′UTR complexes to sucrose gradient centrifugation or to footprinting techniques, assumptions that may not be warranted.

Perhaps the best evidence against a threading mechanism where the thread-hole is a permanent and integral feature of the 40S subunits is the recent work showing that ribosomes can translate a covalently closed, circular RNA, provided the RNA incorporates an IRES element (Chen and Sarnow 1995). This eliminates threading through a hole that is a permanent feature of the 40S subunit, but it does not eliminate threading through a type of temporary "clasp," composed of the 40S subunit and some initiation factor, perhaps eIF3. In internal initiation on a circular mRNA, the clasp could be formed by the 40S subunit and the initiation factor binding independently but adjacently on the mRNA and joining to form the clasp. In the scanning model, a 40S/eIF3 complex could "thread" from the 5′end, but when initiation occurred and eIF3 dissociated, the thread channel would disappear (perhaps to be replaced by the cleft between the 40S and 60S ribosomal subunits). However, even a model in which scanning involves the threading of the mRNA through a temporary clasp formed of the 40S subunit and possibly eIF3 seems incompatible with the phenomenon of ribosome shunting discussed below.

Is It the 40S Subunit That Moves, or the Helicases?

Although there are many intriguing exceptions, such as internal initiation or the translation of mRNAs with many upstream short ORFs, it is clear that for the majority of eukaryotic cellular and viral mRNAs, initiation occurs exclusively or almost exclusively at the 5′-proximal AUG codon (Kozak 1989a). Mutation of this AUG usually leads to initiation at the next AUG codon downstream, whereas insertion of a new AUG upstream of the original 5′-proximal one results in initiation at this new AUG codon. This underlying 5′-3′directionality in the mechanism of initiation site selection is surely beyond dispute. Does this necessarily mean that the directionality results from migration of the 40S ribosomal subunit through the 5′UTR? Evidence in favor of actual subunit migration has been claimed from the effects of the antibiotic edeine in promoting the binding of several 40S subunits to the mRNA (Kozak and Shatkin 1978; Kozak 1979), the effect of ATP depletion in blocking a 40S subunit at or near the cap (Kozak 1980), and evidence for 40S subunits queuing on long 5′UTRs (Kozak 1989b, 1991b). However, evidence for such queuing and proof that the additional 40S subunits are actually located within the 5′UTR are rather meager. To locate a queuing 40S subunit within the 5′UTR, it was necessary to have a stable hairpin in the UTR a little way upstream of the initiation codon (Kozak 1989b). It was argued that without this hairpin, the 40S subunit would have scanned off the mRNA fragments during the nuclease treatment step necessary to map the sites at which the queuing 40S subunit is located.

The requirement for ATP hydrolysis for translation initiation (Kozak 1980; Jackson 1991) and the trapping of a 40S subunit near the very 5′end of the mRNA when ATP is depleted (Kozak 1980) seemed at the time to be highly suggestive, if not definitive proof, that the subunits actually moved. However, the subsequent realization that two of the translation initiation factors (eIF4F holoenzyme and eIF4A) have ATP-dependent RNA helicase activities demands some reassessment of these ideas (Rozen et al. 1990; Jaramillo et al. 1991). Is the ATP hydrolyzed both by the helicases and by the scanning ribosome or is it required only for the helicases? In the latter eventuality, could the apparent 5′-3′ directionality of the initiation site selection mechanism be related *not* to 40S subunit migration, but to a preferred 5′-3′directionality of the helicase action in unwinding the mRNA from the 5′end? Although it is true that the helicases were reported to be capable of unwinding RNA in either direction (Rozen et al. 1990), the need for an unpaired segment to initiate such unwinding might result in the favored direction being 5′-3′, with the reverse 3′-5′unwinding limited to special cases (Sonenberg 1991, 1993).

In an extreme form of this alternative model, it is postulated that there is no obligatory migration of the 40S ribosomal subunit but that such subunits can bind directly at any site internally located within the mRNA, provided the binding site is unstructured (Sonenberg 1991, 1993). The directionality of the initiation site selection process would then be explained by the preferred 5′-3′directionality of the action of the helicase initiation factors in unwinding the mRNA. This model has the merit that both scanning and internal initiation can be accommodated in a single unified model; in internal initiation, it would be a case of the unwinding starting at some internal site rather than at the 5′end. However, this alternative model has problems in explaining why it is usually the first, 5′-proximal, AUG codon that is selected exclusively as the initiation site, unless it invokes what would appear to be an arbitrary postulate that the unwinding by helicases stops at or shortly after the 5′-proximal AUG codon. It also cannot readily explain translation/reinitiation events such as occur in the translation of yeast GCN4 mRNA, whereas the scanning ribosome model can (Hinnebusch 1994 and this volume).

Both models, the migrating 40S subunit or the migrating helicases, lead to the prediction that there should be a finite lag between addition of mRNA to a system and the first initiation events, a lag that should be related to the length of the 5′UTR. The helicase model predicts that if the mRNA is preincubated with ATP in a system that is complete except for the presence of salt-washed ribosomal subunits, the lag should be largely abolished, and certainly any dependency of the lag on the length of the 5′UTR should disappear, since the unwinding would have occurred in the preincubation step before the addition of ribosomes. Another prediction relates to the fact that dATP can substitute for the ATP required for eIF4F and eIF4A helicase activities (Rozen et al. 1990). Can dATP substitute for the ATP requirement for initiation? If the answer is negative, the implication is that initiation involves at least one additional ATP-dependent step distinct from the helicase-dependent events. For the purposes of further discussion, we will assume that it is the 40S subunit that moves and thereby generates the 5′-3′ directionality.

Is Ribosomal Scanning a Systematic Linear Search or a Random Diffusion?

Although perhaps not explicitly stated as such, an implicit feature of the models presented by Kozak would seem to be that ribosome scanning is a systematic and unidirectional linear migration or search for an appropriate initiation codon. Direct evidence on this question is still lacking. It would be useful to know whether there really is ATP hydrolysis coupled to the actual movement of the 40S preinitiation complex, rather than coupled to the action of the helicase initiation factors. If so, then we would further like to know whether there is a defined stoichiometry of a certain number of nucleotides scanned per molecule of ATP hydrolyzed. A defined stoichiometry of ATP hydrolysis for 40S preinitiation complex scanning per se would clearly support the idea of a systematic and unidirectional migration.

An alternative view is that ribosome scanning is merely a random diffusion process, which from time to time may include some movement in the 3′-5′ direction, as well as in the conventional 5′-3′ direction. According to this interpretation, the overall 5′-3′ directionality of the scanning process would merely be a consequence of the fact that the only entry point for the scanning 40S subunit, the point at which the diffusion process starts, is at or very near the 5′ end of the mRNA. A single entry point at the 5′ end coupled with a random diffusion process would generate behavior that appears to be scanning in a 5′-3′ direction, even though diffusion in the opposite 3′-5′ direction may occur periodically.

According to this view, the scanning eukaryotic ribosomes would show some resemblance to the scanning of prokaryotic 30S subunits which evidently occurs following translation termination, as witnessed by the fact that a termination codon allows otherwise silent but nearby initiation sites to be utilized (Adhin and van Duin 1990; Sonenberg 1993). Although limited to a distance of about 40 residues, this scanning by prokaryotic ribosomes is as "systematic" as by eukaryotic ribosomes insofar as the nearest potential initiation site to the termination codon is always selected (even to the extent of preferring a nearby UUG codon to a more distant AUG), but it is clearly bidirectional at the macroscopic level in that initiation can occur at sites either downstream or upstream of the stop codon, with the small qualification that the efficiency of initiation tends to be lower at an upstream site rather than at a downstream site. These observations suggest that following termination, each 30S subunit scans in a series of steps of random directionality. As there are reports that termination codons in eukaryotic mRNAs can likewise potentiate initiation at otherwise silent sites, even if the AUG initiation codon is located upstream of the termination codon (Peabody and Berg 1986; Peabody et al. 1986; Thomas and Capecchi 1986), it seems that scanning by eukaryotic ribosomes can also be essentially a random diffusion process.

One indication that scanning by eukaryotic ribosomes is not a highly systematic stepwise linear search is the failure of the scanning process to discriminate between two very closely spaced AUG codons. The best example is the NA/NB mRNA of the influenza B viruses, where initiation occurs slightly more frequently at the downstream rather than the upstream of the two AUG codons in the sequence ..AAA**AUG**A ACA**AUG**CUA.. (Williams and Lamb 1989); yet a simple interpretation of the effects of context would suggest that the upstream site should be favored. Of the various mutations and manipulations tested in an attempt to obtain the predicted preference for the 5′-proximal AUG codon, the most effective was, significantly, the introduction of additional sequences to increase the separation of the two AUG codons. Although far from conclusive, this result is more easily explained by a random diffusion model than by a systematic stepwise linear search. It has recently been shown that the upstream of two closely spaced AUG codons will be selected exclusively in vitro provided it is surrounded by *every* known favorable context feature: a 5′UTR of sufficient length, a GCCACC sequence immediately upstream of the 5′-proximal AUG codon, a G at +4, and a hairpin loop at the appropriate distance downstream (Kozak 1995). In contrast to this author's interpretation, these results do not seem to me to prove that scanning of the "average" mRNA is invariably a systematic unidirectional migration.

Another related and still unsolved, or indeed seldom posed, question is whether there is a finite off-rate during scanning. Assuming that ribosomes "enter" at the 5′end of the mRNA at a certain rate, and scan in a 5′-3′direction, a model in which there is no off-rate requires that all those ribosomes which enter at the 5′end must initiate translation at some point in the mRNA. If the context of the 5′-proximal AUG codon is good, then virtually all the ribosomes will initiate at this site. However, if the "strength" of the 5′-proximal site is reduced, either by downgrading the context or by changing it to a non-AUG initiator, then the decrease in initiation events at this 5′-proximal site should be precisely counterbalanced by an increase in initiation frequency at the next suitable initiation site(s). Thus, the total number of initiation events on each mRNA per unit time should be uninfluenced by changes in the strength of the 5′-proximal site and should equal the rate of ribosome loading at the actual 5′end. These predictions certainly seem to be upheld quite well in experiments where the first and second initiation sites are in-frame and quite close together (Kozak 1989c, 1990, 1991a). In other cases where the separation of the first and second in-frame AUG codons is much larger, manipulations that decreased the frequency of initiation at the 5′-proximal site, such as alterations in the Mg^{++} concentration or mutating the initiation codon to CUG, did not result in a compensatory increase in the frequency of initiation at the next in-frame AUG codons further downstream (Grünert and Jackson 1994). However, this analysis was complicated by the uncertainties that there were intervening out-of-frame AUG codons, and the frequency of initiation at these sites could not be evaluated.

Another puzzling observation apparently inconsistent with the idea that every 40S subunit that is loaded at the 5′end of an mRNA necessarily initiates translation at some downstream site concerns the human parainfluenza virus-1 P/C mRNA, where the 5′-proximal initiation site for C′protein synthesis is an exceptionally strong GUG codon, and the synthesis of the P and C proteins is initiated at two downstream AUGs. The curious feature is that mutation of the GUG to AUG results in only a modest (<20%) increase in the yield of C′, yet almost complete abrogation of P and C synthesis (Boeck et al. 1992).

In our attempts to interpret these puzzling data, it would seem prudent to allow for the possibility that there is a finite off-rate and that not every 40S subunit loaded at the 5′end necessarily accomplishes an initiation event further downstream.

Ribosome Shunting or Nonlinear Scanning

The scanning ribosome model postulates that when a scanning 40S ribosomal subunit encounters a hairpin loop in the 5′UTR, it does not skip over the loop but unwinds it, probably with the aid of the helicase initiation factors. This is based on the observation that if the 5′-proximal AUG codon is buried in the stem of such a hairpin loop, it is nevertheless still used as the preferred initiation site, rather than the next AUG codon after the stem-loop (Kozak 1986b). Nevertheless, there seem to be some exceptions to this general rule, most notably the 35S cauliflower mosaic virus (CaMV) RNA, where a fair proportion of the scanning ribosomes, but not all of them, seem to skip over a large segment of the approximately 600-nucleotide 5′UTR, which has no less than seven small ORFs, the last of these overlapping the main 5′-proximal gene VII ORF. The skipping or "ribosomal shunt" was inferred from the fact that insertion of an entire β-glucuronidase (GUS) ORF in the middle of the 35S mRNA 5′UTR reduced expression of a reporter cistron substituted in the gene VII position by only 70% (Fütterer et al. 1993). Thus, it was inferred that in the absence of the GUS insertion, up to two thirds of the scanning ribosomes negotiate the approximately 600-nucleotide 5′UTR by linear scanning and reinitiation, but one third of the scanning ribosomes bypass or "shunt" through the central section of the 5′UTR, with a take-off site upstream of the site of insertion of the GUS cistron and a landing site downstream from this insertion site.

A similar scenario seems to apply to the late adenovirus mRNAs, as witnessed by the fact that insertion of a very stable stem-loop structure in the distal half of the 5′UTR had almost no influence on the frequency of initiation at the authentic site (Schneider 1995). This experiment seems to provide as solid an operational criterion for ribosome shunting as is available at present, and as such, it is very timely, since the growing fashionability of ribosome shunting makes it almost predictable that shunting will be invoked, without any real supporting evidence, as an explanation for all results that cannot obviously be accommodated within the most simplistic reading of the scanning ribosome model (see, e.g., Hellen et al. 1994). The results obtained with the adenovirus late mRNAs suggest the following operational criteria for ribosome shunting: Insertion into the putative bypass region of a stem-loop structure, which is so intrinsically stable that it should neither be disrupted by neighboring sequences nor should it, itself, disrupt neighboring secondary structure motifs, ought to have minimal effect on the frequency of initiation at the authentic initiation codon. Moreover, if the position of insertion in the 5′UTR of this intrinsically stable stem-loop structure is varied, then it should be possible to define the "take-off" and "landing" sites of any scanning ribosomal shunt or bypass. A great deal of potential confusion in the field would be avoided if no claims for a ribosomal shunt mechanism were made without positive evidence of this type that such a bypass of a specific segment of the 5′UTR does indeed operate.

"Cap-independent" Initiation by the Scanning Mechanism

For well over a decade, dating back long before the discovery of true internal initiation of translation as exemplified by the picornavirus IRESs, the translation characteristics of different eukaryotic cellular and viral mRNAs have been classified as "cap-dependent" or "cap-independent."

As we have argued in detail elsewhere (Jackson et al. 1995), this is a most inopportune classification, which we believe has hindered the development of ideas in the field, largely because it is seldom clear whether the term is being used as an operational criterion or as a mechanistic explanation; and when used as a mechanistic interpretation, it is often not at all clear whether or not what is implied is true internal initiation.

There are two valid tests for internal initiation: the dicistronic mRNA assay or even better the circularized RNA system recently developed by Chen and Sarnow (1995). Any mRNA that fails this dicistronic mRNA assay must be presumed to be translated by a 5′-end-dependent scanning mechanism, no matter how "cap-independent" the translation of the mRNA may appear to be according to the various operational criteria that have been applied: (1) a comparatively high resistance of translation to inhibition by cap analogs (m^7GTP or m^7GDP) or by antibodies against eIF4E or eIF4G; (2) a relatively small difference in translation efficiency between capped and uncapped versions of the same mRNA species; (3) a relatively high translational efficiency in cell-free extracts of poliovirus-infected cells; and (4) persistence of translation in vivo following the general shutoff of host-cell mRNA translation caused by poliovirus infection. Among the capped mRNAs that best satisfy one or more of these operational criteria, the most closely studied have been alfalfa mosaic virus RNA 4 (AMV RNA 4), the mRNAs coding for the heat shock proteins, and the adenovirus late mRNAs transcribed from the major late promoter.

Close scrutiny of the data suggests, however, that none of these can be regarded as true cap-independent translation. Although the translation of AMV RNA 4 is quite resistant to inhibition by cap analogs (m^7GDP or m^7GTP), it is not totally resistant, especially if high salt concentrations are used (Herson et al. 1979; Fletcher et al. 1990). Although AMV RNA 4 is translated quite efficiently in extracts of poliovirus-infected HeLa cells (Sonenberg et al. 1982), this relative efficiency (40% with respect to control extracts) is less than was observed with satellite tobacco necrosis virus (STNV) RNA (50%) or the internal initiated encephalomyocarditis virus (EMCV) RNA (90%). As for the heat-shock protein mRNAs, it is true that they are translated with increased efficiency when intracellular eIF4F levels are reduced as a consequence of expression of antisense transcripts to eIF4E mRNA (Joshi-Barve et al. 1992; Rhoads et al. 1993) or under heat shock conditions when eIF4F activity is impaired in some way that is not yet fully understood (Lamphear and Panniers 1991; Rhoads and Lamphear 1995). However, their translation in poliovirus-infected cells is not nearly so resistant to the general shutoff of cellular mRNA translation as is that of BiP mRNA (Muñoz et al. 1984; Sarnow 1989), which, as discussed below, is translatable by a true internal initiation mechanism. Late adenovirus mRNAs with the tripartite 5′UTR were once considered to be translated by cap-independent mechanisms, since their translation persists in poliovirus-infected cells (Castrillo and Carrasco 1987; Dolph et al. 1988), when translation of the vast majority of capped cellular mRNAs is shut off as the result of the inactivation of eIF4F by proteolytic cleavage of the eIF4G subunit. However, translation of such mRNAs in vitro is inhibited if the foot-and-mouth disease virus (FMDV) L protease is expressed in the system to inactivate eIF4F (Thomas et al. 1992), and so the fact that their translation is refractory to poliovirus infection in vivo is probably due to incomplete cleavage of eIF4G.

With the adenovirus late mRNAs, not only does the dicistronic mRNA test for true internal initiation give a negative result, but introduction of a new (out-of-frame) AUG triplet into the 5′UTR strongly inhibits expression from the main ORF (Schneider 1995), a result that is highly indicative of translation by a scanning mechanism. It is unfortunate that this type of AUG insertion experiment has not been more widely used, since it provides the best positive test for scanning-dependent initiation and thus complements the dicistronic mRNA assay for true internal initiation. If the inserted AUG codon is in-frame with the main reading frame, we would expect it to result in the synthesis of a longer product (predominantly but not necessarily exclusively) with little change in total product yield. If the new AUG is out-of-frame, then the mutation should very significantly reduce the yield of product from the authentic initiation codon. In contrast, when an AUG is introduced (or is naturally present) just upstream of the actual internal ribosome entry site of an RNA that is translated by true internal initiation, this upstream AUG is not used as an initiation site (Kaminski et al. 1990).

It has become apparent that the concentration of eIF4F required for maximal rates of scanning-dependent initiation of translation varies according to the mRNA species under scrutiny and appears to be related to the degree of secondary structure present in the 5′UTR, particularly in the extreme 5′-proximal region. Thus, overexpression of eIF4E results in increased expression from mRNAs with hairpin loops in the 5′UTR (Koromilas et al. 1992). Even more instructive are the comparisons between different mRNAs in a fractionated wheat-germ system. The concentration of eIF4F required for half-maximal translation (the apparent K_m) varied widely between different mRNA species, and among all capped mRNAs tested, the apparent K_m was lowest with AMV RNA 4

(Fletcher et al. 1990; Timmer et al. 1993a,b). These considerations suggest that AMV RNA 4, the heat shock protein mRNAs, and adenovirus late mRNAs should not be considered as a distinct class on their own. The balance of evidence suggests that their translation is by the conventional scanning ribosome mechanism and does indeed require eIF4F, albeit in unusually low concentrations. Thus, they merely lie at the extreme (lower) end of the whole spectrum of capped mRNAs that vary in the concentration of eIF4F required for efficient translation.

Also pertinent to these issues are questions about the mechanism and factor requirements for the translation of uncapped forms of mRNAs that are normally capped. Since decapping mRNA decreases the efficiency of translation initiation but does not usually change the preference for selection of the 5′-proximal AUG codon (Kozak 1989a), the implication is that such decapped mRNAs are also translated by a scanning mechanism. This is supported by the recent finding that when an uncapped mRNA is produced in vivo by placing the reporter cistron under the control of an RNA polymerase III promoter, it appears to be translated by a mechanism akin to scanning as judged from the negative influence of an upstream AUG triplet on the expression of the reporter (Gunnery and Mathews 1995). However, contrary to beliefs widely held for a long time, the translation of an uncapped derivative of a normally capped mRNA does actually require eIF4F activity. In the fractionated wheat-germ system, the apparent K_m for eIF4F was manyfold higher for uncapped AMV RNA 4 than for the capped version, and if sufficient eIF4F is added, then the uncapped form is translated as efficiently as the capped species (Fletcher et al. 1990; Timmer et al. 1993a,b). Apart from this difference with respect to eIF4F concentrations, another difference is that only with uncapped mRNA can eIF4F proteolytically cleaved in its eIF4G subunit substitute effectively for the intact eIF4F holoenzyme: Prior expression of FMDV L protease in vitro resulted in the expected inhibition of translation of a capped mRNA in the subsequent assay step, but caused an unexpected stimulation of translation of the uncapped version of the same mRNA species (Ohlmann et al. 1995).

Left to the end of this discussion is the case of the naturally uncapped STNV RNA, which represents probably the greatest challenge to any attempt to classify an RNA according to initiation mechanisms. The translation of this RNA in a fractionated wheat-germ system exhibits a lower apparent K_m for eIF4F than any other mRNA species, and unlike other mRNAs (such as AMV RNA 4), the K_m is uninfluenced by capping (Fletcher et al. 1990; Timmer et al. 1993a,b). This property seems to be conferred largely by an approximately 100-nucleotide 3′UTR segment just downstream from the translation termination codon and to a lesser extent by the short 29-nucleotide 5′UTR. Mutations and deletions in either element reduce the efficiency of translation of uncapped STNV RNA, but this effect can be reversed by capping (Timmer et al. 1993b). Nevertheless, there is as yet no evidence that STNV RNA is translated by true internal initiation, and until positive results are obtained from dicistronic mRNA assays, the presumption must be that it is translated by a 5′-end-dependent mechanism, albeit with a requirement for eIF4F only at very low concentrations and no requirement for a 5′-cap, both properties being conferred partly by the unusual nature of the 5′UTR and more especially by the influence of the 3′UTR motif (Danthine et al. 1993; Timmer et al. 1993b).

INTERNAL INITIATION IN EUKARYOTES

The Picornavirus Paradigm

A detailed discussion of internal initiation of translation of picornavirus RNAs is given elsewhere in this volume (Ehrenfeld) and in recent reviews (Jackson et al. 1994, 1995; Kaminski et al. 1994b), whereas Figure 2 depicts current ideas in a diagrammatic form. In summary, all picornavirus IRESs are about 450 nucleotides long, and by the criteria of sequence conservation can be divided into one minor and two major groups: (1) hepatitis A virus (the minor group); (2) the entero- and rhinovirus IRESs; and (3) the cardio- and aphthovirus IRESs. Within each group there is strong conservation of IRES secondary structure and fair conservation of primary sequence, but there is little obvious homology between groups. The phylogenetic comparisons within each group suggest that most of the 450 nucleotides are required by virtue of its secondary, and presumably also tertiary structure, in order to present a number of quite short conserved primary sequence motifs, most of them in unpaired regions, in the correct three-dimensional spatial organization (Jackson et al. 1994, 1995; Kaminski et al. 1994b).

The actual ribosome entry site, defined as the most 5′-proximal point at which initiation can occur, is at an AUG codon located at the 3′end of the IRES, some 25 nucleotides downstream from a pyrimidine-rich tract which is the only extended primary sequence motif common to all picornavirus IRESs (Jackson et al. 1990). If another AUG codon occurs naturally or is introduced by mutation slightly upstream of this 3′-terminal AUG (i.e., closer to the oligopyrimidine tract), it neither is used as a functional initiation site nor influences initiation at the correct site (Kaminski et al. 1990). As the segment between the pyrimidine-rich tract at the AUG entry site is always G-poor but can otherwise be quite vari-

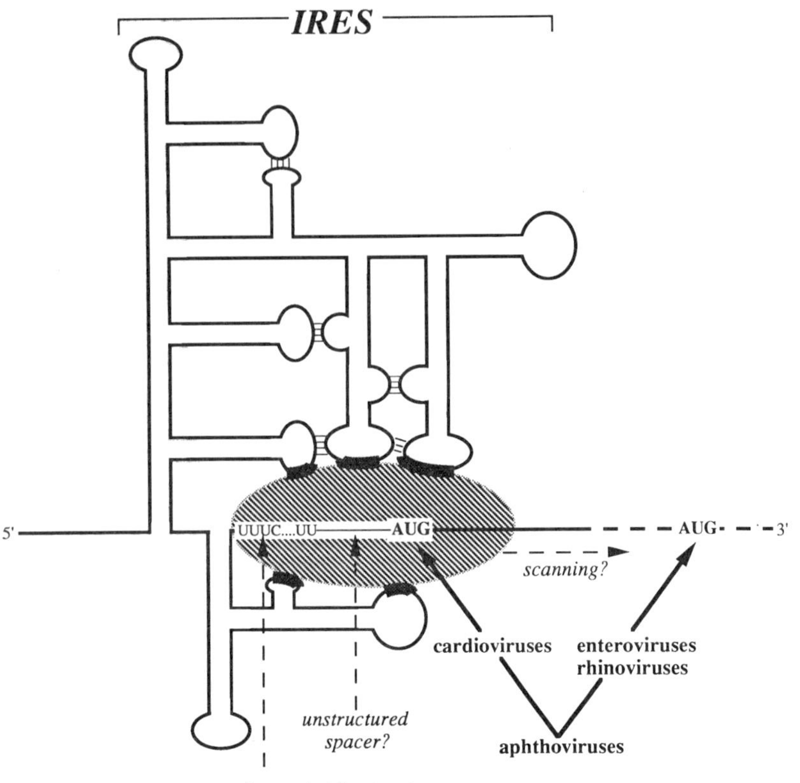

Figure 2 General model for internal initiation of translation on picornavirus IRESs. The IRES consists of a number of base-paired stem-loops, with possible tertiary structure interactions between the loops and bulges. This structure presents several quite short unpaired primary sequence motifs, denoted by thickened lines, in the appropriate three-dimensional spatial organization for internal ribosome entry. These motifs may be binding sites for specific RNA-binding proteins, which are then recognized by the initiating ribosome (*shaded oval*) or they may be recognized directly by the ribosome and associated initiation factors. The actual ribosome entry site is at the AUG triplet located at the 3′end of the IRES, some 25 nucleotides downstream from the start of a conserved pyrimidine-rich tract. The segment between the oligopyrimidine tract and the AUG entry site is thought to serve as an unstructured spacer of defined length. The AUG at the 3′end of the IRES is the authentic initiation site for viral polyprotein synthesis in the cardioviruses. In the case of the entero-/rhino-viruses, this AUG is likewise the ribosome entry site but is not used as an initiation site; the ribosomes are transferred, probably by a scanning mechanism, to the authentic initiation site, which is the next AUG codon further downstream. The aphthoviruses represent an intermediate case, with some of the internally entering ribosomes initiating synthesis at the AUG at the 3′end of the IRES and some utilizing the next downstream AUG.

able in sequence (but not length) between different virus species and strains, it is likely to function as a spacer of defined length. Evidence in support of this idea has come from studying the effects of changing its length (Pilipenko et al. 1992; Gmyl et al. 1993; Kaminski et al. 1994a). The pyrimidine tract itself may also be part of this spacer, since mutation of the pyrimidines in cardiovirus IRESs has comparatively little effect on internal initiation efficiency (Kaminski et al. 1994a; Pilipenko et al. 1994), although in the case of poliovirus, the 5′-proximal UUUC sequence must be retained (Nicholson et al. 1991; Pestova et al. 1991; Pilipenko et al. 1992).

The critical AUG codon located at the 3′end of the IRES and 25 nucleotides downstream from the oligopyrimidine tract is the main initiation site in cardioviruses, although some 15% of the ribosomes may initiate at the next AUG located a short distance downstream (Hunt et al. 1993; Pilipenko et al. 1994). In the aphthoviruses, a slight minority of the internally entering ribosomes initiate translation at this AUG, whereas the rest are transferred, probably by a scanning mechanism, to the next AUG codon downstream (Belsham 1992). In contrast, in the case of the entero-/rhinoviruses, there is virtually no initiation at the AUG codon at the 3′end of the IRES, and it is the next AUG downstream that is the functional initiation site. The sequences between these two AUG codons are hypervariable between different poliovirus strains but never include an additional AUG triplet (Poyry et al. 1992). Moreover, deletions or insertions in this region have little effect on translation efficiency or infectivity unless a hairpin loop or a sequence with an AUG codon is inserted (Kuge and Nomoto 1987; Pelletier and Sonenberg 1988; Kuge et al. 1989). These results suggest that the ribosomes first enter at the silent AUG triplet at the 3′end of the IRES and then scan to the authentic initiation codon, although other models of a more direct transfer have been suggested (Hellen et al. 1994). The question of why ribosomes should show different behavior toward the AUG codon at the ribosome entry site at the 3′end of the IRESs of different picornavirus species is still an enigma, and the reader is referred to recent reviews for more detailed discussion of possible explanations (Jackson et al. 1994, 1995; Kaminski et al. 1994b).

Contrasts between the IRESs of Picornaviruses and Other RNAs

Among cellular mRNAs that have been proven to be translated by internal initiation by the criterion of the dicistronic mRNA assay, the best characterized are the *Antennapedia* mRNA of *Drosophila* and mammalian BiP mRNA (Macejak and Sarnow 1991; OH et al. 1992), which codes for immunoglobulin heavy-chain binding protein, also known as GRP 78 (78-kD glucose-regulated protein). The 5′UTRs of the two forms of the *Antennapedia* mRNA, derived from two different promoters, are exceedingly long (>1500 nucleotides) and have several short ORFs particularly toward the 5′end. However, even though the complete 5′UTR gave maximum IRES activity, 35–40% of this activity was obtained with just exons D and E, the approximately 400 nucleotides of 5′UTR sequence immediately preceding the authentic initiation codon, a segment that is devoid of upstream AUG codons (OH et al. 1992). (Even exon D on its own had about 20% of the IRES activity of the full-length 5′UTR.) The approximately 200-nucleotide BiP mRNA 5′UTR is likewise devoid of AUG codons. In neither case do we know the exact internal ribosome entry point or whether AUG codons inserted upstream of the correct initiation site would be ignored as potential initiation codons.

Of the viral IRESs other than those of picornaviruses, the best studied is that in the hepatitis C virus (HCV) genome. The 5′UTR of HCV is approximately 341 nucleotides long with up to five upstream AUG codons (the length and number of AUG codons depending on the particular strain) and is by far the most conserved region of the genome among different HCV strains and isolates (Bukh et al. 1992). In dicistronic mRNA assays in which the downstream cistron was the 5′-proximal part of the HCV core protein-coding sequence, the 5′boundary of the IRES was mapped to a region which implied that the IRES was the approximately 200 nucleotides immediately preceding the authentic initiation codon (Tsukiyama-Kohara et al. 1992). However, subsequent experiments from other groups using standard reporter ORFs for both cistrons of the dicistronic construct mapped the 5′boundary as lying between nucleotides 30 and 70 of the viral genome, implying a larger IRES of some 300 nucleotides but still significantly shorter than the picornavirus IRESs (Wang et al. 1993). In our hands, there is a very strong requirement for the first approximately 10 codons of the HCV polyprotein coding sequences for efficient internal initiation (Reynolds et al. 1995), in contrast to the picornaviruses where there seems to be little influence of coding sequences on IRES function (Hunt et al. 1993). As a possible correlation with the presence of strong determinants of HCV IRES function lying downstream from the authentic initiation codon, internal initiation driven by the HCV IRES retained high relative efficiency (80–90%) even when the authentic initiation codon was mutated to CUG or AUU (Reynolds et al. 1995), a result that has never been found with any picornavirus IRES. On the other hand, one feature in common with at least the cardiovirus subfamily of the Picornaviridae is that the actual internal ribosome entry site on the HCV genome is at or very close to the authentic initiation codon, since an AUG codon introduced just upstream of the authentic site was not utilized under conditions of IRES-driven initiation but was used if most of the IRES was deleted to convert the RNA to a form translated by the 5′-end-dependent scanning mechanism.

There is no obvious resemblance among the IRESs of *Antennapedia*, BiP, and HCV RNAs, nor do any of these IRESs show clear homology with any picornavirus IRES. None of these three IRESs has a pyrimidine-rich tract located 25 nucleotides upstream of the initiation site or putative ribosome entry site, as in the picornavirus IRESs. Although there is a pyrimidine-rich tract followed by an AUG triplet roughly central in the HCV 5′UTR, mutation of the pyrimidine tract (Wang et al. 1994) or of the AUG triplet had no effect on internal initiation (J.E. Reynolds and R.J. Jackson, unpubl.), and indeed this particular AUG is missing from a minority of field isolates of HCV (Bukh et al. 1992).

If internal initiation is dependent on a number of base-pairing interactions between the IRES element and the small ribosomal subunit rRNA, then it seems possible that it may be a different set of interactions involving different rRNA motifs for the different types of IRES element, in much the same way as different types of enhancers of prokaryotic mRNA translation (e.g., the Epsilon element and Ω) are postulated to base pair with different regions of 16S rRNA. However, as the phylogenetic comparison of the picornavirus IRESs implies that any conserved motifs which might be involved in such base-pairing interactions are highly dispersed in the linear sequence and are each quite short (Kaminski et al. 1994b), it is really rather difficult to identify similar motifs in an entirely unrelated IRES. Thus, it may be premature to rule out the possibility that these non-picornavirus IRESs lack the critical conserved short primary sequence elements found in the picornaviruses.

Trans-acting Factors Required for Internal Initiation

As discussed in detail elsewhere in this volume (Ehrenfeld) and in other recent reviews (Jackson et al. 1994, 1995; Kaminski et al. 1994b), internal initiation dependent on the picornavirus IRESs seems to require the same set of canonical initiation factors in roughly the same concentrations as are needed for translation of capped mRNAs. Perhaps surprisingly, this conclusion also extends to eIF4F (Scheper et al. 1992), although with the qualification that eIF4F which has been cleaved in its largest (eIF4G) subunit by picornavirus proteases is as active, if not more active, in promoting internal initiation than intact eIF4F holoenzyme (Buckley and Ehrenfeld 1987; Thomas et al. 1992; Liebig et al. 1993), whereas only the intact factor is active with respect to capped mRNA translation. It has recently been shown that cleavage of endogenous eIF4F by FMDV L protease also stimulates translation of uncapped mRNAs in reticulocyte lysates (Ohlmann et al. 1995), notwithstanding the fact that such mRNAs are most probably translated by a scanning mechanism. Thus, intact eIF4F is specifically required only for translation of capped mRNAs by the scanning mechanism, and the cleaved factor suffices for translation of uncapped mRNAs irrespective of whether such translation is by scanning or by internal ribosome entry. The cleavage splits eIF4G into an amino-terminal one-third fragment to which eIF4E is bound, and a carboxy-terminal two-thirds fragment associated with eIF4A (Lamphear et al. 1995; Mader et al. 1995). Thus, it is almost certainly the carboxy-terminal cleavage product and its associated eIF4A helicase which is the fragment that is the active entity in promoting internal initiation and the translation of uncapped mRNAs.

Despite similar initiation factor requirements, however, there does seem to be a difference with respect to the ATP requirement for initiation driven by the EMCV IRES as opposed to scanning-dependent initiation. Using sucrose density gradient centrifugation to detect the formation of initiation complexes in ATP-depleted reticulocyte lysates, it was found that complex formation with EMCV virion RNA required either very little hydrolyzable ATP or possibly none at all, the uncertainty being due to technical constraints (Jackson 1991). This has recently been confirmed using an entirely different assay method, and in this case it could be shown that ATP is actually needed but only at very much lower concentrations than required for mRNAs translated by the scanning mechanism (S. Grünert and R.J. Jackson, in prep.). This observation may be correlated with the fact that internal initiation of EMCV RNA translation requires very little, if any, ribosome scanning (Kaminski et al. 1990).

Even though internal initiation seems to require the same set of canonical initiation factors as the conventional scanning mechanism, it may well need additional *trans*-acting factors as witnessed by the fact that no IRES has been reported to function in the wheat-germ extract, although such extracts are capable of translating the vast majority of capped mRNAs that are translated by a scanning mechanism. In addition, poliovirus and rhinovirus RNAs are translated efficiently and accurately in extracts of HeLa cells or (to a lesser extent) Krebs II ascites or L cells, yet are inefficiently and inaccurately translated in rabbit reticulocyte lysates unless HeLa cell (or Krebs II ascites cell) factors are added

(Brown and Ehrenfeld 1979; Dorner et al. 1984; Svitkin et al. 1988; Borman et al. 1993). Thus, reticulocyte lysates evidently contain sufficient quantities of such *trans*-acting factors as may be required for cardio-/aphthovirus IRES function, but they are deficient in factors needed for initiation driven by the entero-/rhinovirus IRESs. Moreover, the translation of hepatitis A virus RNA in reticulocyte lysates was stimulated by liver extracts, but not by a HeLa cell extract that was active in stimulation of poliovirus IRES function (Glass and Summers 1993).

The search for potential *trans*-acting factors is often made by UV cross-linking cellular proteins to radioactively labeled IRESs or by gel-retardation assays (Kaminski et al. 1994b; Ehrenfeld, this volume). However, the problem with such assays is that they offer no evidence that the RNA-binding protein really does have a direct role in the internal initiation mechanism. A salutary warning against automatically assuming that a protein which binds to a specific site within the IRES must necessarily have a role in internal initiation is provided by the example of the specific binding of a 50-kD HeLa cell protein to a particular stem-loop in the poliovirus IRES, which, however, is completely missing in certain viable deletion mutants (Najita and Sarnow 1990).

There is thus really no substitute for functional assays, and so far, only two IRES-binding proteins have been examined in this way. One is the autoantigen La, which is present in HeLa cell extracts at higher concentrations than in rabbit reticulocyte lysates and cross-links to sequences near the 3′end of the poliovirus IRES (Meerovitch et al. 1989, 1993). Addition of recombinant La to rabbit reticulocyte lysates stimulates initiation at the authentic initiation site of poliovirus RNA and suppresses initiation at incorrect sites, and thus mimics the effect of supplementing the lysates with HeLa (or Krebs II ascites) cell extracts (Meerovitch et al. 1993). However, the concentrations of recombinant La required are very high indeed (~100 μg/ml), certainly much higher than would be present when HeLa cell extract is added to reticulocyte lysates to achieve the same outcome. This problem is illustrated by Figure 1 of Svitkin et al. (1994b), where 1 μg of La elicits a distinctly lower response than 15 μg of Krebs II ascites cell ribosomal salt wash; this result implies that La ought to represent 10% of the salt wash proteins, which most clearly is not the case. An additional discrepancy is that the dissociation constant for the binding of La to the poliovirus 5′UTR was reported to be less than 10 nM, which is not consistent with the fact that molar ratios of La/poliovirus RNA in excess of 100-fold had to be used to elicit a response in translation assays (Svitkin et al. 1994b). Moreover, when HeLa cell extracts are fractionated to purify the factor that specifically stimulates translation dependent on the rhinovirus IRES, two activities are resolved, each of which is active on its own (Borman et al. 1993), but neither of which copurifies with La (S.L. Hunt and R.J. Jackson, unpubl.). One of these two activities appears to be identical to PTB (discussed below) in purification characteristics and specific activity. The other remains unidentified, but copurifies, to apparent homogeneity, with a 38-kD protein and a 96-kD protein that is cross-linkable to the rhinovirus IRES. Rhinovirus and poliovirus IRES function in reticulocyte lysates is strongly stimulated by addition of either of the two activities at concentrations in the range 1–5 μg/ml. Thus, although the results show that La in high concentrations can promote initiation driven by the poliovirus IRES, there are reasons to suspect that it may be mimicking some other entity and that it is not the major physiologically relevant factor for efficient poliovirus RNA translation.

The other IRES-binding protein that has been shown to promote internal initiation in a functional assay is PTB (polypyrimidine-tract-binding protein), a protein that was once thought to be a constitutive pre-mRNA splicing factor possibly functioning in conjuction with another such factor, PSF (Patton et al. 1991, 1993), but which is now considered more likely to be a negative regulator of alternative splicing (Lin and Patton 1995; Singh et al. 1995). PTB binds to all picornavirus IRESs but appears to have higher affinity for the IRESs of the cardio-/aphthoviruses (Borman et al. 1993). When HeLa cell extracts were immunodepleted of PTB, the capacity for translation dependent on the EMCV or poliovirus IRES was lost, but translation of mRNAs dependent on the scanning ribosome mechanism was unimpaired (Hellen et al. 1993). However, the capacity for IRES-driven translation could not be restored by addition of recombinant PTB, a result that raised some doubt as to whether PTB was really the activity required for IRES-dependent initiation. This uncertainty has been resolved by using an affinity column method to deplete reticulocyte lysates of PTB. The depleted lysates retained the capacity for translation dependent on the scanning mechanism but lost all capacity for translation dependent on the EMCV IRES, which could, however, be fully restored by addition of recombinant PTB (Kaminski et al. 1995). However, translation dependent on the IRES of another cardiovirus, Theiler's murine encephalomyelitis virus (TMEV), was only slightly impaired in the depleted lysates and could not be improved by addition of PTB, a surprising result in view of the fact that PTB cross-links efficiently to the TMEV IRES. The depleted lysate was also completely unimpaired for translation dependent on the HCV IRES, but this is possibly less surprising since cross-linking of PTB to the HCV IRES is barely above the background levels seen with random RNAs (Kaminski et al. 1995). Moreover, there is no cross-linking of PTB to either of the two well-characterized IRESs found in cellular mRNAs, notably BiP mRNA and the *Antennapedia* mRNA of *Drosophila.* Thus, PTB is certainly required for the function of some IRESs, but by no means all of them; among the picornavirus IRESs, there is no obvious correlation between the requirement for PTB for internal initiation driven by a particular IRES and the labeling of PTB in a standard UV cross-linking assay with that same IRES.

What is striking about the outcome of the searches conducted so far for *trans*-acting factors is that no single factor essential for internal initiation on all IRESs has been found, nor are there even any hints that such a universal factor may exist. This raises some doubt as to whether either PTB or La, or indeed any other IRES-binding protein, can be regarded as an essential catalyst of the internal initiation process in the same way as, say, eIF2 carries out an essential catalytic function in initiation. Moreover, if PTB were to function in internal initiation of translation of eukaryotic mRNAs in much the same way as ribosomal protein S1 in the prokaryotic system, we might expect that there should be a single specific PTB-binding site on the 40S ribosomal subunit, which does not appear to be the case according to current evidence. In addition, except at very high mRNA concentrations, excess S1 is inhibitory in prokaryotic systems (Subramanian 1983; Boni et al. 1991), presumably because the excess free S1 binds to the mRNA and thus prevents the ribosome-associated S1 from making contact with the mRNA. In contrast, there is no evidence (yet) that excess PTB inhibits internal initiation.

One possible explanation for why PTB is required for internal initiation with some IRESs but not others may be that by binding to its specific site(s) on the IRES, it refolds the RNA so as to present the important sequence motifs in the correct orientation; this chaperone-like function has already been proposed in the case of La (Meerovitch and Sonenberg 1993; Svitkin et al. 1994a). Those IRESs whose function is independent of PTB may adopt the appropriate higher-order RNA structure in the absence of bound PTB. Against this view one might wonder why IRESs have not evolved to achieve the optimal conformation without the need for PTB binding. However, if the evolution of the viral IRESs has taken place in an environment in which PTB and La were invariably present, then it follows that if an IRES has evolved to the stage where it can carry out its function optimally provided PTB is bound to it, then there is no selective pressure for further evolution toward a PTB-independent structure.

CONCLUSIONS AND FUTURE PROSPECTS

We are thus faced with three apparently very different modes of translation initiation: the prokaryotic mechanism, highly dependent on base pairing between the SD motif and the 16S rRNA but also influenced by various enhancer sequences that may also work through base pairing with rRNA; the eukaryotic scanning mechanism dependent on the capped 5′end of the mRNA; and the eukaryotic internal initiation mechanism. Those who seek infinite diversity might argue that each of the two modes of eukaryotic initiation can be further subdivided. However, as suggested previously, the distinction between cap-dependent initiation and the so-called cap-independent translation that is still nevertheless dependent on the 5′end of the mRNA, and very probably also on ribosome scanning, is likely to be an artificial subdivision which is merely a reflection of the fact that the apparent K_m for eIF4F differs quite widely between different mRNA species (Fletcher et al. 1990; Timmer et al. 1993a,b).

As for internal initiation, there is some temptation to subdivide this further into different categories, given the limited sequence homology between the IRESs of the cardio-/aphthovirus subgroup of picornaviruses and the entero-/rhinoviruses, and the total lack of homology between any picornavirus IRES and all of the nonpicornavirus IRESs. The difference in the *trans*-acting factor requirements might provide an additional justification and alternative criteria for subdivision. However, it seems extraordinary enough that two different basic mechanisms of initiation, internal initiation and 5′-end-dependent scanning, should have evolved in eukaryotes, and it seems instinctively improbable that there should be two or more entirely different mechanisms of internal initiation. It seems much more likely that all internal initiation follows roughly the same mechanism, same pathway, and same principles but that we are not yet in a position to perceive these common features. One possibility is that all internal initiation depends on base pairing between motifs in the IRES and segments of the 18S rRNA, a hypothesis that would allow for some variation between different IRESs to the extent that not all IRESs might pair with the same set of 18S rRNA motifs. Internal initiation in eukaryotes would thus be seen as similar to prokaryotic initiation, or at least intermediate between prokaryotic initiation and ribosome scanning. By no means the least attractive aspect of this idea is that it is probably untestable, certainly in the immediate future, and thus can be assured of a long half-life while other more immediately testable ideas can be tried out and found wanting. Ultimately, the only way to prove the base-pairing hypothesis would be to generate compensating mutations in the

IRES and in the rRNA, an experiment that can only be contemplated at present with yeast.

Alternatives to the base-pairing hypothesis place an emphasis on *trans*-acting proteins that bind to the IRES. Internal initiation might occur as a result of these proteins binding to specific sites on the IRES and then the primed 40S preinitiation complex recognizing the bound proteins via specific protein/protein interactions. This type of hypothesis runs into the problem of the apparent differences between various IRESs with respect to the *trans*-acting factors required for internal initiation and the proteins that bind specifically to each IRES. There is no indication yet of any universal initiation factor required specifically and exclusively for internal initiation. Moreover, hypotheses that emphasize the importance of protein/protein interactions require that the specific IRES-binding protein, for example, PTB, should also interact specifically either with one of the canonical initiation factors or with a ribosomal protein. There is no indication yet of any such specific protein/protein interactions with any IRES-binding protein. This is another area where a system amenable to genetic analysis could greatly help resolve many outstanding questions, and thus the recent report that internal initiation is operative in yeast, or at least in yeast cell-free systems (Iizuka et al. 1994), takes on great significance, offering the prospect of identifying *trans*-acting factors by genetic screens rather than by the more capricious biochemical assays used at present.

How fundamentally different is internal initiation from initiation by ribosome scanning? The similarities in the initiation factor requirements, albeit with some difference in the required concentration of ATP, suggest strong similarities between the two processes. Moreover, it seems inherently improbable that ribosomes, or rather the primed 40S/Met-$tRNA_i$ preinitiation complex, would behave in an entirely different way when confronted with the two types of mRNAs. A more probable scenario is that these are two different ways of delivering the preinitiation complex to the correct site. In internal initiation, this is perhaps achieved by direct RNA/RNA interaction or RNA/protein interaction at the right site. For the vast majority of cellular mRNAs and capped viral mRNAs lacking such signals for direct internal ribosome entry, the delivery has to be by the scanning mechanism, which can be looked upon as a more efficient analog of the way in which initiation of fd gene VII translation can only be achieved by ribosomes that have been delivered to the proximity of the initiation site as a result of translating the upstream gene V cistron (Ivey-Hoyle and Steege 1992).

Aside from delving further into the minutiae of the two mechanisms of eukaryotic initiation, the conceptual challenges for the immediate future are to perceive the aspects of internal initiation that are common to all IRES elements and to understand the extent to which internal initiation and initiation by ribosome scanning undoubtedly share features in common.

ACKNOWLEDGMENTS

Although I alone bear responsibility for any outlandish misinterpretations, I would like to thank present and recent past members of my group for their numerous inputs toward the development of these ideas: Ann Kaminski, Sarah Hunt, Michael Howell, Andrew Borman, Stefan Grünert, Catherine Gibbs, Joanna Reynolds, Simon Fletcher, Carola Lempke, Paul Crisell, and Annette Lasham. Likewise I acknowledge the valuable contributions of recent collaborators: Kathie Kean, Tim Skern, Bert Semler, Sue Milburn, and John Hershey. Work from our own laboratory described herein was supported by grants from the Wellcome Trust.

REFERENCES

Adhin, M.R. and J. van Duin. 1990. Scanning model for translational reinitiation in eubacteria. *J. Mol. Biol.* **213:** 811–818.

Agol, V.I. 1991. The 5′-untranslated region of picornaviral genomes. *Adv. Virus Res.* **40:** 103–180.

Backendorf, C., G.P. Overbeek, J.H. van Boom, G. van der Marel, G. Veeneman, and J. van Duin. 1980. Role of 16S RNA in ribosome messenger recognition. *Eur. J. Biochem.* **110:** 599–604.

Belsham, G.J. 1992. Dual initiation sites of protein synthesis on foot-and-mouth disease virus RNA are selected following internal entry and scanning of ribosomes in vivo. *EMBO. J.* **11:** 1106–1110.

Benne, R. and J.W.B. Hershey. 1978. The mechanism of action of protein synthesis initiation factors from rabbit reticulocytes. *J. Biol. Chem.* **253:** 3078–3087.

Benne, R., N. Naaktgeboren, J. Gubbens, and H.O. Voorma. 1973. Recycling of initiation factors IF-1, IF-2 and IF-3. *Eur. J. Biochem.* **32:** 372–380.

Berkhout, B. and J. van Duin. 1985. Mechanism of translational coupling between coat protein and replicase genes of RNA bacteriophage MS2. *Nucleic Acids Res.* **13:** 6955–6967.

Blumberg, B.M., T. Nakamoto, and F.J. Kezdy. 1979. Kinetics of initiation of bacterial protein synthesis. *Proc. Natl. Acad. Sci.* **76:** 251–255.

Boeck, R., J. Curran, Y. Matsuoka, R. Compans, and D. Kolakofsky. 1992. The parainfluenza virus type 1 P/C gene uses a very efficient GUG codon to start its C′ protein. *J. Virol.* **66:** 1765–1768.

Boni, I., D. Isaeva, M. Musychenko, and N. Tzareva. 1991. Ribosome-messenger recognition: mRNA target sites for ribosomal protein S1. *Nucleic Acids Res.* **19:** 155–162.

Borman, A., M.T. Howell, J.G. Patton, and R.J. Jackson. 1993. The involvement of a spliceosome component in internal initiation of human rhinovirus RNA translation. *J. Gen. Virol.* **74:** 1775–1788.

Brown, B. and E. Ehrenfeld. 1979. Translation of poliovirus RNA in vitro: changes in cleavage pattern and initiation sites by ribosomal salt wash. *Virology* **97:** 396–405.

Buckley, B. and E. Ehrenfeld. 1987. The cap-binding protein complex in uninfected and poliovirus infected HeLa cells. *J. Biol. Chem.* **262:** 13599–13606.

Bukh, J., R.H. Purcell, and R.H. Miller. 1992. Sequence analysis of the 5′ noncoding region of hepatitis C virus. *Proc. Natl. Acad. Sci.* **89:** 4942–4946.

Calogero, R.A., C.L. Pon, M.A. Canonaco, and C.O. Gualerzi. 1988. Selection of the mRNA translation initiation region by *Escherichia coli* ribosomes. *Proc. Natl. Acad. Sci.* **85:** 6427–6431.

Castrillo, J.L. and L. Carrasco. 1987. Adenovirus late protein synthesis is resistant to the inhibition of translation induced by poliovirus. *J. Biol. Chem.* **262:** 7328–7334.

Chen C.-Y. and P. Sarnow. 1995. Initiation of protein synthesis by the eukaryotic translational apparatus on circular RNAs. *Science* **268:** 415–417.

Danthine, X., J. Seurinck, F. Meulewaeter, M. van Montagu, and M. Cornelissen. 1993. The 3′ untranslated region of satellite tobacco necrosis virus RNA stimulates translation in vitro. *Mol. Cell. Biol.* **13:** 3340–3349.

Darnbrough C.H., S. Legon, T. Hunt, and R.J. Jackson. 1973. Initiation of protein synthesis: Evidence for messenger RNA-independent binding of methionyl-transfer RNA to the 40S ribosomal subunit. *J. Mol. Biol.* **76:** 379–403.

Dasso, M.C., S.C. Milburn, J.W.B. Hershey, and R.J. Jackson. 1990. Selection of the 5′-proximal translation initiation site is influenced by mRNA and eIF-2 concentrations. *Eur. J. Biochem.* **187:** 361–371.

de Smit, M. and J. van Duin. 1990a. Control of prokaryotic translational initiation by mRNA secondary structure. *Prog. Nucleic Acid Res. Mol. Biol.* **38:** 1–35

———. 1990b. Secondary structure of ribosome binding site determines translational efficiency. *Proc. Natl. Acad. Sci.* **87:** 7668–7672.

Dolph, P.J., V. Racaniello, A. Villamarin, F. Palladion, and R.J. Schneider. 1988. The adenovirus tripartite leader eliminates the requirement for cap binding protein during translation inititation. *J. Virol.* **62:** 2059–2066.

Dorner, A.J., B.L. Semler, R.J. Jackson, R. Hanecak, E. Duprey, and E. Wimmer. 1984. In vitro translation of poliovirus RNA: Utilization of internal initiation sites in reticulocyte lysate. *J. Virol.* **50:** 507–514.

Dreyfus, M. 1988. What consitutes the signal for initiation of protein synthesis on *Escherichia coli* mRNAs? *J. Mol. Biol.* **204:** 79–94.

Faxen, M., J. Plumbridge, and L.A. Isaksson. 1991. Codon choice and potential complementarity between mRNA downstream of the initiation codon and bases 1471-1480 in 16S ribosomal RNA affects expression of *glnS*. *Nucleic Acids Res.* **19:** 5247–5251.

Fletcher, L., S.D. Corbin, K.G. Browning, and J.M. Ravel. 1990. The absence of a m^7G cap on β-globin mRNA and alfalfa mosaic virus RNA 4 increases the amounts of initiation factor 4F required for translation. *J. Biol. Chem.* **265:** 19582–19587.

Fütterer, J., Z. Kiss-Laszlo, and T. Hohn. 1993. Nonlinear ribosome migration on cauliflower mosaic virus 35S RNA. *Cell* **73:** 789–802.

Gallie, D.R. and C.I. Kado. 1989. A translational enhancer derived from tobacco mosaic virus is functionally equivalent to a Shine-Dalgarno sequence. *Proc. Natl. Acad. Sci.*

86: 129–132.

Glanville, N., M. Ranki, J. Morser, L Kääriäinen, and A.E. Smith. 1976. Initiation of translation directed by 42S and 26S RNAs from Semliki forest virus in vitro. *Proc. Natl. Acad. Sci.* **73:** 3059–3063.

Glass, M.J. and D.F. Summers. 1993. Identification of a *trans*-acting activity from liver that stimulates hepatitis A virus translation in vitro. *Virology* **193:** 1047–1050.

Gmyl, A.P., E.V. Pilipenko, S.V. Maslova, G.A. Belov, and V.I. Agol. 1993. Functional and genetic plasticities of the poliovirus genome: Quasi-infectious RNAs modified in the 5′- untranslated region yield a variety of pseudo-revertants. *J. Virol.* **67:** 6309–6316.

Goelet, P., G.P. Lommonosoff, P.J.G. Butler, M.E. Akam, M.J. Gait, and J. Karn. 1982. Nucleotide sequence of tobacco mosaic virus RNA. *Proc. Natl. Acad. Sci.* **79:** 5818–5822.

Gold, L. 1988. Post-transcriptional regulatory mechanisms in *Escherichia coli. Annu. Rev. Biochem.* **57:** 199–233.

Gold, L., D. Pribnow, T. Schneider, S. Shinedling, B.S. Singer, and G. Stormo. 1981. Translational initiation in prokaryotes. *Annu. Rev. Microbiol.* **35:** 365–403.

Grant, C.M. and A.G. Hinnebusch. 1994. Effect of sequence context at stop codons on efficiency of reinitiation in GCN4 translational control. *Mol. Cell. Biol.* **14:** 606–618.

Grünert, S. and R.J. Jackson. 1994. The immediate downstream codon strongly influences the efficiency of utilization of eukaryotic translation initiation sites. *EMBO J.* **13:** 3618–3630.

Gualerzi, C.O. and C.L. Pon. 1990. Initiation of mRNA translation in prokaryotes. *Biochemistry* **29:** 5881–5889.

Gualerzi, C.O., G. Risuleo, and C.L. Pon. 1977. Initial rate kinetic analysis of the mechanism of initiation complex formation and the role of initiation factor IF-3. *Biochemistry* **16:** 1684–1689.

Gunnery, S. and M.B. Mathews. 1995. Functional mRNA can be generated by RNA polymerase III. *Mol. Cell. Biol.* **15:** 3597–3607.

Gutell, R.R., B. Weiser, C.R. Woese, and H.F. Noller. 1985. Comparative anatomy of 16-S-like ribosomal RNA. *Prog. Nucleic Acid Res. Mol. Biol.* **32:** 155–216.

Guthrie, C. and M. Nomura. 1968. Initiation of protein synthesis: A critical test of the 30S subunit model. *Nature* **219:** 232–235.

Hartz, D., D.S. McPheeters, and L. Gold. 1989. Selection of the initiator tRNA by *Escherichia coli* initiation factors. *Genes Dev.* **3:** 1899–1912.

Hartz, D., D.S. McPheeters, L. Green, and L. Gold. 1991. Detection of *Escherichia coli* ribosome binding at translation initiation sites in the absence of tRNA. *J. Mol. Biol.* **218:** 99–105.

Hellen, C.U.T., T.V. Pestova, and E. Wimmer. 1994. Effect of mutations downstream of the internal ribosome entry site on initiation of poliovirus protein synthesis. *J. Virol.* **68:** 6312–6322.

Hellen, C.U.T., G.W.Witherell, M. Schmid, S.H. Shin, T.V. Pestova, A. Gil, and E. Wimmer. 1993. The cellular polypeptide p57 that is required for translation of picornavirus RNA by internal ribosome entry is identical to the nuclear pyrimidine-tract binding protein. *Proc. Natl. Acad. Sci.* **90:** 7642–7646.

Herson, D., A. Schmidt, S.N. Seal, A. Marcus, and L. van Vloten-Doting. 1979. Competitive mRNA translation in an in vitro system from wheat germ. *J. Biol. Chem.* **254:** 8245–8249.

Hinnebusch, A.G. 1994. Translational control of GCN4: An in vivo barometer of initiation-factor activity. *Trends Biochem. Sci.* **19:** 409–414.

Hirashima, A. and A. Kaji. 1972. Factor dependent release of ribosome from messenger RNA: Requirement for two heat stable factors. *J. Mol. Biol.* **65:** 43–58.

Hoerz, W. and K.S. McCarty. 1969. Evidence for a proposed initiation complex for protein synthesis in reticulocyte polyribosome profiles. *Proc. Natl. Acad. Sci.* **63:** 1206–1213.

Howard, G.A., S.D. Adamson, and E. Herbert. 1970. Subunit recycling during translation in a reticulocyte cell-free system. *J. Biol. Chem.* **245:** 6237–6239.

Hui, A. and H.A. de Boer. 1987. Specialized ribosome system: Preferential translation of a single mRNA species by a subpopulation of mutated ribosomes in *Escherichia coli*. *Proc. Natl. Acad. Sci.* **84:** 4762–4766.

Hunt, S.L., A. Kaminski, and R.J. Jackson. 1993. The influence of viral coding sequences on the efficiency of internal initiation of translation of cardiovirus RNAs. *Virology* **197:** 801–807.

Hunter, A.R., R.J. Jackson, and T. Hunt. 1977. The role of complexes between the 40S ribosomal subunit and Met-tRNA$_f$ in the initiation of protein synthesis in the wheat-germ system. *Eur. J. Biochem.* **75:** 159–170.

Hunter, A.R., T. Hunt, J. Knowland, and D. Zimmern. 1976. Messenger RNA for the coat protein of tobacco mosaic virus. *Nature* **260:** 759–764.

Ichikawa, S., M. Ryoji, Z. Siegfried, and A. Kaji. 1989. Localisation of the ribosome releasing factor gene in the *Escherichia coli* genome. *J. Bacteriol.* **171:** 3689–3695.

Iizuka, N., L. Najita, A. Franzusoff, and P. Sarnow. 1994. Cap-dependent and cap-independent translation by internal initiation of mRNAs in cell extracts prepared from *Saccharomyces cerevisiae*. *Mol. Cell. Biol.* **14:** 7322–7330.

Ivey-Hoyle, M. and D.A. Steege. 1992. Mutational analysis of an inherently defective translation initiation site. *J. Mol. Biol.* **224:** 1039–1054.

Jackson, R.J. 1991. The ATP requirement for initiation of eukaryotic translation varies according to the mRNA species. *Eur. J. Biochem.* **200:** 285–294.

Jackson, R.J., M.T. Howell, and A. Kaminski. 1990. The novel mechanism of initiation of picornavirus RNA translation. *Trends Biochem. Sci.* **15:** 477–483.

Jackson, R.J., S.L. Hunt, C.L. Gibbs, and A. Kaminski. 1994. Internal initiation of translation of picornavirus RNAs. *Mol. Biol. Rep.* **19:** 147–159.

Jackson, R.J., S.L. Hunt, J.E. Reynolds, and A. Kaminski. 1995. Cap-dependent and cap-independent translation: Operational distinctions and mechanistic interpretations. *Curr. Top. Microbiol. Immunol.* **203:** 1–29.

Jaramillo, M., T.E. Dever, W.C. Merrick, and N. Sonenberg. 1991. RNA unwinding in translation: Assembly of helicase complex intermediates comprising eukaryotic initiation factors eIF-4F and eIF-4B. *Mol. Cell. Biol.* **11:** 5992–5997.

Jay, G. and R. Kaempfer. 1974. Sequence of events in initiation of translation: A role for initiator transfer RNA in the recognition of messenger RNA. *Proc. Natl. Acad. Sci.* **71:** 3199–3203.

———. 1975. Initiation of protein synthesis: binding of messenger RNA. *J. Biol. Chem.* **250:** 5742–5748.

Joshi-Barve, S., A. De Benedetti, and R.E. Rhoads. 1992. Preferential translation of heat shock mRNAs in HeLa cells deficient in protein synthesis initiation factors eIF-4E and eIF-4γ. *J. Biol. Chem.* **267:** 21038–21043.

Kääriäinen, L., K. Takkinen, S. Keranen, and H. Soderlund. 1987. Replication of the genome of alphaviruses. *J. Cell. Sci. Suppl.* **7:** 231–250.

Kaminski, A., G.J. Belsham, and R.J. Jackson. 1994a. Translation of encephalomyocarditis virus RNA: Parameters influencing the selection of the internal initiation site. *EMBO. J.* **13:** 1673–1681.

Kaminski, A., M.T. Howell, and R.J. Jackson. 1990. Initiation of encephalomyocarditis virus RNA translation: The authentic initiation site is not selected by a scanning mechanism. *EMBO J.* **9:** 3753–3759.

Kaminski, A., S.L. Hunt, C.L. Gibbs, and R.J. Jackson. 1994b. Internal initiation of mRNA translation in eukaryotes. In *Genetic engineering: Principles and methods* (ed. J. Setlow), vol. 16, pp. 115–155. Plenum Press, New York.

Kaminski, A., S.L. Hunt, J.G. Patton, and R.J. Jackson. 1995. Direct evidence that polypyrimidine tract binding protein (PTB) is essential for internal initiation of translation of encephalomyocarditis virus RNA. *RNA* **1:** (in press).

Kolakofsky, D., T. Ohta, and R.E. Thach. 1968. Junction of the 50S ribosomal subunit with the 30S initiation complex. *Nature* **220:** 244–247.

Koromilas, A.E., A. Lazaras-Karatzas, and N. Sonenberg. 1992. mRNAs containing extensive secondary structure in their 5′ non-coding region translate efficiently in cells overexpressing initiation factor eIF-4E. *EMBO J.* **11:** 4153–4158.

Kozak, M. 1979. Migration of 40S ribosomal subunits on messenger RNA when initiation is perturbed by lowering magnesium or adding drugs. *J. Biol. Chem.* **254:** 4731–4735.

———. 1980. Role of ATP in binding and migration of 40S ribosomal subunits. *Cell* **22:** 459–467.

———. 1986a. Point mutations define a sequence flanking the AUG initiator codon that modulates translation by eukaryotic ribosomes. *Cell* **44:** 283–292.

———. 1986b. Influences of mRNA secondary structure on initiation by eukaryotic ribosomes. *Proc. Natl. Acad. Sci.* **83:** 2850–2854.

———. 1989a. The scanning model for translation: An update. *J. Cell Biol.* **108:** 229–241.

———. 1989b. Circumstances and mechanism of inhibition of translation by secondary structure in eukaryotic mRNAs. *Mol. Cell. Biol* **9:** 5134–5142.

———. 1989c. Context effects and inefficient initiation at non-AUG codons in eukaryotic cell-free translation systems. *Mol. Cell. Biol.* **9:** 5073–5080.

———. 1990. Downstream secondary structure facilitates recognition of initiator codons by eukaryotic ribosomes. *Proc. Natl. Acad. Sci.* **87:** 8301–8305.

———. 1991a. A short leader sequence impairs the fidelity of initiation by eukaryotic ribosomes in vitro. *Gene Expr.* **1:** 111–116.

———. 1991b. Effects of long 5′ leader sequences on initiation by eukaryotic ribosomes *in vitro*. *Gene Expr.* **1:** 117–125.

———. 1995. Adherence to the first-AUG rule when a second AUG codon follows closely upon the first. *Proc. Natl. Acad. Sci.* **92:** 2662–2666.

Kozak, M. and A.J. Shatkin. 1978. Migration of 40S ribosomal subunits in the presence of edeine. *J. Biol. Chem.* **253:** 6568–6577.

Kuge, S. and A. Nomoto. 1987. Construction of viable deletion and insertion mutants of the Sabin strain of type 1 poliovirus: Function of the 5′ noncoding sequence in viral replication. *J. Virol.* **61:** 1478–1487.

Kuge, S., N. Kawamura, and A. Nomoto. 1989. Strong inclination toward transition mutations in nucleotide substitutions by poliovirus replicase. *J. Mol. Biol.* **207:**

175–182.

Lamphear, B.J. and R. Panniers. 1991. Heat shock impairs the interaction of cap binding protein complex with 5′ mRNA cap. *J. Biol. Chem.* **266:** 2789–2794.

Lamphear, B.J., R. Kirchweger, T. Skern, and R.E. Rhoads. 1995. Mapping of functional domains in eukaryotic protein synthesis initiation factor 4G (eIF4G) with picornaviral proteases. *J. Biol. Chem.* **270:** 21975–21983.

Liebig, H.-D., E. Ziegler, R. Yan, K. Hartmuth, H. Klump, H. Kowlaski, D. Blass, W. Sommergruber, L. Frasel, B. Lamphear, R.E. Rhoads, E. Kuechler, and T. Skern. 1993. Purification of two picornaviral 2A proteinases: Interaction with eIF-4γ and influence on in vitro translation. *Biochemistry* **32:** 7581–7588.

Lin, C.-H. and J.G. Patton. 1995. Regulation of the alternative 3′ splice site selection by constitutive splicing factors. *RNA* **1:** 234–245.

Macejak, D.G. and P. Sarnow. 1991. Internal initiation of translation mediated by the 5′ leader of a cellular mRNA. *Nature* **353:** 90–94.

Mader, S., H. Lee, A. Pause, and N. Sonenberg. 1995. The translation initiation factor eIF-4E binds to a common motif shared by the translation factor eIF-4γ and the translational repressors 4E-binding proteins. *Mol. Cell. Biol.* **15:** 4990–4997.

Martin, J. and R.E. Webster. 1975. The in vitro translation of a termination signal by a single *Escherichia coli* ribosome: The fate of the subunits. *J. Biol. Chem.* **250:** 8132–8139.

McCarthy, J.E.G. and R. Brimacombe. 1994. Prokaryotic translation: The interactive pathway leading to initiation. *Trends Genet.* **10:** 402–407.

McCarthy, J.E.G., H.U. Scaude, and W. Sebald. 1985. Translation initiation frequency of *atp* genes from *Escherichia coli*: Identification of an intercistronic spacer that enhances translation. *EMBO J.* **4:** 519–526.

McCarthy, J.E.G., W. Sebald, G. Gross, and R. Lammers. 1986. Enhancement of translational efficiency by the *Escherichia coli atpE* translational initiation region: Its fusion with two human genes. *Gene* **41:** 201–206.

Meerovitch, K. and N. Sonenberg. 1993. Internal initiation of picornavirus RNA translation. *Semin. Virol.* **4:** 217–227.

Meerovitch, K., J. Pelletier, and N. Sonenberg. 1989. A cellular protein that binds to the 5′-noncoding region of poliovirus RNA: Implications for internal translation initiation. *Genes Dev.* **3:** 1026–1034.

Meerovitch, K., Y.V. Svitkin, H.S. Lee, F. Lejbkowicz, D.J. Kenan, E.K.L. Chan, V.I. Agol, J.D. Keene, and N. Sonenberg. 1993. La autoantigen enhances and corrects aberrant translation of poliovirus RNA in reticulocyte lysate. *J. Virol.* **67:** 3798–3807.

Muñoz, A., M.A. Alonson, and L. Carrasco. 1984. Synthesis of heat-shock proteins in HeLa cells: Inhibition by virus infection. *Virology* **137:** 150–159.

Najita, L. and P. Sarnow. 1990. Oxidation-reduction sensitive interaction of a cellular 50 kDa protein with an RNA hairpin in the 5′ noncoding region of the poliovirus genome. *Proc. Natl. Acad. Sci.* **87:** 5846–5850.

Nicholson, R., J. Pelletier, S.-Y. Le, and N. Sonenberg. 1991. Structural and functional analysis of the ribosome landing pad of poliovirus type 2: In vivo translation studies. *J. Virol.* **65:** 5886–5894.

Nomura, M., R. Gourse, and G. Baughman. 1984. Regulation of the synthesis of ribosomes and ribosomal components. *Annu. Rev. Biochem.* **53:** 75–117.

OH, S.K., M.P. Scott, and P. Sarnow. 1992. Homeotic gene antennapedia messenger RNA contains 5′-noncoding sequences that confer translational initiation by internal ribosome binding. *Genes Dev.* **6:** 1643–1653.

Ohlmann, T., M. Rau, S.J. Morley, and V.M. Pain. 1995. Proteolytic cleavage of initiation factor eIF-4γ in the reticulocyte lysate inhibits translation of capped mRNAs but enhances that of uncapped mRNAs. *Nucleic Acids Res.* **23:** 334–340.

Olins, P.O. and S.H. Rangwala. 1989. A novel sequence derived from bacteriophage T7 mRNA acts as an enhancer of translation of the *lacZ* gene of *Escherichia coli. J. Biol. Chem.* **264:** 16973–16976.

Olins, P.O., C.S. Devine, S.H. Rangwala, and K.S. Kavka. 1988. The T7 phage gene 10 leader RNA, a ribosome-binding site that dramatically enhances the expression of foreign genes in *Escherichia coli. Gene* **173:** 227–235.

Patton, J.G., S.A. Mayer, P. Tempst, and B. Nadal-Ginard. 1991. Characterization and molecular cloning of polypyrimidine tract-binding protein: A component of a complex necessary for pre-mRNA splicing. *Genes Dev.* **5:** 1237–1251.

Patton J.G., E.B. Porro, J. Galceran, P. Tempst, and B. Nadal-Ginard. 1993. Cloning and characterization of PSF, a novel pre-mRNA splicing factor. *Genes Dev.* **7:** 393– 406.

Peabody, D.S. and P. Berg. 1986. Termination-reinitiation occurs in the translation of mammalian cell mRNAs. *Mol. Cell. Biol.* **6:** 2695–2703.

Peabody, D.S., S. Subramani, and P. Berg. 1986. Effect of upstream reading frames on translation efficiency in simian virus 40 recombinants. *Mol. Cell. Biol.* **6:** 2704–2711.

Pelletier, J. and N. Sonenberg. 1988. Internal initiation of translation of eukaryotic mRNA directed by a sequence derived from poliovirus RNA. *Nature* **334:** 320–325.

Pestova, T.V., C.U.T. Hellen, and E. Wimmer. 1991. Translation of poliovirus RNA; role of an essential *cis*-acting oligopyrimidine element within the 5′ nontranslated region and involvement of a cellular 57-kilodalton protein. *J. Virol.* **65:** 6194–6204.

Pilipenko, E.V., A.P. Gmyl, S.V. Maslova, Y.V. Svitkin, A.N. Sinyakov, and V.I. Agol. 1992. Prokaryotic-like *cis* elements in the cap-independent internal initiation of translation on picornavirus RNA. *Cell* **68:** 119–131.

Pilipenko, E.V., A.P. Gmyl, S.V. Maslova, G.A. Belov, A.N. Sinyakov, M. Huang, T.D.K. Brown, and V.I. Agol. 1994. Starting window, a distinct element in the cap-independent internal initiation of translation of picornaviral RNA. *J. Mol. Biol.* **241:** 398–414.

Poyry, T., L. Kinnunen, and T. Hovi. 1992. Genetic variation in vivo and proposed functional domains of the 5′ noncoding region of poliovirus RNA. *J. Virol.* **66:** 5313–5319.

Reynolds, J.E., A. Kaminski, H.J. Kettinen, K. Grace, B.E. Clarke, A.R. Carroll, D.J. Rowlands, and R.J. Jackson. 1995. Unique features of internal initiation of hepatitis C virus RNA translation. *EMBO J.* **14:** (in press).

Rhoads, R.E. and B.J. Lamphear. 1995. Cap-independent translation of heat-shock messenger RNAs. *Curr. Top. Microbiol. Immunol.* **203:** 131–153.

Rhoads, R.E., S. Joshi-Barve, and C. Rinker-Schaeffer. 1993. Mechanism of action and regulation of protein synthesis initiation factor 4E: Effects on mRNA discrimination, cellular growth rate, and oncogenesis. *Prog. Nucleic Acid Res. Mol. Biol.* **46:** 183–219.

Ringquist, S., S. Shinedling, D. Barrick, L. Green, J. Binkley, G.D. Stormo, and L. Gold. 1992. Translation initiation in *Escherichia coli*; sequences within the ribosome-binding site. *Mol. Microbiol.* **6:** 1219–1229.

Rozen, F., I. Edery, K. Meerovitch, T.E. Dever, W.C. Merrick, and N. Sonenberg. 1990. Bidirectional RNA helicase activity of eucaryotic initiation factors 4A and 4F. *Mol. Cell. Biol.* **10:** 1134–1144.

Sarnow, P. 1989. Translation of glucose regulated protein 78/immunoglobulin heavy chain binding protein mRNA is increased in poliovirus-infected cells at a time when cap-dependent translation of cellular mRNAs is inhibited. *Proc. Natl. Acad. Sci.* **86:** 5795–5799.

Scheper, G.C., H.O. Voorma, and A.A.M. Thomas. 1992. Eukaryotic initiation factors-4E and -4F stimulate 5′ cap-dependent as well as internal initiation of protein synthesis. *J. Biol. Chem.* **267:** 7269–7274.

Schneider, R.J. 1995. Cap-independent translation in adenovirus infected cells. *Curr. Top. Microbiol. Immunol.* **203:** 117–129.

Seong, B.L. and U.L. Rajbhandary. 1987. *Escherichia coli* formylmethionine tRNA: Mutations in GGG/CCC sequence conserved in anticodon stem of initiator tRNAs affect initiation of protein synthesis and conformation of anticodon loop. *Proc. Natl. Acad. Sci.* **84:** 334–338.

Shean, C.S. and M.E. Gottesman. 1992. Translation of the prophage λ cI transcript. *Cell* **70:** 513–522.

Singh, R., J. Vilcárel, and M.R. Green. 1995. Distinct binding specificities and functions of higher eukaryotic polypyrimidine tract-binding proteins. *Science* **268:** 1173–1176.

Sogin, M.L., J.H. Gunderson, H.J. Elwood, R.A. Alonso, and D.A. Peattie. 1989. Phylogenetic significance of the kingdom concept: An unusual eukaryotic 16S-like ribosomal RNA from *Giardia lamblia*. *Science* **243:** 75–77.

Sonenberg, N. 1991. Picornavirus RNA translation continues to surprise. *Trends Genet.* **7:** 105–106.

———. 1993. Remarks on the mechanism of ribosome binding to eukaryotic mRNAs. *Gene Expr.* **3:** 317–323.

Sonenberg, N., D. Guertin, and K.A.W. Lee. 1982. Capped mRNAs with reduced secondary structure can function in extracts from poliovirus infected cells. *Mol. Cell. Biol.* **2:** 1633–1638.

Sprengart, M.L., H.P. Fatscher, and E. Fuchs. 1990. The initiation of translation in *Escherichia coli*: Apparent base-pairing between the 16S rRNA and downstream sequences of the mRNA. *Nucleic Acids Res.* **18:** 1719–1723.

Steitz, J.A. and K. Jakes. 1975. How ribosomes select initiator regions in mRNA: Base-pair formation between the 3′ terminus of 16S rRNA and the mRNA during initiation of protein synthesis in *Escherichia coli*. *Proc. Natl. Acad. Sci.* **72:** 4734–4738.

Subramanian, A.R. 1983. Structure and functions of ribosomal protein S1. *Prog. Nucleic Acid Res. Mol. Biol.* **28:** 101–142.

Svitkin, Y.V., A. Pause, and N. Sonenberg. 1994a. La autoantigen alleviates translational repression by the 5′ leader sequence of the human immunodeficiency virus type 1 mRNA. *J. Virol.* **68:** 7001–7007.

Svitkin, Y.V., T.V. Pestova, S.V. Maslova, and V.I. Agol. 1988. Point mutations modify the response of poliovirus RNA to a translation initiation factor: A comparison of neurovirulent and attenuated strains. *Virology* **166:** 394–404.

Svitkin, Y.V., K. Meerovitch, H.S. Lee, J.N. Dholakia, D.J. Kenan, V.I. Agol, and N. Sonenberg. 1994b. Internal translation initiation on poliovirus RNA: Further characterization of La function in poliovirus translated in vitro. *J. Virol.* **68:** 1544–1550.

Thanaraj, T.A. and M.W. Pandit. 1989. An additional ribosome-binding site on mRNA of highly expressed genes and a bifunctional site on the colicin fragment of 16 S rRNA from *Escherichia coli*: Important determinants of the efficiency of translation initiation. *Nucleic Acids Res.* **17:** 2973–2985.

Thomas, A.A.M., G.C. Scheper, M. Kleijn, M. DeBoer, and H.O. Voorma. 1992. Dependence of the adenovirus tripartite leader on the p220 subunit of eukaryotic initiation factor 4F during in vitro translation. *Eur. J. Biochem.* **207:** 471–477.

Thomas, J.O. and W. Szer. 1982. RNA-helix-destabilising proteins. *Prog. Nucleic Acid Res. Mol. Biol.* **27:** 157–187.

Thomas, K.R. and M.R. Capecchi. 1986. Introduction of homologous DNA sequences into mammalian cells induces mutations in the cognate gene. *Nature* **324:** 34–38.

Timmer, R.T., S.R. Lax, D.L. Hughes, W.C. Merrick, J.M. Ravel, and K.S. Browning. 1993a. Characterisation of wheat germ protein synthesis initiation factor eIF-4C and comparison of eIF-4C from wheat germ and rabbit reticulocytes. *J. Biol. Chem.* **268:** 24863–24867.

Timmer, R.T., L.A. Benkowski, D. Schodin, S.R. Lax, A.M. Metz, J.M. Ravel, and K.S. Browning. 1993b. The 5′ and 3′ untranslated regions of satellite tobacco necrosis virus RNA affect translational efficiency and dependence on 5′ cap structure. *J. Biol. Chem.* **268:** 9504–9510.

Trachsel, H., B. Erni, M.H. Schreier, and T. Staehelin. 1977. Initiation of mammalian protein synthesis II. The assembly of the initiation complex with purified initiation factors. *J. Mol. Biol.* **116:** 755–767.

Tsukiyama-Kohara, K., N. Iizuka, M. Kohara, and A. Nomoto. 1992. Internal ribosome entry site within hepatitis C virus RNA. *J. Virol.* **66:** 1476–1483.

van Duin, J., G.P. Overbeek, and C. Backendorf. 1980. Functional recognition of phage RNA by 30S ribosomal subunits in the absence of initiator tRNA. *Eur. J. Biochem.* **110:** 593–597.

Wakao, H., P. Romby, E. Westhof, S. Laalami, M. Grunberg-Manago, J.-P. Ebel, C. Ehresmann, and B. Ehresmann. 1989. The solution structure of the *Escherichia coli* initiator tRNA and its interactions with initiation factor 2 and the ribosomal 30S subunit. *J. Biol. Chem.* **264:** 20363–20371.

Wallaczek, J., R. Albrecht-Ehrlich, G. Stöffler, and M. Stöffler-Meilicke. 1990. Three dimensional localization of the NH_2- and carboxyl-terminal domain of ribosomal protein S1 on the surface of the 30S subunit from *Escherichia coli*. *J. Biol. Chem.* **265:** 11138–11344.

Wang, C., P. Sarnow, and A. Siddiqui. 1993. Translation of human hepatitis C virus RNA in cultured cells is mediated by an internal ribosome-binding mechanism. *J. Virol.* **67:** 3338–3344.

———. 1994. A conserved helical element is essential for internal initiation of translation of hepatitis C virus RNA. *J. Virol.* **68:** 7301–7307.

Williams, M.A. and R.A. Lamb. 1989. Effect of mutations and deletions in a bicistronic mRNA on the synthesis of influenza B virus NB and NA glycoproteins. *J. Virol.* **63:** 28–35.

Yates, J.L., D. Dean, W.A. Strycharz, and M. Nomura. 1981. *E. coli* ribosomal protein L10 inhibits translation of L10 and L7/L12 mRNAs by acting at a single site. *Nature* **294:** 190–192.

4

Binding of Initiator Methionyl-tRNA to Ribosomes

Hans Trachsel
Institute of Biochemistry und Molecular Biology
University of Berne
3012 Berne, Switzerland

Among the important events in the expression of a gene is the selection of the correct codon on the messenger RNA to initiate the synthesis of the encoded protein. This is achieved in the multistep pathway of translation initiation. In these reactions, a ribosome binds the mRNA and the first amino acid to be incorporated into the future polypeptide chain. All organisms use the amino acid methionine to initiate translation and a specific translation initiation factor to carry it in the form of methionyl-tRNA to the ribosome. In eukaryotes, the main steps of the translation initiation pathway include binding of initiator methionyl-tRNA (Met-$tRNA_i$) to the small (40S) ribosomal subunit, selection of an AUG codon by the (40S•Met-$tRNA_i$) complex on mRNA, and joining of a large (60S) ribosomal subunit with the 40S initiation complex to form an 80S ribosome competent for translation elongation (for details, see Merrick and Hershey, this volume). This chapter describes the mechanism and regulation of Met-$tRNA_i$ binding to ribosomes.

COMPONENTS INVOLVED IN MET-tRNA BINDING

Most of the proteins that catalyze specific steps of translation initiation in eukaryotes were originally purified from the rabbit reticulocyte ribosomal salt wash fraction, and their functions were determined by reconstitution of the initiation pathway in vitro. From these early studies, it could be shown that some of the reaction intermediates of the pathway identified in vitro, such as the (40S•Met-$tRNA_i$) complex, also existed in extracts derived from rabbit reticulocytes and wheat germ. They were therefore considered to be physiologically relevant (Safer and Anderson 1978).

Initiator tRNA

Methionine is carried to the ribosome bound to initiator tRNA ($tRNA_i^{Met}$). This tRNA is structurally distinct from the elongator tRNA

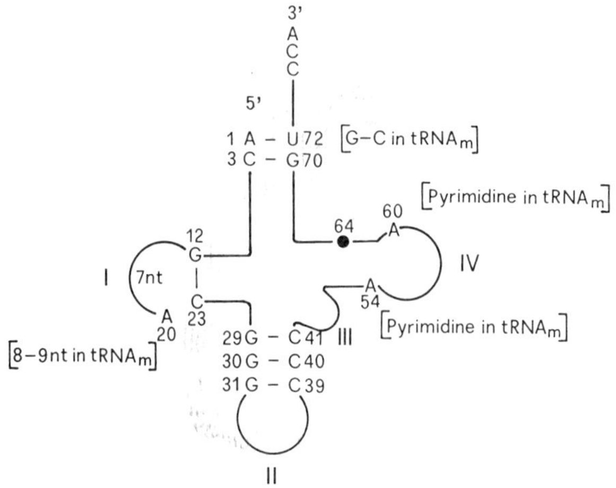

Figure 1 Schematic two-dimensional drawing of $tRNA_i$. A simplified version of yeast $tRNA_i$ is shown (Aström et al. 1993). For a detailed description of sequences, see Sprinzl et al. (1991). (I–IV) Loops; (*closed circle*) purine with 2′-*O*-ribosylphosphate.

($tRNA^{Met}$), which inserts methionine at internal positions in the polypeptide chain. Eukaryotic organisms usually harbor several genes encoding initiator and elongator tRNAs. In the yeast *Saccharomyces cerevisiae*, for example, five genes encode a single type of elongator tRNA and four genes encode a single type of $tRNA_i$ (Aström et al. 1993).

The two tRNAs differ in the first base pair (1/72) of the amino acid acceptor stem, in the size of loop I (7 nucleotides in $tRNA_i$ and 8–9 nucleotides in $tRNA^{Met}$) and in nucleotides in positions 54 and 60 of loop IV (Fig. 1). Furthermore, $tRNA_i$ has conserved G-C base pairs at positions 3/70 and 12/23, three consecutive G-C base pairs in the anticodon stem, an A residue at position 20, and a purine 2′*O*-ribosylphosphate at position 64 (only present in fungal and plant $tRNA_i^{Met}$) (Desgres et al. 1989). Tertiary RNA interactions through hydrogen bonds between loops I and IV in the initiator $tRNA_i^{Met}$ create a special three-dimensional structure in $tRNA_i$ not present in other tRNAs, as revealed by the analysis of crystallized *S. cerevisiae* $tRNA_i^{Met}$ (Basavappa and Sigler 1991). This structure and some of the structural determinants described above are probably responsible for functional differences between initiator and elongator tRNAs, such as binding of the $tRNA_i$ to eIF2 and its exclusion from the elongation process (Drabkin et al. 1993) (see below).

Methionine is activated by reaction with ATP to form the intermediate methionyl-AMP and then ligated to $tRNA_i^{Met}$. Both reactions are catalyzed by the same enzyme, methionyl-tRNA synthetase, which (together with nine other enzymes) belongs to the class I aminoacyl-tRNA synthetases (Eriani et al. 1990). These enzymes have conserved sequence elements involved in ATP binding (Eriani et al. 1990), charge tRNA at the 2′-OH group of the 3′-terminal ribose, and are associated in vivo in a high-molecular-weight complex. The function(s) of the synthetase multienzyme complexes in eukaryotic cells is still unclear (Kisselev and Wolfson 1994). Methionyl-tRNA synthetase is a single polypeptide chain with a molecular mass, in eukaryotes, of approximately 100 kD. It charges both initiator and elongator tRNAs with methionine. As for the majority of synthetases, the most important determinant in the tRNA for discrimination against noncognate tRNAs by the enzyme is the anticodon base sequence (Muramatsu et al. 1988; Schulman and Pelka 1988). In addition to the anticodon base sequence, bases in the anticodon loop and the base A73 in the amino acid acceptor stem also contribute to recognition by the synthetase (Senger et al. 1992).

In one of the first steps of the translation initiation pathway, Met-$tRNA_i$ is recognized and bound by initiation factor eIF2. Recognition is mediated by sequences in the acceptor stem of the $tRNA_i^{Met}$. This was shown in yeast cells by suppressing the growth defect elicited by mutations in the acceptor stem of $tRNA_i^{Met}$ through overexpression of the α and β subunits of eIF2 (Aström et al. 1993). Disruption of the A-U base pair (position 1/72, Fig. 1) may allow the 3′end of the tRNA with the methionyl residue to fold back on the tRNA and to create a surface that is recognized by eIF2 (for discussion, see Basavappa and Sigler 1991).

In contrast to eIF2, which positions Met-$tRNA_i$ at the P-site on the ribosome, the elongation factor eEF1A binds elongator tRNA and carries it to the A-site of the ribosome. Interestingly, discrimination against $tRNA^{Met}$ as a $tRNA_i$ is stronger than for $tRNA_i^{Met}$ as an elongator tRNA. This was deduced from experiments with yeast *S. cerevisiae* strains having all $tRNA_i^{Met}$ and $tRNA^{Met}$ genes deleted and $tRNA_i^{Met}$ or $tRNA^{Met}$ supplied by genes carried on plasmids. Overexpression of $tRNA_i^{Met}$ allowed growth in the absence of $tRNA^{Met}$, whereas cells lacking $tRNA_i^{Met}$ were inviable (Aström et al. 1993). Binding of $tRNA_i^{Met}$ to eEF1A is prevented in vitro by the purine 2′*O*-ribosylphosphate at position 64 in fungal and plant $tRNA_i^{Met}$ (Kiesewetter et al. 1990). The contribution of the 2′*O*-ribosylphosphate modification to discrimination by eEF1A in vivo, however, remains debatable, since a recent report shows that elimination of the 2′*O*-ribosylphosphate has no effect on viability and growth of *S. cerevisiae* cells (Aström and Byström 1994). In conclusion, we know the structure of $tRNA_i^{Met}$ in remarkable detail. On the other hand, detailed knowledge about the interaction of $tRNA_i^{Met}$ with the synthetase and with eIF2 is still lacking.

Initiation Factor eIF2

Met-$tRNA_i$ binding to the ribosome is promoted by eIF2. This protein is able to form a ternary complex with Met-$tRNA_i$ and GTP in the absence of other components (Safer et al. 1975, 1976; Benne et al. 1976; Schreier et al. 1977). Since its discovery (Levin et al. 1973; Schreier and Staehelin 1973), eIF2 has been isolated from many sources, including mammalian and yeast *S. cerevisiae* cells. It consists of three subunits (α, β, γ) with molecular masses in mammalian cells of 36 (α), 38 (β), and 52 (γ) kD. Most workers found that these subunits are stably associated in vitro. However, in the presence of RNA, dissociation of eIF2 into its subunits was reported (Barrieux and Rosenfeld 1978), and in vivo, the free α-subunit was shown to equilibrate quite rapidly with the complex-bound α-subunit in transfected COS cells (Choi et al. 1992). These are interesting observations that require further investigation. At present, the mechanism and biological significance of subunit exchange in eIF2 are not known.

Biochemical and genetic analyses demonstrate that the three subunits of eIF2 have distinct activities. The α-subunit carries the Ser-51 that is phosphorylated by eIF2α kinases and has a central role in the regulation of eIF2 activity (see below). The β-subunit is involved in RNA binding (Donahue et al. 1988; Pathak et el. 1988; Flynn et al. 1994). Mutations in this protein affect AUG initiation codon selection (Donahue et al. 1988) (see below). Furthermore, the β-subunit may contribute to Met-$tRNA_i$ binding by the γ-subunit and to GDP-GTP exchange mediated by initiation factor eIF2B (Flynn et al. 1993). Earlier reports in the literature describe eIF2 preparations that lack the β-subunit but are nevertheless active in all steps of initiation of translation tested at that time (Mitsui et al. 1981). These findings are difficult to reconcile with today's knowledge about eIF2 subunit functions, especially with the finding that deletion of the β-subunit in the yeast *S. cerevisiae* is lethal for these cells (Cigan et al. 1989).

For quite some time, the GTP-binding site of eIF2 was a matter of controversy. It was originally claimed that the α-subunit binds GTP (Barrieux and Rosenfeld 1977), but affinity-labeling experiments with GTP analogs done later by several groups (Bommer and Kurzchalia 1989; Dholakia et al. 1989; Anthony et al. 1990) indicated that both the β and γ subunits of eIF2, but not the α-subunit, are involved in GTP binding. This issue was recently resolved when sequence analysis of the γ-subunit showed a strong relationship with elongation factors EF1A of prokaryotes and eEF1A of eukaryotes (Hannig et al. 1993; Gaspar et al. 1994). The similarity is most evident in the GTP-binding domain, and affinity labeling experiments revealed the ability of the γ-subunit to bind Met-$tRNA_i$ and GTP (Gaspar et al. 1994). The γ-subunit is thus the GTP-binding subunit. However, it is possible that the β-subunit can bind an additional molecule of GTP (Dholakia et al. 1989; Anthony et al. 1990) or ATP (Dholakia et al. 1989; Gonsky et al. 1990) and that this nucleotide regulates eIF2 activity (see below).

In light of the central part eIF2 has in initiation of translation, it is not surprising that its activity is regulated in several ways. Besides regulation of eIF2 activity by GDP and GTP and by phosphorylation of the α-subunit (see below), there is regulation of the expression of eIF2 subunits at the transcriptional and translational levels. Unfortunately, rather little is currently known about the structure of genes encoding eIF2 subunits. In mammalian cells, only the single-copy gene encoding the α-subunit of eIF2 has been characterized (Hümbelin et al. 1989) and its transcription studied (Jacob et al. 1989; Silverman et al. 1992; Efiok et al. 1994). These studies led to the discovery of a novel transcription factor, α-Pal, the identification of its DNA recognition site, and the cloning of a cDNA encoding this transcription factor (Efiok et al. 1994). Transcription factor α-Pal is essential for eIF2α gene transcription.

Interestingly, the α-Pal DNA-binding site occurs also in the eIF2β gene, as well as in a number of other genes whose transcription is stimulated in growing cells. As expected, the expression of eIF2 correlates positively with proliferation of cells. Increased levels of α-subunit mRNA were found in human T lymphocytes stimulated to proliferate (Cohen et al. 1990) and decreased levels were found in aging cells (Kimball et al. 1992). The increased levels of α-subunit mRNA in proliferating T lymphocytes are due to stabilization of transcripts, rather than to enhanced transcription, and result in an approximately 50-fold elevation of the mRNA level over that of resting cells (Cohen et al. 1990).

Further control of the expression of eIF2 subunits is exerted at the level of translation (e.g., in K562 cells). The synthesis of roughly equal amounts of α and β subunits is regulated in these cells at the level of translation elongation: The translation of the more abundant β-chain mRNA is about four times slower than that of the less abundant α-chain mRNA (Chiorini et al. 1993).

TERNARY COMPLEX FORMATION

Met-$tRNA_i$, initiation factor eIF2, and GTP associate in the ternary complex (eIF2•GTP•Met-$tRNA_i$) (Fig. 2, step 2). Since eIF2 forms a binary

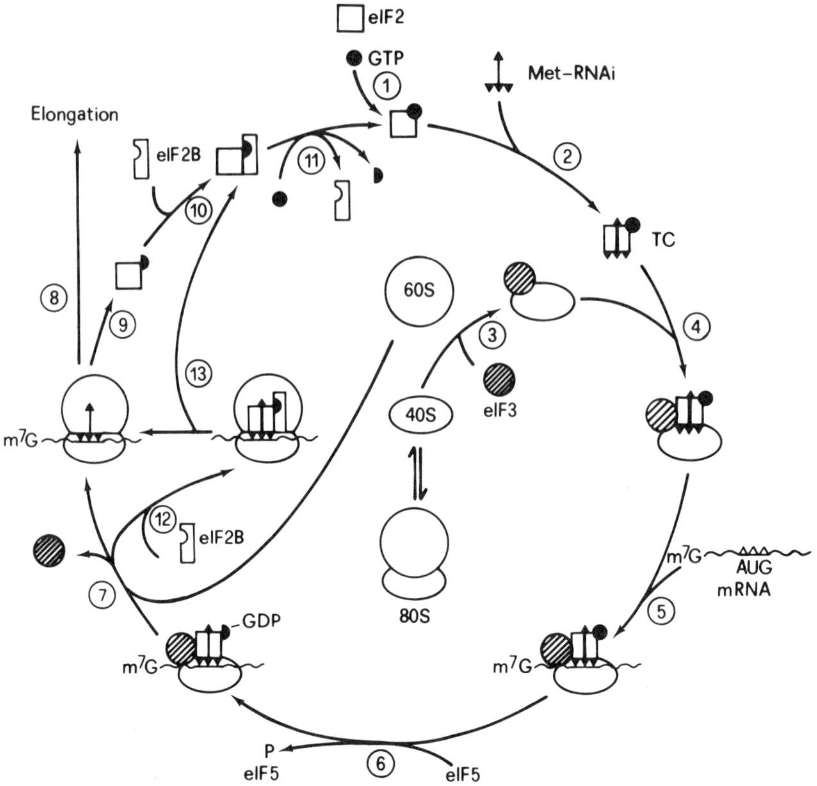

Figure 2 Binding of Met-tRNA$_i$ to the ribosome. The individual steps (1, 2,...) are described in the text. (eIF) Eukaryotic initiation factor; (TC) ternary complex (eIF2•GTP•Met-tRNA$_i$); (m^{7}G) cap structure m^{7}Gppp; (AUG) initiation codon; (P) phosphate.

complex with GTP (Fig. 2, step 1) but not with Met-tRNA$_i$, we have good reason to assume that an (eIF2•GTP) binary complex is formed initially and that Met-tRNA$_i$ is bound subsequently. This reaction proceeds readily in vitro with purified components and is easily monitored by filtering the incubation mixture containing Met-tRNA$_i$, GTP, and eIF2 through nitrocellulose filters and measuring the labeled nucleotide or Met-tRNA$_i$ retained on the filter. The nitrocellulose filter binds eIF2 but not Met-tRNA$_i$ or GTP unless they are bound to the protein.

Ternary complex formation is also supported by nonhydrolyzable GTP analogs showing that GTP hydrolysis is not required at this stage. GTP hydrolysis is actually not required for any of the subsequent steps of initiation involving eIF2 and Met-tRNA$_i$ until ribosomal subunit joining (see below). eIF2 also binds ATP (Dholakia et al. 1989; Gonsky et al.

1990; Flynn et al. 1994). These researchers agree that (1) ATP binds to a site that is different from the GTP-binding site, (2) the site is probably located on the β-subunit, and (3) ATP binding reduces the activity of eIF2 in ternary complex formation in vitro. Both strong binding (Gonsky et al. 1990) and weak binding of ATP to eIF2 were reported (Flynn et al. 1994), and there is no agreement as to whether ATP inhibits ternary complex formation by inhibiting Met-$tRNA_i$ binding to eIF2 (Gonsky et al. 1990) or by interfering with GTP binding to the γ-subunit (Flynn et al. 1994). More experiments are required to answer the question of whether physiological concentrations of ATP have a role in the regulation of ternary complex formation in vivo.

A direct demonstration of the existence of ternary complex in vivo is lacking. Nevertheless, it is reasonable to assume that the ternary complex also forms in crude translation systems and in vivo prior to binding of eIF2 to the ribosome. This pathway is supported by the in vitro finding that eIF2 by itself binds poorly to ribosomes, although it binds well in the presence of charged initiator tRNA (Trachsel and Staehelin 1978).

Ternary complex (eIF2•GTP•Met-$tRNA_i$) formation in vitro depends (among other parameters) on the GTP:GDP ratio (Walton and Gill 1976). The GTP concentration in living cells is on the order of 0.1 mM, and the GTP:GDP ratio is approximately 10:1 (Walton and Gill 1976). Decreasing the GTP:GDP ratio has a strong inhibitory effect on ternary complex formation in vitro and decreasing the ratio from 10:1 to 5:1 inhibits translation initiation in extracts of Ehrlich ascites tumor cells (Hucul et al. 1985). The GTP:GDP ratio depends on the ATP:ADP ratio since ATP is used by the enzyme nucleoside diphosphate kinase to rapidly regenerate GTP from GDP (Walton and Gill 1976; Hucul et al. 1985). The regulatory potential of the ATP:ADP ratio (through the GTP:GDP ratio) is due to the fact that eIF2 has an approximately 100–400 times higher affinity for GDP than for GTP (Barrieux and Rosenfeld 1977; Panniers et al. 1988) and that the (eIF2•GDP) complex is inactive in Met-$tRNA_i$ binding.

Some debate exists about protein factors that influence ternary complex formation and stability. In vitro, at concentrations of eIF2 and Met-$tRNA_i$ considered to correspond to physiological concentrations (a few micromolar) and in the presence of an ATP- and GTP-regenerating system, ternary complex formation proceeds efficiently in the absence of other translation factors (Benne et al. 1979). This changes at low eIF2 concentrations when RNA is added to the reaction mixture. It was shown that eIF2 binds RNA (Barrieux and Rosenfeld 1977; Gonsky et al. 1990; Flynn et al. 1994) and that RNA binding to eIF2 inhibits ternary complex formation in vitro (Barrieux and Rosenfeld 1978; Gonsky et al. 1990). This inhibition can be overcome by ancilliary factors isolated from the rabbit reticulocyte ribosomal salt wash fraction. One of these factors is eIF2C (previously called Co-eIF-2A) (Nasrin et al. 1986; Roy et al. 1988), a single polypeptide chain of 94 kD that stimulates ternary complex formation at eIF2 concentrations in the nanomolar range in the presence of RNA. Stimulation of ternary complex formation in vitro was also reported for a factor previously termed Co-eIF-2C, at that time probably containing the factors eIF3 and eIF2B (Roy et al. 1988). Whereas the role of eIF2B in ternary complex formation is well established (see below), a role of eIF3 in this step of translation remains to be demonstrated. This is in contrast to the well-known function of eIF3 as a stimulator of ternary complex binding to 40S ribosomes (described below).

BINDING OF THE TERNARY COMPLEX TO RIBOSOMES

The reactions of the translation initiation pathway culminate in the identification of the initiation codon on mRNA. This task is fulfilled by the small ribosomal subunit. At physiological Mg^{++} concentrations (~2 mM), 80S ribosomes dissociate spontaneously into 40S and 60S subunits. 40S subunits bind the translation initiation factors eIF1A and eIF3, and eIF3 inhibits the reassociation of ribosomal subunits (Fig. 2, step 3) (Thompson et al. 1977; Trachsel and Staehelin 1979). Initiation factor eIF1A is a single polypeptide chain with a mass of about 17 kD, whereas initiation factor eIF3 is a high-molecular-weight polypeptide complex that, in mammalian cells, contains eight subunits (for details, see Merrick and Hershey, this volume). The 60S ribosomal subunits bind eIF6, a single polypeptide chain with a mass of 23–25 kD, which is by itself able to inhibit ribosomal subunit reassociation (Russell and Spremulli 1979; Raychaudhuri et al. 1984) and therefore contributes to eIF3-mediated anti-association activity (eIF6 is omitted from Fig. 2).

The (40S•eIF3) complex binds the ternary complex (eIF2•GTP•Met-$tRNA_i$) such that the methionyl-tRNA sits in the P-site of the 40S subunit (Fig. 2, step 4). This can be shown in vitro by incubating purified 40S ribosomal subunits with translation initiation factors, GTP and Met-$tRNA_i$, followed by separation of free components from ribosome-associated components by sucrose density gradient centrifugation. The binding site of the ternary complex on the 40S ribosomal subunit is identified by joining a 60S ribosomal subunit to the 40S initiation complex and by reacting the resulting 80S initiation complex with the antibiotic puromycin (Trachsel et al. 1977; Benne and Hershey 1978). Puromycin requires a free A-site on the ribosome for binding, and it reacts with methionyl-tRNA in the P-site to form methionyl-puromycin. Methionyl-puromycin can be extracted from incubation mixtures and measured. The amount of methionyl-puromycin formed is a measure of the amount of Met-$tRNA_i$ in the P-site of the ribosome.

The ternary complex is stabilized on the 40S ribosomal subunit by the initiation factors eIF1A and eIF3 (Trachsel et al. 1977; Benne and Hershey 1978; Trachsel and Staehelin 1979), as first demonstrated in vitro in reconstituted translation systems. An important role for eIF3 in ternary complex binding to the ribosome in vivo can be deduced from experiments in the yeast *S. cerevisiae*: One of the subunits of yeast eIF3 (Naranda et al. 1994) is encoded by the gene *PRT1* (Keierleber et al. 1986; Hanic-Joyce et al. 1987). A temperature-sensitive mutation in this gene causes growth arrest of yeast cells and inhibition of Met-$tRNA_i$ binding to 40S ribosomal subunits in extracts derived from these cells (Feinberg et al. 1982).

Ternary complex binding to 40S ribosomal subunits occurs before mRNA binding. This was convincingly shown in rabbit reticulocyte lysates (Darnbrough et al. 1973) and in reconstituted translation systems (Levin et al. 1973; Schreier and Staehelin 1973; Trachsel et al. 1977; Benne and Hershey 1978). Experiments with rabbit reticulocyte lysate revealed that more than 30% of the free 40S ribosomal subunits carry Met-$tRNA_i$. Fewer than 10% of these free 40S ribosomal subunits, however, are bound to mRNA. This shows that a significant fraction of the 40S ribosomal subunits in rabbit reticulocyte lysate are (40S•Met-$tRNA_i$) complexes devoid of mRNA. When these (40S•Met-$tRNA_i$) complexes were incubated with mRNA, they rapidly formed 80S complexes, demonstrating that they were competent to bind mRNA (Darnbrough et al. 1973) and strongly suggesting that Met-$tRNA_i$ binding precedes mRNA binding in the translation initiation pathway. This view was even more firmly established through experiments in translation systems reconstituted from purified components. In these systems, the binding of labeled mRNA to 40S ribosomal subunits was absolutely dependent on Met-$tRNA_i$ bound to the 40S ribosomal subunit but not vice versa (Trachsel et al. 1977; Benne and Hershey 1978).

Once formed, the (40S•Met-$tRNA_i$) initiation complex binds to mRNA. This appears to occur on most mRNAs at or near the cap structure. From there, the 40S initiation complex moves in the 5′ to 3′ direction along the mRNA searching for an AUG initiation codon (scanning process; see Merrick and Hershey; Jackson; both this volume).

Despite the strong evidence in favor of Met-$tRNA_i$ binding preceding mRNA binding in initiation complex formation, we should not dismiss the possibility that binding in the reverse order may occur in special situations. A candidate case is limited availability of ternary complex, e.g., due to down-regulation of eIF2 activity in the cell. Under these conditions, 40S ribosomes carrying initiation factors but lacking ternary complex might bind to the 5′ end of mRNA and begin the scanning process. They would then traverse AUG codons until the ternary complex (eIF2•GTP•Met-$tRNA_i$) associates with the incomplete initiation complex. After binding of the ternary complex, the resulting 40S initiation complex would select the next downstream AUG codon in favorable sequence context for initiation, join with a 60S ribosomal subunit, and proceed to the elongation phase. Results from experiments carried out by Dasso et al. (1990) could be interpreted in this way. These authors studied initiation of translation on mRNA bearing two in-frame AUG codons and showed that increases in the concentration of eIF2 in the rabbit reticulocyte lysate resulted in preferential recognition of the first of the two AUG codons (Dasso et al. 1990). Another situation in which mRNA binding may precede ternary complex binding to the 40S ribosomal subunit is reinitiation of translation. This is discussed in Open Questions at the end of this chapter.

INITIATOR AUG CODON RECOGNITION

On most eukaryotic mRNAs, ribosomes initiate translation at the first AUG codon they encounter (Kozak 1989). Very rarely, another codon is selected for initiation (Mehdi et al. 1990). The (40S•Met-$tRNA_i$) complex recognizes an AUG codon through codon-anticodon base pairing with the Met-$tRNA_i$ (Fig. 2, step 5). This was nicely illustrated by genetic experiments in the yeast *S. cerevisiae* system. Cigan et al. (1988) showed that initiation from a non-AUG codon is possible when a mutant initiator tRNA is expressed whose anticodon bases can base pair with the mutated initiator codon (for a detailed description of the AUG recognition process, see Merrick and Hershey, this volume). In addition to the AUG codon, bases at positions –3 and +4 relative to the A of AUG on the mRNA are important for the recognition process. In mammalian cells, for instance, purines in the context A/GCCAUGA/G (AUG initiator codon underlined) are optimal for initiation (Kozak 1989). More recently, it became evident that even positions +5 and +6 influence AUG selection (Boeck and Kolakofsky 1994; Grünert and Jackson 1994). In the yeast *S. cerevisiae*, the requirement for specific nucleotides flanking the AUG codon is less stringent than in higher eukaryotes; e.g., an AUG codon flanked by a pyrimidine instead of a purine at position –3 is only about threefold less active as an initiation codon (Baim and Sherman 1988).

It is not yet clear how the AUG codon and its flanking sequences are recognized by the (40S•Met-$tRNA_i$) complex. The α and β subunits of eIF2 are good candidates for proteins involved in this process. Using biochemical methods, several workers demonstrated that eIF2 binds to RNA and perhaps preferentially to ribosome-binding sites on mRNA (Kaempfer et al. 1981; Gonsky et al. 1990). Recently, the eIF2 subunit involved in RNA binding was identified as the β-subunit (Flynn et al. 1994). RNA-binding activity of the β-subunit was earlier inferred from sequence analysis of the mammalian cDNA (Pathak et al. 1988) and the *S. cerevisiae* gene (Donahue et al. 1988) encoding the β-subunit of eIF2: Both contain a zinc finger motif and stretches of basic amino acids that may be involved in interaction with RNA. The most convincing evidence for an involvement of eIF2 in AUG codon recognition, however, is the finding that mutations in the genes encoding the α and β subunits of eIF2 in *S. cerevisiae* allow initiation of translation from non-AUG codons (Donahue et al. 1988).

It is tempting to speculate that the RNA-binding activity of eIF2 is regulated by covalent modification of the protein (see Open Questions at the end of this chapter) or by ligand binding. Concerning the latter, it was reported that ATP stimulates binding of eIF2 to mRNA (Gonsky et al. 1990). More recently, however, this finding could not be reproduced (Flynn et al. 1994). A further candidate protein for AUG initiation codon recognition by the (40S•Met-$tRNA_i$) complex on mRNA is the factor SUI1. This single polypeptide chain with a mass of 12 kD, isolated from *S. cerevisiae*, is essential for growth of yeast cells. A mutation in the gene encoding SUI1 protein, like mutations in the genes encoding the α and β subunits of eIF2, allows initiation from a non-AUG codon by eIF2 (Yoon and Donahue 1992). A SUI1-like protein is present in higher eukaryotic cells as shown by the isolation of a gene from mammalian cells encoding a protein that has sequence homology with the SUI1 protein (Fields and Adams 1994). The SUI1 homolog is identical to eIF1 (Kasperaitis et al. 1995).

A third initiation factor suggested some time ago to be a candidate for AUG recognition is eIF4B. The generally accepted role of this protein in translation initiation is a helper function in eIF4A-mediated RNA secondary structure unwinding (see Merrick and Hershey, this volume). Data from trinucleotide-binding experiments in vitro showed that rabbit reticulocyte eIF4B bound AUG triplets better than any other of a variety of triplets tested (Goss et al. 1987). It was also claimed that eIF4B from wheat germ binds preferentially to RNA fragments containing an AUG codon and its flanking regions derived from a plant viral RNA (Butler and Clark 1984). However, preferential binding of AUG codon-containing oligoribonucleotides to eIF4B could not be confirmed recently using the wheat-germ factor (Sha et al. 1994). Furthermore, the eIF4B preparation used in the study of Butler and Clark (1984) was not homogeneous, and the main 80-kD protein in the preparation was later shown not to be eIF4B (Browning et al. 1989). Therefore, the question of whether eIF4B contributes to the AUG recognition process and how RNA secondary structure unwinding and AUG recognition are possibly linked requires further investigation.

Finally, there remains the interesting possibility that rRNA sequences of the small ribosomal subunit contribute to AUG recognition through interaction, e.g., by base pairing, with mRNA sequences flanking the AUG initiator codon. The anticodon of Met-$tRNA_i$ base pairs with the AUG codon during initiation, but, for steric reasons, it is unlikely that further Met-$tRNA_i$ sequences interact with mRNA sequences flanking the AUG codon. Interestingly, rRNA sequences of the small ribosomal subunit, which are conserved between prokaryotes and eukaryotes, the 530-loop, and the 1400-region (numbering according to the *Escherichia coli* rRNA), were shown to interact with mRNA in *E. coli* (Dontsova et al. 1992). In particular, the 530-loop with the sequence 3′-UGGCGC CGAC-5′ is complementary to nucleotide sequences flanking strong initiator codons in eukaryotes if the AUG codon on the mRNA is put in register with the CGC triplet on the rRNA (Lagunez-Otero 1993). Even though this is a highly speculative model at the moment, it merits further investigation.

RECYCLING OF eIF2

Recognition of an AUG initiation codon by the (40S•Met-$tRNA_i$) complex on mRNA is followed by the joining of a 60S ribosomal subunit with the 40S initiation complex. Ribosomal subunit joining is triggered by binding of initiation factor eIF5 to the 40S initiation complex (Fig. 2, step 6). This factor is a single polypeptide chain with a molecular mass of 45 kD in *S. cerevisiae* (Chakravarti and Maitra 1993) and 49 kD in rat cells (Das et al. 1993). Factor eIF5 promotes the hydrolysis of GTP bound to eIF2 on the 40S initiation complex (Peterson et al. 1979; Raychaudhuri et al. 1985) (Fig. 2, step 6) and the release of bound initiation factors (Fig. 2, step 7; only eIF3 is shown in Fig. 2). This is a prerequisite for the joining of a 60S ribosomal subunit with the 40S initiation complex (Fig. 2, step 7). Whether the GTPase activity resides in eIF2 and eIF5 acts as a GTPase-activating factor or whether eIF5 carries the GTPase activity is still not clear. The resulting 80S initiation complex carries the Met-tRNA in the P-site and is competent for elongation (Fig. 2, step 8).

Ribosomal subunit joining can be studied in vitro in reconstituted translation systems (Trachsel et al. 1977; Benne and Hershey 1978). Incubation mixtures containing 40S and 80S initiation complexes are separated on sucrose density gradients, and the transfer of labeled initiator tRNA and mRNA from the 40S to the 80S ribosome position in the gradient is determined. This transfer serves as a measure for subunit joining. The reaction is absolutely dependent on eIF5 and is inhibited by nonhydrolyzable GTP analogs. The binding site of Met-$tRNA_i$ on the ribosome can be determined by incubation of 80S initiation complexes with puromycin as described above (see Binding of Ternary Complex to Ribosomes).

The hydrolysis of GTP during the subunit joining reaction results in the formation of the (eIF2•GDP) complex. The (eIF2•GDP) complex is then released from the ribosome and GDP is replaced by GTP. These reactions are catalyzed by initiation factor eIF2B and result in the regeneration of eIF2 for a new round of initiation. Factor eIF2B consists of five subunits and was isolated from mammalian (Konieczny and Safer 1983) and *S. cerevisiae* cells (Bushman et al. 1993a,b; Cigan et al. 1993).

Two pathways for the release of the (eIF2•GDP) complex from the ribosome were described. In the first pathway (accepted by the majority of workers in the field), the (eIF2•GDP) complex is spontaneously released from the ribosome (Fig. 2, step 9) and then recycled off the ribosome by initiation factor eIF2B (Fig. 2, steps 10 and 11) (Rowlands et al. 1988; Dholakia and Wahba 1989; Oldfield and Proud 1992). In the second pathway, the (eIF2•GDP) complex remains associated with the 80S initiation complex after the subunit joining step (Fig. 2, step 12) bound to the 60S subunit, and eIF2B stimulates the dissociation of the (eIF2•GDP) complex from the 60S ribosomal subunit (Fig. 2, step 13) (Gross et al. 1985, 1987; Thomas et al. 1985). In this pathway, an (eIF2•GDP•eIF2B) complex is released from the ribosome. As in the first pathway, GDP is then exchanged for GTP on eIF2 (Fig. 2, step 11) if the GTP:GDP ratio is high enough to allow GTP to compete successfully with GDP for rebinding to eIF2 (Rowlands et al. 1988).

The molecular mechanism of the nucleotide exchange reaction (Fig. 2, step 11) is still not known. Two different models were proposed. In the first model (Rowlands et al. 1988), eIF2B binds the (eIF2•GDP) complex and allows dissociation of GDP and binding of GTP through a displacement mechanism. Factor eIF2B then dissociates from the (eIF2•GTP) complex. In the second model (Dholakia and Wahba 1989; Oldfield and Proud 1992), a quarternary (eIF2•GDP•eIF2B•GTP) complex is formed, and GDP and eIF2B are then sequentially released from this complex. In both pathways for (eIF2•GTP) complex release and eIF2 regeneration, the nucleotide exchange reaction is assumed to occur in solution off the ribosome. This is supported by the observation that eIF2B catalyzes nucleotide exchange on eIF2 in vitro in the absence of ribosomes. At present, we cannot dismiss an alternative pathway in which eIF2B-mediated nucleotide exchange on eIF2 occurs, at least in some situations, on the ribosome. This hypothetical pathway and its implications are discussed below (see Open Questions). The nucleotide exchange activity of eIF2B can be conveniently measured by incubating the preformed (eIF2•GDP) complex containing radioactively labeled GDP with cold GDP or GTP and determining the loss of label from eIF2 by filtering the factor-nucleotide complex through nitrocellulose filters (Panniers et al. 1988). This assay can be performed in vitro in reconstituted translation systems or in unfractionated cell extracts. Note that the regulation of the activity of eIF2 by the GTP:GDP ratio at the level of ternary complex (eIF2•GTP•Met-tRNA$_i$) formation through the action of the guanine nucleotide exchange factor eIF2B and at the level of GTP hydrolysis through the action of initiation factor eIF5 is strikingly similar to the regulation of G-protein activities in signal transduction pathways (Bourne 1990).

The recycling of eIF2 by eIF2B is regulated by phosphorylation of eIF2 (Farrell et al. 1977) and, very likely, by modulation of eIF2B activity (see below). Factor eIF2 is phosphorylated on the α-subunit at Ser-51 by eIF2-specific protein kinases (see Clemens, this volume). The phosphorylated factor is still active in binding of Met-tRNA$_i$ to the 40S ribosomal subunit (Trachsel and Staehelin 1978), but it cannot be regenerated for further rounds of initiation. The block was found to be at the level of the GDP-GTP exchange reaction catalyzed by initiation factor eIF2B. Phosphorylated eIF2 has an approximately 150-fold higher affinity for eIF2B than the unphosphorylated factor and acts as a competitive inhibitor in the nucleotide exchange reaction (Fig. 2, step 11) (Rowlands et al. 1988). Tight binding of phosphorylated eIF2 by eIF2B leads to inhibition of nucleotide exchange on eIF2 and consequently to inhibition of translation initiation.

Genetic and biochemical analyses of the subunits of *S. cerevisiae* eIF2B have led to the identification of amino acid sequences in three eIF2B subunits that may be involved in the binding of the α-subunit of eIF2. These sequences probably bind more tightly to the phosphorylated form than to the unphosphorylated form. Perhaps the sole activity of the smallest subunit of yeast eIF2B (GCN3 protein) is binding to the α-subunit of eIF2. This eIF2B subunit is dispensable for normal growth of yeast cells and is only required for the down-regulation of translation in response to eIF2α phosphorylation (Hinnebusch 1994).

Since eIF2B is usually present in cells or cell lysates in concentrations lower than those of eIF2, phosphorylation of a fraction of the eIF2 is sufficient to give very strong inhibition. The ratio of eIF2 to eIF2B in cells may thus determine the sensitivity of translation initiation activity in response to eIF2α phosphorylation. This ratio was shown to vary more than fourfold in different mammalian tissues, mainly due to variations in the amounts of eIF2B (Oldfield et al. 1994). In the rabbit reticulocyte lysate, approximately 30% of eIF2 in the phosphorylated form is sufficient to block initiation because eIF2B is about fivefold less abundant than eIF2.

In the pathway in which eIF2B stimulates the dissociation of the (eIF2•GDP) complex from the 60S ribosomal subunit (Fig. 2, steps 12 and 13), phosphorylation of the α-subunit of eIF2 by eIF2 kinases inhibits the release (Gross et al. 1985, 1987; Thomas et al. 1985). As a consequence, the 60S subunit dissociates from the initiation complex associated with the (eIF2•GDP) complex, whereas the 40S subunit remains attached to the mRNA. Met-$tRNA_i$ associated with these 40S initiation complexes may subsequently undergo deacylation of the Met-$tRNA_i$ (Kaplansky et al. 1982). In this pathway, phosphorylation of eIF2 inhibits translation initiation at two levels: the release of the (eIF2•GDP) complex from 80S initiation complexes (Fig. 2, step 13) and the nucleotide exchange on eIF2 (Fig. 2, step 11).

Down-regulation of eIF2B activity in response to eIF2α phosphorylation is indirect. Several workers also reported regulation of eIF2B activity by mechanisms other than phosphorylation of the α-subunit of eIF2. Enhanced eIF2B activity was observed in rat fibroblasts after treatment with insulin, epidermal growth factor, serum, or phorbol esters. eIF2B activity could be measured in extracts derived from these cells and occurred in the absence of changes in the phosphorylation status of the α-subunit of eIF2 (Welsh and Proud 1992). Two different types of regulation of eIF2B activity were proposed: allosteric regulation and regulation by phosphorylation of the largest subunit of eIF2B (see Clemens, this volume). Allosteric regulation of eIF2B activity was shown in vitro. It is exerted by ligands that bind to eIF2B such as NAD^+,

NADP+ (Dholakia et al. 1986; Oldfield and Proud 1992), and polyamines (Gross and Rubino 1989; Oldfield and Proud 1992). In the rabbit reticulocyte lysate, glucose-6-phosphate was shown to stimulate eIF2B activity independent of eIF2α phosphorylation (Gross et al. 1988), but a direct effect on eIF2B activity could not be demonstrated in vitro (Oldfield and Proud 1992). The molecular mechanism(s) responsible for allosteric regulation by these ligands is not known at present.

The largest subunit of eIF2B is phosphorylated in vitro by the casein kinases I and II (Dholakia and Wahba 1988; Oldfield and Proud 1992). Phosphorylation of eIF2B by casein kinase I does not alter its activity. The effect of phosphorylation by casein kinase II on eIF2B activity is controversial: One group of workers reports that phosphorylation of the largest subunit of eIF2B in vitro by casein kinase II enhances the GDP-GTP exchange activity of eIF2B (Dholakia and Wahba 1988; Singh et al. 1994), whereas another group finds no effect (Oldfield and Proud 1992). The largest subunit of eIF2B is also a target for glycogen synthase kinase-3 in vitro (Welsh and Proud 1993). Phosphorylation of eIF2B by this kinase inactivates eIF2B (Price and Proud 1994). Since all of these effects were measured in vitro, it remains to be demonstrated whether allosteric regulation of eIF2B and regulation of its activity by phosphorylation have a role in vivo.

OPEN QUESTIONS

Many questions concerning the mechanism and regulation of Met-tRNA$_i$ binding to ribosomes remain open. A few of them were selected and are discussed below.

1. Where Does Guanine Nucleotide Exchange on eIF2 by eIF2B Occur, in Solution or on the Ribosome?

Models of translation initiation that involve eIF2B in the release of the (eIF2•GDP) complex from the ribosome (see Recycling of eIF2 above and Fig. 2, steps 12 and 13) are potentially interesting for our understanding of the mechanism of reinitiation of translation. According to this model, the formation of stable 80S initiation complexes requires the release of the (eIF2•GDP) complex from the ribosome by eIF2B. Phosphorylation of the α-subunit of eIF2 slows this process down, which may allow the 60S subunit to dissociate from the 80S initiation complex. The resulting 40S initiation complex could then resume scanning. From such a 40S initiation complex, eIF2B could slowly release the (eIF2•GDP) complex during the scanning process. Alternatively, the α-subunit of eIF2 could be dephosphorylated by a phosphatase, resulting in more efficient (eIF2•GDP) complex release. Such 40S initiation complexes might then rebind the ternary complex and become competent to select an AUG codon for initiation (reinitiation). One might even imagine that the eIF2B-catalyzed guanine nucleotide exchange reaction takes place on the scanning 40S ribosome after dephosphorylation of the α-subunit of eIF2. This might explain the findings of de Benedetti and Baglioni (1983, 1985), who claimed that 40S initiation complexes, which carry phosphorylated eIF2 can be dephosphorylated and are then able to join 60S ribosomal subunits without dissociating from mRNA.

2. *Is There a Contribution to Regulation of Met-tRNA$_i$ Binding to the Ribosome by Phosphorylation of the β-subunit of eIF2?*

The β-subunit of eIF2 was shown to be phosphorylated in vitro by three kinases: casein kinase II, protein kinase C, and cAMP-dependent kinase (Welsh et al. 1994). The two sites on the β-subunit that are phosphorylated by casein kinase II in vitro are also phosphorylated in the rabbit reticulocyte lysate (Welsh et al. 1994) and may be phosphorylated in vivo. Interestingly, changes in the phosphorylation status of the β-subunit in vivo accompany increased translation initiation activity (Duncan and Hershey 1985). An interesting proposal that has not yet been experimentally tested is that phosphorylation of the β-subunit might regulate the RNA-binding activity of eIF2 (Flynn et al 1994).

3. *What Are the Functions of Some Less Well Characterized Translation Initiation Factors in Met-tRNA$_i$ Binding to the Ribosome?*

Among these less well characterized initiation factors are the initiation factors eIF1, eIF2A, and SUI1. Is eIF2 the only factor able to transfer Met-tRNA$_i$ to the 40S ribosome? It is interesting to recall that in the early days of initiation factor analysis, an additional factor with Met-tRNA$_i$-binding activity was purified. This factor was originally called M1 and later eIF2A. It is a single polypeptide chain of 65 kD. Factor eIF2A was found to catalyze Met-tRNA$_i$ binding to the 40S ribosomal subunit in an AUG-dependent but GTP-independent reaction (Merrick and Anderson 1975). It was shown to be less active than eIF2 in catalyzing methionyl-puromycin synthesis in the presence of mRNA (Adams et al. 1975). Its function (if any) in translation was never established. Perhaps it would be worthwhile to test whether this factor has a role in alternative initiation pathways such as reinitiation and internal initiation.

Except for the determination of partial amino acid sequences (Kasperaitis et al. 1995), eIF1 has not been further studied during recent years. This factor is a single polypeptide chain with a molecular mass of 15 kD (Schreier et al. 1977). It stimulates Met-tRNA$_i$ and mRNA binding to the 40S ribosomal subunit in vitro in a reconstituted translation system (Trachsel et al. 1977). However, the effects of eIF1 on these reactions as well as stimulation of overall translation by this factor were only marginal. Since eIF1 is closely realted to SUI1, this suggests that the mammalian initiation factor might also influence initiation codon selection while promoting the initiation pathway.

4. What Is the Function of Met-tRNA$_i$ Deacylase?

The existence of a Met-tRNA$_i$ deacylase was reported many years ago and was postulated to have a role in the shutoff of translation in heme deficiency in rabbit reticulocyte lysates (Gross 1979). The contribution of this enzyme to translational regulation has not been studied further and questions concerning its physiological role therefore remain open.

PERSPECTIVE

In recent years, the role of initiation factor eIF2 in individual steps of the translation initiation pathway was elucidated, and the mechanism by which phosphorylation of the α-subunit of eIF2 leads to inhibition of the recycling of the (eIF2·GDP) complex was unraveled. In the future, we may expect to learn more about the role of eIF2 and eIF2B activity regulation in vivo under certain physiological conditions. We already know from experiments with cells in which eIF2α phosphorylation is depressed in vivo by expression of a dominant negative eIF2α kinase that eIF2 activity levels are important for growth regulation (Koromilas et al. 1992; Meurs et al. 1993). Such cells apparently lose growth control and become transformed, indicating a role for eIF2 in translational regulation of individual mRNAs. Down-regulation of eIF2 activity is expected to result in inhibition of translation of most cellular mRNAs. However, elegant genetic and biochemical experiments in the yeast *S. cerevisiae* system have more recently revealed that reduction of eIF2 activity does not necessarily result in diminished translation of all mRNAs. Translation of the yeast *GCN4* mRNA which encodes a transcription factor is actually stimulated, and the increased synthesis of this protein is of central physiological significance when yeast cells are starved for amino acids (see Hinnebusch, this volume). It is reasonable to expect that there exist more mRNAs, also in higher eukaryotes, whose translation is up-regulated by down-regulation of eIF2 activity. If so, phosphorylation of the α-subunit of eIF2 or direct modulation of eIF2B activity allows cells to simultaneously fine-tune the translational activities of many mRNAs in response to intracellular and extracellular signals.

SUMMARY

The selection of the correct AUG codon on the mRNA for initiation of translation is achieved by translation initiation factor eIF2 and initiator Met-tRNA (Met-$tRNA_i$) and influenced by additional initiation factors. Initiation factor eIF2 catalyzes the formation of the ternary complex (eIF2•GTP•Met-$tRNA_i$). The activity of eIF2 for ternary complex formation is regulated by the GTP:GDP ratio and the degree of phosphorylation of the α-subunit of eIF2 at Ser-51. The ternary complex binds to the 40S ribosomal subunit and the resulting (40S•eIF2•Met-$tRNA_i$) complex selects the initiator AUG codon on mRNA by as yet poorly characterized factor-RNA interaction(s) and by base-pairing between the initiator tRNA and the AUG codon. After this process, eIF2-bound GTP is hydrolyzed to GDP and inorganic phosphate through the action of initiation factor eIF5, and the (eIF2•GDP) complex is released from the ribosome. The GDP molecule bound to eIF2 is replaced by a molecule of GTP in a reaction catalyzed by initiation factor eIF2B. The control of eIF2 activity by the nucleotide GTP, GTPase-activating factor eIF5, and guanine nucleotide exchange factor eIF2B is analogous to the regulation of the activity of G proteins in signal transduction pathways. Finally, an alternative model of (eIF2•GDP) complex release from the ribosome and a number of open questions are discussed.

ACKNOWLEDGMENTS

I thank N. Schmitz and M. Altmann for critical reading of the manuscript, M. König for typing the manuscript, and S. Jungi for drawing the figures.

REFERENCES

Adams, S.L., B. Safer, W.F. Anderson, and W.C. Merrick. 1975. Eukaryotic initiation complex formation: Evidence for two distinct pathways. *J. Biol. Chem.* **250:** 9083–9089.

Anthony, D.D., T.G. Kinzy, and W.C. Merrick. 1990. Affinity labelling of eukaryotic in-

itiation factor 2 and elongation factor 1αβγ with GTP analogs. *Arch. Biochem. Biophys.* **281:** 157–162.
Aström, S.U. and A.S. Byström. 1994. Rit1, a tRNA backbone-modifying enzyme that mediates initiator and elongator tRNA discrimination. *Cell* **79:** 535–546.
Aström, S.U., U. Vonpawelrammingen, and A.S. Byström. 1993. The yeast initiator transfer RNA(Met) can act as an elongator transfer RNA(Met) in vivo. *J. Mol. Biol.* **233:** 43–58.
Baim, S.B. and F. Sherman. 1988. mRNA structures influencing translation in the yeast *Saccharomyces cerevisiae*. *Mol. Cell. Biol.* **8:** 1591–1601.
Barrieux, A. and M.G. Rosenfeld. 1977. Characterization of GTP-dependent Met-$tRNA_f$ binding protein. *J. Biol. Chem.* **252:** 3843–3847.
———. 1978. mRNA induced dissociation of initiation factor 2. *J. Biol. Chem.* **253:** 6311–6314.
Basavappa, R. and P. Sigler. 1991. The 3Å crystal structure of yeast initiator tRNA: Functional implications in initiator/elongator discrimination. *EMBO J.* **10:** 3105–3111.
Benne, R. and J.W.B. Hershey. 1978. The mechanism of action of protein synthesis initiation factors from rabbit reticulocytes. *J. Biol. Chem.* **253:** 3078–3087.
Benne, R., H. Amesz, J.W.B. Hershey, and H.O. Voorma. 1979. The activity of eukaryotic initiation factor eIF-2 in ternary complex formation with GTP and Met-$tRNA_f$. *J. Biol. Chem.* **254:** 3201–3205.
Benne, R., C. Wong, M. Luedi, and J.W.B. Hershey. 1976. Purification and characterization of initiation factor IF-E2 from rabbit reticulocytes. *J. Biol. Chem.* **251:** 7675–7681.
Boeck, R. and D. Kolakofsky. 1994. Positions +5 and +6 can be major determinants of the efficiency of non-AUG initiation codons for protein synthesis. *EMBO J.* **13:** 3608–3617.
Bommer, U.-A. and T.V. Kurzchalia. 1989. GTP interacts with different subunits of the eukaryotic initiation factor eIF-2. *FEBS Lett.* **244:** 323–327.
Bourne, H.R. 1990. The GTPase superfamily: A conserved switch for diverse cell functions (review). *Nature* **348:** 125–132.
Browning, K.S., L. Fletcher, S.R. Lax, and J. Ravel. 1989. Evidence that the 59-kDa protein synthesis initiation factor from wheat germ is functionally similar to the 80-kDa initiation factor 4B from mammalian cells. *J. Biol. Chem.* **264:** 8491–8494.
Bushman, J.L., A.I. Asuru, R.L. Matts, and A.G. Hinnebusch. 1993a. Evidence that GCD6 and GCD7, translational regulators of GCN4, are subunits of the guanine nucleotide exchange factor for eIF-2 in *Saccharomyces cerevisiae*. *Mol. Cell. Biol.* **13:** 1920–1932.
Bushman, J.L., M. Foiani, A.M. Cigan, C.J. Paddon, and A.G. Hinnebusch. 1993b. Guanine nucleotide exchange factor for eukaryotic translation initiation factor-2 in *Saccharomyces cerevisiae* — Interactions between the essential subunits GCD2, GCD6, and GCD7 and the regulatory subunit GCN3. *Mol. Cell. Biol.* **13:** 4618–4631.
Butler, J.S. and J.M. Clark. 1984. Eukaryotic eIF-4B of wheat germ binds to the translation initiation region of a mRNA. *Biochemistry* **23:** 809–815.
Choi, S.-Y., B.J. Scherrer, J. Schnier, M.V. Davies, and R.J. Kaufman. 1992. Stimulation of protein synthesis in COS cells transfected with variants of the α-subunit of initiation factor eIF-2. *J. Biol. Chem.* **267:** 286–293.
Cigan, A.M., L. Feng, and T.F. Donahue. 1988. $tRNA^{met}/_i$ functions in directing the scanning ribosome to the start site of translation. *Science* **242:** 93–96.
Cigan, A.M., J.L. Bushman, T.R. Boal, and A.G. Hinnebusch. 1993. A protein complex

of translational regulators of GCN4 messenger RNA is the guanine nucleotide-exchange factor for translation initiation factor-2 in yeast. *Proc. Natl. Acad. Sci.* **90:** 5350–5354.

Cigan, A.M., E.K. Pabich, L. Feng, and T.F. Donahue. 1989. Yeast translation initiation suppressor sui2 encodes the α-subunit of eukaryotic initiation factor 2 and shares sequence identity with the human α-subunit. *Proc. Natl. Acad. Sci.* **86:** 2784–2788.

Chakravarti, D. and U. Maitra. 1993. Eukaryotic translation initiation factor-5 from *Saccharomyces cerevisiae* — Cloning, characterization, and expression of the gene encoding the 45,346 Da protein. *J. Biol. Chem.* **268:** 10524–10533.

Chiorini, J.A., T.R. Boal, S. Miyamoto, and B. Safer. 1993. A difference in the rate of ribosomal elongation balances the synthesis of eukaryotic translation initiation factor (eIF)-2α and eIF-2β. *J. Biol. Chem.* **268:** 13748–13755.

Cohen, R.B., T.R. Boal, and B. Safer. 1990. Increased eIF-2α expression in mitogen-activated primary T lymphocytes. *EMBO J.* **9:** 3831–3837.

Darnbrough, C., S. Legon, T. Hunt, and R.J. Jackson. 1973. Initiation of protein synthesis: Evidence for messenger RNA-independent binding of methionyl-transfer RNA to the 40S ribosomal subunit. *J. Mol. Biol.* **76:** 379–403.

Das, K., J. Chevesich, and U. Maitra. 1993. Molecular cloning and expression of cDNA for mammalian translation initiation factor-5. *Proc. Natl. Acad. Sci.* **90:** 3058–3062.

Dasso, M.C., S.C. Milburn, J.W.B. Hershey, and R.J. Jackson. 1990. Selection of the 5′ proximal translation initiation site is influenced by mRNA and eIF-2 concentrations. *Eur. J. Biochem.* **187:** 361–371.

de Benedetti, A. and C. Baglioni. 1983. Phosphorylation of initiation factor eIF-2α, binding of mRNA to 48S complexes, and its reutilization in initiation of protein synthesis. *J. Biol. Chem.* **258:** 14556–14562.

———. 1985. Kinetics of dephosphorylation of eIF-2(αP) and reutilization of mRNA. *J. Biol. Chem.* **260:** 3135–3139.

Desgres, J., G. Keith, K.C. Kuo, and C.W. Gehrke. 1989. Presence of phosphorylated O-ribosyl adenosine in T-Ψ-stem of yeast methionine initiator tRNA. *Nucleic Acids Res.* **17:** 862–865.

Dholakia, J.N. and A.J. Wahba. 1988. Phosphorylation of the guanine nucleotide exchange factor from rabbit reticulocytes regulates its activity in polypeptide initiation. *Proc. Natl. Acad. Sci.* **85:** 51–54.

———. 1989. Mechanism of the nucleotide exchange reaction in eukaryotic polypeptide chain initiation. *J. Biol. Chem.* **264:** 546–550.

Dholakia, J.N., B.R. Francis, B.E. Haley, and A.J. Wahba. 1989. Photoaffinity labeling of the rabbit reticulocyte guanine nucleotide exchange factor and eukaryotic initiation factor 2 with 8-azidopurine nucleotides. *J. Biol. Chem.* **264:** 20638–20642.

Dholakia, J.N., T.C. Mueser, C.L. Woodley, L.J. Parkhurst, and A.J. Wahba. 1986. The association of NADPH with the guanine nucleotide exchange factor from rabbit reticulocytes: A role of pyridine dinucleotides in eukaryotic polypeptide chain initiation. *Proc. Natl. Acad. Sci.* **83:** 6746–6750.

Donahue, T.F., A.M. Cigan, E.K. Pabich, and B. Castillo Valavicius. 1988. Mutations at a Zn(II) finger motif in the yeast eIF-2β gene alter ribosomal start-site selection during the scanning process. *Cell* **54:** 621–632.

Dontsova, O., S. Dokudovskaya, A. Kopylov, A. Bogdanov, J. Rinke-Appel, N. Jünke, and R. Brimacombe. 1992. Three widely separated positions in the 16S RNA lie in or close to the ribosomal decoding region; a site-directed cross-linking study with mRNA analogues. *EMBO J.* **11:** 3105–3116.

Drabkin, H.J., B. Helk, and U.L. RajBhandary. 1993. The role of nucleotides conserved in eukaryotic initiator methionine transfer RNAs in initiation of protein synthesis. *J. Biol. Chem.* **268:** 25221–25228.

Duncan, R. and J.W.B. Hershey. 1985. Regulation of initiation factors during translational repression caused by serum depletion. *J. Biol. Chem.* **260:** 5493–5497.

Efiok, B.J.S., J.A. Chiorini, and B. Safer. 1994. A key transcription factor for eukaryotic initiation factor-2 alpha is strongly homologous to developmental transcription factors and may link metabolic genes to cellular growth and development. *J. Biol. Chem.* **269:** 18921–18930.

Eriani, G., M. Delarue, O. Poch, J. Gangloff, and D. Moras. 1990. Partition of tRNA synthetases in two classes based on mutually exclusive sets of sequence motifs. *Nature* **347:** 203–206.

Farrell, P.J., K. Balkow, T. Hunt, R.J. Jackson, and H. Trachsel. 1977. Phosphorylation of initiation factor eIF-2 and the control of reticulocyte protein synthesis. *Cell* **11:** 187–200.

Feinberg, B., C.S. McLaughlin, and K. Moldave. 1982. Analysis of temperature-sensitive mutant ts 187 of *Saccharomyces cerevisiae* altered in a component required for the initiation of protein synthesis. *J. Biol. Chem.* **257:** 10846–10851.

Fields, C. and M.D. Adams. 1994. Expressed sequence tags identify a human isolog of the Sui1 translation initiation factor. *Biochem. Biophys. Res. Commun.* **198:** 288–291.

Flynn, A., S. Oldfield, and C.G. Proud. 1993. The role of the beta-subunit of initiation factor-eIF-2 in initiation complex formation. *Biochim. Biophys. Acta* **1174:** 117–121.

Flynn, A., I.N. Shatsky, C.G. Proud, and A. Kaminsky. 1994. The RNA-binding properties of protein synthesis initiation factor eIF-2. *Biochim. Biophys. Acta* **1219:** 293–301.

Gaspar, N.J., T.G. Kinzy, B.J. Scherer, M. Hümbelin, J.W.B. Hershey, and W.C. Merrick. 1994. Translation initiation factor-eIF-2. Cloning and expression of the human cDNA encoding the gamma-subunit. *J. Biol. Chem.* **269:** 3415–3422.

Gonsky, R., M.A. Lebendiker, R. Harary, Y. Bonai, and R. Kaempfer. 1990. Binding of ATP to eukaryotic initiation factor 2. *J. Biol. Chem.* **265:** 9083–9089.

Goss, D.J., C.L. Woodley, and A.J. Wahba. 1987. A fluorescence study of the binding of eukaryotic initiation factors to messenger RNA and messenger RNA analogues. *Biochemistry* **26:** 1551–1556.

Gross, M. 1979. Control of protein synthesis by hemin: Evidence that the hemin-controlled translational repressor inhibits formation of 80S initiation complexes from 48S intermediate initiation complexes. *J. Biol. Chem.* **254:** 2370–2377.

Gross, M. and M.S. Rubino. 1989. Regulation of eukaryotic initiation factor 2B activity by polyamines and amino acid starvation in rabbit reticulocyte lysate. *J. Biol. Chem.* **264:** 21879–21884.

Gross, M., R. Redman, and D.A. Kaplansky. 1985. Evidence that the primary effect of phosphorylation of eukaryotic initiation factor 2α in rabbit reticulocyte lysate is inhibition of the release of eukaryotic initiation factor-2-GDP from 60 S ribosomal subunits. *J. Biol. Chem.* **260:** 9491–9500.

Gross, M., M.S. Rubino, and T.K. Starn. 1988. Regulation of protein synthesis in rabbit reticulocyte lysate. Glucose-6-phosphate is required to maintain the activity of eukaryotic initiation factor eIF-2B by a mechanism that is independent of the phosphorylation of eIF-2α. *J. Biol. Chem.* **263:** 12486– 12492.

Gross, M., M. Wing, C. Rundquist, and M.S. Rubino. 1987. Evidence that phosphorylation of eIF-2α prevents the eIF-2B-mediated dissociation of eIF-2-GDP from the 60 S subunit of complete initiation complexes. *J. Biol. Chem.* **262:** 6899–6907.

Grünert, S. and R.J. Jackson. 1994. The immediate downstream codon strongly influences the efficiency of utilization of eukaryotic translation initiation codons. *EMBO J.* **13:** 3618–3630.

Hanic-Joyce, P.J., R.A. Singer, and G.C. Johnston. 1987. Molecular characterization of the yeast *PRT1* gene in which mutations affect translation initiation and regulation of cell proliferaton. *J. Biol. Chem.* **262:** 2845–2851.

Hannig, E.M., A.M. Cigan, B.A. Freeman, and T.G. Kinzy. 1993. *GCD11*, a negative regulator of *GCN4* expression, encodes the γ-subunit of eIF-2 in *Saccharomyces cerevisiae. Mol. Cell. Biol.* **13:** 506–520.

Hinnebusch, A.G. 1994. Translational control of GCN4: An *in vivo* barometer of initiation factor activity. *Trends Biochem. Sci.* **19:** 409–414.

Hucul, J.A., E.C. Henshaw, and D.A. Young. 1985. Nucleoside diphosphate regulation of overall rates of protein biosynthesis acting at the level of initiation. *J. Biol. Chem.* **260:** 15585–15591.

Hümbelin, M., B. Safer, J.A. Chiorini, J.W.B. Hershey, and R.B. Cohen. 1989. Isolation and characterization of the promoter and flanking regions of the gene encoding the human protein synthesis initiation factor 2α. *Gene* **81:** 315–324.

Jacob, W.F., T.A. Silverman, R.B. Cohen, and B. Safer. 1989. Identification and characterization of a novel transcription factor participating in the expression of eukaryotic initiation factor 2 alpha. *J. Biol. Chem.* **264:** 20372–20384.

Kaempfer, R., J. van Emmolo, and W. Fiers. 1981. Specific binding of eukaryotic initiation factor 2 to satellite tobacco necrosis virus RNA at a 5′-terminal sequence comprising the ribosome binding site. *Proc. Natl. Acad. Sci.* **78:** 1542–1546.

Kaplansky, D.A., A. Kwan, and M. Gross. 1982. Effect of Met-$tRNA_f$ deacylase on polypeptide chain initiation in rabbit reticulocyte lysate. *J. Biol. Chem.* **257:** 5722–5729.

Kasperaitis, M.A.M., H.O. Voorma, and A.A.M. Thomas. 1995. The amino acid sequence of eukaryotic translation initiation factor 1 and its similarity to yeast initiation factor SUI1. *FEBS Lett.* **365:** 47–50.

Keierleber, C., M. Wittekind, S. Qin, and C.S. McLaughlin. 1986. Isolation and characterization of *PRT1*, a gene required for the initiation of protein biosynthesis in *Saccharomyces cerevisiae. Mol. Cell. Biol.* **6:** 4419–4424.

Kiesewetter, S., G. Ott, and M. Sprinzl. 1990. The role of modified purine 64 in initiator/elongator discrimination of tRNA-Met_i from yeast and wheat germ. *Nucleic Acids Res.* **18:** 4677–4682.

Kimball, S.R., T.C. Vary, and L.S. Jefferson. 1992. Age-dependent decrease in the amount of eukaryotic initiation factor-2 in various rat tissues. *Biochem. J.* **286:** 263–268.

Kisselev, L.L. and A.D. Wolfson. 1994. Aminoacyl-tRNA synthetases from higher eukaryotes. *Prog. Nucleic Acid Res.* **48:** 83–142.

Konieczny, A. and B. Safer. 1983. Purification of the eIF-2•eIF-2B complex and characterization of its guanine nucleotide exchange activity during protein synthesis initiation. *J. Biol. Chem.* **258:** 3402–3408.

Koromilas, A.E., S. Roy, G.N. Barber, M.G. Katze, and N. Sonenberg. 1992. Malignant transformation by a mutant of the IFN-inducible dsRNA-dependent protein kinase.

Science **257:** 1685–1689.

Kozak, M. 1989. The scanning model for translation: An update. *J. Cell Biol.* **108:** 229–241.

Lagunez-Otero, J. 1993. rRNA-mRNA complementarity: Implications for translation initiation. *Trends Biochem. Sci.* **18:** 406–408.

Levin, D.H., D. Kyner, and G. Acs. 1973. Protein synthesis initiation in eukaryotes: Characterization of ribosomal factors from mouse fibroblasts. *J. Biol. Chem.* **248:** 6416–6425.

Mehdi, H., E. Ono, and K.C. Gupta. 1990. Initiation of translation at CUG, GUG and ACG codons in mammalian cells. *Gene* **91:** 173–178.

Merrick, W.C. and W.F. Anderson. 1975. Purification and characterization of homogeneous protein synthesis initiation factor M1 from rabbit reticulocytes. *J. Biol. Chem.* **250:** 1107–1111.

Meurs, E.F., J. Galabru, G.N. Barber, M.G. Katze, and A.G. Hovanessian. 1993. Tumor suppressor function of the interferon-induced double-stranded RNA-activated protein kinase. *Proc. Natl. Acad. Sci.* **90:** 232–236.

Mitsui, K.-I., A. Datta, and S. Ochoa. 1981. Removal of the β-subunit of the eukaryotic polypeptide chain initiation factor 2 by limited proteolysis. *Proc. Natl. Acad. Sci.* **78:** 4128–4132.

Muramatsu, T., K. Nishikawa, F. Nemoto, Y. Kuchino, S. Nishimura, T. Miyazawa, and S. Yokoyama. 1988. Codon and amino-acid specificities of a transfer RNA are both converted by a single post-transcriptional modification. *Nature* **336:** 179–181.

Naranda, T., S.E. MacMillan, and J.W.B. Hershey. 1994. Purified yeast translational initiation factor eIF-3 is an RNA-binding protein complex that contains the PRT1 protein. *J. Biol. Chem.* **269:** 32286–32292.

Nasrin, N., M.F. Ahmad, P. Tarburton, and N.K. Gupta. 1986. Protein synthesis in yeast *Saccharomyces cerevisiae*. Purification of Co-eIF-2A and mRNA-binding factor(s) and studies of their roles in Met-tRNA-40S-mRNA complex formation. *Eur. J. Biochem.* **161:** 1–6.

Oldfield, S. and C.G. Proud. 1992. Purification, phosphorylation and control of the guanine-nucleotide-exchange factor from rabbit reticulocyte lysates. *Eur. J. Biochem.* **208:** 73–81.

Oldfield, S., B.L. Jones, D. Tanton, and C.G. Proud. 1994. Use of monoclonal antibodies to study the structure and function of eukaryotic protein synthesis initiation factor eIF-2B. *Eur. J. Biochem.* **221:** 399–410.

Panniers, R., A.G. Rowlands, and E.C. Henshaw. 1988. The effect of Mg^{2+} and guanine nucleotide exchange factor on the binding of guanine nucleotides to eukaryotic initiation factor 2. *J. Biol. Chem.* **263:** 5519–5525.

Pathak, V.K., P.J. Nielsen, H. Trachsel, and J.W.B. Hershey. 1988. Structure of the beta subunit of translational initiation factor eIF-2. *Cell* **54:** 633–639.

Peterson, D.T., B. Safer, and W.C. Merrick. 1979. Role of eukaryotic initiation factor 5 in the formation of 80S initiation complexes. *J. Biol. Chem.* **254:** 7730–7735.

Price, N.T. and C.G. Proud. 1994. The guanine nucleotide exchange factor, eIF-2B. *Biochimie* **76:** 748–760.

Raychaudhuri, P., A. Chaudhuri, and U. Maitra. 1985. Formation and release of eukaryotic initiation factor 2•GDP complex during eukaryotic ribosomal polypeptide chain initiation complex formation. *J. Biol. Chem.* **260:** 2140–2145.

Raychaudhuri, P., E.A. Stringer, D.M. Valenzuela, and U. Maitra. 1984. Ribosomal subunit anti-association activity in rabbit reticulocyte lysates: Evidence for a low molecular weight ribosomal subunit anti-association protein factor (M_r = 25,000). *J. Biol. Chem.* **259:** 11930–11935.

Rowlands, A.G., R. Panniers, and E.C. Henshaw. 1988. The catalytic mechanism of guanine nucleotide exchange factor action and competitive inhibition by phosphorylated initiation factor 2. *J. Biol. Chem.* **263:** 5526–5533.

Roy, A.L., D. Chakrabarti, B. Datta, R.E. Hileman, and N.K. Gupta. 1988. Natural mRNA is required for directing Met-tRNA binding to 40S ribosomal subunits in animal cells: Involvement of Co-eIF-2A in natural mRNA-directed initiation complex formation. *Biochemistry* **27:** 8203–8209.

Russell, D.W. and L.L. Spremulli. 1979. Purification and characterization of a ribosome dissociation factor (eukaryotic initiation factor) from wheat germ. *J. Biol. Chem.* **254:** 8796–8800.

Safer, B. and W.F. Anderson. 1978. The molecular mechanism of hemoglobin synthesis and its regulation in the reticulocyte. *CRC Crit. Rev. Biochem.* 261–289.

Safer, B., W.C. Anderson, and W.C. Merrick. 1975. Purification and physical properties of homogeneous initiation factor MP from rabbit reticulocytes. *J. Biol. Chem.* **250:** 9067–9075.

Safer, B., S.L. Adams, W.M. Kemper, K.W. Berry, M. Lloyd, and W.C. Merrick. 1976. Purification and characterization of two initiation factors required for maximal activity of a highly fractionated globin mRNA translation system. *Proc. Natl. Acad. Sci.* **73:** 2584–2588.

Schreier, M.H. and T. Staehelin. 1973. Initiation of eukaryotic protein synthesis: (Met-$tRNA_f$-40S ribosome) initiation complex catalyzed by purified initiation factors in the absence of mRNA. *Nat. New Biol.* **242:** 35–38.

Schreier, M.H., B. Erni, and T. Staehelin. 1977. Initiation of mammalian protein synthesis. I. Purification and characterization of seven initiation factors. *J. Mol. Biol.* **116:** 727–754.

Schulman, L.H. and H. Pelka. 1988. Anticodon switching changes the identity of methionine and valine transfer RNAs. *Science* **242:** 765–768.

Senger, B., L. Despons, P. Walter, and F. Fasiolo. 1992. The anticodon triplet is not sufficient to confer methionine acceptance to a transfer RNA. *Proc. Natl. Acad. Sci.* **89:** 10768–10771.

Sha, M., M.L. Balasta, and D.J. Goss. 1994. An interaction of wheat germ initiation factor 4B with oligoribonucleotides. *J. Biol. Chem.* **269:** 14872–14877.

Silverman, T.A., M. Noguchi, and B. Safer. 1992. Role of sequences within the 1st intron in the regulation of expression of eukaryotic initiation factor-2 alpha. *J. Biol. Chem.* **267:** 9738–9742.

Singh, L.P., A.R. Aroor, and A.J. Wahba. 1994. Phosphorylation of the guanine nucleotide exchange factor and eukaryotic initiation factor 2 by casein kinase II regulates guanine nucleotide binding and GDP/GTP exchange. *Biochemistry* **33:** 9152–9157.

Sprinzl, M., N. Dank, S. Nock, and A. Schön. 1991. Compilation of tRNA sequences and sequences of tRNA genes. *Nucleic Acids Res.* **19:** 2127–2171.

Thomas, N.S.B., R.L. Matts, D.H. Levin, and I.M. London. 1985. The 60S ribosomal subunit as a carrier of eukaryotic initiation factor 2 and the site of reversing factor activity during protein synthesis. *J. Biol. Chem.* **260:** 9860–9866.

Thompson, H.A., I. Sadnik, I. Scheinbuks, and K. Moldave. 1977. Studies on native ribosomal subunits from rat liver. Purification and characterization of a ribosome dissociation factor. *Biochemistry* **16:** 2221–2230.

Trachsel, H. and T. Staehelin. 1978. Binding and release of eukaryotic initiation factor eIF-2 and GTP during protein synthesis initiation. *Proc. Natl. Acad. Sci.* **75:** 204–208.

———. 1979. Initiation of mammalian protein synthesis. The multiple functions of the initiation factor eIF-3. *Biochim. Biophys. Acta* **565:** 305–314.

Trachsel, H., B. Erni, M.H. Schreier, and T. Staehelin. 1977. Initiation of mammalian protein synthesis. II. The assembly of the initiation complex with purified initiation factors. *J. Mol. Biol.* **116:** 755–768.

Walton, G.M. and G.N. Gill. 1976. Preferential regulation of protein synthesis initiation complex formation by purine nucleotides. *Biochim. Biophys. Acta* **447:** 11–19.

Welsh, G.I. and C.G. Proud. 1992. Regulation of protein synthesis in Swiss 3T3 fibroblasts — Rapid activation of the guanine-nucleotide-exchange factor by insulin and growth factors. *Biochem. J.* **284:** 19–23.

———. 1993. Glycogen synthase kinase-3 is rapidly inactivated in response to insulin and phosphorylates eukaryotic initiation factor eIF-2B. *Biochem. J.* **294:** 625–629.

Welsh, G.I., N.T. Price, B.A. Bladergroen, G. Bloomberg, and C.G. Proud. 1994. Identification of novel phosphorylation sites in the β-subunit of translation initiation factor eIF-2. *Biochem. Biophys. Res. Commun.* **201:** 1279–1288.

Yoon, H. and T.F. Donahue. 1992. The *sui1* suppressor locus in *Saccharomyces cerevisiae* encodes a translation factor that functions during $tRNA_i^{Met}$ recognition of the start codon. *Mol. Cell. Biol.* **12:** 248–260.

5

Protein Kinases That Phosphorylate eIF2 and eIF2B, and Their Role in Eukaryotic Cell Translational Control

Michael J. Clemens
Division of Biochemistry, Department of Cellular and Molecular Sciences
St. George's Hospital Medical School
Cranmer Terrace, London SW17 0RE, United Kingdom

Translational initiation in eukaryotes is a complex process involving many components, described in detail in other chapters of this volume (see, e.g., Merrick and Hershey; Trachsel; Jackson). This chapter presents a review of the current state of knowledge concerning the regulation of initiation, concentrating on the mechanisms by which the activity of initiation factor eIF2 is controlled by reversible protein phosphorylation. In recent years, it has become clear that the phosphorylation of the smallest (α)-subunit of eIF2 is a widely used mechanism of translational control in many organisms, and we now know of several physiologically important situations where eIF2α kinases are activated or inhibited.

In most situations, in vivo translation is limited by the rate of initiation, although the precise stage at which this limitation occurs probably varies depending on the circumstances. A rate-limiting step in initiation when protein synthesis is relatively rapid occurs at the level of binding of messenger RNAs to 43S preinitiation complexes (Sonenberg 1993). This process involves the least abundant initiation factor in the cell, the cap-binding protein eIF4E, as well as other components of the eIF4F complex; its regulation is discussed by Sonenberg (this volume). In contrast, under a variety of conditions of cellular stress, the binding of initiator Met-tRNA$_i$ to the 40S ribosomal subunit, catalyzed by initiation factor eIF2, can become the rate-limiting step. Examples of stresses that have this effect are hemin deprivation of reticulocytes (London et al. 1976; Chen et al. 1994), nutrient limitation (Pain 1994; Hinnebusch, this volume), heat shock (Panniers 1994; Duncan, this volume), and viral infection (Katze; Mathews; Schneider; all this volume). Inhibition of eIF2 activity is often a consequence of changes in the phosphorylation state of the α-subunit of this factor, although other mechanisms can also control eIF2, as indicated below.

eIF2 AND eIF2B IN REGULATION OF INITIATION

The principal roles of eIF2 and its guanine nucleotide exchange factor eIF2B (also known as GEF) in the mechanism and regulation of protein synthesis are now reasonably well understood. The three subunits of eIF2 and the five subunits of eIF2B have been cloned from a number of eukaryotic species, and their sequences show several conserved features (Merrick and Hershey, this volume). In the case of eIF2, identification of sites that can be phosphorylated has enabled studies of site-directed mutagenesis to establish the functional significance of such phosphorylation events. Both the α and β subunits of eIF2 can be phosphorylated in vivo and in vitro, but the physiological significance of such modifications has only been established in the case of a single site on the α-subunit (Ser-51 in mammalian species). There is evidence that the phosphorylation of eIF2α at this position leads to an increased affinity of the initiation factor for eIF2B and thus increases the proportion of the latter that is trapped as an inactive complex with phosphorylated eIF2 and GDP (Rowlands et al. 1988a). Since eIF2B is usually present in vivo at less than stoichiometric levels with respect to eIF2 (Rowlands et al. 1988b; Kimball et al. 1994; Oldfield et al. 1994), the reduction in free eIF2B results in a fall in the overall rate of guanine nucleotide exchange on the remaining unphosphorylated eIF2 (Trachsel, this volume). In situations where the eIF2/2B step in initiation is (or can become) rate-limiting, changes in the state of phosphorylation of eIF2α in the cell thus lead to overall changes in protein synthesis. (Note, however, that if the eIF4F-catalyzed binding of mRNA to ribosomes is rate-limiting, it is at least theoretically possible that some change in eIF2 phosphorylation could be tolerated by the cell without a concomitant alteration of overall translation.)

It seems likely that direct control of eIF2B activity, by mechanisms not involving eIF2α phosphorylation, can also regulate the initiation rate at the level of guanine nucleotide exchange on eIF2 (Akkaraju et al. 1991). It has been suggested, on the basis of in vitro stimulatory effects of casein kinase II (CKII) (Dholakia and Wahba 1988; Singh et al. 1994) and inhibitory effects of glycogen synthase kinase-3 (GSK-3) (C. Proud, pers. comm.), that the activity of eIF2B may be rapidly modulated by

Table 1 Properties of eIF2α protein kinases

Name	M_r	Properties	Conditions for activation	Conditions for inhibition
HCR[a]	70 kD	heme-regulated protein kinase	deficiencies of iron or hemin; heat shock	adequate hemin levels + normal physiological temperature
PKR[b]	62–65 kD	dsRNA-regulated protein kinase	low concentrations of dsRNA; treatment with heparin; others?	high concentrations of dsRNA; small RNAs; p58 inhibitor protein dsRNA-sequestration by other proteins
GCN2[c]	182 kD	essential for translational induction of GCN4 in yeast	nitrogen starvation	adequate availability of amino acids

[a]Reviewed in Chen et al. (1994).
[b]Reviewed in Mathews (this volume).
[c]Reviewed in Hinnebusch (this volume).

direct phosphorylation of this factor in response to certain environmental stimuli. One example of such an effect in vivo may be the changes in protein synthesis in skeletal muscle associated with diabetes and insulin treatment. These changes, as well as the effects of insulin on other cell types, involve direct effects on eIF2B rather than an altered phosphorylation state of eIF2 (Jeffrey et al. 1990; Welsh and Proud 1992, 1993; Karinch et al. 1993; Kimball and Jefferson 1994; Price and Proud 1994). Whether phosphorylation of eIF2B itself is responsible remains unclear, but GSK-3 is known to be rapidly inhibited by insulin treatment of cells (Welsh and Proud 1993), in parallel with a stimulation of guanine nucleotide exchange activity toward eIF2 (Welsh and Proud 1992). GSK-3-dependent (and/or CKII-dependent) regulation of eIF2B may also be of importance for the control of protein synthesis by other agents such as epidermal growth factor and phorbol esters (Welsh and Proud 1992).

THE PROTEIN KINASES THAT PHOSPHORYLATE eIF2 AND eIF2B

Three distinct enzymes that all phosphorylate the same site on the α-subunit of eIF2 (Ser-51) have been characterized in eukaryotic cells (Table 1). These protein kinases are the "hemin-controlled repressor" (HCR, also known as HRI and PK_h), the double-stranded RNA-activated

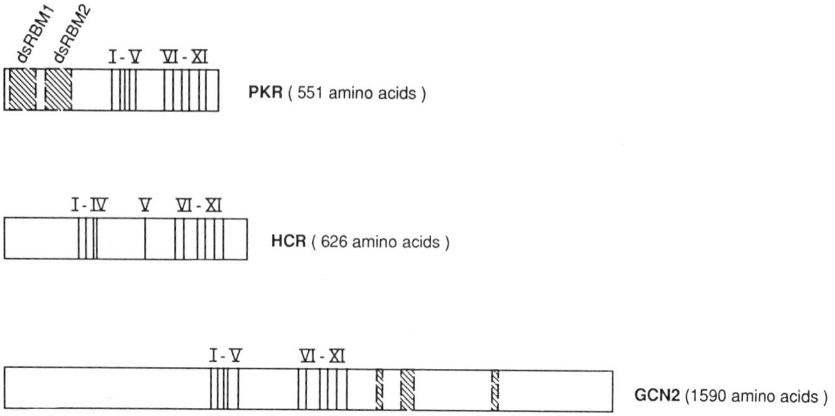

Figure 1 Domain structure of the eIF2α-specific protein kinases PKR, HCR, and GCN2. An alignment is shown of PKR, HCR, and GCN2 in which the protein kinase domains I–XI, as defined by Hanks et al. (1988), are indicated by vertical continuous lines. Shaded areas show additional domains in PKR and GCN2, which are thought to be involved in recognition of dsRNA and tRNA, respectively. In the case of PKR, two imperfect repeats occur near the amino terminus (labeled dsRBM1 and dsRBM2) that are each able to interact with dsRNA. The amino acid sequences of human PKR (GenBank accession number M35663), rabbit HCR (GenBank accession number M69035), and yeast GCN2 (GenBank accession numbers M20487 and M27082) are described in Meurs et al. (1990), Chen et al. (1991), and Wek et al. (1989), respectively.

kinase PKR (also called DAI, dsI, p68, P1, or PK_{ds}), and the *Saccharomyces cerevisiae* protein kinase GCN2. Cloning of cDNAs for the three enzymes has revealed a number of interesting sequence similarities between them, mainly in the protein kinase domains (see Fig. 1) (Wek et al. 1989; Meurs et al. 1990; Chen et al. 1991; Icely et al. 1991; Chong et al. 1992; Mellor et al. 1994). However, it is not yet clear which features account for their common substrate specificity toward eIF2α. When expressed in yeast, mammalian HCR and PKR can exhibit functional equivalence to GCN2 (Dever et al. 1993), indicating strong conservation of the properties of these kinases over a long period of evolution. However, it is uncertain whether the known properties of HCR, PKR, and GCN2 can account for all of the changes in eIF2α phosphorylation that have been reported in various physiological systems, and it is possible that other eIF2α kinases still await identification or full characterization. So far, such enzymes have eluded attempts to clone them (Mellor et al. 1994).

The Heme-regulated Protein Kinase HCR

HCR is a soluble enzyme that occurs predominantly, although perhaps not exclusively, in erythroid cells (Crosby et al. 1994; Mellor et al. 1994). Complementary DNAs for both the rabbit and rat proteins have been cloned and sequenced (Chen et al. 1991; Mellor et al. 1994). Northern blotting analyses of RNAs from various tissues have produced conflicting conclusions concerning the tissue specificity of HCR expression. Crosby et al. (1994) were unable to see expression of the HCR gene in nonerythroid cell types, a result that was consistent with measurements of HCR protein by Western blotting (Pal et al. 1991; Crosby et al. 1994). On the other hand, Mellor et al. (1994) cloned rat HCR from a brain cDNA library and subsequently found that the corresponding mRNA was abundant not only in reticulocytes, but also in rat psoas muscle. Lower levels were also present in RNA preparations from lung, heart, liver, and kidney, whereas globin mRNA was not detectable in these samples. If these findings are confirmed, it will be important to establish the physiological significance of the expression of HCR and the mode of regulation of this protein kinase in nonerythroid cell types. HCR is inhibited by hemin in vivo and in vitro and is activated when reticulocytes become deficient for iron or heme, as well as in response to many other stimuli (Jackson 1991). The sequence of the protein does not provide many clues to the mechanism of activation, but reduction of disulfide bonds appears to be an important step in the process (see later in this chapter). The sequence of HCR shows similarities to a number of other protein kinases, with the closest relationships being to GCN2 and PKR (Chen et al. 1991). There are 11 conserved protein kinase domains, with a large insert separating these domains into two groups (Fig. 1).

The Double-stranded RNA-regulated Protein Kinase PKR

In recent years, PKR has become the most widely studied eIF2α kinase. Complementary DNAs for PKR have been cloned from human and mouse cells (Meurs et al. 1990; Icely et al. 1991; Patel and Sen 1992a; Thomis et al. 1992; Baier et al. 1993), and the mouse PKR gene has also been fully sequenced (Tanaka and Samuel 1994). Comparison of the cDNA and genomic sequences indicates that the mouse PKR gene consists of 16 exons (Tanaka and Samuel 1995). The human and murine enzymes, with masses of 62 kD and 65 kD, respectively, show high amino acid sequence homology, although there are also a number of differences (see below).

The sequences of the cDNAs for human and mouse PKR reveal several features characteristic of protein kinases in general (Fig. 1) (Meurs et al. 1990; Icely et al. 1991). As with HCR, 11 protein kinase subdomains are identifiable, the first of which contains the consensus ATP-binding sequence GXGXXG. Another conserved ATP-binding region, the sequence DFG, occurs in subdomain VII. Subdomain II contains an invariant lysine residue at position 296; this amino acid is probably directly involved in phosphate transfer catalyzed by PKR, and its mutation to arginine completely abrogates all kinase activity (Katze et al. 1991). The equivalent residue in HCR is Lys-199 (Chen et al. 1991). A "kinase insertion sequence" is found in PKR between subdomains V and VI; this contains a conserved sequence, LXIQMXXC, which is also found in HCR and GCN2 (Chong et al. 1992). The function of this sequence is unknown, but it could be involved in substrate recognition. Protein kinase subdomain V of mouse PKR is three amino acids shorter than its human counterpart and there is also a further deletion just carboxy-terminal to this region (Patel and Sen 1992a). Kinase subdomain X is also less well conserved than the other subdomains between the human and mouse enzymes.

The predicted isoelectric point of the unmodified PKR protein is quite high (pI = 8.57 for the human enzyme), with basic amino acids occurring in a number of distinct dense clusters. However, PKR exists in multiple forms with much more acidic isoelectric points in vivo, suggesting that the protein is phosphorylated at several positions (Krust et al. 1984; Jeffrey et al. 1995). It is likely that this reflects the autophosphorylation of the enzyme that occurs during its activation. However, the possibility that PKR is also a substrate for other protein kinases cannot be ruled out. The sites of phosphorylation on PKR have not yet been reported. Much of the original interest in PKR came from its involvement in some of the antiviral effects of the interferons (Pestka et al. 1987; Hovanessian 1989; Samuel 1991; Meurs et al. 1992). In addition, there is now the exciting prospect that PKR may act as a tumor suppressor gene product (Clemens 1992; Koromilas et al. 1992; Lengyel 1993; Meurs et al. 1993) and may be involved in the regulation of apoptosis (Lee and Esteban 1994). PKR occurs at a low level in most mammalian cells, but its expression can be induced severalfold by interferon treatment. Curiously, it is present not only in the cytoplasm as a ribosome-associated protein, but also in the nucleus (Jiménez-García et al. 1993). Nuclear PKR is estimated to account for approximately 20% of the total in cells not induced with interferons (Jeffrey et al. 1995). Both these studies have shown that the majority of this nuclear PKR is in fact located in the nucleolus, where it may be associated with nascent ribosomes, but the significance of this is unknown.

PKR binds double-stranded RNA (dsRNA) with high affinity and is normally dependent on low concentrations of dsRNA for its activation. This leads to inhibition of protein synthesis in the presence of nanogram per milliliter levels of dsRNA in systems where PKR is present, such as rabbit reticulocyte lysates (Jackson 1991) or extracts from interferon-treated cells (Roberts et al. 1976). Other activators of PKR include polyanions such as heparin (Patel et al. 1994) and viral RNAs with extensive secondary structure such as reovirus S1 mRNA (Henry et al. 1994), hepatitis δ RNA (Robertson et al. 1996), and transcripts from the *Bam*HI W repeat region of the Epstein-Barr virus genome (A. Elia and M.J. Clemens, unpubl.). It is of interest that the translation of PKR itself is apparently impaired by activation of the protein kinase, leading to a self-limiting rate of synthesis of the enzyme (Thomis and Samuel 1992; Barber et al. 1993b; Lee et al. 1993). This suggests that the mRNA for PKR may be able to activate its own translation product.

Structural features responsible for the interaction between PKR and dsRNA have been closely investigated by several laboratories. Mutational analysis has identified two dsRNA-binding motifs near the amino terminus of the protein (amino acids 11–77 and 101–167 in human PKR) that are both necessary and sufficient for dsRNA binding (Barber et al. 1991; Katze et al. 1991; Green and Mathews 1992; McCormack et al. 1992, 1994; Patel and Sen 1992b; Patel et al. 1994; Clarke and Mathews 1995). Sequences in several other dsRNA-binding proteins show features in common with these domains of PKR (St Johnston et al. 1992; Chang and Jacobs 1993; Gatignol et al. 1993). A number of site-directed mutagenesis studies have been performed, and the roles of the various residues in these sequences in the recognition and binding of dsRNA by PKR are beginning to be delineated (Green et al. 1995; Romano et al. 1995). In vitro, PKR can also be activated by heparin; curiously, deletions near the amino terminus of the kinase that block PKR activation by dsRNA do not inhibit this alternative mode of stimulation, suggesting that another mechanism must be involved (Patel et al. 1994). A recent report indicating that mutant forms of PKR which cannot bind dsRNA can still be activated in vivo (S.B. Lee et al. 1994) adds weight to the possibility of alternative modes of PKR activation.

Although PKR appears to be a reasonably stable protein in vivo, it is susceptible to cleavage at a specific site during purification, giving rise to a fragment with an apparent size of approximately 46 kD that is recognized by a monoclonal antibody with specificity for the amino-terminal part of the protein (Clarke et al. 1991; Jeffrey et al. 1995). This fragment can still bind dsRNA and other small RNA ligands (Galabru and Hovanessian 1987; Clarke et al. 1991). It is not known whether PKR undergoes selective proteolytic cleavage at this site in vivo, but if so such a process would generate a product that has no protein kinase activity but can still bind dsRNA, perhaps in competition with the full-length active protein.

The Nutritionally Regulated Protein Kinase GCN2

The *S. cerevisiae* eIF2α protein kinase GCN2 was identified as a result of extensive genetic and biochemical investigations into the translational derepression of the transcription factor GCN4 in cells subjected to amino acid starvation (Hinnebusch, this volume). GCN2 is a 182-kD protein that is associated with ribosomes, and its activity can be regulated to enhance the synthesis of GCN4 at the translational level during nitrogen depletion. The mechanism of synthesis of GCN4, which involves a paradoxical stimulation of translation of this protein under conditions where overall protein synthesis is depressed, is described in detail by Hinnebusch elsewhere in this volume and will not be considered further here. It is of interest that GCN2 has some sequence similarities to mammalian PKR (Fig. 1) (Chong et al. 1992), but it is not clear whether these features reflect related roles for the two protein kinases in vivo. One feature that GCN2 may share with PKR is its activation by RNA. However, whereas dsRNA is the activator in the case of PKR, the most likely activator of GCN2 is uncharged tRNA (Ramirez et al. 1992) (see Fig. 1 for sites in GCN2 which, by virtue of their similarities to histidyl-tRNA synthetases, may be involved in binding tRNA). It is also possible that the similarities between GCN2 and PKR may be concerned with recognition of eIF2 or with binding to ribosomes, rather than with the mechanisms of activation of the two enzymes.

Are There Other eIF2α Kinases?

The presently known eIF2α-specific protein kinases have several structural and functional features in common, and it has been shown that both PKR and HCR can functionally substitute for GCN2 in phosphorylating eIF2α in yeast (Dever et al. 1993). Nevertheless, the question exists whether other members of this family remain to be discovered. A number of papers have described kinases that are potential additions to the list (Hiddinga et al. 1988; Sarre 1989; Barber et al. 1992; Feldhoff et al.

1993; Mellor et al. 1993; Olmsted et al. 1993), but none of these enzymes have yet been fully purified or their cDNA cloned. Certain features of the mammalian cellular response to amino acid or glucose starvation are reminiscent of the regulation of initiation in the yeast GCN4 system (see later in this chapter), raising the question of whether a GCN2 homolog in mammalian cells mediates the translational effects of these nutritional changes. At present, however, the data do not entirely rule out the possibility that PKR or HCR is the enzyme responsible.

Protein Phosphatases Acting on eIF2α

In contrast to the extensive characterization of eIF2α-specific protein kinases, relatively little is known about phosphatases that may act on this protein substrate. Both type-1 and -2A protein phosphatases have been suggested as potentially able to fulfill this function (Pato et al. 1983; Redpath and Proud 1990). The best-characterized candidate for a specific eIF2α phosphatase so far described is an enzyme involved in the yeast GCN4 regulatory pathway. This protein, the product of the *GLC7* gene, is a type-1 protein phosphatase that is also involved in the regulation of glycogen metabolism (Wek et al. 1992). Mutant forms of the phosphatase with a dominant negative effect on the wild-type protein counteract the phenotypic effect of partially defective GCN2 kinase, suggesting the involvement of GLC7 in the dephosphorylation of yeast eIF2α in vivo. Indeed, overexpression of the dominant negative mutant phosphatase does lead to increased phosphorylation of eIF2α (Wek et al. 1992). However, it is not clear whether this is a consequence of a direct or indirect action of the wild-type GLC7 protein on the initiation factor. A type-1 protein phosphatase has also been shown to dephosphorylate (and thus potentially inactivate) the eIF2α kinase, PKR (Szyszka et al. 1989b).

Protein Kinases That Phosphorylate eIF2β

Although a large amount of evidence indicates that it is the phosphorylation of the α-subunit of eIF2 that regulates the function of this initiation factor in vivo, there are many reports that the β-subunit can also be phosphorylated, both by purified protein kinases and in intact cells. CKII and protein kinase C (PKC) are capable of phosphorylating eIF2β. The former enzyme phosphorylates Ser-2 and Ser-67, whereas a rat brain PKC preparation phosphorylates Ser-13, at least in vitro (Clark et al. 1989; Welsh et al. 1994). cAMP-dependent protein kinase also has the ability to phosphorylate eIF2β in vitro on Ser-218 (Welsh et al. 1994). The state of phosphorylation of eIF2β does change under some circumstances where protein synthesis rates are modified in vivo, but the regulatory significance of this, if any, has not been established. However, Singh et al. (1994) recently reported that dephosphorylation of eIF2β with alkaline phosphatase had a modest (approximately twofold) stimulatory effect on the extent of GDP binding to the factor, and this effect could be reversed by phosphorylation in vitro with CKII.

Protein Kinases That Phosphorylate eIF2B

The strongest candidates for protein kinases that can phosphorylate eIF2B with consequent effects on the activity of this guanine nucleotide exchange factor are CKII and GSK-3. In vitro, CKII phosphorylates the 82-kD ε-subunit of eIF2B on a number of serine residues (Dholakia and Wahba 1988; Aroor et al. 1994), and this is associated with a stimulation of the guanine nucleotide exchange activity of the factor of up to fivefold (Singh et al. 1994). The stimulatory effect may be due to an enhancement of the ability of eIF2B to bind GTP. CKII is a ubiquitous protein kinase of broad specificity that, as indicated above, also phosphorylates the β-subunit of eIF2 itself. The physiological mechanisms of regulation of CKII have not been established, and the extent to which this enzyme has a role in the control of the eIF2/2B system in vivo remains to be clarified.

The regulation of GSK-3, in contrast to that of CKII, is well-characterized. This enzyme phosphorylates glycogen synthase and inhibits its activity. GSK-3 is inactivated in cells in response to insulin treatment, consistent with the stimulatory effect of this hormone on glycogen synthesis in vivo. GSK-3 also phosphorylates the transcription factor c-*jun* and inhibits its activity (Boyle et al. 1991). In vitro, eIF2B is phosphorylated by GSK-3 on its ε-subunit (Welsh and Proud 1993), and addition of GSK-3 to the reticulocyte lysate system causes a decrease in guanine nucleotide exchange activity catalyzed by eIF2B (G. Welsh and C. Proud, pers. comm.). It will be of interest to see in the future under what circumstances GSK-3 utilizes eIF2B as a substrate, and whether eIF2B activity is controlled in vivo by the balance between CKII-dependent (stimulatory) and GSK-3-dependent (inhibitory) modifications.

ACTIVATION OF eIF2α KINASES

Mechanism of Activation of HCR

The mechanism of activation of HCR has been the subject of some disagreement in the literature, although it is clear that hemin deficiency provides the signal to initiate the process. Physiologically, iron or hemin deficiency in erythroid cells leads to a general inhibition of protein synthesis. This provides a means for the coordination of heme and globin production in cells specialized for the synthesis and accumulation of large amounts of hemoglobin during erythropoiesis (London et al. 1976; Chen et al. 1994). It is to be noted, however, that the effects of HCR are not specific for globin synthesis, and in other systems where HCR is expressed, translation is again compromised in the absence of hemin. Thus, for example, insect cells infected with a recombinant baculovirus expressing mammalian HCR produce much greater yields of both the kinase itself and other co-expressed proteins if exogenous hemin is supplied (Chefalo et al. 1994). Several lines of evidence suggest that hemin and certain other porphyrin compounds (Méndez et al. 1992; Yang et al. 1992) promote the formation of intermolecular and/or intramolecular disulfide bonds involved in HCR regulation. Models in which hemin favors either the covalent homodimerization of HCR itself (Chen et al. 1989) or heterodimerization between HCR and the heat shock protein HSP90 (Matts and Hurst 1989; Méndez et al. 1992; Méndez and de Haro 1994) have been put forward as a basis for the ability of hemin to prevent the activation of the kinase (Fig. 2). However, despite the fact that the activation of HCR that occurs in the absence of hemin involves the reduction of disulfide bonds, it is not clear whether subsequent dissociation of the protein subunits is necessary for protein kinase activity (Yang et al. 1992; Méndez and de Haro 1994). Antibodies against HSP90 can adsorb HCR from hemin-supplemented but not from hemin-deficient reticulocyte lysates and can also prevent the reformation of inactive HCR when hemin is added (Matts and Hurst 1989; Matts et al. 1992). These results are in accord with the proposal that (unphosphorylated) HSP90 is associated with the inactive form but not the active form of HCR. HSP90 has been reported to inhibit translation in hemin-supplemented lysates (Rose et al. 1989).

Activation of HCR is accompanied by phosphorylation of the kinase protein itself, at sites that have not yet been identified. Additional phosphorylation of the associated HSP90 protein has also been described by Méndez and de Haro (1994). These authors concluded that phosphorylation of HSP90 (or of both HSP90 and HCR) by CKII is necessary for HCR activation and that this occurs when dissociation takes place in the absence of hemin. Previously, the phosphorylated form of HSP90 was reported to enhance the kinase activity of HCR against eIF2α (Szyszka et al. 1989a). Additional autophosphorylation of HCR may also occur during its activation, and the relative contributions of this process and of CKII-mediated phosphorylation have not been clarified.

In addition to the hemin-dependent mechanism described above, HCR can be activated by heat shock or by conditions that mimic this stress, such as exposure to heavy metals, in reticulocyte lysates (Matts et al. 1991, 1993; Matts and Hurst 1992; Duncan, this volume). These phenomena are consistent with the disulfide-bond-dependent interaction of HCR with heat shock proteins and with the ability of heavy metals to affect disulfide bond formation via interference with the thioredoxin/thioredoxin reductase system (Matts et al. 1991). In addition to HSP90, HCR can also bind to other proteins involved in cellular stress responses, such as members of the HSP70 family, p56, and EC1 antigen (Matts and Hurst 1992; Matts et al. 1992), and these may also regulate its activation during heat shock. Gross et al. (1994) have demonstrated that addition of

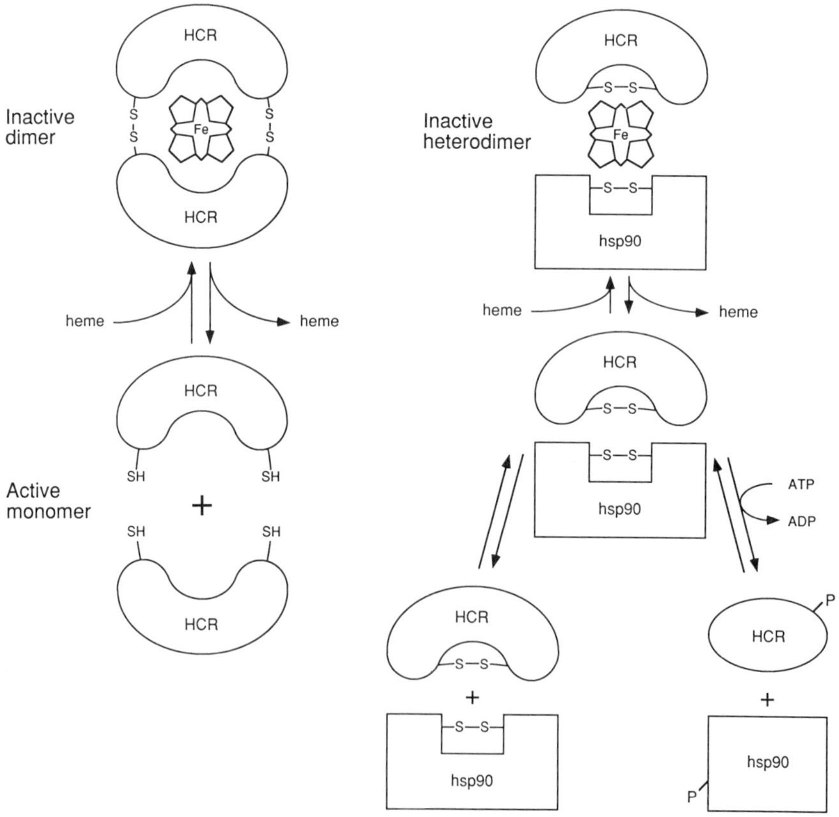

Figure 2 (*See facing page for legend.*)

the 72-kD heat shock protein HSP70(R) to hemin-deficient reticulocyte lysates reverses the inhibition of protein synthesis. As a general mechanism for translational control in response to cellular stress, it has been suggested that proteins denatured in response to heat or other stresses may bind to HSP70 proteins in competition with HCR, thus leading to dissociation and activation of the latter (Matts and Hurst 1992; Matts et al. 1993). However, HSP70 appears to remain associated with HCR when the latter is activated by hemin deficiency.

Heat shock causes increased phosphorylation of eIF2α and inhibits eIF2 activity in other cell types besides reticulocytes (Duncan and Hershey 1984, 1989; Scorsone et al. 1987; Hu et al. 1993; Panniers 1994; Sierra 1994), suggesting that activation of the same or related protein kinase(s) is involved in these systems as well. In general, the enzymes responsible for eIF2α phosphorylation under these circumstances have not been well characterized, although De Benedetti and Baglioni (1986) provided evidence for the involvement of an HCR-like kinase in the inhibition of protein synthesis in heat-shocked HeLa cells.

Figure 2 Two models for the activation of the heme-regulated protein kinase HCR. (*Left*) Model proposed by Chen et al. (1989) in which the inactive form of HCR exists as a homodimer with two 90-kD monomers linked by disulfide bridges. In the absence of hemin, reduction of the disulfide bonds occurs. This is accompanied by autophosphorylation and activation of the protein, which can then dissociate into the monomeric form. Bifunctional thiol-reactive crosslinking agents can restore covalent dimerization and this inhibits protein kinase activity. In contrast, monofunctional thiol-reactive agents activate the kinase by favoring the monomeric form. Dephosphorylation catalyzed by protein phosphatases inhibits HCR activity. Under denaturing conditions, the dimeric and monomeric forms behave as 180-kD and 90-kD proteins, respectively. Under native conditions, the dimeric form of the kinase behaves as a 290-kD complex, suggesting an elongated shape for the molecule. (*Right*) Alternative model proposed by Méndez and de Haro (1994) in which HCR is held in an inactive state in the presence of hemin by association with the heat shock protein HSP90. In the absence of heme, spontaneous dissociation of the HSP90 subunit promotes formation of a heme-reversible monomeric form of HCR, whereas phosphorylation of HCR and/or HSP90 by CKII causes dissociation and activation of the kinase in a heme-irreversible manner. Only dephosphorylation of this form of HCR can allow heterodimerization and restoration of an inactive form of the kinase (for further details, see Chen et al. [1989] and Méndez and de Haro [1994], from which these models have been redrawn).

Mechanism of Activation of PKR

The activation of PKR by dsRNA is also accompanied by autophosphorylation of the enzyme (Galabru and Hovanessian 1987). The mechanism by which dsRNA binding brings this about is not yet clear, but protein dimerization and/or conformational changes have been suggested to be required (Kostura and Mathews 1989; Langland and Jacobs 1992; Manche et al. 1992; Romano et al. 1995). It is likely that one PKR molecule can phosphorylate its neighbor, perhaps while both are bound to a single stretch of dsRNA, as indicated schematically in Figure 3. However, additional truly intramolecular autophosphorylation has not

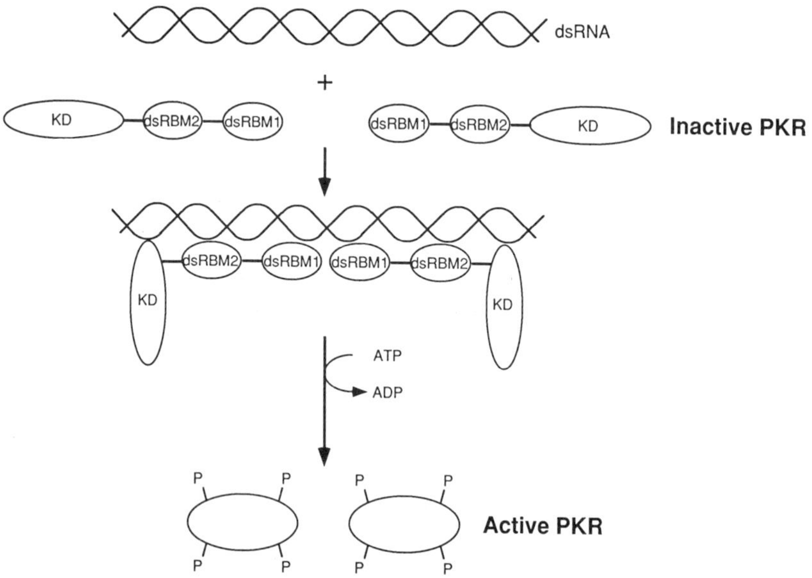

Figure 3 A model for the activation of the dsRNA-regulated protein kinase PKR. Analysis of the structure of PKR has shown that the protein kinase domain (labeled KD) occurs in the carboxy-terminal half of the enzyme, whereas the two dsRNA-binding domains (labeled dsRBM1 and dsRBM2) are at the amino terminus. In the absence of dsRNA, it is proposed that PKR exists in an inactive monomeric form (*top*). Activation of PKR occurs when two or more molecules of the protein kinase bind to a dsRNA molecule (through their dsRNA-binding domains) in close proximity to each other and interact in a way that allows transphosphorylation and/or autophosphorylation at multiple sites on the enzyme (*center*). PKR dimerization may occur not only through binding to a common dsRNA molecule, but also by direct protein-protein interaction, at the amino terminus of the protein (Patel et al. 1995). Phosphorylation of PKR is believed to result in a conformational change (*bottom*) that allows the kinase to phosphorylate eIF2α and other specific substrates in a dsRNA-independent manner.

been excluded. Once PKR is phosphorylated, its protein kinase activity against other substrates becomes independent of dsRNA. Consistent with the likely need for initial dsRNA-mediated homodimerization, high concentrations of dsRNA prevent PKR activation (Manche et al. 1992), probably because the formation of monomeric [PKR.dsRNA] complexes rather than [PKR.dsRNA.PKR] dimers is favored when the dsRNA concentration is high. An alternative model (Galabru et al. 1989), proposing the existence of both high-affinity (activating) and low-affinity (inactivating) dsRNA-binding sites on PKR, has not been supported by more recent kinetic investigations and RNA-binding studies (Kostura and Mathews 1989; Manche et al. 1992; Sharp et al. 1993b). It is of interest that homodimerization of both HCR and PKR has been implicated in the regulation of their activation (albeit with opposite effects), and this possibility may also extend to the control of the functionally related yeast GCN2 kinase (Diallinas and Thireos 1994).

PHYSIOLOGICAL CONDITIONS INVOLVING REGULATION OF eIF2α KINASES

As indicated earlier, the existence of protein kinases that can phosphorylate eIF2α at amino acid Ser-51 is a widespread phenomenon that has been observed from yeast to mammals. Given such strong evolutionary conservation, it comes as no surprise that a large variety of physiological regulatory pathways that affect protein synthesis involve changes in eIF2 phosphorylation (Table 2). In general, increases in eIF2α phosphorylation are observed in cells responding to various environmental stresses (Rowlands et al. 1988b). Cellular stresses such as heat shock (Duncan and Hershey 1984, 1989; Scorsone et al. 1987), viral infection (Chinchar and Dholakia 1989; DeStefano et al. 1990; Huang and Schneider 1990), nutrient deprivation (Scorsone et al. 1987), and ischemia (Hu and

Table 2 Physiological conditions promoting phosphorylation of eIF2α

Condition	Reference[a]
Heat shock	Panniers (1994)
Viral infection	Mathews and Shenk (1991)
Amino acid, glucose, or serum starvation	Pain (1994)
Iron or heme deficiency (erythroid cells)	Chen et al. (1994)
Growth factor deprivation	Proud (1992)
Calcium mobilization	Kimball and Jefferson (1994)

[a]The references given here are recent reviews in which much of the primary literature is listed. For further details, see text.

Wieloch 1993; Burda et al. 1994) all stimulate the phosphorylation of eIF2α in various systems. In reticulocytes and their precursors, the phosphorylation of the factor in response to deficiencies of iron or heme can also be regarded as a kind of specialized stress response (Farrell et al. 1977; Leroux and London 1982). It is also likely that changes in eIF2α phosphorylation are involved in the normal regulation of cell growth (Mundschau and Faller 1991; Chong et al. 1992). Control of the phosphorylation state of eIF2, or of the activity of eIF2α-specific kinases, occurs in response to stimulation of cell proliferation by mitogens and growth factors (Duncan and Hershey 1985; Montine and Henshaw 1989; Ito et al. 1994), as well as the induction of cell differentiation (Petryshyn et al. 1984, 1988).

A growing body of evidence indicates that changes in both intracellular and extracellular calcium levels can control eIF2 activity, and thus the rate of polypeptide chain initiation, in mammalian cells (for review, see Brostrom and Brostrom 1990; Kimball and Jefferson 1994; Nairn and Palfrey, this volume). When calcium is removed from the growth medium of cells in culture or from cell extracts, by use of specific chelating agents, polypeptide initiation becomes inhibited at the level of 43S complex formation (Kumar et al. 1989). In at least some systems, these effects cannot be attributed wholly to phosphorylation of eIF2α, since the 43S complexes are also depleted of eIF3 when calcium is chelated. However, mobilization of sequestered intracellular calcium also inhibits protein synthesis, and in this case, there is clear evidence for involvement of changes in the state of phosphorylation of eIF2α. It appears to be the concentration of calcium within intracellular stores, rather than the free cytosolic level of the cation, that is critical for these effects on initiation (Brostrom and Brostrom 1990; Brostrom et al. 1990). This may have relevance for the effects of physiological agents such as vasopressin, epinephrine, or angiotensin II that cause mobilization of sequestered calcium in the liver through production of inositol trisphosphate (Kimball and Jefferson 1992, 1994). When calcium is mobilized by exposure of perfused livers or cultured cells to these and other agents, the ensuing inhibition of translation is associated with increased phosphorylation of eIF2α and an inhibition of eIF2B activity (Kimball and Jefferson 1990, 1994; Prostko et al. 1992, 1993). Recent evidence suggests that the protein kinase responsible for these effects is PKR (Prostko et al. 1995; Srivastava et al. 1995), but further work is required to determine the mechanism by which calcium mobilization or other changes in endoplasmic reticulum (ER) function lead to activation of this enzyme.

The stress of calcium mobilization also induces longer-term cellular responses that include induction of proteins such as GRP78/BiP. The latter, an ER-associated protein, appears able to induce subsequent tolerance of protein synthesis to the initial inhibitory stress (Brostrom et al. 1990). This chronic adjustment of the initiation rate is associated with a decrease in the level of phosphorylated eIF2α (Prostko et al. 1992). However, a recent report (Brostrom et al. 1995) suggests that GRP78/BiP induction and the phosphorylation of eIF2α are separate events, with the former being more sensitive to perturbation of ER function and occurring earlier than the modification of the initiation factor. Interestingly, the longer-term adaptation to the effects of calcium mobilization is also characterized by resistance of protein synthesis to inhibition by other cellular stresses such as elevated temperature and sodium arsenite.

It is not always clear whether the alterations in the phosphorylation state of eIF2α, and hence the activity of the eIF2/eIF2B system, are causes or consequences of the various physiological responses that have been studied in different systems. Nevertheless, in some cases, eIF2 phosphorylation has been shown to be required for mediation of the inhibitory effects on protein synthesis of various cell treatments. The most convincing evidence comes from studies where the phosphorylation site in eIF2α (Ser-51) has been mutated to alanine, resulting in at least a partial protection of protein synthesis in intact cells from inhibition by stresses such as heat shock, adenovirus infection, or transfection with certain plasmids (Davies et al. 1989; Kaufman et al. 1989; Choi et al. 1992; Murtha-Riel et al. 1993). A similar protective effect of this mutation against the action of HCR has been demonstrated in vitro (Ramaiah et al. 1994).

Roles of PKR in the Cell

It is likely that the participation of PKR in the interferon-induced antiviral response reflects its ability to be activated in infected cells by viral dsRNA molecules or viral mRNAs containing extensive secondary structure (Maran and Mathews 1988; Henry et al. 1994), thus resulting in an impairment of the overall rate of protein synthesis and a slowing of virus replication (Meurs et al. 1992; Lee and Esteban 1993; Mathews, this volume). In uninfected cells, and cells not exposed to interferons, the situation is much less clear. It is possible that some cellular RNAs with extensive regions of secondary structure can function as activators of PKR to regulate protein synthesis and cell growth; cellular RNA preparations from a variety of sources have been shown to activate the kinase in cell-free systems (Pratt et al. 1988; Li and Petryshyn 1991). However, there has been little characterization of individual RNA species that may have this ability. Curiously, the mRNA for PKR itself could be one candidate since translation of this message leads to activation of the protein kinase in vivo and in vitro (Thomis and Samuel 1992; Lee et al. 1993; Barber et al. 1993b).

A role for PKR in cell growth regulation (for review, see Clemens 1992; Lengyel 1993) is supported by a recent demonstration that interleukin-3 inactivates the kinase in cells dependent on this cytokine for growth (Ito et al. 1994). Expression of wild-type PKR in yeast inhibits cell proliferation (Chong et al. 1992) and the kinase can induce apoptosis in HeLa cells (Lee and Esteban 1994). On the other hand, there is evidence that PKR activity is *required* for some aspects of signal transduction by the mitogenic platelet-derived growth factor (PDGF) (Mundschau and Faller 1995).

Expression of catalytically inactive mutant forms of the kinase in 3T3 cells results in a tumorigenic phenotype (Koromilas et al. 1992; Meurs et al. 1993). It has been proposed that the latter effect is caused by a dominant negative effect of the PKR mutants on the activity of the wild-type kinase. The suggestion that cell transformation in these experiments is due to inactivation of PKR is supported by the recent finding that overexpression of a mutant form of the α-subunit of eIF2 in which Ser-51 has been replaced by alanine can result in a tumorigenic phenotype (Donzé et al. 1995). Such a form of eIF2α cannot be phosphorylated by PKR (or other kinases) at this position and might be expected to mimic the effects of dominant negative regulators of PKR activity.

A number of mechanisms for inhibition of wild-type PKR by inactive mutant forms of the enzyme can be suggested, including competition between mutant and wild-type proteins for limiting amounts of dsRNA activators. In vitro evidence in favor of such a mechanism has been obtained (Sharp et al. 1993a, 1995), but other recent data suggest that heterodimerization between mutant and wild-type forms of PKR may be involved (Romano et al. 1995). Indeed, some inactive PKR mutants that can no longer bind dsRNA can still be tumorigenic (Barber et al. 1995). This suggests that mutant forms of the kinase may exert dominant negative effects and/or transform cells by more than one pathway. Whatever the mechanism, however, it is important to note that there is so far no direct evidence that PKR truly functions as a tumor suppressor gene product (i.e., as an enzyme that has a normal role in negatively regulating cell cycle progression in normal cells and that leads to an increased probability of cell transformation when it is inactivated). To ascertain the role of PKR in tumorigenesis, it will be necessary to establish whether any rearrangements or other disruptions of the PKR gene (on chromosome 2p21-p22 in human cells; Barber et al. 1993a; Hanash et al. 1993; Squire et al. 1993) are associated with naturally occurring cancers.

Phosphorylation of eIF2α during Nutritional Deprivation of Mammalian Cells

Changes in the supply of essential cellular nutrients such as amino acids or glucose bring about rapid alterations in the rate of initiation of protein synthesis by mechanisms involving eIF2α phosphorylation (Scorsone et al. 1987; Pain 1994). Such effects can also be mimicked by subjecting mammalian cells containing temperature-sensitive aminoacyl-tRNA synthetases to a nonpermissive temperature (Clemens et al. 1987; Pollard et al. 1989). The nature of the signal initiating the phosphorylation of eIF2α under these conditions is not known. In the case of amino acid starvation, current evidence does not favor the accumulation of uncharged tRNA as a basis for the activation of a specific kinase; rather, it is possible that a direct regulatory connection exists between the aminoacyl-tRNA synthetases themselves (perhaps in the form of a multienzyme complex) and eIF2 phosphorylation (Clemens 1990). It has not proved possible thus far to identify the protein kinase(s) responsible for the regulation of polypeptide chain initiation under these conditions. Kimball et al. (1991) suggest that amino acid deprivation may inhibit eIF2α phosphatase activity in liver, but the nature of the enzyme involved or its mode of regulation by changes in nutritional conditions has not been further investigated.

In mammalian cells, it has not been established whether environmental changes such as alterations in nutrient availability can exert specific effects on the expression of individual genes at the translational level, as is seen in the case of the induction of GCN4 in *S. cerevisiae* (Hinnebusch 1990 and this volume; Dever et al. 1992; Altmann and Trachsel 1993). It is mechanistically quite feasible that under conditions where eIF2 phosphorylation is increased, a GCN4-type induction process can operate in higher cells, since several examples are now known of polycistronic mRNAs in which short upstream open reading frames precede the main coding sequence (Geballe, this volume). The strong similarities in the control of eIF2 phosphorylation by nutrient starvation in yeast and higher organisms suggest that this is a real possibility. There are indications that eIF2 can influence the selection of initiation codons in artificially constructed mRNAs (Dasso et al. 1990).

INHIBITION OF eIF2α PROTEIN KINASE ACTIVATION AND ACTIVITY

Given the widespread occurrence and importance of mechanisms for regulation of initiation involving the phosphorylation of eIF2, it is not surprising to find that a number of pathways exist for the negative control of the eIF2α-specific kinases. Several such mechanisms have been well characterized, particularly in the case of PKR. Consistent with the importance of this protein kinase in the strategy by which the interferons exert an antiviral effect, several viruses have evolved ways of blocking the activation of PKR (Mathews, this volume). A particularly interesting example of such a mechanism is the production of small viral RNAs that bind to the kinase in competition with the dsRNA activator (Mathews and Shenk 1991; Clemens et al. 1994; Mathews; Schneider, both this volume).

As well as the viral pathways, many cellular mechanisms exist for the negative regulation of PKR (summarized in Table 3). A well-characterized cell-encoded product that acts as a PKR inhibitor is the protein p58 (Katze, this volume), which was originally identified as being involved in the influenza-virus-induced down-regulation of PKR activity in cells infected with this virus (Lee et al. 1990, 1992). The cDNA for this protein has been cloned and has revealed a number of interesting features, including regions of homology with the DnaJ family of proteins and nine copies of a 34-residue repeat characteristic of the tetratricopeptide family of proteins (T.G. Lee et al. 1994). In keeping with the potential tumor suppressor role of PKR, when p58 is overexpressed in stably transfected cell lines, it induces a tumorigenic phenotype (Barber et al. 1994). A heat-sensitive inhibitor of PKR has also been described that may be responsible for the lack of activity of the kinase in cells transformed with the *ras* oncogene (Mundschau and Faller 1992). Thus, it is possible that loss of PKR activity could contribute to the tumorigenic phenotype of *ras*-expressing cells. Other cell-encoded inhibitory proteins and RNAs have been identified (Judware and Petryshyn 1991, 1992; Saito and Kawakita 1991), and one interesting class may be proteins that are themselves dsRNA-binding proteins. For example, both the cellular TAR-binding protein, which interacts with the TAR region of human immunodeficiency virus type-1 (HIV-1) RNAs, and the La antigen, which binds double-stranded as well as single-stranded RNAs (and can unwind the former), have been reported to inhibit the activation of PKR in vitro (Park et al. 1994; Xiao et al. 1994).

A completely separate mechanism for protecting eIF2α from phosphorylation by the protein kinases discussed here has been described by

Table 3 Potential cellular mechanisms for inhibition of eIF2α-specific protein kinases

Adequate hemin concentration (HCR)
Association with heat shock proteins (HCR)
Sequestration or unwinding of dsRNA by other proteins (PKR)
p58 protein (PKR)
p67 protein (HCR and PKR)

For details, see text.

Gupta and colleagues (Ray et al. 1992, 1993; Wu et al. 1993). This involves a 67-kD protein that associates with the initiation factor and prevents it being phosphorylated by either HCR or PKR (Chakraborty et al. 1994). This protein, cDNAs for which have been cloned from both rat and human sources (Wu et al. 1993; Li and Chang 1995), bears some limited sequence similarities in its amino terminus to the β-subunit of human and yeast eIF2. Notably, all three proteins contain blocks of polylysine. It is possible that p67 binds to eIF2 by mimicking the region of the β-subunit that interacts with one or both of the other two subunits of the factor. In doing so, it may mask the phosphorylation site on eIF2α and protect it from the protein kinases. Additionally, the association of p67 with eIF2 may somehow lessen the effect of phosphorylation of the α-subunit on protein synthesis. Such mimicry of an eIF2 protein sequence is reminiscent of that seen in the case of a virally encoded protein, the vaccinia virus K3L gene product (Schneider; Mathews, both this volume). However, the K3L protein resembles part of the α-subunit rather than the β-subunit of eIF2 and acts as a "decoy" to block the activity of PKR and HCR against their normal substrate.

Evidence that p67 may provide an alternative means to regulate the effects of the eIF2α kinases comes from in vitro studies in which the protein was able to permit translation to take place in reticulocyte lysates in the absence of hemin (Datta et al. 1988). It is not yet clear under what circumstances p67 may work in vivo or how its effects are regulated, but the protein is modified by an unusual O-linked glycosylation and it is possible that the state of glycosylation may be important for its function (Chakraborty et al. 1994).

Effects of Sequence Changes at the Site of Phosphorylation of eIF2α

Although it is widely accepted that Ser-51, or the equivalent residue in other species, is the amino acid that is modified when eIF2α is phosphorylated by HCR, PKR, or GCN2 (Colthurst et al. 1987), the exact part played by phosphorylated Ser-51 remains unclear. There is evidence indicating that a nearby serine residue, Ser-48, may also influence the behavior of eIF2 in translation, although this site is probably not itself phosphorylated. As might be predicted, replacement of Ser-51 by an alanine allows eIF2 containing this mutant α-subunit to continue to function in the presence of activated PKR or GCN2 in transfected cells (Kaufman et al. 1989; Choi et al. 1992; Dever et al. 1992; Murtha-Riel et al. 1993). On the other hand, substitution of an aspartic acid residue at this position mimics the effect of phosphorylation (Choi et al. 1992), presumably by putting a negative charge in a position that increases the affinity of eIF2 for eIF2B. Changing Ser-48 to alanine is partially protective against the inhibitory effect of either phosphorylation of Ser-51 or substitution of the latter by aspartic acid (Kaufman et al. 1989; Choi et al. 1992). It appears that other mutations downstream from Ser-51 can also ameliorate the effects of eIF2α phosphorylation in yeast (Vazquez de Aldana et al. 1993). Taken together, these results suggest that this region of eIF2α makes phosphorylation-sensitive contacts with eIF2B, probably involving the α, β, and γ subunits of the latter protein (Hinnebusch, this volume). There are no known naturally occurring mutants of eIF2α in which Ser-51 is replaced by alanine, aspartic acid, or other amino acids, suggesting that no evolutionary advantage attaches to loss of phosphorylation at this position. Indeed, it is possible that loss of translational regulation due to such a mutation may be harmful to the cell under conditions of natural selection. This conclusion is supported by the perfect conservation of the amino acid sequence NIEGMILLSEL<u>S</u>-RRRIRSI around the site of phosphorylation (underlined) in human, *Drosophila*, and *S. cerevisiae* eIF2α (Qu and Cavener 1994).

CONCLUDING REMARKS

This review has addressed a number of issues concerning the structure, regulation, and physiological significance of the protein kinases that phosphorylate initiation factors eIF2 and eIF2B. These enzymes clearly have great importance for the control of protein synthesis at the translational level in a variety of systems.

What does the future hold for this field? There are a number of areas in which we can expect more information to emerge regarding the mechanisms of action of the kinases and the regulation of eIF2 and eIF2B function by protein phosphorylation. In addition to the question of whether other eIF2α kinases exist, relatively little is known about the protein phosphatases that may reverse the effects of the kinases on the initiation of protein synthesis. Enzymes that act to remove phosphate groups from HCR or PKR might be expected to counteract the autophosphorylation-dependent activation of these protein kinases, as has been demonstrated in vitro with the relatively nonspecific bacterial alkaline phosphatase (Katze et al. 1987). It is also possible that further novel protein phosphatases that are active against phosphorylated eIF2 remain to be discovered and these may have substrate preferences or regulatory properties that are more specific than those of the enzymes studied so far (Pato et al. 1983; Redpath and Proud 1990). Of particular interest is the possible existence of a mammalian equivalent of the yeast *GLC7*-encoded protein which may be involved in regulating the state of phosphorylation of eIF2α in vivo (Wek et al. 1992).

Another important question is whether the eIF2α kinases have other substrates and may regulate pathways other than translation in the cell. This is a particularly interesting possibility in the case of PKR since this enzyme has recently been shown to phosphorylate the transcriptional regulatory factor I-κB, an inhibitor of members of the NF-κB/c-*rel* family (Kumar et al. 1994). Activation of NF-κB stimulates the transcription of a wide range of genes associated with growth regulation, differentiation, immune, and inflammatory responses. It is possible that the PKR-catalyzed phosphorylation of I-κB constitutes one means of inducing some of these genes, many of which are transcriptionally activated following exposure of cells to dsRNA. Abrogation of PKR expression using a targeted antisense approach has been demonstrated to prevent the dsRNA-dependent activation of NF-κB (Maran et al. 1994). Such a mechanism may be of great importance for the ability of PKR (and perhaps other related kinases) to regulate cell growth and other functions.

ACKNOWLEDGMENTS

I am grateful to Drs. Glen Barber, Jane-Jane Chen, Chris Prostko, and Chris Proud for discussion of their work prior to publication and to Dr. Jenny Pain for critical reading of the manuscript. Work in my laboratory is supported by project grants from the Cancer Research Campaign and the Leukaemia Research Fund, with additional funding from the Wellcome Trust and the Sylvia Reed Cancer Fund.

REFERENCES

Akkaraju, G.R., L.J. Hansen, and R. Jagus. 1991. Increase in eukaryotic initiation factor 2B activity following fertilization reflects changes in redox potential. *J. Biol. Chem.* **266:** 24451–24459.

Altmann, M. and H. Trachsel. 1993. Regulation of translation initiation and modulation of cellular physiology. *Trends Biochem. Sci.* **18:** 429–432.

Aroor, A.R., N.D. Denslow, L.P. Singh, T.W. O'Brien, and A.J. Wahba. 1994.

Phosphorylation of rabbit reticulocyte guanine nucleotide exchange factor in vivo: Identification of putative casein kinase II phosphorylation sites. *Biochemistry* **33:** 3350–3357.

Baier, L.J., T. Shors, S.T. Shors, and B.L. Jacobs. 1993. The mouse antiphosphotyrosine immunoreactive kinase, TIK, is indistinguishable from the double-stranded RNA-dependent interferon-induced protein kinase, PKR. *Nucleic Acids Res.* **21:** 4830–4835.

Barber, G.N., S. Edelhoff, M.G. Katze, and C.M. Disteche. 1993a. Chromosomal assignment of the interferon-inducible double-stranded RNA-dependent protein kinase (PRKR) to human chromosome 2p21-p22 and mouse chromosome 17 E2. *Genomics* **16:** 765–767.

Barber G.N., J. Tomita, A.G. Hovanessian, E. Meurs, and M.G. Katze. 1991. Functional expression and characterization of the interferon-induced double-stranded RNA activated P68 protein kinase from *Escherichia coli. Biochemistry* **30:** 10356–10361.

Barber, G.N., M. Wambach, S. Thompson, R. Jagus, and M.G. Katze. 1995. Mutants of the RNA-dependent proteins kinase (PKR) lacking double-stranded RNA binding domain I can act as transdominant inhibitors and induce malignant transformation. *Mol. Cell. Biol.* **15:** 3138–3146.

Barber, G.N., J. Tomita, M.S. Garfinkel, E. Meurs, A. Hovanessian, and M.G. Katze. 1992. Detection of protein kinase homologs and viral RNA-binding domains utilizing polyclonal antiserum prepared against a baculovirus-expressed ds RNA-activated 68,000-Da protein kinase. *Virology* **191:** 670–679.

Barber, G.N., M. Wambach, M.-L. Wong, T.E. Dever, A.G. Hinnebusch, and M.G. Katze. 1993b. Translational regulation by the interferon-induced double-stranded-RNA-activated 68-kDa protein kinase. *Proc. Natl. Acad. Sci.* **90:** 4621–4625.

Barber, G.N., S. Thompson, T.G. Lee, T. Strom, R. Jagus, A. Darveau, and M.G. Katze. 1994. The 58-kilodalton inhibitor of the interferon-induced double-stranded RNA-activated protein kinase is a tetratricopeptide repeat protein with oncogenic properties. *Proc. Natl. Acad. Sci.* **91:** 4278–4282.

Boyle, W.J, T. Smeal, L.H. Defize, P. Angel, J.R. Woodgett, M. Karin, and T. Hunter. 1991. Activation of protein kinase C decreases phosphorylation of c-Jun at sites that negatively regulate its DNA-binding activity. *Cell* **64:** 573–584.

Brostrom, C.O. and M.A. Brostrom. 1990. Calcium-dependent regulation of protein synthesis in intact mammalian Cells. *Annu. Rev. Physiol.* **52:** 577–590.

Brostrom, M.A., C.R. Prostko, D. Gmitter, and C.O. Brostrom. 1995. Independent signaling of *grp78* gene transcription and phosphorylation of eukaryotic initiation factor 2α by the stressed endoplasmic reticulum. *J. Biol. Chem.* **270:** 4127–4132.

Brostrom, M.A., C. Cade, C.R. Prostko, D. Gmitter-Yellen, and C.O. Brostrom. 1990. Accommodation of protein synthesis to chronic deprivation of intracellular sequestered calcium. A putative role for GRP78. *J. Biol. Chem.* **265:** 20539–20546.

Burda, J., M.E. Martín, A. García, A. Alcázar, J.L. Fando, and M. Salinas. 1994. Phosphorylation of the α subunit of initiation factor 2 correlates with the inhibition of translation following transient cerebral ischaemia in the rat. *Biochem J.* **302:** 335–338.

Chakraborty, A., D. Saha, A. Bose, M. Chatterjee, and N.K. Gupta. 1994. Regulation of eIF-2 α-subunit phosphorylation in reticulocyte lysate. *Biochemistry* **33:** 6700–6706.

Chang, H.-W. and B.L. Jacobs. 1993. Identification of a conserved motif that is necessary for binding of the vaccinia virus E3L gene products to double-stranded RNA. *Virology* **194:** 537–547.

Chefalo, P.J., J.M. Yang, K.V.A. Ramaiah, L. Gehrke, and J.-J. Chen. 1994. Inhibition of protein synthesis in insect cells by baculovirus-expressed heme-regulated eIF-2α kinase. *J. Biol. Chem.* **269:** 25788–25794.

Chen, J.-J., J.S. Crosby, and I.M. London. 1994. Regulation of heme-regulated eIF2α kinase and its expression in erythroid cells. *Biochimie* **76:** 761–769.

Chen, J.-J., J.M. Yang, R. Petryshyn, N. Kosower, and I.M. London. 1989. Disulfide bond formation in the regulation of eIF-2α kinase by heme. *J. Biol. Chem.* **264:** 9559–9564.

Chen, J.-J., M.S. Throop, L. Gehrke, I. Kuo, J.K. Pal, M. Brodsky, and I.M. London. 1991. Cloning of the cDNA of the heme-regulated eukaryotic initiation factor 2α (eIF-2α) kinase of rabbit reticulocytes: Homology to yeast GCN2 protein kinase and human double-stranded-RNA-dependent eIF-2α kinase. *Proc. Natl. Acad. Sci.* **88:** 7729–7733.

Chinchar, V.G. and J.N. Dholakia. 1989. Frog virus 3-induced translational shut-off: Activation of an eIF-2 kinase in virus-infected cells. *Virus Res.* **14:** 207–224.

Choi, S.-Y., B.J. Scherer, J. Schnier, M.V. Davies, R.J. Kaufman, and J.W.B. Hershey. 1992. Stimulation of protein synthesis in COS cells transfected with variants of the α-subunit of initiation factor eIF-2. *J. Biol. Chem.* **267:** 286–293.

Chong, K.L., L. Feng, K. Schappert, E. Meurs, T.F. Donahue, J.D. Friesen, A.G. Hovanessian, and B.R.G. Williams. 1992. Human p68 kinase exhibits growth suppression in yeast and homology to the translational regulator *GCN2*. *EMBO J.* **11:** 1553–1562.

Clark, S.J., A.J. Ashford, N.T. Price, and C.G. Proud. 1989. Casein kinase-2 phosphorylates serine-2 in the β-subunit of initiation factor-2. *Biochim. Biophys. Acta* **1010:** 377–380.

Clarke, P.A. and M.B. Mathews. 1995. Interactions between the double-stranded RNA binding motif and RNA: Definition of the binding site for the interferon-induced protein kinase DAI (PKR) on adenovirus VA RNA. *RNA* **1:** 7–20.

Clarke, P.A., M. Schwemmle, J. Schickinger, K. Hilse, and M.J. Clemens. 1991. Binding of Epstein-Barr virus small RNA EBER-1 to the double-stranded RNA-activated protein kinase DAI. *Nucleic Acids Res.* **19:** 243–248.

Clemens, M.J. 1990. Does protein phosphorylation play a role in translational control by eukaryotic aminoacyl-tRNA synthetases? *Trends Biochem. Sci.* **15:** 172–175.

———. 1992. Suppression with a difference. *Nature* **360:** 210–211.

Clemens, M.J., A. Galpine, S.A. Austin, R. Panniers, E.C. Henshaw, R. Duncan, J.W.B. Hershey, and J.W. Pollard. 1987. Regulation of polypeptide chain initiation in Chinese hamster ovary cells with a temperature-sensitive leucyl-tRNA synthetase. *J. Biol. Chem.* **262:** 767–771.

Clemens, M.J., K. Laing, I.W. Jeffrey, A. Schofield, T.V. Sharp, A. Elia, V. Matys, M.C. James, and V.J. Tilleray. 1994. Regulation of the interferon-inducible eIF-2α protein kinase by small RNAs. *Biochimie* **76:** 770–778.

Colthurst, D.R., D.G. Campbell, and C.G. Proud. 1987. Structure and regulation of eukaryotic initiation factor 2: Sequence of the site in the α subunit phosphorylated by the heme-controlled repressor and by the double-stranded RNA-activated inhibitor. *Eur. J. Biochem.* **166:** 357–363.

Crosby, J.S., K. Lee, I.M. London, and J.-J. Chen. 1994. Erythroid expression of the heme-regulated eIF-2α kinase. *Mol. Cell. Biol.* **14:** 3906–3914.

Dasso, M.C., S.C. Milburn, J.W.B. Hershey, and R.J. Jackson. 1990. Selection of the 5′-proximal translation initiation site is influenced by mRNA and eIF-2 concentrations. *Eur. J. Biochem.* **187:** 361–371.

Datta, B., D. Chakrabarti, A.L. Roy, and N.K. Gupta. 1988. Roles of a 67-kDa polypeptide in reversal of protein synthesis inhibition in heme-deficient reticulocyte lysate. *Proc. Natl. Acad. Sci.* **85:** 3324–3328.

Davies, M.V., M. Furtado, J.W.B. Hershey, B. Thimmappaya, and R.J. Kaufman. 1989. Complementation of adenovirus virus-associated RNA I gene deletion by expression of a mutant eukaryotic translation initiation factor. *Proc. Natl. Acad. Sci.* **86:** 9163–9167.

De Benedetti, A. and C. Baglioni. 1986. Activation of hemin-regulated initiation factor-2 kinase in heat-shocked HeLa cells. *J. Biol. Chem.* **261:** 338–342.

DeStefano, J., E. Olmsted, R. Panniers, and J. Lucas-Lenard. 1990. The α subunit of eucaryotic initiation factor 2 is phosphorylated in mengovirus-infected mouse L cells. *J. Virol.* **64:** 4445–4453.

Dever, T.E., L. Feng, R.C. Wek, A.M. Cigan, T.F. Donahue, and A.G. Hinnebusch. 1992. Phosphorylation of initiation factor 2α by protein kinase GCN2 mediates gene-specific translational control of *GCN4* in yeast. *Cell* **68:** 585–596.

Dever, T.E., J.-J. Chen, G.N. Barber, A.M. Cigan, L. Feng, T.F. Donahue, I.M. London, M.G. Katze, and A.G. Hinnebusch. 1993. Mammalian eukaryotic initiation factor 2α kinases functionally substitute for GCN2 protein kinase in the GCN4 translational control mechanism of yeast. *Proc. Natl. Acad. Sci.* **90:** 4616–4620.

Dholakia, J.N. and A.J. Wahba. 1988. Phosphorylation of the guanine nucleotide exchange factor from rabbit reticulocytes regulates its activity in polypeptide chain initiation. *Proc. Natl. Acad. Sci.* **85:** 51–54.

Diallinas, G. and G. Thireos. 1994. Genetic and biochemical evidence for yeast GCN2 protein kinase polymerization. *Gene* **143:** 21–27.

Donzé, O., R. Jagus, A.E. Koromilas, J.W.B. Hershey, and N. Sonenberg. 1995. Abrogation of translation initiation factor eIF-2 phosphorylation causes malignant transformation of NIH 3T3 cells. *EMBO J.* **14:** 3828–3834.

Duncan, R.F. and J.W.B. Hershey. 1984. Heat shock-induced translational alterations in HeLa cells. *J. Biol. Chem.* **259:** 11882–11889.

———. 1985. Regulation of initiation factors during translational repression caused by serum depletion. Covalent modification. *J. Biol. Chem.* **260:** 5493–5497.

———. 1989. Protein synthesis and protein phosphorylation during heat stress, recovery and adaptation. *J. Cell Biol.* **109:** 1467–1481.

Farrell, P.J., K. Balkow, T. Hunt, R.J. Jackson, and H. Trachsel. 1977. Phosphorylation of initiation factor eIF-2 and control of reticulocyte protein synthesis. *Cell* **11:** 187–200.

Feldhoff, R.C., A.M. Karinch, S.R. Kimball, and L.S. Jefferson. 1993. Purification of eukaryotic initiation factors eIF-2, eIF-2B and eIF-2α kinase from bovine liver. *Prep. Biochem.* **23:** 363–374.

Galabru, J. and A.G. Hovanessian. 1987. Autophosphorylation of the protein kinase dependent on double-stranded RNA. *J. Biol. Chem.* **262:** 15538–15544.

Galabru, J., M.G. Katze, N. Robert, and A.G. Hovanessian. 1989. The binding of double-stranded RNA and adenovirus VAI RNA to the interferon-induced protein kinase. *Eur. J. Biochem.* **178:** 581–589.

Gatignol, A., C. Buckler, and K.-T. Jeang. 1993. Relatedness of an RNA-binding motif in human immunodeficiency virus type 1 TAR RNA-binding protein TRBP to human P1/dsI kinase and *Drosophila Staufen*. *Mol. Cell. Biol.* **13:** 2193–2202.

Green, S.R. and M.B. Mathews. 1992. Two RNA-binding motifs in the double-stranded RNA-activated protein kinase, DAI. *Genes Dev.* **6:** 2478–2490.

Green, S.R., L. Manche, and M.B. Mathews. 1995. Two functionally distinct RNA-binding motifs in the regulatory domain of the protein kinase DAI. *Mol. Cell. Biol.* **15:** 358–364.

Gross, M., A. Olin, S. Hessefort, and S. Bender. 1994. Control of protein synthesis by hemin. Purification of a rabbit reticulocyte hsp 70 and characterization of its regulation of the activation of the hemin-controlled eIF2(α) kinase. *J. Biol. Chem.* **269:** 22738–22748.

Hanash, S.M., L. Beretta, C.L. Barcroft, S. Sheldon, T.W. Glover, D. Ungar, and N. Sonenberg. 1993. Mapping of the gene for interferon-inducible dsRNA-dependent protein kinase to chromosome region 2p21-22: A site of rearrangements in myeloproliferative disorders. *Genes Chromosomes Cancer* **8:** 34–37.

Hanks, S.K., A.M. Quinn, and T. Hunter. 1988. The protein kinase family: Conserved features and deduced phylogeny of the catalytic domains. *Science* **241:** 42–52.

Henry, G.L., S.J. McCormack, D.C. Thomis, and C.E. Samuel. 1994. Mechanism of interferon action. Translational control and the RNA-dependent protein kinase (PKR): Antagonists of PKR enhance the translational activity of mRNAs that include a 161 nucleotide region from reovirus S1 mRNA. *J. Biol. Regul. Homeostatic Agents* **8:** 15–24.

Hiddinga, H.J., C.J. Crum, J. Hu, and D.A. Roth. 1988. Viroid-induced phosphorylation of a host protein related to a dsRNA-dependent protein kinase. *Science* **241:** 451–453.

Hinnebusch, A.G. 1990. Involvement of an initiation factor and protein phosphorylation in translational control of *GCN4* mRNA. *Trends Biochem. Sci.* **15:** 148–152.

Hovanessian, A.G. 1989. The double-stranded RNA-activated protein kinase induced by interferon: dsRNA-PK. *J. Interferon Res.* **9:** 641–647.

Hu, B.-R. and T. Wieloch. 1993. Stress-induced inhibition of protein synthesis initiation: Modulation of initiation factor 2 and guanine nucleotide exchange factor activities following transient cerebral ischemia in the rat. *J. Neurosci.* **13:** 1830–1838.

Hu, B.-R., Y.-B.O. Yang, and T. Wieloch. 1993. Heat-shock inhibits protein synthesis and eIF-2 activity in cultured cortical neurons. *Neurochem. Res.* **18:** 1003–1007.

Huang, J. and R.J. Schneider. 1990. Adenovirus inhibition of cellular protein synthesis is prevented by the drug 2-aminopurine. *Proc. Natl. Acad. Sci.* **87:** 7115–7119.

Icely, P.L., P. Gros, J.J.M. Bergeron, A. Devault, D.E.H. Afar, and J.C. Bell. 1991. TIK, a novel serine/threonine kinase, is recognized by antibodies directed against phosphotyrosine. *J. Biol. Chem.* **266:** 16073–16077.

Ito, T., R. Jagus, and W.S. May. 1994. Interleukin 3 stimulates protein synthesis by regulating double-stranded RNA-dependent protein kinase. *Proc. Natl. Acad. Sci.* **91:** 7455–7459.

Jackson, R.J. 1991. Binding of Met-tRNA. In *Translation in eukaryotes* (ed. H. Trachsel), pp. 193–229, CRC Press, Boca Raton, Florida.

Jeffrey, I.W., F.J. Kelly, R. Duncan, J.W.B. Hershey, and V.M. Pain. 1990. Effect of starvation and diabetes on the activity of the eukaryotic initiation factor eIF-2 in rat skeletal muscle. *Biochimie* **72:** 751–757.

Jeffrey, I.W., S. Kadereit, E.F. Meurs, T. Metzger, M. Bachmann, M. Schwemmle, A.G. Hovanessian, and M.J. Clemens. 1995. Nuclear localization of the interferon-inducible protein kinase PKR in human cells and transfected mouse cells. *Exp. Cell Res.* **218:** 17–27.

Jiménez-García, L.F., S.R. Green, M.B. Mathews, and D.L. Spector. 1993. Organization of the double-stranded RNA-activated protein kinase DAI and virus-associated VA

RNA_I in adenovirus-2-infected HeLa cells. *J. Cell Sci.* **106:** 11–22.

Judware, R. and R. Petryshyn. 1991. Partial characterization of a cellular factor that regulates the double-stranded RNA-dependent eIF-2α kinase in 3T3-F442A fibroblasts. *Mol. Cell. Biol.* **11:** 3259–3267.

———. 1992. Mechanism of action of a cellular inhibitor of the dsRNA-dependent protein kinase from 3T3-F442A cells. *J. Biol. Chem.* **267:** 21685–21690.

Karinch, A.M., S.R. Kimball, T.C. Vary, and L.S. Jefferson. 1993. Regulation of eukaryotic initiation factor-2B activity in muscle of diabetic rats. *Am. J. Physiol. (Endocrinol. Metab. 27)* **264:** E101–E108.

Katze, M.G., D. DeCorato, B. Safer, J. Galabru, and A.G. Hovanessian. 1987. The adenovirus VAI RNA complexes with the p68 protein kinase to regulate its autophosphorylation and activity. *EMBO J.* **6:** 689–697.

Katze, M.G., M. Wambach, M.-L. Wong, M. Garfinkel, E. Meurs, K. Chong, B.R.G. Williams, A.G. Hovanessian, and G.N. Barber. 1991. Functional expression and RNA binding analysis of the interferon-induced, double-stranded RNA-activated, 68,000-M_r protein kinase in a cell-free system. *Mol. Cell. Biol.* **11:** 5497–5505.

Kaufman, R.J., M.V. Davies, V.K. Pathak, and J.W.B. Hershey. 1989. The phosphorylation state of eucaryotic initiation factor 2 alters translational efficiency of specific mRNAs. *Mol. Cell. Biol.* **9:** 946–958.

Kimball, S.R. and L.S. Jefferson. 1990. Mechanism of inhibition of protein synthesis by vasopressin in rat liver. *J. Biol. Chem.* **265:** 16794–16798.

———. 1992. Regulation of protein synthesis by modulation of intracellular calcium in rat liver. *Am. J. Physiol. (Endocrinol. Metab. 26)* **263:** E958–E964.

———. 1994. Mechanisms of translational control in liver and skeletal muscle. *Biochimie* **76:** 729–736.

Kimball, S.R., D.A. Antonetti, R.M. Brawley, and L.S. Jefferson. 1991. Mechanism of inhibition of peptide chain initiation by amino acid deprivation in perfused rat liver. Regulation involving inhibition of eukaryotic initiation factor 2α phosphatase activity. *J. Biol. Chem.* **266:** 1969–1976.

Kimball, S.R., A.M. Karinch, R.C. Feldhoff, H. Mellor, and L.S. Jefferson. 1994. Purification and characterization of eukaryotic translational initiation factor eIF2B from liver. *Biochim. Biophys. Acta* **1201:** 473–481.

Koromilas, A.E., S. Roy, G.N. Barber, M.G. Katze, and N. Sonenberg. 1992. Malignant transformation by a mutant of the IFN-inducible dsRNA-dependent protein kinase. *Science* **257:** 1685–1689.

Kostura, M. and M.B. Mathews. 1989. Purification and activation of the double-stranded RNA-dependent eIF-2 kinase DAI. *Mol. Cell. Biol.* **9:** 1576–1586.

Krust, B., J. Galabru, and A.G. Hovanessian. 1984. Further characterization of the protein kinase activity mediated by interferon in mouse and human cells. *J. Biol. Chem.* **259:** 8494–8496.

Kumar, A., J. Haque, J. Lacoste, J. Hiscott, and B.R.G. Williams. 1994. Double-stranded RNA-dependent protein kinase activates transcription factor NF-κB by phosphorylating IκB. *Proc. Natl. Acad. Sci.* **91:** 6288–6292.

Kumar, R.V., A. Wolfman, R. Panniers, and E.C. Henshaw. 1989. Mechanism of inhibition of polypeptide chain initiation in calcium-depleted Ehrlich ascites tumor cells. *J. Cell Biol.* **108:** 2107–2115.

Langland, J.O. and B.L. Jacobs. 1992. Cytosolic double-stranded RNA-dependent protein kinase is likely a dimer of partially phosphorylated M_r = 66,000 subunits. *J. Biol.*

Chem. **267:** 10729–10736.

Lee, S.B. and M. Esteban. 1993. The interferon-induced double-stranded RNA-activated human p68 protein kinase inhibits the replication of vaccinia virus. *Virology* **193:** 1037–1041.

———. 1994. The interferon-induced double-stranded RNA-activated protein kinase induces apoptosis. *Virology* **199:** 491–496.

Lee, S.B., Z. Melkova, W. Yan, B.R.G. Williams, A.G. Hovanessian, and M. Esteban. 1993. The interferon-induced double-stranded RNA-activated human p68 protein kinase potently inhibits protein synthesis in cultured cells. *Virology* **192:** 380–385.

Lee, S.B., S.R. Green, M.B. Mathews, and M. Esteban. 1994. Activation of the double-stranded RNA (dsRNA)-activated human protein kinase *in vivo* in the absence of its dsRNA binding domain. *Proc. Natl. Acad. Sci.* **91:** 10551–10555.

Lee, T.G., N. Tang, S. Thompson, J. Miller, and M.G. Katze. 1994. The 58,000-dalton cellular inhibitor of the interferon-induced double-stranded RNA-activated protein kinase (PKR) is a member of the tetratricopeptide repeat family of proteins. *Mol. Cell. Biol.* **14:** 2331–2342.

Lee, T.G., J. Tomita, A.G. Hovanessian, and M.G. Katze. 1990. Purification and partial characterization of a cellular inhibitor of the interferon-induced protein kinase of M_r 68,000 from influenza virus-infected cells. *Proc. Natl. Acad. Sci.* **87:** 6208–6212.

———. 1992. Characterization and regulation of the 58,000-dalton cellular inhibitor of the interferon-induced, dsRNA-activated protein kinase. *J. Biol. Chem.* **267:** 14238–14243.

Lengyel, P. 1993. Tumor-suppressor genes: News about the interferon connection. *Proc. Natl. Acad. Sci.* **90:** 5893–5895.

Leroux, A. and I.M. London. 1982. Regulation of protein synthesis by phosphorylation of eukaryotic initiation factor 2α in intact reticulocytes and reticulocyte lysates. *Proc. Natl. Acad. Sci.* **79:** 2147–2151.

Li, J. and R.A. Petryshyn. 1991. Activation of the double-stranded RNA-dependent eIF-2α kinase by cellular RNA from 3T3-F442A cells. *Eur. J. Biochem.* **195:** 41–48.

Li, X. and Y.-H. Chang. 1995. Molecular cloning of a human complementary DNA encoding an initiation factor 2-associated protein (p^{67}). *Biochim. Biophys. Acta* **1260:** 333–336.

London, I.M., M.J. Clemens, R.S. Ranu, D.H. Levin, L.F. Cherbas, and V. Ernst. 1976. The role of hemin in the regulation of protein synthesis in erythroid cells. *Fed. Proc.* **35:** 2218–2222.

Manche, L., S.R. Green, C. Schmedt, and M.B. Mathews. 1992. Interactions between double-stranded RNA regulators and the protein kinase DAI. *Mol. Cell. Biol.* **12:** 5238–5248.

Maran, A. and M.B. Mathews. 1988. Characterization of the double-stranded RNA implicated in the inhibition of protein synthesis in cells infected with a mutant adenovirus defective for VA RNA_I. *Virology* **164:** 106–113.

Maran, A., R.K. Maitra, A. Kumar, B. Dong, W. Xiao, G. Li, B.R.G. Williams, P.F. Torrence, and R.H. Silverman. 1994. Blockage of NF-κB signaling by selective ablation of an mRNA target by 2-5A antisense chimeras. *Science* **265:** 789–792.

Mathews, M.B. and T. Shenk. 1991. Adenovirus virus-associated RNA and translation control. *J. Virol.* **65:** 5657–5662.

Matts, R.L. and R. Hurst. 1989. Evidence for the association of the heme-regulated eIF2α kinase with the 90-kDa heat shock protein in rabbit reticulocyte lysate *in situ*. *J. Biol.*

Chem. **264:** 15542–15547.

———. 1992. The relationship between protein synthesis and heat shock proteins levels in rabbit reticulocyte lysates. *J. Biol. Chem.* **267:** 18168–18174.

Matts, R.L., R. Hurst, and Z. Xu. 1993. Denatured proteins inhibit translation in hemin-supplemented rabbit reticulocyte lysate by inducing the activation of the heme-regulated eIF-2α kinase. *Biochemistry* **32:** 7323–7328.

Matts, R.L., Z. Xu, J.K. Pal, and J.-J. Chen. 1992. Interactions of the heme-regulated eIF-2α kinase with heat shock proteins in rabbit reticulocyte lysates. *J. Biol. Chem.* **267:** 18160–18167.

Matts, R.L., J.R. Schatz, R. Hurst, and R. Kagen. 1991. Toxic heavy metal ions activate the heme-regulated eukaryotic initiation factor-2α kinase by inhibiting the capacity of hemin-supplemented reticulocyte lysates to reduce disulfide bonds. *J. Biol. Chem.* **266:** 12695–12702.

McCormack, S.J., D.C. Thomis, and C.E. Samuel. 1992. Mechanism of interferon action: Identification of a RNA binding domain within the N-terminal region of the human RNA-dependent P1/eIF-2α protein kinase. *Virology* **188:** 47–56.

McCormack, S.J., L.G. Ortega, J.P. Doohan, and C.E. Samuel. 1994. Mechanism of interferon action: Motif I of the interferon-induced, RNA-dependent protein kinase (PKR) is sufficient to mediate RNA-binding activity. *Virology* **198:** 92–99.

Mellor, H., K.M. Flowers, S.R. Kimball, and L.S. Jefferson. 1994. Cloning and characterization of cDNA encoding rat hemin-sensitive initiation factor-2α (eIF-2α) kinase. Evidence for multitissue expression. *J. Biol. Chem.* **269:** 10201–10204.

Mellor, H., N.T. Price, S. Oldfield, T.F. Sarre, and C.G. Proud. 1993. Purification and characterization of an initiation-factor-2 kinase from uninduced mouse erythroleukaemia cells. *Eur. J. Biochem.* **211:** 529–538.

Méndez, R. and C. de Haro. 1994. Casein kinase II is implicated in the regulation of heme-controlled translational inhibitor of reticulocyte lysates. *J. Biol. Chem.* **269:** 6170–6176.

Méndez, R., A. Moreno, and C. de Haro. 1992. Regulation of heme-controlled eukaryotic polypeptide chain initiation factor 2 α-subunit kinase of reticulocyte lysates. *J. Biol. Chem.* **267:** 11500–11507.

Meurs, E.F., J. Galabru, G.N. Barber, M.G. Katze, and A.G. Hovanessian. 1993. Tumor suppressor function of the interferon-induced double-stranded RNA-activated protein kinase. *Proc. Natl. Acad. Sci.* **90:** 232–236.

Meurs, E., K. Chong, J. Galabru, N.S.B. Thomas, I.M. Kerr, B.R.G. Williams, and A.G. Hovanessian. 1990. Molecular cloning and characterization of the human double-stranded RNA-activated protein kinase induced by interferon. *Cell* **62:** 379–390.

Meurs, E.F., Y. Watanabe, S. Kadereit, G.N. Barber, M.G. Katze, K. Chong, B.R.G. Williams, and A.G. Hovanessian. 1992. Constitutive expression of human double-stranded RNA-activated p68 kinase in murine cells mediates phosphorylation of eukaryotic initiation factor 2 and partial resistance to encephalomyocarditis virus growth. *J. Virol.* **66:** 5805–5814.

Montine, K.S. and E.C. Henshaw. 1989. Serum growth factors cause rapid stimulation of protein synthesis and dephosphorylation of eIF-2 in serum deprived Ehrlich cells. *Biochim. Biophys. Acta* **1014:** 282–288.

Mundschau, L.J. and D.V. Faller. 1991. BALB/c-3T3 fibroblasts resistant to growth inhibition by beta interferon exhibit aberrant platelet-derived growth factor, epidermal growth factor, and fibroblast growth factor signal transduction. *Mol. Cell. Biol.* **11:**

3148–3154.

———. 1992. Oncogenic *ras* induces an inhibitor of double-stranded RNA-dependent eukaryotic initiation factor 2α-kinase activation. *J. Biol. Chem.* **267:** 23092–23098.

———. 1995. Platelet-derived growth factor signal transduction through the interferon-inducible kinase PKR. Immediate early gene induction. *J. Biol. Chem.* **270:** 3100–3106.

Murtha-Riel, P., M.V. Davies, B.J. Scherer, S.-Y. Choi, J.W.B. Hershey, and R.J. Kaufman. 1993. Expression of a phosphorylation-resistant eukaryotic initiation factor 2 α-subunit mitigates heat shock inhibition of protein synthesis. *J. Biol. Chem.* **268:** 12946–12951.

Oldfield, S., B.L. Jones, D. Tanton, and C.G. Proud. 1994. Use of monoclonal antibodies to study the structure and function of eukaryotic protein synthesis initiation factor eIF2B. *Eur. J. Biochem.* **221:** 399–410.

Olmsted, E.A., L. O'Brien, E.C. Henshaw, and R. Panniers. 1993. Purification and characterization of eukaryotic initiation factor (eIF)-2α kinases from Ehrlich ascites tumor cells. *J. Biol. Chem.* **268:** 12552–12559.

Pain, V.M. 1994. Translational control during amino acid starvation. *Biochimie* **76:** 718–728.

Pal, J.K., J.-J. Chen, and I.M. London. 1991. Tissue distribution and immunoreactivity of heme-regulated eIF-2α kinase determined by monoclonal antibodies. *Biochemistry* **30:** 2555–2562.

Panniers, R. 1994. Translational control during heat shock. *Biochimie* **76:** 737–747.

Park, H., M.V. Davies, J.O. Langland, H.-W. Chang, Y.S. Nam, J. Tartaglia, E. Paoletti, B.L. Jacobs, R.J. Kaufman, and S. Venkatesan. 1994. TAR RNA-binding protein is an inhibitor of the interferon-induced protein kinase PKR. *Proc. Natl. Acad. Sci.* **91:** 4713–4717.

Patel, R.C. and G.C. Sen. 1992a. Construction and expression of an enzymatically active human-mouse chimeric double-stranded RNA-dependent protein kinase. *J. Interferon Res.* **12:** 389–393.

———. 1992b. Identification of the double-stranded RNA-binding domain of the human interferon-inducible protein kinase. *J. Biol. Chem.* **267:** 7671–7676.

Patel, R.C., P. Stanton, and G.C. Sen. 1994. Role of the amino-terminal residues of the interferon-induced protein kinase in its activation by double-stranded RNA and heparin. *J. Biol. Chem.* **269:** 18593–18598.

Patel, R.C., P. Stanton, N.M.J. McMillan, B.R.G. Williams, and G.C. Sen. 1995. The interferon-inducible double-stranded RNA-activated protein kinase self-associates *in vitro* and *in vivo*. *Proc. Natl. Acad. Sci.* **92:** 8283–8287.

Pato, M.D., R.S. Adelstein, D. Crouch, B. Safer, T.S. Ingebritsen, and P. Cohen. 1983. The protein phosphatases involved in cellular regulation. *Eur. J. Biochem.* **132:** 283–287.

Pestka, S., J.A. Langer, K.C. Zoon, and C.E. Samuel. 1987. Interferons and their actions. *Annu. Rev. Biochem.* **56:** 727–777.

Petryshyn, R., J.-J. Chen, and I.M. London. 1984. Growth-related expression of a double-stranded RNA-dependent protein kinase in 3T3 cells. *J. Biol. Chem.* **259:** 14736–14742.

———. 1988. Detection of activated double-stranded RNA-dependent protein kinase in 3T3-F442A cells. *Proc. Natl. Acad. Sci.* **85:** 1427–1431.

Pollard, J.W., A.R. Galpine, and M.J. Clemens. 1989. A novel role for aminoacyl-tRNA synthetases in the regulation of polypeptide chain initiation. *Eur. J. Biochem.* **182:** 1–9.

Pratt, G., A. Galpine, N. Sharp, S. Palmer, and M.J. Clemens. 1988. Regulation of in vitro translation by double-stranded RNA in mammalian mRNA preparations. *Nucleic Acids Res.* **16:** 3497–3510.

Price, N. and C. Proud. 1994. The guanine nucleotide-exchange factor, eIF-2B. *Biochimie* **76:** 748–760.

Prostko, C.R., M.A. Brostrom, and C.O. Brostrom. 1993. Reversible phosphorylation of eukaryotic initiation factor 2α in response to endoplasmic reticular signaling. *Mol. Cell. Biochem.* **127-128:** 255–265.

Prostko, C.R., M.A. Brostrom, E.M. Malara, and C.O. Brostrom. 1992. Phosphorylation of eukaryotic initiation factor (eIF) 2α and inhibition of eIF-2B in GH_3 pituitary cells by perturbants of early protein processing that induce GRP78. *J. Biol. Chem.* **267:** 16751–16754.

Prostko, C.R., J.N. Dholakia, M.A. Brostrom, and C.O. Brostrom. 1995. Activation of the double-stranded RNA-regulated protein kinase by depletion of endoplasmic reticular calcium stores. *J. Biol. Chem.* **270:** 6211–6215.

Proud, C.G. 1992. Protein phosphorylation in translational control. *Curr. Top. Cell. Regul.* **32:** 243–369.

Qu, S. and D.R. Cavener. 1994. Isolation and characterization of the *Drosophila melanogaster eIF-2α* gene encoding the alpha subunit of translation initiation factor eIF-2. *Gene* **140:** 239–242.

Ramaiah, K.V.A., M.V. Davies, J.-J. Chen, and R.J. Kaufman. 1994. Expression of mutant eukaryotic initiation factor 2 α subunit (eIF-2α) reduces inhibition of guanine nucleotide exchange activity of eIF-2B mediated by eIF-2α phosphorylation. *Mol. Cell. Biol.* **14:** 4546–4553.

Ramirez, M., R.C. Wek, C.R. Vazquez de Aldana, B.M. Jackson, B. Freeman, and A.G. Hinnebusch. 1992. Mutations activating the yeast eIF2α kinase GCN2: Isolation of alleles altering the domain related to histidyl-tRNA synthetases. *Mol. Cell. Biol.* **12:** 5801–5815.

Ray, M.K., B. Datta, A. Chakraborty, A. Chattopadhyay, S. Meza-Keuthen, and N.K. Gupta. 1992. The eukaryotic initiation factor 2-associated 67-kDa polypeptide (p^{67}) plays a critical role in regulation of protein synthesis initiation in animal cells. *Proc. Natl. Acad. Sci.* **89:** 539–543.

Ray, M.K., A. Chakraborty, B. Datta, A. Chattopadhyay, D. Saha, A. Bose, T.G. Kinzy, S. Wu, R.E. Hileman, W.C. Merrick, and N.K. Gupta. 1993. Characteristics of the eukaryotic initiation factor 2 associated 67-kDa polypeptide. *Biochemistry* **32:** 5151–5159.

Redpath, N.T. and C.G. Proud. 1990. Activity of protein phosphatases against initiation factor-2 and elongation factor-2. *Eur. J. Biochem.* **272:** 175–180.

Roberts, W.K., M.J. Clemens, and I.M. Kerr. 1976. Interferon-induced inhibition of protein synthesis in L-cell extracts: An ATP-dependent step in the activation of an inhibitor by double-stranded RNA. *Proc. Natl. Acad. Sci.* **73:** 3136–3140.

Robertson, H.D., L. Manche, and M.B. Mathews. 1996. Paradoxical interactions between human delta hepatitis agent RNA and the cellular protein kinase DAI (PKR). *J. Virol.* (in press).

Romano, P.R., S.R. Green, G.N. Barber, M.B. Mathews, and A.G. Hinnebusch. 1995. Structural requirements for double-stranded RNA binding, dimerization, and activation of the human eIF-2α kinase DAI in *Saccharomyces cerevisiae*. *Mol. Cell. Biol.* **15:**

365–378.

Rose, D.W., W.J. Welch, G. Kramer, and B. Hardesty. 1989. Possible involvement of the 90 kDa heat shock protein in the regulation of protein synthesis. *J. Biol. Chem.* **264:** 6239–6244.

Rowlands, A.G., R. Panniers, and E.C. Henshaw. 1988a. The catalytic mechanism of guanine nucleotide exchange factor action and competititve inhibition by phosphorylated eucaryotic initiation factor 2. *J. Biol. Chem.* **263:** 5526–5533.

Rowlands, A.G., K.S. Montine, E.C. Henshaw, and R. Panniers. 1988b. Physiological stresses inhibit guanine-nucleotide-exchange factor in Ehrlich cells. *Eur. J. Biochem.* **175:** 93–99.

Saito, S. and M. Kawakita. 1991. Inhibitor of interferon-induced double-stranded RNA-dependent protein kinase and its relevance to alteration of cellular protein kinase activity level in response to external stimuli. *Microbiol. Immunol.* **35:** 1105–1114.

Samuel, C.E. 1991. Antiviral actions of interferon: Interferon-regulated cellular proteins and their surprisingly selective antiviral activities. *Virology* **183:** 1–11.

Sarre, T.F. 1989. Presence of haemin-controlled eIF-2α kinases in both undifferentiated and differentiating mouse erythroleukaemia cells. *Biochem J.* **262:** 569–574.

Scorsone, K.A., R. Panniers, A.G. Rowlands, and E.C. Henshaw. 1987. Phosphorylation of eukaryotic initiation factor 2 during physiological stresses which affect protein synthesis. *J. Biol. Chem.* **262:** 14538–14543.

Sharp, T.V., Q. Xiao, I. Jeffrey, D.R. Gewert, and M.J. Clemens. 1993a. Reversal of the double-stranded-RNA-induced inhibition of protein synthesis by a catalytically inactive mutant of the protein kinase PKR. *Eur. J. Biochem.* **214:** 945–948.

Sharp, T.V., Q. Xiao, J. Justesen, D.R. Gewert, and M.J. Clemens. 1995. Regulation of the interferon-inducible protein kinase PKR and 2′5′ oligoadenylate synthetase by a catalytically inactive PKR mutant through competition for double-stranded RNA binding. *Eur. J. Biochem.* **230:** 97–103.

Sharp, T.V., M. Schwemmle, I. Jeffrey, K. Laing, H. Mellor, C.G. Proud, K. Hilse, and M.J. Clemens. 1993b. Comparative analysis of the regulation of the interferon-inducible protein kinase PKR by Epstein-Barr virus RNAs EBER-1 and EBER-2 and adenovirus VA_1 RNA. *Nucleic Acids Res.* **21:** 4483–4490.

Sierra, J.M. 1994. Translational regulation of the heat shock response. *Mol. Biol. Rep.* **19:** 211–220.

Singh, L.P., A.R. Aroor, and A.J. Wahba. 1994. Phosphorylation of the guanine nucleotide exchange factor and eukaryotic initiation factor 2 by casein kinase II regulates guanine nucleotide binding and GDP/GTP exchange. *Biochemistry* **33:** 9152–9157.

Sonenberg, N. 1993. Translation factors as effectors of cell growth and tumorigenesis. *Curr. Opin. Cell Biol.* **5:** 955–960.

Squire, J., E.F. Meurs, K.L. Chong, N.A.J. McMillan, A.G. Hovanessian, and B.R.G. Williams. 1993. Localization of the human interferon-induced, ds-RNA activated p68 kinase gene (PRKR) to chromosome 2p21-p22. *Genomics* **16:** 768–770.

Srivastava, S.P., M.V. Davies, and R.J. Kaufman. 1995. Calcium depletion from the endoplasmic reticulum activates the double-stranded RNA-dependent protein kinase (PKR) to inhibit protein synthesis. *J. Biol. Chem.* **270:** 16619–16624.

St Johnston, D., N.H. Brown, J.G. Gall, and M. Jantsch. 1992. A conserved double-stranded RNA-binding domain. *Proc. Natl. Acad. Sci.* **89:** 10979–10983.

Szyszka, R., G. Kramer, and B. Hardesty. 1989a. The phosphorylation state of reticulocyte 90 kDa heat shock protein affects its ability to increase phosphorylation of peptide initiation factor 2α subunit by the heme-sensitive kinase. *Biochemistry* **28:** 1435–1438.

Szyszka, R., W. Kudlick, G. Kramer, B. Hardesty, J. Galabru, and A. Hovanessian. 1989b. A type 1 phosphoprotein phosphatase active with phosphorylated M_r 68,000 initiation factor 2 kinase. *J. Biol. Chem.* **264:** 3827–3831.

Tanaka, H. and C.E. Samuel. 1994. Mechanism of interferon action: Structure of the mouse PKR gene encoding the interferon-inducible RNA-dependent protein kinase. *Proc. Natl. Acad. Sci.* **91:** 7995–7999.

———. 1995. Sequence of the murine interferon-inducible RNA-dependent protein kinase (PKR) deduced from genomic clones. *Gene* **153:** 283–284.

Thomis, D.C. and C.E. Samuel. 1992. Mechanism of interferon action: Autoregulation of RNA-dependent P1/eIF-2α protein kinase (PKR) expression in transfected mammalian cells. *Proc. Natl. Acad. Sci.* **89:** 10837–10841.

Thomis, D.C., J.P. Doohan, and C.E. Samuel. 1992. Mechanism of interferon action: cDNA structure, expression, and regulation of the interferon-induced, RNA-dependent P1/eIF-2α protein kinase from human cells. *Virology* **188:** 33–46.

Vazquez de Aldana, C.R., T.E. Dever, and A.G. Hinnebusch. 1993. Mutations in the α subunit of eukaryotic translation initiation factor 2 (eIF-2α) that overcome the inhibitory effect of eIF-2α phosphorylation on translation initiation. *Proc. Natl. Acad. Sci.* **90:** 7215–7219.

Wek, R.C., B.M. Jackson, and A.G. Hinnebusch. 1989. Juxtaposition of domains homologous to protein kinases and histidyl-tRNA synthetases in GCN2 protein suggests a mechanism for coupling GCN4 expression to amino acid availability. *Proc. Natl. Acad. Sci.* **86:** 4579–4583.

Wek, R.C., J.F. Cannon, T.E. Dever, and A.G. Hinnebusch. 1992. Truncated protein phosphatase GLC7 restores translational activation of *GCN4* expression in yeast mutants defective for the eIF2α kinase GCN2. *Mol. Cell. Biol.* **12:** 5700–5710.

Welsh, G.I. and C.G. Proud. 1992. Regulation of protein synthesis in Swiss 3T3 fibroblasts. Rapid activation of the guanine-nucleotide-exchange factor by insulin and growth factors. *Biochem J.* **284:** 19–23.

———. 1993. Glycogen synthase kinase-3 is rapidly inactivated in response to insulin and phosphorylates eukaryotic initiation factor eIF-2B. *Biochem J.* **294:** 625–629.

Welsh, G.I., N.T. Price, B.A. Bladergroen, G. Bloomberg, and C.G. Proud. 1994. Identification of novel phosphorylation sites in the β-subunit of translation initiation factor eIF-2. *Biochem. Biophys. Res. Commun.* **201:** 1279–1288.

Wu, S., S. Gupta, N. Chatterjee, R.E. Hileman, T.G. Kinzy, N.D. Denslow, W.C. Merrick, D. Chakrabarti, J.C. Osterman, and N.K. Gupta. 1993. Cloning and characterization of complementary DNA encoding the eukaryotic initiation factor 2-associated 67-kDa protein (p^{67}). *J. Biol. Chem.* **268:** 10796–10801.

Xiao, Q., T.V. Sharp, I.W. Jeffrey, M.C. James, G.J.M. Pruijn, W.J. Van Venrooij, and M.J. Clemens. 1994. The La antigen inhibits the activation of the interferon-inducible protein kinase PKR by sequestering and unwinding double-stranded RNA. *Nucleic Acids Res.* **22:** 2512–2518.

Yang, J.M., I.M. London, and J.-J. Chen. 1992. Effects of hemin and porphyrin compounds on intersubunit disulfide formation of heme-regulated eIF-2α kinase and the regulation of protein synthesis in reticulocyte lysates. *J. Biol. Chem.* **267:** 20519–20524.

6

Translational Control Mediated by Upstream AUG Codons

Adam P. Geballe
Department of Molecular Medicine and Division of Clinical Research
Fred Hutchinson Cancer Research Center, and
Departments of Medicine and Microbiology
University of Washington, Seattle, Washington 98104

A commonly held, although not wholly accurate conception of eukaryotic messenger RNA structure envisions one major open reading frame (ORF) encoding a single polypeptide product. Flanking the ORF are sequences that may regulate properties such as the stability, subcellular distribution, or translational efficiency of the mRNA. The RNA sequence between the methyl guanosine cap and the initiation codon of the ORF typically consists of 20–100 nucleotides (Kozak 1987a) and is conventionally known as the 5′-untranslated region (5′UTR), even though this term is a misnomer when applied to transcript leader sequences containing upstream AUG (uAUG) codons that may function as translation initiation sites.

According to the "scanning model" of eukaryotic translation (Kozak 1989), the 40S ribosomal subunit with its associated factors engages the mRNA at or near the cap and then scans in a 3′direction. Upon encountering the first initiation codon, the 60S subunit joins the 40S subunit to form a complete 80S ribosome, and polypeptide synthesis commences. This model predicts that sequence elements within 5′UTRs have the potential to control the access of ribosomes to the downstream coding ORF. Indeed, studies of genes from many systems have revealed a variety of regulatory elements within 5′UTRs. This chapter reviews our current understanding of the regulatory effects of uAUG codons and associated upstream ORFs (uORFs). Although the scanning model provides a useful conceptual framework for considering the effects of uAUG codons and uORFs, the model remains unproven, and alternative mechanisms are clearly involved in translation of some eukaryotic mRNAs (see Jackson, this volume). Information on other regulatory elements within the 5′UTR and 3′of the major ORF can be found elsewhere in this volume (see Spirin; Rouault et al.; Meyuhas et al.; Jacobson).

PREVALENCE OF UPSTREAM INITIATION CODONS IN EUKARYOTIC GENES

Overall, approximately 10% of eukaryotic gene transcripts contain uAUG codons (Kozak 1987a). However, the frequency of uAUG codons is substantially higher in certain subsets of genes. For example, 42% of *Drosophila* genes express transcripts containing uAUG codons (Cavener and Cavener 1993). There are many examples of uAUG codons expressed in the 5′UTRs of viral and cellular genes involved in control of growth and differentiation, including transcription factors, translation factors, tumor suppressors, growth factors, cellular receptors, and genes involved in development (Kozak 1991). The presence of uAUG codons in two thirds of proto-oncogene transcripts suggests the hypothesis that deregulation of oncoprotein translation may be a pathway to oncogenic transformation. In fact, mutations of regulatory uAUG codons can augment the transformation efficiencies of growth factor genes (Bates et al. 1991), although no naturally occurring tumors are known to result from uAUG codon mutations. Given the diversity of genes in which they reside, uAUG codons are likely involved in a broad range of biological processes.

Despite the accumulation of an enormous amount of nucleotide sequence data during the last few years, no recent compilation of uAUG codons among eukaryotic genes has been reported. One impediment to preparing such an inventory is that entries in sequence databases, even those reported to be "cDNAs," do not consistently include the entire or precise 5′UTR sequences. Another problem is that numerous genes express more than one transcript, as a result of either alternative transcript initiation sites or posttranscriptional processing events. Finally, non-AUG codons sometimes function as initiation codons. Because the vast majority of known initiation codons are AUG codons, the designation "uAUG codon" is used throughout most of this chapter. However, since we are only beginning to understand the rules with which to identify these non-AUG initiation codons (Boeck and Kolakofsky 1994; Grünert and Jackson 1994), we should bear in mind that upstream non-AUG initiation codons may be more widespread than is currently appreciated. Eukaryotic transcripts that are now considered simple monocistronic mRNAs may in fact have a more complex polycistronic structure.

ORGANIZATION OF ORFS IN TRANSCRIPTS WITH UPSTREAM INITIATION CODONS

For purposes of discussion, it is useful to distinguish three potential configurations that an ORF following an uAUG codon may assume, relative

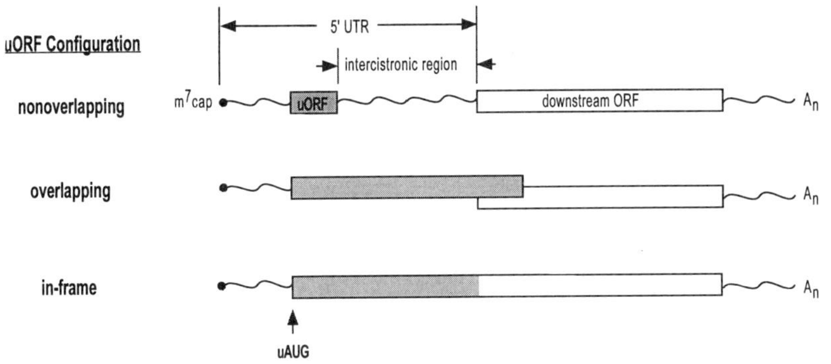

Figure 1 Three possible configurations of ORFs following uAUG codons. ORFs that initiate at uAUG codons may be nonoverlapping, overlapping, or in-frame with the downstream ORF.

to the downstream ORF (Fig. 1). The term uORF as used here refers to an initiation codon within the 5′UTR of an mRNA, followed by any number of internal codons and a termination codon different from that of the major coding ORF. This definition does not specify any minimal length of the uORF since there are no known mechanisms by which a ribosome, while initiating at the AUG codon, can detect the overall length of the ORF. Nonoverlapping and overlapping uORFs are distinguished by the position of the termination codon relative to the initiation codon of the downstream ORF. In both cases, the potential peptide product of the uORF will be entirely different from that of the downstream ORF. Finally, translation initiation at alternative in-frame initiation codons on a single mRNA may produce two proteins differing only in their amino termini. The more 5′-proximal AUG codon in such transcripts can be viewed as an uAUG codon in the 5′UTR of the ORF starting at the more 3′ initiation codon.

Multiple uAUG codons and complex combinations of the three configurations of uORFs are not uncommon in naturally occurring transcripts. The herpesvirus ateles thymidylate synthase transcript contains an extraordinary 29 uAUG codons (Richter et al. 1988). SV40 expresses RNAs with ORFs representing all three possible configurations (for review, see Sedman et al. 1989).

PREDICTED EFFECTS OF UPSTREAM INITIATION CODONS ON TRANSLATION OF THE DOWNSTREAM ORF

In the simplest version of the scanning model, 40S ribosomal subunits identify only one AUG codon as an initiation site, and those that do so at the 5′-proximal AUG codon will be unavailable to translate the down-

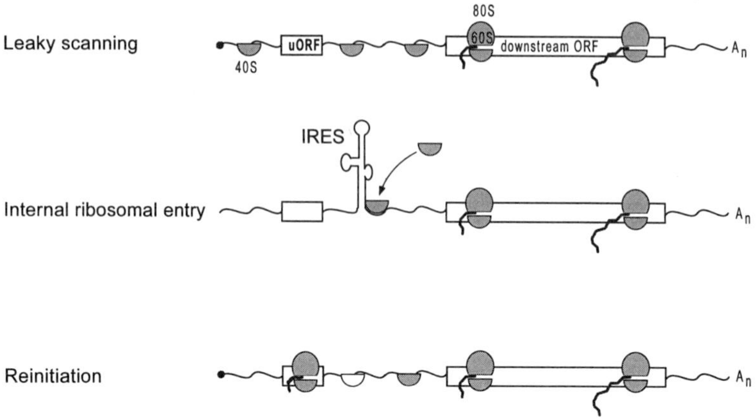

Figure 2 Translational mechanisms of avoiding the inhibitory effects of an uAUG codon. Ribosomes may (1) scan past an uAUG codon without initiating until reaching the downstream AUG codon (leaky scanning), (2) load on the mRNA downstream from an uAUG codon, as at an internal ribosomal entry site (IRES), or (3) initiate at an uAUG codon and translate the uORF, but then reinitiate at a downstream AUG codon.

stream ORF (Kozak 1989). This view predicts that expression of a downstream ORF will vary inversely with initiation at the uAUG codon. However, several refinements and additions to the scanning model describe mechanisms that counter the expected inhibitory impact of uAUG codons (Fig. 2).

By using the process of "leaky scanning," ribosomes may ignore the 5′-proximal AUG codon. If the nucleotides flanking the uAUG codon are suboptimal for initiation or if the uAUG codon is very close to the 5′ end of the mRNA (Kozak 1989; Sedman et al. 1990), the 40S ribosomal subunit may fail to recognize the uAUG codon and continue scanning to the next AUG codon. Parameters other than the primary sequence influence the efficiency of uAUG codon recognition by scanning 40S subunits. For example, secondary structure in the RNA immediately downstream from a poor context uAUG codon may retard ribosomal scanning and thereby facilitate initiation at the uAUG codon (Kozak 1990).

In several viral and a few cellular transcripts, ribosomes bypass uAUG codons due to the presence of an internal ribosomal entry site (IRES; see Jackson, this volume). Ribosomes apparently load onto the mRNA near the 3′end of the IRES and thus never encounter uAUG codons present further upstream.

In some transcripts, ribosomes reinitiate at a downstream AUG codon after having translated a more 5′-proximal ORF. Regulation of *GCN4* expression, the most thoroughly characterized example of reinitiation, is discussed by Hinnebusch (this volume). In this and other systems, the spacing between the two cistrons may influence the efficiency of reinitiation. Although increasing the intercistronic spacing from 11 to 79 nucleotides increased the frequency of reinitiation from 10% to 100%, compared to a construct with no uAUG codon in one study (Kozak 1987b), other studies cast doubt on the generality of these numbers (Sedman et al. 1989). For example, efficient reinitiation can occur following an intercistronic distance of only 30–40 nucleotides (Grant et al. 1994; Cao and Geballe 1995). The identity of the downstream cistron may be an important determinant overriding the requirement for a long intercistronic spacing (Grant et al. 1994). In some prokaryotic genes, translation of a downstream cistron is facilitated by or even dependent on translation of the uORF (Normark et al. 1983; Ivey-Hoyle and Steege 1989; Adhin and van Duin 1990). In these examples of translational coupling, the termination codon of the uORF and the initiation codon of the downstream ORF are often close together or overlapping. Most data in eukaryotes suggest that reinitiation at the downstream AUG codon in transcripts with a short intercistronic spacing between two ORFs is inefficient (Kozak 1987b; Bates et al. 1991), although it may in some cases be sufficient for low but significant expression of the downstream ORF (Johansen et al. 1984; Hormath et al. 1990). Two special cases of reinitiation of translation, backward scanning and ribosomal shunting, are discussed below.

A gene may avoid the regulatory effects of an uAUG codon by expressing an alternate mRNA that lacks the uAUG codon, either by using a different transcription initiation site or by posttranscriptional mRNA processing. In practice, this issue confounds analyses of uORFs (Kozak 1989), in part because 5′UTRs are often heterogeneous or their precise 5′ends are unknown. When multiple mRNAs have the potential to express the same downstream ORF, elucidation of the regulatory impact of the uAUG codons requires a careful evaluation of the translational efficiencies of the various wild-type transcripts and of derivatives containing uAUG codon mutations (Barkan and Mertz 1984; Sedman et al. 1989).

The presence of multiple uORFs in a single 5′UTR, as occurs in many natural transcripts, can greatly complicate analyses of uORF regulation. For example, a point mutation in a termination codon of one uORF sometimes also alters the initiation codon of a second uORF in the same 5′UTR, thereby confounding the dissection of the regulatory roles of the individual uORFs (Zimmer et al. 1994; Cao and Geballe 1995). Potential interactions between ribosomes translating different ORFs on the same transcript further complicate the design and interpretation of these studies (Barkan and Mertz 1984; Sedman et al. 1989; Fajardo and Shatkin 1990; Hill and Morris 1993; Cao and Geballe 1995). Nevertheless, as well demonstrated in the case of *GCN4*, understanding the composite effects of multiple uORFs can be the key to unraveling novel translational regulatory mechanisms.

NONOVERLAPPING uORFs

Only a small subset of the numerous 5′UTRs that contain uAUG codons have been analyzed in sufficient detail to ascertain whether the uAUG codons affect translation of the downstream ORF. 5′UTR deletion mutants are inadequate proof of regulation by uAUG codons since other features of the 5′UTR, such as its overall length, secondary structure, and repressor binding sites, may be responsible for the effects. For example, the 5′UTR of the platelet-derived growth factor, B-chain mRNA, is a potent inhibitor of downstream translation (Rao et al. 1988). However, deletion of all three uAUG codons does not reverse the translational effects of the 5′UTR, indicating that the uORFs are not essential components of the inhibitory sequences. To implicate an uAUG codon in the regulation of downstream expression requires, at a minimum, a demonstration that an uAUG point mutation alters the effects of the 5′UTR. Because the AUG nucleotides could be components of another regulatory sequence (e.g., a transcription factor binding site), proof of translational regulation by the uAUG codon also requires verification that the uAUG mutation affects translation, rather than another pretranslational event. In most cases, fulfillment of this criterion has been achieved by demonstrating that the transcript abundance of the wild-type and uAUG codon mutants is similar (see, e.g., Marth et al. 1988; Imataka et al. 1994; Zimmer et al. 1994) or that the AUG codon mutant mRNA associates with larger polysomes than does the wild type (Schleiss et al. 1991; Hill and Morris 1992).

Even the demonstration that mutation of an uAUG codon eliminates the translational effects of a 5′UTR does not completely prove that the mechanism of inhibition depends on translation initiation function of the uAUG codon. Instead, the AUG nucleotides might be an essential component of another translational sequence or structure in the 5′UTR. For example, although mutation of the seventh uAUG codon in the poliovirus 5′UTR reduces downstream translation, this result is likely not due to any effect on translation initiation at this uAUG codon (Pelletier et al. 1988).

Table 1 shows examples of eukaryotic genes that have uORFs for which there is strong in vivo evidence that one or more uAUG codons affect downstream translation. Most of these uORFs are nonoverlapping with the downstream ORF and inhibit its expression. Genes for which a phenotype associated with an uAUG codon has been documented only in cell-free translation assays are not included because of the inconsistent fidelity of cell-free systems in reproducing in vivo translational phenomena (Kozak 1989; Roner et al. 1993; Michelet et al. 1994). Mutations in the uAUG codons in the genes listed in Table 1 do not necessarily completely eliminate the 5′UTR effects on downstream translation. For example, mutation of the second of three uAUG codons in the *lck* proto-oncogene 5′UTR increases downstream translation three- to fourfold, but the translational efficiency of a 5′UTR deletion mutant is even greater, suggesting that other uAUG codons or 5′UTR sequences are also inhibitory (Marth et al. 1988).

The effects of some uAUG codons are specific for a particular cell type or tissue. The sixth of seven uAUG codons in the 5′UTR of the basic transcription element-binding protein (BTEB) transcription factor contributes to inhibition of downstream translation in HeLa cells but not in neuroblastoma cells (Imataka et al. 1994). Studies of transgenic mice revealed that uORFs in the retinoic acid receptor-β_2 5′UTR repress translational expression in the heart and brain but not in other tissues (Zimmer et al. 1994). These examples of cell specificity predict that *trans*-acting factors, such as translation initiation factors that mediate recognition of uAUG codons or that are needed for reinitiation, are differentially expressed in various cells and tissues. Although the abundance of the cap-binding complex, eIF4F, influences the differential recognition of alternative AUG codons in vitro (Tahara et al. 1991), little is known about in vivo differences in activities of translation factors that might explain cell-type-specific effects of uORFs.

In contrast to the examples shown in Table 1, other 5′UTRs have no apparent impact on downstream translation despite containing uAUG codons. Since few detailed analyses of uAUG codons in translationally neutral 5′UTRs have been reported, the frequency of neutral uAUG codons is unknown. Mechanisms such as leaky scanning or reinitiation do not easily account for the apparent translational neutrality of the uORF in all cases. For example, an optimal context uAUG codon in the human cytomegalovirus (CMV) pp150 gene 5′UTR has no effect on

Table 1 Nonoverlapping or overlapping uORFs that affect downstream translation in vivo

	Gene	References
RNA viruses	Rous sarcoma virus	Donzé and Spahr (1992); Moustakas et al. (1993b)
	reovirus S1 RNA	Fajardo and Shatkin (1990); Belli and Samuel (1993)
DNA viruses	cytomegalovirus gp48	Schleiss et al. (1991); Degnin et al. (1993); Cao and Geballe (1994, 1995)
	SV40	
	16S RNA	Grass and Manley (1987); Perez et al. (1987); Sedman et al. (1989)
	19S RNA	Sedman and Mertz (1988); Sedman et al. (1989)
Yeast	*GCN4*	Hinnebusch (this volume)
	CPA1	Werner et al. (1987); Delbecq et al. (1994)
Plants	*Opaque*-2	Lohmer et al. (1993)
	Lc	Damiani and Wessler (1993)
	CaMV 35S RNA	Fütterer and Hohn (1992); Fütterer et al. (1993)
	pma-1	Michelet et al. (1994)
Mammals	transforming growth factor β3	Arrick et al. (1991)
	S-adenosylmethionine decarboxylase (AdoMetDC)	Hill and Morris (1992, 1993); Ruan et al. (1994)
	β_2-adrenergic receptor	Parola and Kobilka (1994)
	retinoic acid receptor-β_2	Zimmer et al. (1994)
	lck	Marth et al. (1988)
	BTEB	Imataka et al. (1994)
	erythrocyte carbonic anhydrase inhibitor	Bergenhem et al. (1992)
	fibroblast growth factor 5	Bates et al. (1991)

Analyses of the translational impact of mutating auAUG codon in each of these genes demonstrate that the uORF affects downstream translation in vivo.

downstream translation, regardless of the intercistronic spacing (Biegalke and Geballe 1990). Observations such as these highlight limitations in our understanding of parameters that determine translation initiation and reinitiation and suggest that in some cases, additional *cis*-acting sequences are involved in the regulation.

CODING SEQUENCE-DEPENDENT uORFs

Attenuation Control in Prokaryotes

Before considering the role of uORF-coding sequences in eukaryotic mRNAs, it is instructive to consider two examples in prokaryotes. Attenuation control of transcription of the *Escherichia coli trp* operon is mediated by translation of a nonoverlapping uORF containing two critical tryptophan codons (Yanofsky 1988). When $tRNA^{trp}$ is scarce, ribosomes pause upon encountering the tryptophan codons of the uORF. This ribosomal pausing induces a change in the structure of the nascent RNA that results in transcription of the full-length mRNA. It is noteworthy that even though the uORF-coding information is essential, the peptide product of the uORF seems to be only a by-product, not a direct mediator, of this regulatory mechanism. Translational attenuation controls expression of the bacterial antibiotic resistance genes *cat* or *erm*. In the presence of the relevant antibiotic, the nascent peptide product of a nonoverlapping uORF causes ribosomes to stall while translating the uORF. The resulting change in mRNA structure unmasks a ribosomal binding site enabling translation of the downstream ORF (Lovett 1990). Mechanisms analogous to transcriptional or translational attenuation control have not been reported in eukaryotes. Indeed, the physical separation of transcription and translation in eukaryotes makes a mechanism similar to transcriptional attenuation implausible in eukaryotes, except perhaps for RNA viruses that replicate (transcribe) and translate their genomes in the cytoplasm. However, the notion that translation of uORFs can have significant effects on RNA structure may be germane to understanding the roles of some eukaryotic uORFs.

Eukaryotic Sequence-dependent uORFs

In five eukaryotic genes (Table 2), inhibition of downstream translation appears to depend on the coding sequence of the uORF. Among these genes, data supporting the amino acid sequence dependence of the uORF are strongest for mammalian *S*-adenosylmethionine decarboxylase

Table 2 Amino acid sequence dependent uORFs

Gene; source; function	uORF sequence[a]	uORF length
AdoMetDC; mammals; polyamine biosynthesis	MAG**DIS**	6
CPA1; yeast; arginine and pyrimidine biosynthesis	MFSLSNSQYT**C**Q**D**YISDHIWKTSSH	25
β_2-adrenergic receptor; mammals; receptor	MKLPGVRPRPAAP**RRR**CTR	19
gp48; human cytomegalovirus; virion glycoprotein	MQPLVLSAKKLSSLLTC**KYIPP**	22
Lc; plants; transcription factor	MEV**L**ALLRCFSSFFLLRLSSIR**M**PLVRRF**T**RHRL**M**ISR	38

For references, see Table 1.
[a]Single missense mutations (bold) or mutations of several adjacent codons eliminate the inhibitory effect of the uORF.

(AdoMetDC), yeast *CPA1*, and CMV gp48 (or gpUL4). Some, but not all, missense mutations of these uORFs eliminate the inhibitory effect of the 5′UTRs. For example, mutations of the 6-codon AdoMetDC uORF revealed that the three carboxy-terminal codons, but not codons 2 or 3, are essential for inhibition of downstream translation (Hill and Morris 1993). Frameshift mutations in the 22-codon gp48 uORF (Schleiss et al. 1991; Cao and Geballe 1994) and certain missense mutations, such as those affecting either of two carboxy-terminal proline codons (positions 21 and 22), eliminate the inhibitory effect, whereas changes in codons 10, 11, and 17 retain the wild-type phenotype (Degnin et al. 1993). Although carboxy-terminal codons appear to be necessary, they are not sufficient for inhibition by the AdoMetDC and gp48 uORFs (Degnin et al. 1993; Hill and Morris 1993).

Analyses of the *CPA1* 5′UTR provided elegant confirmation of the regulatory significance of this uORF (Werner et al. 1987). In the presence of arginine and the product of the *trans*-acting gene *CPAR*, the *CPA1* 5′UTR inhibits downstream expression. All mutations in strains selected in vivo for constitutive expression of the downstream cistron mapped to the uORF. In addition to nonsense and uAUG codon mutations, missense mutations of codons 11 and 13 were isolated. An analogous study of gp48 5′UTR mutants derived by in vivo selection supported the role of the gp48 uORF-coding content (Cao and Geballe 1994).

Synonymous mutations of the AdoMetDC, gp48, and *CPA1* uORFs uniformly retain the inhibitory phenotype, supporting the conclusion that the amino acid sequence, not the nucleotide sequence per se, mediates the effect (Schleiss et al. 1991; Degnin et al. 1993; Hill and Morris 1993; Delbecq et al. 1994). In contrast, regulation by uORFs in the *GCN4* transcript depends on the nucleotide sequence, not on the amino-acid-coding information (see Hinnebusch, this volume).

Analyses suggest that 5′UTRs of the mammalian β_2-adrenergic receptor and the plant *Lc* transcription factor contain sequence-dependent uORFs (Damiani and Wessler 1993; Parola and Kobilka 1994). However, additional studies are needed since only a few mutations have been analyzed, and in the case of *Lc*, a synonymous mutation in the uORF eliminated most of the inhibitory effect of the 5′UTR.

Role of Initiation and Termination of uORF Translation

Not surprisingly, mutations that influence translation initiation frequency at the uAUG codons of sequence-dependent uORFs affect the magnitude of the inhibitory effect. The AdoMetDC uORF inhibits downstream translation in T-cell lines, but in other cell types, the short length of the 5′UTR between the cap and the uAUG codon causes ribosomes to bypass the uAUG codon (Ruan et al. 1994). The inhibitory gp48 uORF is surrounded by a suboptimal context of nucleotides, resulting in only approximately 10% of ribosomes initiating at the uAUG codon (Cao and Geballe 1995). Mutation of the nucleotides surrounding the uAUG codon to an optimal context increases the inhibitory effect of the uORF tenfold. The suboptimal initiation at the uAUG codons of AdoMetDC and gp48 may serve to assure that expression of the downstream cistron is very sensitive to relatively minor changes in the efficiency of initiation (Ruan et al. 1994).

In addition to the uAUG codon and codons within the uORF, the authentic position of the termination codon is required for inhibition by the uORFs of AdoMetDC and gp48. Mutations of the termination codon resulting in carboxy-terminal extension of the uORF, even by only a single codon, uniformly eliminate the inhibitory effect, although which of the three termination codons ends the uORF does not seem to be a determinant in the regulation (Degnin et al. 1993; Hill and Morris 1993). This requirement for the position of the termination codon argues against a mechanism involving ribosomal stalling at a particular uORF codon such as might occur if the corresponding charged tRNA is limiting. Termination rather than elongation of translation appears to be the critical step mediating inhibition by these uORFs.

In contrast, the *CPA1* uORF termination codon does not appear to be necessary for the uORF-mediated inhibition since expression of an inframe fusion of the uORF to a downstream reporter ORF, eliminating the uORF termination codon, was inhibited (Delbecq et al. 1994). These data may reflect a significant difference in the mechanisms of regulation by the *CPA1* uORF compared to the AdoMetDC and gp48 uORFs.

5′UTR sequences flanking the sequence-dependent uORFs do not appear to be involved in the translational regulation. For example, the gp48 inhibitory uORF still functions after removal of two other uORFs in the 5′UTR, after deletion of the entire 5′UTR 3′of the uORF, or when embedded in a retroviral vector 5′UTR that is more than 1 kb in length and contains multiple additional AUG codons and uORFs (Schleiss et al. 1991; Cao and Geballe 1994). The *CPA1* uORF also functions independently of flanking sequences (Delbecq et al. 1994). For AdoMetDC, the short 5′UTR sequence upstream of the uAUG codon is responsible for cell-type-specific inhibition, but the length and sequence of the 5′UTR downstream from the uORF do not seem to be important (Hill and Morris 1993). Again in contrast to these genes, *GCN4* regulation depends on the intercistronic spacing and on sequences following the termination codons (see Hinnebusch, this volume). The absence of intercistronic spacing or sequence requirements by the sequence-dependent uORFs suggests that, unlike *GCN4*, reinitiation of translation does not have a central role in the inhibitory mechanism.

Possible Mechanisms of Inhibition by Sequence-dependent uORFs

A reasonable hypothesis, based on the coding sequence dependence of these uORFs, is that the peptide products of the uORFs mediate the inhibitory effects. However, none of these uORF peptide products have been identified thus far. If the peptides do mediate the inhibition, then the observation that the uORFs of all five genes appear to function only in *cis*, at least in vivo, is perplexing (Werner et al. 1987; Damiani and Wessler 1993; Degnin et al. 1993; Hill and Morris 1993; Parola and Kobilka 1994). The β_2-adrenergic receptor uORF peptide inhibits translation in *trans*, but it does so only in cell-free extracts and only at high concentrations (Parola and Kobilka 1994), suggesting that the effect may be nonspecific.

Several models can be considered which reconcile the coding sequence dependence of these uORFs with their exclusively *cis*-acting effects. The peptide product of the uORF might be synthesized and released but only achieve a high enough concentration in the local microenvironment to inhibit translation. Alternatively, the nascent peptide, still attached to the translational machinery, might mediate repression. For example, the peptide might interact with the ribosome or a ribosome-associated translation factor to delay or prevent termination of translation. Although very little is known about termination, especially in eukaryotes, it is thought to be a relatively slow step and thus one that might easily be subject to regulation (Wolin and Walter 1988). In fact, recent studies demonstrate that ribosomes arrest during termination of translation of the gp48 uORF (J. Cao and A.P. Geballe, in prep.). Finally, it is conceivable that repression might result from the act of translating particular codons of the uORF, and the peptide is only a by-product, analogous to transcriptional attenuation in bacteria. However, the requirement for the authentic position of the termination codon for inhibition of AdoMetDC and gp48 expression argues against this last model.

The gp48 uORF exerts a potent inhibitory effect on the downstream ORF despite paradoxically inefficient utilization of the uAUG codon. One model accounting for these results is that even though few ribosomes initiate at the uAUG codon, those that do so synthesize the uORF peptide product, which in turn prolongs termination of translation or release of the ribosome from the mRNA. As a result, no additional scanning ribosomal subunits can gain access to the AUG codon of the downstream uORF (Cao and Geballe 1995). Regardless of the mechanism, studies of gp48 illustrate that even a weak uAUG codon can have a significant impact on downstream translation (Barkan and Mertz 1984; Fajardo and Shatkin 1990).

Two of the sequence-dependent uORFs have been shown to regulate downstream translation in response to environmental conditions (Werner et al. 1987; Hill and Morris 1992, 1993). The *CPA1* uORF represses downstream translation only in the presence of arginine. Translation of AdoMetDC is inhibited in resting T cells. In principle, derepression of *CPA1* and AdoMetDC translation by arginine depletion (Werner et al. 1987) and T-cell activation (Mach et al. 1986), respectively, could result either from reduced initiation at the uAUG codon or from a loss of the inhibitory influence of the uORF-coding sequences. Efficient translation of AdoMetDC in non-T-cell lines results from inefficient use of the uAUG codon (Ruan et al. 1994), but it remains to be established whether the translational effects of T-cell activation result from reduction in uAUG codon recognition.

Although sequence-dependent uORFs have been reported for only five eukaryotic genes, few other genes containing uORF have been analyzed in sufficient detail to assess the role of the uORF-coding sequences. Comparisons of the sequences the five uORFs and of the genes in which they reside illuminate no common features. Alignments of the uORFs reveal no consensus codons (see Table 2). Although AdoMetDC, *CPA1*, and *Lc* function in cellular biosynthetic activities, no link has been recognized between uORF codons and the functions of the downstream ORFs. Nevertheless, the occurrence of these uORFs in evolutionarily divergent eukaryotes, including yeast, plants, mammals, and animal viruses, suggests that regulation by such uORFs is more common than is currently recognized.

OVERLAPPING uORFs

Although most parameters that determine the translational effects of a nonoverlapping uORF also apply to overlapping uORFs, additional constraints are imposed by the overlapping configuration. A ribosome that has translated a nonoverlapping uORF may resume scanning, in the conventional 5′ to 3′ direction, and reinitiate at a downstream AUG codon. In contrast, with an overlapping uORF, the ribosome must scan in a reverse direction to encounter the AUG codon of the downstream ORF. Reasonably strong evidence in support of "backward scanning" as an explanation of translation of overlapping ORFs has been reported for a few transcripts, especially if the region of overlap is no more than approximately 50 nucleotides (Johansen et al. 1984; Peabody and Berg 1986; Peabody et al. 1986; Thomas and Capecchi 1986). However, studies of other genes demonstrate that backward scanning and reinitiation do not always enable efficient downstream translation, even with relatively short overlapping regions (Michelet et al. 1994; Cao and Geballe 1995).

The overlapping configuration of ORFs also provides the opportunity for an 80S ribosome that has paused during translation of one ORF to block a trailing ribosome engaged in translation of the overlapping ORF. The reovirus S1 transcript contains overlapping ORFs encoding the σ1 protein, translated by initiation at the most 5′ AUG codon, and the p14 (or σ1NS) protein, initiating at a downstream AUG codon within the σ1 ORF. Surprisingly, the level of σ1 protein expression is independent of its AUG codon context, suggesting that a translational step other than initiation is rate-limiting (Fajardo and Shatkin 1990; Belli and Samuel 1993). One interpretation is that elongation of σ1 translation is a slow step and that ribosomes translating the σ1 uORF impede ribosomes translating the p14 ORF. However, conflicting data and interpretation of studies of reovirus S1 RNA translation obscure any clear conclusions about the mechanism of regulation by this cistronic configuration.

IN-FRAME UPSTREAM INITIATORS

Several eukaryotic genes express two or more proteins by use of alternative in-frame initiation codons. Recognition that the larger, amino-terminally extended protein results from initiation at the more 5′ site was delayed in several cases because the upstream site is a non-AUG initiation codon. For example, the c-*myc* AUG initiation codon was shown to initiate the c-*myc* 2 protein before the discovery that the larger c-*myc* 1 protein results from initiation at an upstream CUG codon (Hann et al. 1988). In other genes, such as *erbA*α, protein initiation at the 5′ initiation codon was demonstrated first and only subsequent analyses revealed that shorter proteins arose from initiation at downstream AUG codons (Bigler and Eisenman 1988). Regardless of the order of discovery, quite a few eukaryotic genes have now been shown to express multiple proteins by use of alternative initiation codons (Table 3).

Differential use of the 5′-proximal versus alternative downstream initiation codons can have significant consequences for gene function, since the two proteins may have alternative, cooperative, or even opposing activities. The CUG and AUG initiated forms of basic fibroblast growth factor cooperate in transformation of cells in culture (Couderc et al. 1991). Initiation at the different AUG codons in the *lap* mRNA, encoding a liver-enriched transcriptional factor, results in synthesis of the activator protein LAP and an inhibitor of LAP called LIP (Descombes and Schibler 1991).

Differential use of in-frame initiation codons can also direct related proteins to different subcellular compartments. Forms of the *int-2* and basic fibroblast growth factor resulting from initiation at 5′-proximal CUG codons localize to the nucleus, whereas initiation at the downstream in-frame AUG codons produces cytoplasmic or secreted proteins (Acland et al. 1990; Bugler et al. 1991). The larger CUG-initiated form of the *pim-1* oncogene is relatively stable and associates with a complex unlike the smaller AUG-initiated form (Saris et al. 1991).

Differential selection of initiation codons is regulated in some genes with in-frame ORFs. For example, methionine depletion increases the relative utilization of the c-*myc* CUG initiation codon (Hann et al. 1992). A change in relative use of alternative AUG codons during development results in a fivefold increase in the LAP to LIP ratio, dramatically altering the overall transcriptional effects of this gene (Descombes and Schibler 1991). AUG codon selection in C/EBP α transcripts also appears to be developmentally regulated (Lin et al. 1993). The factors and mechanisms involved in the regulation of initiation codon selection are unknown.

STIMULATORY uORFs

In contrast to most uORFs that are either neutral or repressive with respect to translation of the downstream ORF, a few uORFs have been shown to act as positive effectors of downstream translation. Translational coupling in prokaryotes is an example in which an uORF is needed for efficient translation of the downstream ORF (Normark et al. 1983). An example among eukaryotes is the 5′-proximal uORF in the *GCN4* 5′UTR which enables ribosomes to bypass the strongly repressive fourth uORF under derepressing conditions (see Hinnebusch, this volume). The uORFs in the Rous sarcoma virus 5′UTR may enhance downstream translation, although the data are conflicting (Donzé and Spahr 1992; Moustakas et al. 1993b).

A fascinating example of an uORF acting as a positive effector of downstream translation occurs in plant viruses. The 5′UTR of the polycistronic 35S cauliflower mosaic virus (CaMV) RNA is highly struc-

Table 3 Use of alternative in-frame initiation codons in eukaryotic mRNAs

	Gene	References
Viruses	respiratory syncytial virus–G gene	Roberts et al. (1994)
	hepatitis B virus–X gene	Kwee et al. (1992)
	SV40 19S RNA	Sedman and Mertz (1988)
Yeast	*MOD5*	Gillman et al. (1991); Slusher et al. 1991
Vertebrates	basic fibroblast growth factor[a]	Florkiewicz and Sommer (1989); Prats et al. (1989); Bugler et al. (1991); Couderc et al. (1991); Vagner et al. (1995)
	cAMP response element modulator τ	Delmas et al. (1992)
	CCAAT/ enhancer binding protein α	Lin et al. (1993)
	c-*myc*[a]	Hann et al. (1988, 1992)
	erbA	Bigler and Eisenman (1988); Bigler et al. (1992)
	hck1	Lock et al. (1991)
	Ia antigen-associated invariant chain	Strubin et al. (1986)
	int-2[a]	Ackland et al. (1990)
	lap	Descombes and Schibler (1991)
	pim-1[a]	Saris et al. (1991)
	progesterone receptor A and B	Conneely et al. (1989)

[a] Contains upstream CUG initiation codon(s).

tured (600 nucleotides long) and contains seven short uORFs. Expression of the downstream cistrons in CaMV and related viruses requires an uORF in the 5′ UTR and expression of a viral *trans*-activator protein that is the product of ORF VI (Bonneville et al. 1989; Gowda et al. 1991; Fütterer and Hohn 1992; Fütterer et al. 1993). The uORF must be approximately 30 codons long, but it need not have a particular coding sequence for efficient translation of the downstream ORF (Fütterer and Hohn 1992). Remarkably, translation of the downstream ORF occurs even if the region between the uORF and the downstream ORF contains sequences, such as a whole cistron or stable stem-loop structure, that normally preclude downstream translation. In fact, the CaMV leader and the downstream ORF can be on different RNA molecules as long as the two transcripts are tethered through a region of complementary nucleotide sequence (Fütterer et al. 1993). One model explaining these data proposes that, following translation of the first uORF, an RNA secondary structure forms that causes shunting of the ribosome to a far downstream ORF. Elucidation of the mechanism of ribosomal shunting and its dependence on the uORFs and the *trans*-acting factor promises dramatically new insights into eukaryotic translation.

uORFs INVOLVED IN VIRAL RNA PACKAGING

For some viruses, transcripts serve both as genomes and as mRNAs. The mechanisms that determine whether a particular RNA molecule is packaged into new virions or is translated are largely unknown, nor is it clear that these are mutually incompatible fates. Because the 5′ends of viral RNAs are often important in virus replication as well as translation and packaging, experimental dissection of the various functions of these sequences is complex.

In a few viruses, translation initiating at uAUG codons affects encapsidation of the viral genome. The hepatitis B virus (HBV) full-length genomic transcript is both packaged and used for expression of its core antigen. A transcript initiating a short distance upstream from the genomic transcript contains the e antigen ORF, which is an amino-terminal in-frame extension of the core ORF. Unlike the genomic transcript, the e antigen transcript is not packaged into virions, even though both contain the *cis*-acting sequences required for packaging. Initiation of translation at the e antigen AUG codon and elongation for at least 13 codons prevent encapsidation of e antigen transcripts (Nassal et al. 1990). Apparently, 80S ribosomes translating up to or through the HBV packaging signal disrupt an interaction between the RNA and encapsidation proteins.

The 5′UTRs of avian sarcoma-leukosis retroviruses contain three uORFs that are highly conserved in length and position within the 5′ UTR, but not in amino-acid-coding content (Hackett et al. 1991), and that seem to be vital for virus replication (Petersen et al. 1989; Donzé and Spahr 1992; Moustakas et al. 1993a,b). Ribosomes initiate translation at the first uAUG (uAUG1), and at least in cell-free translation assays, they synthesize the seven-amino-acid peptide product of uORF1 (Hackett et al. 1986). In contrast to conflicting data regarding the impact of the various uORFs on downstream translation (Donzé and Spahr 1992; Moustakas et al. 1993b), a consistent finding is that mutations of AUG1 or AUG3 interfere with the production of infectious virus, suggesting that translation of uORF1 and uORF3 is required for packaging of viral RNA. For example, translation of uORF1 and uORF3 might disrupt an RNA secondary structure that otherwise inhibits packaging.

PEPTIDE PRODUCTS RESULTING FROM INITIATION AT uAUG CODONS

In vivo translation of multiple proteins from a single mRNA has been demonstrated most often in viral systems, where evolutionary pressures are believed to favor compact genomes (Kozak 1986). In theory, all transcripts containing uAUG codons have the potential to function as templates for translation of multiple polypeptides. However, the only example of synthesis of the peptide product of a short uORF is the demonstration, using in vitro translation assays, of the unstable peptide encoded by the avian retroviral uORF (Hackett et al. 1986). Despite strong circumstantial evidence, no data directly verify synthesis of the peptides encoded by sequence-dependent uORFs. More sensitive methods may be necessary to detect these peptides, which may be scarce and unstable.

Although use of alternative in-frame initiation codons has been reported predominantly for genes involved in the control of cellular growth (see Table 3), other less intensely scrutinized genes likely exploit the same mechanism to express multiple proteins. The fact that the AUG codons initiating most proteins deviate at least in part from the optimum sequence (Kozak 1987a) suggests that a fraction of ribosomes loading on most mRNAs may bypass the usual initiation codon, initiate at a downstream AUG codon, and thus produce an alternative protein.

CONCLUSIONS

The recognition of regulatory uAUG codons and uORFs in genes from widely divergent biological systems forecasts the discovery of many new examples. Progress in identifying the *cis*-acting sequences and *trans*-acting factors that influence translation initiation, termination, and reinitiation should help clarify the roles and mechanisms of regulation by these elements.

The control of translation initiation is one key to understanding the effects of uAUG codons. Several eukaryotic genes utilize multiple initiation codons to express alternative proteins, often with related but distinct functions, from a single mRNA template. Although little is known about the *trans*-acting cellular factors that mediate selection of alternative translation initiation codons, developmental and environmental cues may modulate gene expression by affecting initiation at the uAUG codons of uORFs.

Our knowledge of mRNA features that influence translation initiation at uAUG codons is insufficient for predicting the effects of uORFs based solely on sequence inspection. Some optimal uAUG codons have no evident regulatory effects, whereas other weak uAUG codons profoundly inhibit downstream expression. Elements other than those affecting translation initiation are clearly important determinants of the effects of some uORFs as illustrated by studies of amino-acid-sequence-dependent uORFs. A better understanding the mechanisms of, and factors required for, termination and reinitiation and of the interactions between ribosomes translating different ORFs on the same mRNA should help reveal the mechanism of regulation by these uORFs.

Although most reports have highlighted inhibitory effects of uORFs on downstream translation, some uORFs instead augment downstream translation or have a role in nontranslational events such encapsidation of viral RNAs. The diversity of effects of uORFs illustrated in this chapter should serve as a guide for the formidable but important task of dissecting the composite effects of the multiple uORFs that are found in many natural mRNAs.

ACKNOWLEDGMENTS

I thank David Morris and Jianhong Cao for critical review of this chapter. Research in the author's laboratory was supported by U.S. Public Health Service grant AI-26672.

REFERENCES

Acland, P., M. Dixon, G. Peters, and C. Dickson. 1990. Subcellular fate of the Int-2 oncoprotein is determined by choice of initiation codon. *Nature* **343:** 662–665.

Adhin, M.R. and J. van Duin. 1990. Scanning model for translational reinitiation in eubacteria. *J. Mol. Biol.* **213:** 811–818.

Arrick, B.A., A.L. Lee, R.L. Grendell, and R. Derynck. 1991. Inhibition of translation of transforming growth factor-β3 mRNA by its 5′ untranslated region. *Mol. Cell. Biol.* **11:** 4306–4313.

Barkan, A. and J.E. Mertz. 1984. The number of ribosomes on simian virus 40 late 16S mRNA is determined in part by the nucleotide sequence of its leader. *Mol. Cell. Biol.* **4:**

813–816.

Bates, B., J. Hardin, X. Zhan, K. Drickamer, and M. Goldfarb. 1991. Biosynthesis of human fibroblast growth factor-5. *Mol. Cell. Biol.* **11:** 1840–1845.

Belli, B.A. and C.E. Samuel. 1993. Biosynthesis of reovirus-specified polypeptides: Identification of regions of the bicistronic reovirus S1 mRNA that affect the efficiency of translation in animal cells. *Virology* **193:** 16–27.

Bergenhem, N.C.H., P.J. Venta, P.J. Hopkins, H.J. Kim, and R.E. Tashian. 1992. Mutation creates an open reading frame within the 5′ untranslated region of macaque erythrocyte carbonic anhydrase (CA) I mRNA that suppresses CA I expression and supports the scanning model for translation. *Proc. Natl. Acad. Sci.* **89:** 8798–8802.

Biegalke, B.J. and A.P. Geballe. 1990. Translational inhibition by cytomegalovirus transcript leaders. *Virology* **177:** 657–667.

Bigler, J. and R.N. Eisenman. 1988. c-*erbA* encodes multiple proteins in chicken erythroid cells. *Mol. Cell. Biol.* **8:** 4155–4161.

Bigler, J., W. Hokanson, and R.N. Eisenman. 1992. Thyroid hormone receptor transcriptional activity is potentially autoregulated by truncated forms of the receptor. *Mol. Cell. Biol.* **12:** 2406–2417.

Boeck, R. and D. Kolakofsky. 1994. Positions +5 and +6 can be major determinants of the efficiency of non-AUG initiation codons for protein synthesis. *EMBO J.* **13:** 3608–3617.

Bonneville, J.M., H. Sanfacon, J. Fütterer, and T. Hohn. 1989. Posttranscriptional *trans*-activation in cauliflower mosaic virus. *Cell* **59:** 1135–1143.

Bugler, B., F. Amalric, and H. Prats. 1991. Alternative initiation of translation determines cytoplasmic or nuclear localization of basic fibroblast growth factor. *Mol. Cell. Biol.* **11:** 573–577.

Cao, J. and A.P. Geballe. 1994. Mutational analysis of the translational signal in the human cytomegalovirus gpUL4 (gp48) transcript leader by retroviral infection. *Virology* **205:** 151–160.

———. 1995. Translational inhibition by a human cytomegalovirus upstream open reading frame despite inefficient utilization of its AUG codon. *J. Virol.* **69:** 1030–1036.

Cavener, D.R. and B.A. Cavener. 1993. Translation start sites and mRNA leaders. In *An atlas of* Drosophila *genes* (ed. G. Maroni), pp. 359–377. Oxford University Press, New York.

Conneely, O.M., D.M. Kettelberger, M.-J. Tsai, W.T. Schrader, and B.W. O'Malley. 1989. The chicken progesterone receptor A and B isoforms are products of an alternate translation initiation event. *J. Biol. Chem.* **264:** 14062–14064.

Couderc, B., H. Prats, F. Bayard, and F. Amalric. 1991. Potential oncogenic effects of basic fibroblast growth factor requires cooperation between CUG and AUG-initiated forms. *Cell Regul.* **2:** 709–718.

Damiani, R.D. Jr. and S.R. Wessler. 1993. An upstream open reading frame represses expression of *Lc*, a member of the *R/B* family of maize transcriptional activators. *Proc. Natl. Acad. Sci.* **90:** 8244–8248.

Degnin, C.R., M.R. Schleiss, J. Cao, and A.P. Geballe. 1993. Translational inhibition mediated by a short upstream open reading frame in the human cytomegalorivus gpUL4 (gp48) transcript. *J. Virol.* **67:** 5514–5521.

Delbecq, P., M. Werner, A. Feller, R.K. Filipkowski, F. Messenguy, and A. Piérard. 1994. A segment of mRNA encoding the leader peptide of the *CPA1* gene confers repression by arginine on a heterologous yeast gene transcript. *Mol. Cell. Biol.* **14:**

2378–2390.

Delmas, V., B.M. Laoide, D. Masquilier, R.P. deGroot, N.S. Foulkes, and P. Sassone-Corsi. 1992. Alternative usage of initiation codons in mRNA encoding the cAMP-responsive-element modulator generates regulators with opposite functions. *Proc. Natl. Acad. Sci.* **89:** 4226–4230.

Descombes, P. and U. Schibler. 1991. A liver-enriched transcriptional activator protein, LAP, and a transcriptional inhibitory protein, LIP, are translated from the same mRNA. *Cell* **67:** 569–579.

Donzé, O. and P.-F. Spahr. 1992. Role of the open reading frames of Rous sarcoma virus leader RNA in translation and genome packaging. *EMBO J.* **11:** 3747–3757.

Fajardo, J.E. and A.J. Shatkin. 1990. Translation of bicistronic viral mRNA in transfected cells: Regulation at the level of elongation. *Proc. Natl. Acad. Sci.* **87:** 328–332.

Florkiewicz, R.Z. and A. Sommer. 1989. Human basic fibroblast growth factor gene encodes four polypeptides: Three initiate translation from non-AUG codons. *Proc. Natl. Acad. Sci.* **86:** 3978–3981.

Fütterer, J. and T. Hohn. 1992. Role of an upstream open reading frame in the translation of polycistronic mRNAs in plant cells. *Nucleic Acids Res.* **20:** 3851–3857.

Fütterer, J., Z. Kiss-László, and T. Hohn. 1993. Nonlinear ribosome migration on cauliflower mosaic virus 35S RNA. *Cell* **73:** 789–802.

Gillman, E.C., L.B. Slusher, N.C. Martin, and A.K. Hopper. 1991. *MOD5* translation initiation sites determine N^6-isopentenyladenosine modification of mitochondrial and cytoplasmic tRNA. *Mol. Cell. Biol.* **11:** 2382–2390.

Gowda S., H.S. Scholthof, F.C. Wu, and R.J. Shepherd. 1991. Requirement for gene VII in *cis* for the expression of downstream genes on the major transcript of figwort mosaic virus. *Virology* **185:** 867–871.

Grant, C.M., P.F. Miller, and A.G. Hinnebusch. 1994. Requirements for intercistronic distance and level of eukaryotic initiation factor 2 activity in reinitiation of *GCN4* mRNA vary with the downstream cistron. *Mol. Cell. Biol.* **14:** 2616–2628.

Grass, D.S. and J.L. Manley. 1987. Selective translation initiation on bicistronic simian virus 40 late mRNA. *J. Virol.* **61:** 2331–2335.

Grünert, S. and R.J. Jackson. 1994. The immediate downstream codon strongly influences the efficiency of utilization of eukaryotic translation initiation codons. *EMBO J.* **13:** 3618–3630.

Hackett, P.B., M.W. Dalton, D.P. Johnson, and R.B. Petersen. 1991. Phylogenetic and physical analysis of the 5′ leader RNA sequences of avian retroviruses. *Nucleic Acids Res.* **19:** 6929–6934.

Hackett, P.B., R.B. Petersen, C.H. Hensel, F. Albericio, S.I. Gunderson, A.C. Palmenberg, and G. Barany. 1986. Synthesis *in vitro* of a seven amino acid peptide encoded in the leader RNA of Rous sarcoma virus. *J. Mol. Biol.* **190:** 45–57.

Hann, S.R., K. Sloan-Brown, and G.D. Spotts. 1992. Translational activiation of the non-AUG-initiated c-*myc* 1 protein at high cell densities due to methionine deprivation. *Genes Dev.* **6:** 1229–1240.

Hann, S.R., M.W. King, D.L. Bentley, C.W. Anderson, and R.N. Eisenman. 1988. A non-AUG translational initiation in c-*myc* exon 1 generates an N-terminally distinct protein whose synthesis is disrupted in Burkitt's lymphoma. *Cell* **52:** 185–195.

Hill, J.R. and D.R. Morris. 1992. Cell-specific translation of S-adenosylmethionine decarboxylase mRNA. *J. Biol. Chem.* **267:** 21886–21893.

———. 1993. Cell-specific translational regulation of S-adenosylmethionine decarboxylase mRNA. *J. Biol. Chem.* **268:** 726–731.

Hormath, C.M., M.A. Williams, and R.A. Lamb. 1990. Eukaryotic coupled translation of tandem cistrons: Identification of the influenza B virus BM2 polypeptide. *EMBO J.* **9:** 2639–2647.

Imataka, H., K. Nakayama, K.-I. Yasumoto, A. Mizuno, Y. Fujii-Kuriyama, and M. Hayami. 1994. Cell-specific translational control of transcription factor BTEB expression. *J. Biol. Chem.* **269:** 20668–20673.

Ivey-Hoyle, M. and D.A. Steege. 1989. Translation of phage f1 gene VII occurs from an inherently defective initiation site made functional by coupling. *J. Mol. Biol.* **208:** 233–244.

Johansen, H., D. Schümperli, and M. Rosenberg. 1984. Affecting gene expression by altering the length and sequence of the 5′ leader. *Proc. Natl. Acad. Sci.* **81:** 7698–7702.

Kozak, M. 1986. Bifunctional messenger RNAs in eukaryotes. *Cell* **47:** 481–483.

———. 1987a. An analysis of 5′-noncoding sequences from 699 vertebrate messenger RNAs. *Nucleic Acids Res.* **15:** 8125–8148.

———. 1987b. Effects of intercistronic length on the efficiency of reinitiation by eucaryotic ribosomes. *Mol. Cell. Biol.* **7:** 3438–3445.

———. 1989. The scanning model for translation: An update. *J. Cell Biol.* **108:** 229-241.

———. 1990. Downstream secondary structure facilitates recognition of initiator codons by eukaryotic ribosomes. *Proc. Natl. Acad. Sci.* **87:** 8301–8305.

———. 1991. An analysis of vertebrate mRNA sequences: Intimations of translational control. *J. Cell Biol.* **115:** 887–903.

Kwee, L., R. Lucito, B. Aufiero, and R.J. Schneider. 1992. Alternate translation initiation on hepatitis B virus X mRNA produces multiple polypeptides that differentially transactivate class II and III promoters. *J. Virol.* **66:** 4382–4389.

Lin, F.-T., O.A. MacDougald, A.M. Diehl, and M.D. Lane. 1993. A 30-kDa alternative translation product of the CCAAT/enhancer binding protein α message: Transcriptional activator lacking antimitotic activity. *Proc. Natl. Acad. Sci.* **90:** 9606–9610.

Lock, P., S. Ralph, E. Stanley, I. Boulet, R. Ramsay, and A.R. Dunn. 1991. Two isoforms of murine *hck*, generated by utilization of alternative translational initiation codons, exhibit different patterns of subcellular localization. *Mol. Cell. Biol.* **11:** 4363–4370.

Lohmer, S., M. Maddaloni, M. Motto, F. Salamini, and R.D. Thompson. 1993. Translation of the mRNA of the maize transcriptional activator *opaque-2* is inhibited by upstream open reading frames present in the leader sequence. *Plant Cell* **5:** 65–73.

Lovett, P.S. 1990. Translational attenuation as the regulator of inducible *cat* genes. *J. Bacteriol.* **172:** 1–6.

Mach, M., M.W. White, M. Neubauer, J.L. Degen, and D.R. Morris. 1986. Isolation of a cDNA clone encoding *S*-adenosylmethionine decarboxylase: Expression of the gene in mitogen-activated lymphocytes. *J. Biol. Chem.* **261:** 11697–11703.

Marth, J.D., R.W. Overell, K.E. Meier, E.G. Krebs, and R.M. Perlmutter. 1988. Translational activation of the *lck* proto-oncogene. *Nature* **332:** 171–173.

Michelet, B., M. Lukaszewicz, V. Dupriez, and M. Boutry. 1994. A plant plasma membrane proton-ATPase gene is regulated by development and environment and shows signs of a translational regulation. *Plant Cell* **6:** 1375–1389.

Moustakas, A., T.S. Sonstegard, and P.B. Hackett. 1993a. Alterations of the three short open reading frames in the Rous sarcoma virus leader RNA modulate viral replication and gene expression. *J. Virol.* **67:** 4337–4349.

———. 1993b. Effects of the open reading frames in the Rous sarcoma virus leader RNA on translation. *J. Virol.* **67:** 4350–4357.

Nassal, M., M. Junker-Niepmann, and H. Schaller. 1990. Translational inactiviation of RNA function: Discrimination against a subset of genomic transcripts during HBV nucleocapsid assembly. *Cell* **63:** 1357–1363.

Normark, S., S. Bergström, T. Edlund, T. Grundström, B. Jaurin, F.P. Lindberg, and O. Olsson. 1983. Overlapping genes. *Annu. Rev. Genet.* **17:** 499–525.

Parola, A.L. and B.K. Kobilka. 1994. The peptide product of a 5′ leader cistron in the β_2 adrenergic receptor mRNA inhibits receptor synthesis. *J. Biol. Chem.* **269:** 4497–4505.

Peabody, D.S. and P. Berg. 1986. Termination-reinitiation occurs in the translation of mammalian cell mRNA. *Mol. Cell. Biol.* **6:** 2695–2703.

Peabody, D.S., S. Subramani, and P. Berg. 1986. Effect of upstream reading frames on translation efficiency in simian virus 40 recombinants. *Mol. Cell. Biol.* **6:** 2704–2711.

Pelletier, J., M.E. Flynn, G. Kaplan, V. Racaniello, and N. Sonenberg. 1988. Mutational analysis of upstream AUG codons of poliovirus RNA. *J. Virol.* **62:** 4486–4492.

Perez, L., J.W. Wills, and E. Hunter. 1987. Expression of the Rous sarcoma virus *env* gene from a simian virus 40 late-region replacement vector: Effects of upstream initiation codons. *J. Virol.* **61:** 1276–1281.

Petersen, R.B., A. Moustakas, and P.B. Hackett. 1989. A mutation in the short 5′-proximal open reading frame on Rous sarcoma virus RNA alters virus production. *J. Virol.* **63:** 4748–4796.

Prats, H., M. Kaghad, A.C. Prats, M. Klagsbrun, J.M. Lélias, P. Liauzun, P. Chalon, J.P. Tauber, F. Amalric, J.A. Smith, and D. Caput. 1989. High molecular mass forms of basic fibroblast growth factor are initiated by alternative CUG codons. *Proc. Natl. Acad. Sci.* **86:** 1836–1840.

Rao, C.D., M. Pech, K.C. Robbins, and S.A. Aaronson. 1988. The 5′ untranslated sequence of the c-*sis*/platelet-derived growth factor 2 transcript is a potent translational inhibitor. *Mol. Cell. Biol.* **8:** 284–292.

Richter, J., I. Puchtler, and B. Fleckenstein. 1988. Thymidylate synthase gene of herpesvirus ateles. *J. Virol.* **62:** 3530–3535.

Roberts, S.R., D. Lichtenstein, L.A. Ball, and G.W. Wertz. 1994. The membrane-associated and secreted forms of the respiratory syncytial virus attachment glycoprotein G are synthesized from alternative initation codons. *J. Virol.* **68:** 4538–4546.

Roner, M.R., L.A. Roner, and W.K. Joklik. 1993. Translation of reovirus RNA species m1 can initiate at either of the first two in-frame initiation codons. *Proc. Natl. Acad. Sci.* **90:** 8947–8951.

Ruan, H., J.R. Hill, S. Fatemie-Nainie, and D.R. Morris. 1994. Cell-specific translational regulation of *S*-adenosylmethionine decarboxylase mRNA. *J. Biol. Chem.* **269:** 17905–17910.

Saris, C.J.M., J. Domen, and A. Berns. 1991. The *pim*-1 oncogene encodes two related protein-serine/threonine kinases by alternative initiation at AUG and CUG. *EMBO J.* **10:** 655–664.

Schleiss, M.R., C.R. Degnin, and A.P. Geballe. 1991. Translational control of human cytomegalovirus gp48 expression. *J. Virol.* **65:** 6782–6789.

Sedman, S.A. and J.E. Mertz. 1988. Mechanisms of synthesis of virion proteins from the functionally bigenic late mRNAs of simian virus 40. *J. Virol.* **62:** 954–961.

Sedman, S.A., G.W. Gelembiuk, and J.E. Mertz. 1990. Translation initiation at a downstream AUG occurs with increased efficiency when the upstream AUG is located very close to the 5′ cap. *J. Virol.* **64:** 453–457.

Sedman, S.A., P.J. Good, and J.E. Mertz. 1989. Leader-encoded open reading frames modulate both the absolute and relative rates of synthesis of the virion proteins of simian virus 40. *J. Virol.* **63:** 3884–3893.

Slusher, L.B., E.C. Gillman, N.C. Martin, and A.K. Hopper. 1991. mRNA leader length and initiation codon context determine alternative AUG selection for the yeast gene *MOD5*. *Proc. Natl. Acad. Sci.* **88:** 9789–9793.

Strubin, M., E.O. Long, and B. Mach. 1986. Two forms of the Ia antigen-associated invariant chain result from alternative initiations at two in-phase AUGs. *Cell* **47:** 619–625.

Tahara, S.M., T.A. Dietlin, T.E. Dever, W.C. Merrick, and L.M. Worrilow. 1991. Effect of eukaryotic initiation factor 4F on AUG selection in a bicistronic mRNA. *J. Biol. Chem.* **266:** 3594–3601.

Thomas, K.R. and M.R. Capecchi. 1986. Introduction of homologous DNA sequences into mammalian cells induces mutations in the cognate gene. *Nature* **324:** 34–38.

Vagner, S., M.-C. Gensac, A. Maret, F. Bayard, F. Amalric, H. Prats, and A.-C. Prats. 1995. Alternative translation of human fibroblast growth factor 2 mRNA occurs by internal entry of ribosomes. *Mol. Cell. Biol.* **15:** 35–44.

Werner, M., A. Feller, F. Messenguy, and A. Piérard. 1987. The leader peptide of yeast gene CPA1 is essential for the translational repression of its expression. *Cell* **49:** 805–813.

Wolin, S.L. and P. Walter. 1988. Ribosome pausing and stacking during translation of a eukaryotic mRNA. *EMBO J.* **7:** 3559–3569.

Yanofsky, C. 1988. Transcription attenuation. *J. Biol. Chem.* **263:** 609–612.

Zimmer, A., A.M. Zimmer, and K. Reynolds. 1994. Tissue specific expression of the retinoic acid receptor-β_2: Regulation by short open reading frames in the 5′-noncoding region. *J. Cell Biol.* **127:** 1111–1119.

7

Translational Control of *GCN4*: Gene-specific Regulation by Phosphorylation of eIF2

Alan G. Hinnebusch
Laboratory of Eukaryotic Gene Regulation
National Institute of Child Health and Human Development
Bethesda, Maryland 20892

OVERVIEW OF *GCN4* TRANSLATIONAL CONTROL

When subjected to various kinds of starvation, stress, or certain viral infections, mammalian cells respond by reducing the overall rate of protein synthesis by phosphorylating the α-subunit of translation initiation factor-2 (eIF2) (see Clemens; Mathews; Katze; Duncan; Schneider; all this volume). This down-regulation of protein synthesis is presumably a means of conserving resources and limiting cell division under adverse growth conditions or of preventing virus multiplication. Studies on the yeast *Saccharomyces cerevisiae* have shown that eIF2α becomes phosphorylated when cells are deprived of an amino acid or purine and that this event leads to an inhibition of translation by the same mechanism that operates in mammalian cells. Interestingly, it also regulates the translation of a specific mRNA encoding the transcription factor GCN4, causing increased synthesis of GCN4 protein under conditions in which the translation of other yeast messenger RNAs is being reduced. GCN4 activates transcription of at least 40 different genes encoding amino acid biosynthetic enzymes (Hinnebusch 1988); thus, the induction of GCN4 alleviates the limitation for nutrients that triggers phosphorylation of eIF2 in yeast. The extent of eIF2 phosphorylation required to induce *GCN4* translation is too low to cause a significant reduction in the rate of general protein synthesis. Consequently, *GCN4* expression is a very sensitive indicator of the activity of eIF2 and associated translation initiation factors.

The unique induction of *GCN4* translation under starvation conditions is mediated by four short upstream open reading frames (uORFs) in the leader of *GCN4* mRNA, located between 150 and 360 nucleotides upstream of the AUG start codon. These sequences act in *cis* to prevent ribosomes from initiating translation at the *GCN4* start site when nutrients are abundant, but they are much less effective in repressing *GCN4* translation when eIF2 is phosphorylated in response to nutrient deprivation. To understand how the inhibitory effect of the uORFs on *GCN4* translation can be coupled to the level of eIF2 activity in the cell, we must first summarize some fundamental aspects of translation initiation and the mechanism for down-regulating protein synthesis by phosphorylation of eIF2 that have been elucidated for mammalian cells.

In mammals, the first step in translation initiation is the formation of a ternary complex composed of eIF2 (made up of three nonidentical subunits in 1:1:1 stoichiometry), GTP, and charged initiator $tRNA^{Met}$ (Met-$tRNA_i$) (see Trachsel, this volume). This ternary complex associates with the small ribosomal subunit, preloaded with initiation factors eIF3 and eIF1A, to form a 43S preinitiation complex. In the translation of most mRNAs, the 43S complex binds to the mRNA near the capped 5′end, migrates downstream, and recognizes the first AUG codon present in a suitable sequence context, whereupon an 80S initiation complex is formed and translation begins (see Merrick and Hershey, this volume). This "scanning" mechanism for translation initiation probably applies to the translation of most mRNAs in eukaryotic cells (Kozak 1989; for review, see Jackson, this volume). Following AUG recognition, the GTP bound to eIF2 is hydrolyzed to GDP and eIF2 is released as an eIF2•GDP binary complex. To re-form the ternary complex, the GDP bound to eIF2 must be replaced by GTP, and this nucleotide exchange reaction is catalyzed by a factor known as eIF2B (Merrick and Hershey, this volume).

Phosphorylation of the α-subunit of eIF2 on a serine residue at position 51 inhibits guanine nucleotide exchange on eIF2. The binary complex eIF2[αP]•GDP containing phosphorylated eIF2α has a higher affinity than nonphosphorylated eIF2•GDP for eIF2B; in addition, it is generally considered that eIF2B cannot catalyze the exchange of GDP for GTP on phosphorylated eIF2. Consequently, GDP-GTP exchange on both phosphorylated and nonphosphorylated eIF2 is impaired, and the formation of new ternary complexes is diminished by phosphorylation of eIF2α. Phosphorylation of only a fraction of eIF2α is sufficient to inactivate all of the eIF2B, presumably because eIF2 is present at considerably higher levels than eIF2B (see Clemens; Merrick and Hershey; both this volume).

Translational control of *GCN4* by phosphorylation of eIF2 is dependent on the four uORFs in the mRNA leader; however, the first and fourth of these small uORFs (counting from the 5′end) are sufficient for nearly wild-type regulation. These two uORFs have very different functions in regulating *GCN4* translation. When present alone in a *GCN4* construct in which ORFs 1–3 have been eliminated by point mutations, uORF4 reduces *GCN4* translation to only 1% of the value seen in the absence of all four uORFs, whether or not cells are starved for an amino acid (Mueller and Hinnebusch 1986). According to the scanning mechanism described above, essentially all ribosomes that bind to the 5′end of *GCN4* mRNA are expected to translate uORF4 when no other uORFs are present in the leader. To account for the strong inhibitory properties of uORF4, we proposed that none of these ribosomes can reinitiate downstream at *GCN4*, presumably because they dissociate from the mRNA following peptide chain termination at uORF4 (Mueller and Hinnebusch 1986). The first uORF must be present upstream of uORF4 for the induction of *GCN4* translation that occurs in response to eIF2 phosphorylation (Mueller and Hinnebusch 1986). This stimulatory function of uORF1 can be explained by proposing that translation of uORF1 allows ribosomes to bypass the start sites at uORFs 2–4 and reinitiate at *GCN4* instead when phosphorylation of eIF2 reaches a critical level (Abastado et al. 1991b; Dever et al. 1992).

According to the model shown in Figure 1, essentially all ribosomes that bind to the 5′end of *GCN4* mRNA will translate uORF1, under both starvation and nonstarvation conditions. About 50% of these ribosomes will remain attached to the mRNA and resume scanning downstream, presumably as 40S subunits. Under nonstarvation conditions, when the active form of eIF2 is abundant, these 40S subunits will rapidly rebind the eIF2•GTP•Met-$tRNA_i$ ternary complex and regain the ability to recognize an AUG codon as a translational start site. Consequently, most of these ribosomes will reinitiate at uORFs 2, 3, or 4, dissociate from the mRNA following chain termination, and thus fail to reach the *GCN4* start codon. Under starvation conditions, eIF2α will be phosphorylated on Ser-51 by the protein kinase GCN2, and this will reduce the level of eIF2•GTP•Met-$tRNA_i$ ternary complexes in the cell due to inhibition of the recycling factor eIF2B. Following translation of uORF1, many ribosomes will scan the entire distance between uORF1 and uORF4 without rebinding ternary complexes. Lacking the initiator $tRNA^{Met}$, they cannot recognize the AUG start codons at uORFs 2, 3, and 4 (Cigan et al. 1988) and will continue scanning downstream. While traversing the remaining leader segment between uORF4 and *GCN4*, most of these ribosomes will bind the ternary complex and reinitiate translation at *GCN4*. Thus, reducing the level of ternary complexes by phosphorylation of eIF2α will decrease the rate at which ribosomes become competent to reinitiate following translation of uORF1. This will enable ribosomes to bypass

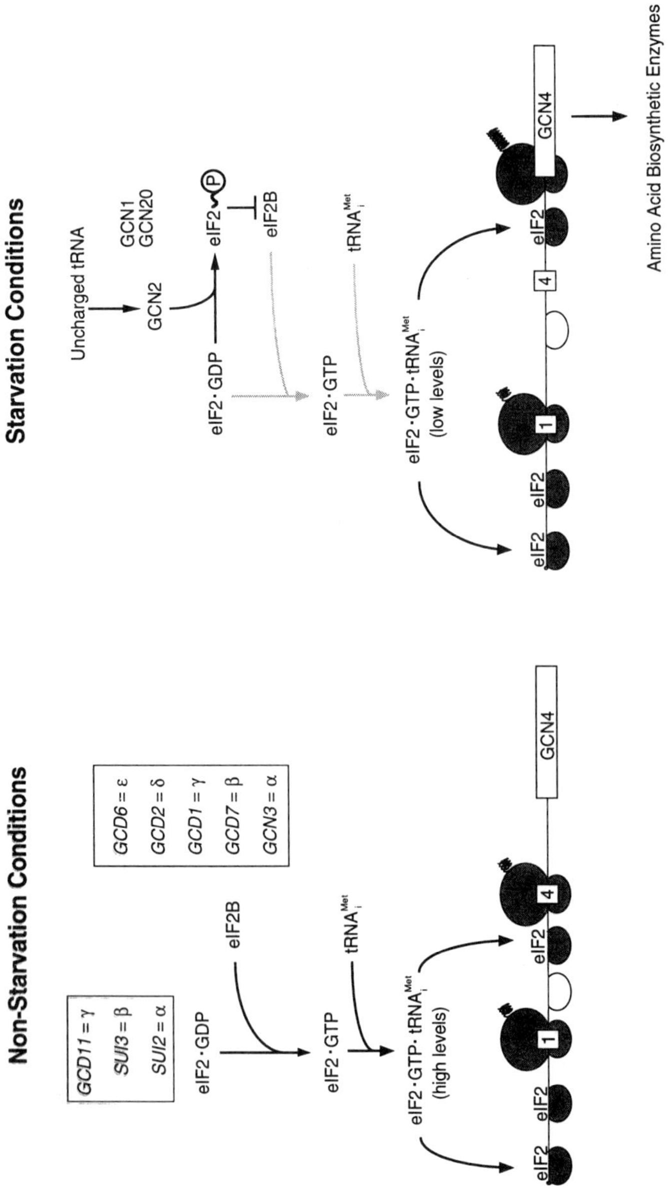
Non-Starvation Conditions
GCD11 = γ
SUI3 = β
SUI2 = α
eIF2·GDP
eIF2B
eIF2·GTP
tRNA$_i^{Met}$
eIF2·GTP·tRNA$_i^{Met}$
(high levels)
GCD6 = ε
GCD2 = δ
GCD1 = γ
GCD7 = β
GCN3 = α
eIF2
eIF2
eIF2
1
4
GCN4
Starvation Conditions
Uncharged tRNA
GCN2
GCN1
GCN20
eIF2·GDP
eIF2
P
eIF2B
eIF2·GTP
tRNA$_i^{Met}$
eIF2·GTP·tRNA$_i^{Met}$
(low levels)
eIF2
eIF2
eIF2
1
4
GCN4
Amino Acid Biosynthetic Enzymes

Figure 1 A model for translational control of yeast *GCN4* by phosphorylation of eIF2α by the protein kinase GCN2. *GCN4* mRNA is shown with uORFs 1 and 4 and the *GCN4*-coding sequences, all indicated as boxes. 40S ribosomal subunits are shaded when they are associated with the ternary complex composed of eIF2, GTP, and Met-$tRNA_i$ and are thus competent to reinitiate translation; unshaded 40S subunits lack the ternary complex and therefore cannot reinitiate. The α, β, and γ subunits of eIF2 in yeast are encoded by *SUI2*, *SUI3*, and *GCD11*, respectively. The subunits of eIF2B are encoded by *GCD6*, *GCD2*, *GCD1*, *GCD7*, and *GCN3*, as shown. Under nonstarvation conditions, eIF2•GDP is readily recycled to eIF2•GTP by eIF2B, leading to high levels of eIF2•GTP and ternary complex formation. The ternary complexes thus formed reassemble with 40S ribosomes scanning downstream from uORF1, causing reinitiation to occur at uORF4. Under starvation conditions, uncharged tRNA accumulates and stimulates the activity of the protein kinase GCN2. GCN2 phosphorylates eIF2α and the phosphorylated eIF2 inhibits eIF2B, reducing the recycling of eIF2•GDP to eIF2•GTP. The resulting low level of eIF2•GTP and ternary complex formation diminishes the rate at which initiation complexes are reassembled on 40S subunits following translation of uORF1. Consequently, many 40S subunits scanning downstream from uORF1 are not competent to initiate at uORF4; these subunits acquire ternary complex while scanning the interval between uORF4 and *GCN4* and reinitiate at the *GCN4* start site instead. GCN1 and GCN20 are proteins required for increased eIF2α phosphorylation by GCN2 under conditions of amino acid deprivation (see text for details and references). (Adapted from Abastado et al. [1991a] and Dever et al. [1992].)

uORFs 2–4 and reinitiate further downstream at *GCN4* instead (Abastado et al. 1991b; Dever et al. 1992).

The remainder of this chapter reviews the evidence supporting the model shown in Figure 1, beginning with the genetic results showing that *GCN4* expression is regulated by a translational mechanism involving the uORFs and a hierarchy of positive and negative *trans*-acting factors. Following this is a summary of the experiments that identified the distinct regulatory functions of each uORF and those that established a reinitiation mechanism for *GCN4* translation. We then consider the evidence that phosphorylation of eIF2 by GCN2 is critical for the induction of *GCN4* translation and that the phosphorylated form of eIF2 regulates translation in yeast by inhibiting the recycling factor eIF2B. Next will follow a discussion of the mechanism for activation of the kinase GCN2 in nutrient-deprived cells, and finally, the identification of a GCN2-independent pathway for inducing *GCN4* translation.

A HIERARCHY OF *TRANS*-ACTING FACTORS REGULATES *GCN4* TRANSLATION BY MODULATING THE INHIBITORY EFFECTS OF THE uORFs ACCORDING TO THE AVAILABILITY OF AMINO ACIDS

Mutations have been isolated in numerous unlinked genes that alter the expression of *GCN4* at the translational level, with corresponding effects on the 40 or more amino acid biosynthetic genes whose transcription depends on GCN4. On the basis of their phenotypes, these mutations fall into two classes. A recessive mutation in *GCN1*, *GCN2*, or *GCN3* prevents the induction of *GCN4* translation that normally occurs in wild-type strains in response to amino acid deprivation (Table 1). This Gcn$^-$ (*g*eneral *c*ontrol *n*on-derepressible) phenotype indicates that the *GCN* products are positive regulators of *GCN4* expression (for review, see Hinnebusch 1988, 1992). Because of their inability to derepress *GCN4*, the *gcn* mutants are more sensitive than wild-type strains to various compounds that inhibit amino acid biosynthesis, such as 3-aminotriazole.

A recessive mutation in any one of the genes *GCD1*, *GCD2*, *GCD5*, *GCD6*, *GCD7*, *GCD10*, *GCD11*, *GCD13*, *SUI2*, or *SUI3* results in constitutive high-level expression of *GCN4* and its target genes (Table 1) (Hinnebusch 1988, 1992). This Gcd$^-$ (*g*eneral *c*ontrol *d*erepressed) phenotype indicates that the *GCD* and *SUI* gene products are negative effectors of *GCN4* expression. Because they produce high levels of GCN4 protein constitutively, *gcd* mutants are more resistant than wild-type strains to toxic amino acid analogs that become incorporated into proteins during translation, e.g., 5-fluorotryptophan. All of the *GCD* and

SUI genes listed in Table 1 encode essential components of the translational machinery; consequently, the recessive reduced-function mutations in these genes lead to slow growth or temperature-sensitive growth on rich medium in addition to affecting *GCN4* expression. The observation that *gcn gcd* double mutants generally exhibit the constitutively derepressed phenotype of *gcd* single mutants led to the idea that GCN1, GCN2, and GCN3 stimulate *GCN4* expression by antagonizing one or more negatively acting *GCD* or *SUI* gene products (Hinnebusch 1988, 1992). As described in detail below, this interpretation of the genetic data has been borne out for each of these GCN factors.

Each of the positive and negative regulators listed in Table 1 regulate *GCN4* expression at the translational level via the four uORFs in the leader of *GCN4* mRNA. An early indication that these factors function at the translational level was that the effects of various *gcn* and *gcd* mutations on expression of a GCN4-LacZ fusion protein could not be accounted for by changes in the steady-state level of the corresponding *GCN4-lacZ* mRNA (Hinnebusch 1984, 1985). A second important finding was that regulation of *GCN4* expression by the *trans*-acting factors remained intact when the *GCN4* promoter was exchanged with that of the galactose-inducible *GAL1* gene (Hinnebusch 1985). The third critical piece of evidence was that mutations in the *trans*-acting factors had little or no effect on expression of a *GCN4-lacZ* reporter lacking the four uORFs in the *GCN4* leader. Eliminating all four uORFs by a deletion (Hinnebusch 1984; Thireos et al. 1984) or by point mutations in the four ATG start codons (Mueller and Hinnebusch 1986) results in high-level *GCN4* expression under both repressing and derepressing conditions (Fig. 2A). These mutations do not alter the 5′ end or abundance of *GCN4* mRNA, implying that the uORFs inhibit *GCN4* expression at the translational level, presumably by restricting the flow of scanning ribosomes from the 5′ end of the transcript to the *GCN4* start codon. The mutations in the *GCN*, *GCD*, and *SUI* genes that impair regulation of wild-type *GCN4* have little or no effect on expression of this and other uORF-less *GCN4-lacZ* constructs (Hinnebusch 1985; Mueller and Hinnebusch 1986; Mueller et al. 1987; Williams et al. 1989; Bushman et al. 1993a; Marton et al. 1993), strongly suggesting that these *trans*-acting factors regulate *GCN4* translation through the uORFs. In this view, the *GCD* and *SUI* products would be required for the inhibitory effects of the uORFs on *GCN4* translation, whereas GCN1, GCN2, and GCN3 would act to overcome the translational barrier to *GCN4* translation imposed by the uORFs. These conclusions were supported by the fact that insertion of various segments containing the four uORFs into the leader of a *GAL1*-

Table 1 Genetic properties of *trans*-acting regulators of *GCN4* translation

Gene	Product	M_r (kD)[1]	Allele: current name	Allele: previous name	Nature of mutation[2]	Phenotype[3]: *GCN4* expression	Phenotype[3]: cell growth
GCD1	γ-subunit, eIF2B (1,2)	66 (3,1)	*gcd1-101*	*tra3-1* (4)		Cdr (5)	Slg⁻, Tsm⁻ (4)
			*gcd1*Δ (3)		deletion		lethal (3)
			gcd1-2 (5a)	*cdr2-1* (6)			Slg⁻ (6)
GCD2	δ-subunit, eIF2B (1,2)	71 (7)	*gcd2-1*	*cdr1-1* (6)	Δ 631–651 (8)	Cdr (9)	Slg⁻, Tsm⁻ (7)
			gcd2-503	*gcd12-503* (10)	R334S (8)	Cdr (10)	Slg⁻, Tsm⁻ (10)
			*gcd2*Δ (11)		deletion		lethal (11)
GCD5/ *KRS1*	lysyl-tRNA synthetase (13)	68 (14)	*gcd5-1* (15)		R488 (13)	Cdr (13)	Slg⁻ (15)
			*gcd5*Δ (13a)		deletion		lethal (13a)
GCD6	ε-subunit, eIF2B (16,2,8)	81 (16)	*gcd6-1* (15)			Cdr (16)	Slg⁻, (15,16)
			*gcd6*Δ (16)		deletion		lethal (16)
GCD7	β-subunit, eIF2B (16,2,8)	43 (16)	*gcd7-201*	*gcd3-201* (17)		Cdr (17)	Slg⁻ (17,16)
			*gcd7*Δ (16)		deletion		lethal (16)
			GCD7-I118T (12)		I188T	Ndr (12)	WT (12)
GCD11	γ-subunit, eIF2 (18)	58 (18)	*GCD11-501* (10)			Cdr (10)	Slg⁻ (10)
			*gcd11*Δ (18)		deletion		lethal (18)
SUI2	α-subunit, eIF2 (19)	36 (19)	*sui2-1* (20)		P14S (19)	Cdr (21)	Slg⁻, Tsm⁻ (20)
			SUI2-S51A(22)		S51A (22)	Ndr (22)	WT (22)
			SUI2-L84F (23)		L84F (23)	Ndr (23)	WT (22)
			*sui2*Δ (22)		deletion		lethal (22)

SUI3	β-subunit, eIF2 (24)	32 (24)	*SUI3-2* (20)		S268Y (24)	Cdr (21)	WT (20)
			SUI3-104 (25)		C236L, Δ237-261, C262D (25)	Cdr[a] (25)	lethal (25)
GCN1	positive regulator of Gcn2 (26)	297 (26)	*gcn1-1*	*ndr1-1* (27)		Ndr[a] (27)	WT (27)
			gcn1Δ (26)		deletion	Ndr (26)	WT (26)
GCN2	eIF2α kinase (22)	182 (28)	*gcn2-1*	*ndr2-1* (27)		Ndr (5)	WT (27)
			gcn2-K559R (29)		K559R	Ndr (29)	WT (29)
			gcn2Δ (30)		deletion	Ndr (30,29)	WT (30)
			GCN2[c]-515 (22)		E1537G, E532K (22)	Cdr (22)	Slg (22)
			GCN2-ar[2] (31)		E1456K (31)	Ndr[b] (31)	Slg^-, Tsm^- (31)
GCN3	α-subunit, eIF2B (1,2,8)	34 (22)	*gcn3-101*	*aas2-1* (4)		Ndr (33)	WT (4)
			gcn3Δ (32)		deletion	Ndr (33)	WT (32)
			gcn3[c]R104K (34)		R104K (34)	Cdr (34)	Slg^- (34)

This is not an exhaustive list. Alleles were included if they were the first mutation described in that gene with a particular phenotype; alleles isolated subsequently with the same phenotype generally were not listed. A deletion allele was included for each gene whenever available.

[1]The molecular weights of the protein products are deduced from the DNA sequences of the corresponding genes.

[2]Amino acid substitutions are indicated by listing the wild-type residue (in single-letter code), the position in the protein, and the substituting amino acid in that order;

[3](Cdr) Constitutively derepressed *GCN4-lacZ* expression; Gcd^- phenotype; (Ndr) nonderepressible *GCN4-lacZ* expression; Gcn^- phenotype.; (Slg^-) slow growth on rich medium; (Tsm^-) temperature-sensitive growth on rich medium.

[a]This phenotype is deduced from the effect of the mutation on one or more amino acid biosynthetic genes under the control of Gcn4.

[b]This phenotype seems to result from the extreme inhibition of translation initiation that occurs in this strain.

References: (1) Cigan et al. 1991; (2) Cigan et al. 1993; (3) Hill and Struhl 1988; (4) Wolfner et al. 1975; (5) Hinnebusch 1985; (5a) J.L. Bushman and A.G. Hinnebusch, unpubl.; (6) Miozzari et al. 1978; (7) Paddon et al. 1989; (8) Bushman et al. 1993b; (9) Williams et al. 1989; (10) Harashima and Hinnebusch 1986; (11) Paddon and Hinnebusch 1989; (12) Vazquez de Aldana and Hinnebusch 1994; (13) Lanker et al. 1992; (13a) Martinez et al. 1991; (14) Mirande and Waller 1988; (15) Niederberger et al. 1986; (16) Bushman et al. 1993a; (17) Myers et al. 1986; (18) Hannig et al. 1992; (19) Cigan et al. 1989; (20) Castilho-Valavicius et al. 1990; (21) Williams et al. 1989; (22) Dever et al. 1992; (23) Vazquez de Aldana et al. 1993; (24) Donahue et al. 1988; (25) Castilho-Valavicius et al. 1992; (26) Marton et al. 1993; (27) Schurch et al. 1974; (28) Wek et al. 1989; (29) Wek et al. 1990; (30) Roussou et al. 1988; (31) Diallinas and Thireos 1994; (32) Hannig and Hinnebusch 1988; (33) Harashima et al. 1987; (34) Hannig et al. 1990.

Figure 2 Effects of selected mutations in the *GCN4* mRNA leader on expression of β-galactosidase from a *GCN4-lacZ* fusion. The schematics on the left depict the *GCN4* leader sequences with the uORFs (boxes labeled 1–4) and the beginning of *GCN4*-coding sequences (box labeled *GCN4*) in the various constructs. An "X" replacing a box indicates point mutations in the uORF start codons. The nucleotide positions of the start codons for uORFs 1 and 4 and GCN4 are indicated above the first construct shown in panel *A*. In panel *B*, uORF5 in construct 6 is a 43-codon uORF created by inserting an ATG codon at a position approximately 20 nucleotides 3′ to uORF4 in a construct lacking uORFs 2–4. The P uORF in construct 8 is a synthetic uORF inserted into a construct lacking the segment of the leader containing uORFs 1 and 2. The P′ element in construct 9 contains a point mutation in the ATG start codon of the P uORF. The wavy lines indicate the locations of deletion junctions. Construct 11 in panel *C* contains a deletion of all but 32 nucleotides normally found between uORF1 and *GCN4* (pG67; Grant et al. 1994). In construct 12, the distance between uORF1 and *GCN4* is 140 nucleotides (pM199; Grant et al. 1994). The 1-el and 4-el constructs in panels *C* and *E* are elongated 93-codon versions of uORFs 1 and 4, respectively, that overlap the beginning of *GCN4* by 130 nucleotides; they were produced by removing the uORF stop codon and several additional stop codons located downstream. For the constructs in panel *D*, the 16 nucleotides normally located upstream, and the 25 nucleotides normally located downstream from uORF1 or uORF4, respectively, are indicated by rectangles attached to the uORFs. (*Open rectangles*) Sequences derived from uORF1; (*closed rectangles*) sequences derived from uORF4. In the case of the 1/4 construct (no. 15), uORF1 and the 25 nucleotides normally 3′ to it are replaced by the corresponding sequences found at uORF4; for the 4/1 construct (no. 16), only the 16 nucleotides 5′ to uORF1 have been replaced with the corresponding sequence found at uORF4. Construct 17 in panel *E* is identical to construct 5 in panel *A*. Construct 18 contains an insertion capable of forming a secondary structure with an 8-bp stem and 5-nucleotide loop (pA46; Abastado et al. 1991b) Constructs 20 and 21 contain insertions of 146 and 144 nucleotides at the indicated positions. Construct 22 has a deletion of 104 nucleotides between uORF4 and *GCN4* (construct B/X; Williams et al. 1988). uORF8 in construct 23 consists of the first 11 codons of *GCN4*, followed by a stop codon, inserted at the exact position of the *GCN4* start codon; uORF8 is followed by the 11 nucleotides normally found 5′ to *GCN4* and then the wild-type *GCN4*-coding region (pG17; Grant et al. 1994). In all constructs, *GCN4*-coding sequences are fused to *lacZ* sequences at codon 55 of *GCN4* (Hinnebusch 1984). All of the constructs were assayed by measuring β-galactosidase activity in the constitutively repressed *gcn2-1* mutant strain H15 (column labeled R) and the constitutively derepressed *gcd1-101* strain F98 (column labeled DR) after growth in minimal medium. The data were taken from the following references indicated to the right of each construct: (1) Mueller and Hinnebusch 1986; (2) Abastado et al. 1991b; (3) Hinnebusch et al. 1988; (4) Mueller et al. 1988; (5) Williams et al. 1988; (6) Grant et al. 1994; (7) Miller and Hinnebusch 1989.

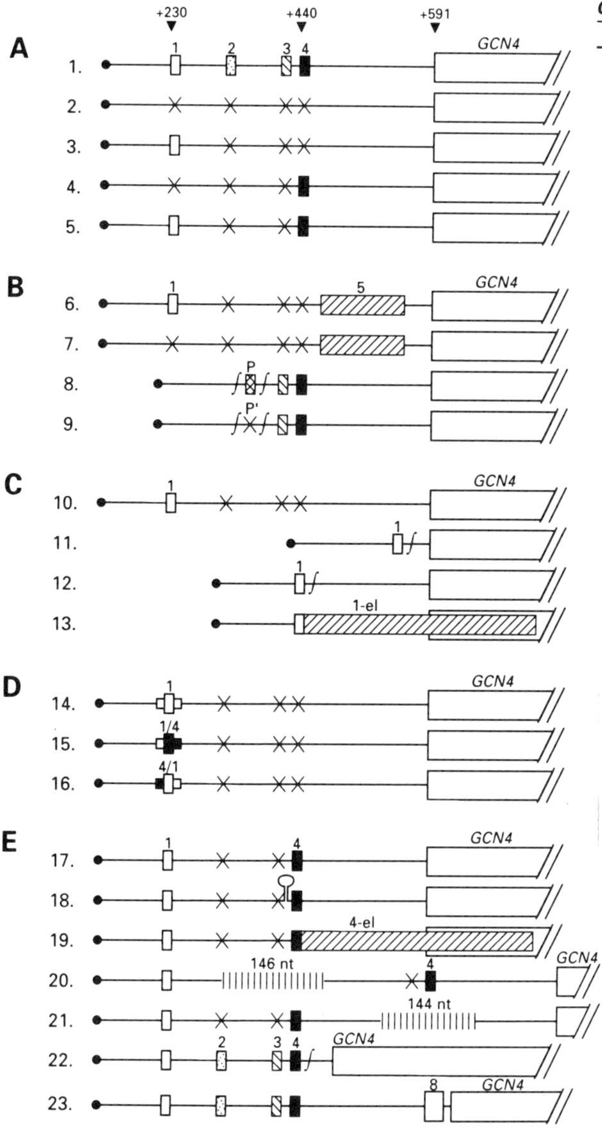

	***GCN4-lacZ* Expression (U)**		
	R	**DR**	**Ref.**
A 1.	6	310	1
2.	1300	1100	1
3.	710	690	1,2
4.	11	21	1,2
5.	16	390	1,3
B 6.	40	220	4
7.	41	65	4
8.	6	58	5
9.	4	10	5
C 10.	570	680	1,6
11.	55	90	6
12.	340	500	6
13.	12	23	6
D 14.	710	690	7
15.	73	83	7
16.	640	780	7
E 17.	16	390	1,3
18.	5	12	2
19.	22	280	2
20.	28	67	2
21.	32	370	2
22.	2	61	5
23.	3	6	6

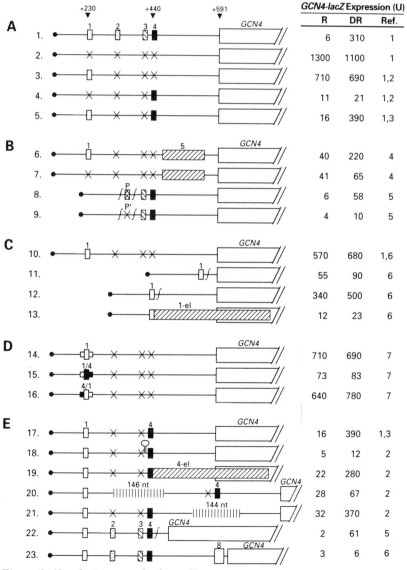

Figure 2 (*See facing page for legend.*)

lacZ transcript caused expression of a GAL1-LacZ fusion protein to be regulated by GCN2, GCN3, and GCD1 in much the same fashion observed for the *GCN4-lacZ* fusion bearing the uORFs (Mueller et al. 1987).

The individual contributions of the uORFs to the translational control of *GCN4* expression were evaluated by making different combinations of point mutations in the ATG start codons of the uORFs. When present alone in the *GCN4* mRNA leader, uORFs 1, 3, and 4 each represses *GCN4* translation; however, solitary uORF1 lowers *GCN4* expression by only approximately 50%, whereas the presence of uORF3 or uORF4 alone reduces *GCN4* translation to only a few percent of that seen in the absence of all four uORFs. In addition, removal of uORF1 in the presence of uORF 3 or 4 leads to a reduction, not an increase, in *GCN4* expression (Fig. 2A) (Mueller and Hinnebusch 1986; Tzamarias et al. 1986). Thus, in wild-type *GCN4*, uORF1 is a positive control element. Because it stimulates *GCN4* expression only in the presence of uORFs 3 and 4, it appears that uORF1 acts positively by allowing ribosomes to overcome the strong translational barrier imposed by uORFs 3 and 4. A *GCN4* allele containing only uORFs 1 and 4 shows nearly wild-type regulation, indicating that uORFs 3 and 4 are redundant negative elements (Fig. 2A) (Mueller and Hinnebusch 1986).

The distinctive functions of uORF1 and uORF4 in translational control appear to be completely unrelated to the small peptides they are predicted to encode. Numerous missense mutations in uORF1 have no effect on *GCN4* expression (Miller and Hinnebusch 1989; Grant and Hinnebusch 1994). In addition, it was shown that the authentic uORFs could be replaced with heterologous small uORFs without destroying *GCN4* translational control (Fig. 2B). The heterologous uORFs did not function as effectively as did the authentic uORF sequences, and conferred derepression ratios that were substantially reduced in comparison to constructs containing wild-type uORFs. Nevertheless, the qualitative hallmarks of the regulatory mechanism were retained by constructs containing heterologous uORFs. These include the requirements for uORF1 and GCN2 for induction of *GCN4* expression in response to starvation and the requirement for GCD1 in repressing *GCN4* under nonstarvation conditions (Fig. 2B) (Mueller et al. 1988; Tzamarias and Thireos 1988; Williams et al. 1988). Although the authentic uORFs could be substituted individually with heterologous uORFs, switching the order of uORFs 1 and 4 (Williams et al. 1988), or substituting uORF1 and its immediately flanking nucleotides with the corresponding sequences from uORF4 (Miller and Hinnebusch 1989), abolished induction of *GCN4* translation in response to starvation. This led to the idea that uORFs 1 and 4 are optimized for opposing roles in the regulatory mechanism and are therefore not functionally interchangeable.

THE REQUIREMENTS FOR EFFICIENT REINITIATION AT *GCN4* FOLLOWING TRANSLATION OF uORF1

A possible explanation for the fact that solitary uORF1 has a relatively small inhibitory effect on *GCN4* translation could be that most ribosomes "leaky scan" past uORF1 and initiate at *GCN4* instead. Inconsistent with this idea was the observation that moving uORF1 to within 50 or 32 nucleotides of *GCN4* decreased *GCN4* translation by about 90% (Fig. 2C) (Grant et al. 1994). If most ribosomes skipped uORF1, then moving it closer to *GCN4* should not drastically reduce *GCN4* translation. Such alterations would be inhibitory, however, if all ribosomes must first translate uORF1 and then reinitiate at *GCN4*, given that the efficiency of reinitiation seems to increase with the distance from an uORF (Kozak 1987). A second indication that all ribosomes translate uORF1 en route to *GCN4* was that elongating uORF1 by fusing it to an extended ORF that overlaps the beginning of *GCN4* by 130 nucleotides abolished *GCN4* expression (Fig. 2C) (Grant et al. 1994). This result suggests that essentially all ribosomes translate uORF1, resume scanning, and translate *GCN4* via reinitiation events. With the construct containing elongated uORF1, ribosomes cannot reinitiate at *GCN4* after terminating uORF1 translation, presumably because they encounter four other AUG codons present in the overlap region while scanning backward to the *GCN4* start site (Grant et al. 1994).

Genetic analysis has shown that the functional differences between uORF1 and uORF4 are determined in large part by the different nucleotide sequences surrounding these two uORFs. The nucleotides immediately flanking the stop codon of uORF1 appear to be critical in determining its modest inhibitory effect on *GCN4* translation when uORF1 is present singly in the leader, and also the ability of uORF1 to stimulate *GCN4* translation in the presence of uORFs 2–4. Replacing the last codon and ten nucleotides immediately 3′ to the uORF1 stop codon with the corresponding nucleotides from uORF4 converted uORF1 into a strong translational barrier and destroyed its ability to stimulate *GCN4* translation when situated upstream of uORF4 (Fig. 2D) (Miller and Hinnebusch 1989). The same results were obtained for mutations in the uORF1 stop codon that cause uORF1 to terminate at various stop codons downstream from uORF1, thus changing the sequence of the uORF1 termination region (Mueller et al. 1988; Miller and Hinnebusch 1989). In contrast, several missense mutations in the second codon of uORF1, or exchanging the 16 nucleotides immediately upstream of uORF1 with the sequence present at uORF4, had no effect on *GCN4* expression (Fig. 2D) (Miller and Hinnebusch 1989). These results led to the idea that the high efficiency of reinitiation following uORF1 translation is dependent on particular nucleotides flanking the termination codon at uORF1. In this view, the corresponding sequence at uORF4 is incompatible with the resumption of scanning required for reinitiation downstream; thus, swapping the termination region of uORF1 with that present at uORF4 abolishes reinitiation at *GCN4*. Because this alteration also eliminates the ability of uORF1 to stimulate *GCN4* translation when uORF4 is present downstream, it was proposed that ribosomes must translate uORF1 and resume scanning in order to subsequently traverse uORFs 2–4 and reinitiate at *GCN4* in nutrient-deprived cells (Mueller et al. 1988; Tzamarias and Thireos 1988; Williams et al. 1988). It was suggested that the sequence surrounding the uORF4 stop codon prevents reinitiation by promoting ribosome dissociation from the mRNA following peptide chain release (Miller and Hinnebusch 1989).

An extensive mutational study of the last codon of uORF1 and ten nucleotides 3′ to its stop codon revealed that a large number of AU-rich sequences at these positions can promote high-level reinitiation at *GCN4*. Twenty-two different A+U-rich triplets were found to confer efficient reinitiation at *GCN4* even when the uORF4 termination sequence was present immediately 3′ to the uORF1 stop codon. Thus, these 22 triplets promote reinitiation more effectively than does the third codon of native uORF1 (TGC). Numerous triplets of a higher G+C content were examined that encode the same amino acids, and in some cases are decoded by the same tRNAs, as one of the 22 permissive A+U-rich triplets. These G+C-rich triplets conferred lower levels of reinitiation than did the corresponding A+U-rich triplets (Grant and Hinnebusch 1994). It was also discovered that a large number of A+U-rich sequences introduced at the ten nucleotides immediately 3′ to uORF1 were compatible with efficient reinitiation. By replacing blocks of three to four nucleotides in the uORF4 termination region with a consensus A+U-rich sequence, it was found that the three bases immediately 3′ to the uORF4 stop codon (CGG) have the greatest negative effect on reinitiation at *GCN4* (Grant and Hinnebusch 1994).

The fact that numerous A+U-rich sequences surrounding the uORF1 stop codon can promote reinitiation dowstream, whereas G+C-rich sequences appear to be incompatible with reinitiation, led to the suggestion that stable RNA-RNA interactions involving the mRNA sequences immediately surrounding the stop codon of an uORF promote ribosome dissociation from the mRNA following peptide chain release. One possibility is that stable base-pairing between the sequences surrounding the uORF stop codon and sites on the rRNA lengthens the time spent by the ribosome in the termination region, and thus increases the probability of binding a factor analogous to *Escherichia coli* RF4 (formerly RRF) that catalyzes ribosome release from deacylated tRNA and mRNA following peptide termination (Ryoji et al. 1981; Ichikawa and Kaji 1989). In contrast, A+U-rich sequences surrounding the uORF stop codon would allow ribosomes to resume scanning more rapidly and exit the termination region before an RF4-like factor can bind (Grant and Hinnebusch 1994).

TRANSLATION INITIATION AT *GCN4* IS STIMULATED IN NUTRIENT-DEPRIVED CELLS BY SUPPRESSING REINITIATION EVENTS AT uORFs 2–4

Following termination at uORF1, ribosomes reach the *GCN4* AUG codon by scanning past the start sites at uORFs 2, 3, and 4 without initiating translation, rather than by translating these uORFs and reinitiating again at *GCN4*, or by being shunted directly to the *GCN4* start codon. The latter possibility is inconsistent with the observation that small insertions with the potential to form stem-loop structures in the vicinity of uORF4 completely abolish *GCN4* translation (Fig. 2E), strongly suggesting that ribosomes must traverse uORF4 sequences to reach the *GCN4* start codon rather than physically circumventing uORF4 (Abastado et al. 1991b). A ribosome shunting model would also have difficulty explaining why moving uORF1 progressively closer to *GCN4* leads to a continuous decline in *GCN4* translation (Fig. 2C) (Grant et al. 1994). It is much easier to account for this observation by a scanning-reinitiation mechanism in which the probability of reinitiation increases with the distance scanned between uORF1 and *GCN4*.

An important result indicating that ribosomes ignore the start sites at uORFs 2–4 en route to *GCN4* was that mutations which remove the uORF4 stop codon and lengthen uORF4 extensively so that it overlaps the beginning of *GCN4* by 130 nucleotides have no significant effect on *GCN4* expression (Fig. 2E) (Abastado et al. 1991b). If ribosomes had to translate uORF4 before they could reinitiate at *GCN4*, they would in the case of the elongated-uORF4 construct have to scan upstream 130 nucleotides and ignore four other AUG codons in the region where uORF4 and *GCN4* overlap before reaching the *GCN4* start codon. A much simpler explanation is that the ribosomes which initiate at GCN4 have not previously translated uORF4; thus, the location of the uORF4 stop codon is of little consequence (Abastado et al. 1991b).

The idea that ribosomes scan past uORFs 2–4 in order to reinitiate at *GCN4* gained support from measurements of the initiation rate at an uORF4-*lacZ* fusion that was constructed by altering the elongated uORF4 sequence just described to place it in the same reading frame as the *GCN4-lacZ*-coding sequence. The resulting uORF4-*lacZ* fusion showed a pattern of translation dramatically different from that of *GCN4-lacZ* fusion. The latter is translated at low levels in a constitutively repressed *gcn2* mutant and at 10–50-fold higher levels in a constitutively derepressed *gcd1* strain. In contrast, translation of the uORF4-*lacZ* fusion occurred at high levels in the *gcn2* mutant, presumably reflecting efficient reinitiation at uORF4, and at 50% lower levels in the *gcd1* strain, indicative of reduced reinitiation at uORF4. The absolute amount by which uORF4-*lacZ* translation was decreased in the *gcd1* strain versus the *gcn2* mutant was approximately the same as the amount that *GCN4-lacZ* translation increased under the same conditions. Moreover, comparing *GCN4-lacZ* expression in the presence and absence of uORF4 leads to the prediction that only about 50% of the ribosomes that translate uORF1 and resume scanning are able to bypass uORF4 and translate *GCN4* in the *gcd1* mutant. Thus, several pieces of evidence indicate that approximately one half of the ribosomes that scan from uORF1 to uORF4 skip over the uORF4 start site and reinitiate at *GCN4* instead under derepressing conditions (Abastado et al. 1991b).

Several additional *lacZ* fusions made to uORF3 and uORF4 did not show the reduction in expression under the derepressing conditions described above for the fusion made to elongated uORF4 (Hinnebusch et al. 1988; Tzamarias and Thireos 1988). In these other fusions, *lacZ* was inserted only a few codons 3′ to the uORF start codons, and it is possible that this greatly increased the secondary structure near the beginning of the uORFs. Such increased structure could restore high-level reinitiation at the uORF3-*lacZ* or uORF4-*lacZ* fusions under derepressing conditions by increasing the time required for ribosomes to scan past the uORF start sites. In the *lacZ* fusion made to elongated uORF4, in contrast, the *lacZ* sequences were inserted approximately 300 nucleotides downstream from the uORF4 start site, and the sequences between the uORF4 AUG codon and the fusion point differ by only a few nucleotides from the wild-type *GCN4* sequence in this extended interval (Abastado et al. 1991b). Consequently, this elongated-uORF4-*lacZ* fusion should not contain any novel secondary structure in the first 300 nucleotides of the uORF4-*lacZ* sequence, making it a more accurate indicator of initiation rates at authentic uORF4.

Another critical piece of evidence that ribosomes reach *GCN4* by scanning past the start sites at uORFs 2–4 was the finding that *GCN4* translation is strongly dependent on the intercistronic distances separating uORF1 from uORF4 and uORF1 from *GCN4*. When the uORF1-uORF4 interval was gradually increased by inserting multiple copies of a spacer sequence originating downstream from uORF4, *GCN4* translation under derepressing conditions was progressively decreased (Abastado et al. 1991b). It is noteworthy that when the uORF1-uORF4 spacing was increased to roughly the same size as the wild-type uORF1-*GCN4* interval, derepression of *GCN4* was substantially reduced (Fig. 2E). Importantly, these insertions did not decrease *GCN4* translation when uORF1 was present alone in the leader, ruling out the possibility that the spacer insertions simply blocked the scanning process. Instead, they seemed to interfere specifically with the ability of ribosomes to bypass uORF4 under derepressing conditions.

These findings can be explained by proposing that the time it takes to scan the uORF1-uORF4 interval is sufficient under repressing conditions to allow the reassembly of competent initiation complexes by essentially all the 40S subunits scanning downstream from uORF1 before they reach uORF4; consequently, these subunits all reinitiate at uORF4 and fail to translate *GCN4*. Under derepressing conditions, only about 50% of the 40S subunits scanning from uORF1 become competent to reinitiate before reaching uORF4; the remaining 50% continue scanning downstream and recover the ability to reinitiate while scanning the uORF4-*GCN4* interval, allowing them to translate *GCN4* (see Fig. 1). When the uORF1-uORF4 interval was expanded to approximately the wild-type distance between uORF1 and *GCN4*, most of the 40S subunits were competent to reinitiate by the time they reached uORF4, even under derepressing conditions; consequently, none could bypass uORF4 and reinitiate downstream at *GCN4*. Thus, translational control of *GCN4* exploits the increased scanning time needed for efficient reinitiation in amino-acid-starved cells versus unstarved cells (Abastado et al. 1991a,b).

It was shown that lengthening the interval between uORF4 and *GCN4* does not decrease *GCN4* translation (Fig. 2E). This result is consistent with the foregoing model, since increasing the uORF4-*GCN4* interval should only increase the probability of reinitiation at *GCN4* by ribosomes that have bypassed uORF4 (Abastado et al. 1991a,b). The fact that *GCN4* translation does not increase substantially when the uORF4-*GCN4* spacing is expanded indicates that the wild-type distance is large enough to ensure that nearly all ribosomes that bypass uORF4 will be competent to reinitiate after scanning the uORF4-*GCN4* interval. As expected, a large deletion between uORF4 and *GCN4* significantly reduced *GCN4* translation (Fig. 2E) (Williams et al. 1988), since it shortens the interval in which ribosomes must recover the ability to reinitiate translation after bypassing uORF4 and before reaching *GCN4*. The idea that the distance from uORF1 is a critical determinant of the efficiency of reinitiation received additional support from the fact that a small heterologous uORF placed midway between uORF4 and *GCN4* appeared to be skipped less often than was uORF4; moreover, this heterologous uORF was not skipped at all when it was placed even further downstream at the normal position of the *GCN4* start codon (Grant et al. 1994). The latter result confirms the notion that virtually all ribosomes that scan past uORF4 under derepressing conditions will be competent to reinitiate by the time they reach *GCN4*.

Analysis of the numerous constructs available in which the distance between uORF1 and uORF4 differs from the wild-type spacing led to the prediction that reinitiation at uORF4 under derepressing conditions increases in roughly linear fashion with the spacing between the two sequences (Abastado et al. 1991a; Grant et al. 1994). Reinitiation at uORF4 is nearly 100% when the uORF1-uORF4 spacing is 150 nucleotides larger than the wild-type distance, 50–70% at the wild-type spacing of 200 nucleotides, and only 20% when the spacing is decreased to 32 nucleotides. Reinitiation at uORF4 is considerably more efficient under repressing conditions for uORF1-uORF4 intervals of 200 nucleotides or less. This difference in the intercistronic length dependence for reinitiation at uORF4 between repressing and derepressing conditions forms the basis for *GCN4* translational control.

A final issue to be addressed regarding the *GCN4* leader sequences is the role of uORFs 2 and 3 in the regulatory mechanism. uORF3 functions similarly to uORF4, although it does not repress *GCN4* translation as effectively as uORF4 does. Consequently, a construct containing uORFs 1 and 3 has a smaller derepression ratio than a construct containing uORFs 1 and 4. uORF2 can partially substitute for uORF1; however, it is less effective than uORF1 in stimulating *GCN4* translation when situated upstream of uORFs 3–4. Constructs lacking uORF3 exhibit slightly higher levels of *GCN4-lacZ* expression under repressing conditions compared to constructs containing uORFs 3 and 4 (Mueller and Hinnebusch 1986). Thus, it seems likely that uORF3 (and perhaps uORF2) is present in the wild-type leader to provide multiple sites for reinitiation and thereby ensure very low levels of *GCN4* translation under repressing conditions. The presence of these two negative elements between uORF1 and uORF4 does not interfere with the efficiency of derepressing *GCN4* because all of those ribosomes that failed to rebind the ternary complex by the time they reached uORF4 would also have lacked this critical initiation factor when they encountered uORFs 2 and 3 further upstream, causing them to be skipped as well as uORF4.

GENETIC EVIDENCE THAT THE PROTEIN KINASE GCN2 STIMULATES *GCN4* TRANSLATION BY DOWN-REGULATING THE RECYCLING OF eIF2 BY eIF2B

The large increase in *GCN4* translation that occurs when yeast cells are starved for an amino acid or for purines is dependent on the protein kinase GCN2 (Hinnebusch 1988; Roussou et al. 1988; Rolfes and Hinnebusch 1993). *GCN2* encodes a 180-kD protein containing a segment of approximately 425 amino acids that exhibits strong similarity with the catalytic subunit of eukaryotic protein kinases (Roussou et al. 1988; Wek et al. 1989). Substitutions of the invariant lysine residue (Lys-559) in the predicted ATP-binding domain of the GCN2 kinase moiety abolish both its ability to stimulate *GCN4* expression in vivo (Wek et al. 1989) and its autokinase activity in vitro (Wek et al. 1990). These findings demonstrated that the protein kinase activity of GCN2 is required for its role as an activator of *GCN4* translation. The only obvious effect of deleting *GCN2* is to impair the derepression of *GCN4* translation that normally occurs in response to nutrient deprivation (Table 1) (Hinnebusch 1988; Roussou et al. 1988); thus, GCN2 appears to be dedicated to the regulation of *GCN4* expression. The sequences of cloned cDNAs encoding the mammalian eIF2α kinases HRI (Chen et al. 1991) and PKR (Meurs et al. 1990) exhibit much greater similarity to the GCN2 sequence (36% and 42%, respectively) than they do to other known protein kinase sequences (Chen et al. 1991; Ramirez et al. 1991, 1992; Chong et al. 1992), suggesting that these three enzymes represent a subfamily of phylogenetically related proteins. This observation and the sequence identity between yeast (Cigan et al. 1989) and human eIF2α (Ernst et al. 1987) in the vicinity of the site of phosphorylation on eIF2α by PKR and HRI (Ser-51) (Colthurst et al. 1987; Pathak et al. 1988) were important clues that GCN2 regulates translation by phosphorylating eIF2α.

Equally important in establishing this conclusion were genetic and biochemical observations indicating that GCN2 modulates *GCN4* translation by down-regulating the activities of eIF2 and a general initiation factor composed of GCD proteins and the positive regulator GCN3. Most mutations in *GCD* genes were identified on the basis of causing constitutive derepression of *GCN4* translation, and this Gcd⁻ phenotype was observed whether or not a functional copy of *GCN2* was present in the strain. The fact that recessive *gcd* mutations can overcome the requirement for GCN2 in attaining high-level *GCN4* expression was interpreted to indicate that GCN2 antagonizes one or more GCD factors as the means of stimulating *GCN4* translation (Hinnebusch 1988). In this view, the *gcd* mutations mimic the negative effect that GCN2 exerts on the corresponding wild-type GCD proteins under amino acid starvation conditions.

All known *gcd* mutations lead to temperature-sensitive growth or unconditional slow growth on nutrient-rich medium. In addition, deletions of *GCD1* and *GCD2* are lethal (Table 1), indicating that these factors are required for an essential function in addition to controlling *GCN4* expression. Given that GCD1 and GCD2 regulate *GCN4* at the translational level, it was suggested (Harashima and Hinnebusch 1986) that these factors have general functions in protein synthesis. In support of this idea, it was found that the in vivo rate of total protein synthesis dropped rapidly when *gcd1* (Wolfner et al. 1975; Hill and Struhl 1988) or *gcd2* (Foiani et al. 1991) mutants were shifted to the restrictive temperature. Analysis of polysomes in these mutants revealed aberrant profiles indicative of a reduced rate of translation initiation (Cigan et al. 1991; Foiani et al. 1991). These findings, together with the genetic relationships between *gcn2* and *gcd* mutations, led to the idea that GCN2 stimulates *GCN4* translation by down-regulating the activity of a general translation initiation factor composed of one or more GCD proteins.

Certain *GCD* genes were identified as being more directly involved in *GCN4* translational control than others, because mutations in the former class showed very specific interactions with mutations affecting the positive regulator *GCN3*. Deletion of *GCN3* impairs derepression of *GCN4* translation in response to amino acid starvation; however, *GCN4* expression is constitutively derepressed in *gcd gcn3* double mutants (Hinnebusch 1988, 1992). Thus, like GCN2, GCN3 appears to function as a positive regulator of *GCN4* by down-regulating one or more GCD factors. Inactivating *GCN3* in otherwise wild-type cells has no effect on growth under nonstarvation conditions (Hannig and Hinnebusch 1988); however, deletion of *GCN3* is lethal in combination with the *gcd1-101* mutation (Hannig and Hinnebusch 1988) and exacerbates the growth defects associated with various other *gcd1* and *gcd2* mutations (Harashima et al. 1987; Paddon and Hinnebusch 1989). In contrast, the *gcn3-102* allele suppresses the growth defect of these same *gcd* alleles (Hinnebusch and Fink 1983; Harashima et al. 1987). Moreover, the growth defects of certain mutations in *GCD11* were suppressed by the *gcn3-101* mutation (Harashima et al. 1987). These genetic interactions suggested that GCN3 is physically associated with GCD1, GCD2, and GCD11. In contrast with the genetic interactions just described, the phenotypes of *gcd10* and *gcd13* mutations were unaffected by the presence or absence of *GCN3* (Harashima et al. 1987), suggesting that GCD10 and GCD13 affect some other aspect of the initiation pathway.

Implicit in the idea that GCN3 directly interacts with certain GCD proteins was the notion that GCN3 functions downstream from GCN2 in the regulatory pathway. This idea received additional support from the finding that a different class of mutations in *GCN3* could be obtained in which *GCN4* is constitutively derepressed (Gcd$^-$ phenotype), in contrast to the uninducible (Gcn$^-$) phenotype of a *GCN3* deletion. Like the *gcd* mutations, many such *gcn3*^{c} alleles lead to high-level *GCN4* expression in the absence of *GCN2* (Table 1) (Hannig et al. 1990). Thus, it appeared that GCN3 could be mutationally altered to antagonize GCD1, GCD2, or GCD11 in a way that substitutes for the regulatory functions of wild-type GCN2 and GCN3 in nutrient-deprived cells. It has also been possible to isolate dominant mutations in *GCN2* that lead to constitutive derepression of *GCN4* translation (a Gcd$^-$ phenotype) (Table 1) (Wek et al. 1990; Ramirez et al. 1992). These *GCN2*^{c} mutations were found to be wholly or partially dependent on *GCN3* function for their derepressed phenotype (Hannig et al. 1990; Ramirez et al. 1992). The simplest explanation for all the available genetic data was that GCN3 is a regulatory subunit of a complex containing GCD1, GCD2, and GCD11, which performs an essential function in translation initiation. Under amino acid starvation conditions, GCN3 would mediate the inhibitory effect of GCN2 on the function of this complex as the means of stimulating translation of *GCN4* mRNA.

A functional link between the proposed GCD/GCN3 complex and eIF2 was first provided by the isolation of mutations with a Gcd$^-$ phenotype in the genes encoding the α and β subunits of eIF2 (*SUI2* and *SUI3*, respectively). Similar to the *gcd* mutations, the *sui2-1* and *SUI3-2* alleles lead to constitutive derepression of *GCN4* expression, independently of GCN2, and only when uORFs 1 and 4 are present together in the *GCN4* mRNA leader (Williams et al. 1989; Castilho-Valavicius et al. 1990). In addition, the *sui2-1* mutation, which causes temperature-sensitive growth in otherwise wild-type strains, is lethal when combined with a deletion of *GCN3* (Williams et al. 1989). Subsequently, it was discovered that *GCD11* encodes the γ-subunit of eIF2 (Hannig et al. 1992). Thus, mutations in all three subunits of eIF2 were identified whose deleterious effects on cell growth were exacerbated by inactivation of *GCN3*. The *sui2-1* and *SUI3-2* mutations reduce the ability of eIF2 to form ternary complexes in vitro (Donahue et al. 1988; Cigan et al. 1989). These results placed *SUI2* and *SUI3* together with the *GCD* genes in the *GCN4* regulatory pathway (see Fig. 1) and led to the suggestion that a reduction in eIF2 function mediated by GCN2 and GCN3 is responsible for stimulating *GCN4* translation in amino-acid-starved cells (Williams et al. 1989). In accord with these genetic observations, it was shown that binding of Met-$tRNA_i$ to 40S subunits, a function mediated by eIF2 in mammalian cells, was reduced in yeast cells in response to several conditions in which *GCN4* translation is derepressed, including a *gcd1* mutation, overexpression of GCN2, and a nutritional shift-down from amino-acid-rich medium to minimal medium (Tzamarias et al. 1989). These biochemical data suggested that reducing ternary complex formation is required for increased translation initiation at *GCN4* in response to amino acid starvation and further indicated that GCD1 protein is required for high levels of ternary complex formation in vivo.

The prediction that GCD1, GCD2, GCN3, and the subunits of eIF2 are components of the same protein complex was confirmed by purifying these proteins from yeast. Using antibodies specific for GCD1, GCD2, and GCN3, it was shown that all three proteins copurified in a high-molecular-mass complex of about 600 kD that also contained about 10–20% of the total eIF2α (SUI2) and eIF2β (SUI3) proteins present in the cell. Furthermore, it was demonstrated that the majority of GCD1, GCD2, and GCN3, and a fraction of eIF2α and eIF2β, could be specifically coimmunoprecipitated from a ribosomal salt wash using antibodies against GCD1 or GCD2 (Cigan et al. 1991). Subsequently, it was established that the complex containing GCD1, GCD2, and GCN3 is composed of eight subunits and contains the γ-subunit of eIF2 (GCD11) and the products of *GCD6* and *GCD7* in addition to the α and β subunits of eIF2 (Bushman et al. 1993a,b; Cigan et al. 1993).

The physical properties of the GCD/GCN3 complex are very reminiscent of mammalian eIF2B, the guanine nucleotide exchange factor that catalyzes the conversion of eIF2•GDP to eIF•GTP (see Merrick and Hershey, this volume). This initiation factor has five subunits with molecular weights similar to those of GCD1, GCD2, GCD6, GCD7, and GCN3 (Table 1) (Konieczny and Safer 1983; Bushman et al. 1993a) and copurifies with a fraction of the eIF2 present in mammalian cells. In addition, the largest subunit of the yeast complex encoded by *GCD6* is 30% identical in sequence to the largest (ε) subunit of rabbit eIF2B (Bushman et al. 1993a). Similarly, GCD2 is 36% identical to the δ-subunit of eIF2B from rabbit (Price et al. 1994) and mouse (Henderson et al. 1994), and GCN3 is 42% identical to the α-subunit of rat eIF2B (Flowers et al. 1995). Direct evidence that the GCD/GCN3 complex is the yeast equivalent of eIF2B was provided by demonstrating that a purified preparation of the complex could catalyze release of radioactively labeled GDP from a purified eIF2•GDP binary complex in the presence of excess GDP or GTP but not in the presence of ADP or ATP. In addition, the purified GCD/GCN3 complex could reverse the inhibitory effect of GDP on the formation of eIF2•GTP•Met-$tRNA_i$ ternary complexes. Finally, the GDP-GTP exchange activity was shown to cosediment and coimmunoprecipitate with the subunits of the GCD/GCN3/eIF2 complex (Cigan et al. 1993). The realization that the GCD/GCN3/eIF2 complex was the yeast equivalent of eIF2B•eIF2 provided a simple reason for why mutations in any of the three subunits of eIF2 or the GCD factors comprising eIF2B would have a derepressing effect on *GCN4* translation, as all would be expected to decrease the level of eIF2 activity in the cell.

GCN2 STIMULATES *GCN4* TRANSLATION BY PHOSPHORYLATING THE α-SUBUNIT OF eIF2 ON SER-51

The realization that the GCD/GCN3 complex is the yeast equivalent of eIF2B gave strong impetus to the idea that GCN2 stimulates *GCN4* translation by phosphorylating eIF2, as the ability of eIF2B to recycle eIF2 is impaired in mammalian cells by phosphorylation of the α-subunit of eIF2 on Ser-51. To examine this possibility directly, the in vivo phosphorylation state of yeast eIF2α was measured using isoelectric focusing gel electrophoresis to separate differentially phosphorylated forms of the protein. These studies revealed that a hyperphosphorylated isoform of eIF2α increased in abundance when cells were starved for histidine and that its presence was completely dependent on GCN2, being undetectable in a *gcn2Δ* strain. In addition, the hyperphosphorylated form of eIF2α was found to be predominant in the absence of amino acid starvation in strains containing constitutively activated *GCN2*^{c} alleles. Substitution of Ser-51 with alanine in the eIF2α gene (the *SUI2-S51A* allele) completely eliminated the hyperphosphorylated form of the protein in *GCN2* cells under starvation conditions and in derepressed *GCN2*^{c} mutants. In addition to these in vivo results, it was shown that GCN2 protein isolated in immune complexes specifically phosphorylated the α-subunit of eIF2 purified from rabbit or yeast and did not phosphorylate the mutant form of yeast eIF2α containing the Ala-51 substitution (Dever et al. 1992). These findings strongly indicated that GCN2 phosphorylates eIF2α on

Ser-51 in vivo and that the level of eIF2α phosphorylation increases in response to amino acid starvation.

The Ala-51 substitution in eIF2α that prevents phosphorylation by GCN2 also impairs derepression of *GCN4* under starvation conditions to the same extent that occurs when *GCN2* is deleted. As expected, this eIF2α mutation also reverses the derepressing effect of a *GCN2*^{c} mutation (Fig. 3A). Conversely, substitution of Ser-51 with an aspartate residue, which resembles phosphoserine, leads to partial derepression of *GCN4* translation in the absence of amino acid starvation or GCN2 function (Dever et al. 1992). These results provide strong evidence that GCN2-catalyzed phosphorylation of eIF2α on Ser-51 is required for increased *GCN4* expression in response to amino acid starvation. Additional evidence supporting this conclusion was provided by the fact that expression of the mammalian eIF2α kinases PKR and HRI in yeast leads to induction of *GCN4* translation in the absence of *GCN2*. Importantly, expression of these mammalian enzymes has no effect on *GCN4* expression in strains harboring the mutant form of eIF2α with alanine at position 51 (Dever et al. 1993).

The mutational analysis of the uORFs in *GCN4* mRNA led to the conclusion that derepression of *GCN4* occurs when ribosomes scanning downstream from uORF1 cannot reassemble an initiation complex rapidly enough to ensure reinitiation at uORFs 2–4. The genetic and biochemical analyses of GCN2, eIF2, and eIF2B in yeast led to the prediction that the eIF2•GTP•Met-tRNA$_{i}$ ternary complex is the key component of the initiation complex that is rate-limiting for reinitiation at uORFs 2–4 in amino-acid-starved cells. These concepts were combined with the biochemical mechanism for the inhibition of translation initiation by phosphorylated eIF2 elucidated in mammalian systems to produce the model for *GCN4* translational control shown in Figure 1.

One of the interesting features of the *GCN4* translational control mechanism is that the level of eIF2α phosphorylation achieved under amino acid starvation conditions is sufficient to induce *GCN4* translation by 10–50-fold without substantially affecting the rate of initiation at uORF1 or at the AUG start codons in most other mRNAs. To understand the gene-specific characteristics of this response, it is important to note first that the large induction of *GCN4* translation that occurs under derepressing conditions results from only about a 2-fold decrease in reinitiation efficiency at uORFs 2–4. Because only about 1% of the ribosomes bypass uORFs 2–4 and reach *GCN4* under nonstarvation conditions, when the probability of bypass increases to 50% under starvation conditions, a 50-fold increase is produced in *GCN4* translation.

A.

	***GCN4-lacZ* Expression (U)**			
	GCN2		*GCN2*^{C}	
Genotype	**R**	**DR**	**R**	**DR**
WT	16	100	140	170
SUI2-S51A	8	20	7	17
SUI2-L84F	12	30	19	32
GCD7-I118T,D178Y	13	40	16	45

B.

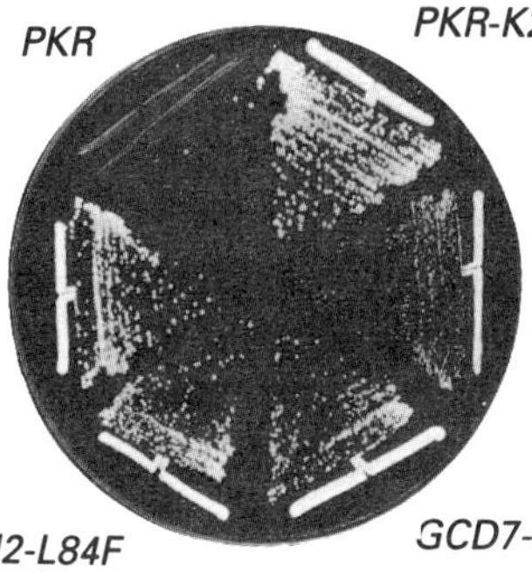

Figure 3 Regulatory mutations in eIF2α and the GCN3 (α) and GCD7 (β) subunits of eIF2B that overcome the in vivo effects of eIF2 hyperphosphorylation. (*A*) *GCN4-lacZ* expression (in units of β-galactosidase activity) measured in strains with the indicated alterations in eIF2α or GCD7 (listed at the left) in strains containing either *GCN2* or a constitutively activated *GCN2*^{c} allele under nonstarvation conditions (R, repressing) or histidine starvation conditions (DR, derepressing). An isogenic *gcn2Δ* strain that is otherwise wild-type shows *GCN4-lacZ* expression of 10 and 34 units under repressing and derepressing conditions, respectively. (*B*) Yeast strains containing the indicated *GCD7*, *SUI2*, or *GCN3* mutations and also bearing the cDNA for wild-type human PKR or the catalytically inactive mutant K296R (*PKR-K296R*) under the control of a galactose-inducible promoter were streaked on medium containing galactose to induce expression of the PKR constructs and incubated for 10 days. (Data taken from Vazquez de Aldana and Hinnebusch 1994.)

A second important consideration is that reinitiation at uORFs 2–4 has kinetic constraints that probably do not apply to primary initiation events occurring at the first (5′-proximal) AUG codons on most other mRNAs. It is thought that ribosomes generally bind the ternary complex before interacting with mRNA (see Merrick and Hershey, this volume).

Thus, a reduction in the level of ternary complexes may decrease the pool of 43S preinitiation complexes that are competent to bind at the capped ends of mRNAs and initiate the scanning process; however, it should not cause leaky scanning past the first AUG codons on these mRNAs if ribosomes bind the ternary complex before interacting with the transcript. In contrast, a reduction in the rate of ternary complex binding to ribosomes should lead to considerable leaky scanning at uORFs 2–4 (with attendant stimulation of *GCN4* translation), because ribosomes must rebind the ternary complex in the time it takes to scan from uORF1 to the start sites at these uORFs. Apart from the unique kinetic constraints on reinitiation, it is possible that after translating uORF1, ribosomes lack an initiation factor that normally facilitates binding of the ternary complex (e.g., eIF3; see Merrick and Hershey, this volume). Consequently, ternary complexes may rebind less efficiently to ribosomes scanning on *GCN4* mRNA downstream from uORF1 than they would to free 43S complexes.

GENETIC ANALYSIS OF THE MECHANISM OF eIF2B INHIBITION BY PHOSPHORYLATED eIF2

Reduced-function mutations in the subunits of yeast eIF2B were first identified by their effects on *GCN4* translation. This was a strong indication that phosphorylation of eIF2 stimulates *GCN4* translation by down-regulating eIF2B. In support of this interpretation, deletion of *GCN3* encoding the α-subunit of eIF2B greatly diminishes *GCN4* translation in response to eIF2 phosphorylation by GCN2 (Hinnebusch 1988; Dever et al. 1993). This last finding also indicated that GCN3 functions primarily as a regulatory subunit of eIF2B, mediating the inhibitory effect of eIF2(αP) on eIF2B activity. Genetic evidence additionally implicating the β and δ subunits of eIF2B (GCD7 and GCD2, respectively) in this regulatory function was provided by the isolation of mutations affecting these two proteins that uncouple *GCN4* translation from phosphorylation of eIF2. These mutations were isolated as suppressors of a hyperactivated *GCN2*^{c} allele that produces higher levels of eIF2α phosphorylation than occurs when wild-type GCN2 is activated in amino-acid-starved cells. Thus, this *GCN2*^{c} allele reduces the rate of general translation initiation and thereby inhibits cell growth (Ramirez et al. 1992) in addition to causing derepression of *GCN4* in the absence of amino acid starvation. As would be expected, both phenotypes are completely eliminated by the Ala-51 substitution in eIF2α (Dever et al. 1992; Ramirez et al. 1992). Spontaneous revertants of the slow-growth phenotype of

GCN2^{c} alleles were found to contain single-amino-acid substitutions in eIF2α at positions other than Ser-51, or in the eIF2B subunits encoded by *GCN3*, *GCD7*, or *GCD2*. Mutations in *GCN3* were expected to arise in this genetic selection based on the previous finding that the slow-growth phenotype associated with *GCN2*^{c} alleles (Dever et al. 1992; Ramirez et al. 1992) or resulting from expression of PKR or HRI in yeast (Dever et al. 1993) was either abolished or greatly diminished by mutations in *GCN3*. Four suppressors mapping in *GCD7* and one mapping in *GCD2* were found to have the same phenotype as a deletion of *GCN3*, impairing derepression of *GCN4* translation in response to eIF2α phosphorylation by GCN2 (Gcn^{-} phenotype).

In otherwise wild-type strains, the *GCD7* and *GCD2* suppressor mutations have no effect on cell growth in rich medium where *GCN4* expression is repressed. This indicates that these mutations have little or no effect on the catalytic function of eIF2B and act primarily to make eIF2B insensitive to inhibition by eIF2(αP). It is also noteworthy that an allele of *GCD7* constructed by combining amino acid substitutions found in two *GCD7* suppressors (*GCD7-I118T, D178Y*) exceeds a deletion of *GCN3* in making eIF2B insensitive to eIF2(αP) (Fig. 3B) (Vazquez de Aldana and Hinnebusch 1994). Even high-level induction of PKR, which is normally lethal, has no detectable effect on the cellular growth rate in strains bearing the *GCD7-I118T, D178Y* allele. These last results indicate that GCD7 and GCN3 each make critical contributions to the inhibitory effects of eIF2(αP) on eIF2B function and eliminate the possibility that the *GCD7* suppressor mutations simply cause GCN3 to be excluded from the eIF2B complex.

The suppressors of *GCN2*^{c} alleles that alter eIF2α (exemplified by *SUI2-L84F*, Fig. 3) all map in the amino-terminal one third of the protein within 38 residues of the phosphorylation site; however, only one of the four alleles seems to reduce the level of eIF2α phosphorylation. In fact, the remaining four exhibit higher levels of eIF2(αP) compared to the parental strain (Vazquez de Aldana et al. 1993). This latter phenomenon has also been described for suppressors of *GCN2*^{c} alleles mapping in *GCN3* (Dever et al. 1993), *GCD7*, and *GCD2* (Vazquez de Aldana and Hinnebusch 1994) and probably reflects a negative feedback loop involving eIF2α kinases and eIF2(αP). Thus, these mutations render eIF2(αP) impotent as an inhibitor of eIF2B, indicating that specific amino acids in the vicinity of the phosphorylation site on eIF2α contribute to the inhibitory interaction between eIF2(αP) and eIF2B. Ser-48 of eIF2α appears to have a similar role both in yeast (Dever et al. 1992) and in mammalian cells (Choi et al. 1992).

It is interesting that GCD7, GCN3, and the carboxy-terminal half of GCD2 share regions of high sequence similarity (Hinnebusch 1992; Bushman et al. 1993a), particularly at their extreme carboxyl termini. This sequence relatedness is particularly interesting in light of the above described genetic data indicating that all three proteins contribute to the inhibition of eIF2B by eIF2(αP). One attractive possibility is that GCD2, GCD7, and GCN3 are juxtaposed in the eIF2B complex and that their related carboxy-terminal segments form a surface that directly interacts with the α-subunit of eIF2 in the region surrounding Ser-51 (Fig. 4) (Vazquez de Aldana et al. 1993; Hinnebusch 1994; Vazquez de Aldana and Hinnebusch 1994). Likewise, the *SUI2* suppressors may identify surface contacts between eIF2α and the regulatory domain in eIF2B. The effects of the point mutations in GCN3, GCD7, GCD2, and eIF2α that overcome the toxicity of eIF2(αP) could be explained by proposing that they decrease the affinity of eIF2B for eIF2(αP), allowing nucleotide exchange to proceed on the nonphosphorylated fraction of eIF2 (Fig. 4). Alternatively, the mutations may permit eIF2B to catalyze nucleotide exchange on phosphorylated and nonphosphorylated eIF2 with similar efficiencies.

The amino acid substitutions in GCD7 and eIF2α that diminish the effects of eIF2 phosphorylation also suppress a subset of *gcn3*^{c} alleles (Vazquez de Aldana et al. 1993; Vazquez de Aldana and Hinnebusch 1994). As mentioned above, the *gcn3*^{c} mutations mimic the deleterious effects of eIF2 phosphorylation on translation initiation. One way to explain how both *gcn3*^{c} mutations and eIF2α hyperphosphorylation could be suppressed by the same mutations in eIF2α or GCD7 would be to propose that the *gcn3*^{c} mutations decrease the dissociation rate of nonphosphorylated eIF2 from eIF2B and thereby prevent eIF2B from functioning catalytically in the recycling of eIF2. The suppressor mutations affecting eIF2α and GCD7 would weaken interactions between nonphosphorylated eIF2α and mutant eIF2B containing a *gcn3*^{c} subunit, and also between phosphorylated eIF2α and wild-type eIF2B, by eliminating different contacts between GCN3 or GCD7 and eIF2α (Hinnebusch 1994).

REGULATION OF GCN2 KINASE ACTIVITY BY NUTRIENT AVAILABILITY

When GCN2 was replaced by the mammalian eIF2α kinases DAI or HRI, *GCN4* translation was induced independently of amino acid availability (Dever et al. 1993). This implied that the increased phosphorylation of eIF2α that occurs when yeast cells are deprived of amino acids reflects an increase in GCN2 kinase activity rather than inhibition of eIF2α phosphatase activity. Measurements of *GCN2* mRNA and protein levels have shown little or no increase in GCN2 abundance in response to amino acid starvation (Wek et al. 1990), suggesting that the catalytic activity of GCN2, or its access to the substrate eIF2α, is stimulated in response to starvation. In accord with this conclusion, the *GCN2*ᶜ mutations described above derepress *GCN4* translation without increasing the steady-state level of GCN2 protein (Wek et al. 1990). Six of the known *GCN2*ᶜ alleles alter single amino acids in the protein kinase domain, suggesting a direct effect on GCN2 catalytic activity. A large number of *GCN2*ᶜ alleles contain single-amino-acid substitutions in the approximately 670 residues carboxy-terminal to the protein kinase moiety, identifying this region as a regulatory domain (Wek et al. 1990; Ramirez et al. 1992; Diallinas and Thireos 1994). Several two-codon insertions and various in-frame deletions of these carboxy-terminal sequences impair the ability of GCN2 to stimulate *GCN4* translation in vivo (Fig. 5), and in several cases, these mutations were shown not to affect the autophosphorylation activity of GCN2 in vitro (Wek et al. 1990). Thus, the carboxyl terminus of GCN2 appears to be required in vivo to stimulate its protein kinase function in response to amino acid starvation.

It has been proposed that uncharged tRNA is an activating ligand for GCN2 because mutations in two different aminoacyl-tRNA synthetases lead to increased expression of *GCN4* (Lanker et al. 1992), or of amino acid biosynthetic genes under GCN4 control (Messenguy and Delforge 1976; Vazquez de Aldana et al. 1994), without any limitation for the cognate amino acids. In the case of the lysyl-tRNA synthetase encoded by *KRS1*, there is strong genetic evidence that accumulation of uncharged $tRNA^{Lys}$ in mutant strains containing a defective *KRS1* product (*gcd5-1* mutants) leads to translational derepression of *GCN4* that is dependent on GCN2 (Lanker et al. 1992). The idea that uncharged tRNA is an activator of GCN2 is also supported by the finding that 530 residues in the carboxy-terminal regulatory domain of GCN2 show 22% identity and 45% similarity with the entire sequence of yeast histidyl-tRNA synthetase (HisRS) (Wek et al. 1989). Sequence similarities between GCN2 and the *E. coli* and human HisRS sequences are also evident. The portion of HisRS that is most highly conserved in GCN2 corresponds to the core region shared by all class II aminoacyl-tRNA synthetases (Ramirez et al. 1992), containing the binding sites for amino acid, ATP, and the acceptor stem of tRNA (Cusack et al. 1991). Two of the two-codon insertions mentioned above that inactivate the positive regulatory function of GCN2 map in the HisRS-like region of the protein (Wek et al. 1989).

Given that aminoacyl-tRNA synthetases bind uncharged tRNA as a substrate and distinguish between charged and uncharged forms of tRNA (Schimmel and Soll 1979) and that accumulation of uncharged tRNA is thought to trigger the derepression of *GCN4* expression, it was proposed that the HisRS-related domain of GCN2 functions to detect an increase in the level of uncharged tRNA under amino acid starvation conditions. Binding of uncharged tRNA to this domain would produce an allosteric change in the adjacent protein kinase moiety that increases the ability of GCN2 to phosphorylate eIF2α. Because *GCN4* translation increases in response to starvation for numerous amino acids, GCN2 would have

Figure 4 Hypothetical model for inhibition of the guanine nucleotide exchange activity of eIF2B by phosphorylated eIF2. The heterotrimeric eIF2 complex is shown shaded with a binding site for GDP or GTP on the γ-subunit. The five subunits of the eIF2B complex are labeled by their gene designations in yeast. GCN3, GCD7, and the carboxy-terminal half of GCD2 are identically shaded to reflect similarities in their amino acid sequences. They are grouped together in eIF2B and shown interacting with the α-subunit of eIF2 because of the fact that mutations in *GCN3*, *GCD7*, and *GCD2* have been obtained that render eIF2B insensitive to eIF2(αP). GCD1 and GCD6 are shown interacting with each other because of sequence similarities between the two (Bushman et al. 1993a); the notion that they comprise the active site on eIF2B is purely hypothetical. The fact that the γ-subunit of eIF2 (encoded by *GCD11*) contains sequence motifs involved in GTP binding and hydrolysis that are conserved among GTP-binding translation factors (e.g., EF1α) leads to the prediction that it contains the GTP-binding site on eIF2 (Hannig et al. 1992); therefore, the GCD11 subunit is shown interacting directly with the active site of eIF2B. (Panel I) Exchange of GDP for GTP on eIF2 being catalyzed by eIF2B; (panel II) the α-subunit has been phosphorylated on Ser-51 by GCN2. This leads to a stronger interaction between eIF2α and the GCN3, GCD7, and GCD2 subunits of eIF2B and also a structural alteration in the GCD6 and GCD1 subunits that prevents GDP/GTP exchange on eIF2. The latter feature of the model is included to explain observations that eIF2B cannot catalyze nucleotide exchange on phosphorylated eIF2. As a result of the greater affinity of eIF2B for phosphorylated versus nonphosphorylated eIF2, eIF2B is not available to catalyze nucleotide exchange on nonphosphorylated eIF2. (Panel III) A mutation in the GCD7 subunit overcomes the inhibitory effect of eIF2α phosphorylation by weakening the interaction between the phosphorylated α-subunit of eIF2 and eIF2B. This allows nucleotide exchange to occur on nonphosphorylated eIF2 even in the presence of high levels of the phosphorylated form of the protein. An alternative possibility not depicted is that the mutation in GCD7 allows eIF2B to catalyze nucleotide exchange on eIF2(αP) rather than decreasing the affinity of eIF2B for eIF2(αP). (Adapted from Hinnebusch 1994.)

diverged sufficiently from HisRS that it now lacks the ability to discriminate between different uncharged tRNA species (Wek et al. 1989). Consistent with this hypothesis is the fact that those regions in authentic HisRS that are least conserved in GCN2 are the extreme amino- and carboxy-terminal segments that are thought to have functions specific to HisRS, such as binding of histidine or $tRNA^{His}$ (Cusack et al. 1991). In addition, the *GCN2*[c] mutations that alter the HisRS-like domain all map in or near the most highly conserved regions shared among class II synthetases (Ramirez et al. 1992). In particular, two of these mutations

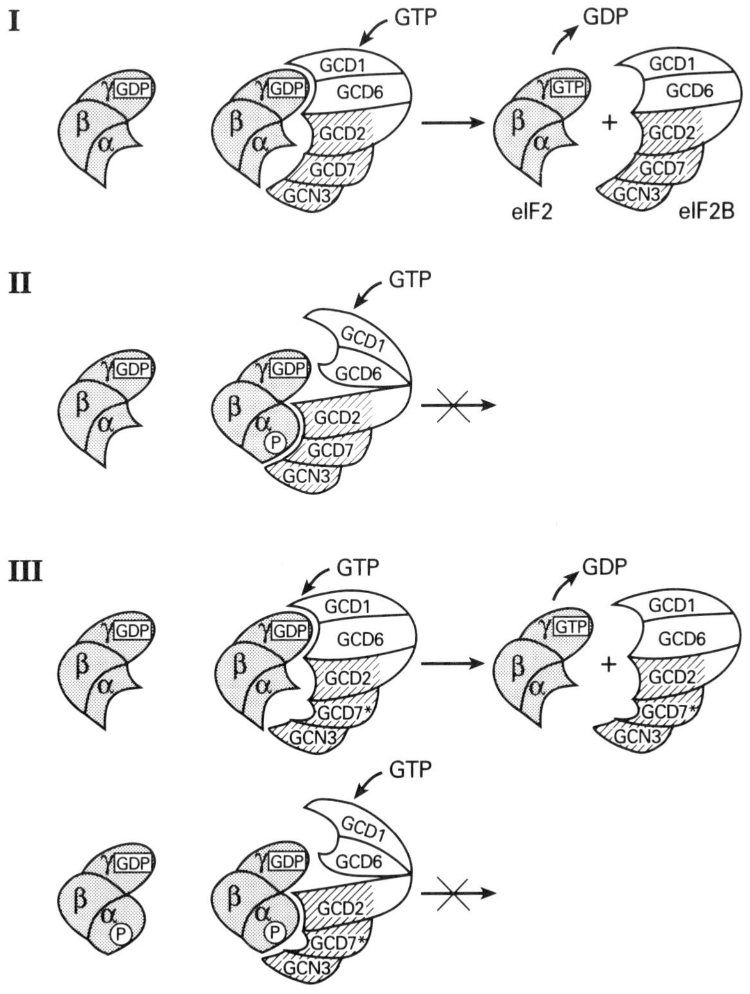

Figure 4 (*See facing page for legend.*)

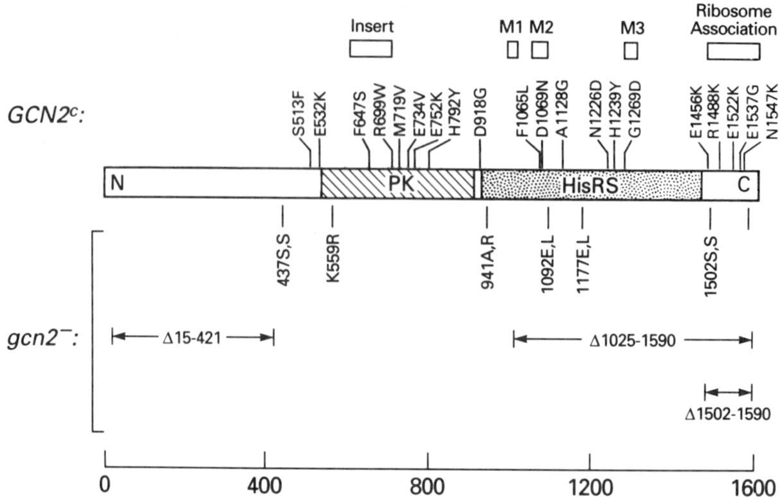

Figure 5 Genetics of the eIF2α kinase GCN2. The 1590-amino-acid polypeptide chain encoded by *GCN2* is depicted from amino- to carboxyl terminus. (*Hatched area*) The protein kinase domain (PK); (*stippled area*) the histidyl-tRNA synthetase (HisRS)-like domain. Indicated by boxes above the sequence are the positions of a 136-amino-acid insert in kinase subdomain IV (Wek et al. 1989), three short sequence motifs 1–3 (M1–M3) conserved between class II tRNA synthetases (Cusack et al. 1991), and a carboxy-terminal segment that is required for ribosome-association by GCN2 in vivo (Ramirez et al. 1991). Recessive mutations that inactivate GCN2 regulatory function in vivo are shown below the coding sequence. These include deletions (Δ), two codon insertions (designated with single-letter amino acid code), and the Lys-559 substitution with arginine (K559R) in kinase subdomain II. Only the K559R mutation abolishes kinase function in vitro; the other mutations are believed to impair the activation of kinase function in amino-acid-starved cells (Wek et al. 1990). Immediately above the coding region are shown dominant mutations, introducing single-amino-acid substitutions, that lead to phosphorylation of eIF2α and derepression of *GCN4* translation under nonstarvation conditions (*GCN2*ᶜ alleles) (Wek et al. 1990; Ramirez et al. 1992; Diallinas and Thireos 1994).

make conservative substitutions in sequence motif II, which in the crystal structure of aspartyl-tRNA synthetase forms a loop that directly contacts the acceptor stem of tRNA (Cusack et al. 1991). Perhaps these *GCN2*ᶜ mutations increase the affinity of GCN2 for uncharged tRNA, allowing activation to occur in the face of less severe amino acid starvation. Alternatively, they might allow both charged and uncharged tRNAs to function as an activator of GCN2.

There is biochemical evidence that much of the GCN2 in the cell is bound to translating ribosomes and free ribosomal subunits (Ramirez et al. 1991). In exponentially growing cells, a substantial fraction of the GCN2 protein was found to cosediment with polysomes, 80S ribosomes, or free subunits. However, when polysomes were run off in vitro, most of the GCN2 cosedimented with free 40S and 60S ribosomal subunits or was found unassociated with ribosomes, despite the accumulation of inactive 80S ribosomal couples (80S particles unbound to mRNA). Moreover, during exponential growth, the amount of GCN2 per ribosome was considerably greater for free 40S and 60S subunits than for 80S ribosomes or polysomes (Ramirez et al. 1991). These latter observations may indicate that GCN2 interacts preferentially with ribosomal subunits engaged in initiation. This might provide GCN2 with access to its substrate, as there is evidence that eIF2 interacts with free ribosomal subunits (Gross et al. 1985; Thomas et al. 1985).

The association between GCN2 and ribosomal subunits is dependent on the carboxy-terminal 120 amino acids of GCN2 (Ramirez et al. 1991). Interestingly, among the *GCN2*^{c} alleles containing single-amino-acid substitutions, those that activate GCN2 function in vivo to the greatest extent are located in the carboxy-terminal ribosome-association domain (Wek et al. 1990; Ramirez et al. 1992; Diallinas and Thireos 1994). These observations raise the possibility that activation by uncharged tRNA occurs while GCN2 is ribosome-bound, perhaps when uncharged tRNA enters the acceptor (A) site of the ribosome during the translation elongation cycle (Ramirez et al. 1992). This would be akin to the activation of the relA protein of *E. coli* by uncharged tRNA during the stringent response (Cashel and Rudd 1987). In this view, the extreme carboxyl terminus of GCN2 would function both in anchoring GCN2 near the A-site of the ribosome and in the recognition of uncharged tRNA by the HisRS-like domain. Deprivation of cells for purines also leads to activation of GCN2 kinase function, even in the presence of all 20 amino acids in the culture medium; however, it is not known whether purine starvation signals GCN2 via uncharged tRNA or by some other activating molecule (Rolfes and Hinnebusch 1993).

There is some genetic and biochemical evidence that oligomerization of GCN2 is important for some aspect of its activation or catalytic function. The *GCN2*^{c}*-E1456K* allele (Fig. 5) leads to an inhibition of translation initiation that is so severe, apparently, that *GCN4* translation cannot be induced. The slow-growth phenotype of this allele was reversed by overexpression of eIF2α, indicating that it probably arises from eIF2 hyperphosphorylation. Interestingly, when wild-type *GCN2* was present in the same strain, the slow-growth phenotype of *GCN2*^{c}*-E1456K* was completely overcome and *GCN4* expression was constitutively derepressed.

These last results suggest that eIF2α phosphorylation catalyzed by the *GCN2*^{c}*-E1456K* product is diminished (but not eliminated) by the presence of wild-type GCN2. To account for this phenomenon, it was proposed that the hyperactivated mutant and wild-type GCN2 proteins form oligomers that are less active for eIF2α phosphorylation than the corresponding oligomers containing only *GCN2*^{c}-encoded proteins. Several other *GCN2*^{c} alleles with a less severe phenotype isolated in the same study appeared to be completely dominant over wild-type *GCN2* for derepression of *GCN4* (Diallinas and Thireos 1994). The differences among *GCN2*^{c} alleles with respect to dominance could be related to different levels of expression of the mutant proteins, to a varying capacity to oligomerize with wild-type GCN2, or to different mechanisms of kinase activation.

A second indication that GCN2 may form oligomers was that much of the protein eluted from a sizing column with an apparent molecular weight in excess of monomeric GCN2, with peak fractions corresponding to the molecular weights of dimers or tetramers of GCN2 (Diallinas and Thireos 1994). Presumably, the protein-protein interactions responsible for these high-molecular-weight complexes are relatively weak, because GCN2 is found exclusively in a monomeric form when fractionated on a sizing column in 0.5 M KCl (Ramirez et al. 1991). It is noteworthy that the HisRS-like domain of GCN2 contains a relatively good match to motif I, a sequence that mediates dimer formation in class II tRNA synthetases (Cusack et al. 1991). It has been proposed that a GCN2 homodimer might bind to a single tRNA molecule and that intermolecular autophosphorylation by the two GCN2 protomers would activate the eIF2α kinase activity of GCN2 (Diallinas and Thireos 1994). Further experiments are required to determine whether the recessive nature of *GCN2*^{c}*-E1456K* and the high-molecular-weight forms of GCN2 detected in cell extracts reflect oligomerization of the protein or interactions with other accessory factors required for activation or localization of GCN2 in the cell.

GCN1 is a 297-kD protein that is required for the stimulation of GCN2 kinase function in amino-acid- or purine-starved cells (Hinnebusch 1988; Marton et al. 1993; Rolfes and Hinnebusch 1993). Deletion of *GCN1* also reduces the extent of eIF2α phosphorylation in strains bearing an activated *GCN2*^{c} allele. GCN1 is not required for expression of *GCN2* or for GCN2 kinase function in vitro, suggesting that it is required in vivo for the stimulation of GCN2 kinase activity by uncharged tRNA. It is interesting that GCN1 contains a large domain with sequence similarity to the fungus-specific translation elongation factor EF3

(Marton et al. 1993). Perhaps this homology indicates that GCN1 interacts with ribosomes or tRNA in performing its regulatory function. Interestingly, another factor, called GCN20, is required in addition to GCN1 for full activity of $GCN2^c$ kinases (Vazquez de Aldana and Hinnebusch 1994). It is conceivable that GCN1 and GCN20 regulate GCN2 indirectly by influencing the aminoacylation of tRNAs rather than by mediating the stimulatory effect of uncharged tRNA. Another open question concerns the role of the approximately 510 amino acids in GCN2 located amino-terminal to the protein kinase domain (Fig. 5). Mutational analysis has shown that at least the portion of this amino-terminal domain closest to the protein kinase moiety is essential for GCN2 regulatory function in vivo (Wek et al. 1989).

GCN2-INDEPENDENT DEREPRESSION OF *GCN4* TRANSLATION

The regulatory mechanism described above applies to the derepression of *GCN4* translation that occurs under conditions of prolonged amino-acid-limited growth. It appears that a related, but distinct, mechanism is responsible for a transient derepression of *GCN4* that accompanies a "shiftdown" from amino-acid-rich medium to minimal medium. This increase in *GCN4* expression occurs at the translational level and requires uORF1; however, it was observed in a *gcn2* mutant (Tzamarias et al. 1989), suggesting that either a different eIF2α kinase is being activated transiently during a nutritional shiftdown or that a different mechanism operates under these conditions for suppressing ribosomal reinitiation at uORFs 2–4.

This GCN2-independent mechanism may be responsible for the uORF1-dependent derepression of *GCN4* translation that occurred when a yeast cell-free translation system was limited for amino acids or supplemented with nonacylated tRNAs. Under these conditions, binding of eIF2•GTP•Met-$tRNA_i$ ternary complexes to 40S ribosomes was reduced (Krupitza and Thireos 1990), just as occurs in vivo during a nutritional shiftdown (Tzamarias et al. 1989). Moreover, the increased translation of *GCN4* seen in vitro was independent of GCN2, again implying the existence of a second mechanism for reducing eIF2 activity that stimulates translation of *GCN4* relative to mRNAs lacking uORFs (Krupitza and Thireos 1990).

The molecular details of this alternative mechanism for down-regulating eIF2 function remain to be identified; however, recent work implicates the cAMP-dependent protein kinase (PKA) in the response. The rapid induction of *HIS3* transcription that occurs following a nutritional shiftdown was not observed in a mutant strain containing a constitutively low level of PKA activity (*tpk1*^{w} strain). This suggests that the rapid induction of *GCN4* translation that occurs under these conditions, with attendant derepression of *HIS3* transcription, requires high-level PKA function. (An assumption of this interpretation was that the rapid induction of *HIS3* mRNA following nutritional shiftdown is completely attributable to GCN4 [see Driscoll-Penn et al. 1984].) A substantial induction of *HIS3* transcription during prolonged histidine-limited growth still occurred in the *tpk1*^{w} strain, indicating that the GCN2-dependent derepression of *GCN4* does not require elevated PKA function (Engelberg et al. 1994).

In accord with the idea that PKA is an activator of *GCN4* translation, partial derepression of *GCN4* translation occurred during growth on rich medium in mutant strains with constitutively high levels of PKA function (*RAS2*^{Val19} and *bcy1*Δ mutants). As expected, expression of *HIS3* and *HIS4*, two targets of GCN4 transcriptional activation, was also induced in these mutants. A GCN4-dependent induction of *HIS3* transcription in response to the *bcy1* mutation was observed in a *gcn2*Δ strain and also in one containing the Ala-51 substitution in eIF2α, suggesting that PKA-mediated induction of *GCN4* translation is independent of phosphorylation on Ser-51 by GCN2 (Engelberg et al. 1994). This last conclusion was not established directly, however, by assaying *GCN4* expression in *bcy1 gcn2* double mutants; moreover, it was unclear whether the derepression of *HIS3* transcription associated with constitutive activation of PKA was completely independent of GCN2. Nevertheless, the available data strongly suggest that a GCN2-independent mechanism for derepressing *GCN4* translation requires activation of PKA. There is also evidence that UV-irradiation of yeast cells increases *GCN4* translation by a PKA-dependent mechanism (Engelberg et al. 1994); however, the magnitude of this effect on *GCN4* expression and the involvement of GCN2 in the response remain to be determined.

Another case of GCN2-independent derepression of *GCN4* translation was observed in strains overexpressing tRNAs under conditions in which the excess tRNA could not be aminoacylated. The most complete derepression of *GCN4* expression was observed in strains overexpressing a form of tRNA^{Val} (AAC) containing a mutation in the acceptor stem that almost certainly impairs aminoacylation of this tRNA. Overexpression of this unchargeable mutant tRNA^{Val} derepresses *GCN4* translation under nonstarvation conditions in a manner that depends on the uORFs in *GCN4* mRNA but is independent of GCN2 and Ser-51 on eIF2α. Overexpression of the mutant tRNA^{Val} did not affect cellular growth under nonstarvation conditions in an otherwise wild-type strain; however, it did specifically exacerbate the slow-growth phenotype of a *GCN2*^{c} mutant. These findings suggest that the presence of excess unchargeable tRNA^{Val} does lower the level of eIF2 activity, even though it involves a mechanism that is independent of eIF2α phosphorylation. Much weaker GCN2-independent derepression was observed in *gcn2Δ* strains overexpressing wild-type tRNA^{His} under histidine starvation conditions, indicating that this GCN2-independent mechanism can also be activated by an excess of uncharged wild-type tRNA (Vazquez de Aldana et al. 1994). It is not known whether activation of PKA is involved in the GCN2-independent response to the overexpressed tRNAs. In any event, it is intriguing that yeast contains two separate pathways for down-regulating eIF2 activity that both respond to uncharged tRNA. Whereas starvation for histidine or valine increases eIF2α phosphorylation in strains containing wild-type GCN2, surprisingly, overexpression of the mutant tRNA^{Val} did not. Moreover, derepression of *GCN4* in response to steady-state deprivation of histidine, tryptophan, or valine is almost completely dependent on GCN2, whereas the derepression response to the mutant tRNA^{Val} is largely independent of GCN2 (Vazquez de Aldana et al. 1994). One way to explain these results would be to propose that GCN2 cannot be activated by this particular tRNA^{Val} isoacceptor, even though it is probably the major isoacceptor for valine (Hinnebusch and Liebman 1991). It is also possible that GCN2 does not have access to the mutant tRNA^{Val}.

DEPHOSPHORYLATION OF eIF2α IN YEAST

To reverse the effects of amino acid starvation on *GCN4* translation when cells are no longer limited for amino acids requires the dephosphorylation of eIF2(αP). Genetic evidence indicates that the type I protein phosphatase encoded by *GLC7* is a major eIF2α phosphatase in yeast. A truncated allele of *GLC7* lacking the carboxy-terminal 104 codons was isolated by its ability to restore derepression of *GCN4* translation in a strain containing a partially functional allele of *GCN2*. The same phenotype was seen for a recessive reduced-function allele known as *glc7-1*, suggesting that overexpression of the truncated protein interferes with the function of the native *GLC7*-encoded phosphatase (dominant-negative phenotype). Consistent with this interpretation, the presence of additional copies of wild-type *GLC7* in an otherwise wild-type strain impaired derepression of *GCN4*, thus mimicking a *gcn2* mutation. These results indicate that GCN2 and GLC7 have opposing functions in regulating *GCN4*, as would be expected if GLC7 dephosphorylates eIF2α. Importantly, the dominant-negative *glc7* allele has no effect on *GCN4* expression in strains lacking GCN2 or Ser-51 on eIF2α. This shows that GLC7 reverses the GCN2-mediated pathway for inducing *GCN4* translation but seems to have no effect on the GCN2-independent derepression mechanism (Wek et al. 1992). Unlike *GCN2* and *GCN4*, deletion of *GLC7* is lethal (Feng et al. 1991), suggesting that this phosphatase is involved in regulating one or more essential functions in yeast, beyond its involvement in *GCN4* translational control. It is known that GLC7 function is required for normal levels of glycogen accumulation in yeast cells, although glycogen synthesis is not essential in yeast (Feng et al. 1991). Recent results indicate that GLC7 function is required for proper chromosome segregation (Francisco et al. 1994) and for cell cycle progression at the G_2/M boundary (Hisamoto et al. 1994).

CONCLUSIONS

Phosphorylation of eIF2α on Ser-51 is a highly conserved mechanism for decreasing the rate of total protein synthesis under conditions of starvation and stress. In yeast cells, it also leads to increased production of GCN4, an important regulator of amino acid biosynthetic pathways. The induction of *GCN4* translation depends on the interplay between uORFs in the *GCN4* mRNA leader, wherein recognition of the particular uORFs that block ribosomal scanning to *GCN4* (uORFs 2–4) can be suppressed by decreasing the level of eIF2 activity in the cell. Either because of kinetic constraints on reinitiation events or because the process of reinitiation is inherently inefficient, reductions in eIF2 activity sufficient for full induction of *GCN4* have little effect on translation of bulk mRNAs lacking uORFs. Presumably, this allows yeast to induce a key activator of anabolic pathways long before nutritional deprivation becomes so severe that it blocks protein synthesis.

Currently, *GCN4* is the only known mRNA whose translation is coupled to eIF2 activity by multiple uORFs. The best candidate for a second gene governed by this control mechanism is *cpc-1* of *Neurospora crassa* that encodes the functional homolog of GCN4 in this organism (Paluh et al. 1988). Several other eukaryotic mRNAs are found in yeast and mammals in which uORFs regulate translation; however, these mostly contain a single uORF whose peptide product has a role in preventing ribosomes from reaching the authentic start codon downstream (see Geballe, this volume). Probably no aspect of *GCN4* translational control is unique to yeast, however, making it possible that this mechanism, or one related to it, will surface elsewhere. The best place to look for it would be under conditions of starvation or stress where the cell is down-regulating protein synthesis and decreasing its growth rate in response to a diminishing supply of nutrients or a harmful agent in the environment.

The *GCN4* regulatory mechanism provides a powerful tool for elucidating the molecular details of many different aspects of protein synthesis. As indicated in this chapter, these include the functions of eIF2 and its recycling factor eIF2B and the manner in which these factors are regulated by eIF2α kinases, and possibly by PKA as well. Recent results strongly suggest that the GCD10 protein is a subunit of yeast eIF3 (Garcia-Barrio et al. 1995), a multisubunit complex implicated in several aspects of the initiation process, including the stimulation of ternary complex binding to the small ribosomal subunit (see Merrick and Hershey, this volume). If *gcd10* mutations impair this particular function of yeast eIF3, it would explain why they lead to derepression of *GCN4* translation in the absence of eIF2α phosphorylation (Harashima and Hinnebusch 1986; Mueller et al. 1987). It is likely that other initiation factors besides eIF2, eIF2B, and eIF3 contribute to the process of reinitiation on *GCN4* mRNA, and mutations in these other factors may be found on the basis of having a Gcn^- or Gcd^- phenotype.

Although the eIF2•GTP•Met-$tRNA_i$ ternary complex seems to be the limiting factor for reinitiation at uORFs 2–4, recent findings suggest that another factor is rate-limiting for reinitiation at the *GCN4* start codon. In constructs containing only uORF1 and *GCN4*, rebinding of this hypothetical factor to all ribosomes scanning downstream from uORF1 seems to require the entire approximately 350-nucleotide interval separating uORF1 from *GCN4*, even when ternary complexes are abundant. In contrast, under the same conditions, nearly all ribosomes can reinitiate at uORF4 after scanning less than half this distance. Either the hypothetical factor limiting reinitiation at *GCN4* is dispensable for reinitiation at uORF4 or some feature of uORF4 facilitates rebinding of this factor to ribosomes once they reach uORF4 (Grant et al. 1994). At present, there are no strong clues about the factor that is rate-limiting for reinitiation at *GCN4*, but it may be possible to identify it by isolating mutations that alter the scanning distance required for efficient reinitiation at *GCN4*.

In addition to the intricate controls over initiation events, *GCN4* regulation depends on the fact that roughly half of the ribosomes remain attached to the mRNA after terminating translation at uORF1, but essentially all dissociate from the transcript after completing uORF4 translation. Thus, it is probable that *GCN4* expression can be used as a sensitive indicator of the functions of rRNA sequences and protein factors in the reactions involved in translation termination and ribosome release. Perhaps the yeast equivalent of the ribosome release factor (RF4) can be identified genetically by its involvement in the termination aspects of *GCN4* translational control.

In addition to exploiting the *GCN4* system to probe the structure and regulation of the machinery for protein synthesis, it will be interesting to identify the full range of environmental conditions that stimulate GCN2 kinase activity and to determine how signals for different types of nutritional deprivation are transduced by GCN1 and other proteins involved in regulating GCN2 activity. Recent results indicate that GCN1 and GCN20 are components of a heteromeric protein complex and that GCN20 is a member of the ATP-binding cassette (ABC) family of proteins. GCN20 shows strong sequence similarity to the ABC domains of EF3; thus, GCN1 and GCN20 are closely related to different segments of EF3. Interestingly, GCN20 is even more highly related to other ABC proteins of unknown function in plants and animals, raising the possibility that higher eukaryotes contain a mechanism for coupling eIF2 phosphorylation to levels of uncharged tRNA, which is similar to that involving GCN2, GCN1, and GCN20 (Vazquez de Aldana et al. 1995). In addition to learning more about how GCN2 is regulated, we must also identify the proteins involved in the GCN2-independent mechanism for inducing *GCN4* translation and the part played by PKA in this alternative response to starvation. Finally, it is intriguing to consider whether eIF2α phosphorylation has a role in cell cycle control or cellular differentiation (Gimeno et al. 1992) in response to nutrient deprivation in yeast.

ACKNOWLEDGMENTS

I thank Carlos Vazquez de Aldana for helpful comments on the manuscript and Kathy Shoobridge for help in its preparation.

REFERENCES

Abastado, J.-P., P.F. Miller, and A.G. Hinnebusch. 1991a. A quantitative model for translational control of the *GCN4* gene of *Saccharomyces cerevisiae*. *New Biol.* **3:** 511–524.

Abastado, J.P., P.F. Miller, B.M. Jackson, and A.G. Hinnebusch. 1991b. Suppression of ribosomal reinitiation at upstream open reading frames in amino acid-starved cells forms the basis for *GCN4* translational control. *Mol. Cell. Biol.* **11:** 486–496.

Bushman, J.L., A.I. Asuru, R.L. Matts, and A.G. Hinnebusch. 1993a. Evidence that GCD6 and GCD7, translational regulators of *GCN4*, are subunits of the guanine nucleotide exchange factor for eIF-2 in *Saccharomyces cerevisiae*. *Mol. Cell. Biol.* **13:**

1920–1932.

Bushman, J.L., M. Foiani, A.M. Cigan, C.J. Paddon, and A.G. Hinnebusch. 1993b. Guanine nucleotide exchange factor for eIF-2 in yeast: genetic and biochemical analysis of interactions between essential subunits GCD2, GCD6 and GCD7 and regulatory subunit GCN3. *Mol. Cell. Biol.* **13:** 4618–4631.

Cashel, M. and K.E. Rudd. 1987. The stringent response. In Escherichia coli *and* Salmonella typhimurium: *Cellular and molecular biology* (ed. F.C. Neidhardt et al.), pp. 1410–1438. American Society for Microbiology, Washington, D.C.

Castilho-Valavicius, B., G.M. Thompson, and T.F. Donahue. 1992. Mutation analysis of the Cys-X_2-Cys-X_{19}-Cys-X_2-Cys motif in the β subunit of eukaryotic translation factor 2. *Gene Expr.* **2:** 297–309.

Castilho-Valavicius, B., H. Yoon, and T.F. Donahue. 1990. Genetic characterization of the *Saccharomyces cerevisiae* translational initiation suppressors *sui1, sui2* and *SUI3* and their effects on *HIS4* expression. *Genetics* **124:** 483–495.

Chen, J.-J., M.S. Throop, L. Gehrke, I. Kuo, J.K. Pal, M. Brodsky, and I.M. London. 1991. Cloning of the cDNA of the heme-regulated eukaryotic initiation factor 2α (eIF-2α) kinase of rabbit reticulocytes: Homology to yeast GCN2 protein kinase and human double-stranded-RNA-dependent eIF-2α kinase. *Proc. Natl. Acad. Sci.* **88:** 7729–7733.

Choi, S.-Y., B.J. Scherer, J. Schnier, M.V. Davies, and R.J. Kaufman. 1992. Stimulation of protein synthesis in COS cells transfected with variants of the α-subunit of initiation factor eIF-2. *J. Biol. Chem.* **267:** 286–293.

Chong, K.L., L. Feng, K. Schappert, E. Meurs, T.F. Donahue, J.D. Friesen, A.G. Hovanessian, and B.R.G. Williams. 1992. Human p68 kinase exhibits growth suppression in yeast and homology to the translational regulator GCN2. *EMBO J.* **11:** 1553–1562.

Cigan, A.M., L. Feng, and T.F. Donahue. 1988. $tRNA_i^{Met}$ functions in directing the scanning ribosome to the start site of translation. *Science* **242:** 93–97.

Cigan, A.M., J.L. Bushman, T.R. Boal, and A.G. Hinnebusch. 1993. A protein complex of translational regulators of *GCN4* is the guanine nucleotide exchange factor for eIF-2 in yeast. *Proc. Natl. Acad. Sci.* **90:** 5350–5354.

Cigan, A.M., M. Foiani, E.M. Hannig, and A.G. Hinnebusch. 1991. Complex formation by positive and negative translational regulators of *GCN4*. *Mol. Cell. Biol.* **11:** 3217–3228.

Cigan, A.M., E.K. Pabich, L. Feng, and T.F. Donahue. 1989. Yeast translation initiation suppressor *sui2* encodes the α subunit of eukaryotic initiation factor 2 and shares identity with the human α subunit. *Proc. Natl. Acad. Sci.* **86:** 2784–2788.

Colthurst, D.R., D.G. Campbell, and C.G. Proud. 1987. Structure and regulation of eukaryotic initiation factor eIF-2: Sequence of the site in the α subunit phosphorylated by the haem-controlled repressor and by the double-stranded RNA-activated inhibitor. *Eur. J. Biochem.* **166:** 357–363.

Cusack, S., M. Hartlein, and R. Leberman. 1991. Sequence, structural and evolutionary relationships between class 2 aminoacyl-tRNA synthetases. *Nucleic Acids Res.* **19:** 3489–3498.

Dever, T.E., L. Feng, R.C. Wek, A.M. Cigan, T.D. Donahue, and A.G. Hinnebusch. 1992. Phosphorylation of initiation factor 2α by protein kinase GCN2 mediates gene-specific translational control of *GCN4* in yeast. *Cell* **68:** 585–596.

Dever, T.E., J.-J. Chen, G.N. Barber, A.M. Cigan, L. Feng, T.F. Donahue, I.M. London,

M.G. Katze, and A.G. Hinnebusch. 1993. Mammalian eukaryotic initiation factor 2α kinases functionally substitute for GCN2 in the *GCN4* translational control mechanism of yeast. *Proc. Natl. Acad. Sci.* **90:** 4616–4620.

Diallinas, G. and G. Thireos. 1994. Genetic and biochemical evidence for yeast GCN2 protein kinase polymerization. *Gene* **143:** 21–27.

Donahue, T.F., A.M. Cigan, E.K. Pabich, and B. Castilho-Valavicius. 1988. Mutations at a Zn(II) finger motif in the yeast eIF-2β gene alter ribosomal start-site selection during the scanning process. *Cell* **54:** 621–632.

Driscoll-Penn, M., G. Thireos, and H. Greer. 1984. Temporal analysis of general control of amino acid biosynthesis in *Saccharomyces cerevisiae*: Role of positive regulatory genes in initiation and maintenance of mRNA derepression. *Mol. Cell. Biol.* **4:** 520–528.

Engelberg, D., C. Klein, H. Martinetto, K. Struhl, and M. Karin. 1994. The UV response involving the ras signaling pathway and AP-1 transcription factors is conserved between yeast and mammals. *Cell* **77:** 381–390.

Ernst, H., R.F. Duncan, and J.W.B. Hershey. 1987. Cloning and sequencing of complementary DNAs encoding the α-subunit of translational initiation factor eIF-2. *J. Biol. Chem.* **262:** 1206–1212.

Feng, Z., S.E. Wilson, Z.-Y. Peng, K.K. Schlender, E.M. Reimann, and R.J. Trumbly. 1991. The yeast *GLC7* gene required for glycogen accumulation encodes a type 1 protein phosphatase. *J. Biol. Chem.* **226:** 23796–23801.

Flowers, K.M., S.R. Kimball, R.C. Feldhoff, A.G. Hinnebusch, and L.S. Jefferson. 1995. Molecular cloning and characterization of cDNA encoding the α-subunit of the rat protein synthesis initiation factor eIF-2B. *Proc. Natl. Acad. Sci.* **92:** 4274–4278.

Foiani, M., A.M. Cigan, C.J. Paddon, S. Harashima, and A.G. Hinnebusch. 1991. GCD2, a translational repressor of the *GCN4* gene, has a general function in the initiation of protein synthesis in *Saccharomyces cerevisiae*. *Mol. Cell. Biol.* **11:** 3203–3216.

Francisco, L., W. Wang, and C.S. Chan. 1994. Type 1 protein phosphatase acts in opposition to IpL1 protein kinase in regulating yeast chromosome segregation. *Mol. Cell. Biol.* **14:** 4731–4740.

Garcia-Barrio, M.T., T. Naranda, C.R. Vazquez de Aldana, R. Cuesta, A.G. Hinnebusch, J.W.B. Hershey, and M. Tamame. 1995. GCD10, a translational repressor of *GCN4*, is the RNA-binding subunit of eukaryotic translation initiation factor-3. *Genes Dev.* **9:** 1781–1796.

Gimeno, C.J., P.O. Ljungdahl, C.A. Styles, and G.R. Fink. 1992. Unipolar cell divisions in the yeast *S. cerevisiae* lead to filamentous growth: Regulation by starvation and RAS. *Cell* **68:** 1077–1090.

Grant, C.M. and A.G. Hinnebusch. 1994. Effect of sequence context at stop codons on efficiency of reinitiation in *GCN4* translational control. *Mol. Cell. Biol.* **14:** 606–618.

Grant, C., P. Miller, and A. Hinnebusch. 1994. Requirements for intercistronic distance and level of eIF-2 activity in reinitiation on GCN4 mRNA varies with the downstream cistron. *Mol. Cell. Biol.* **14:** 2616–2628.

Gross, M., R. Redman, and D.A. Kaplansky. 1985. Evidence that the primary effect of phosphorylation of eukaryotic initiation factor 2α in rabbit reticulocyte lysate is inhibition of the release of eukaryotic initiation factor-2•GDP from 60S ribosomal subunits. *J. Biol. Chem.* **260:** 9491–9500.

Hannig, E.M. and A.G. Hinnebusch. 1988. Molecular analysis of *GCN3*, a translational activator of *GCN4*: Evidence for posttranslational control of *GCN3* regulatory function.

Mol. Cell. Biol. **8:** 4808–4820.

Hannig, E.M., A.M. Cigan, B.A. Freeman, and T.G. Kinzy. 1992. *GCD11*, a negative regulator of *GCN4* expression, encodes the gamma subunit of eIF-2 in *Saccharomyces cerevisiae. Mol. Cell. Biol.* **13:** 506–520.

Hannig, E.H., N.P. Williams, R.C. Wek, and A.G. Hinnebusch. 1990. The translational activator GCN3 functions downstream from GCN1 and GCN2 in the regulatory pathway that couples *GCN4* expression to amino acid availability in *Saccharomyces cerevisiae. Genetics* **126:** 549–562.

Harashima, S. and A.G. Hinnebusch. 1986. Multiple *GCD* genes required for repression of *GCN4*, a transcriptional activator of amino acid biosynthetic genes in *Saccharomyces cerevisiae. Mol. Cell. Biol.* **6:** 3990–3998.

Harashima, S., E.M. Hannig, and A.G. Hinnebusch. 1987. Interactions between positive and negative regulators of *GCN4* controlling gene expression and entry into the yeast cell cycle. *Genetics* **117:** 409–419.

Henderson, R.A., G.W. Krissansen, R.Y. Yong, E. Leung, and J.D. Watson. 1994. The delta-subunit of murine guanine nucleotide exchange factor eIF-2B. Characterization of cDNAs predicts isoforms differing at the amino-terminal end. *J. Biol. Chem.* **269:** 30517–30523.

Hill, D.E. and K.Struhl. 1988. Molecular characterization of *GCD1*, a yeast gene rquired for general control of amino acid biosynthesis and cell-cycle initiation. *Nucleic Acids Res.* **16:** 9253–9265.

Hinnebusch, A.G. 1984. Evidence for translational regulation of the activator of general amino acid control in yeast. *Proc. Natl. Acad. Sci.* **81:** 6442–6446.

———. 1985. A hierarchy of *trans*-acting factors modulate translation of an activator of amino acid biosynthetic genes in *Saccharomyces cerevisiae. Mol. Cell. Biol.* **5:** 2349–2360.

———. 1988. Mechanisms of gene regulation in the general control of amino acid biosynthesis in *Saccharomyces cerevisiae. Microbiol. Rev.* **52:** 248–273.

———. 1992. General and pathway-specific regulatory mechanisms controlling the synthesis of amino acid biosynthetic enzymes in *Saccharomyces cerevisiae.* In *The molecular and cellular biology of the yeast* Saccharomyces: *Gene expression* (ed. E.W. Jones et al.), vol. 2, pp. 319–414. Cold Spring Harbor Laboratory Press, Cold Spring Harbor, New York.

———. 1994. Translational control of *GCN4*: An *in vivo* barometer of initiation factor activity. *Trends Biochem. Sci.* **19:** 409–414.

Hinnebusch, A.G. and G.R. Fink. 1983. Positive regulation in the general amino acid control of *Saccharomyces cerevisiae. Proc. Natl. Acad. Sci.* **80:** 5374–5378.

Hinnebusch, A.G. and S.W. Liebman. 1991. Protein synthesis and translational control in *Saccharomyces cerevisiae.* In *The molecular and cellular biology of the yeast* Saccharomyces: *Genome dynamics, protein synthesis, and energetics* (ed. J.R. Broach et al.), vol. 1, pp. 627–735. Cold Spring Harbor Laboratory Press, Cold Spring Harbor, New York.

Hinnebusch, A.G., B.M. Jackson, and P.P. Mueller. 1988. Evidence for regulation of reinitiation in translational control of *GCN4* mRNA. *Proc. Natl. Acad. Sci.* **85:** 7279–7283.

Hisamoto, N., K. Sugimoto, and K. Matsumoto. 1994. The Glc7 type 1 protein phosphatase of *Saccharomyces cerevisiae* is required for cell cycle progression in G2/M. *Mol. Cell. Biol.* **14:** 3158–3165.

Ichikawa, S. and A. Kaji. 1989. Molecular cloning and expression of ribosome releasing factor. *J. Biol. Chem.* **264:** 20054–20059.

Konieczny, A. and B. Safer. 1983. Purification of the eukaryotic initiation factor 2-eukaryotic initiation factor 2B complex and characterization of its guanine nucleotide exchange activity during protein synthesis initiation. *J. Biol. Chem.* **258:** 3402–3408.

Kozak, M. 1987. Effects of intercistronic length on the efficiency of reinitiation by eucaryotic ribosomes. *Mol. Cell. Biol.* **7:** 3438–3445.

———. 1989. The scanning model for translation: An update. *J. Cell Biol.* **108:** 229–241.

Krupitza, G. and G. Thireos. 1990. Translational activation of GCN4 mRNA in a cell-free system is triggered by uncharged tRNAs. *Mol. Cell. Biol.* **10:** 4375–4378.

Lanker, S., J.L. Bushman, A.G. Hinnebusch, H. Trachsel, and P.P. Mueller. 1992. Autoregulation of the yeast lysyl-tRNA synthetase gene *GCD5/KRS1* by translational and transcriptional control mechanisms. *Cell* **70:** 647–657.

Martinez, R., M.-T. Latreille, and M. Mirande. 1991. A *PMR2* tandem repeat with a modified C-terminus is located downstream from the *KRS1* gene encoding lysyl-tRNA synthetase in *Saccharomyces cerevisiae*. *Mol. Gen. Genet.* **227:** 149–154.

Marton, M. J., D. Crouch, and A. G. Hinnebusch. 1993. GCN1, a translational activator of *GCN4* in *S. cerevisiae*, is required for phosphorylation of eukaryotic translation initiation factor 2 by protein kinase GCN2. *Mol. Cell. Biol.* **13:** 3541–3556.

Messenguy, F. and J. Delforge. 1976. Role of transfer ribonucleic acids in the regulation of several biosyntheses in *Saccharomyces cerevisiae*. *Eur. J. Biochem.* **67:** 335–339.

Meurs, E., K. Chong, J. Galabru, N.S.B. Thomas, I.M. Kerr, B.R.G. Williams, and A.G. Hovanessian. 1990. Molecular cloning and characterization of the human double-stranded RNA-activated protein kinase induced by interferon. *Cell* **62:** 379–390.

Miller, P.F. and A.G. Hinnebusch. 1989. Sequences that surround the stop codons of upstream open reading frames in *GCN4* mRNA determine their distinct functions in translational control. *Genes Dev.* **3:** 1217–1225.

Miozzari, G., P. Niederberger, and R. Huetter. 1978. Tryptophan biosynthesis in *Saccharomyces cerevisiae*: Control of the flux through the pathway. *J. Bacteriol.* **134:** 48–59.

Mirande, M. and J.-P. Waller. 1988. The yeast lysyl-tRNA synthetase gene: Evidence for general amino acid control of its expression and domain structure of the encoded protein. *J. Biol. Chem.* **263:** 18443–18451.

Mueller, P.P. and A.G. Hinnebusch. 1986. Multiple upstream AUG codons mediate translational control of *GCN4*. *Cell* **45:** 201–207.

Mueller, P.P., S. Harashima, and A.G. Hinnebusch. 1987. A segment of *GCN4* mRNA containing the upstream AUG codons confers translational control upon a heterologous yeast transcript. *Proc. Natl. Acad. Sci.* **84:** 2863–2867.

Mueller, P.P., B.M. Jackson, P.F. Miller, and A.G. Hinnebusch. 1988. The first and fourth upstream open reading frames in *GCN4* mRNA have similar initiation efficiencies but respond differently in translational control to changes in length and sequence. *Mol. Cell. Biol.* **8:** 5439–5447.

Myers, P.L., R.C. Skvirsky, M.L. Greenberg, and H. Greer. 1986. Negative regulatory gene for general control of amino acid biosynthesis in *Saccharomyces cerevisiae*. *Mol. Cell. Biol.* **6:** 3150–3155.

Niederberger, P., M. Aebi, and R. Huetter. 1986. Identification and characterization of four new *GCD* genes in *Saccharomyces cerevisiae*. *Curr. Genet.* **10:** 657–664.

Paddon, C.J. and A.G. Hinnebusch. 1989. *gcd12* mutations are *gcn3*-dependent alleles of *GCD2*, a negative regulator of *GCN4* in the general amino acid control of *Saccharomyces cerevisiae*. *Genetics* **122:** 543–550.

Paddon, C.J., E.M. Hannig, and A.G. Hinnebusch. 1989. Amino acid sequence similarity between GCN3 and GCD2, positive and negative translational regulators of *GCN4*: Evidence for antagonism by competition. *Genetics* **122:** 551–559.

Paluh, J.L., M.J. Orbach, T.L. Legerton, and C. Yanofsky. 1988. The cross-pathway control gene of *Neurospora crassa, cpc-1*, encodes a protein similar to GCN4 of yeast and the DNA-binding domain of the oncogene v-*jun*-encoded protein. *Proc. Natl. Acad. Sci.* **85:** 3728–3732.

Pathak, V.K., D. Schindler, and J.W.B. Hershey. 1988. Generation of a mutant form of protein synthesis initiation factor eIF-2 lacking the site of phosphorylation by eIF-2 kinases. *Mol. Cell. Biol.* **8:** 993–995.

Price, N.T., G. Francia, L. Hall, and C.G. Proud. 1994. Guanine nucleotide exchange factor for eukaryotic initiation factor-2. Cloning of cDNA for the δ-subunit of rabbit translation initiation factor-2B. *Biochim. Biophys. Acta* **1217:** 207–210.

Ramirez, M., R.C. Wek, and A.G. Hinnebusch. 1991. Ribosome-association of GCN2 protein kinase, a translational activator of the *GCN4* gene of *Saccharomyces cerevisiae*. *Mol. Cell. Biol.* **11:** 3027–3036.

Ramirez, M., R.C. Wek, C.R. Vazquez de Aldana, B.M. Jackson, B. Freeman, and A.G. Hinnebusch. 1992. Mutations activating the yeast eIF-2α kinase GCN2: Isolation of alleles altering the domain related to histidyl-tRNA synthetases. *Mol. Cell. Biol.* **12:** 5801–5815.

Rolfes, R.J. and A.G. Hinnebusch. 1993. Translation of the yeast transcriptional activator GCN4 is stimulated by purine limitation: Implications for activation of the protein kinase GCN2. *Mol. Cell. Biol.* **13:** 5099–5111.

Roussou, I., G. Thireos, and B.M. Hauge. 1988. Transcriptional-translational regulatory circuit in *Saccharomyces cerevisiae* which involves the *GCN4* transcriptional activator and the GCN2 protein kinase. *Mol. Cell. Biol.* **8:** 2132–2139.

Ryoji, M., J.W. Karpen, and A. Kaji. 1981. Further characterization of ribosome releasing factor and evidence that it prevents ribosomes from reading though a termination codon. *J. Biol. Chem.* **256:** 5798–5801.

Schimmel, P.R. and D. Soll. 1979. Aminoacyl-tRNA synthetases: General features and recognition of transfer RNAs. *Annu. Rev. Biochem.* **48:** 601–648.

Schurch, A., J. Miozzari, and R. Huetter. 1974. Regulation of tryptophan biosynthesis in *Saccharomyces cerevisiae*: Mode of action of 5-methyltryptophan- and 5-methyltryptophan-sensitive mutants. *J. Bacteriol.* **117:** 1131–1140.

Thireos, G., M. Driscoll-Penn, and H. Greer. 1984. 5′ untranslated sequences are required for the translational control of a yeast regulatory gene. *Proc. Natl. Acad. Sci.* **81:** 5096–5100.

Thomas, N.S.B., R.L. Matts, D.H. Levin, and I.M. London. 1985. The 60S ribosomal subunit as a carrier of eukaryotic initiation factor 2 and the site of reversing factor activity during protein synthesis. *J. Biol. Chem.* **260:** 9860–9866.

Tzamarias, D. and G. Thireos. 1988. Evidence that the *GCN2* protein kinase regulates reinitiation by yeast ribosomes. *EMBO J.* **7:** 3547–3551.

Tzamarias, D., D. Alexandraki, and G. Thireos. 1986. Multiple *cis*-acting elements modulate the translational efficiency of GCN4 mRNA in yeast. *Proc. Natl. Acad. Sci.*

83: 4849–4853.

Tzamarias, D., I. Roussou, and G. Thireos. 1989. Coupling of GCN4 mRNA translational activation with decreased rates of polypeptide chain initiation. *Cell* **57:** 947–954.

Vazquez de Aldana, C.R. and A.G. Hinnebusch. 1994. Mutations in the GCD7 subunit of yeast guanine nucleotide exchange factor eIF-2B overcome the inhibitory effects of phosphorylated eIF-2 on translation initiation. *Mol. Cell. Biol.* **14:** 3208–3222.

Vazquez de Aldana, C.R., T.E. Dever, and A.G. Hinnebusch. 1993. Mutations in the α subunit of eukaryotic translation initiation factor 2 (eIF-2α) that overcome the inhibitory effects of eIF-2α phosphorylation on translation initiation. *Proc. Natl. Acad. Sci.* **90:** 7215–7219.

Vazquez de Aldana, C.R., M.J. Marton, and A.G. Hinnebusch. 1995. GCN20, a novel ABC protein, and GCN1 reside in a protein complex that mediates activation of the eIF2α kinase GCN2 in amino acid-starved cells. *EMBO J.* **14:** 3184–3199.

Vazquez de Aldana, C.R., R.C. Wek, P. San Segundo, A.G. Truesdell, and A.G. Hinnebusch. 1994. Multicopy tRNA genes functionally suppress mutations in yeast eIF-2α kinase GCN2: Evidence for separate pathways coupling *GCN4* expression to uncharged tRNA. *Mol. Cell. Biol.* **14:** 7920–7932.

Wek, R.C., B.M. Jackson, and A.G. Hinnebusch. 1989. Juxtaposition of domains homologous to protein kinases and histidyl-tRNA synthetases in GCN2 protein suggests a mechanism for coupling *GCN4* expression to amino acid availability. *Proc. Natl. Acad. Sci.* **86:** 4579–4583.

Wek, R.C., J.F. Cannon, T.E. Dever, and A.G. Hinnebusch. 1992. Truncated protein phosphatase GLC7 restores translational activation of *GCN4* expression in yeast mutants defective for the eIF-2α kinase GCN2. *Mol. Cell. Biol.* **12:** 5700–5710.

Wek, R.C., M. Ramirez, B.M. Jackson, and A.G. Hinnebusch. 1990. Identification of positive-acting domains in GCN2 protein kinase required for translational activation of *GCN4* expression. *Mol. Cell. Biol.* **10:** 2820–2831.

Williams, N.P., A.G. Hinnebusch, and T.F. Donahue. 1989. Mutations in the structural genes for eukaryotic initiation factors 2α and 2β of *Saccharomyces cerevisiae* disrupt translational control of *GCN4* mRNA. *Proc. Natl. Acad. Sci.* **86:** 7515–7519.

Williams, N.P., P.P. Mueller, and A.G. Hinnebusch. 1988. The positive regulatory function of the 5′-proximal open reading frames in *GCN4* mRNA can be mimicked by heterologous, short coding sequences. *Mol. Cell. Biol.* **8:** 3827–3836.

Wolfner, M., D. Yep, F. Messenguy, and G.R. Fink. 1975. Integration of amino acid biosynthesis into the cell cycle of *Saccharomyces cerevisiae*. *J. Mol. Biol.* **96:** 273–290.

8

mRNA 5′ Cap-binding Protein eIF4E and Control of Cell Growth

Nahum Sonenberg
Department of Biochemistry and McGill Cancer Center
McGill University, Montreal, Quebec
Canada, H3G 1Y6

Modulation of translation rates accompanies major biological processes including proliferation, differentiation, and development. A general increase in the rate of translation is required for both entry into and transit through the cell cycle (Brooks 1977). Reentry of cells from the resting state, G_0, into the cell cycle is effected by several extracellular stimuli. These stimuli include growth factors and hormones that activate multiple signaling pathways that enhance transcription and translation rates and stimulate cell proliferation. Several major signal transduction pathways have been elucidated, and many of the components that link extracellular receptors to the translational and transcriptional machineries have been characterized. Many of these components function in phosphorylation cascades.

Translation rates increase in response to treatment with growth factors, cytokines, hormones, and mitogens (for reviews, see Rhoads 1991; Frederickson and Sonenberg 1993; Sonenberg 1993; Morris 1995). The list of translation components, whose involvement in control of cell growth has been documented, is large and expanding. These include the translation initiation factors (eIF, for eukaryotic initiation factor) eIF2, eIF2B, eIF4E, possibly eIF4B, eIF4G, and eIF3; the elongation factors eEF1 and eEF2; and the ribosomal protein S6. Evidence also exists in yeast that translation initiation controls G_1 progression, and mutations in several eIFs cause an early G_1 arrest (Brenner et al. 1988; Hartwell and McLaughlin 1968). The large number of eIFs implicated in control of cell growth indicates that the mechanisms by which translation rates modulate cell growth are highly complex and integrate the activities of numerous elements. Most of the control of translation occurs at the initiation step, which is usually the rate-limiting step in translation (see Mathews et al., this volume). This regulation involves the reversible phosphorylation of key initiation factors, such as eIF2 and eIF4E. These translation components are phosphorylated in response to a wide variety of extracellular stimuli. Phosphorylation of several initiation factors (eIF2B, eIF4B, eIF4E, eIF4G) positively correlates with increased translation rates and cell proliferation, whereas phosphorylation of another eIF (eIF2) results in inhibition of translation and suppression of cell growth.

Of particular importance to the understanding of the regulation of cell growth by initiation factors are eIF4E, the messenger RNA 5′cap-binding protein, and eIF2. This is underscored by the finding that overexpression of eIF4E in rodent fibroblasts results in malignant transformation, which is consistent with the idea that it is an important transducer of growth signals (De Benedetti et al. 1994; Lazaris-Karatzas et al. 1990, 1992; Lazaris-Karatzas and Sonenberg 1992). On the other hand, expression of a dominant negative mutant of PKR (double-stranded RNA-activated kinase), the kinase that phosphorylates and inactivates eIF2, causes malignant transformation in NIH-3T3 cells (see Clemens, this volume). Furthermore, expression of a nonphosphorylatable mutant of eIF2 also causes transformation in NIH-3T3 cells (Donzé et al. 1995). These findings demonstrate that aberrant expression or regulation of translation factors can cause malignancy. Inasmuch as phosphorylation of eIF2 and eIF4E modulates their activity, it is important to elucidate the signaling pathways responsible for the phosphorylation of these initiation factors. This should increase our understanding of how extracellular stimuli and oncogenes effect cellular proliferation and the part that translation plays in this process. This chapter deals chiefly with the role of eIF4E in the control of cell growth, as the evidence for such a role is compelling. The function of other mRNA-binding initiation factors in the control of cell growth is summarized briefly as the available information is still limited. A number of chapters in this volume describe the involvement of additional translation components in growth control: eIF2 (Clemens; Hinnebusch), eEF2 (Nairn and Palfrey), and S6 (Jefferies and Thomas; Meyuhas).

MECHANISM OF mRNA BINDING TO RIBOSOMES IN EUKARYOTES

Several alternative models for the mechanism of mRNA binding to the ribosome are reviewed by Merrick and Hershey (this volume). One of these models is shown in Figure 1. Ribosome binding to mRNA is an energy-dependent process, which requires ATP hydrolysis. Three initiation factors, eIF4A, eIF4B, and eIF4F, are required for mRNA ribosome

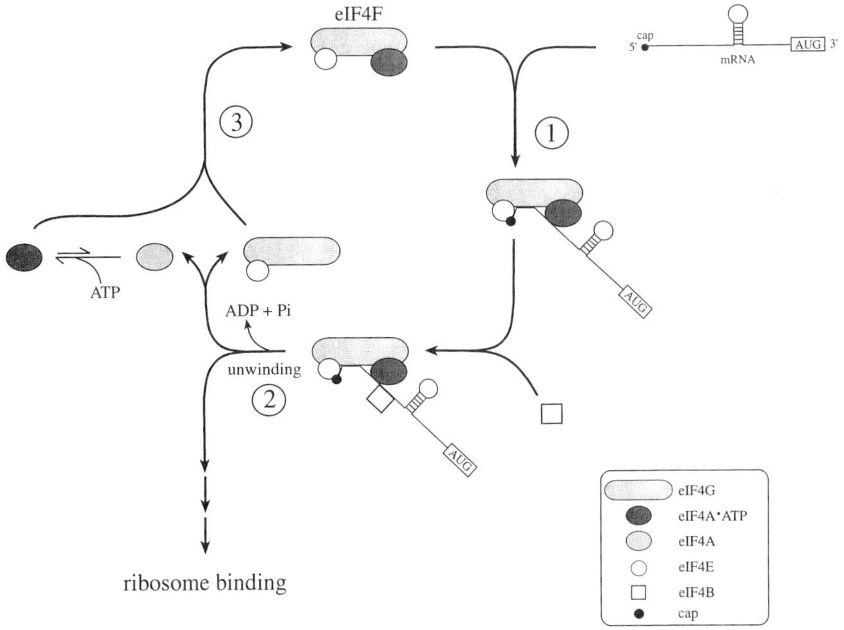

Figure 1 Model for the mechanism of action of mRNA-binding initiation factors. The steps leading to ribosome binding are (1) binding of eIF4F to the cap structure of the mRNA, followed by the binding of eIF4B, (2) unwinding of the proximal mRNA 5′secondary structure, in an ATP-dependent manner, and (3) recycling of eIF4A through the eIF4F complex. It is possible that eIF4A cycles through the eIF4F complex on the mRNA. It is also possible that eIF4F (perhaps through eIF4G) mediates ribosome association prior to its release from the mRNA. (Modified from Pause et al. 1994a.)

binding. eIF4A, which in mammals exists as two different but very similar gene products (Nielsen and Trachsel 1989), is an RNA-dependent ATPase, and in conjunction with eIF4B exhibits RNA helicase activity (Abramson et al. 1987; Rozen et al. 1990). eIF4F in mammals is a three-subunit complex (Tahara et al. 1981; Edery et al. 1983; Grifo et al. 1983) that is composed of (1) eIF4A, (2) eIF4E, a 24-kD polypeptide that specifically interacts with the cap structure (Sonenberg et al. 1978, 1979), and (3) eIF4G, a high-molecular-weight subunit, earlier referred to as p220, that binds to both eIF4E and eIF4A. Recent experiments identified the binding sites of eIF4E and eIF4A on eIF4G; eIF4E binds to the amino-terminal half of eIF4G (Lamphear et al. 1995; Mader et al. 1995), and eIF4A binds to the carboxy-terminal half of eIF4G (Lamphear et al. 1995). In addition, eIF4G can also interact with eIF3 (Lamphear et al.

1995), suggesting that one function of eIF4G is to serve as a scaffold to assemble a complex protein machinery that directs the ribosome to the mRNA. eIF4F also exhibits sequence-nonspecific RNA-binding activity (Jaramillo et al. 1991). This activity could explain the finding that eIF4F is approximately 20 times more potent as an RNA helicase than its catalytic subunit, eIF4A, in its free form (Rozen et al. 1990). These results are consistent with more recent observations that eIF4A does not function as a separate helicase entity, but rather cycles through eIF4F (Yoder-Hill et al. 1993; Pause et al. 1994a). Taken together, these features of eIF4F support the hypothesis that eIF4F functions together with eIF4B in the unwinding of the mRNA 5′secondary structure to promote ribosome binding (Edery et al. 1984; Ray et al. 1985; Abramson et al. 1987; Jaramillo et al. 1991).

A large body of evidence supports the "mRNA helicase hypothesis." For example, components of the helicase machinery can suppress the inhibition of translation caused by secondary structure in the 5′-untranslated region (5′UTR). This was shown in NIH-3T3 cells for eIF4E, which is the limiting component in eIF4F (Koromilas et al. 1992), and for eIF4B in yeast (Altmann et al. 1993). In addition, mRNAs with reduced secondary structure in their 5′UTR function in extracts from poliovirus-infected cells, in which eIF4F is rendered inactive for translation of cap-dependent mRNAs (Sonenberg et al. 1982). However, eIF4A is required for the translation of all mRNAs, including those that have reduced secondary structure in their 5′UTR, inasmuch as translation of all mRNAs is abolished in extracts prepared from yeast in which the eIF4A gene was disrupted (Blum et al. 1992). Moreover, dominant negative mutants of the mammalian eIF4A inhibit translation of all mRNAs when added to translation extracts (Pause et al. 1994a). This is consistent with the suggestion that eIF4A is also involved in additional RNA helicase activities, possibly melting mRNA-rRNA interactions during translation initiation (Blum et al. 1992).

According to the model shown in Figure 1, eIF4F binds first to the mRNA cap structure. The hypothesis that the eIF4F complex, rather than the eIF4E subunit alone (Rhoads 1988; Joshi et al. 1994), binds first to the mRNA cap structure is based on several observations. eIF4E as a subunit of eIF4F cross-links more efficiently to the cap than eIF4E alone (Lee et al. 1985). This is consistent with the finding that eIF4E in extracts prepared from poliovirus-infected cells cross-links inefficiently to the cap, although it is not itself modified or inactivated (Lee and Sonenberg 1982). Rather, the eIF4G subunit is cleaved, resulting in the separation of eIF4E from eIF4A (Ehrenfeld; Mathews; both this volume). The binding affinity of eIF4F to the cap is partly determined by the interaction of the cap-binding subunit, eIF4E, with the cap structure. However, binding of eIF4F to the mRNA is likely to be dependent not only on the cap, but also on eIF4G, as this subunit binds strongly to RNA (Jaramillo et al. 1991). Indeed, eIF4F binds much more avidly to RNA than either eIF4E or eIF4A in a gel-shift assay (Jaramillo et al. 1991). Binding of eIF4F to the mRNA is postulated to be followed by the association of eIF4B, and melting of secondary structure (Jaramillo et al. 1991), thus creating a single-stranded RNA region that serves as a ribosome-binding site. It is not clear from this model to what extent the 5′UTR must be melted prior to ribosome binding. The model also does not address the question of where and when the recycling of eIF4A occurs—either on the mRNA or in solution. It should be emphasized that ribosome-binding models which are different from that described here have been proposed (Merrick and Hershey, this volume). For example, it was suggested that eIF4E alone binds first to the mRNA, which then binds to the 43S ribosome complex that is already associated with eIF4G (Joshi et al. 1994; Rhoads et al. 1994). A different model posits that ribosomes plus associated factors bind directly to the cap structure followed by melting of secondary structure and scanning (Kozak 1989).

eIF4E PHOSPHORYLATION AND REGULATION OF TRANSLATION AND CELL GROWTH

mRNA binding to ribosomes is regulated under different physiological conditions, and regulation is tightly correlated with the growth status of the cell. The activity of several mRNA-binding initiation factors and ribosomal protein S6 is modulated under these conditions. eIF4E is one of the best characterized of these components. A priori, two features of eIF4E render it an excellent candidate as a key player in regulation of translation and cell growth. First, it is present in limiting amounts relative to other initiation factors, 0.01–0.2 molecules per ribosome, as compared to 0.5–3 molecules per ribosome for other initiation factors (Hiremath et al. 1985; Duncan et al. 1987). Second, there is an excellent correlation between the phosphorylation state of eIF4E and translation rates in vivo. For example, during mitosis (Bonneau and Sonenberg 1987; Huang and Schneider 1991) or following heat shock (Duncan et al. 1987; Duncan and Hershey 1989; Lamphear and Panniers 1990; Duncan, this volume), translation is reduced concomitantly with a reduction in eIF4E phosphorylation. In addition, a decrease in eIF4E phosphorylation correlates with the shutoff of host-protein synthesis following infection with several viruses (Feigenblum and Schneider 1993; Huang and Schneider 1991; Schneider, this volume). Conversely, eIF4E is phosphorylated in response to treatment of cells with numerous growth factors and mitogens, as well as upon activation of T and B lymphocytes (for reviews, see Rhoads 1991; Frederickson and Sonenberg 1993; Sonenberg 1993).

Posttranslational modification of eIF4E was initially suggested by the finding that the purified protein from rabbit reticulocyte lysate could be resolved into two isoelectric variants (Sonenberg et al. 1979). Additional minor forms have been detected in human erythrocytes and HeLa cells, but at least some of them may have originated during the preparation of the samples for analysis (Buckley and Ehrenfeld 1986; Rychlik et al. 1986). Rychlik et al. (1986) demonstrated that the two major isoelectric variants of eIF4E in reticulocytes have isoelectric points of 5.9 and 6.3 and that the more acidic form is phosphorylated. Using tryptic peptide analyses, phosphorylation was detected primarily on a single peptide (Frederickson et al. 1991, 1992; Morley and Traugh 1990). The phosphorylated amino acid was initially identified as Ser-53 (Rychlik et al. 1987). However, in a more recent analysis, the major phosphorylation site was reassigned to Ser-209 (Joshi et al. 1995). In addition, an insulin-stimulated protamine kinase was reported to phosphorylate both Ser-209 and Thr-210 (Makkinje et al. 1995). Protein kinase C (PKC) also phosphorylates Ser-209 preferentially, but it can also phosphorylate Thr-210 when used in large amounts or after a long incubation (A.-C. Gingras and S. Whalen, unpubl.). Furthermore, phosphorylation of Thr-210 was observed in vivo when a Ser-209 to alanine mutant of eIF4E was transiently expressed in the 293 cells (A.-C. Gingras and S. Whalen, unpubl.).

Stimuli that induce eIF4E phosphorylation include various peptide growth factors, such as epidermal growth factor (EGF), platelet-derived growth factor (PDGF), and tumor necrosis factor-α (TNF-α); hormones, such as insulin and angiotensin II; and a differentiation-promoting factor, nerve growth factor (NGF) (Rhoads 1991; Frederickson and Sonenberg 1993). Phosphorylation of eIF4E also increases after mitogenic stimulation of human T cells by phytohemagglutinin (Boal et al. 1993) or of porcine peripheral blood mononuclear cells by either phorbol ester or concanavalin A (Con A) (Morley et al. 1993), concomitant with an increase in the rate of protein synthesis. Extracellular stimuli exert their effects on cell growth and differentiation through binding and activation of specific cell surface receptors, many of which are phosphotyrosine kinases. Tyrosine-phosphorylated receptors recruit SH2-containing proteins, leading to the activation of Ras and downstream serine/threonine/tyrosine protein kinases to transmit both mitogenic and differentiation signals to relevant effector molecules. Thus, oncogenic variants of components of these signaling pathways would be predicted to elevate eIF4E phosphorylation in the absence of serum or growth factors. Indeed, expression of v-*ras*, pp60v-*src*, or a transforming variant of cellular pp60*src* in fibroblasts results in a significant increase in steady-state levels of phosphorylated eIF4E in serum-deprived cells (Frederickson et al. 1991; Rinker-Schaeffer et al. 1992). Phosphorylation of eIF4E is also implicated in the regulation of development as fertilization of sea urchin eggs and starfish oocytes results in a dramatic increase in translation rates coincident with the increase in phosphorylation of eIF4E (Jagus et al. 1993; Xu et al. 1993). eIF4E, presumably in its phosphorylated form, was also shown to have an important role in a later stage of *Xenopus* development by activating an inductive signal required for mesoderm formation (Klein and Melton 1994). Injection of mouse eIF4E mRNA into early *Xenopus* embryos led to mesoderm induction in ectodermal explants, in a Ras-dependent manner. Furthermore, eIF4E stimulated the synthesis of activin, a member of the transforming growth factor-β (TGF-β) family, in *Xenopus* embryos. Since TGF-β activates Ras, it was postulated that eIF4E establishes a positive feedback autocrine loop that results in Ras activation and induction of mesoderm formation. Mechanistically, this model is similar to the induction of the Ras pathway in mammalian cells by overexpression of eIF4E (Lazaris-Karatzas et al. 1992), which has a role in malignant transformation (see below).

It is not certain what the physiological kinase of eIF4E is, although PKC appears to be the strongest candidate. Phorbol esters, which bind to and activate various PKC subtypes (Nishizuka 1988), induce eIF4E phosphorylation in several cell lines and in T cells (Morley and Traugh 1989, 1990; Frederickson et al. 1992; Boal et al. 1993). Significantly, insulin-induced phosphorylation of eIF4E in 3T3-L1 cells is highly dependent on PKC as down-regulation of PKC through long-term exposure of cells to phorbol ester prevents phosphorylation (Morley and Traugh 1990). Similarly, angiotensin-II-induced phosphorylation of eIF4E is mediated by PKC (Rao et al. 1994). Moreover, coinjection of eIF4E and PKC potentiates the mitogenic activity of eIF4E in quiescent 3T3 cells (Smith et al. 1991). The significance of these studies is strengthened by the findings that PKC phosphorylates eIF4E in vitro (Smith et al. 1991; Tuazon et al. 1989, 1990) and that this phosphorylation occurs on Ser-209 (A.-C. Gingras and S. Whalen, unpubl.). However, there is also evidence for eIF4E phosphorylation that is independent of the conventional isoforms of PKC, as the NGF-induced response in PC12 cells is unaffected by PKC down-regulation (Frederickson et al. 1992). Moreover, in B lymphocytes stimulated by lipopolysaccharide (LPS), eIF4E phosphorylation is insensitive to H7 and HA1004, two inhibitors of PKC and cyclic nucleotide-dependent kinases (Rychlik et al. 1990). However, it is important to note that there are PKC isoforms, such as PKC ζ (Ways et al. 1992), that are not activated by phorbol esters and are therefore not down-regulated by prolonged treatment with phorbol esters. It is also conceivable that other kinases mediate eIF4E phosphorylation depending on the cell and the effector in question. One reported candidate is an insulin-stimulated protamine kinase that also phosphorylates eIF4E on Ser-209 at an even lower K_m than PKC (Makkinje et al. 1995).

The regulation of the eIF4E phosphorylation state might involve modulation of both eIF4E kinases and phosphatases. Treatment of cells in culture with the phosphatase inhibitor okadaic acid stimulated the EGF-induced phosphorylation of eIF4E (Donaldson et al. 1991). However, it should be noted that okadaic acid could inhibit a phosphatase that acts upstream in the phosphorylation cascade, rather than directly on eIF4E. B lymphoid cells that were activated with LPS or phorbol ester showed an increased rate of phosphate turnover, as the increase in the amount of the phosphorylated form (~2–3-fold) could not fully account for the enhanced labeling (~50-fold) (Rychlik et al. 1990). Similar changes in phosphate turnover rates were noted in NGF-stimulated PC12 cells (Frederickson et al. 1992), Src-transformed NIH-3T3 cells (Frederickson et al. 1991), and Ras-transformed cloned rat embryo fibroblasts (CREFs) (Rinker-Schaeffer et al. 1992). Such an increased turnover of the phosphate moiety would be expected if it were consumed during each translation initiation cycle. However, there is no current biochemical evidence to support this notion. Nevertheless, these findings suggest that an eIF4E phosphatase activity is a potential target for regulating the phosphorylation state of eIF4E and its activity.

The increase of eIF4E phosphorylation by such a wide variety of growth-promoting signals is consistent with an important role for eIF4E in the mitogenic response that is effected by an increase in translational efficiency. Phosphorylation appears to enhance eIF4E activity, since only the phosphorylated form of eIF4E is present in the 48S mRNA ribosome complex (Joshi-Barve et al. 1990). In addition, phosphorylation of eIF4F (in this case phosphorylation of both eIF4E and eIF4G) by PKC increases its activity in mRNA translation (Morley et al. 1991). How does phosphorylation of eIF4E increase translation rates? eIF4E phosphorylation following mitogenic stimulation of T cells with phorbol esters or

Con A resulted in an increase in its association with eIF4G and eIF4A to assemble the cap-binding complex, eIF4F (Morley et al. 1993). Consistent with these findings, heat shock treatment resulted in reduced phosphorylation of eIF4E and reduced recovery of the eIF4F complex by cap-column chromatography (Lamphear and Panniers 1991). These changes are concomitant with the shutoff of cap-dependent host-protein translation after heat shock treatment (Lamphear and Panniers 1991). As noted above, there is evidence that eIF4F binds with higher affinity to the mRNA cap structure than eIF4E alone, as determined by cross-linking studies (Lee et al. 1985). In addition, eIF4F has been reported to bind better to a matrix-linked cap structure relative to eIF4E (Bu et al. 1993). Taken together, these results are consistent with the idea that the phosphorylation of eIF4E increases the amount of eIF4F complex in the cell with subsequent increased binding to the mRNA. Another explanation is that eIF4E phosphorylation directly increases its affinity to the cap structure, since phosphorylated eIF4E exhibits a threefold higher binding affinity to a cap-column relative to the unphosphorylated form (Minich et al. 1994). These two possibilities are not mutually exclusive.

PROTEINS THAT LINK TRANSLATION INITIATION AND GROWTH-PROMOTING SIGNAL TRANSDUCTION PATHWAYS

The understanding of the mechanism by which growth-promoting signals are relayed to the translation machinery has been greatly enhanced by the recent characterization of two proteins that link the insulin pathway and the translation machinery. Pause et al. (1994b) have described two low-molecular-weight (118 and 120 amino acids) homologous human proteins that can complex with eIF4E and inhibit cap-dependent, but not cap-independent, translation. These proteins were termed 4E-BP1 and 4E-BP2 (for 4E-binding protein; a recent survey of the databanks revealed the existence of a third member of this family). A rat cDNA homolog of 4E-BP1 was independently cloned and found to encode a protein termed PHAS-I (for *p*hosphorylated *h*eat- and *a*cid-*s*table protein regulated by *i*nsulin) (Hu et al. 1994). This protein had been characterized earlier in several laboratories as a major phosphorylation target in response to treatment of cells with insulin and growth factors (Belsham and Denton 1980; Belsham et al. 1982; Blackshear et al. 1982, 1983). The major phosphorylation site of 4E-BP1 following activation is Ser-64, which is phosphorylated in vitro by the *m*itogen-*a*ctivated *p*rotein (MAP) kinases ERK1 (*e*xtracellular *r*egulated protein *k*inase-1) and ERK2 (Haystead et al. 1994; Lin et al. 1994). Stimulation of 3T3-L1 adipocytes with

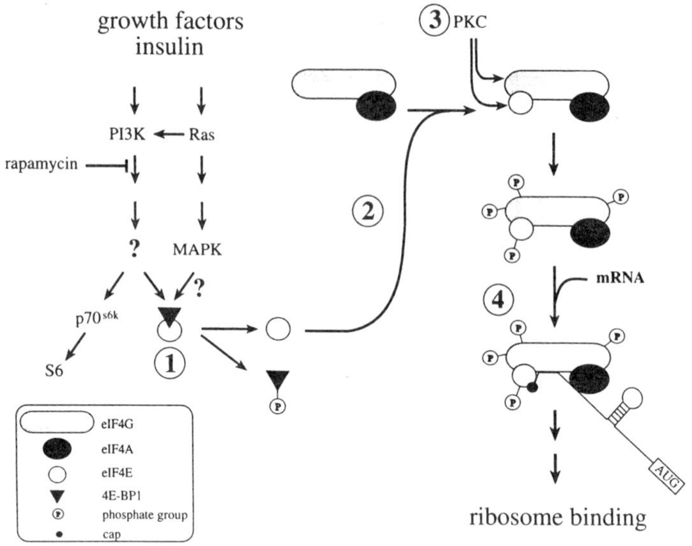

Figure 2 Mechanism of stimulation of translation by growth factors and hormones, such as insulin, through the activation of eIF4E. Binding of 4E-BP1 to eIF4E prevents the interaction of eIF4E with eIF4G to form the complex eIF4F. In step 1, growth factors and insulin cause the phosphorylation of 4E-BP1 by a rapamycin-sensitive kinase, which has not been identified. The involvement of MAPK was reported, but is now questionable (see text). Phosphorylation of 4E-BP1 results in its dissociation from eIF4E, which is available to form an active cap-binding complex (eIF4F) with eIF4A and eIF4G (step 2). In step 3, PKC phosphorylates eIF4E and eIF4G (the number of phosphates added is not known). eIF4F that contains a phosphorylated eIF4E has a higher affinity for the mRNA cap structure, thus enhancing binding to the mRNA (step 4). Insulin and growth factors were reported to stimulate the phosphorylation of eIF4E through a PKC-dependent pathway.

EGF or insulin led to an increase in MAP kinase activity. This was the only detectable kinase activity, and it cofractionated with the kinase activity that phosphorylated PHAS-I (4E-BP1) (Lin et al. 1994). ERK1 and ERK2 become activated by numerous extracellular stimulators of cell growth, primarily via the Ras signal transduction pathway (Blenis 1993; Davis 1994; Myers et al. 1994).

Consistent with a role for MAP kinase in regulating 4E-BP1 association with eIF4E, phosphorylation of 4E-BP1 by MAP kinase renders the protein incapable of interacting with eIF4E (Lin et al. 1994). This finding provides an attractive mechanism for the stimulation of translation in response to treatment of cells with hormones, such as insulin, and growth factors (Fig. 2). Such a mechanism predicts that only cap-dependent translation would be stimulated and that mRNAs which translate poorly because of extensive secondary structure in their 5′UTR would be preferentially stimulated. This is indeed the case: Insulin stimulates specifically cap-dependent translation (Gallie and Traugh 1994). In addition, translation of ornithine decarboxylase (ODC) mRNA, which contains a high amount of secondary structure in its 5′UTR, is increased approximately 30-fold after insulin stimulation, whereas general translation is increased only approximately 2-fold (Manzella et al. 1991).

The involvement of MAP kinases in phosphorylation of 4E-BP1 in vivo has been seriously questioned, however, for several reasons. (1) In recent experiments, it was shown that phosphorylation of 4E-BP1 is inhibited by the immunosuppressant drug rapamycin in smooth muscle cells and several cell lines, including fibroblasts (3T3-L1 and NIH-3T3) and Jurkat T cells (Graves et al. 1995; Lin et al. 1995; Beretta et al. 1996). Rapamycin is a specific inhibitor of $p70^{s6k}$ and does not inhibit the MAP kinase pathway (see Jefferies and Thomas, this volume). (2) An inhibitor of MAP kinase kinase (MAPKK or MEK), which phosphorylates MAP kinase, PD 058059, drastically inhibited insulin-stimulated MAP kinase activity in 3T3-L1 cells but had no effect on the phosphorylation of 4E-BP1 (Lin et al. 1995). (3) The insulin growth factor, IGF-1, increased the phosphorylation of 4E-BP1 in aortic smooth muscle cells, without an increase in MAP kinase activity (Graves et al. 1995). (4) Stimulation of 4E-BP1 phosphorylation by insulin was demonstrated to occur in the absence of MAP kinase activation in 293 cells (von Manteuffel et al. 1996). (5) Angiotensin II stimulated the phosphorylation of 4E-BP1 in vascular smooth muscle cells by a MAP kinase-independent mechanism (M. Fleurent et al., unpubl.). Taken together, these results indicate that the major 4E-BP1 phosphorylation pathway is rapamycin-sensitive and not the MAP kinase pathway initially reported (Haystead et al. 1994). However, the identity of the rapamycin-sensitive kinase that phosphorylates the 4E-BPs in vivo is not known. The rapamycin-sensitive S6 kinase, $p70^{s6k}$, does not directly phosphorylate 4E-BP1 and 4E-BP2 in vitro (Haystead et al. 1994; A.-C. Gingras and N. Sonenberg, unpubl.). 4E-BP1 can be phosphorylated in vitro by casein kinase II (CKII) and PKC (Haystead et al. 1994); however, the physiological significance of this is not known.

How do 4E-BPs inhibit translation? Binding of either 4E-BP1 or 4E-BP2 to eIF4E does not prevent eIF4E interaction with the cap structure (Lin et al. 1994; Pause et al. 1994b). It was therefore hypothesized that 4E-BPs compete with eIF4G for association to eIF4E (see Fig. 2). This hypothesis has recently been verified by showing that 4E-BPs and eIF4G cannot bind simultaneously to eIF4E (Haghighat et al. 1995). The binding site for eIF4E on eIF4G has been recently characterized (Lamphear et al. 1995; Mader et al. 1995). Interestingly, a similar binding site is also present in the two 4E-BPs (Mader et al. 1995). Mutations in the common sequence abrogate the binding of 4E-BPs and eIF4G to eIF4E. Thus, 4E-BPs and eIF4G share a common motif for binding to eIF4E. As discussed earlier, eIF4F binds better than eIF4E to the mRNA cap structure (Lee et al. 1985), and therefore the prevention of eIF4F assembly by 4E-BPs would interfere with efficient cap binding.

The existence of 4E-BP1 and 4E-BP2 and the regulation of their binding to eIF4E by extracellular signals might explain, at least in part, some of the biological activities of eIF4E that were reported earlier (Lazaris-Karatzas et al. 1992). For example, inhibition of Ras activity by overexpression of GAP (GTPase-activating factor) reversed the malignantly transformed phenotype of eIF4E-overexpressing cells (Lazaris-Karatzas et al. 1992). Likewise, microinjection of antibodies against Ras strongly diminished the mitogenic activity of eIF4E (Lazaris-Karatzas et al. 1992). Since several signaling pathways lie downstream from Ras, the possibility cannot be excluded that the inhibition of signaling by the Ras antagonists results in the dephosphorylation of 4E-BPs, enhanced binding to eIF4E, and reduction in eIF4E function.

A pertinent question concerns the effect of 4E-BP1 and 4E-BP2 on eIF4E phosphorylation. For example, what is the temporal relationship between eIF4E binding to 4E-BPs and eIF4E phosphorylation? Can phosphorylation of eIF4E occur when complexed to 4E-BPs? Recent experiments demonstrated that phosphorylation of eIF4E by PKC in vitro is prevented by the interaction of 4E-BP1 with eIF4E (A.-C. Gingras and S. Whalen, unpubl.). These results might also explain earlier findings showing that PKC phosphorylates eIF4E poorly, but eIF4F efficiently (Tuazon et al. 1990). It is plausible that the eIF4E preparation that was prepared from reticulocytes was complexed with the 4E-BPs, which prevented phosphorylation by PKC. Thus, a likely scenario for the activation of eIF4E is the initial dissociation of 4E-BPs from eIF4E, followed by eIF4E phosphorylation and the assembly of eIF4F (see Fig. 2)

eIF4E AND MALIGNANT TRANSFORMATION

The correlation between the regulation of eIF4E activity and cell growth underlies the findings that deregulation of eIF4E activity transforms rodent cells to malignancy (Lazaris-Karatzas et al. 1990, 1992; Lazaris-Karatzas and Sonenberg 1992; De Benedetti et al. 1994) or causes aberrant growth of HeLa cells (De Benedetti and Rhoads 1990). Consistent with these findings, eIF4E is mitogenic as microinjection of eIF4E into quiescent 3T3 fibroblasts induces DNA synthesis (Smith et al. 1990). The initial report on oncogenic activity of eIF4E showed that overexpression of wild-type eIF4E, but not of a mutant containing a change of Ser-53 to alanine, transforms NIH-3T3 cells to malignancy (Lazaris-Karatzas et al. 1990). The Ser-53 mutation was chosen, because this Ser-53 was assigned as the major phosphorylation site of eIF4E (Rychlik et al. 1987). However, as noted above, the phosphorylation site has been reassigned to Ser-209 (Lamphear et al. 1995; Makkinje et al. 1995). Nevertheless, the Ser-53 mutation inactivates a biochemical property of eIF4E, as it prevented the association of eIF4E with the 43S preinitiation complexes (Joshi-Barve et al. 1990). These findings could explain the lack of activity of mutant Ser-53 in several biological assays, including transformation, aberrant cell growth, and activation of DNA synthesis (see below). Subsequent experiments showed that eIF4E also transforms nonimmortalized rat embryo fibroblasts, in collaboration with immortalizing genes such as E1A or v-*myc* (Lazaris-Karatzas and Sonenberg 1992). Thus, eIF4E can replace Ras in the two-oncogene transformation assay. This is consistent with the finding that Ras becomes activated in eIF4E-transformed cells, as the ratio of GTP•Ras to GDP•Ras increased dramatically (Lazaris-Karatzas et al. 1992). The constitutive activity of Ras in these cells could explain the transformation phenotype. The central role that Ras has in control of the eIF4E-mediated cell growth and transformation was demonstrated by several experiments. Overexpression of GAP, which promotes the conversion of GTP•Ras to GDP•Ras and thus inhibits Ras function, in cells transformed by eIF4E, reverses the transformation phenotype (Lazaris-Karatzas et al. 1992). In addition, neutralizing anti-Ras antibodies and a dominant negative mutant of Ras inhibited the activation of DNA synthesis following microinjection of eIF4E (Smith et al. 1991; Lazaris-Karatzas et al. 1992).

One possible pathway underlying the activation of Ras and transformation by eIF4E is through an autocrine positive feedback loop mechanism (Fig. 3). According to this model, eIF4E enhances preferentially the translation of mRNAs encoding growth-promoting proteins, such as growth factors, growth factor receptors, and other growth-promoting proteins. Enhanced expression of any of these proteins would result in the increased secretion of growth factors with subsequent elevation of growth factor receptor occupancy. Most if not all of the growth factor receptors signal by activating Ras (Maruta and Burgess 1994). Ras activation by eIF4E overexpression results in increased phosphorylation of

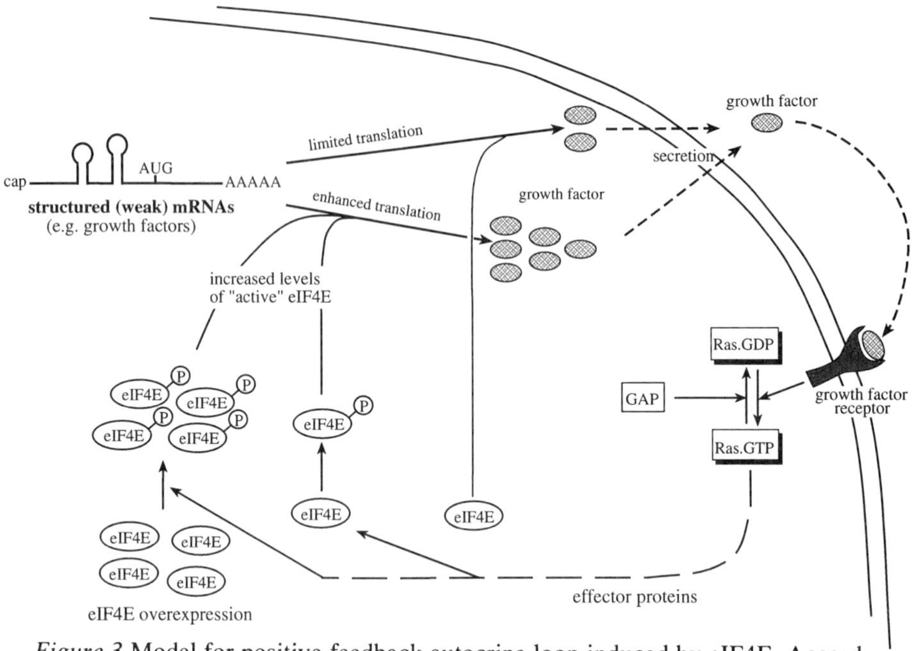

Figure 3 Model for positive feedback autocrine loop induced by eIF4E. According to this model, growth factors that activate the Ras pathway induce the phosphorylation of eIF4E, which results in its activation. Active eIF4E preferentially stimulates the translation of weak mRNAs, such as those encoding growth factors, which when secreted will bind to their receptors and further activate the Ras pathway. This establishes an autocrine positive feedback loop. Overexpression of eIF4E results in constitutive activation of the Ras pathway and leads to malignant transformation. (Adapted from Lazaris-Karatzas et al. 1992.)

eIF4E and translational activation. Thus, a positive feedback loop is established. A similar model could explain the eIF4E inductive signal during development in *Xenopus*, as described above.

What is the molecular basis for the specificity of translational enhancement by eIF4E? A large proportion of growth-related mRNAs contain long 5′UTRs with the potential to form extensive secondary structures that reduce translation efficiency (Pelletier and Sonenberg 1985). It is thought that because the mRNA helicase activity of eIF4F is limiting in cells, these mRNAs are discriminated against by the translational machinery. By overexpressing eIF4E, which is the limiting component of the helicase complex, eIF4F, these mRNAs would be translated more efficiently. Consistent with this hypothesis, insertion of an increased amount of secondary structure into the 5′UTR of mRNAs, which normally inhibits translation, had no such effect in eIF4E-overexpressing cells (Koromilas et al. 1992).

A most pertinent question is the identity of the target mRNA(s) whose translation is activated in eIF4E-overexpressing cells. There is likely to be more than one mRNA species in this class as the transformed phenotype elicited by overexpression of eIF4E is very striking. The latency period of tumor formation in nude mice is very short, 10–15 days, which is comparable to classical oncogenes, such as Ha-*ras* (Lazaris-Karatzas et al. 1990; Lazaris-Karatzas and Sonenberg 1992). Several target mRNAs were described in eIF4E-overexpressing NIH-3T3 cells: Increased expression of cyclin D1 (Rosenwald et al. 1993a), ODC (Shantz and Pegg 1994), and P23 (Bommer et al. 1995) was demonstrated. Some of these proteins participate in control of cell growth: Cyclin D1 is required for cell cycle progression from G_1 to S (Xiong et al. 1991). ODC is a rate-limiting enzyme in polyamine synthesis, which has an essential role in cell growth and differentiation (Pegg 1988). It is highly significant that overexpression of cyclin D1 and ODC proteins may result in malignant transformation in various systems (Auvinen et al. 1992; Hinds et al. 1994; Lovec et al. 1994). Additional potential target mRNAs were described in other cell lines: Increased expression of c-Myc and several forms of human fibroblast growth factor-2 (FGF-2) were demonstrated in Chinese hamster ovary cells overexpressing eIF4E (De Benedetti et al. 1994; A. De Benedetti, pers. comm.). The translation of ornithine aminotransferase is stimulated by overexpression of eIF4E in a retinoblastoma cell line (Fagan et al. 1991), and transcription factor NF-AT (nuclear factor of activated T cells) expression is increased by eIF4E in the $CD4^+$ T_H2 subset of T cells (Barve et al. 1994). The effects of the 5′UTRs of some of the potential target mRNAs on translation have been studied. Human FGF-5 (Bates et al. 1991), ODC (Manzella et al. 1991), and c-Myc (Darveau et al. 1985) mRNAs contain long and complex 5′UTRs and are translationally regulated. Thus, it appears that eIF4E overexpression activates multiple parallel pathways that engender dramatic changes in cell growth control.

eIF4E might also have an important role in the transformation process mediated by established oncogenes. c-Myc has a critical role in growth control as it is an immediate early gene product that is expressed after mitogenic stimulation and encodes a transcriptional activator (Marcu et al. 1992). Furthermore, its transcriptional activation by translocation or amplification is associated with several tumors (Marcu et al. 1992). Rosenwald et al. (1993b) made use of an estrogen receptor–c-Myc fusion protein, whose expression is regulated by dexamethasone, and showed a correlation between expression of c-Myc and an increase in eIF4E mRNA level. Moreover, the eIF4E promoter that has been recently cloned contains Myc consensus binding sites (E. Schmidt, pers. comm.). Thus, the eIF4E gene is a possible target for transcriptional activation by c-Myc. On the basis of its temporal expression after mitogenic induction and activation by c-Myc, eIF4E belongs to the delayed early-growth genes (Lanahan et al. 1992). This class of genes is required for progression through the cell cycle. It is therefore plausible that the oncogenic activity of c-Myc is partly mediated through eIF4E.

Several other reports suggest that translation has an important role in tumorigenesis by established oncogenes. Rinker-Schaeffer et al. (1993) demonstrated that expression of antisense RNA to eIF4E in T24-*ras*-transformed cloned rat embryo fibroblasts partially reversed the transformed phenotype of these cells. My laboratory has shown that overexpression of 4E-BP1 and 4E-BP2 in NIH-3T3 cells transformed by v-*ras* or v-*src* or eIF4E partially reverts their transformed phenotype (D. Rousseau, unpubl.). Finally, two reports were recently published showing an increase of eIF4E mRNA or protein in transformed cell lines and in tumors in situ. Elevated levels (7–15-fold) of eIF4E mRNA were reported in a variety of chemically, virally, and oncogenically transformed tumor cell lines (Miyagi et al. 1995). Enhanced expression of eIF4E was also detected in breast carcinomas by immunohistological staining in situ (Kerekatte et al. 1995). It is imperative to determine whether there is a causal relationship between the increase in eIF4E levels and cancer formation. Notwithstanding this reservation, these results raise the possibility that eIF4E can serve as a target for chemotherapy. Inhibitors of eIF4E function would be effective as anticancer drugs if they could be manipulated to diminish the levels or activity of eIF4E in cancer cells to physiological levels.

PROSPECTS AND CONCLUDING REMARKS

In addition to eIF4E, phosphorylation of eIF4A, eIF4B, and eIF4G also appears to regulate mRNA binding to ribosomes and translation during growth stimulation, differentiation, and development.

Phosphorylation of eIF4A occurs in maize under conditions of oxygen starvation and correlates with inhibition of translation (Webster et al. 1991). A similar situation occurs during *Drosophila* development: eIF4A is phosphorylated in *Drosophila* oocytes and becomes dephosphorylated following fertilization (S. Lavoie and P. Lasko, pers. comm.). Translational activity in *Drosophila* oocytes is low, but not completely blocked as in sea urchins, and is stimulated approximately twofold following fertilization. In contrast to these examples, although two isoelectric forms have been observed for mammalian eIF4A, there is no evidence for phosphorylation of eIF4A in mammals or in yeast.

Phosphorylation of the eIF4G subunit of eIF4F also correlates with changes in cell growth. It becomes multiply phosphorylated in response to treatment of quiescent cells with phorbol esters (Morley and Traugh 1989), insulin (Morley and Traugh 1990), and EGF (Bu and Hagedorn 1991). eIF4B, like eIF4G, is phosphorylated at multiple sites. The phosphorylation state of eIF4B correlates with translational activity as serum deprivation and heat shock treatment cause dephosphorylation of eIF4B (Duncan and Hershey 1984, 1987). In contrast, the phosphorylation of eIF4B is enhanced in response to treatment of cells with phorbol esters or serum (Morley and Traugh 1990). Both eIF4B and eIF4G are phosphorylated by several kinases in vitro, notably by S6 kinase and PKC (Tuazon et al. 1989; Morley et al. 1991). eIF4B is also phosphorylated by PKA and CKII (Tuazon et al. 1989). Although there is evidence that eIF4G phosphorylation either by PKC (together with the phosphorylation of eIF4E) or by S6 kinase increases the translational activity of eIF4F (Morley et al. 1991), there is not yet any such evidence for eIF4B.

The accumulated data point to an important role of the mRNA-binding translation factor, eIF4E, in the regulation of translation rates and control of cell growth. However, it is abundantly clear that other initiation factors, in addition to elongation factors, and ribosomal proteins (particularly S6) collaborate with eIF4E in regulating protein synthesis (Hershey 1991; Morley and Thomas 1991; Redpath and Proud 1994). Elucidation of the interactions between the components of the translational machinery, and the effects of their modifications by phosphorylation on translational efficiency, is critical for a full understanding of the control of protein synthesis.

The ability of two initiation factors, a nonphosphorylatable mutant of eIF2α and wild-type eIF4E, to transform cells to malignancy raises the pertinent question as to whether overexpression of other initiation factors would also result in transformation. Such experiments were undertaken in the author's laboratory. Overexpression of the α, β, and γ subunits of eIF2, separately, eIF4A, or eIF5A (the last two are very abundant proteins) in NIH-3T3 cells had no effect on cell growth and failed to transform the cells (A. Lazaris-Karatzas, unpubl.). Because the outcome of these experiments is negative, they should be interpreted with caution.

Nevertheless, the findings that two translation initiation factors, which have key roles in translational control, regulate cell growth, differentiation, and development, and transform cells to malignancy underscore the importance of translation in regulating cell physiology.

ACKNOWLEDGMENTS

I am grateful to A.-C. Gingras for the preparation of the figures and to G. Belsham and the members of my laboratory for their contributions to the work, and for critical comments and suggestions on the manuscript. The author's work is supported by grants from the National Cancer Institute and the Medical Research Council of Canada.

REFERENCES

Abramson, R.D., T.E. Dever, T.G. Lawson, B.K. Ray, R.E. Thach, and W.C. Merrick. 1987. The ATP-dependent interaction of eukaryotic initiation factors with mRNA. *J. Biol. Chem.* **262:** 3826–3832.

Altmann, M., P.P. Müller, B. Wittmer, F. Ruchti, S. Lanker, and H. Trachsel. 1993. A *Saccharomyces cerevisiae* homologue of mammalian translation initiation factor 4B contributes to RNA helicase activity. *EMBO J.* **12:** 3997–4004.

Auvinen, M., A. Paasinen, L.C. Andersson, and E. Holtta. 1992. Ornithine decarboxylase activity is critical for cell transformation. *Nature* **360:** 355–358.

Barve, S.S., D.A. Cohen, A. De Benedetti, R.E. Rhoads, and A.M. Kaplan. 1994. Mechanism of differential regulation of IL-2 in murine Th1 and Th2 T cell subsets. 1. Induction of IL-2 transcription in Th2 cells by up-regulation of transcription factors with the protein synthesis initiation factor 4E. *J. Immunol.* **152:** 1171–1181.

Bates, B., J. Hardin, X. Zhan, K. Drickamer, and M. Goldfarb. 1991. Biosynthesis of human fibroblast growth factor-5. *Mol. Cell. Biol.* **11:** 1840–1845.

Belsham, G.J. and R.M. Denton. 1980. The effect of insulin and adrenaline on the phosphorylation of a 22 000-molecular weight protein within isolated fat cells; possible identification as the inhibitor-1 of the "general phosphatase." *Biochem. Soc. Trans.* **8:** 382–383.

Belsham, G.J., R.W. Brownsey, and R.M. Denton. 1982. Reversibility of the insulin-stimulated phosphorylation of ATP citrate lyase and a cytoplasmic protein of subunit M_r 22000 in adipose tissue. *Biochem. J.* **204:** 345–352.

Beretta, L., A.-G. Gingras, Y. Svitkin, M.N. Hall, and N. Sonenberg. 1996. Rapamycin blocks the phosphorylation of 4E-BP1 and inhibits cap-dependent initiation of translation. *EMBO J.* (in press).

Blackshear, P.J., R.A. Nemenoff, and J. Avruch. 1982. Preliminary characterization of a heat-stable protein from rat adipose tissue whose phosphorylation is stimulated by insulin. *Biochem. J.* **204:** 817–824.

———. 1983. Insulin and growth factors stimulate the phosphorylation of a M_r-22000 protein in 3T3-L1 adipocytes. *Biochem. J.* **214:** 11–19.

Blenis, J. 1993. Signal transduction via the MAP kinases: Proceed at your own RSK. *Proc. Natl. Acad. Sci.* **90:** 5889–5892.

Blum, S., S.R. Schmid, A. Pause, P. Buser, P. Linder, N. Sonenberg, and H. Trachsel. 1992. ATP hydrolysis by initiation factor 4A is required for translation initiation in *Saccharomyces cerevisiae*. *Proc. Natl. Acad. Sci.* **89:** 7664–7668.

Boal, T.R., J.A. Chiorini, R.B. Cohen, S. Miyamoto, R.M. Frederickson, N. Sonenberg, and B. Safer. 1993. Regulation of eukaryotic translation initiation factor expression during T-cell activation. *Biochim. Biophys. Acta* **1176:** 257–264.

Bommer, U.-A., A. Lazaris-Karatzas, A. De Benedetti, P. Nürnberg, R. Benndorf, H. Bielka, and N. Sonenberg. 1995. Translational regulation of the mammalian growth-related protein P23: Involvementof eIF-4E. *Cell. Mol. Biol. Res.* **40:** 633–641.

Bonneau, A.M. and N. Sonenberg. 1987. Involvement of the 24-kDa cap-binding protein in regulation of protein synthesis in mitosis. *J. Biol. Chem.* **262:** 11134–11139.

Brenner, C., N. Nakayama, M. Goebl, K. Tanaka, A. Toh-e, and K. Matsumoto. 1988. CDC33 encodes mRNA cap-binding protein eIF-4E of *Saccharomyces cerevisiae*. *Mol. Cell. Biol.* **8:** 3556–3559.

Brooks, R.F. 1977. Continuous protein synthesis is required to maintain the probability of entry into S phase. *Cell* **12:** 311–317.

Bu, X. and C.H. Hagedorn. 1991. Platelet-derived growth factor stimulates phosphorylation of the 25 kDa mRNA cap binding protein (eIF-4E) in human lung fibroblasts. *FEBS Lett.* **283:** 219–222.

Bu, X., D.W. Haas, and C.H. Hagedorn. 1993. Novel phosphorylation sites of eukaryotic initiation factor-4F and evidence that phosphorylation stabilizes interactions of the p25 and p220 subunits. *J. Biol. Chem.* **268:** 4975–4978.

Buckley, B. and E. Ehrenfeld. 1986. Two-dimensional gel analyses of the 24-kDa cap binding protein from poliovirus-infected and uninfected HeLa cells. *Virology* **152:** 497–501.

Darveau, A., J. Pelletier, and N. Sonenberg. 1985. Differential efficiencies of in vitro translation of mouse c-*myc* transcripts differing in the 5′ untranslated region. *Proc. Natl. Acad. Sci.* **82:** 2315–2319.

Davis, R.J. 1994. MAPKs: New JNK expands the group. *Trends Biochem. Sci.* **19:** 470–473.

De Benedetti, A. and R.E. Rhoads. 1990. Overexpression of eukaryotic protein synthesis initiation factor 4E in HeLa cells results in aberrant growth and morphology. *Proc. Natl. Acad. Sci.* **87:** 8212–8216.

De Benedetti, A., B. Joshi, J.R. Graff, and S.G. Zimmer. 1994. CHO cells transformed by the initiation factor eIF-4E display increased c-*myc* expression but require overexpression of Max for tumorigenicity. *Mol. Cell. Differ.* **2:** 347–371.

Donaldson, R.W., C.H. Hegedorn, and S. Cohen. 1991. Epidermal growth factor or okadaic acid stimulates phosphorylation of eukaryotic initiation factor 4F. *J. Biol. Chem.* **266:** 3162–3166.

Donzé, O., R. Jagus, A.E. Koromilas, J.W.B. Hershey, and N. Sonenberg. 1995. Abrogation of translation initiation factor eIF-2 phosphorylation causes malignant transformation of NIH-3T3 cells. *EMBO J.* **14:** 3828–3834.

Duncan, R. and J.W.B. Hershey. 1984. Heat shock-induced translational alterations in HeLa cells: Initiation factor modifications and the inhibition of translation. *J. Biol. Chem.* **259:** 11882–11889.

———. 1987. Initiation factor protein modifications and inhibition of protein synthesis. *Mol. Cell. Biol.* **7:** 1293–1295.

———. 1989. Protein synthesis and protein phosphorylation during heat stress, recovery, and adaptation. *J. Cell Biol.* **109:** 1467–1481.

Duncan, R., S.C. Milburn, and J.W.B. Hershey. 1987. Regulated phosphorylation and low abundance of HeLa cell initiation factor eIF-4F suggest a role in translational control. Heat shock effects on eIF-4F. *J. Biol. Chem.* **262:** 380–388.

Edery, I., K.A. Lee, and N. Sonenberg. 1984. Functional characterization of eukaryotic mRNA cap binding protein complex: Effects on translation of capped and naturally uncapped RNAs. *Biochemistry* **23:** 2456–2462.

Edery, I., M. Humbelin, A. Darveau, K.A. Lee, S. Milburn, J.W. Hershey, H. Trachsel, and N. Sonenberg. 1983. Involvement of eukaryotic initiation factor 4A in the cap recognition process. *J. Biol. Chem.* **258:** 11398–11403.

Fagan, R.J., A. Lazaris-Karatzas, N. Sonenberg, and R. Rozen. 1991. Translational control of ornithine aminotransferase. Modulation by initiation factor eIF-4E. *J. Biol. Chem.* **266:** 16518–16523.

Feigenblum, D. and R.J. Schneider. 1993. Modification of eukaryotic initiation factor 4F during infection by influenza virus. *J. Virol.* **67:** 3027–3035.

Frederickson, R.M. and N. Sonenberg. 1993. eIF-4E phosphorylation and the regulation of protein synthesis. In *Translational regulation of gene expression* (ed. J. Ilan), pp. 143–162. Plenum Press, New York.

Frederickson, R.M., K.S. Montine, and N. Sonenberg. 1991. Phosphorylation of eukaryotic translation initiation factor 4E is increased in Src-transformed cell lines. *Mol. Cell. Biol.* **11:** 2896–2900.

Frederickson, R.M., W.E. Mushynski, and N. Sonenberg. 1992. Phosphorylation of translation initiation factor eIF-4E is induced in a ras-dependent manner during nerve growth factor-mediated PC12 cell differentiation. *Mol. Cell. Biol.* **12:** 1239–1247.

Gallie, D.R. and J.A. Traugh. 1994. Serum and insulin regulate cap function in 3T3-L1 cells. *J. Biol. Chem.* **269:** 7174–7179.

Graves, L.M., K.E. Bornfeldt, G.M. Argast, E.G. Krebs, X.M. Kong, T.A. Lin, and J.C. Lawrence. 1995. cAMP- and rapamycin-sensitive regulation of the association of eukaryotic initiation factor 4E and the translational regulator PHAS-I in aortic smooth muscle cells. *Proc. Natl. Acad. Sci.* **92:** 7222–7226.

Grifo, J.A., S.M. Tahara, M.A. Morgan, A.J. Shatkin, and W.C. Merrick. 1983. New initiation factor activity required for globin mRNA translation. *J. Biol. Chem.* **258:** 5804–5810.

Haghighat, A., S. Mader, A. Pause, and N. Sonenberg. 1995. Repression of cap dependent translation by 4E-BP1: Competition with p220 for binding to eIF-4E. *EMBO J.* (in press).

Hartwell, L.H. and C.S. McLaughlin. 1968. Temperature-sensitive mutants of yeast exhibiting a rapid inhibition of protein synthesis. *J. Bacteriol.* **93:** 1662–1670.

Haystead, T.A., C.M. Haystead, C. Hu, T.A. Lin, and J. Lawrence, Jr. 1994. Phosphorylation of PHAS-I by mitogen-activated protein (MAP) kinase. Identification of a site phosphorylated by MAP kinase in vitro and in response to insulin in rat adipocytes. *J. Biol. Chem.* **269:** 23185–23191.

Hershey, J.W.B. 1991. Translational control in mammalian cells. *Annu. Rev. Biochem.* **60:** 715–755.

Hinds, P.W., S.F. Dowdy, E.N. Eaton, A. Arnold, and R.A. Weinberg. 1994. Function of a human cyclin gene as an oncogene. *Proc. Natl. Acad. Sci.* **91:** 709–713.

Hiremath, L.S., N.R. Webb, and R.E. Rhoads. 1985. Immunological detection of the messenger RNA cap-binding protein. *J. Biol. Chem.* **260:** 7843–7849.

Hu, C., S. Pang, X. Kong, M. Velleca, and J. Lawrence, Jr. 1994. Molecular cloning and tissue distribution of PHAS-I, an intracellular target for insulin and growth factors. *Proc. Natl. Acad. Sci.* **91:** 3730–3734.

Huang, J.T. and R.J. Schneider. 1991. Adenovirus inhibition of cellular protein synthesis involves inactivation of cap-binding protein. *Cell* **65:** 271–280.

Jagus, R., W. Huang, L.S. Hiremath, B.D. Stern, and R.E. Rhoads. 1993. Mechanism of action of developmentally regulated sea urchin inhibitor of eIF-4. *Dev. Genet.* **14:** 412–423.

Jaramillo, M., T.E. Dever, W.C. Merrick, and N. Sonenberg. 1991. RNA unwinding in translation: Assembly of helicase complex intermediates comprising eukaryotic initiation factors eIF-4F and eIF-4B. *Mol. Cell. Biol.* **11:** 5992–5997.

Joshi, B., R. Yan, and R.E. Rhoads. 1994. In vitro synthesis of human protein synthesis initiation factor 4γ and its localization on 43 and 48 S initiation complexes. *J. Biol. Chem.* **269:** 2048–2055.

Joshi, B., A.L. Cai, B.D. Keiper, W.B. Minich, R. Mendez, C.M. Beach, J. Stepinski, R. Stolarski, E. Darzynkiewicz, and R.E. Rhoads. 1995. Phosphorylation of eukaryotic protein synthesis initiation factor 4E at ser-209. *J. Biol. Chem.* **270:** 14597–14603.

Joshi-Barve, S., W. Rychlik, and R.E. Rhoads. 1990. Alteration of the major phosphorylation site of eukaryotic protein synthesis initiation factor 4E prevents its association with the 48 S initiation complex. *J. Biol. Chem.* **265:** 2979–2983.

Kerekatte, V., K. Smiley, B. Hu, A. Smith, F. Gelder, and A. de Benedetti. 1995. The proto-oncogene translation factor eIF-4E—A survey of its expression in breast carcinomas. *Int. J. Cancer* **64:** 27–31.

Klein, P.S. and D.A. Melton. 1994. Induction of mesoderm in *Xenopus laevis* embryos by translation initiation factor 4E. *Science* **265:** 803–806.

Koromilas, A.E., A. Lazaris-Karatzas, and N. Sonenberg. 1992. mRNAs containing extensive secondary structure in their 5′ non-coding region translate efficiently in cells overexpressing initiation factor eIF-4E. *EMBO J.* **11:** 4153–4158.

Kozak, M. 1989. The scanning model for translation: An update. *J. Cell Biol.* **108:** 229–241.

Lamphear, B.J. and R. Panniers. 1990. Cap binding protein complex that restores protein synthesis in heat-shocked Ehrlich cell lysates contains highly phosphorylated eIF-4E. *J. Biol. Chem.* **265:** 5333–5336.

———. 1991. Heat shock impairs the interaction of cap-binding protein complex with 5′ mRNA cap. *J. Biol. Chem.* **266:** 2789–2794.

Lamphear, B.J., R. Kirchweger, T. Skern, and R.E. Rhoads. 1995. Mapping of functional domains in eIF-4ψ with picornaviral proteases. Implications for cap-dependent and cap-independent translational initiation. *J. Biol. Chem.* **270:** 21975–21983.

Lanahan, A., J.B. Williams, L.K. Sanders, and D. Nathans. 1992. Growth factor-induced delayed early response genes. *Mol. Cell. Biol.* **12:** 3919–3929.

Lazaris-Karatzas, A. and N. Sonenberg. 1992. The mRNA 5′ cap-binding protein, eIF-4E, cooperates with v-*myc* or E1A in the transformation of primary rodent fibroblasts. *Mol. Cell. Biol.* **12:** 1234–1238.

Lazaris-Karatzas, A., K.S. Montine, and N. Sonenberg. 1990. Malignant transformation by a eukaryotic initiation factor subunit that binds to mRNA 5′ cap. *Nature* **345:** 544–547.

Lazaris-Karatzas, A., M.R. Smith, R.M. Frederickson, M.L. Jaramillo, Y.L. Liu, H.F. Kung, and N. Sonenberg. 1992. Ras mediates translation initiation factor 4E-induced malignant transformation. *Genes Dev.* **6:** 1631–1642.

Lee, K.A. and N. Sonenberg. 1982. Inactivation of cap-binding proteins accompanies the shut-off of host protein synthesis by poliovirus. *Proc. Natl. Acad. Sci.* **79:** 3447–3451.

Lee, K.A., I. Edery, and N. Sonenberg. 1985. Isolation and structural characterization of cap-binding proteins from poliovirus-infected HeLa cells. *J. Virol.* **54:** 515–524.

Lin, T.A., X.M. Kong, A.R. Saltiel, P.J. Blackshear, and J.C. Lawrence. 1995. Control of PHAS-I by insulin in 3T3-L1 adipocytes—Synthesis, degradation, and phosphorylation by a rapamycin-sensitive and mitogen-activated protein kinase-independent pathway. *J. Biol. Chem.* **270:** 18531–18538.

Lin, T.A., X. Kong, T.A. Haystead, A. Pause, G. Belsham, N. Sonenberg, and J. Lawrence, Jr. 1994. PHAS-I as a link between mitogen-activated protein kinase and translation initiation. *Science* **266:** 653–656.

Lovec, H., A. Sewing, F.C. Lucibello, R. Muller, and T. Moroy. 1994. Oncogenic activity of cyclin D1 revealed through cooperation with Ha-*ras*: Link between cell cycle control and malignant transformation. *Oncogene* **9:** 323–326.

Mader, S., H. Lee, A. Pause, and N. Sonenberg. 1995. The translation initiation factor eIF-4E binds to a common motif shared by the translation factor eIF-4γ and the translational repressors 4E-binding proteins. *Mol. Cell. Biol.* **15:** 4990–4997.

Makkinje, A., H.S. Xiong, M. Li, and Z. Damuni. 1995. Phosphorylation of eukaryotic protein synthesis initiation factor 4E by insulin-stimulated protamine kinase. *J. Biol. Chem.* **270:** 14824–14828.

Manzella, J.M., W. Rychlik, R.E. Rhoads, J.W. Hershey, and P.J. Blackshear. 1991. Insulin induction of ornithine decarboxylase. Importance of mRNA secondary structure and phosphorylation of eucaryotic initiation factors eIF-4B and eIF-4E. *J. Biol. Chem.* **266:** 2383–2389.

Marcu, K.B., S.A. Bossone, and A.J. Patel. 1992. *myc* function and regulation. *Annu. Rev. Biochem.* **61:** 809–860.

Maruta, H. and A.W. Burgess. 1994. Regulation of Ras signalling networking. *BioEssays* **16:** 489–496

Minich, W.B., M.L. Balasta, D.J. Goss, and R.E. Rhoads. 1994. Chromatographic resolution of in vivo phosphorylated and nonphosphorylated eukaryotic translation initiation factor eIF-4E: Increased cap affinity of the phosphorylated form. *Proc. Natl. Acad. Sci.* **91:** 7668–7672.

Miyagi, Y., A. Sugiyama, A. Asai, T. Okazaki, Y. Kuchino, and S.J. Kerr. 1995. Elevated levels of eukaryotic translation initiation factor eIF-4E, mRNA in a broad spectrum of transformed cell lines. *Cancer Lett.* **91:** 247–252.

Morley, S.J. and G. Thomas. 1991. Intracellular messengers and the control of protein synthesis. *Pharmacol. Ther.* **50:** 291–319.

Morley, S.J. and J.A. Traugh. 1989. Phorbol esters stimulate phosphorylation of eukaryotic initiation factors 3, 4B, and 4F. *J. Biol. Chem.* **264:** 2401–2404.

———. 1990. Differential stimulation of phosphorylation of initiation factors eIF-4F, eIF-4B, eIF-3, and ribosomal protein S6 by insulin and phorbol esters. *J. Biol. Chem.* **265:** 10611–10616.

Morley, S.J., T.E. Dever, D. Etchison, and J.A. Traugh. 1991. Phosphorylation of eIF-4F by protein kinase C or multipotential S6 kinase stimulates protein synthesis at initiation. *J. Biol. Chem.* **266:** 4669–4672.

Morley, S.J., M. Rau, J.E. Kay, and V.M. Pain. 1993. Increased phosphorylation of eukaryotic initiation factor 4α during early activation of T lymphocytes correlates with increased initiation factor 4F complex formation. *Eur. J. Biochem.* **218:** 39–48.

Morris, D. 1995. Growth control of translation in mammalian cells. *Prog. Nucleic Acids Res. Mol. Biol.* **51:** 339–363.

Myers, M.G.J., X.J. Sun, and M.F. White. 1994. The IRS-1 signaling system. *Trends Biochem. Sci.* **19:** 289–295.

Nielsen, P.J. and H. Trachsel. 1989. The mouse protein synthesis initiation factor 4A gene family includes two related functional genes which are differentially expressed. *EMBO J.* **7:** 2097–2105.

Nishizuka, Y. 1988. The molecular heterogeneity of protein kinase C and its implications for cellular regulation. *Nature* **334:** 661–665.

Pause, A., N. Methot, Y. Svitkin, W.C. Merrick, and N. Sonenberg. 1994a. Dominant negative mutants of mammalian translation initiation factor eIF-4A define a critical role for eIF-4F in cap-dependent and cap-independent initiation of translation. *EMBO J.* **13:** 1205–1215.

Pause, A., G.J. Belsham, A.C. Gingras, O. Donze, T.A. Lin, J. Lawrence, Jr., and N. Sonenberg. 1994b. Insulin-dependent stimulation of protein synthesis by phosphorylation of a regulator of 5 ′-cap function. *Nature* **371:** 762–767.

Pegg, A.E. 1988. Polyamine metabolism and its importance in neoplastic growth and as a target for chemotherapy. *Cancer Res.* **48:** 759–774.

Pelletier, J. and N. Sonenberg. 1985. Insertion mutagenesis to increase secondary structure within the 5′ noncoding region of a eukaryotic mRNA reduces translational efficiency. *Cell* **40:** 515–526.

Rao, G.N., K.K. Griendling, R.M. Frederickson, N. Sonenberg, and R.W. Alexander. 1994. Angiotensin II induces phosphorylation of eukaryotic protein synthesis initiation factor 4E in vascular smooth muscle cells. *J. Biol. Chem.* **269:** 7180–7184.

Ray, B.K., T.G. Lawson, J.C. Kramer, M.H. Cladaras, J.A. Grifo, R.D. Abramson, W.C. Merrick, and R.E. Thach. 1985. ATP-dependent unwinding of messenger RNA structure by eukaryotic initiation factors. *J. Biol. Chem.* **260:** 7651–7658.

Redpath, N.T. and C.G. Proud. 1994. Molecular mechanisms in the control of translation by hormones and growth factors. *Biochim. Biophys. Acta* **1220:** 147–162.

Rhoads, R.E. 1988. Cap recognition and the entry of mRNA into the protein synthesis initiation cycle. *Trends Biochem. Sci.* **13:** 52–56.

———. 1991. Protein synthesis, cell growth, and oncogenesis. *Curr. Opin. Cell Biol.* **3:** 1019–1024.

Rhoads, R.E., B. Joshi, and W.B. Minich. 1994. Participation of initiation factors in the recruitment of mRNA to ribosomes. *Biochimie* **76:** 831–838.

Rinker-Schaeffer, C.W., V. Austin, S. Zimmer, and R.E. Rhoads. 1992. *ras*-Transformation of cloned rat embryo fibroblasts results in increased rates of protein synthesis and phosphorylation of eukaryotic initiation factor 4E. *J. Biol. Chem.* **267:** 10659–10664.

Rinker-Schaeffer, C.W., J.R. Graff, A. De Benedetti, S.G. Zimmer, and R.E. Rhoads. 1993. Decreasing the level of translation initiation factor 4E with antisense RNA causes reversal of *ras*-mediated transformation and tumorigenesis of cloned rat embryo fibroblasts. *Int. J. Cancer* **55:** 841–847.

Rosenwald, I.B., A. Lazaris-Karatzas, N. Sonenberg, and E.V. Schmidt. 1993a. Elevated levels of cyclin D1 protein in response to increased expression of eukaryotic initiation factor 4E. *Mol. Cell. Biol.* **13:** 7358–7363.

Rosenwald, I.B., D.B. Rhoads, L.D. Callanan, K.J. Isselbacher, and E.V. Schmidt. 1993b.

Increased expression of eukaryotic translation initiation factors eIF-4E and eIF-2α in response to growth induction by c-myc. *Proc. Natl. Acad. Sci.* **90:** 6175–6178.

Rozen, F., I. Edery, K. Meerovitch, T.E. Dever, W.C. Merrick, and N. Sonenberg. 1990. Bidirectional RNA helicase activity of eucaryotic translation initiation factors 4A and 4F. *Mol. Cell. Biol.* **10:** 1134–1144.

Rychlik, W., M.S. Russ, and R.E. Rhoads. 1987. Phosphorylation site of eukaryotic initiation factor 4E. *J. Biol. Chem.* **262:** 10434–10437.

Rychlik, W., P.R. Gardner, T.C. Vanaman, and R.E. Rhoads. 1986. Structural analysis of the messenger RNA cap-binding protein. Presence of phosphate, sulfhydryl, and disulfide groups. *J. Biol. Chem.* **261:** 71–75.

Rychlik, W., J.S. Rush, R.E. Rhoads, and C.J. Waechter. 1990. Increased rate of phosphorylation-dephosphorylation of the translational initiation factor eIF-4E correlates with the induction of protein and glycoprotein biosynthesis in activated B lymphocytes. *J. Biol. Chem.* **265:** 19467–19471.

Shantz, L.M. and A.E. Pegg. 1994. Overproduction of ornithine decarboxylase caused by relief of translational repression is associated with neoplastic transformation. *Cancer Res.* **54:** 2313–2316.

Smith, M.R., M. Jaramillo, Y.L. Liu, T.E. Dever, W.C. Merrick, H.F. Kung, and N. Sonenberg. 1990. Translation initiation factors induce DNA synthesis and transform NIH-3T3 cells. *New Biol.* **2:** 648–654.

Smith, M.R., M. Jaramillo, P.T. Tuazon, J.A. Traugh, Y.L. Liu, N. Sonenberg, and H.F. Kung. 1991. Modulation of the mitogenic activity of eukaryotic translation initiation factor-4E by protein kinase C. *New Biol.* **3:** 601–607.

Sonenberg, N. 1993. Translation factors as effectors of cell growth and tumorigenesis. *Curr. Opin. Cell Biol.* **5:** 955–960.

Sonenberg, N., D. Guertin, and K.A.W. Lee. 1982. Capped mRNAs with reduced secondary structure can function in extracts from poliovirus-infected cells. *Mol. Cell. Biol.* **2:** 1633–1638.

Sonenberg, N., M.A. Morgan, W.C. Merrick, and A.J. Shatkin. 1978. A polypeptide in eukaryotic initiation factors that crosslinks specifically to the 5′-terminal cap in mRNA. *Proc. Natl. Acad. Sci.* **75:** 4843–4847.

Sonenberg, N., K.M. Rupprecht, S.M. Hecht, and A.J. Shatkin. 1979. Eukaryotic mRNA cap binding protein: Purification by affinity chromatography on sepharose-coupled m^7GDP. *Proc. Natl. Acad. Sci.* **76:** 4345–4349.

Tahara, S.M., M.A. Morgan, and A.J. Shatkin. 1981. Two forms of purified m7G-cap binding protein with different effects on capped mRNA translation in extracts of uninfected and poliovirus-infected HeLa cells. *J. Biol. Chem.* **256:** 7691–7694.

Tuazon, P.T., W.C. Merrick, and J.A. Traugh. 1989. Comparative analysis of phosphorylation of translational initiation and elongation factors by seven protein kinases. *J. Biol. Chem.* **264:** 2773–2777.

Tuazon, P.T., S.J. Morley, T.E. Dever, W.C. Merrick, R.E. Rhoads, and J.A. Traugh. 1990. Association of initiation factor eIF-4E in a cap binding protein complex (eIF-4F) is critical for and enhances phosphorylation by protein kinase C. *J. Biol. Chem.* **265:** 10617–10621.

von Manteuffel, S.R., A.-C. Gingeras, N. Sonenberg, and G. Thomas. 1996. 4E-BP1 phosphorylation is mediated by the FRAP-$p70^{s6k}$ pathway and is MAP kinase independent. *Proc. Natl. Acad. Sci.* (in press).

Ways, D.K., P.P. Cook, C. Webster, and P.J. Parker. 1992. Effect of phorbol esters on protein kinase C-z. *J. Biol. Chem.* **267:** 4799–4805.

Webster, C., R.L. Gaut, K.S. Browning, J.M. Ravel, and J.K.M. Roberts. 1991. Hypoxia enhances phosphorylation of eukaryotic initiation factor 4A in maize root tips. *J. Biol. Chem.* **266:** 23341–23346.

Xiong, Y., T. Connolly, B. Futcher, and D. Beach. 1991. Human D-type cyclin. *Cell* **65:** 691–699.

Xu, Z., J.N. Dholakia, and M.B. Hille. 1993. Maturation hormone induced an increase in the translational activity of starfish oocytes coincident with the phosphorylation of the mRNA cap binding protein, eIF-4E, and the activation of several kinases. *Dev. Genet.* **14:** 424–439.

Yoder-Hill, J., A. Pause, N. Sonenberg, and W.C. Merrick. 1993. The p46 subunit of eukaryotic initiation factor (eIF)-4F exchanges with eIF-4A. *J. Biol. Chem.* **268:** 5566–5573.

9

Translational Control during Heat Shock

Roger F. Duncan
University of Southern California School of Pharmacy
Department of Molecular Pharmacology and Toxicology
and School of Medicine, Department of Molecular
Microbiology and Immunology
Los Angeles, California 90033

Heat shock or heat stress, i.e., raising the temperature of cells or organisms 5–10°C above their normal growth temperature, is the principal stress system from which most of our understanding of stress-induced translational control has emerged. There are two principal events comprising translational control during heat shock: (1) a general repression of the translation of non-heat shock protein messenger RNAs (referred to here as normal mRNAs) and (2) the preferential translation of a small class of mRNAs, the heat shock protein (HSP) mRNAs. Current knowledge about the molecular basis for each is reviewed in this chapter. Substantial progress has been made in recent years, and there is reason for optimism that where uncertainty remains, molecular details will be forthcoming shortly.

A consequence of heat-shock-induced translational reprogramming is the production of massive amounts of the heat shock proteins. These enable cells to survive what would otherwise be lethal heat shock and greatly accelerate recovery of normal cell function following heat shock. This process, termed acquired thermotolerance, has been extensively reviewed elsewhere (Parsell and Lindquist 1993). At the translational level, heat shock proteins are proposed to counteract the *repressive* influences of heat shock, resulting in less inhibition during heat shock and a more rapid restoration of normal protein synthesis rate after heat shock. This process, termed translational thermotolerance, which is a subset of acquired thermotolerance, is described in more detail below.

The preferential translation of HSP mRNAs provides a mechanism for the rapid, massive accumulation of these beneficial proteins under adverse circumstances. The repression of general translation, which may be either transient or prolonged, likely serves to restrict the production of missense, incomplete, or malfolded proteins that might subsequently have deleterious effects on cell growth. The magnitude of preferential translation and general repression differs substantially between cell types, species, and specific circumstances of heat shock. Whenever one considers the translational heat shock response, it is important to bear in mind this variation.

TRANSLATIONAL CONTROL OF NON-HEAT SHOCK mRNA TRANSLATION BY HEAT SHOCK

Two primary translational events affecting non-heat shock (normal) mRNAs are set in motion by heat shock: the repression of their translation at the beginning of the stressful episode, followed by subsequent translational restoration when cells are returned to their normal growth temperature. Detailed descriptions of these two events and their molecular bases are considered in the following sections.

Repression of Normal Translational Activity

Repression of normal translational activity is virtually always observed following heat shock, in all organisms. The characteristics and molecular basis for the phenomenon have been intensively investigated in *Drosophila* and mammalian cells. There are numerous similarities in their responses, as well as some significant differences. Heat-shock-induced repression of translation and its subsequent restoration in *Drosophila* are depicted in Figure 1 (panels A and B), showing protein synthetic rate and type by pulse-labeling at intervals after heat shock. The extent of repression is directly related to the heat shock severity. In *Drosophila* cells, inhibition becomes detectable at around 32–33°C (~10°C above their normal growth temperature), reaches approximately 50% inhibition at 34°C, and exceeds 90% inhibition at 37°C (the commonly employed "heat shock" temperature for these cells). In mammalian cells, protein synthesis is more than 80% inhibited at 41°C, only 4°C above their normal growth temperature. In most cell types, the normal mRNAs are neither degraded nor substantially modified. The clearest evidence comes from in vitro translation of mRNAs extracted from heat-shocked cells, which reveals no loss of mRNA activity. The pattern of proteins synthesized resembles the nonstressed in vivo pattern, with heat shock proteins superimposed (Fig. 1C, compare lane 3 [in vitro] to lane 1 [in vivo]; see also Kruger and Benecke 1981; Lindquist 1981; Petersen and Mitchell 1981). Additionally, full translational recovery can occur using the preexisting mRNAs when new RNA syn-

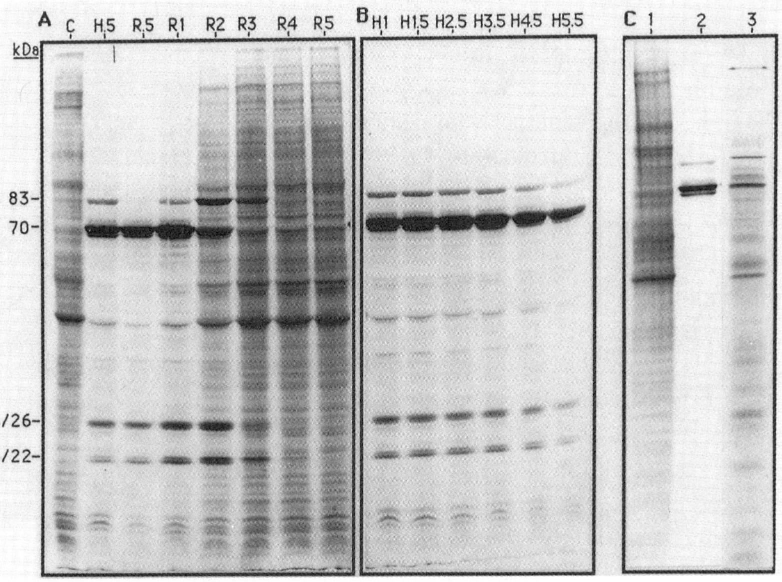

Figure 1 Protein synthesis patterns in *Drosophila* cells during heat shock. (*A,B*) *Drosophila* S2 cultured cells were heat-shocked, and aliquots were pulse-labeled with [^{35}S]methionine for 10 min, ending at the indicated times. Proteins were extracted and equal amounts of protein were analyzed by SDS-PAGE and autoradiography. The cells in panel *A* were returned to their normal growth temperature after 30-min heat shock at 37°C (recovery conditions), whereas in panel *B*, they were maintained at heat shock temperature. The times of sampling and cell status are indicated above the lanes as R (recovery duration) or H (heating duration). Heat shock causes the repression of virtually all non-heat shock mRNAs, but transcriptionally induced HSP mRNAs translate efficiently. During recovery, normal translation gradually recovers to the pre-heat shock levels over 4–6 hr. HSP70 translation is repressed, whereas the other HSP mRNA's translation persists for a longer interval. During continuous heat shock, the preferential translation pattern is maintained with no evidence for adaptation. The molecular weights of the major fly heat shock proteins are indicated to the left of panel *A*. (*C*) The translation pattern in *Drosophila* is compared between non-heated cells (lane *1*), heat-shocked cells (lane *2*), and mRNAs extracted from heat-shocked cells and translated in vitro in the rabbit reticulocyte lysate (lane *3*).

thesis is prevented by treatment with inhibitors preceding return to normal temperature (Lindquist 1981; DiDomenico et al. 1982a; Duncan and Hershey 1989). In a few cell types, such as yeast (Lindquist 1981), heat-induced mRNA degradation is observed and accounts, at least in part, for the decreased normal mRNA translation.

What intracellular events signal translational repression? The accumulation of improperly folded proteins caused by, for example, incorporation of amino acid analogs signals *transcriptional* activation (DiDomenico et al. 1982a; Thomas and Mathews 1984; Ananthan et al. 1986; Welch and Suhan 1986; Russel et al. 1987). Denatured proteins also cause *translational* repression upon addition to the rabbit reticulocyte lysate, due to the titration of HSP70 off the heme-regulated eIF2α kinase, resulting in its activation (Matts et al. 1993). Treatment with amino acid analogs partially inhibits translation rate in intact cells (~50%) (Duncan and Hershey 1987), but this extent is far less than provoked by heat shock and does not appear to involve eIF2α kinase activation (Duncan and Hershey 1987). The significance of eIF2α kinase activation by heat shock is detailed below.

Other clues about the proximal event signaling heat shock translational repression can be provided by identifying agents that mitigate repression. Intriguingly, free radical scavengers reduce repression, as do calcium chelators (Burdon 1987). This suggests that radical-mediated protein damage, including but not restricted to protein denaturation, and calcium-dependent signaling pathways may be involved. Conversely, agents that increase intracellular free calcium (A23187) or promote free radical formation such as thiol depletion agents (diethylmaleate) cause greater heat-shock-induced translational repression (Burdon 1987). Substantial transient increases in intracellular free calcium (~10-fold) occur during heat shock (Drummond et al. 1986); their significance in protein synthesis repression has not been directly determined. The recent characterization of free radical-activated MAP-family kinases (Derijard et al. 1994; Kyriakis et al. 1994) may provide an entree to molecular signaling pathways affecting heat shock translation. Numerous other agents that affect the heat shock response have been identified, but in most instances, their specific impact on protein synthesis has not been examined.

Heat-shock-induced translational repression can be reduced by certain *pretreatments*, in a process termed translational thermotolerance. The clearest and paradigmatic example involves preheating cells (a "priming" heat shock) followed by a recovery interval. During a subsequent heat shock (the "challenge" heat shock), the extent of inhibition is reduced (Hahn and Shiu 1985; Mizzen and Welch 1988; Ciavarra and Simeone 1990; Liu et al. 1992; Li and Duncan 1995). Acquired thermotolerance due to priming treatments resulting in enhanced cell viability is correlated with translational thermotolerance, suggesting that translational events are key determinants of general thermotolerance (Schamhart et al. 1984a,b). However, enhanced survival correlates best with enhanced *rates of recovery following heat shock.* This aspect, as well as the molecular basis for translational tolerance, is further discussed in the next section.

Restoration of Normal mRNA Translation

Following heat shock, cells recover full translational activity, over a relatively long interval. Whereas repression occurs within minutes, recovery requires several hours. Resumption of normal, non-heat shock mRNA translation can be first detected after about 1 hour following a moderate to severe heat shock, increasing progressively over the next 4–6 hours until full activity is restored. An example drawn from *Drosophila* cells is presented in Figure 1A (recovery lanes); mammalian cells respond with similar kinetics of restoration. Careful analysis of the recovery kinetics for individual mRNAs suggests that all recover as a cohort (DiDomenico et al. 1982b). When more severe heat shocks are administered, the recovery kinetics are delayed (DiDomenico et al. 1982a; Lindquist 1993), although non-heat shock mRNAs again appear to recover synchronously.

The prolonged recovery interval suggests new synthesis may be required, and several lines of evidence suggest that HSP70 synthesis and accumulation in specific regulate recovery. On the basis of careful quantitations in *Drosophila* cells, DiDomenico et al. (1982a,b) determined that the synchronous recovery of non-heat shock mRNA translation began when a threshold level of HSP70 was attained. As discussed below, the same threshold signals the repression of HSP70 mRNA translation (DiDomenico et al. 1982b). Reducing the synthesis of functional HSP70 by transcriptional inhibition or by incorporating amino acid analogs delays recovery until the threshold level of active HSP70 is reached. The HSP70 threshold model for translational recovery has been reproduced in mammalian cells (Mizzen and Welch 1988), suggesting that HSP70-based translational reactivation is an ancient, conserved strategy. The specific function of HSP70 in regulating restoration likely relates to its chaperone-like activities in protein disaggregation and refolding (Gething and Sambrook 1992).

To demonstrate unambiguously that HSP70 in particular was a molecular rheostat, antisense-mediated inhibition was carried out to specifically reduce its accumulation. Recovery rate was reduced concomitantly (Solomon et al. 1991). Conversely, specific HSP70 overexpression accelerates recovery in mammalian and *Drosophila* cells (Liu et al. 1992; Li and Duncan 1995). Reduction of HSP26 accumulation, on the other hand, does not reduce translational recovery rate (McGarry and Lindquist 1986). However, results described below suggest that other non-HSP70 elements also promote recovery.

Mammalian cells are capable of another, distinct form of "recovery" during *continuous* heat shock, often termed "adaptation." For example, cells continuously heat shocked at 41–42°C initially show ≥80% reduction in translation, but after approximately 2 hours, the translation rate begins to rise, and after 4–6 hours, it reaches a rate only slightly less than normal (McCormick and Penman 1969; Hickey and Weber 1982; Burdon 1987; Duncan and Hershey 1989). Adaptation is not observed during more severe heat shock (>42°C). Similar adaptation occurs to "calcium stress" caused by ionophore A23187, metalloendoproteinase inhibitor, thapsigargin or EGTA treatment, or chronic treatment with reducing agents such as DTT (Brostrom et al. 1990, 1991; Wong et al. 1993). Continuous exposure to modest amounts of certain translational inhibitors also results in adaptation (Hallberg 1986). Thus, cells clearly possess one or more adaptive mechanisms that can "desensitize" the translational machinery. Adaptation in mammalian cells differs from recovery: Following a mild 41°C heat shock, cells can recover normal translation at normal temperature, but cannot adapt to 41°C, in the presence of the RNA synthesis inhibitor DRB (Duncan and Hershey 1989). These results are consistent with adaptation but not recovery, requiring new HSP synthesis for restoration of activity. Adaptation is not observed in *Drosophila* heat shocked at 37°C; the heat-specific translation pattern persists for more than 6 hours (Fig. 1B; Delavalle et al. 1994). However, adaptation clearly occurs in other organisms, including *Tetrahymena* (Fung et al. 1995) and prokaryotes (Lemaux et al. 1978).

Translational thermotolerance can also be observed as an enhanced rate of recovery of normal mRNA translation at the normal growth temperature (Petersen and Mitchell 1981; DiDomenico et al. 1982a; Sciandra and Subjeck 1984; Mizzen and Welch 1988; De Maio et al. 1993a,b; Li and Duncan 1995), paralleling translational thermotolerance measured *during* heat shock as described previously. As noted above, HSP70 overexpression accelerates translational recovery (Liu et al. 1992; Li and Duncan 1995), although this effect was not observed in other investigations (Solomon et al. 1991). HSP70 colocalizes with ribosomes during recovery (Welch and Suhan 1986; Beck and De Maio 1994), suggesting a direct role in ribosome reactivation, and accelerates the reactivation of heat-shock-inhibited protein synthesis initiation factors (Chang et al. 1994). Cell lines selected for heat-resistant growth frequently show substantially increased constitutive expression of HSP70 isoforms (see, e.g., Laszlo and Li 1985; Anderson et al. 1991). Translational thermotolerance, although not examined in these cell lines, likely contributes to their acquired thermotolerance. However, it also appears that overexpressed HSPs, either alone or in combination, provide only partial translational thermotolerance, and other protective events result from the priming treatment. For example, specific overexpression of HSP70 protects translation, but less than in thermotolerant cell lines or in those induced by priming heat treatment (Liu et al. 1992; Li and Duncan 1995). Several cell lines with high intrinsic heat resistance have no detectable elevation of any HSP (see, e.g., Fisher et al. 1991); conversely, certain mutant cells incapable of developing heat-primed translational thermotolerance nevertheless produce normal amounts of HSPs following the priming treatment (Kraus et al. 1986). Therefore, non-HSP elements appear to be involved.

Other candidate mediators of translational thermotolerance have been identified and include other heat shock proteins and ribosome-associated components. Perturbation of calcium homeostasis, by A23187 treatment, results in a transient repression of protein synthesis, which is potentiated by reducing glucose-regulated protein 78 (GRP78) induction with antisense RNA (Brostrom et al. 1990). Similarly, translation restoration and subsequent insensitivity to calcium-perturbing agents is correlated with the accumulation of GRP78 (Brostrom et al. 1990, 1991) and is elevated in a GRP78 overexpressing cell line (Brostrom et al. 1990). A 26-kD ribosome-associated protein reversibly dephosphorylated by calcium stress may influence translational regulation as well (Fawell et al. 1989). Both a 22-kD cytoplasmic protein and a small RNA polymerase-III-transcribed RNA, termed G8, specifically and strongly associate with ribosomes during heat shock in *Tetrahymena* (McMullin and Hallberg 1986; Kraus et al. 1987). Deletion of G8 blocks translational and viability thermotolerance (Fung et al. 1995).

Pretreatment with one form of priming stress often confers translational tolerance to a different type of stress. For example, priming treatment with sodium arsenite provides translational thermotolerance to a subsequent challenge heat shock (Mizzen and Welch 1988). Cross-tolerance can also be induced by a priming treatment with a modest dose of several protein synthesis inhibitors (cycloheximide, emetine) (Hallberg et al. 1985). This provides indirect evidence that modifications of ribosome-associated elements are involved in translational thermotolerance. A current challenge is to identify common molecular alterations induced by arsenite, cycloheximide, and heat leading to translational thermotolerance.

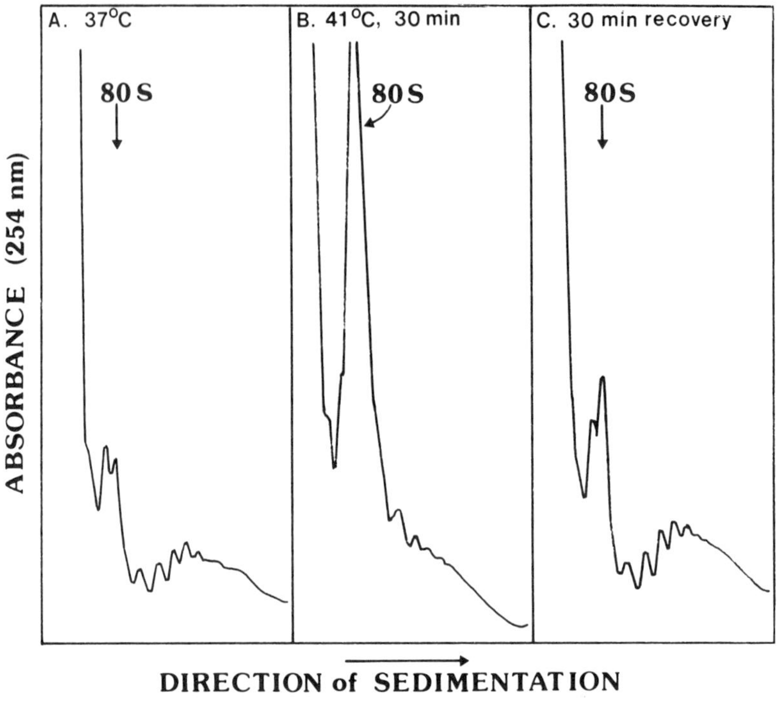

Figure 2 Heat-induced disaggregation of ribosomes. HeLa cells were either sampled at normal growth temperature (37°C) (*A*), heat-shocked for 30 min at 42°C (*B*), or heat-shocked at 42°C, recovered for 30 min at 37°C, and then sampled (*C*). Cell lysates were analyzed by sucrose density centrifugation to display ribosomes as polysomes (active) or monosomes (inactive) (measured by A_{254} absorbance; the location of the 80S monosome peak is indicated). The disaggregation of the polysomal ribosomes by heat shock is characteristic of an initiation block to translation.

Mechanism and Molecular Events Underlying Repression and Restoration

Initiation Is Principally Inhibited, but Elongation May Be Affected as well

The initiation phase of protein synthesis is the principal repressed step following a heat shock, as evidenced by a disaggregation of polyribosomes into inactive monosomes and ribosomal subunits (see Fig. 2 for an example). Initiation-based control will be discussed at length in the following paragraphs. Primary control at the elongation step can be excluded; in that situation, an *accumulation* of ribosomes on polysomes would be observed. However, a few reports suggest that alterations in elongation rate may also accompany heat shock. Ballinger and Pardue (1983) determined that normal mRNAs sediment *as if* polysome associated during heat shock in *Drosophila*; however, there is not a corresponding amount of polyribosomes, suggesting that mRNAs are packaged into fast-sedimenting messenger ribonucleoproteins (mRNPs) during heat shock (see, e.g., McKenzie et al. 1975; Sirkin and Lindquist 1985). Fast-sedimenting nontranslated mRNAs also occur in amino acid analog-stressed mammalian cells (Thomas and Mathews 1984). Conversely, Theodorakis et al. (1988) determined that HSP70 mRNA was poorly translated in *nonheated* cells due to elongation arrest but released from its repression by heat shock. HSP70 mRNA elongation repression may represent a unique form of autoregulation. In non-heat shock circumstances, HSP70 would bind to its own nascent protein (or mRNA) to repress translation, but following heat shock, it would be transferred to other, preferred substrates such as denatured proteins, resulting in the release of the elongation repression. A similar model has been proposed for autogenous regulation of HSP70 *transcription* (Morimoto 1993), as well as for translational regulation of poly(A)-binding protein (PABP) mRNA via PABP binding to its mRNA's 5′-untranslated region (5′UTR) when 3′poly(A) tracts are limiting (Sachs and Davis 1989).

43S Preinitiation Complex Formation Is Inhibited

A distinct deficit in 43S preinitiation complex formation occurs in heat-shocked cells (Panniers and Henshaw 1984), suggesting that binding of the ternary complex eIF2•GTP•Met-$tRNA_i$ is impaired. This suggests an eIF2 lesion, and this has been confirmed by analyses described below. Because a very early stage in the pathway is substantially impaired, it is difficult to ascertain whether or not other subsequent steps, such as mRNA "activation" and binding, function normally or are also impaired. Other approaches are required to address this issue.

Initiation Factors Including eIF2 and eIF4F Show Reduced Activity

A preponderance of evidence suggests that the initiation phase limitation results from inactivation of one or more initiation factors (Duncan and Hershey 1984; Panniers et al. 1985; Sanders et al. 1986). Quantitation of initiation factor activities using a reconstituted system approach indicates that eIF2 and eIF3/eIF4F are significantly inhibited by heat shock (Duncan and Hershey 1984). Supplementation analyses in unfractionated lysates indicate that eIF4F (defined as a complex of eIF4E, eIF4A, and eIF4G) can restore non-heat shock mRNA translation in heat shock cell-free systems from mammalian and *Drosophila* cells (Panniers et al. 1985; Zapata et al. 1991), implicating a heat-induced lesion in this factor. Similar restoration experiments using eIF2 have met with mixed success; Panniers et al. (1985) detected little reactivation of translation, whereas Burdon et al. (1987) observed marked stimulatory activity. The differences likely result from subtle variations in the preparation and status of the in vitro extracts each used. Alternative strategies to quantify eIF2's involvement, described below, provide convincing evidence that its inactivation contributes to heat shock translational repression. Ribosome-associated alterations occur during heat shock, for example, the dephosphorylation of ribosomal protein S6 (Glover 1982; Kennedy et al. 1984), but their consequence for heat shock repression is probably secondary, at most. Suggestions that the principal heat-shock-induced lesion is ribosomal (see, e.g., Scott and Pardue 1981) likely result from copurification of initiation factors with ribosomes under certain fractionation protocols.

Molecular Alterations That Cause Translational Repression

eIF2's activity in promoting Met-tRNA$_i$ binding to the 40S ribosomal subunit is regulated through reversible phosphorylation of the α-subunit at Ser-51 (in the mammalian protein), which inhibits its recycling/reactivation by eIF2B (see Merrick and Hershey; Trachsel; Clemens; all this volume). Heat shock increases eIF2α phosphorylation in a temperature-dependent manner (Duncan and Hershey 1984, 1989; De Benedetti and Baglioni 1986; Rowlands et al. 1988). Further compelling evidence implicating eIF2α phosphorylation comes from transfection of a "dominant-negative" S51A mutant of eIF2α that suppresses eIF2α-phosphorylation-based inhibition (Kaufman et al. 1989). Overexpression of S51A reduces the extent of heat shock (43.5°C)-induced translational repression from more than 80% to approximately 50% (Murtha-Riel et al. 1993), indicating that eIF2α phosphorylation has a significant role in heat-shock-induced repression under these conditions. In other circumstances, the significance of eIF2α phosphorylation is less clear. Mild heat shock (41–42°C) of mammalian cells results characteristically in 70–90% inhibition of protein synthesis, but little if any heat-shock-induced eIF2α phosphorylation can be detected (Duncan and Hershey 1989). Heat-shocked (45°C) thermotolerant Rat-1 cells exhibit more than 90% inhibition of protein synthesis, but again show no increased eIF2α phosphorylation (Chang et al. 1994). Finally, in 37°C heat-shocked *Drosophila* cells, there is little if any increased eIF2α phosphorylation

(Duncan et al. 1995). Thus, eIF2α phosphorylation constitutes an important regulatory event under certain heat shock circumstances, whereas in other cases, other molecular effectors predominate.

The function of eIF4E, like that of eIF2α, is thought to be regulated by phosphorylation. Phosphorylation of eIF4F on the eIF4E and eIF4G subunits increases its initiation activity (Morley et al. 1991). eIF4E phosphorylation decreases substantially during moderate to severe heat shock (>43°C) (Duncan et al. 1987; Duncan and Hershey 1989; Lamphear and Panniers 1990, 1991), implicating eIF4F inactivation in heat-shock-induced translational repression. However, during milder heating (41–42°C), no detectable dephosphorylation occurs (Duncan and Hershey 1989). Similarly, in *Drosophila* at 37°C, eIF4E phosphorylation is not substantially reduced (Duncan et al. 1995). Thus, inactivation of eIF4F via eIF4E dephosphorylation occurs in some but not all heat shock circumstances, paralleling conclusions discussed above for eIF2α phosphorylation.

eIF4E activity is also regulated by reversible protein-protein interactions to form the heterotrimeric eIF4F complex. At moderate to severe heat shock, the mammalian eIF4F complex dissociates, releasing "free" eIF4E (Duncan et al. 1987; Lamphear and Panniers 1990, 1991); during mild heat shock (41–42°C), complex disruption is not observed (Duncan and Hershey 1989). Recent evidence that eIF4E incorporation into the heterotrimeric complex is regulated by an eIF4E-binding protein(s) suggests a possible basis for complex dissociation during heat shock (Lin et al. 1994; Pause et al. 1994). Phosphorylation of eIF4E may be the molecular event that regulates eIF4F heterotrimer association/dissociation (Lamphear and Panniers 1991), and which may be controlled by the eIF4E-binding protein. eIF4F subunit dissociation has also been reported in 37°C heat-shocked *Drosophila* cells (Zapata et al. 1991). Another phosphoprotein initiation factor, eIF4B, is loosely associated with eIF4F. It dissociates during moderate to severe heat shock and is substantially dephosphorylated in both mammalian and *Drosophila* cells (Duncan et al. 1987, 1995).

In summary, the two key initiation factor protein regulators of translation, eIF2α and eIF4E (eIF4F), are converted into their inactive forms during moderate to severe heat shock in mammalian cells. However, in other heat shock circumstances, such as mild febrile heat shock, protein synthesis inhibition appears to be caused by other, as yet uncharacterized, molecular alterations. In 37°C heat-shocked *Drosophila* cells, which retain the capacity to translate HSP mRNAs very efficiently (detailed below), neither factor appears to be substantially covalently modified, although eIF4F complex dissociation may occur. A current goal is to determine the identity of these other heat-shock-affected translational regulators. Potential candidates include (1) molecules affecting polyadenylation, since poly(A) changes have been reported in heat-shocked *Drosophila* (Storti et al. 1980) and provide the basis for translational control in other circumstances (see Richter, this volume) and (2) an hsrω-like transcript; this molecule has thus far only been identified in *Drosophila* species (Garbe et al. 1989), where its transcription and abundance are markedly stimulated by heat shock and other stresses and functionally is proposed to serve as a rheostat that regulates protein synthesis rates (Bendena et al. 1989; Fini et al. 1989).

PREFERENTIAL TRANSLATION OF HEAT SHOCK PROTEIN mRNAs

An intriguing facet of the heat shock translational response is the ability of HSP mRNAs to be very efficiently translated when virtually all of the non-heat shock mRNAs are severely repressed (Lindquist 1981). This likely results from unique structural features in heat shock mRNAs that enable them to bypass the initiation factor lesions cited in the previous section. Before proceeding, a few salient points should be mentioned. First, "exclusive" translation of HSP mRNAs during heat shock is *not* observed in all cell types. For example, in *Drosophila*, there *is* a striking discrimination between HSP and normal mRNAs, but in mammalian cells, the translational efficiency of HSP mRNAs is similar to or only a fewfold greater than non-heat shock mRNAs during heat shock (Hickey and Weber 1982; Duncan and Hershey 1989; M. Hess et al., unpubl.). Second, and likely related, preferential translation may depend on the heat shock severity. For example, in *Drosophila*, HSP mRNAs are induced at 31–32°C; overall translation is not significantly repressed at this temperature and hence there is no preferential translation. Between 36°C and 37°C, marked preferential translation is observed; at temperatures above 39°C, even HSP mRNAs cannot be translated (DiDomenico et al. 1982a). Thus, preferential HSP mRNA translation may be restricted to a temperature window, which can range from broad (e.g., in *Drosophila*) to very narrow (e.g., as may apply to mammalian cells).

Finally, a plausible model for preferential translation of HSP mRNAs is that mRNAs transcribed during heat shock are "marked" for activity, but this has been disproved in two ways: a non-heat shock mRNA expressed from a heat shock promoter during heat shock cannot be translated (Klementz et al. 1985), and conversely, an HSP mRNA expressed at normal temperature, such as the constitutively synthesized HSP83 mRNA, is translated efficiently during heat shock (Lindquist 1981).

Results described in the following sections convincingly document that special sequence elements or features in HSP mRNAs confer preferential translation. These results derive almost exclusively from investigations of *Drosophila* because of its marked capacity to translate HSP mRNAs during heat shock.

HSP mRNA Primary Structure Confers Preferential Translation

The 5′UTR of HSP mRNAs contains sufficient primary sequence information for preferential translation. In early studies, gene fusions linking the 5′UTR of *Drosophila* HSP70 mRNA to the 5′UTR and coding region of a non-heat shock mRNA were produced (Di Nocera and Dawid 1983). These chimeric RNAs were translated during heat shock (Di Nocera and Dawid 1983). Subsequent fine-structure analysis has determined that the precise amounts and organization of the HSP and normal portions of the 5′UTR can significantly influence the preferential translation extent of the chimeric transcript, as detailed in the following sections.

Salient Sequence Features of HSP mRNAs 5′ UTRs

There are at least seven distinct HSP mRNAs in *Drosophila melanogaster*. Several common features occur: (1) relatively long (200–250 nucleotides) 5′UTRs; (2) conserved sequence blocks in similar positions relative to the cap site can be identified (Holmgren et al. 1981; Hultmark et al. 1986; Lindquist 1987), suggestive of an important regulatory function; (3) very high adenosine content (~45–50%) (Ingolia and Craig 1981); and (4) correspondingly, low potential for forming secondary structure (compared to most non-HSP mRNA 5′UTRs) (Lindquist and Petersen 1990). Which, if any, of these features are critical determinants for HSP mRNA preferential translation?

First, progressive 5′-end deletions of the HSP70 5′UTR result in progressive decreases in heat shock translation (Lindquist and Petersen 1990). This suggests that length influences preferential translation capacity. However, a different large (~170) nucleotide deletion encompassing nucleotides that retain the first 38 nucleotides translates relatively well (70% of wild type) (Lindquist 1987; Lindquist and Petersen 1990), indicating that long length per se is not required. Similarly, deleting approximately 130 nucleotides from the HSP22 5′UTR while retaining the cap-proximal 27 nucleotides has no detrimental effect on its preferential translation (Hultmark et al. 1986). These results suggest that the cap-proximal portion of the leader constitutes a primary determinant.

Second, the conserved features at the 5′terminus of the 5′UTR are not *required* for heat shock translatability, nor is the internal conserved region: Deletion mutants lacking either segment, or both conjointly, are efficiently translated during heat shock (Lindquist 1987). Although these results do not rigorously exclude a functional role for these regions, they indicate at the very least that functional, and perhaps redundant, "elements" are dispersed throughout the 5′UTR. The 5′-terminal conserved segment has been suggested to be a transcriptional regulatory element (Hultmark et al. 1986). To address whether essential sequence elements exist at any point in the HSP70 5′UTR, a series of overlapping small, 25–60-long nucleotide deletion mutants covering the entire 5′UTR were prepared. All translated relatively well during heat shock (Lindquist 1987), excluding this possibility.

Third, a high adenosine content is *not sufficient* to confer preferential translation: Synthetic 5′UTRs that retain the authentic base composition of the HSP70 5′UTR but scramble the nucleotide order fail to translate during heat shock (Lindquist and Petersen 1990).

Finally, the paucity of secondary structure may constitute a critical feature. This remains to be tested. However, it is clear that lack of structure per se cannot constitute the only required feature, since the A-rich scrambled leaders described in the last paragraph fail to translate even though they recreate a relatively unstructured 5′UTR (based on computer folding algorithms). Thus, there appears to be some dependence on specific sequences within the HSP 5′UTR. What these sequences are and why they are functionally significant remain absorbing problems and may provide clues to the molecular mechanism regulating preferential translation.

A Cap-independent Mode of Preferential Translation?

Several lines of evidence strongly suggest that preferential translation is based on a capacity of HSP mRNAs to bypass an eIF4F lesion. First, by several criteria, eIF4F is inhibited by heat shock (as detailed above). Second, in heat-shocked cell-free translation lysates that reproduce the in vivo HSP mRNA discrimination, supplementation with eIF4F partially restores normal mRNA translation (Panniers et al. 1985; Zapata et al. 1991). Third, in vitro eIF4F limitation caused by treatment with the m^7G cap analog, with eIF4E or eIF4G antibody, or with the foot-and-mouth disease virus (FMDV) L protease, results in normal mRNA translation being substantially more inhibited than HSP mRNA translation (Maroto and Sierra 1988; Zapata et al. 1991, 1994; Song et al. 1995). Similarly, in vivo diminution of eIF4F abundance by expressing antisense eIF4E RNA reduces normal mRNA translation, but HSP mRNAs are induced and translate relatively efficiently (Joshi-Barve et al. 1992).

A critical issue is whether HSP mRNA translation can occur by cap-independent, internal initiation, like poliovirus mRNA (see Ehrenfeld, this volume), or whether it can competitively "sequester" a minimal amount of active eIF4F, resulting in its preferential, efficient translation. Suggestive evidence for the former possibility comes from studies showing that GRP78/BiP, the endoplasmic-reticulum-localized HSP70 variant, can translate via a cap-independent, internal initiation mechanism (Sarnow 1989; Macejak and Sarnow 1991). However, evidence disfavoring this view comes from experiments in which approximately 30 nucleotides were appended to the 5′end of the HSP70 5′UTR. These severely repressed its heat shock translation (McGarry and Lindquist 1985), which suggests a cap-independent, internal initiation pathway does not occur. This has been confirmed using other 5′UTR extensions (M. Hess and R. Duncan, unpubl.). Determination of the capacity of the HSP 5′UTR to promote internal initiation in a bicistronic construct will conclusively resolve this issue.

Regulation of Translation by HSP70

HSP70 regulates its own translation, in a process termed autoregulation (DiDomenico et al. 1982a,b). During heat shock followed by recovery, HSP70 initially translates very efficiently and then its translation is rapidly repressed (DiDomenico et al. 1982a,b) due to rapid HSP70 mRNA degradation (Petersen and Lindquist 1988, 1989). The duration of HSP70 mRNA translation is extended when the accumulation rate of functional HSP70 protein is limited in *Drosophila* and mammalian cells (DiDomenico et al. 1982a; Mizzen and Welch 1988; Solomon et al. 1991), consistent with a fixed threshold level of HSP70 accumulation being the trigger that initiates its down-regulation. Conversely, cells in which HSP70 has been preinduced repress HSP70 translation more rapidly (DiDomenico et al. 1982a; R. Duncan, unpubl.).

HSP70 accumulation also regulates the recovery of normal mRNA translation in *Drosophila* (DiDomenico et al. 1982b). The molecular basis is currently unclear. However, tantalizing clues from mammalian cells implicate HSP90 and HSP70 as *trans*-acting factors regulating eIF2α phosphorylation, and hence some instances of protein synthesis inhibition and recovery. HSP90 and HSP70 interact with the eIF2α kinase HCR to regulate its activity in reticulocytes (Rose et al. 1989; Matts and Hurst 1992; Matts et al. 1992). The accumulation of HSP70 following heat shock may help reverse the heat-shock-induced eIF2α phosphorylation, since cell lines overexpressing HSP70 dephosphorylate eIF2 more rapidly during recovery (Chang et al. 1994). A unifying hypothesis is that HSP70 shuttles between binding to normal cellular proteins and to denatured proteins. Heat shock increases the amount of denatured proteins and causes HSP70 to redistribute to these preferred substrates. Its dissociation from non-heat shock proteins such as HCR results in their activities being changed, which are only restored to their non-heat shock state when all denatured protein substrates have been titrated by HSP70. Furthermore, in HSP70-overexpressing cells, denatured proteins would be titrated more rapidly, resulting in more rapid restoration of normal cell function. This model can account for both autoregulation and HSP70-mediated translational control if protein complex formation with HSP70 is required to maintain the activities of the mRNA degradatory and translational machineries.

Interaction of Trans-*acting Factors with the HSP70 mRNA 5′ UTR*

Many sequence or structural features of mRNAs that regulate translation are recognized by *trans*-acting protein factors. For example, the internal initiation sites in viral mRNAs and the iron-responsive regulatory site in ferritin mRNA require specific protein binding to elicit their translational effect (see Ehrenfeld; Rouault et al.; both this volume).

Preferential translation of the HSP mRNAs does not seem to require such an mRNA/protein interaction. An extensive investigation of RNA/protein interactions between the *Drosophila* HSP70 5′ UTR and lysate proteins, using gel retardation, UV-light cross-linking, and biotin-labeling affinity purification, failed to detect any marked changes in sequence-specific protein binding upon heat shock—either as increases or as decreases (Hess and Duncan 1994). This result, although negative, is consistent with preferential HSP mRNA translation being primarily determined by its minimal secondary structure alleviating a requirement for unwinding factors.

TRANSLATIONAL REPROGRAMMING DURING HEAT SHOCK: SUMMARY AND CONCLUDING THOUGHTS

Partial inactivation of eIF4F in concert with eIF2α phosphorylation is sufficient to account for the repression of non-heat shock mRNA translation during heat shock. Heat shock mRNA preferential translation can be explained based on eIF4F inactivation if HSP mRNAs have a reduced requirement for this factor. However, there are several aspects of heat shock translation still in need of resolution. What is the molecular basis for heat-shock-induced inhibition during febrile temperatures and in *Drosophila*? If eIF2α phosphorylation is involved in inhibition under any heat shock circumstances, how do the HSP mRNAs evade the eIF2 lesion?

Primary sequence and deletion analyses of HSP mRNA 5′UTRs have established some general features that are required for preferential HSP mRNA translation during heat shock, yet uncertainties remain. If specific 5′UTR sequences are required, then what is their identity, are they reiterated throughout the 5′UTR, and how do they act? Can a cap-independent, internal initiation pathway account for the characteristics of HSP mRNA translation during heat shock? Alternatively, can minimal secondary structure combined with a specific cap-proximal sequence efficiently recruit the small ribosomal subunit, perhaps by a Shine-Dalgarno-like sequence complementarity, under conditions of limiting eIF4F? The resolution of these questions, which should be forthcoming soon, holds the promise not only of clarifying the molecular basis for heat shock translational control, but also of uncovering a possible new mode for discriminatory mRNA translation.

REFERENCES

Ananthan, J., A.L. Goldberg, and R. Voellmy. 1986. Abnormal proteins serve as eukaryotic stress signals and trigger the activation of heat shock genes. *Science* **232:** 252–254.

Anderson, R.L., K.J. Fong, T. Gabriele, P. Lavagnini, G.M. Hahn, J.W. Evans, C.A. Waldren, T.D. Stamato, and A.J. Giaccia. 1991. Loss of the intrinsic heat resistance of human cells and changes in *M*r 70,000 heat shock protein expression in human X hamster hybrids. *Cancer Res.* **51:** 2636–2641.

Ballinger, D.G. and M.L. Pardue. 1983. The control of protein synthesis during heat shock in *Drosophila* cells involves altered polypeptide elongation rates. *Cell* **33:** 103–114.

Beck, S.C., and A. De Maio. 1994. Stabilization of protein synthesis in thermotolerant cells during heat shock: Association of heat shock protein-72 with ribosomal subunits of polysomes. *J. Biol. Chem.* **269:** 21803–21811.

Bendena, W.G., J.C. Garbe, K.L. Traverse, S.C. Lakhotia, and M.L. Pardue. 1989. Multiple inducers of the *Drosophila* heat shock locus 93D (hsro): Inducer-specific patterns of the three transcripts. *J. Cell Biol.* **108:** 2017–2028.

Brostrom, M.A., C. Cade, C.R. Prostko, D. Gmitter-Yellen, and C.O. Brostrom. 1990.. Accommodation of protein synthesis to chronic deprivation of intracellular sequestered calcium: A putative role for GRP78. *J. Biol. Chem.* **265:** 20539–20546.

Brostrom, M.A., C.R. Prostko, D. Gmitter-Yellen, L.J. Grandison, G. Kuznetsov, W.L. Wong, and C.O. Brostrom. 1991. Inhibition of translational initiation by metalloprotease inhibitors. *J. Biol. Chem.* **266:** 7037–7043.

Burdon, R.H. 1987. Temperature and animal cell protein synthesis. *Symp. Soc. Exp. Biol.* **41:** 113–133.

Burdon, R.H., V. Gill, and C.A. Rice-Evans. 1987. Oxidative stress and heat shock protein induction in human cells. *Free Radical Res. Commun.* **3:** 129–139.

Chang, G.C., R. Liu, R. Panniers, and G.C. Li. 1994. Rat fibroblasts transfected with the human 70-kDa heat shock gene exhibit altered translation and eukaryotic initiation factor 2α phosphorylation following heat shock. *Int. J. Hyperthermia* **10:** 325–337.

Ciavarra, R.P. and A. Simeone. 1990. T lymphocyte stress response. II. Protection of translation and DNA replication against some forms of stress by prior hyperthermic stress. *Cell. Immunol.* **131:** 11–26.

De Benedetti, A. and C. Baglioni. 1986. Activation of hemin-regulated initiation factor-2 kinase in heat-shocked HeLa cells. *J. Biol. Chem.* **261:** 338–342.

Dellavalle, R.P., R. Petersen, and S.L. Lindquist. 1994. Preferential deadenylation of *Hsp70* mRNA plays a key role in regulating Hsp70 expression in *Drosophila melanogaster*. *Mol. Cell. Biol* **14:** 3646–3659.

De Maio, A., S.C. Beck, and T.G. Buchman. 1993a. Heat shock gene expression and development of translational thermotolerance in human hepatoblastoma cells. *Circ. Shock* **40:** 177–186.

———. 1993b. Induction of translational thermotolerance in liver of thermally stressed rats. *Eur. J. Biochem.* **278:** 413–420.

Derijard, B., M. Hibi, I.-H. Wu, T. Barrett, B. Su, T. Deng, M. Karin, and R.J. Davis. 1994. JNK-1: A protein kinase stimulated by UV light and Ha-*ras* that binds and phosphorylates the c-*jun* activation domain. *Cell* **76:** 1025–1037.

DiDomenico, B.J., G.E. Bugaisky, and S. Lindquist. 1982a. The heat shock response is self-regulated at both the transcriptional and post-transcriptional levels. *Cell* **31:** 593–603.

———. 1982b. Heat shock and recovery are mediated by different translational mechanisms. *Proc. Natl. Acad. Sci.* **79:** 6181–6185.

Di Nocera, P.P. and I. Dawid. 1983. Transient expression of genes introduced into cultured cells of *Drosophila*. *Proc. Natl. Acad. Sci.* **80:** 7095–7098.

Drummond, I.A.S., S.A. McClure, M. Poenie, R.Y. Tsien, and R.A. Steinhardt. 1986. Large changes in intracellular pH and calcium observed during heat shock are not responsible for the induction of heat shock proteins in *Drosophila melanogaster*. *Mol. Cell. Biol.* **6:** 1767–1775.

Duncan, R. and J.W.B. Hershey. 1984. Heat-shock induced translational alterations in HeLa cells: Initiation factor modifications and the inhibition of translation. *J. Biol. Chem.* **259:** 11882–11889.

———. 1987. Translation repression by chemical inducers of the stress response occurs by different pathways. *Arch. Biochem. Biophys.* **256:** 651–661.

———. 1989. Protein synthesis and protein phosphorylation during heat stress, recovery, and adaptation. *J. Cell Biol.* **109:** 1467–1481.

Duncan, R.F., D.R. Cavener, and S. Qu. 1995. Heat shock effects on phosphorylation of protein synthesis initiation factor proteins eIF-4E and eIF-2α. *Biochemistry* **34:** 2985–2997.

Duncan, R., S.C. Milburn, and J.W.B. Hershey. 1987. Regulated phosphorylation and low abundance of HeLa cell initiation factor eIF4F suggest a role in translational control. *J. Biol. Chem.* **262:** 380–388.

Fawell, E.H., I.J. Boyer, M.A. Brostrom, and C.O. Brostrom. 1989. A novel calcium-dependent phosphorylation of a ribosome-associated protein. *J. Biol. Chem.* **26a4:** 1650–1655.

Fini, M.E., W.G. Bendena, and M.L. Pardue. 1989. Unusual behavior of the cytoplasmic transcript of hsrω: An abundant, stress-inducible RNA that is translated but yields no detectable protein product. *J. Cell Biol.* **108:** 2045–2057.

Fisher, B., P. Kraft, G.M. Hahn, and R.L. Anderson. 1991. Thermotolerance in the absence of induced heat shock proteins in a murine lymphoma. *Cancer Res.* **52:** 2854–2861.

Fung, P.A., J. Gaertig, M.A. Gorovsky, and R.L. Hallberg. 1995. Requirement of a small cytoplasmic RNA for the establishment of thermotolerance. *Science* **268:** 1036–1039.

Garbe, J.C., W.G. Bendena, and M.L. Pardue. 1989. Sequence evolution of the *Drosophila* heat shock locus hsrω. I. The nonrepeated portion of the gene. *Genetics* **122:** 403–415.

Gething, M.J. and J. Sambrook. 1992. Protein folding in the cell. *Nature* **355:** 33–45.

Glover, C.V.C. 1982. Heat shock induces rapid dephosphorylation of a ribosomal protein in *Drosophila. Proc. Natl. Acad. Sci.* **79:** 1781–1785.

Hahn, G.M., and E.C. Shiu. 1985. Protein synthesis, thermotolerance and step down heating. *Int. J. Radiat. Oncol.* **11:** 159–164.

Hallberg, R.L. 1986. No heat shock protein synthesis is required for induced thermostabilization of translational machinery. *Mol. Cell. Biol.* **6:** 2267–2270.

Hallberg, R.L., K.W. Krauss, and E.M. Hallberg. 1985. Induction of acquired thermotolerance in *Tetrahymena thermophila*: Effects of protein synthesis inhibitors. *Mol. Cell. Biol.* **5:** 2061–2069.

Hess, M.A. and R.F. Duncan. 1994. RNA/protein interactions in the 5′untranslated leader of HSP70 mRNA in *Drosophila* lysates: Lack of evidence for specific protein binding. *J. Biol. Chem.* **269:** 10913–10922.

Hickey, E.D. and L.A. Weber. 1982. Modulation of heat-shock polypeptide synthesis in HeLa cells during hyperthermia and recovery. *Biochemistry* **21:** 1513–1521.

Holmgren, R., V. Corces, R. Morimoto, R. Blackman, and M. Meselson. 1981. Sequence homologies in the 5′regions of four *Drosophila* heat-shock genes. *Proc. Natl. Acad. Sci.* **78:** 3775–3778.

Hultmark, D., R. Klemenz, and W.J. Gehring. 1986. Translational and transcriptional control elements in the untranslated leader of the heat shock gene hsp22. *Cell* **44:** 429–438.

Ingolia, T.D. and E.A. Craig. 1981. Primary sequence of the 5′-flanking region of the *Drosophila* heat shock genes in chromosome subdivision 67B. *Nucleic Acids Res.* **9:** 1627–1642.

Joshi-Barve, S., A. De Benedetti, and R.E. Rhoads. 1992. Preferential translation of heat shock mRNAs in HeLa cells deficient in protein synthesis initiation factors eIF-4E and eIF-4Fγ. *J. Biol. Chem.* **267:** 20338–20343.

Kaufman, R.J., M.V. Davies, V.K. Pathak, and J.W.B. Hershey. 1989. The phosphorylation state of eucaryotic initiation factor 2 alters translational efficiency of specific mRNAs. *Mol. Cell. Biol.* **9:** 946–958.

Kennedy, I., R.H. Burdon, and D.P. Leader. 1984. Heat shock causes diverse changes in the phosphorylation of ribosomal proteins of mammalian cells. *FEBS Lett.* **169:** 267–273.

Klementz, R., D. Hultmark, and W.J. Gehring. 1985. Selective translation of heat shock mRNA in *Drosophila melanogaster* depends on sequence information in the leader. *EMBO J.* **4:** 2053–2060.

Kraus, K.W., P.J. Good, and R.L. Hallberg. 1987. A heat shock-induced, polymerase III-transcribed RNA selectively associates with polysomal ribosomes in *Tetrahymena thermophila*. *Proc. Natl. Acad. Sci.* **84:** 383–387.

Kraus, K.W., E.M. Hallberg, and R.L. Hallberg. 1986. Characterization of a *Tetrahymena thermophila* mutant strain unable to develop normal thermotolerance. *Mol. Cell. Biol.* **6:** 3854–3861.

Kruger, C. and B.-J. Benecke. 1981. In vitro translation of *Drosophila* heat-shock and non-heat-shock mRNAs in heterologous and homologous cell-free systems. *Cell* **23:** 595–603.

Kyriakis, J.M., P. Banerjee, E. Nikolakaki, T. Dal, E.A. Rubie, M.F. Ahmad, J. Avruch, and J.R. Woodgett. 1994. The stress-activated protein kinase subfamily of c-Jun kinases. *Nature* **369:** 156–160.

Lamphear, B.J. and R. Panniers. 1990. Cap binding protein complex that restores protein synthesis in heat-shocked Ehrlich cell lysates contains highly phosphorylated eIF-4E. *J. Biol. Chem.* **265:** 5333–5336.

———. 1991. Heat shock impairs the interaction of cap-binding protein complex with 5′mRNA cap. *J. Biol. Chem.* **266:** 2789–2794.

Laszlo, A. and G.C. Li. 1985. Heat-resistant variants of Chinese hamster fibroblasts altered in expression of heat shock proteins. *Proc. Natl. Acad. Sci.* **82:** 8029–8033.

Lemaux, P.G., S.L. Herendeen, P.L. Bloch, F.C. Neidhardt. 1978. Transient rates of synthesis of individual polypeptides in *E. coli* following temperature shifts. *Cell* **13:** 427–434.

Li, D. and R.F. Duncan. 1995. Transient acquired thermotolerance in *Drosophila*, correlated with rapid degradation of Hsp70 during recovery. *Eur. J. Biochem.* **231:** 454–465.

Lin, T.-A., X. Kong, T.A.J. Haystead, A. Pause, G. Belsham, N. Sonenberg, and J.C. Lawrence Jr. 1994. Insulin-dependent stimulation of protein synthesis by phosphorylation of a regulator of 5′-cap function. *Science* **266:** 653–656.

Lindquist, S. 1981. Regulation of protein synthesis during heat shock. *Nature* **293:** 311–314.

———. 1987. Translational regulation in the heat shock response of *Drosophila* cells. In *Translational regulation of gene expression* (ed. J. Ilan), pp. 187–207. Plenum Press, New York.

———. 1993. Translational regulation in the heat shock response of *Drosophila* cells. In *Translational regulation of gene expression* (ed. J. Ilan), vol. II, pp. 187–207. Plenum Press, New York.

Lindquist, S. and R. Petersen. 1990. Selective translation and degradation of heat-shock messenger RNAs in *Drosophila*. *Enzyme* **44:** 147–166.

Liu, R.Y., X. Li, L. Li, and G.C. Li. 1992. Expression of human hsp70 in rat fibroblasts enhances cell survival and facilitates recovery from translational and transcriptional inhibition following heat shock. *Cancer Res.* **52:** 3667–3673.

Macejak, D.G. and P. Sarnow. 1991. Translational regulation of the immunoglobulin heavy-chain binding protein mRNA. *Nature* **353:** 90–94.

Maroto, F.G. and J.M. Sierra. 1988. Translational control in heat-shocked *Drosophila* embryos. *J. Biol. Chem.* **263:** 15720–15725.

Matts, R.L. and R. Hurst. 1992. The relationship between protein synthesis and heat shock protein levels in rabbit reticulocyte lysates. *J. Biol. Chem.* **267:** 18169–18174.

Matts, R.L., R. Hurst, and Z. Xu. 1993. Denatured proteins inhibit translation in hemin-supplemented rabbit reticulocyte lysate by inducing the activation of the heme-regulated eIF-2α kinase. *Biochemistry* **32:** 7323–7328.

Matts, R.L., Z. Xu, J.K. Pal, and J.-J. Chen. 1992. Interactions of the heme-regulated eIF-2α kinase with heat shock proteins in rabbit reticulocyte lysates. *J. Biol. Chem.* **267:** 18160–18167.

McCormick, W. and S. Penman. 1969. Regulation of protein synthesis in HeLa cells: Translation at elevated temperatures. *J. Mol. Biol.* **39:** 315–333.

McGarry, T.J. and S. Lindquist. 1985. The preferential translation of *Drosophila* hsp70 mRNA requires sequences in the untranslated leader. *Cell* **42:** 903–911.

———. 1986. Inhibition of heat shock protein synthesis by heat-inducible antisense RNA. *Proc. Natl. Acad. Sci.* **83:** 399–403.

McKenzie, S.L., S. Henikoff, and M. Meselson. 1975. Localization of RNA from heat-induced polysomes at puff sites in *Drosophila* melanogaster. *Proc. Natl. Acad. Sci.* **72:** 1117–1121.

McMullin, T.W. and R.L. Hallberg. 1986. Effect of heat shock on ribosome structure: Appearance of a new ribosome-associated protein. *Mol. Cell. Biol.* **6:** 2527–2535.

Mizzen, L.A. and W.J. Welch. 1988. Characterization of the thermotolerant cell. I. Effects on protein synthesis activity and the regulation of heat-shock protein 70 expression. *J. Cell Biol.* **106:** 1105–1116.

Morimoto, R.I. 1993. Cells in stress: Transcriptional activation of heat shock genes. *Science* **259:** 1409–1410.

Morley, S.J., T.E. Dever, D.E. Etchison, and J.A. Traugh. 1991. Phosphorylation of eIF-4F by protein kinase C or multipotential S6 kinase stimulates protein synthesis at initiation. *J. Biol. Chem.* **266:** 4669–4672.

Murtha-Riel, P., M. Davies, B.J. Scherer, S.-Y. Choi, J.W.B. Hershey, and R.J. Kaufman. 1993. Expression of a phosphorylation-resistant eukaryotic initiation factor 2 α-subunit mitigates heat shock inhibition of protein synthesis. *J. Biol. Chem.* **268:** 12946–12951.

Panniers, R. and E.C. Henshaw. 1984. Mechanism of inhibition of polypeptide chain initiation in heat-shocked Ehrlich ascites tumour cells. *Eur. J. Biochem.* **140:** 209–214.

Panniers, R., E.B. Stewart, W.C. Merrick, and E.C. Henshaw. 1985. Mechanism of inhibition of polypeptide chain initiation in heat-shocked Ehrlich cells involves reduction of eukaryotic initiation factor 4F activity. *J. Biol. Chem.* **260:** 9648–9653.

Parsell, D.A. and S. Lindquist. 1993. The function of heat-shock proteins in stress tolerance: Degradation and reactivation of damaged proteins. *Annu. Rev. Genet.* **27:** 437–496.

Pause, A., G.J. Belsham, A.-C. Gingas, O. Donzé, T.-A. Lin, J.C. Lawrence Jr., and N. Sonenberg. 1994. Insulin-dependent stimulation of protein synthesis by phosphorylation of a regulator of 5′-cap function. *Nature* **371:** 762–767.

Petersen, N.S. and H.K. Mitchell. 1981. Recovery of protein synthesis after heat shock: Prior treatment affects the ability of cells to translate mRNA. *Proc. Natl. Acad. Sci.* **78:** 1708–1711.

Petersen, R. and S. Lindquist. 1988. The *Drosophila* hsp70 message is rapidly degraded at normal temperatures and stabilized by heat shock. *Gene* **72:** 161–168.

———. 1989. Regulation of hsp70 synthesis by messenger RNA degradation. *Cell Regul.* **1:** 135–149.

Rose, D.W., W.J. Welch, G. Kramer, and B. Hardesty. 1989. Possible involvement of the 90-kDa heat shock protein in the regulation of protein synthesis. *J. Biol. Chem.* **264:**

6239–6244.

Rowlands, A.G., K.S. Montine, E.C. Henshaw, and R. Panniers. 1988. Physiological stresses inhibit guanine-nucleotide-exchange factor in Ehrlich cells. *Eur. J. Biochem.* **175:** 93–99.

Russell, J., E.C. Stow, N.D. Stow, and C.M. Preston. 1987. Abnormal forms of the herpes simplex virus immediate early polypeptide Vmw175 induce the cellular stress response. *J. Gen. Virol.* **68:** 2397–2406.

Sachs, A.B. and R.W. Davis. 1989. The poly(A)-binding protein is required for poly(A) shortening and 60S ribosomal-subunit dependent translational initiation. *Cell* **58:** 857–867.

Sanders, M.M., D.F. Triemer, and A.S. Olsen. 1986. Regulation of protein synthesis in heat-shocked *Drosophila* cells: Soluble factors control translation *in vitro*. *J. Biol. Chem.* **261:** 2189–2196.

Sarnow, P. 1989. Translation of glucose-regulated protein 78/immunoglobulin heavy-chain binding protein mRNA is increased in poliovirus-infected cells at a time when cap-dependent translation of cellular mRNAs is inhibited. *Proc. Natl. Acad. Sci.* **86:** 5795–5799.

Schamhart, D.H., V. Berendson, J. van Rijn, and R. van Wijk. 1984a. Comparative studies of heat sensitivity of several rat hepatoma cell lines and hepatocytes in primary culture. *Cancer Res.* **44:** 4507–4516.

Schamhart, D.H., H.S. van Walraven, F.A.C. Wiegant, W.A.M. Linnemans, J. van Rijn, J. van den Berg, and R. van Wijk. 1984b. Thermotolerance in cultured rat hepatoma cells: Cell viability, cell morphology, protein synthesis and heat shock proteins. *Radiat. Res.* **98:** 82–95.

Sciandra, J.J. and J.R. Subjeck. 1984. Heat shock proteins and protection of proliferation and translation in mammalian cells. *Cancer Res.* **44:** 5188–5194.

Scott, M.P. and M.L. Pardue. 1981. Translational control in lysates of *Drosophila melanogaster* cells. *Proc. Natl. Acad. Sci.* **78:** 3353–3357.

Sirkin, E.R. and S. Lindquist. 1985. Translation regulation in the *Drosophila* heat shock response. In *Sequence specificity in transcription and translation* (ed. R. Calendar and L. Gold), pp. 669–679. Alan R. Liss, New York.

Solomon, J.M., J.M. Rossi, K. Golic, T. McGarry, and S. Lindquist. 1991. Changes in hsp70 alter thermotolerance and heat-shock regulation in *Drosophila*. *New Biol.* **3:** 1106–1120.

Song, H.-J., D.R. Gallie, and R.F. Duncan. 1995. m(7)GpppG cap dependence for efficient translation of *Drosphila* 70-kDa heat-shock-protein (Hsp70) mRNA. *Eur. J. Biochem.* **232:** 778–788.

Storti, R.V., M.P. Scott, A. Rich, and M.L. Pardue. 1980. Translational control of protein synthesis in response to heat shock in *D. melanogaster* cells. *Cell* **22:** 825–834.

Theodorakis, N.G., S.S. Banerji, and R.I. Morimoto. 1988. HSP 70 mRNA translation in chicken reticulocytes is regulated at the level of elongation. *J. Biol. Chem.* **263:** 14579–14585.

Thomas, G.P. and M.B. Mathews. 1984. Alterations of transcription and translation in HeLa cells exposed to amino acid analogues. *Mol. Cell. Biol.* **4:** 1063–1072.

Welch, W.J. and J.P. Suhan. 1986. Cellular and biochemical events in mammalian cells during and after recovery from physiological stress. *J. Cell Biol.* **103:** 2035–2052.

Wong, W.L., M.A. Brostrom, G. Kuznetsov, D. Gmitter-Yellen, and C.O. Brostrom. 1993. Inhibition of protein synthesis and early protein processing by thapsigargin in cultured cells. *Biochem. J.* **289:** 71–79.

Zapata, J.M., F.G. Maroto, and J.M. Sierra. 1991. Inactivation of mRNA cap-binding protein complex in *Drosophila melanogaster* embryos during heat shock. *J. Biol. Chem.* **266:** 16007–16014.

Zapata, J.M., M.A. Martinez, and J.M. Sierra. 1994. Purification and characterization of eukaryotic polypeptide chain initiation factor 4F from *Drosophila melanogaster* embryos. *J. Biol. Chem.* **269:** 18047–18052.

10

Regulation of Protein Synthesis by Calcium

Angus C. Nairn
The Laboratory of Molecular and Cellular Neuroscience
Rockefeller University, New York, New York 10021

H. Clive Palfrey
Department of Pharmacological and Physiological Sciences, University of Chicago
Chicago, Illinois 60637

Eukaryotic protein synthesis is highly regulated by a variety of acute and chronic processes. Paramount among these processes is the reversible phosphorylation of protein synthetic components. Many of the factors involved in translation are phosphoproteins, and a certain amount is known about the kinases and phosphatases responsible for the phosphorylation and dephosphorylation of these factors. Considerably less is known about the overall physiological role of phosphorylation of translational components. In particular, two aspects remain to be fully understood, namely, the general involvement of Ca^{++} in the regulation of protein synthesis and specifically, the role of Ca^{++}-dependent phosphorylation of elongation factor-2 (eEF2). This chapter reviews these two topics. For more detailed reviews of the mechanistic details of protein synthesis and the role of protein phosphorylation in protein synthesis, see Merrick and Hershey, Trachsel, Clemens, and Hinnebusch (all this volume).

REGULATION OF PROTEIN SYNTHESIS ELONGATION BY CA^{++}

Identification of eEF2 Phosphorylation

It has generally been assumed that initiation is the key regulatory step in protein synthesis and that changes in elongation rate could only affect global rates of translation without any type of specificity. However, increasing evidence suggests that elongation rates are subject to several forms of control and that this may result in selective effects on the translation of certain messenger RNAs (for further discussion, see below). An important factor in this renewed interest in the regulation of elongation is the discovery that both of the factors involved in polypeptide elongation, namely, eEF1 and eEF2 (see Merrick and Hershey, this volume), are physiological targets for protein phosphorylation (for further discussion of eEF1 phosphorylation, see Venema et al. 1991a,b; Mulner-Lorillon et al. 1994).

eEF2 was first identified as a phosphoprotein in two independent studies (Nairn and Palfrey 1987; Ryazanov 1987). The starting point for one study was the description of a widely distributed 100-kD protein that was phosphorylated in a Ca^{++}/calmodulin-dependent mannner in extracts from a number of rat tissues (Palfrey 1983). Amino acid sequencing of the 100-kD protein indicated that it was eEF2, and this was confirmed by the fact that the protein was a target for diphtheria-toxin-mediated ADP-ribosylation and photolabeling with a GTP analog (Nairn and Palfrey 1987). The basis for the other study was that a 100-kD polypeptide was phosphorylated in rabbit reticulocyte and rat liver lysates and that this polypeptide appeared to be identical to eEF2 (Ryazanov 1987).

At the time of the initial characterization of the phosphorylation of the 100-kD protein by a distinct Ca^{++}/calmodulin-dependent (CaM) kinase (Nairn et al. 1985), four other CaM kinase family members had been well characterized. These included two highly specific enzymes: phosphorylase kinase and myosin-light-chain kinase; an enzyme of somewhat broader specificity, CaM kinase I; and the multifunctional enzyme, CaM kinase II (Hanson and Schulman 1992; Nairn and Picciotto 1994). Thus, the 100-kD kinase was termed CaM kinase III. Subsequently, an additional, multifunctional enzyme, termed CaM kinase IV, was also characterized. However, from the initial characterization and from later, more extensive studies, it is likely that eEF2 is the sole substrate for CaM kinase III and that eEF2 is not a physiological substrate for any other kinase (Mitsui et al. 1993; Redpath and Proud 1993b). Thus, by analogy to the highly specific phosphorylase and myosin-light-chain kinases, CaM kinase III is also referred to as eEF2 kinase.

Phosphorylation of eEF2 In Vitro

Initial studies indicated that at least one threonine residue contained in a tryptic peptide derived from amino acids 51–60 of eEF2 was phosphorylated by eEF2 kinase (Fig. 1) (Nairn and Palfrey 1987). Subsequent studies indicated that all three threonine residues within this region (at positions 53, 56, and 58) could be phosphorylated in vitro (Ovchinnikov et al. 1990; Price et al. 1991; Redpath et al. 1993). However, Thr-56 is the major site for eEF2 kinase, with Thr-58 being phosphorylated to a lesser extent and Thr-53 being very poorly phosphorylated. It has been

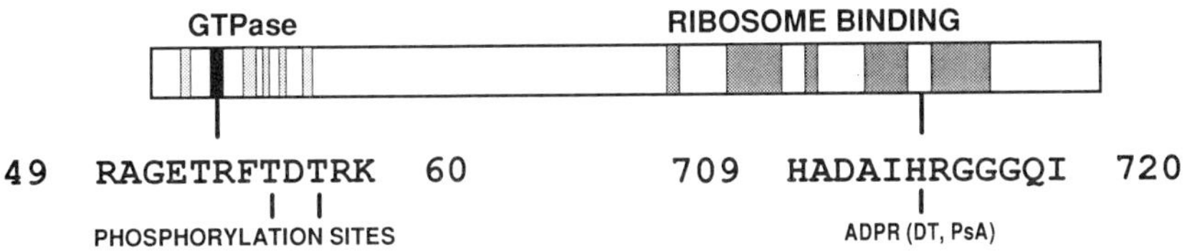

Figure 1 Domain structure of eEF2. The position of the sites phosphorylated by eEF2 kinase (Thr-56 and Thr-58). The amino-terminal shaded regions indicate amino acid sequences homologous to those found in the GTPase domain of other GTP-binding proteins. The carboxy-terminal shaded regions indicate amino acid sequences homologous to those found in other elongation factors and believed to be involved in ribosome binding. Other ribosomal binding sites may also exist within the GTPase domain. The crystal structure of the bacterial elongation factor, EF1A, has been recently determined in both the GTP and GDP forms (Berchtold et al. 1993; Sprinzl 1994). On the basis of homology between the GTPase domains of eEF2 and EF1A, it has been possible by analogy to model the structure of this region of eEF2 (Perentesis et al. 1992). The carboxy-terminal region of eEF2 also contains an unusual amino acid that results from posttranslational modification of a histidine residue (His-714) (Ueda and Hayaishi 1985). This amino acid is the site of ADP-ribosylation catalyzed by diphtheria toxin as well as *Pseudomonas* toxin A and as a consequence has been termed "diphthamide" (Ueda and Hayaishi 1985). ADP ribosylation by these exogenous toxins renders eEF2 totally inactive, resulting in cell death. A number of toxin-resistant cell lines have been isolated in which mutation of Arg-716 of eEF2 to glycine is found (Kohno and Uchida 1987), resulting in disruption of ADP ribosylation of the mutant eEF2 by the toxin (Foley et al. 1992).

suggested that phosphorylation of Thr-58 is dependent on prior phosphorylation of Thr-56 (Redpath et al. 1993), although recent preliminary studies of mutants of eEF2 in which Thr-56 has been changed to alanine indicate that Thr-58 is phosphorylated well by eEF2 kinase, suggesting that phosphorylation of Thr-56 may act as a negative, rather than a positive, determinant (K. Nastiuk and A. Nairn, unpubl.).

The phosphorylatable threonines in eEF2, Thr-56 and Thr-58 (but not Thr-53), are highly conserved throughout eukaryotic cell evolution (Proud 1992). In addition, yeast has been shown to contain a Ca^{++}/CaM-dependent activity that possesses properties in common with mammalian eEF2 kinase (Donovan and Bodley 1991; K. Mitsui and A. Nairn, unpubl.). The evolutionary conservation of this modification system strongly suggests an important physiological role for eEF2 phosphorylation. Although eEF2 kinase phosphorylates eEF2 with high affinity and specificity, the enzyme does not phosphorylate short synthetic peptides encompassing residues 49–60 (Redpath et al. 1993; K. Mitsui and A. Nairn, unpubl.), suggesting that elements of secondary or tertiary structure of eEF2 may be critical for its phosphorylation by eEF2 kinase.

Regulation of eEF2 by Phosphorylation In Vitro

Phosphorylation of eEF2 by eEF2 kinase was initially shown to result in almost complete inhibition of polypeptide synthesis as demonstrated in vitro by measurement of poly(U)-directed poly(Phe) synthesis (Nairn and Palfrey 1987; Ryazanov et al. 1988). The inhibition was readily reversed by dephosphorylation of the factor with protein phosphatase-2A, the likely physiological eEF2 phosphatase (Nairn and Palfrey 1987) (see below). An inhibitory effect of eEF2 phosphorylation was also apparent in the reticulocyte lysate system, where it was found that inhibition of eEF2 dephosphorylation by the protein phosphatase inhibitor, okadaic acid, was associated with elongation block (Redpath and Proud 1989, 1991). Studies using either paradigm indicated that phosphorylation of Thr-56 alone was sufficient to inactivate eEF2 (Nairn and Palfrey 1987; Carlberg et al. 1990; Redpath et al. 1993), raising the question of whether phosphorylation of Thr-58 makes any significant contribution (Redpath et al. 1993).

Although these initial studies clearly demonstrated a dramatic effect of phosphorylation, the precise mechanism whereby this modification abrogates the function of eEF2 is still controversial. eEF2 function can be broken down into several steps: binding of GTP to eEF2, attatchment of eEF2 to the ribosome, performance of the translocation reaction, ribosome-dependent GTP hydrolysis, and dissociation of eEF2-GDP from the ribosome. In theory, any one or more of these reactions could be impaired by phosphorylation. Notably, Thr-56 and Thr-58 are located close to the GTP-binding domain of eEF2, raising the obvious possibility that effects of phosphorylation on GTP binding or hydrolysis might explain the inhibition of function. Several studies have directly addressed this possibility and reported no obvious effect of phosphorylation (Carlberg et al. 1990; A. Nairn and H.C. Palfrey, unpubl.), but a recent report using fluorescence methods has suggested that phosphorylation of eEF2 results in a decreased affinity for GTP but not for GDP (Dumont-Miscopein et al. 1994).

There is a more general consensus that phosphorylation of eEF2 results in a decreased affinity of the eEF2-GTP complex for the ribosome. Unphosphorylated eEF2 bound to ribosomes is not accessible to eEF2 kinase (Nilsson and Nygård 1991; Dumont-Miscopein et al. 1994). Studies in reticulocyte lysates also indicate that phospho-eEF2 has a much lower affinity for the ribosome and that ribosome-dependent GTPase activity is consequently reduced (Carlberg et al. 1990). Furthermore, the ability of phosphorylated eEF2 to form a complex with ribosomes in the presence of a nonhydrolyzable GTP analog was significantly decreased (Dumont-Miscopein et al. 1994).

Together, these results suggest that phosphorylation of eEF2 may inhibit both GTP binding and the ability of the eEF2-GTP complex to bind to the ribosome. However, other studies in which poly(U)-directed poly(Phe) synthesis was measured have suggested that phospho-eEF2 is able to compete with dephospho-eEF2, implying that the phosphorylated protein can interact with ribosomes (Ryazanov et al. 1988; Redpath et al. 1993; A. Nairn and H.C. Palfrey, unpubl.). If correct, this result would be important since it might indicate that phospho-eEF2 could exert a "dominant negative" effect on translation. In contrast, studies with the more physiologically relevant reticulocyte lysate system do not support this possibility. Analysis of polysomes following elongation block failed to detect phospho-eEF2 (Redpath and Proud 1989). In addition, even when approximately 97% of the eEF2 in a reticulocyte lysate is apparently phosphorylated, protein synthetic rates only decrease by 50% (Redpath et al. 1993). Although this result is somewhat surprising, given that the ratio of eEF2:ribosomes in reticulocyte lysates is approximately 1:1 (Nygård and Nilsson 1984), the fact that a ratio of phospho-eEF2:eEF2 of greater than 30 only halved translation rates argues against the dominant negative hypothesis, at least in this translation system.

Phosphorylation of eEF2 and the Regulation of Elongation in Intact Cells

eEF2 phosphorylation has been observed in many different types of cells grown in culture or acutely dissociated from tissue (for a listing of many of the systems that have been studied, see Proud 1992; Ryazanov and Spirin 1993). For example, stimulation of quiescent fibroblasts with serum or bradykinin (Palfrey et al. 1987) or endothelial cells with histamine (Mackie et al. 1989) leads to a rapid increase in eEF2 phosphorylation. Increased eEF2 phosphorylation has also been found in acinar cells (isolated from rat parotid gland) treated with a variety of secretagogues such as carbachol, substance P, and ATP (Hincke and Nairn 1992), as well as in chromaffin cells treated with acetylcholine (Haycock et al. 1987). In a number of different cell types where appropriate studies

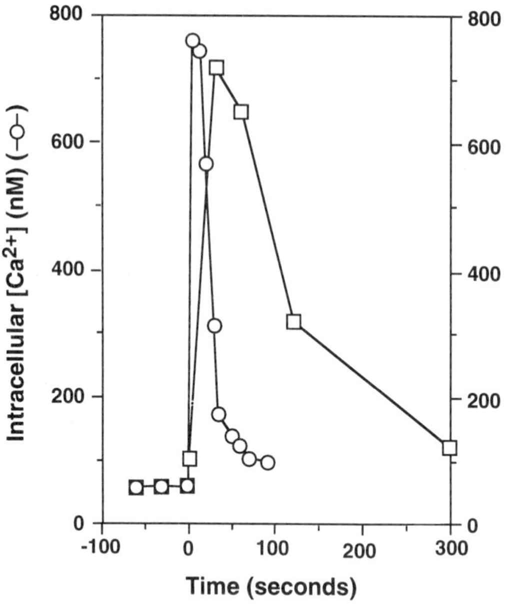

Figure 2 Effect of bradykinin on intracellular [Ca^{++}] and eEF2 phosphorylation in fibroblasts. Bradykinin (10 ng/ml) was added to both types of assays at zero time. Intracellular [Ca^{++}] was measured by using the Ca^{++}-binding fluorescent probe, Fura. eEF2 phosphorylation was analyzed by immunoprecipitation from ^{32}P-labeled cells. (Adapted from Palfrey et al. 1987.)

have been carried out, the increased eEF2 phosphorylation qualitatively reflects the intracellular [Ca^{++}] transients produced by the various ligands used (Fig. 2). Thus, a general feature of eEF2 phosphorylation is that it is proportional to the status of physiologically important cytoplasmic Ca^{++} concentrations in cells. Another important feature is that in most cases the phosphorylation of eEF2 is relatively transient. This dynamic phosphorylation and dephosphorylation presumably reflects the transient nature of the increase in Ca^{++} concentration, coupled with the rapid dephosphorylation of eEF2 by cellular phosphatases (see below).

Although transient in nature, the phosphorylation of eEF2 can reach high stoichiometry in intact cells stimulated with physiological agonists. For example, in endothelial cells, isoelectric focusing methods have indicated that the proportion of phosphorylated eEF2 rises from approximately 15% in the basal state to about 90% after incubation with thrombin for 2 minutes (Mackie et al. 1989). Two-dimensional phosphopeptide mapping studies indicate that Thr-56 is the major site phosphorylated in endothelial cells as well as in other cells stimulated with physiological agonists (Palfrey et al. 1987; Mackie et al. 1989).

Studies in reticulocytes treated with the calcium ionophore A23187 have shown that eEF2 is rapidly and stably phosphorylated (A. Nairn and H.C. Palfrey, unpubl.), and this correlates with a buildup of polysomes reflecting the ionophore-induced elongation block noted earlier in the same system (Wong et al. 1991). Studies have also been carried out in which dephosphorylation of eEF2 in intact cells was inhibited using the protein phosphatase inhibitors okadaic acid or calyculin A. Under these circumstances, it has been possible to observe an agonist-dependent and sustained inhibition of protein synthesis that parallels the phosphorylation of eEF2 (K. Mitsui et al., in prep.). In addition, these studies have demonstrated that the inhibition of protein synthesis is associated with an increased ribosomal transit time (K. Mitsui et al., in prep.). Thus, although it is possible that Ca^{++} and/or protein phosphatase inhibition affects the phosphorylation of other factors involved in protein synthesis, the demonstration that eEF2 is phosphorylated and elongation is inhibited in intact cells, coupled with in vitro studies clearly demonstrating that phosphorylation of eEF2 by eEF2 kinase is inhibitory, suggests that inhibition of elongation by eEF2 phosphorylation occurs in vivo.

An area receiving increased attention is the role of Ca^{++} in neuronal excitotoxicity. In particular, Ca^{++} influx into neurons and inhibition of protein synthesis may be involved in ischemia-induced cell death. The molecular basis for the inhibition of protein synthesis remains to be established, but recent studies have indicated that treatment of cortical neurons in culture, with the glutamate agonist, NMDA, results in increased phosphorylation of eEF2 (P. Marin et al., unpubl.). Furthermore, NMDA has also been found to increase the phosphorylation of a number of proteins in developing frog tectum (Scheetz and Constantine-Paton 1994), one of which has recently been shown to be eEF2 (A. Scheetz et al., unpubl.). Thus, these preliminary studies raise the possibility that phosphorylation of eEF2 has some role in glutamate-dependent changes in neurons.

eEF2 Phosphorylation and Cell Cycle Regulation

There is substantial evidence that both Ca^{++} and CaM have important roles in the cell cycle (for review, see Lu and Means 1993; Lu et al. 1993). For example, Ca^{++} appears to be important for cell cycle progression in sea urchin embryos (Whitaker and Patel 1990). Although Ca^{++} transients have been found in some mammalian cells during G_1/S and M phases, consensus has not been reached on the timing or magnitude of Ca^{++} changes in cycling mammalian cells (Poenie et al. 1985, 1986;

Ratan et al. 1986; Hepler 1989; Kao et al. 1990). A number of studies have shown that protein synthetic rates vary throughout the cell cycle. There is a significant suppression of protein synthesis during mitosis that may involve polypeptide initiation, since polysomes were found to decrease while ribosomal transit times did not change (Fan and Penman 1970; for studies in *Xenopus* embryos, see also Kanki and Newport 1991). However, polysomal profiles in other cell types are not so drastically affected during the cell cycle (Eremenko and Volpe 1975), suggesting that inhibition of elongation might also be important.

Several studies have analyzed both the phosphorylation of eEF2 and the level of eEF2 kinase during the cell cycle; however, the results obtained are somewhat contradictory. As discussed above, transient eEF2 phosphorylation has been demonstrated in response to a number of different mitogens, at a time when the previously quiescent cells begin to move into the G_1 phase of the cell cycle. Other studies indicate that both eEF2 kinase activity and eEF2 phosphorylation are regulated during the cell cycle (Celis et al. 1990; Carlberg et al. 1991). For example, the ratio of phospho-eEF2:eEF2 increases from approximately 0.1 in interphase amnion cells to about 0.25 in cells synchronized by mitotic shake-off (Celis et al. 1990). Studies in intact COS cells overexpressing insulin receptors exhibited eEF2 kinase activity (and consequently eEF2 phosphorylation) that was at its highest in quiescent cells, and this was reduced following treatment with serum or insulin (N.T. Redpath et al., unpubl.). Notably, the ribosomal transit time was decreased about two-fold following treatment with serum or insulin, possibly as a result of eEF2 dephosphorylation. In addition, the effect of serum or insulin was blocked by rapamycin, an inhibitor of mitogen-activated 70-kD S6 kinase (see Jefferies and Thomas, this volume), suggesting that eEF2 kinase may be regulated by this pathway. Other studies in intact HeLa cells, measured using an antibody specific for phospho-Thr-56, indicated that eEF2 phosphorylation was highest during the G_0/G_1 and G_1/S phases of the cell cycle, whereas the levels of eEF2 kinase were highest in extracts prepared from quiescent cells (K. Mitsui et al., unpubl.). In contrast, however, in Ehrlich ascites cells, the level of eEF2 kinase has been found to be highest in lysates prepared from cells in S phase (Carlberg et al. 1991).

Although these various studies are suggestive of a role for eEF2 phosphorylation in some aspect of the cell cycle, it is clear that since the regulation of the cell cycle is extremely complex, ascribing a specific role for eEF2 phosphorylation is not straightforward. (eEF1 is also known to be phosphorylated by cdc kinase, raising the possibility that other mechanisms are involved in cell-cycle-dependent regulation of elongation [Mulner-Lorillon et al. 1994].) Furthermore, it is important to distinguish measurements of eEF2 phosphorylation from those of the levels of eEF2 kinase analyzed in cell extracts. In situ, this kinase would be expected to be completely dependent on Ca^{++} for activity, and thus alterations in the absolute levels of the enzyme may not necessarily be correlated with changes in eEF2 phosphorylation in the absence of changes in cytosolic Ca^{++}.

In an attempt to define further the role of eEF2 phosphorylation in cell function, recent studies have used site-directed mutagenesis of the eEF2 phosphorylation sites to render eEF2 nonphosphorylatable in cells (K. Nastiuk et al., unpubl.). A potential problem in studies of eEF2 is the analysis of any eEF2 mutants in the presence of the high endogenous levels of eEF2. To overcome this problem, variant eEF2 molecules that not only lack Thr-56 and/or Thr-58, but are also resistant to toxin-mediated ADP-ribosylation have been engineered. As described previously (Kohno and Uchida 1987), transfection with eEF2[G717R] resulted in viable cells when grown in the presence of normally lethal levels of toxin. When cells were transfected with eEF2[T56A,G717R] or eEF2[T58A,G717R], toxin-resistant cells were also obtained, although these cells exhibited a slower growth rate. However, cells transfected with eEF2[T56A,T58A,G717R] were nonviable when grown in the presence of toxin. Thus, while preliminary and subject to a number of different interpretations, these results suggest that phosphorylation of eEF2 may have an important role in the cell cycle.

Characterization of eEF2 Kinase

eEF2 kinase has been purified to homogeneity from rabbit reticulocytes and rat pancreas (Mitsui et al. 1993; Redpath and Proud 1993b). The properties of the enzyme from either source were essentially identical. eEF2 kinase has an apparent molecular mass of 140 kD as measured by gel filtration and a subunit size of 95 kD as measured by SDS-PAGE. These results, together with the sedimentation coefficient, indicate that the enzyme is an elongated monomer. A report claiming that eEF2 kinase is identical to the heat shock protein, HSP90, is incorrect (Nilsson et al. 1991; Nygård et al. 1991). However, it appears that HSP90 (a very abundant protein in reticulocytes and many other mammalian cells) may interact with eEF2 kinase to some extent (Palmquist et al. 1994), although the physiological significance of this is unclear.

As discussed above, eEF2 kinase is highly specific for eEF2. Enzyme activity is completely dependent on Ca^{++}/CaM at neutral pH and above, but some Ca^{++}/CaM-independent activity is evident at pH 6–6.5 (Nairn et al. 1987; Mitsui et al. 1993). In addition, it appears that eEF2 kinase binds a single CaM molecule with high affinity, as do the catalytic subunits of other CaM kinases, except phosphorylase kinase (Nairn and Picciotto 1994).

eEF2 kinase is autophosphorylated in vitro through an intramolecular mechanism in the presence of Ca^{++}/CaM and ATP (Mitsui et al. 1993; Redpath and Proud 1993b). Autophosphorylation occurs at multiple sites (possibly as many as five) on both serine and threonine residues and is associated with the partial generation of Ca^{++}/CaM-independent eEF2 kinase activity. In this respect, autophosphorylation of eEF2 kinase resembles CaM kinase II, which can be rendered Ca^{++}/CaM-independent both in vitro and in intact cells (Hanson and Schulman 1992). However, the autophosphorylation of eEF2 kinase has yet to be demonstrated in intact cells, and thus its physiological significance remains to be established.

Purified eEF2 kinase is also an excellent substrate for cAMP-dependent protein kinase (PKA) in either the absence or presence of Ca^{++}/CaM (Redpath and Proud 1993a,b). Phosphorylation by PKA is also associated with the generation of partial Ca^{++}/CaM-independent eEF2 kinase activity, and it is possible that Ca^{++}/CaM-dependent autophosphorylation and PKA phosphorylation occur at overlapping sites (Redpath and Proud 1993a,b). Again, the physiological significance of phosphorylation of eEF2 kinase by PKA is unknown, since treatments that elevate cellular cAMP do not alter the phosphorylation of eEF2 (Palfrey et al. 1987; Mackie et al. 1989; M. Brady and H.C. Palfrey, unpubl.). Potentially, phosphorylation of eEF2 kinase by PKA may be involved in regulating its expression or degradation (see below).

Regulation of eEF2 Kinase Expression

eEF2 kinase appears to be an exclusively cytoplasmic species that is present in all eukaryotic cells and tissues examined (Palfrey 1983; Nairn et al. 1985). However, significant changes in the levels of eEF2 kinase measured in cytosolic extracts have been found in various tissues and cell types (see also above). High levels of enzyme activity are found in extracts prepared from proliferating cells in culture and from tissues exhibiting high levels of protein synthesis such as pancreas. In contrast, brain contains the lowest level of the enzyme. The level of expression of eEF2 kinase has also been found to be subject to significant regulation in a number of different cell systems. The best characterized of these is in PC12 cells where high eEF2 kinase levels measured in cell extracts are dramatically down-regulated over a period of hours by treatment with nerve growth factor (NGF) or agents that increase cAMP levels (End et al. 1982; Nairn et al. 1987; Brady et al. 1990). In PC12 cells, treatment with epidermal growth factor (EGF) causes this effect, although to a lesser extent, and down-regulation of eEF2 kinase is found in response to cAMP in other cells of nervous system origin (H.C. Palfrey, unpubl.). In addition, eEF2 kinase activity is decreased in extracts prepared from stationary phase C_6 gliomal cells compared to that in extracts prepared from cells in logarithmic phase (Bagaglio et al. 1993). Little or no eEF2 kinase activity was detected in white matter from rat forebrain, supporting the hypothesis that high levels of enzyme were specifically associated with the proliferating malignant cells. Finally, down-regulation of eEF2 kinase activity has been observed during the final stages of *Xenopus laevis* oogenesis (Severinov et al. 1990). The significance of these changes is presently unclear, since as discussed above, the absolute levels of eEF2 kinase may not necessarily be correlated with the levels of eEF2 phosphorylation in intact cells in the absence of changes in cytosolic Ca^{++}. However, the amount of active eEF2 kinase would be expected to determine the rate of eEF2 phosphorylation during periods of elevated $[Ca^{++}]_{cyt}$.

The down-regulation of eEF2 kinase is prominent in cells of neuronal origin as they differentiate into sympathetic neurons. This phenomenon might therefore be important in developing neurons where Ca^{++} influx via voltage-sensitive Ca^{++} channels controls many processes from growth-cone migration to neurotransmitter release. An additional feature is that high levels of eEF2 kinase are often found in cytoplasmic extracts prepared from cells growing in culture, whereas low levels of kinase activity are found in extracts prepared from terminally differentiated neurons or in certain nonproliferating cells. An obvious implication is that inhibition of polypeptide elongation would be less sensitive to Ca^{++} after down-regulation of eEF2 kinase; however, the physiological role of such a process in developing neurons or in nonproliferating cells is unclear.

De novo protein synthesis is necessary for recovery of enzyme levels after NGF treatment of PC12 cells (Nairn et al. 1987). PKA is clearly involved in the down-regulation of eEF2 kinase in PC12 cells exposed to agents that elevate cAMP, and it also seems to be involved in the response to NGF (Nairn et al. 1987; Brady et al. 1990). PC12 cells that are deficient in PKA fail to down-regulate eEF2 kinase in response to NGF or forskolin (Brady et al. 1990). However, the down-regulation observed in response to EGF is not affected in PKA-deficient cells. Other data suggest that the activity of the high-affinity NGF receptor, *Trk*, and $p21^{ras}$ is necessary for down-regulation (M. Brady and H.C. Palfrey, unpubl.), but the connection between these signaling elements and PKA has not been defined. As discussed above, eEF2 kinase can be phosphorylated by PKA in vitro, although this results in an apparent activation, rather than an inhibition, of enzyme activity. Furthermore, down-regulation in intact cells is a significantly slower process than that of activation of PKA by cAMP. However, given that eEF2 kinase is an excellent substrate for PKA, it seems possible that loss of eEF2 kinase activity may still be a consequence of PKA-mediated phosphorylation of the enzyme. Autophosphorylation of CaM kinase II, which leads to the generation of Ca^{++}/CaM-independent activity, also results in increased thermal instability (Lai et al. 1986; Ishida and Fujisawa 1995). By analogy, autophosphorylation or phosphorylation of eEF2 kinase by PKA might also result in down-regulation by a similar mechanism.

Dephosphorylation of eEF2

Initial studies indicated that dephosphorylation of phosphorylated eEF2 by protein phosphatase-2A (PP2A) was able to reverse fully the inhibitory effect of phosphorylation on protein synthesis (Nairn and Palfrey 1987). More extensive analysis in vitro has indicated that of the four major serine/threonine protein phosphatases (Shenolikar and Nairn 1991), PP2A is the most active toward phospho-eEF2, although PP2C displays significant activity, and PP1 and PP2B (calcineurin) are essentially inactive (Redpath and Proud 1989, 1990, 1991; Redpath et al. 1993; A. Nairn, unpubl.). Notably, PP2A appeared to be able to dephosphorylate Thr-56 and Thr-58 equally well, and PP2C appeared to dephosphorylate Thr-56 preferentially (Redpath et al. 1993). Furthermore, phosphorylation of Thr-58 appeared to inhibit the ability of PP2C to dephosphorylate Thr-56. Studies with inhibitors of protein phosphatases, such as okadaic acid, microcystin, and calyculin A, indicate that PP2A is the major eEF2 phosphatase in cell extracts (Redpath and Proud 1989, 1991). In addition, studies using these inhibitors in intact cells (K. Mitsui et al., in prep.) are consistent with PP2A being the most likely phosphatase responsible for phospho-eEF2 dephosphorylation in vivo.

Regulation of eEF2 Synthesis and the Role of Ribosomal Protein S6 Phosphorylation

eEF2 levels in most cells seem to be approximately stoichiometric with the number of ribosomes (Gill and Dinius 1973; Nygård and Nilsson 1984). However, eEF2 biosynthesis is subject to regulation (Levenson et al. 1989; Rattan 1991; Vary et al. 1994). Recent studies have demonstrated that eEF2 mRNA is a member of a class of polypyrimidine tract mRNAs whose translation is regulated through phosphorylation of ribosomal S6 protein (Jefferies et al. 1994; Terada et al. 1994; see Jefferies and Thomas, this volume). In these studies, the 70-kD S6 kinase inhibitor, rapamycin, was found to inhibit specifically the synthesis of only a few proteins, prominent among which was eEF2. Other proteins whose synthesis was inhibited included eEF1A and a number of ribosomal proteins. In contrast, in other studies, stimulation of quiescent lymphocytes with mitogens resulted in a coordinated pattern of increased synthesis of the same mRNAs, which was correlated with increased phosphorylation of ribosomal S6 protein (N. Terada, unpubl.). Together, these studies suggest the possibility that the regulation of expression of eEF1A, eEF2, and other ribosomal proteins may be coordinated with the control of cell growth and proliferation.

REGULATION OF PROTEIN SYNTHESIS INITIATION BY CA^{++}

Regulation of eIF2α Phosphorylation by Ca^{++}

The regulation of eIF2 function results from phosphorylation of a single serine residue, Ser-51, in the α-subunit by a number of distict protein kinases that are themselves subject to various novel forms of regulation (see Trachsel; Clemens; both this volume). On the basis of studies in several cell types, but particularly in reticulocytes, at least two enzymes, the "heme-controlled repressor" (HCR; so-called because of its inhibition by addition of hemin to a reticulocyte lysate) and a double-stranded RNA-activated enzyme called dsI or DAI (now renamed PKR), are able to phosphorylate Ser-51. Phosphorylation of eIF2α is rapidly modulated in a number of situations, among which are viral infection, heat shock, and serum deprivation of cultured cells (for a summary, see Proud 1992; Pain and Clemens 1991; Clemens; Duncan; Mathews; Katze; Schneider; all this volume.). eIF2α phosphorylation is also regulated by a number of stimuli and treatments that affect cellular Ca^{++} levels. For example, serum treatment of quiescent cells rapidly stimulates protein synthesis, and this may involve, at least in part, the dephosphorylation of eIF2α (Rowlands et al. 1988; Montine and Henshaw 1989; Hershey 1991).

Therefore, since one of the early effects of serum treatment is to raise $[Ca^{++}]_{cyt}$ (Villereal and Byron 1992), a potential connection exists between eIF2α dephosphorylation and $[Ca^{++}]_{cyt}$.

The most compelling evidence for a link between Ca^{++} and eIF2α phosphorylation resulted from studies of cellular Ca^{++} depletion (for review, see Brostrom and Brostrom 1990; Palfrey and Nairn 1995). Initial studies of this phenomenon indicated that treatment of C6 rat glioma cells with a medium containing EGTA in place of Ca^{++} reduced protein synthetic rates to about 15–25% of control values within 30 minutes (Brostrom et al. 1983). Notably, the effect was rapidly reversible following readdition of Ca^{++} to the external medium, although there was a perceptible lag. In other cells, the effect of EGTA has been shown to be augmented by addition of Ca^{++}-mobilizing hormones, and the effect of Ca^{++}-mobilizing hormones also has been shown to be effective under conditions of normal external Ca^{++} (Thomas et al. 1984; Brostrom et al. 1986; Menaya et al. 1988; Chin et al. 1988; Kimball and Jefferson 1990, 1992). Subsequent studies with Ca^{++} ionophores have obtained results similar to those found with EGTA (Brostrom et al. 1989). Significantly, certain cells thought to be "resistant" to the effects of extracellular EGTA (such as HeLa) were found to be susceptible to ionophore treatment, leading to the hypothesis that this form of Ca^{++} regulation of protein synthesis is a widespread phenomenon (Brostrom and Brostrom 1990).

Further analysis supports the idea that it is the removal of Ca^{++} from some intracellular store, rather than changes in $[Ca^{++}]_{cyt}$, that is responsible for the inhibition of protein synthesis (Takuma et al. 1984). Although such conclusions based on studies with Ca^{++} ionophores were subject to a number of interpretations, more recent studies using the relatively specific microsomal Ca^{++} pump inhibitor thapsigargin (Thomas and Hanley 1994) support the model that Ca^{++} depletion from intracellular stores is involved in the regulation of protein synthesis (Preston and Berlin 1992; Wong et al. 1993). Notably, in HeLa cells, treatment with histamine, which provoked a large Ca^{++} transient via an $InsP_3$-mediated mechanism, failed to inhibit protein synthesis even with EGTA in the external medium (Preston and Berlin 1992). Since the decreases in total cell Ca^{++} were equivalent following treatment of either thapsigargin or histamine, it has been suggested that only Ca^{++} depletion from a non-$InsP_3$-sensitive store is effective in inhibiting protein synthesis (Preston and Berlin 1992; cf. Kimball and Jefferson 1991).

Interestingly, studies with several cell types have shown that they can "accommodate" to treatment with EGTA or ionophores, resulting in protein synthesis at rates close to controls (Brostrom et al. 1990; for review, see Brostrom and Brostrom 1990). The mechanism for this accommodation is not understood, but it appears to correlate with the induction of the ER-resident stress protein or "chaperone," termed BiP or GRP-78. Like other chaperones, BiP helps to ensure correct folding of proteins as they are synthesized, suggesting that the effect of Ca^{++} depletion on protein synthesis inhibition is related to protein folding.

A key observation common to many of the studies cited above is that polysome profiles exhibit a breakdown to monosomes with Ca^{++} depletion (Chin et al. 1987; Kimball and Jefferson 1990; Perkins and Pandol 1992). Furthermore, analysis of ribosome transit times has indicated that average elongation rates are generally unaffected by Ca^{++}-depletion (Chin et al. 1987; Kimball and Jefferson 1990). Consequently, the locus of the regulatory event involved in the inhibition of protein synthesis by Ca^{++} depletion has been postulated to be at some step(s) in the initiation cycle (Chin et al. 1987; Kumar et al. 1989; Brostrom and Brostrom 1990; Kimball and Jefferson 1990). Given the importance of eIF2α in regulation of protein synthesis initiation, a number of studies have examined the potential phosphorylation of the factor following depletion of Ca^{++} (Kimball and Jefferson 1990, 1991, 1992; Prostko et al. 1992). In these studies, the phosphorylation of eIF2α has been shown to increase as a consequence of Ca^{++} depletion. A recent study has also shown that overexpression of a mutant eIF2α, in which Ser-51 was changed to alanine, partially protected NIH-3T3 cells from the normally inhibitory effect of Ca^{++} ionophore treatment (Srivastava et al. 1995). A shortcoming of several of these studies is that a precise temporal relationship between eIF2α phosphorylation/eIF2B activity and inhibition/recovery of protein synthesis has not always been clearly established. In addition, in other studies, no relationship was found between eIF2α phosphorylation and inhibition of protein synthesis caused by Ca^{++} depletion, and it has been suggested that the function of other initiation factors such as eIF3 may be affected (Kumar et al. 1989).

In contrast to the studies of eEF2 phosphorylation where the Ca^{++}/CaM dependence of eEF2 kinase has been clearly established, a missing link in the Ca^{++} depletion/initiation block hypothesis is the precise biochemical nature of the Ca^{++}-dependent event(s). It has been suggested that CaM may be involved (Kumar et al. 1991), although neither HCR nor PKR has been shown to be directly regulated by Ca^{++} or CaM (Hinnebusch 1990; Kramer et al. 1993). Furthermore, eIF2α is dephosphorylated by PP1, a Ca^{++}-independent enzyme (Redpath and Proud 1989, 1990). Recent studies have indicated that PKR protein levels are not apparently altered by Ca^{++} depletion and that basal PKR activity in cell lysates is unchanged. Rather, addition of double-stranded RNA to cell lysates results in increased autophosphorylation of PKR and higher enzyme activity (Prostko et al. 1995). In addition, overexpression of a catalytically inactive form of PKR partially protected NIH-3T3 cells from the inhibitory effects of the Ca^{++} ionophore (Srivastava et al. 1995). Previous studies have indicated that PKR is activated by binding of double-stranded RNA to an amino-terminal domain of the enzyme, leading to intersubunit autophosphorylation and stimulation of kinase activity (see Clemens, this volume). Therefore, binding of double-stranded RNA, subunit dimerization, or autophosphorylation might be sensitive to Ca^{++}, possibly involving Ca^{++}-dependent phosphorylation or dephosphorylation of a regulatory site on PKR.

SUMMARY AND SPECULATION

The results reviewed above establish that cellular mechanisms exist that allow the regulation of both initiation and elongation by Ca^{++}. However, although biochemical results indicate that specific initiation and elongation factors are phosphorylated in response to changes in cellular Ca^{++} concentrations, the physiological significance of these Ca^{++}-dependent events remains to be elucidated. For example, cells are generally not exposed to Ca^{++} chelators or ionophores in situ, and thus the effects of drastic intracellular store depletion on protein synthesis may represent a pathological response rather than a physiological response. Indeed, Ca^{++} mobilization by ionophore treatment induces proteins typical of the stress response, reinforcing the view that inhibition of protein synthesis initiation is a general response to stress (Lee 1987). However, conditions exist whereby Ca^{++}-mobilizing hormones cause a reduction in protein synthesis under conditions of normal extracellular Ca^{++}, and convincing evidence indicates that phosphorylation of eIF2α and inhibition of initiation is involved, at least in part, in the reduction in protein synthesis.

Although much is known about the biochemical steps involved in the regulation of eEF2 phosphorylation, it is only possible to speculate as to the physiological significance of this process. Notably, the phosphorylation of eEF2 appears to be highly dynamic. One possible consequence of a transient Ca^{++}-dependent elongation block would be the conservation of cellular energy stores at a time when other Ca^{++}-dependent processes such as secretion or contraction are stimulated by hormones or neurotransmitters. In this respect, it is worth noting that such an economy in energy might involve availability of guanine as well as adenine nucleotides. An intriguing possibility is that eEF2 phosphorylation is the cellular equivalent of treatment with the elongation-blocking drug cycloheximide. Cycloheximide has several effects on gene expression, the most prominent of which is the phenomenon of "superinduction" of the expression of immediate-early gene mRNAs such as c-*fos*, c-*jun*, and c-*myc* which normally have very short half-lives (Greenberg et al. 1986; Cleveland and Yen 1989; Rivera and Greenberg 1990). The mechanisms responsible for regulating mRNA stability are presently not fully established; however, one model suggests that short-lived proteins are involved in the process. Inhibition of protein synthesis would then lead to the rapid disappearance of these proteins and consequently a decrease in the degradation of the normally unstable mRNA. Thus, inhibition of elongation by phosphorylation of eEF2 may affect the stability of this class of mRNAs. In a similar fashion, many transcriptional factors are themselves turned over rapidly. This has led to the speculation that eEF2 phosphorylation may have a "reprogramming" role by transiently regulating the expression of such protein factors and in turn influencing the specific pattern of gene expression (Ryazanov and Spirin 1993; Palfrey and Nairn 1995). However, the effects of cycloheximide and other protein synthesis inhibitors require relatively long exposures that do not mimic the known kinetics of eEF2 phosphorylation observed in response to physiological stimulation.

A number of studies have indicated that eEF2 phosphorylation increases in response to a variety of mitogens, and recent studies have suggested that phosphorylation may be regulated in a cell-cycle-dependent manner. Thus, it is possible that eEF2 phosphorylation early in the mitogenic response or at other phases of the cell cycle serves a synchronization function. It would seem paradoxical that cells would wish to block elongation in response to mitogens when they need to synthesize more protein on emerging from quiescence. However, a consequence of such an elongation block would be to stall polypeptide extension, forcing free ribosomal subunits to engage other, perhaps less efficiently translated, mRNAs. Again, an analogy with the effects of cycloheximide can be drawn. As shown in a number of studies with this elongation inhibitor, the translation of mRNAs with low initiation rates is actually favored by inhibition of elongation because of a reduction in the competition with mRNAs having high initiation rates (Brendler et al. 1981a,b; Godefroy-Colburn and Thach 1981; Walden et al. 1981; Walden and Thach 1986). Thus, the phosphorylation of eEF2 could possibly lead to a diversification of translated messages that may optimize early protein expression. It is also notable from recent studies using rapamycin that an association exists between phosphorylation of ribosomal protein S6, the coordinated biosynthesis of a number of polypyrimidine tract mRNAs that include eEF2, and cell growth and proliferation. In this respect, it is worth considering that phosphorylation of eEF2 is functionally equivalent to decreasing the levels of the factor, albeit in a more transient fashion.

In conclusion, both initiation and elongation phases of protein synthesis can be regulated by distinct Ca^{++}-dependent mechanisms. Elongation block by eEF2 phosphorylation is probably a transient event reflecting $[Ca^{++}]_{cyt}$, whereas the effect of Ca^{++} store depletion may occur over a much longer time-scale and seems independent of prevailing $[Ca^{++}]_{cyt}$. There is no a priori reason for believing that these two sets of controls are mutually exclusive, although one or the other may predominate under certain physiological conditions, in a temporally sequential manner, or in different cellular compartments. Clearly, analysis of the phosphorylation of both eIF2α and eEF2 will be necessary in studies designed to elucidate the contribution made by inhibition of initiation or elongation to the overall Ca^{++}-dependent regulation of protein synthesis.

ACKNOWLEDGMENTS

The authors thank Drs. M.A. and C.O. Brostrom for providing unpublished results. Work from the authors' laboratories was supported by U.S. Public Health Service grants GM-42715 (H.C.P.) and GM-50402 (A.C.N), and the Human Frontiers Program (H.C.P.).

REFERENCES

Bagaglio, D.M., E.H.C. Cheng, F.S. Gorelick, K. Mitsui, A.C. Nairn, and W.N. Hait. 1993. Phosphorylation of elongation factor 2 in normal and malignant rat glial cells. *Cancer Res.* **53:** 2260–2264.

Berchtold, H., L. Reshetnikova, C.O.A. Reiser, N.K. Schirmer, M. Sprinzl, and R. Hilgenfeld. 1993. Crystal structure of active elongation factor Tu reveals major domain rearrangements. *Nature* **365:** 126–132.

Brady, M.J., A.C. Nairn, J.A. Wagner, and H.C. Palfrey. 1990. Nerve growth factor-induced down-regulation of calmodulin-dependent protein kinase III in PC12 cells involves cAMP-dependent protein kinase. *J. Neurochem.* **54:** 1034–1039.

Brendler, T., T. Godefroy-Colburn, R.D. Carlill, and R.E. Thach. 1981a. The role of mRNA competition in regulating translation. II. Development of a quantitative in vitro assay. *J. Biol. Chem.* **256:** 11747–11754.

Brendler, T., T. Godefroy-Colburn, S. Yu, and R.E. Thach. 1981b. The role of mRNA competition in regulating translation. III. Comparison of in vitro and in vivo results. *J. Biol. Chem.* **256:** 11755–11761.

Brostrom, C.O. and M.A. Brostrom. 1990. Calcium-dependent regulation of protein synthesis in intact mammalian cells. *Annu. Rev. Physiol.* **52:** 577–590.

Brostrom, C.O., S.B. Bocckino, and M.A. Brostrom. 1983. Identification of Ca^{2+} requirement for protein synthesis in eukaryotic cells. *J. Biol. Chem.* **258:** 14390–14399.

Brostrom, C.O., S.B. Bocckino, M.A. Brostrom, and E.M. Galuszka. 1986. Regulation of protein synthesis in isolated hepatocytes by Ca^{2+}-mobiiizing hormones. *Mol. Pharmacol.* **29:** 104–111.

Brostrom, C.O., K.V. Chin, W.L. Wong, C. Cade, and M.A. Brostrom. 1989. Inhibition of translational initiation in eukaryotic cells by calcium ionophore. *J. Biol. Chem.* **264:** 1644–1649.

Brostrom, M.A., C. Cade, C.R. Prostko, D. Gmitter-Yellen, and C.O. Brostrom. 1990. Accommodation of protein synthesis to chronic deprivation of intracellular sequestered calcium. A putative role for GRP78. *J. Biol. Chem.* **265:** 20539–20546.

Carlberg, U., A. Nilsson, and O. Nygård. 1990. Functional properties of phosphorylated elongation factor 2. *Eur. J. Biochem.* **191:** 639–645.

Carlberg, U., A. Nilsson, S. Skog, K. Palmquist, and O. Nygård. 1991. Increased activity of the eEF-2 specific, Ca^{2+} and calmodulin dependent protein kinase III during the S-phase in Ehrlich ascites cells. *Biochem. Biophys. Res. Commun.* **180:** 1372–1376.

Celis, J.E., P. Madsen, and A.G. Ryazanov. 1990. Increased phosphorylation of elongation factor 2 during mitosis in transformed human amnion cells correlates with a decreased rate of protein synthesis. *Proc. Natl. Acad. Sci.* **87:** 4231–4235.

Chin, K.V., C. Cade, M.A. Brostrom, and C.O. Brostrom. 1988. Regulation of protein synthesis in intact rat liver by Ca^{2+} mobilizing agents. *Int. J. Biochem.* **20:** 1313–1319.

Chin, K.V., C. Cade, C.O. Brostrom, E.M. Galuska, and M.A. Brostrom. 1987. Calcium-dependent regulation of protein synthesis at translational initiation in eukaryotic cells. *J. Biol. Chem.* **262:** 16509–16514.

Cleveland, D.W. and T.J. Yen. 1989. Multiple determinants of eukaryotic mRNA stability. *New Biol.* **1:** 121–126.

Donovan, M.G. and J.W. Bodley. 1991. *Saccharomyces cerevisiae* elongation factor 2 is phosphorylated by an endogenous kinase. *FEBS Lett.* **291:** 303–306.

Dumont-Miscopein, A., J.-P. Lavergne, D. Guillot, B. Sontag, and J.-P. Reboud. 1994. Interaction of phosphorylated elongation factor EF-2 with nucleotides and ribosomes. *FEBS Lett.* **356:** 283–286.

End, D., M. Hanson, S. Hashimoto, and G. Guroff. 1982. Inhibition of the phosphorylation of a 100,000 dalton soluble protein in whole cells and cell-free extracts of PC12 cells following treatment with NGF. *J. Biol. Chem.* **257:** 9223–9225.

Eremenko, T. and P. Volpe. 1975. Polysome translational state during the cell cycle. *Eur. J. Biochem.* **52:** 203–210.

Fan, H. and S. Penman. 1970. Regulation of protein synthesis in mammalian cells. *J. Mol. Biol.* **50:** 655–670.

Foley, B.T., J.M. Moehring, and T.J. Moehring. 1992. A mutation in codon 717 of the CHO-K1 elongation factor 2 gene prevents the first step in the biosynthesis of diphthamide. *Somatic Cell Mol. Genet.* **18:** 227–231.

Gill, D.M. and L.L. Dinius. 1973. The EF-2 content of mammalian cells. *J. Biol. Chem.* **248:** 654–658.

Godefroy-Colburn, T. and R.E. Thach. 1981. The role of mRNA competition in regulating translation. IV. Kinetic model. *J. Biol. Chem.* **256:** 11762–11773.

Greenberg, M.E., A.L. Hermanowski, and E.A. Ziff. 1986. Effects of protein synthesis inhibitors on growth factor activation of c-*fos*, c-*myc* and actin gene transcription. *Mol. Cell. Biol.* **6:** 1050–1057.

Hanson, P.I. and H. Schulman. 1992. Neuronal Ca^{2+}/calmodulin-dependent protein kinases. *Annu. Rev. Biochem.* **61:** 559–601.

Haycock, J.W., M.D. Browning, and P. Greengard. 1987. Cholinergic regulation of protein phosphorylation in bovine adrenal chromaffin cells. *Proc. Natl. Acad. Sci.* **85:** 1677–1681.

Hepler, P.K. 1989. Calcium transients during mitosis: Observations in flux. *J. Cell Biol.* **109:** 2567–2573.

Hershey, J.W.B. 1991. Translational control in mammalian cells. *Annu. Rev. Biochem.* **60:** 717–755.

Hincke, M.T. and A.C. Nairn. 1992. Phosphorylation of elongation factor 2 during Ca^{2+}-mediated secretion from rat parotid acini. *Biochem. J.* **282:** 877–882.

Hinnebusch, A.G. 1990. Involvement of an initiation factor and protein phosphorylation in translational control of *GCN4* mRNA. *Trends Biochem. Sci.* **15:** 148–152.

Ishida, A. and H. Fujisawa. 1995. Stabilization of calmodulin-dependent protein kinase II through the autoinhibitory domain. *J. Biol. Chem.* **270:** 2163–2170.

Jefferies, H.B.J., C. Reinhard, S.C. Kozma, and G. Thomas. 1994. Rapamycin selectively represses translation of the "polypyrimidine tract" mRNA family. *Proc. Natl. Acad. Sci.* **91:** 4441–4445.

Kanki, J.P. and J.W. Newport. 1991. The cell cycle dependence of protein synthesis during *Xenopus laevis* development. *Dev. Biol.* **146:** 198–213.

Kao, J.P.Y., J.M. Alderton, R.Y. Tsien, and R.A. Steinhardt. 1990. Active involvement of Ca^{2+} in mitogenic progression of Swiss 3T3 cells. *J. Cell Biol.* **111:** 183–196.

Kimball, S.R. and L.S. Jefferson. 1990. Mechanism of the inhibition of protein synthesis by vasopressin in rat liver. *J. Biol. Chem.* **265:** 16794–16798.

———. 1991. Inhibition of microsomal Ca^{2+} sequestration causes an impairment of initiation of protein synthesis in perfused rat liver. *Biochem. Biophys Res. Commun.* **177:** 1082–1086.

———. 1992. Regulation of protein synthesis by modulation on intracellular Ca^{2+} in rat liver. *Am. J. Physiol.* **263:** E958–E964.

Kohno, K. and T. Uchida. 1987. Highly frequent single amino acid substitution in mammalian elongation factor 2 (EF-2) results in expression of resistance to EF-2-ADP-ribosylating toxins. *J. Biol. Chem.* **262:** 12298–12305.

Kramer, G., W. Kudlicki, and B. Hardesty. 1993. Regulation of eIF-2α kinases by phosphorylation. In *Translational regulation of gene expression* (ed. J. Ilan), vol. 2, pp. 373–390. Plenum Press, New York.

Kumar, R.V., R. Panniers, A. Wolfman, and E.C. Henshaw. 1991. Inhibition of protein synthesis by antagonists of calmodulin in Ehrlich ascites tumor cells. *Eur. J. Biochem.* **195:** 313–319.

Kumar, R.V., A. Wolfman, R. Panniers, and E.C. Henshaw. 1989. Mechanism of inhibition of polypeptide chain initiation in calcium-depleted Ehrlich ascites tumor cells. *J. Cell Biol.* **108:** 2107–2115.

Lai, Y., A.C. Nairn, and P. Greengard. 1986. Autophosphorylation reversibly regulates the Ca^{2+}/calmodulin-dependence of Ca^{2+}/calmodulin-dependent protein kinase II. *Proc. Natl. Acad. Sci.* **83:** 4253–4257.

Lee, A.S. 1987. Coordinated regulation of a set of genes by glucose and calcium-ionophore in mammalian cells. *Trends Biochem. Sci.* **12:** 20–24.

Levenson, R.M., A.C. Nairn, and P.J. Blackshear. 1989. Insulin rapidly induces the biosynthesis of elongation factor 2. *J. Biol. Chem.* **264:** 11904–11911.

Lu, K.P. and A.R. Means. 1993. Regulation of the cell cycle by calcium and calmodulin.

Endocr. Rev. **14:** 40–58.

Lu, K.P., S.A. Osmani, A.H. Osmani, and A.R. Means. 1993. Essential roles for calcium and calmodulin in G2/M progression in *Aspergillus nidulans. J. Cell Biol.* **121:** 621–630.

Mackie, K.P., A.C. Nairn, G. Hampel, G. Lam, and E.A. Jaffe. 1989. Thrombin and histamine stimulate the phosphorylation of elongation factor 2 in endothelial cells. *J. Biol. Chem.* **264:** 1748–1753.

Menaya, J., R. Parrilla, and M.S. Ayuso. 1988. Effect of vasopressin on the regulation of protein synthesis initiation in liver cells. *Biochem. J.* **254:** 773–779.

Mitsui, K., M. Brady, H.C. Palfrey, and A.C. Nairn. 1993. Purification and characterization of calmodulin-dependent protein kinase III from rabbit reticulocytes and rat pancreas. *J. Biol. Chem.* **268:** 13422–13433.

Montine, K.S. and E.C. Henshaw. 1989. Serum growth factors cause rapid stimulation of protein synthesis and dephosphorylation of eIF-2α in serum-deprived Ehrlich cells. *Biochim. Biophys. Acta* **1014:** 282–288.

Mulner-Lorillon, O., O. Minella, P. Cormier, J.-P. Capony, J.-C. Cavadore, J. Morales, R. Poulhe, and R. Bellé. 1994. Elongation factor EF-1δ, a new target for maturation-promoting factor in *Xenopus* oocytes. *J. Biol. Chem.* **269:** 20201–20207.

Nairn, A.C. and H.C. Palfrey. 1987. Identification of the major M_r 100,000 substrate for calmodulin-dependent protein kinase III in mammalian cells as elongation factor-2. *J. Biol. Chem.* **262:** 17299–17303.

Nairn, A.C. and M.R. Picciotto. 1994. Calcium/calmodulin-dependent protein kinases. *Semin. Cancer Biol.* **5:** 295–303.

Nairn, A.C., B. Bhagat, and H.C. Palfrey. 1985. Identification of calmodulin-dependent protein kinase III and its major 100,000-molecular-weight substrate in mammalian tissues. *Proc. Natl. Acad. Sci.* **82:** 7939–7943.

Nairn, A.C., R.A. Nichols, M.J. Brady, and H.C. Palfrey. 1987. Nerve growth factor treatment or cyclic AMP elevation reduces calcium-calmodulin-dependent protein kinase III activity in PC12 cells. *J. Biol. Chem.* **262:** 14265–14272.

Nilsson, A., U. Carlberg, and O. Nygård. 1991. Kinetic characterisation of the enzymatic activity of the eEF-2-specific Ca^{2+}- and calmodulin-dependent protein kinase III purified from rabbit reticulocytes. *Eur. J. Biochem.* **195:** 377–383.

Nilsson, L. and O. Nygård. 1991. Altered sensitivity of eukaryotic elongation factor 2 for trypsin after phosphorylation and ribosomal binding. *J. Biol. Chem.* **266:** 10578–10582.

Nygård, O. and L. Nilsson. 1984. Quantification of the different ribosomal phases during the translational elongation cycle in rabbit reticulocyte lysates. *Eur. J. Biochem.* **145:** 345–350.

Nygård, O., A. Nilsson, U. Carlberg, L. Nilsson, and R. Amons. 1991. Phosphorylation regulates the activity of the eEF-2-specific Ca^{2+}- and calmodulin-dependent protein kinase III. *J. Biol. Chem.* **266:** 16425–16430.

Ovchinnikov, L.P., L.P. Motuz, P.G. Natapov, L.J. Averbuch, R.E.H. Wettenhall, R. Szyszka, G. Kramer, and B. Hardesty. 1990. Three phosphorylation sites in elongation factor 2. *FEBS Lett.* **275:** 209–212.

Pain, V.M. and M. Clemens. 1991. Adjustment of translation to special physiological conditions. In *Translational control in eukaryotes* (ed. H. Trachsel), pp. 293–324. CRC Press, Boca Raton, Florida.

Palfrey, H.C. 1983. Presence in many mammalian tissues of an identical major cytosolic substrate (M_r 100 000) for calmodulin-dependent protein kinase. *FEBS. Lett.* **157:**

183–190.

Palfrey, H.C. and A.C. Nairn. 1995. Calcium-dependent regulation of protein synthesis. *Adv. Second Messenger Phosphoprotein Res.* **30:** 191–223.

Palfrey, H.C., A.C. Nairn, L.L. Muldoon, and M.L. Villereal. 1987. Rapid activation of calmodulin-dependent protein kinase III in mitogen-stimulated human fibroblasts: Correlation with intracellular Ca^{2+} transients. *J. Biol. Chem.* **262:** 9785–9792.

Palmquist, K., B. Riis, A. Nilsson, and O. Nygård. 1994. Interaction of the calcium and calmodulin regulated eEF- 2 kinase with heat shock protein 90. *FEBS Lett.* **349:** 239–242.

Perentesis, J.P., L.D. Phan, W.B. Gleason, D.C. LaPorte, D.M. Livingston, and J.W. Bodley. 1992. *Saccharomyces cerevisiae* elongation factor 2. Genetic cloning, characterization of expression, and G-domain modeling. *J. Biol. Chem.* **267:** 1190–1197.

Perkins, P.S. and S.J. Pandol. 1992. Cholecystokinin-induced changes in polysome structure regulate protein synthesis in pancreas. *Biochim. Biophys. Acta Mol. Cell Res.* **1136:** 265–271.

Poenie, M., J. Alderton, R. Steinhardt, and R. Tsien. 1986. Calcium rises abruptly and briefly throughout the cell at the onset of anaphase. *Science* **233:** 886–889.

Poenie, M., J. Alderton, R.Y. Tsien, and R.A. Steinhardt. 1985. Changes of free calcium levels with stages of the cell division cycle. *Nature* **315:** 147–149.

Preston, S.F. and R.D. Berlin. 1992. An intracellular calcium store regulates protein synthesis in HeLa cells, but it is not the hormone-sensitive store. *Cell Calcium* **13:** 303–312.

Price, N.T., N.T. Redpath, K.V. Severinov, D.G. Campbell, J.M. Russell, and C.G. Proud. 1991. Identification of the phosphorylation sites in elongation factor-2 from rabbit reticulocytes. *FEBS Lett.* **282:** 253–258.

Prostko, C.R., J.N. Dholakia, M.A. Brostrom, and C.O. Brostrom. 1995. Activation of the double-stranded RNA-regulated protein kinase by depletion of endoplasmic reticular calcium stores. *J. Biol. Chem.* **270:** 6211–6215.

Prostko, C.R., M.A. Brostrom, E.M. Malara, and C.O. Brostrom. 1992. Phosphorylation of eukaryotic initiation factor (eIF) 2α and inhibition of eIF-2B in GH_3 pituitary cells by perturbants of early protein processing that induce GRP78. *J. Biol. Chem.* **267:** 16751–16754.

Proud, C.G. 1992. Protein phosphorylation in translational control. *Curr. Top. Cell. Regul.* **32:** 243–369.

Ratan, R.R., M.L. Shelanski, and F.R. Maxfield. 1986. Transition from metaphase to anaphase is accompanied by local changes in cytoplasmic free calcium in Pt K2 kidney epithelial cells. *Proc. Natl. Acad. Sci.* **83:** 5136–5140.

Rattan, S.I. 1991. Protein synthesis and the components of the protein synthetic machinery during cellular aging. *Mut. Res.* **256:** 115–125.

Redpath, N.T. and C.G. Proud. 1989. The tumour promoter okadaic acid inhibits reticulocyte-lysate protein synthesis by increasing the net phosphorylation of elongation factor 2. *Biochem. J.* **262:** 69–75.

———. 1990. Activity of protein phosphatases against initiation factor- 2 and elongation factor-2. *Biochem. J.* **272:** 175–180.

———. 1991. Differing effects of the protein phosphatase inhibitors okadaic acid and microcystin on translation in reticulocyte lysates. *Biochim. Biophys. Acta Mol. Cell Res.* **1093:** 36–41.

———. 1993a. Cyclic AMP-dependent protein kinase phosphorylates rabbit reticulocyte elongation factor-2 kinase and induces calcium-independent activity. *Biochem. J.* **293:** 31–34.

———. 1993b. Purification and phosphorylation of elongation factor-2 kinase from rabbit reticulocytes. *Eur. J. Biochem.* **212:** 511–520.

Redpath, N.T., N.T. Price, K.V. Severinov, and C.G. Proud. 1993. Regulation of elongation factor-2 by multisite phosphorylation. *Eur. J. Biochem.* **213:** 689–699.

Rivera, V.M. and M.E. Greenberg. 1990. Growth factor-induced gene expresson: The ups and downs of c-*fos* regulation. *New Biol.* **2:** 751–758.

Rowlands, A.G., K.S. Montine, E.C. Henshaw, and R. Panniers. 1988. Physiological stresses inhibit guanine nucleotide exchange factor in Ehrlich cells. *Eur. J. Biochem.* **175:** 93–99.

Ryazanov, A.G. 1987. Ca^{2+}/calmodulin-dependent phosphorylation of elongation factor 2. *FEBS. Lett.* **214:** 331–334.

Ryazanov, A.G. and A.S. Spirin 1993 Phosphorylation of EF-2: A mechanism to shut off protein syntheis for reprogramming gene expression. In *Translational regulation of gene expression* (ed. J. Ilan), pp. 433-455. Plenum Press, New York.

Ryazanov, A.G., E.A. Shestakova, and P.G. Natapov. 1988. Phosphorylation of elongation factor 2 by EF-2 kinase affects rate of translation. *Nature* **334:** 170–173.

Scheetz, A.J. and M. Constantine-Paton. 1994. NMDA receptor activation regulated phosphoproteins in the optic tectum. *Soc. Neurosci. Abstr.* **20:** 437.

Severinov, K.V., I.G. Melnikova, and A.G. Ryazanov. 1990. Down-regulation of the translational EF-2 kinase in *Xenopus laevis* oocytes at the final stages of oogenesis. *New Biol.* **2:** 887–893.

Shenolikar, S. and A.C. Nairn. 1991. Protein phosphatases: Recent progress. *Adv. Second Messengers Phosphoprotein Res.* **23:** 1–121.

Sprinzl, M. 1994. Elongation factor Tu: A regulatory GTPase with an integrated effector. *Trends Biochem. Sci.* **19:** 245–250.

Srivastava, S.P., M.V. Davies, and R.J. Kaufman. 1995. Calcium depletion from the endoplasmic reticulum activates the double-stranded protein kinase (PKR) to inhibit protein synthesis. *J. Biol. Chem.* **270:** 16619–16624.

Takuma, T., B.L. Kuyatt, and B.M. Baum. 1984. α1-adrenergic inhibition of protein synthesis in rat submandibular gland cells. *Am. J. Physiol.* **247:** G284–G289.

Terada, N., H.R. Patel, K. Takase, K. Kohno, A.C. Nairn, and E.W. Gelfand. 1994. Rapamycin selectively inhibits translation of mRNAs encoding elongation factors and ribosomal proteins. *Proc. Natl. Acad. Sci.* **91:** 11477–11481.

Thomas, A.P., J. Alexander, and J.R. Williamson. 1984. Relationship between inositol polyphosphate production and the increase of cytosolic free Ca^{2+} induced by vasopressin in isolated hepatocytes. *J. Biol. Chem.* **259:** 5574–5584.

Thomas, D. and M.R. Hanley. 1994. Pharmacological tools for perturbing intracellular calcium storage. *Methods Cell Biol.* **40:** 65–89.

Ueda, K. and O. Hayaishi. 1985. ADP-ribosylation. *Annu. Rev. Biochem.* **54:** 73–100.

Vary, T.C., A. Nairn, and C.J. Lynch. 1994. Role of elongation factor 2 in regulating peptide-chain elongation in the heart. *Am. J. Physiol.* **266:** E628–E634.

Venema, R.C., H.I. Peters, and J.A. Traugh. 1991a. Phosphorylation of valyl-tRNA synthetase and elongation factor 1 in response to phorbol esters is associated with stimulation of both activities. *J. Biol. Chem.* **266:** 11993–11998.

———. 1991b. Phosphorylation of elongation factor 1 (EF-1) and valyl-tRNA synthetase by protein kinase C and stimulation of EF-1 activity. *J. Biol. Chem.* **266:**

12574–12580.

Villereal, M.L. and K.L. Byron. 1992. Calcium signals in growth factor signal transduction. *Rev. Physiol. Biochem. Pharmacol.* **119:** 68–121.

Walden, W.E. and R.E. Thach. 1986. Translational control of gene expression in a normal fibroblast. Characterization of a subclass of mRNAs with unusual kinetic properties. *Biochemistry* **25:** 2033–2041.

Walden, W.E., T. Godefroy-Colburn, and R.E. Thach. 1981. The role of mRNA competition in regulating translation. I. Demonstration of competition in vivo. *J. Biol. Chem.* **256:** 11739–11746.

Whitaker, M. and R. Patel. 1990. Calcium and cell cycle control. *Development* **108:** 525–542.

Wong, W.L., M.A. Brostrom, and C.O. Brostrom. 1991. Effects of Ca^{2+} and ionophore A23187 on protein synthesis in intact rabbit reticulocytes. *Int. J. Biochem.* **23:** 605–608.

Wong, W.L., M.A. Brostrom, G. Kuznetsov, D. Gmitter-Yellen, and C.O. Brostrom. 1993. Inhibition of protein synthesis and early protein processing by thapsigargin in cultured cells. *Biochem. J.* **289:** 71–79.

11

Masked and Translatable Messenger Ribonucleoproteins in Higher Eukaryotes

Alexander S. Spirin
Institute of Protein Research
Russian Academy of Sciences
142292 Pushchino, Moscow Region
Russia

In eukaryotes, production of messenger RNA is not automatically followed by its expression. The cytoplasmic control of mRNA expression seems to be at least as important as the transcriptional and nuclear post-transcriptional levels of regulation. As with DNA in the nucleus, individual mRNA species or whole classes of cytoplasmic mRNA can be in an inactive ("repressed" or "masked") form or can be active in translation (in polyribosomes). The status of an mRNA presumably is determined by interactions with special cytoplasmic proteins that regulate its translation. Such mRNA-binding proteins include (1) the mRNA-binding initiation factors, (2) specific translational repressors, (3) translational activators, (4) proteins that affect mRNA stability, (5) mRNA masking proteins, and (6) proteins responsible for the structural organization of mRNPs. This chapter focuses mainly on the latter two categories: the masking and structural proteins. Extensive reviews of other mRNA-binding proteins are found elsewhere in this volume: for the translation initiation factors, see Merrick and Hershey and Sonenberg; for factors that promote initiation by binding to IRES elements, see both Jackson and Ehrenfeld; for specific repressors that bind to the 5′UTR, see Rouault et al. and Meyuhas et al.; for effectors of mRNA stability, see Theodorakis and Cleveland; and for specific regulatory proteins that bind to the 3′UTR, see Wickens et al., Richter, and Jacobson.

The mRNAs inactive in translation are present in the form of free cytoplasmic mRNP particles. In considering these particles, it is useful to postulate two distinct types of nonactive mRNAs. In the one case, initiation of protein synthesis is inhibited by the binding of a specific repressor, usually at the 5′UTR, as exemplified by ferritin mRNA. Such mRNAs may be susceptible to changes in metabolism of their poly(A) tails and to nucleolytic degradation. In the second case, the mRNA is masked in the sense that it becomes inaccessible both to the translational apparatus and to processing and degradative enzymes such as polyadenylate polymerase and nucleases. Such mRNAs are fully inactive and stable (stored mRNA). The masking is caused by interactions of one or more factors with different regions of the mRNA, including the 3′-untranslated region (3′UTR), and seems to involve a structural reorganization of the mRNP.

In general terms, we believe that different, well-defined parts of mRNA are responsible for the binding of the factors with the different functions discussed above (Fig. 1). The 5′-untranslated region (5′UTR) binds initiation factors, other activators of initiation, and translational repressors. Specific sequences within the 3′UTR are recognized by mRNA stabilization and destabilization factors, mRNA polyadenylation and deadenylation factors, and mRNA masking factors. The extreme 3′-terminal portions, such as tRNA-like structures of plant virus RNAs or poly(A) tails, can serve as translational enhancers with the participation of their bound proteins. All sequences of mRNA molecules, including coding sequences, are capable of interacting strongly with core proteins, such as p50, to form mRNPs, as well as displaying weaker affinities for elongation factors, aminoacyl-tRNA synthetases, protein kinases, and some energy-producing enzymes in higher eukaryotes.

MESSENGER RNP PARTICLES

Messenger ribonucleoproteins (mRNP particles or "informosomes") were first discovered in cytoplasmic extracts of early fish embryos (Spirin et al. 1964) and sea urchin embryos (Spirin and Nemer 1965). The discovery led to the hypothesis that the mRNPs represent a masked form of

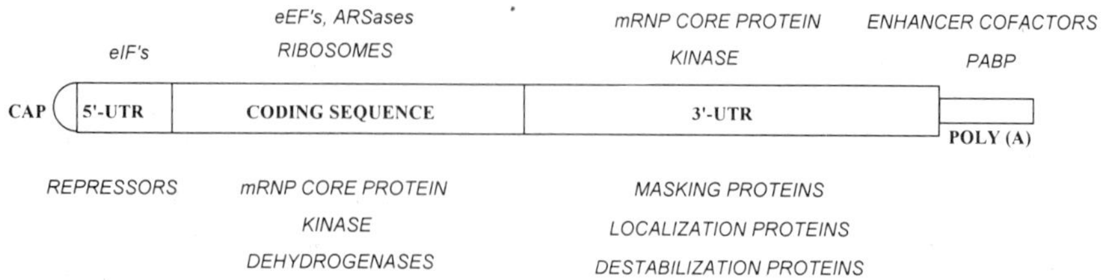

Figure 1 Schematic representation of the distribution of mRNA-binding proteins among different regions of mRNA.

mRNA (Spirin 1966). It was later found that the mRNA released from translating polyribosomes is also complexed with proteins (Cartouzou et al. 1968; Henshaw 1968; Perry and Kelley 1968). Moreover, in a number of cases, a nonselective distribution was reported of various mRNA species between polyribosome-bound mRNPs and free mRNPs (Bag and Sells 1991). Thus, it became evident that the complexing of mRNA with proteins does not automatically result in its masking and storage.

Several classes of mRNA-protein complexes in the cytoplasm may be distinguished on an operational or theoretical basis: (1) polyribosomal mRNPs, i.e., the mRNA-protein complexes within translating polyribosomes; (2) free mRNP particles that are translatable in principle, but either are in transit to polyribosomes, represent a pool of excess mRNA for translation, or are not capable of efficiently competing with other, stronger mRNAs for initiation factors ("weak" mRNAs); (3) nontranslatable mRNP particles where initiation of translation is blocked by specific 5′UTR-bound repressors (the particle with ferritin mRNA and its repressor in the absence of iron is a typical example); and (4) masked mRNP particles that are inactive in translation, stable, and stored in the cytoplasm until receiving a signal for unmasking (typical of free mRNPs from germ cells and cells in other dormant states).

All of the cytoplasmic mRNPs mentioned above have characteristic features in common (for review, see Preobrazhensky and Spirin 1978). They have a significantly higher proportion of protein as compared with ribosomes and, correspondingly, a relatively low buoyant density in CsCl: The ratio of protein to RNA is about 3:1 to 4:1 in free mRNPs (density = 1.4 g/cm) and somewhat lower, down to 2:1, in polyribosomal mRNPs (density is within the range of 1.4–1.5 g/cm). At least two major families of proteins are present in stoichiometries of more than one protein per RNA. One is represented by a protein or a few closely related proteins with an apparent molecular mass (estimated from SDS-electrophoretic mobility) of approximately 50–60 kD in different species; this protein(s) (called p50 in somatic mammalian cells) possesses a high affinity for various heterologous mRNA sequences and much lower affinity for poly(A). The other is represented by a protein with a molecular mass of approximately 70–80 kD (p70 or PABP, poly(A)-binding protein) and a predominant affinity to poly(A) sequences. A great variety of minor protein species are also bound within the mRNP particles. The mRNP particles can be sequentially depleted of their proteins by stepwise increases in salt concentration; the particles are also rather resistant to removal of magnesium, in contrast to ribosomal particles.

mRNP CORE PROTEIN

Among the major proteins bound within free and polysome-associated mRNPs, the so-called p50 component is found to be predominant (Preobrazhensky and Spirin 1978; Minich et al. 1993). It is also the most tightly bound protein component of the mRNPs as demonstrated by experiments using high salt to dissociate proteins from RNA. Thus, p50 can be designated as an mRNP core protein.

Study of the properties of the p50 isolated from free mRNPs of rabbit reticulocytes (Minich et al. 1993) demonstrated that it possesses strong, but nonspecific, RNA-binding activity, can be phosphorylated both in vitro and in vivo, and is characterized by a high glycine and proline content and high isoelectric point (pI ~9.5). In cell-free translation systems, p50 can exert a dual effect: At low concentrations (up to several molecules of the protein per mRNA), it can stimulate translation, whereas at higher concentrations, it is inhibitory (Minich and Ovchinnikov 1992). The cloning and sequence analysis of rabbit reticulocyte p50 have revealed that the protein is highly homologous to the DNA-binding proteins known as Y-box-binding transcription factors (Evdokimova et al. 1995). It is interesting that p50 binding to the DNA Y-box is sequence-specific, but the binding to RNA is not, although the affinity is high. The actual molecular mass of the protein calculated from the amino acid sequence is approximately 35 kD.

One of the most remarkable features of the rabbit reticulocyte p50 is its influence on RNA structure: Instead of protection of RNA against ribonucleases and stabilization of secondary structure, the protein melts up to 60% of the total secondary structure of mRNA upon binding at saturating amounts and increases the sensitivity of the RNA toward nucleolytic degradation (Evdokimova et al. 1995). The same protein is detected as a major mRNP constituent both in nontranslated free mRNP particles and in translating polyribosomes, and thus it is unlikely that it has a defining role in masking, repression, or activation of translation.

A characteristic feature of the protein, at least when in its free state in solution, is that it forms large 18S homomultimeric complexes of about 800 kD (Evdokimova et al. 1995). It is not known if the complexes dissociate into subunits upon interaction with mRNA or if such big protein "globules" can directly accommodate mRNA on their surfaces.

Earlier, two major closely related proteins with apparent molecular masses (by SDS-electrophoretic mobility) of approximately 50–60 kD were shown to be abundant in masked mRNPs of the *Xenopus* oocyte (with a molecular ratio of each polypeptide to mRNA up to 10:1). These proteins are capable of binding nonspecifically to any heterogeneous mRNA sequence, are characterized by a high content of glycine and proline and a high isoelectric point (pI >9), are present in a phosphorylated form, and are oocyte-specific (Darnbrough and Ford 1981; Dearsly et al. 1985; Deschamps et al. 1991, 1992; Murray et al. 1991, 1992; Marello et al. 1992; Wolffe et al. 1992). The actual molecular mass of the proteins determined by sedimentation equilibrium or deduced from their amino acid sequences is 36–38 kD. They also form large multimeric complexes in solution. One of these proteins was identified as the oocyte-specific DNA-binding transcription factor FRGY2 which stimulates the synthesis of mRNAs from Y-box-containing promoters. The other protein was found to be very homologous (85% sequence identity) to FRGY2. A similar Y-box-recognizing protein (MSY1) was isolated from masked (stored) mRNP particles of mouse spermatocytes (Kwon et al. 1993; Tafuri et al. 1993).

When mRNP particles were reconstituted in vitro from mRNA and the protein fraction containing the oocyte-specific Y-box proteins, translation of the mRNA was found to be inhibited (Richter and Smith 1984). In vivo, when the Y-box protein (FRGY2) was expressed in cultured somatic cells, it induced the accumulation of an mRNA transcribed from a Y-box-containing promoter and inhibited translation of the same mRNA (Ranjan et al. 1993). However, it did not protect the mRNA from degradation. FRGY2 was detected in ribosome-containing fractions of a sucrose gradient, suggesting that its presence did not prevent association of mRNAs with ribosomes. In a more recent study of FRGY2, it was confirmed that the protein associates with all mRNAs of the *Xenopus* oocyte, exhibiting no apparent sequence specificity. Moreover, both translationally active (unmasked) and translationally repressed (masked) mRNAs were found to be associated with the protein, being in the form of mRNP particles (Tafuri and Wolffe 1993).

More recently, Bouvet and Wolffe (1994) reported that translation of mRNA synthesized in vivo in *Xenopus* oocytes is inhibited, presumably by FRGY2, more efficiently than that of mRNA presynthesized in vitro and injected into the cytoplasm or the nucleus. At the same time, both the strongly inhibited mRNA and the weakly inhibited mRNA have been found in association with FRGY2. Hence, the effective inhibition of translation seems to be somehow coupled with transcription. The inhibition did not depend on specific sequences in mRNAs. Several interpretations of this interesting observation are possible (see Bouvet and Wolffe 1994). The authors believe that FRGY2 masks mRNA directly, and the nascent mRNA is either better loaded with the protein or capable of better conserving its "repressive secondary structure" upon the loading, as compared with the presynthesized and injected mRNA. The participation of a minor, tightly bound masking protein accompanying nascent mRNA from the transcription site to the cytoplasm is also plausible. It is not clear, however, that the translational inhibition observed is a real masking phenomenon. The inhibition may be the result of a kinetic triggering: The injected mRNA can be immediately involved in the interaction with initiation factors and translation, whereas the nascent mRNA first associates with FRGY2, these two processes being mutually obstructing.

As mentioned above, FRGY2 and the other *Xenopus* Y-box-binding factors are oocyte-specific and are absent from somatic cells. At the same time, there is another Y-box transcription factor, FRGY1, that is present in all somatic cells (Wolffe et al. 1992). It is not known whether FRGY1 is predominantly in the cytoplasm (like FRGY2 in oocytes) or whether it is a component of both free mRNPs and polysomes of *Xenopus* somatic cells.

The central (80-amino-acid-long) domain of the Y-box-binding proteins displays significant homology with the major cold-shock protein of bacteria (Goldstein et al. 1990; Wistow 1990), a single-stranded DNA- and RNA-binding protein. The cold-shock protein-like domain of the Y-box-binding proteins is highly conserved between different species. There are reasons to believe that the domain directly participates in DNA and RNA binding (Tafuri and Wolffe 1992; Wolffe 1994). Immediately adjacent to this domain is a less conserved hydrophilic domain with alternating modules of basic and acidic amino acids that extends to the carboxyl terminus of the Y-box-binding proteins (Ozer et al. 1990; Tafuri and Wolffe 1990). It seems that the carboxy-terminal domain contributes mainly to protein-protein interactions (multimerization) and probably stabilizes protein–nucleic acid interactions (Tafuri and Wolffe 1992; Wolffe 1994). The involvement of the carboxy-terminal domain in the interaction specifically with RNA, but not DNA, has been also reported (Murray 1994).

The members of the Y-box-binding protein family (e.g., FRGY2 and p50) thus seem to be major mRNP proteins both in dormant germ cells and in actively translating somatic cells, such as reticulocytes. They are detected both in free mRNPs and in polyribosomal mRNPs. It seems likely that the p50-like proteins are the major protein component responsible for physically forming cytoplasmic mRNPs, akin to histones forming nucleosomes (Tafuri and Wolffe 1993; Spirin 1994). We can speculate that the protein may serve some kind of structural organization and sequence-nonspecific packaging of mRNA into mRNP particles. This possibly universal form of mRNAs may be available for intra-

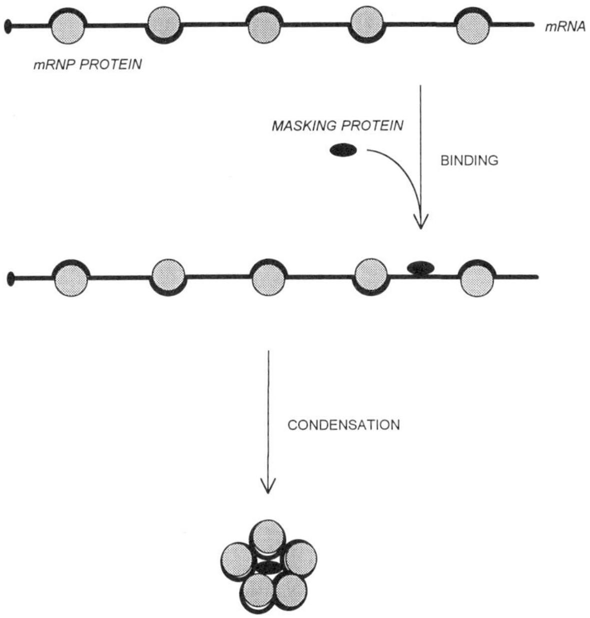

Figure 2 Possible mechanism of mRNA masking. The eukaryotic mRNA is complexed with a large amount of protein, thus forming mRNPs. Binding of a masking protein to a defined section within the 3′UTR may induce a structural rearrangement of the mRNP particle into a condensed form that may be inaccessible to initiation factors, ribosomes, ribonucleases, and poly(A) polymerase (masked mRNP). Possibly the sequence of events can vary; for example, first a masking protein is bound to a nascent mRNA and then the loading of a large number of core proteins results in a compact packing of the mRNP.

cellular transport, translation, masking, degradation, etc., depending on the involvement of other protein components. With participation of a specific masking protein, p50 may be mainly responsible for the proposed condensation of mRNPs (see the subsequent section and Fig. 2).

MASKING PROTEINS AND THE HYPOTHESIS OF INDUCED STRUCTURAL REARRANGEMENTS OF mRNP PARTICLES

The most typical examples of masked mRNAs are mRNAs stored in oocytes and spermatocytes. There are also striking examples of long-term storage of mRNAs in somatic cells (for reviews, see Standart 1992; Spirin 1994). The stored mRNA is unreactive with respect to essentially all processes, including translation, degradation, and polyadenylation/

deadenylation. This is proposed as a definition of masking, in contrast to repression where just one function, translation, is blocked. The most remarkable discovery during recent years was the finding that the masking of mRNA involves primarily the 3′UTRs of mRNAs and that the specific interactions of proteins ("masking proteins") with defined regions within the 3′UTRs are responsible for switching off the functional activities of the respective mRNAs.

The pioneering report in this field was made by Standart et al. (1990; see also Standart and Hunt 1990), who found that a specific region in the middle of the 3′UTR of the masked ribonucleotide reductase mRNA of *Spisula solidissima* oocytes is responsible for binding a masking protein(s). An 82-kD oocyte protein was reported to bind specifically to the "masking element" within the 3′UTR, and its presence in the masked mRNPs correlated with translational inactivity of the corresponding mRNAs (Standart 1992). It has been proposed that the unmasking of maternal mRNAs is due to a maturation-dependent kinase that phosphorylates the protein (see subsequent section).

The best-studied case of mRNA masking/unmasking during the final stages of cell differentiation, rather than at germ cell maturation and activation or during early embryonic development, is the fate of the mRNA encoding erythroid 15-lipoxygenase (LOX). LOX mRNA is synthesized at the early stages of erythropoiesis and becomes masked, i.e., stored in the form of untranslatable and stable cytoplasmic mRNPs, for all subsequent stages, until the late stage of peripheral reticulocytes (Thiele et al. 1982). The unmasking of the LOX mRNA and the synthesis of the enzyme take place during maturation of reticulocytes into erythrocytes. The long 3′UTR of the reticulocyte LOX mRNA contains a characteristic sequence in which a pyrimidine-rich 19-nucleotide motif is repeated ten times (Fleming et al. 1989). It is the 3′UTR repeat region that was found to be responsible for the masking of the LOX mRNA through specific binding of a 48-kD protein (Ostareck-Lederer et al. 1994). The 48-kD protein (LOX-BP) is a part of the translationally inactive (masked) LOX-mRNP in bone marrow cells and reticulocytes. The minimal binding site for LOX-BP sufficient for the effect is two to four repeats. It is noteworthy that the LOX-BP seems to act independently of the 5′cap and 5′UTR, so that the involvement of a "cross-talk" between the two ends of mRNA (see below) is unlikely in this case.

There are also other examples of the involvement of the 3′UTR in the formation of masked mRNPs. Creatine kinase B mRNA is regulated by its 3′UTR, and a specific RNase-resistant, gel-retarded complex was demonstrated to be formed from this mRNA or its 3′UTR and some components (presumably proteins) of the cell extract (Ch'ng et al. 1990). Translation of cytokine, lymphokine, and proto-oncogene mRNAs was also reported to be dependent on their 3′UTRs (Beutler et al. 1986; Kruys et al. 1987, 1989, 1990; Han et al. 1990). The sequences responsible for translational inhibition were found to be rich in uridines and adenosines and to contain several AUUUA repeats (Kruys et al. 1988). Similar AU-rich elements (ARE) in the 3′UTRs were reported earlier to be recognition signals for selective rapid degradation of these mRNAs. The translational control imparted by the presence of the AU-rich element in the 3′UTR seems to be regulated by inducing stimuli. For example, the tumor necrosis factor (TNF) mRNA exists within macrophages in a translationally inactive, masked form, seemingly due to the presence of the AU-rich element in the 3′UTR. Endotoxin (LPS) specifically induces its translation. Just as it is important in suppressing translation, the AU-rich element has been found to be critical for the response to endotoxin (Han et al. 1990). All this suggests the existence of cytoplasmic protein factors that can recognize the AU-rich elements in the 3′UTR, thus inducing mRNA masking, and also respond to a signal to release their masking activity. Indeed, several groups reported the identification of a number of cytoplasmic RNA-binding proteins specifically recognizing the AU-rich and U-rich sequences in the 3′UTRs of cytokine, lymphokine, and oncogene mRNAs (Wickens et al.; Theodorakis and Cleveland; both this volume). The proteins identified by different groups of researchers vary in size from 15 kD to 70 kD. Some of these cytoplasmic proteins, e.g., 36-kD and 43-kD proteins, were reported to be identical to components of nuclear ribonucleoproteins, such as proteins A1 and C, respectively (Hamilton et al. 1993). It is not clear, however, which are mRNA-destabilizing factors and which participate in mRNA masking.

Other special sequences in the 3′UTR may be required for unmasking of mRNA. Thus, removal of the 3′-proximal 100 nucleotides of the tissue plasminogen activator (t-PA) mRNA 3′UTR prevented both translational activation and destabilization of the masked mRNA, as well as its polyadenylation (Strickland et al. 1988). The existence of protein factors that recognize specific structures of the 3′UTRs and thus induce mRNP unmasking has been suspected. Such a *trans*-acting unmasking factor or activator relieving the masking effect of the TNF mRNA 3′UTR has been recently detected in the human cell line, L-929, but not in other cell lines (HeLa, NIH-3T3) (Kruys et al. 1992). The factor seems to abolish the effect of the 3′UTR-binding masking factor discussed above. It is not known if the unmasking factor directly or indirectly competes with the masking factor and displaces it from mRNA.

In virtually all cases of mRNA masking, the effect of the 3′UTR and the 3′UTR-binding protein(s) is displayed as a block at the initiation step of translation. The question arises: How can the 3′-located events affect the 5′-located processes? One hypothesis is that the 3′and 5′parts of an mRNA are in protein-mediated contact with each other, providing some kind of "cross-talk" or even a noncovalent "circularization." This possibility has been claimed and is discussed in detail by Jacobson (this volume). An alternative model is based on the fact that throughout its lifetime, mRNA is complexed with a large amount of protein and thus is organized in mRNP structures. Global structural reorganizations of mRNPs (e.g., like condensation/decondensation of chromatin) could result in mRNA masking/unmasking. Interactions of the 3′UTR with regulatory proteins may induce structural reorganization of mRNP particles. Masked mRNA may be considered as a "condensed" form of mRNP (see Fig. 2) where the RNA is not available for functional interactions with other macromolecules, including ribosomes and/or translational initiation factors, poly(A) polymerase, and ribonucleases. This "structural masking theory" (Spirin 1994) does not necessarily exclude the idea of the 3′part being in proximity to the 5′part of masked mRNP. In any case, the major mRNP core protein (p50/Y-box-binding protein) seems to be requisite, but not sufficient, for mRNA masking.

mRNA-BOUND PROTEIN KINASE AND POSSIBLE REGULATION BY PHOSPHORYLATION

Many mRNP proteins including p50 and FRGY2 are phosphorylated. The processes of phosphorylation/dephosphorylation affect their affinity for RNA (Stepanov and Kandror 1984; Kick et al. 1987; Murray et al. 1991). Several groups found that protein kinase activity is detected among the cytoplasmic RNA-binding proteins and is associated with cytoplasmic mRNPs (Bag and Sells 1979, 1980; Rittschof and Traugh 1982; Stepanov et al. 1982; Thoen et al. 1986; Cummings and Sommerville 1988). Specifically, one of the prominent kinases of the eukaryotic cytoplasm, casein kinase II, was identified as a nonspecific RNA-binding protein and an integral part of mRNPs (Rittschof and Traugh 1982; Kandror and Stepanov 1984; Thoen et al. 1984, 1986). The mRNP proteins were found to be targets of this enzyme. When a highly purified preparation of casein kinase II from frog oocytes was injected back into oocytes, it was reincorporated into cytoplasmic mRNPs (Kandror et al. 1989).

Casein kinase II and its mRNA-binding protein targets are thus compartmentalized within the same macromolecular complex. It is therefore tempting to believe that some kind of triggering of the kinase activity within the mRNP particles determines the level and the pattern of phosphorylation of mRNA-bound proteins and hence regulates translation of the mRNA. Phosphorylation of mRNP proteins, particularly p50 or FRGY2, was reported to be important for mRNA masking, and dephosphorylation at oocyte maturation and during early embryogenesis was discussed as a possible mechanism for unmasking mRNAs (Kick et al. 1987; Sommerville 1990, 1992; Murray et al. 1991; Braddock et al. 1994). In contrast, phosphorylation of FRGY2 by casein kinase II in vitro did not alter its nucleic-acid-binding properties (Tafuri and Wolffe 1993). It is possible that the phosphorylation of p50-like proteins may affect their interactions with other proteins, including masking proteins complexed to mRNA (Wolffe 1994).

On the other hand, recent findings (Standart 1992) have demonstrated that the clam oocyte 82-kD protein which binds the masking element in the 3′UTR of ribonucleotide reductase mRNA (see above) is phosphorylated concomitantly with unmasking. Since the 82-kD protein has all the hallmarks of a masking protein, it has been proposed that unmasking of maternal mRNAs is due to a maturation-dependent kinase. Another example is the 18-kD protein implicated in repression of mouse protamine 2 mRNA by binding to its 3′UTR (Kwon and Hecht 1993). The 18-kD protein no longer binds at a later stage of male gamete differentiation, likely due to its dephosphorylation, and protamine 2 mRNA is actively translated.

CONCLUSIONS

We propose that the function of the binding of various proteins to eukaryotic mRNAs is (1) to govern the conformation of the mRNA, including local or progressive melting of mRNA structure, stabilization of hairpins or tertiary folds, induction or selection of an alternative mRNA conformation, and global conformational rearrangements or packaging of mRNP; (2) to attract other proteins and to assemble larger protein complexes on mRNA (due to protein-protein interactions), or to displace other proteins from mRNA (by competition); (3) to compartmentalize translation-serving proteins and enzymatic activities on mRNA (not covered here).

It should be remembered that eukaryotic mRNA does not exist as a free polyribonucleotide in the cell. Pre-mRNAs, masked mRNAs, and translated mRNAs are present only in the form of RNPs, heavily loaded with various proteins, with fully or partially different protein constituents. The functions of most of the proteins are not known, and the list of the mRNA-bound proteins is not complete. Even less is known about the principles of structural organization of mRNPs.

The massive loading of mRNA with proteins suggests that the following points may be very important in considering mRNA interactions with the translation machinery. (1) The binding of proteins may modify, melt, induce, or switch structural elements in mRNA. There is a suspicion, based on the demonstrated melting activity of p50, that a significant part of moderately stable secondary structure generated by RNA self-folding may not exist in mRNPs. (2) Initiation factors, repressors, ribosomes, etc., interact with mRNPs, rather than with mRNAs. Thus, they may have to compete with proteins that already occupy their binding sites. When considering the "strength" of an mRNA, possible counteracting effects of other mRNA-binding proteins should therefore be taken into account. For example, the "strength" of an mRNA during initiation of translation may depend on the occupancy of the 5′UTR with mRNP proteins and not only on the intrinsic affinity of the 5′UTR for eIF4F. The relative proportions of initiation factors and some mRNA-binding proteins may be of decisive importance. In other cases, synergistic effects of mRNA-bound proteins on the binding of initiation factors (and other positive regulators) may take place. (3) Numerous protein-protein interactions within mRNPs are very likely. This must create additional possibilities for three-dimensional folding and packaging of mRNPs. Masking and enhancing effects of the 3′UTRs and the 3′UTR-binding proteins suggest that the three-dimensional arrangement of mRNPs has an important functional role. Protein-mediated circularization of mRNA and condensation of mRNPs are possible. Hence, the "one-dimensional approach" that is routinely used in almost all analyses of mRNA structure and function is not sufficient, especially in the case of eukaryotic mRNAs, and must be supplemented with a "three-dimensional approach." It is highly appropriate to start serious studies of the three-dimensional structure of mRNP particles.

Regulation of translation is only a part of the functions of mRNA-binding proteins. Intracellular transport, localization, and conservation and degradation of mRNAs are governed by mRNA-binding proteins. Polyribosomes and free mRNP particles form a special functional compartment of the eukaryotic cytoplasm where many factors and enzymes can reside due to their nonspecific RNA-binding activity. All this requires further investigation.

REFERENCES

Bag, J. and B.H. Sells. 1979. The presence of protein kinase activity and acceptors of phosphate groups in nonpolysomal cytoplasmic messenger ribonucleoprotein complexes of embryonic chicken muscle. *J. Biol. Chem.* **254:** 3137–3140.

———. 1980. Presence of cyclic-AMP-independent protein kinase activity in RNA-binding proteins of embryonic chicken muscle. *Eur. J. Biochem.* **106:** 411–424.

———. 1991. mRNA and mRNP. In *Translation in eukaryotes* (ed. H. Trachsel), pp. 71–95. CRC Press, Boca Raton, Florida.

Beutler, B., N. Krochin, I.W. Milsark, C. Luedke, and A. Cerami. 1986. Control of cachectin (tumor necrosis factor) synthesis: Mechanism of endotoxin resistance. *Science* **232:** 977–980.

Bouvet, P. and A.P. Wolffe. 1994. A role for transcription and FRGY2 in masking maternal mRNA within *Xenopus* oocytes. *Cell* **77:** 931–941.

Braddock, M., M. Muckenthaler, M.R.H. White, A.M. Thornburn, J. Sommerville, A.J. Kingsman, and S.M Kingsman. 1994. Intron-less RNA injected into the nucleus of *Xenopus* oocytes accesses a regulated translation control pathway. *Nucleic Acids Res.* **22:** 5255–5264.

Cartouzou, G., J.C. Attali, and S. Lissitzky. 1968. Thyroid messenger RNAs. I. Nuclear and polysomal rapidly labelled RNAs. *Eur. J. Biochem.* **4:** 41–54.

Ch'ng, J.L.C., D.L. Shoemaker, P. Schimmel, and E.W. Holmes. 1990. Reversal of creatine kinase translational repression by 3′ untranslated sequences. *Science* **248:** 1003–1006.

Cummings, A. and J. Sommerville. 1988. Protein kinase activity associated with stored messenger ribonucleoprotein particles of *Xenopus* oocytes. *J. Cell Biol.* **107:** 45–56.

Darnbrough, C.H. and P.J. Ford. 1981. Identification in *Xenopus laevis* of a class of oocyte-specific proteins bound to messenger RNA. *Eur. J. Biochem.* **113:** 415–424.

Dearsly, A.L., R.M. Johnson, P. Barrett, and J. Sommerville. 1985. Identification of a 60-kDa phosphoprotein that binds stored messenger RNA of *Xenopus* oocytes. *Eur. J. Biochem.* **150:** 95–103.

Deschamps, S., A. Viel, H. Denis, and M. le Maire. 1991. Purification of two thermostable components of messenger ribonucleoprotein particles (mRNPs) from *Xenopus laevis* oocytes, belonging to a novel class of RNA-binding proteins. *FEBS Lett.* **282:** 110–114.

Deschamps, S., A. Viel, M. Garrigos, H. Denis, and M. le Maire. 1992. mRNP4, a major mRNA-binding protein from *Xenopus* oocytes is identical to transcription factor FRG Y2. *J. Biol. Chem.* **267:** 13799–13802.

Evdokimova, V.M., C.-L. Wei, A.S. Sitikov, P.N. Simonenko, O.A. Lazarev, K.S. Vasilenko, V.A. Ustinov, J.W.B. Hershey, and L.P. Ovchinnikov. 1995. The major protein of messenger ribonucleoprotein particles in somatic cells is a member of the Ybox binding transcription factor family. *J. Biol. Chem.* **270:** 3186–3192.

Fleming, J., B.J. Thiele, J. Chester, J. O'Prey, S. Janetzki, A. Aitken, I.A. Anton, S.M. Rapoport, and P.R. Harrison. 1989. The complete sequence of the rabbit erythroid cell-specific 15-lipoxygenase mRNA: Comparison of the predicted amino acid sequence of the erythrocyte lipoxygenase with other lipoxygenases. *Gene* **79:** 181–188.

Goldstein, J., N.S. Pollitt, and M. Inouye. 1990. Major cold shock protein of *Escherichia coli*. *Proc. Natl. Acad. Sci.* **87:** 283–287.

Hamilton, B.J., E. Nagy, J.S. Malters, B.A. Arrick, and W.F.C. Rigby. 1993. Association of heterogeneous nuclear ribonucleoprotein A1 and C proteins with reiterated AUUUA sequences. *J. Biol. Chem.* **268:** 8881–8887.

Han, J., T. Brown, and B. Beutler. 1990. Endotoxin-responsive sequences control cachectin/tumor necrosis factor biosynthesis at the translational level. *J. Exp. Med.* **171:** 465–475.

Henshaw, E.C. 1968. Messenger RNA in rat liver polyribosomes: Evidence that it exists as ribonucleoprotein particles. *J. Mol. Biol.* **36:** 401–411.

Kandror, K.V. and A.S. Stepanov. 1984. RNA-binding protein kinase from amphibian oocytes is a casein kinase I. *FEBS Lett.* **170:** 33–37.

Kandror, K.V., A.O. Benumov, and A.S. Stepanov. 1989. Casein kinase II from *Rana temporaria* oocytes. Intracellular localization and activity during progesterone-induced maturation. *Eur. J. Biochem.* **180:** 441–448.

Kick, D., P. Barrett, A. Cummings, and J. Sommerville. 1987. Phosphorylation of a 60 kDa polypeptide from *Xenopus* oocytes blocks messenger RNA translation. *Nucleic Acids Res.* **15:** 4099–4109.

Kruys, V., B. Beutler, and G. Huez. 1990. Translational control mediated by UA-rich sequences. *Enzyme* **44:** 193–202.

Kruys, V.I., M.G. Wathelet, and G.A. Huez. 1988. Identification of a translation inhibitory element (TIE) in the 3′ untranslated region of the human interferon-mRNA. *Gene* **72:** 191–200.

Kruys, V., K. Kemmer, A. Shakhov, V. Jongeneel, and B. Beutler. 1992. Constitutive activity of the tumor necrosis factor promoter is canceled by the 3′ untranslated region in nonmacrophage cell lines; a *trans*-dominant factor overcomes this suppressive effect. *Proc. Natl. Acad. Sci.* **89:** 673–677.

Kruys, V., O. Marinx, G. Shaw, J. Deschamps, and G. Huez. 1989. Translational blockade imposed by cytokine-derived UA-rich sequences. *Science* **245:** 852–855.

Kruys, V., M. Wathelet, P. Poupart, R. Contreras, W. Fiers, J. Content, and G. Huez. 1987. The 3′ untranslated region of the human interferon-mRNA has an inhibitory effect on translation. *Proc. Natl. Acad. Sci.* **84:** 6030–6034.

Kwon, Y.K. and N.B. Hecht. 1993. Binding of a phosphoprotein to the 3′ untranslated region of the mouse protamine 2 mRNA temporally represses its translation. *Mol. Cell. Biol.* **13:** 6547–6557.

Kwon, Y.K., M.T. Murray, and N.B. Hecht. 1993. Proteins homologous to the *Xenopus* germ cell-specific RNA-binding proteins p54/p56 are temporally expressed in mouse male germ cells. *Dev. Biol.* **158:** 90–100.

Marello, K., J. LaRovere, and J. Sommerville. 1992. Binding of *Xenopus* oocyte masking proteins to mRNA sequences. *Nucleic Acids Res.* **20:** 5593–5600.

Minich, W.B. and L.P. Ovchinnikov. 1992. Role of cytoplasmic mRNP proteins in translation. *Biochimie* **74:** 477–483.

Minich, W.B., I.P. Maidebura, and L.P. Ovchinnikov. 1993. Purification and characterization of the major 50-kDa repressor protein from cytoplasmic mRNP of rabbit reticulocytes. *Eur. J. Biochem.* **212:** 633–638.

Murray, M.T. 1994. Nucleic acid-binding properties of the *Xenopus* oocyte Y box protein mRNP. *Biochemistry* **33:** 13910–13917.

Murray, M.T., G. Krohne, and W.W. Franke. 1991. Different forms of soluble cytoplasmic mRNA binding proteins and particles in *Xenopus laevis* oocytes and embryos. *J. Cell Biol.* **112:** 1–11.

Murray, M.T., D.L. Schiller, and W.W. Franke. 1992. Sequence analysis of cytoplasmic mRNA-binding proteins of *Xenopus* oocytes identifies a family of RNA-binding proteins. *Proc. Natl. Acad. Sci.* **89:** 11–15.

Ostareck-Lederer, A., D.H. Ostareck, N. Standart, and B.J. Thiele. 1994. Translation of 15-lipoxygenase mRNA is inhibited by a protein that binds to a repeated sequence in the 3′ untranslated region. *EMBO J.* **13:** 1476–1481.

Ozer, J., M. Faber, R. Chalkley, and L. Sealy. 1990. Isolation and characterization of a cDNA clone for the CCAAT transcription factor EFIA reveals a novel structural motif. *J. Biol. Chem.* **265:** 22143–22152.

Perry, R.P. and D.E. Kelley. 1968. Messenger RNA-protein complexes and newly synthesized ribosomal subunits: Analysis of free particles and components of polyribosomes. *J. Mol. Biol.* **35:** 37–59.

Preobrazhensky, A.A. and A.S. Spirin. 1978. Informosomes and their protein components: The present state of knowledge. *Prog. Nucleic Acid Res. Mol. Biol.* **21:** 1–38.

Ranjan, M., S.R. Tafuri, and A.P. Wolffe. 1993. Masking mRNA from translation in somatic cells. *Genes Dev.* **7:** 1725–1736.

Richter, J.D. and L.D. Smith. 1984. Reversible inhibition of translation by *Xenopus* oocyte-specific proteins. *Nature* **309:** 378–380.

Rittschof, D. and J.A. Traugh. 1982. Identification of casein kinase II and phosphorylated proteins associated with messenger ribonucleoprotein particles from reticulocytes. *Eur. J. Biochem.* **123:** 333–336.

Sommerville, J. 1990. RNA-binding phosphoproteins and the regulation of maternal mRNA in *Xenopus*. *J. Reprod. Fertil.* **42:** 225–233.

———. 1992. RNA-binding proteins: Masking proteins revealed. *BioEssays* **14:** 337–339.

Spirin, A.S. 1966. On 'masked' forms of messenger RNA in early embryogenesis and in other differentiating systems. *Curr. Top. Dev. Biol.* **1:** 1–38.

———. 1994. Storage of messenger RNA in eukaryotes: Envelopment with protein, translational barrier at 5′ side, or conformational masking by 3′ side? *Mol. Reprod. Dev.* **38:** 107–117.

Spirin, A.S. and M. Nemer. 1965. Messenger RNA in early sea-urchin embryos: Cytoplasmic particles. *Science* **150:** 214–217.

Spirin, A.S., N.V. Belitsina, and M.A. Ajtkhozhin. 1964. Messenger RNA in early embryogenesis (*Zh. Obshch. Biol.* **25:** 321–338 [Russian].) *Fed. Proc.* (1965) **24:** T907–T915 (English).

Standart, N. 1992. Masking and unmasking of maternal mRNA. *Semin. Dev. Biol.* **3:** 367–379.

Standart, N. and T. Hunt. 1990. Control of translation of masked mRNAs in clam oocytes. *Enzyme* **44:** 106–119.

Standart, N., M. Dale, E. Stewart, and T. Hunt. 1990. Maternal mRNA from clam oocytes can be specifically unmasked *in vitro* by antisense RNA complementary to the 3′-untranslated region. *Genes Dev.* **4:** 2157–2168.

Stepanov, A.S. and K.V. Kandror. 1984. Effect of self-phosphorylation of RNA-binding proteins on their RNA-binding activity. *Dokl. Acad. Nauk. SSSR* **275:** 1227–1230.

Stepanov, A.S., K.V. Kandror, and S.M. Elizarov. 1982. Protein kinase activity in RNA-binding proteins of amphibia oocytes. *FEBS Lett.* **141:** 157–160.

Strickland, S., J. Huarte, D. Belin, A. Vassalli, R.J. Rickles, and J.-D. Vassalli. 1988. Antisense RNA directed against the 3′ noncoding region prevents dormant mRNA activation in mouse oocytes. *Science* **241:** 680–684.

Tafuri, S.R. and A.P. Wolffe. 1990. *Xenopus* Y-box transcription factors: Molecular cloning, functional analysis, and developmental regulation. *Proc. Natl. Acad. Sci.* **87:** 9028–9032.

———. 1992. DNA binding, multimerization, and transcription stimulation by the *Xenopus* Y box proteins *in vitro*. *New Biol.* **4:** 349–359.

———. 1993. Selective recruitment of masked maternal mRNA from messenger ribonucleoprotein particles containing FRGY2 (mRNP4). *J. Biol. Chem.* **268:** 24255–24261.

Tafuri, S.R., M. Familari, and A.P. Wolffe. 1993. A mouse Y box protein, MSY1, is associated with paternal mRNA in spermatocytes. *J. Biol. Chem.* **268:** 12213–12220.

Thiele, B.J., H. Andree, M. Höhne, and S.M. Rapoport. 1982. Lipoxygenase mRNA in rabbit reticulocytes: Its isolation, characterization and translational repression. *Eur. J. Biochem.* **129:** 133–141.

Thoen, C., E. Deherdt, and H. Slegers. 1986. Identification of the ribosomal proteins phosphorylated by the ribosome-associated casein kinase type II from cryptobiotic gastrulae of the brine shrimp *Artemia* sp. *Biochem. Biophys. Res. Commun.* **135:** 347–354.

Thoen, C., L. Van Hove, E. Piot, and H. Slegers. 1984. Purification and characterization of the messenger ribonucleoprotein-associated casein kinase II of *Artemia salina* cryptobiotic gastrulae. *Biochim. Biophys. Acta* **783:** 105–113.

Wistow, G. 1990. Cold shock and DNA binding. *Nature* **344:** 823–824.

Wolffe, A.P. 1994. Structural and functional properties of the evolutionarily ancient Y-box family of nucleic acid binding proteins. *BioEssays* **16:** 245–251.

Wolffe, A.P., S. Tafuri, M. Ranjan, and M. Familari. 1992. The Y-box factors: A family of nucleic acid binding proteins conserved from *Escherichia coli* to man. *New Biol.* **4:** 290–298

12
Translational Control of Ferritin

Tracey A. Rouault and Richard D. Klausner
Cell Biology and Metabolism Branch
National Institutes of Child Health and Human
Development, National Institutes of Health
Bethesda, Maryland 20892

Joe B. Harford
RiboGene, Inc.
Hayward, California 94545

The translation of messenger RNAs encoding ferritin is highly regulated by iron. A single copy of an RNA motif known as an iron-responsive element or IRE (see Fig. 1) is found within the 5′-untranslated region (5′UTR) of the mRNAs that encode all known vertebrate ferritins. The control of translation of ferritin mRNA by iron is mediated by regulation of an interaction between the IRE and a cytosolic protein. This protein, which binds to the IRE with high affinity and specificity when cells are iron-depleted, has been called the IRE-binding protein (IRE-BP) (Leibold and Munro 1988; Rouault et al. 1988), iron regulatory factor (IRF) (Müllner et al. 1989), ferritin repressor protein (FRP) (Walden et al. 1989), and P90 (Harrell et al. 1991). Recently, a consensus has emerged that favors the term iron regulatory protein (IRP), and we adhere to this convention here. There are now two relatively well-characterized iron regulatory proteins that are described in this chapter. The protein originally called IRE-BP is henceforth referred to as IRP1, and a second IRP that is less abundant in the majority of tissues tested is referred to as IRP2.

In addition to the role of the IRE in mediating the regulation of ferritin biosynthesis, an IRE that functions to produce IRP1-mediated translational regulation is present in the 5′UTR of the mRNA for the erythroid form of δ-aminolevulinic acid (ALA) synthase, the rate-limiting step in the heme biosynthetic pathway (Dierks 1990; Cox et al. 1991; Dandekar et al. 1991; Bhasker et al. 1993; Melefors et al. 1993). IREs also exist in the mRNA encoding the transferrin receptor (TfR) where the IRE/IRP interaction serves to modulate the half-life of the mRNA (Harford 1993). Several recent reviews have appeared in which

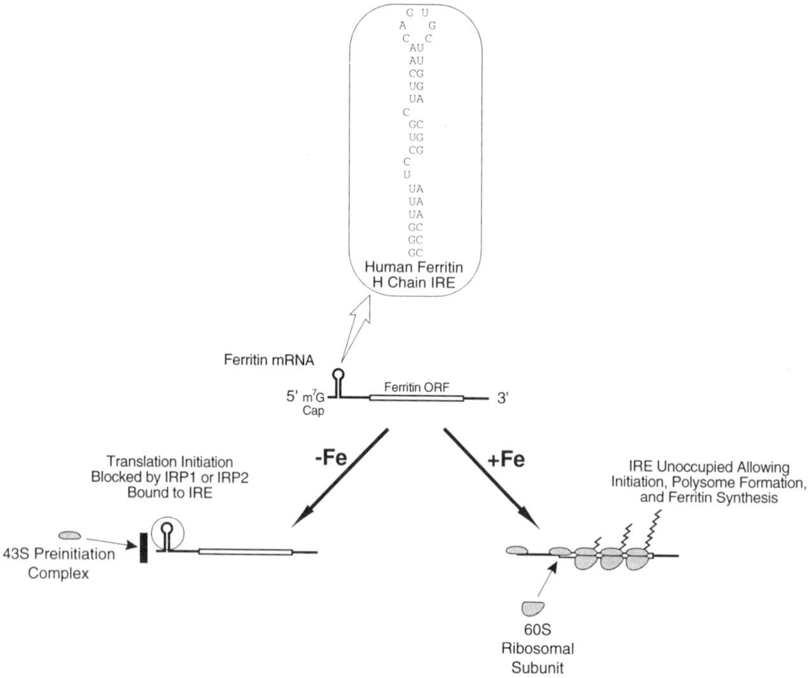

Figure 1 Translational control of ferritin synthesis. The 5′-untranslated region of ferritin mRNA contains a single iron-responsive element (IRE). A representative IRE, that of human ferritin H-chain mRNA, is shown as its predicted secondary structure in the inset. When iron is scarce, a cytoplasmic, high-affinity iron regulatory protein (IRP) interacts with the IRE. In the case of the ferritin mRNA, this interaction serves to reduce the translation of the mRNA by blocking translation initiation through preventing the 43S preinitiation complex from associating with the cap structure of the ferritin mRNA. This block results in reduced ferritin synthesis with no alteration in the level of ferritin mRNA. When iron is abundant, the IRP dissociates from IRE in ferritin mRNA, allowing ferritin mRNA-containing polysomes to form and ferritin to be synthesized.

the coordinate posttranscriptional regulation of ferritin and the TfR have been described (Kühn and Hentze 1992; Leibold and Guo 1992; Munro 1993; Klausner et al. 1993; Theil 1993; Walden 1993; Harford et al. 1994). Given this wealth of reviews, it is not our intention here to provide another exhaustive review of research in this area. Instead, we focus on newer findings, particularly those related to the IRPs and their role in ferritin translational control. We will attempt to restrict our citation of background material to that needed as a context for this focus.

THE NEED FOR REGULATION OF IRON

Iron is an essential component of cellular metabolism but is highly toxic when present in excess. Accordingly, organisms require means both to obtain iron from the environment and to maintain iron intracellularly in a nontoxic state. A large number of critical enzymatic reactions utilize iron, especially those involving electron transfer (Crichton 1991). Numerous proteins utilize iron as a cofactor because of the facility with which it moves between its two oxidation states, ferrous (Fe2+) and ferric (Fe3+). Iron exists in several cellular proteins (e.g., hemoglobin and cytochrome C) in the porphyrin ring structure of heme. Iron-sulfur clusters are also contained in a number of critical cellular enzymes. The Krebs' cycle enzyme aconitase is representative of the iron-sulfur enzymes and is discussed further below because of its relationship to IRP1. A number of other cellular enzymes contain neither heme nor an iron-sulfur cluster but have iron associated in another form or require iron for enzymatic activity. Ribonucleotide reductase, which catalyzes the rate-limiting step in DNA replication, is one example. The essentiality of iron to so many of life's chemical reactions means that iron deficiency can result in serious medical problems (Baynes 1994). However, despite the essential nature of iron, iron in conjunction with oxygen becomes a generator of hydroxyl radicals that have a variety of toxic effects on cells (Gutteridge 1989). The medical aspects of iron toxicity have also been reviewed (Britton et al. 1994).

IRON SEQUESTRATION IN FERRITIN

Given that iron is both indispensable and potentially highly destructive, it is not surprising that organisms possess means to regulate cellular iron uptake from the environment. In certain bacteria, fungi, and plants, iron carrier proteins known as siderophores facilitate uptake of iron. In vertebrates, iron uptake is a regulated process in which the intracellular iron levels determine how much iron will be taken up from the extracellular milieu. This regulation is achieved through regulation of the expression of the TfR, which is responsible for mediating endocytosis of the plasma iron-carrying protein transferrin. Control of TfR biosynthesis is accomplished through iron-dependent regulation of the TfR mRNA turnover (Harford 1993).

In addition to regulation of uptake, there is an intracellular protein that protects the cell from accumulation of iron. The protein, known as ferritin, is uniquely suited to the task of storage and sequestration of iron.

Ferritin exists in the cytoplasm as a hollow, sphere-like structure composed of 24 subunits. The subunits are of two types, the H subunit (H being derived from heart or heavy) and the L subunit (L being derived from liver or light) (for review, see Theil 1987; Andrews et al. 1992). In humans, there is only one copy of expressed H subunit gene, found on chromosome 11 (Hentze et al. 1986), and a single expressed gene encoding the L subunit is found on chromosome 19 (Caskey et al. 1983). The ferritin H and L genes are derived from distinct but related multigene families that each contain pseudogenes which map to a number of human chromosomes (Jain et al. 1985).

Ferritins have the capacity to store up to 4500 molecules of iron within the core of the 24-subunit sphere. There, iron is deposited as an insoluble ferric-oxyhydroxyphosphate (Theil 1987; Andrews et al. 1992). Iron gains access to the inside of the sphere through channels in the sphere. Once deposited as a precipitate within the ferritin molecule, the iron is effectively sequestered and no longer poses a toxic threat to the cell. Few specifics are known about the fate of iron once it has been deposited within ferritin, although it seems likely that the cell can recapture iron from ferritin when the ferritin is degraded to form hemosiderin (Weir et al. 1984). Thus, ferritin serves in the short term to sequester iron and prevent it from facilitating harmful oxidations in the cell. Over the longer term, it may serve as a repository of iron that can be recycled when the ferritin itself is degraded.

TRANSLATIONAL REGULATION OF FERRITIN BIOSYNTHESIS

In the 1970s, Munro and colleagues observed the distribution of ferritin mRNA within sucrose gradients indirectly by assessing the amount of ferritin synthesis that resulted from mixing mRNA isolated from various fractions into an in vitro translation system. They observed that when messenger ribonucleoproteins (mRNPs) were examined from cells that were iron-replete, the majority of the ferritin mRNA was found associated with ribosomes in the polysome fraction. When mRNPs were examined from cells that were iron-depleted, ferritin mRNA was not associated with ribosomes, and it therefore appeared that the level of translation was itself regulated (Zahringer et al. 1976). In the mid-1980s, the cDNAs for ferritins were cloned (Boyd et al. 1984; Santoro et al. 1986), and it became possible to perform Northern hybridization analyses and compare mRNA levels with rates of protein biosynthesis. Studies of this type revealed that mRNA levels remained unaltered when cellular iron status changed (Aziz and Munro 1987; Rouault et al. 1987). The finding that ferritin synthesis responded to iron without a corresponding change in ferritin mRNA levels confirmed that ferritin was posttranscriptionally regulated.

IDENTIFICATION OF IRON RESPONSIVE ELEMENTS

A conserved sequence element was identified in the 5′UTR of ferritin H and L chains that could fold into a moderately stable stem-loop structure. The sequence element was termed the iron-responsive element, or IRE (Hentze et al. 1987), since deletion of the conserved sequence element eliminated iron-mediated regulation of ferritin biosynthesis, and the same element, when ligated to reporter genes, was sufficient to confer iron regulation upon unrelated genes. As with ferritin itself, regulation of the synthesis of the reporter protein was accomplished with no change in the level of the corresponding mRNA, indicating that the level of mRNA translation was regulated.

IREs from the ferritin mRNAs from a number of species have been compared among themselves and with the IREs found in TfR mRNAs for which sequence data are available (Harford 1993). The IRE appears to interact with the IRP1 as a stem-loop structure having a six-membered loop and at least one unpaired base interrupting the stem (see Fig. 1). Studies comparing the native IRE with related but altered RNAs have revealed that the interaction between the IRE and the IRP1 is highly dependent on both RNA sequence and RNA secondary structure (Leibold et al. 1990; Bettany et al. 1992; Jaffrey et al. 1993). In the consensus structure/sequence, which incorporates the conserved characteristics of naturally occurring IREs, there is a base-paired stem interrupted by an unpaired cytosine five base pairs removed from a six-membered loop (Klausner et al. 1993). The sequence of the loop is almost always CAGUGX, where X can be any base but G. The nucleotide composition of the base pairs in the stem can vary as long as base pairing is maintained. Additions, deletions, or changes of bases in the loop all decrease the affinity of IRP1 for the RNA as do changes in the unpaired cytosine residue of the stem or changes in the position of this cytosine relative to the loop (Jaffrey et al. 1993). Results of structure-dependent cleavage studies of the IRE are basically consistent with conclusions reached based on mutagenesis (Theil 1994).

Recently, an in vitro selection technique using purified IRP1 was used to select IRE-binding sequences from among 16,384 sequences in which nucleotides at the position of the bulged C and the CAGUGX sequence of the loop were randomized (Henderson et al. 1994). Two major classes of RNA ligands were selected which differed in the optimal loop sequence.

One class included the sequence CAGUGX found in native IREs, and the second contained the loop sequence UAGUAX. It may be noteworthy that the alternative sequence detected by the randomization of nucleotides in the loop, which binds in vitro with high specificity and affinity to IRP1, has the theoretical capacity for base pairing between positions 1 and 5 of the loop, as does the CAGUGX sequence found in native IREs. Whether the alternative internally base-paired structure of this six-membered "loop" exists in either the free or protein-bound state remains to be determined. In addition, it is not yet clear whether stem-loop structures containing the UAGUAX sequence are important in function or regulation of the naturally occurring transcripts.

THE RELATIONSHIP BETWEEN IRP1 AND MITOCHONDRIAL ACONITASE

Once a *cis*-acting element responsible for translation was identified, it was logical to attempt to identify a *trans*-acting factor that could sense iron levels and regulate translation. Gel-retardation techniques and UV cross-linking techniques revealed a cytosolic protein that bound to isolated IRE sequences with high affinity and specificity (Leibold and Munro 1988; Rouault et al. 1988). Furthermore, the lysates from cells that were iron-depleted prior to lysis showed markedly increased binding, suggesting that the regulatory protein bound to IREs when cells were iron-depleted, and probably served as a translational repressor. In vitro translation of ferritin mRNAs revealed similar information. Reticulocyte lysates were found to translate ferritin mRNA inefficiently due to a specific repressor protein, whereas in vitro translation of ferritin mRNA in wheat-germ lysates was much more efficient because wheat-germ lysates lacked the repressor protein (Dickey et al. 1988; Walden et al. 1988).

A protein that bound to IREs was purified using RNA affinity chromatography (Rouault et al. 1989; Neupert et al. 1990). The protein responsible for repression of ferritin mRNA translation was also purified from rabbit liver using multiple traditional steps of protein purification in conjunction with ferritin translation repression assays (Walden et al. 1989). Purification was followed by cloning of the human (Rouault et al. 1990; Hirling et al. 1992), murine (Philpott et al. 1991), rabbit (Patino and Walden 1992), and rat (Yu et al. 1992) forms of the regulatory protein referred to as IRP1. IRE-binding activity has also been detected

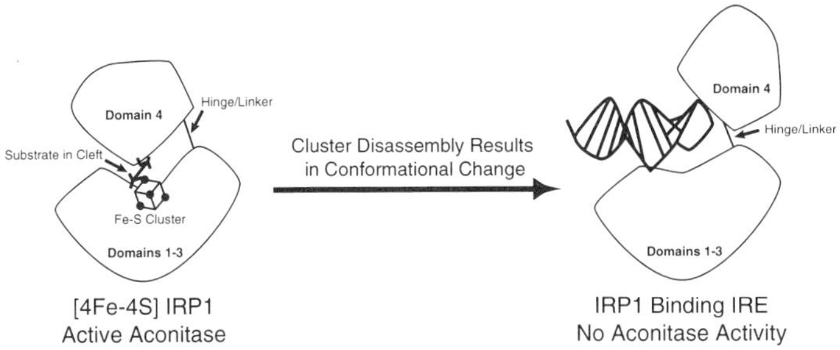

Figure 2 A model for the regulation of the IRP1. The [4Fe-4S] form of the IRP1 is an active aconitase. The substrate is envisioned as contributing to maintenance of the "closed" state of the protein. Cluster disassembly and loss of substrate are seen as allowing the protein to adopt an "open" conformation presumably because of the flexible hinge-linker that connects domains 1–3 to the fourth domain of the protein. In the open form, critical contact points for RNA binding are available. The depiction of the IRP1 is based on the X-ray crystal structure of mitochondrial aconitase, and the RNA is represented as a stem-loop drawn approximately to scale, but the depiction of the protein-RNA complex is merely diagrammatic. Cluster assembly/disassembly may also be involved in iron regulation of the related protein IRP2. However, in most cell types explored to date, the dominant mode of regulation of IRP2 appears to be iron-mediated destabilization of the protein.

in cell lysates from other species including birds, amphibians, fish, and insects (Rothenberger et al. 1990).

Cloning and sequencing of the cDNA for human IRP1 (Rouault et al. 1990) revealed that its predicted amino acid sequence was 30% identical to sequences of porcine mitochondrial aconitase and the mitochondrial aconitase of *Saccharomyces cerevisiae* (Hentze and Argos 1991; Rouault et al. 1991). There was already a great deal of structural and functional information known about mitochondrial aconitase, a Krebs' cycle enzyme found in the inner mitochondrial matrix that catalyzes the stereospecific interconversion of citrate and isocitrate through the intermediate *cis*-aconitate (Beinert and Kennedy 1993). Mitochondrial aconitase is a member of a class of dehydratases that utilize Fe-S clusters in non-redox catalytic mechanisms. A high-resolution solution of the X-ray structure of porcine mitochondrial aconitase (Lauble et al. 1992) revealed that the 83-kD protein folds into four domains. The amino-terminal three domains form a globular structure that is connected to the fourth domain through a potentially flexible hinge-linker peptide (see Fig. 2). Structural analyses, mutagenesis, and functional analyses (Lauble et al. 1992;

Zheng et al. 1992) have defined 23 active-site residues that are conserved in all known aconitases from bacteria to humans. The active-site residues include residues that bind substrate, function in catalysis, support catalytic and binding residues, and ligate the [4Fe-4S] cluster.

Alignment of IRP1 with the yeast and porcine mitochondrial aconitases revealed striking similarities, with the most remarkable feature of the alignment being that all of the aconitase active-site residues were aligned and present in all three proteins (Rouault et al. 1991). This finding suggested that the IRP1 might also have the potential to function enzymatically as an aconitase. Subsequent studies confirmed that the protein which was purified and cloned because of its ability to bind IRE could also catalyze the aconitase reaction (Kaptain et al. 1991) and was apparently the gene for cytosolic aconitase, an enzyme that had long been known to be present in cells but that had eluded purification and cloning. That the IRP1 accounted for all cytosolic aconitase activity was confirmed when the protein was purified using aconitase assays and EPR spectra to follow purification of the Fe-S-cluster-containing form of the protein (Kennedy et al. 1992). The bovine cytosolic aconitase was found to be more than 98% identical to human IRP1. Other enzymes of the Krebs' cycle also exist in the cytoplasm as well as in the mitochondria (for discussion, see Klausner et al. 1993), but the function of these activities in cellular metabolism remains obscure.

MODE OF INHIBITION OF TRANSLATION BY IRP1

To mediate translational repression, IREs must be located within the 5′UTR; IREs in the 3′UTR of TfR mRNA do not affect translation of the TfR transcript (Casey et al. 1988). Furthermore, studies in which the positioning of the IRE with respect to the cap site is changed have revealed that the ability of the IRE to mediate ferritin-like translational control is lost when the IRE is moved beyond about 70 nucleotides 3′of the mRNA cap (Goossen et al. 1990; Goossen and Hentze 1992). Although highly stable stem-loop structures positioned within the 5′UTR can inhibit translation (Pelletier and Sonenberg 1985; Kozak 1986), stabilities higher than those calculated for IREs are required, and IREs do not interfere with translation unless the IRP1 is tightly bound. Recent studies of the translation process have established that IRP1 interferes with translational initiation by inhibiting the association of the 43S complex with the mRNA cap (Gray and Hentze 1994). Furthermore, when binding sites for unrelated RNA-binding proteins are inserted into cap-proximal positions in the 5′UTR and the cognate binding proteins are expressed, inhibition of translation is mediated by the RNA-protein interaction (Stripecke et al. 1994). Thus, it appears that steric inhibition of initiation is the mode by which IRP1 represses translation (see Fig. 1) and furthermore that other RNA protein-binding interactions can function in the same fashion. It should be noted that one report has appeared in which the authors conclude that insertion of the IRE into a heterologous context not only results in translational regulation of the recombinant mRNAs, but also affects cytoplasmic mRNA levels. The authors suggested that there may be iron regulation of synthesis/transport of nuclear ferritin mRNAs in order to explain their observations (Coulson and Cleveland 1993).

THE IRP1 IRON-SULFUR CLUSTER SERVES AS A NOVEL MEDIATOR OF IRON SENSING

The relationship between IRP1 and mitochondrial aconitase led to a hypothesis that the Fe-S cluster was the key to the iron-sensing capability of the protein (Klausner and Rouault 1993). Early experiments demonstrated that latent IRP could be activated for RNA binding by addition of reducing agents such as mercaptoethanol to cell lysates (Haile et al. 1989; Hentze et al. 1989). Since this activation occurred without new protein synthesis, it raised the possibility that changes in IRP activity that occurred in response to changes in cellular iron levels might reflect a stable posttranslational modification of the protein. To test this hypothesis directly, a murine cell line was established that stably expressed the *myc* epitope-tagged human IRP1 (Haile et al. 1992a). By band shift activity, this cell line expressed approximately equal amounts of endogenous murine and exogenous human recombinant protein. Treatment of these cells with an iron source resulted in the loss of RNA-binding activity when measured in cell lysates. This loss of both the murine and human RNA-binding activity could be recovered by the addition of high concentrations of reducing agents to the lysate. No change in total protein level was observed by immunoblotting, and metabolic pulse-chase experiments demonstrated that there was no change in the rate of synthesis or in the rate of degradation of the human IRP1 that accompanied the regulation of RNA-binding activity (Tang et al. 1992). The lack of change of protein levels accompanying the iron-dependent change in IRE-binding activity was reproduced in a number of laboratories examining this activity in cells in culture and in tissues from a variety of sources (see, e.g., Müllner et al. 1992). These results contrasted with reports from Thach and colleagues (Goessling et al. 1992, 1994) who proposed that treatment of cells with hemin resulted in the irreversible loss of IRP through protein degradation. A possible reconciliation of these opposing findings is discussed below.

These results left uncharacterized the nature of the stable posttranslational modification, responsive to oxidation reduction, which could underlie the control of the RNA-binding activity (Hentze et al. 1989). The presence of the Fe-S cluster in mitochondrial aconitase became a possible candidate for such a posttranslational modification. Work with aconitase had demonstrated that because only three of the four irons in the cluster were covalently bound to cysteinyl ligands in the protein, the purified mitochondrial aconitase readily lost its fourth iron, leaving an inactive enzyme with a [3Fe-4S] cluster (Beinert and Kennedy 1989). The cluster could be reassembled in vitro with full recovery of enzyme activity. If this cluster interconversion represented the iron-dependent switch underlying the reversible loss or gain of RNA-binding activity, then the RNA-binding form of the protein in cells would be predicted to have no enzyme activity. Conversely, in iron-replete cells, the [4Fe-4S] cluster would be present; the protein would have full enzyme activity and would not bind RNA. When this prediction was tested in a cell line in which recombinant IRP1 was stably expressed (Haile et al. 1992a,b), reciprocal regulation of the RNA binding and aconitase enzyme activity of the recombinant immunoprecipitated protein was observed in response to iron manipulations. When cells were iron-replete, they had little or no RNA-binding activity and full aconitase activity. When cells were iron-starved, RNA-binding activity was activated and aconitase activity was lost. Similar results were also seen when IRP1 was overexpressed in insect cells and the cellular iron supply was manipulated (Emery-Goodman et al. 1993) and in IRP1 overexpressed from vaccinia virus vectors (Gray et al. 1993). A full understanding of the benefits of coordinate regulation of IRE binding and aconitase activity awaits a more thorough understanding of the interplay between intracellular iron and cellular metabolism.

A more detailed characterization of the status of the cluster during in vivo regulation was done by comparing the activities of recombinant IRP1 isolated from iron-replete cells with traditionally purified cytosolic aconitase from beef heart. As expected, neither of these proteins had RNA-binding activity. When these cluster-containing proteins were titrated with ferricyanide, the enzyme activity was lost at approximately equimolar concentrations of the oxidant and the protein (Haile et al. 1992b). This could be shown by spectroscopic means to result in the production of a [3Fe-4S] cluster. However, IRP1 containing the [3Fe-4S] cluster did not have RNA-binding activity.

Complete oxidative destruction of the cluster was required to activate RNA-binding activity, indicating that the RNA-binding form of the protein is iron-free IRP1. Additional evidence that apoprotein was the RNA-binding form was obtained using the effect of the aconitase substrate on RNA binding as another signature of the status of the cluster. Citrate binds with high affinity to either the 3Fe or 4Fe form of the protein but has essentially no affinity for the apoprotein (Kennedy et al. 1992). When the holoprotein, containing no RNA-binding activity, was treated with high concentrations of 2-mercaptoethanol, a manipulation that recruits RNA-binding activity in the holoprotein (Haile et al. 1989; Hentze et al. 1989), this recovery could be antagonized by the inclusion of substrate for the enzyme. Conversion of the 4Fe cluster to the 3Fe cluster produced protein that still required high concentrations of reducing agents to achieve RNA-binding activity. This activation of RNA binding by 2-mercaptoethanol was also inhibited by the inclusion of enzyme substrates. In contrast, the apoprotein was incapable of binding substrate and the substrate did not inhibit RNA binding. When the active RNA-binding protein was isolated from iron-depleted cells, addition of substrate did not antagonize RNA binding. These observations revealed that the interconversion within cells of the protein from an active enzyme to an RNA-binding entity involved, most likely, complete assembly or disassembly of the Fe-S cluster, and the two active forms represented [4Fe-4S] holoprotein and apoprotein (see Fig. 2) (Haile et al. 1992b).

MUTATIONAL ANALYSIS OF IRP1

Although this reciprocal regulation of aconitase and RNA-binding activity of IRP1 via the assembly/disassembly of the Fe-S cluster was compatible with available data, additional genetic evidence served to further support this mechanism of regulation (Philpott et al. 1994). On the basis of the sequence homology with mitochondrial aconitase, three cysteine residues (C437, C503, and C506 of IRP1) were predicted to ligate the Fe-S cluster. If cluster assembly inhibited RNA-binding activity in the presence of adequate iron levels, mutation of any one of these cluster-ligating cysteines would be predicted to render the protein incapable of sensing iron levels and produce a protein that was permanently in the RNA-binding form. To test these predictions, stable cell lines were made in which wild-type or mutant forms of IRP1 were expressed using a steroid inducible promoter in an Epstein-Barr-virus-based episomal vector. The creation of stably transformed cell lines was essential because overexpression of the wild-type IRP1 in transient expression systems consistently produced apoprotein that was not regulated by iron manipulations of the cells during the course of the transfection. It is possible, although this has not been proven, that this inability to incorporate an Fe-S cluster is a manifestation of substantial protein overexpression from the small percentage of cells transfected in the course of a transient transfection assay. RNA-binding activity of IRP containing single-point mutations of each of the cysteines predicted to ligate the cluster was evaluated by performing supershift assays in which anti-*myc* antibody recognition of the *myc* epitope shifted the complex formed by the recombinant protein to a different position in the gel. Mutation of the cysteines to serines did not affect the specificity or affinity of IRE binding. In the stable cell lines, the wild-type IRP1 was subject to the full range of regulation of RNA binding and aconitase activity in response to iron manipulations, as seen for endogenous protein. However, none of the three mutants (C437, C503, or C506) acquired aconitase activity in response to the addition of iron sources, and the mutant proteins were always active for RNA binding. In contrast, mutation of cysteine C300, not predicted to be involved in the ligation of an Fe-S cluster, produced an IRP that was normally regulated. Similar results were seen when cysteines C437, C503, or C506 were mutated to serines, and the mutant proteins were assayed after manipulations of iron during transient transfections (Hirling et al. 1994). These findings strongly supported the role of the Fe-S cluster assembly in mediating the in vivo regulation of the IRP1.

Little is known about the mechanisms of Fe-S cluster assembly or disassembly in eukaryotic cells. Recently, a prokaryotic enzyme important in de novo assembly of Fe-S clusters has been cloned (Zheng et al. 1993), although, to date, proteins involved in cluster assembly or disassembly have not been identified in eukaryotic cells. Since aconitase activity has been specifically maintained in the cytosol, the role of the enzymatic activity was evaluated in order to determine whether it was linked to cluster assembly/disassembly. Serine S778, which has been shown by mutagenesis to be indispensable in enzymatic function (Zheng et al. 1992), was mutated to alanine (Philpott et al. 1994). Although enzymatic activity was eliminated, regulation of IRE binding of the recombinant protein was maintained, indicating that the enzyme activity was not required for assembly/disassembly of the cluster. Thus, it appears that the cell coordinately regulates both cytosolic citrate/isocitrate metabolism and the fate of IRE-containing RNA transcripts in response to alterations in iron levels but that the two functions are not interdependent. Iron-mediated regulation of RNA binding by IRP1 does not depend on aconitase activity. Neither does aconitase activity require a protein capable of IRE binding since the mitochondrial aconitase lacks IRE-binding activity (Kaptain et al. 1991).

THE RNA-BINDING SITE OF IRP1

Although there is as yet no crystal structure of IRP1, a number of inferences have been drawn from the crystal structure of mitochondrial aconitase. These inferences are justified on the basis of the sequence similarity between IRP1 and mitochondrial aconitase and the fact that both function enzymatically as aconitases. In the crystal structure, the juxtaposition of domain 4 and domains 1–3 forms the active cleft for citrate-isocitrate interconversion (Lauble et al. 1992). To gain some insight into the location of the RNA-binding site, the RNA-binding form of IRP1 (apoprotein) was overexpressed and purified. Radioactively labeled IRE was mixed with the IRP1 and the mixture was subjected to cross-linking by UV light. After UV cross-linking, the IRP1 was subjected to proteolysis, and a single radioactively labeled peptide was sequenced. This peptide, which must be very close to the RNA to be cross-linked, represented amino acids 121–130 of the IRP1 primary sequence containing residues D125 and H126 corresponding to two active-site residues of mitochondrial aconitase (Basilion et al. 1994). Using molecular dimensions of the IRE hairpin and the structure of mitochondrial aconitase as a guide, it is apparent that the fourth domain must move with respect to the first three domains in order to accommodate the IRE structure in proximity to the peptide implicated by the cross-linking experiment (see Fig. 2). In an independent set of cross-linking experiments, the analysis of fragments of IRP1 liberated by limited proteolysis with chymotrypsin was again consistent with the conclusion that peptides in the active site cleft are important in RNA binding (Swenson and Walden 1994).

Because the UV cross-linking had pinpointed a contact site of the IRE with protein and the contact was revealed to be close to the aconitase active site, additional potential residues essential for RNA binding within the enzyme active-site cleft were analyzed in a series of point mutagenesis experiments. Four of the active-site residues of aconitase are arginines that interact with tricarboxylic acid enzyme substrates in mitochondrial aconitase (Lauble et al. 1992; Zheng et al. 1992). The potential role of arginines in RNA recognition has been highlighted by the studies on the residues in the human immunodeficiency virus (HIV)

Tat RNA-binding protein (Calnan et al. 1991; Puglisi et al. 1992, 1993). The guanidinium group of arginine can serve as a potential donor of five hydrogen bonds, enabling it to bind simultaneously to guanine and to neighboring phosphate groups (Puglisi et al. 1993). Four arginine residues of IRP1 (R536, R541, R699, and R780) corresponding to active-site arginine residues of aconitase were each singly replaced with either lysine or glutamine (Philpott et al. 1994). In the case of two of these amino acids (R780, R541), substitution resulted in at least a 20–100-fold drop in RNA-binding affinity even when the substitution was relatively conservative (lysine for arginine). When either R780 or R541 was mutated to glutamine, there was essentially no specific IRE-binding activity. In contrast, R536, when changed to a lysine, demonstrated only a twofold loss of binding affinity for the consensus IRE. When R536 was changed to a glutamine, the affinity drop was 20–100-fold. Finally, when R699 was changed to either a lysine or a glutamine, there was no loss in binding affinity. Thus, it appears that two active-site, substrate-binding arginines are required for the high-affinity IRE recognition. An additional arginine also contributes to binding affinity, but substitution with lysine does not significantly impair binding.

It is noteworthy that R780 corresponds to a domain-4 active-site residue in mitochondrial aconitase. Together with the cross-linking data described above implicating domains 1–3 in RNA binding, the impact of R780 substitution suggests that the IRE makes contact with residues on both sides of the active-site cleft. These data confirm that the conformation of the protein must change substantially to accommodate RNA binding, since an RNA hairpin the size of an IRE would not fit between domain 4 and domains 1–3 without such a change. Multiple contacts between the RNA and IRP1 suggest that the IRE may be enveloped by the protein. This model is consistent with data that the protein-bound IRE is extensively protected from nuclease and chemical attack (Wang et al. 1990; Harrell et al. 1992; Dix et al. 1993).

The involvement of arginine residues in IRE binding by IRP1 suggests an analogy to the Tat-TAR system where the high-affinity binding specificity depends on a single arginine, but surrounding positively charged residues also contribute (Calnan et al. 1991; Puglisi et al. 1992, 1993). The affinity of the IRP-IRE interaction is unusually high (Haile et al. 1989; Barton et al. 1990). It is tempting to speculate that the unusually high affinity of IRP1 for its cognate RNA (10–30 pM vs. 1 nM for the Tat-TAR interaction) might reflect contributions to affinity from at least two arginine residues that participate in formation of multiple hydrogen bonds and at least one residue where charge has an important role.

IRP2: A SECOND IRON REGULATORY PROTEIN

Numerous groups have observed that in lysates from cells and tissues, there are often two specific high-affinity band-shift complexes seen with radioactively labeled IREs (Leibold and Munro 1988; Henderson et al. 1993; Samaniego et al. 1994). UV cross-linking indicated that the second complex contained a protein with an apparent mobility on SDS-PAGE of 105 kD and that an antipeptide antibody raised against the sequence of IRP1 did not immunodeplete the second complex (Henderson et al. 1993). A second highly related clone had been identified when the human IRP1 was initially cloned (Rouault et al. 1990), and expression of the full coding sequence of this cDNA gave rise to a second IRE-binding protein (Rouault et al. 1992; Samaniego et al. 1994). This second IRE-binding protein, henceforth referred to as IRP2, was 57% identical in amino acid sequence to IRP1 and 79% similar, with these similarities extending over the entire length of the protein. The most striking difference noted between the predicted amino acid sequences of IRP1 and IRP2 was a 73-amino-acid insertion in IRP2 between two amino acids that are adjacent residues in IRP1. The extra 73-amino-acid segment found in domain 1 of IRP2 results from insertion of an additional exon within the gene for IRP2 (Samaniego et al. 1994). IRP2 differs from IRP1 in several respects, including the absence of aconitase activity in IRP2 (Guo et al. 1994; Samaniego et al. 1994).

The identification of IRP2 as a distinct gene product suggested that it might be responsible for the second complex that had previously been observed. To characterize endogenous IRP2 in cells, antipeptide antibodies were prepared that were capable of recognizing both murine and human recombinant IRP2 (Samaniego et al. 1994) or rat IRP2 (Guo et al. 1994). The antibodies were raised against a peptide within the 73-amino-acid segment unique to IRP2, and immunoblotting with this antibody demonstrated a protein of 105 kD in all cells tested. Furthermore, in cells in which two specific IRE band-shift complexes are well resolved on gels, only the complex that comigrated with the complex produced by recombinant IRP2 was eliminated from the gel or supershifted by incubation with the IRP2-specific antipeptide antibody (Samaniego et al. 1994; Guo et al. 1994).

REGULATION OF IRP2 BY IRON

When a variety of murine, monkey, and human cells were subjected to iron manipulation, there was significant regulation of IRP2 protein levels. Iron deprivation caused an increase in the level of IRP2, and iron provision caused a marked decrease in levels of immunologically detectable IRP2 (Guo et al. 1994; Samaniego et al. 1994). This is in marked contrast to what has been reported for IRP1 (Tang et al. 1992; Hirling et al. 1994), although in rabbit cells in culture, under certain conditions, Thach and colleagues have reported a similar phenomenon for a protein they believed to be IRP1 (Goessling et al. 1992, 1994). In RD4 cells stably expressing epitope-tagged IRP2, iron regulation of IRP2 protein levels has been observed, and the nature of this regulation has been explored (Samaniego et al. 1994). In this study, no iron-dependent change in the rate of synthesis of IRP2 or in the level of its mRNA was detected. Rather, the major effect of iron can be explained by a change in protein half-life. Under conditions of iron deprivation, there is no detectable turnover of the protein over 24 hours. In contrast, in cells that are iron-replete, the half-life of the protein is approximately 6 hours. Thus, regulation of IRP2 is in the same direction as that of IRP1, but the mechanism of regulation of the two proteins is quite distinct, a conclusion recently confirmed by other investigators (Pantopoulos et al. 1995). Whereas the RNA-binding activity and aconitase activity of IRP1 are coordinately regulated by assembly/disassembly of the Fe-S cluster without change in protein level, IRP2 appears to have its half-life regulated by iron. It is conceivable that the observations of Goessling et al. (1992, 1994) described above are related to these differences in the mode of regulation of IRP1 and IRP2.

THE ROLE OF THE TWO IRPs

The discovery of IRP2 as a second IRE-binding protein coupled with its unique mechanism of regulation raises numerous questions about this protein and how the overall regulation of iron metabolism relates to the functions of both IRP1 and IRP2. The interplay of the two proteins in the overall physiology of iron metabolism will undoubtedly form the basis of many future studies. No cell line or tissue has been identified in which IRP1 is expressed in the absence of IRP2 or vice versa. In tissues that have been characterized to date, the predominant form of IRP as determined at the level of mRNA expression has been IRP1, with the exception of brain where more IRP2 mRNA was found (Samaniego et al. 1994). At the level of protein expression, IRP2, migrating as a distinct complex in gel-retardation assays, was readily detectable in brain and intestinal tissue, whereas only IRP1 was readily detected in other tissues assayed (Henderson et al. 1993).

Since cell-specific expression does not offer an obvious explanation for the existence of two IRPs, other possibilities have been considered to explain why two IRPs are expressed in cells. One possibility is that the two proteins actually have different effector functions. This possibility has been evaluated in a comparison of the efficacy of recombinant IRP1 and IRP2 in mediating ferritin translational repression in an in vitro translation system. Equimolar amounts of purified recombinant IRP1 or IRP2 were added to the in vitro translation system, and the ability to repress translation was compared. In the in vitro setting, the two proteins were equally efficacious in effecting translational regulation (Kim et al. 1995).

Another possible reason for the retention of two IRP activities would be if there were binding targets within transcripts that are recognized by only one of the two proteins. There is evidence to suggest that the binding sites are not interchangeable, in that IRP1 can recognize and bind a sequence in which the sequence of the loop is UAGUAX rather than the more usual CAGUGX, whereas endogenous IRP2 does not bind the former sequence (Henderson et al. 1994). Recombinant IRP2 also binds poorly to this alternative sequence (J.P. Basilion et al., unpubl.). Although endogenous transcripts have not been defined in which the IRE uniquely recognized by IRP1 is present, it remains possible that the two IRPs can bind common targets but that each has, in addition, a set of binding sites which are specific to the individual IRP.

Other targets of IRP regulation, other than ferritin, include the TfR, which is regulated by the IRP by a mechanism other than translational regulation (Harford 1993). The TfR mRNA has five IREs in its 3′UTR, and these sites are bound by IRPs when the cells are iron-depleted. The IRPs prevent degradation of the mRNA by protecting the UTR from endonucleolytic attack (Binder et al. 1994). TfR mRNA levels increase significantly at the onset of rapid cell proliferation following a partial hepatectomy (Cairo and Pietrangelo 1994). RNA band-shift experiments performed as a part of this study indicated an increase in the amount of the faster-migrating IRE-protein complex, which likely corresponds to an IRP2-IRE complex. Although liver regeneration occurs without gross changes in liver iron content, it is not clear whether the observed changes in IRP2 are due to a decrease in iron in the regulatory pool or occur via an iron-independent mechanism.

As was noted earlier, the mRNA for the erythroid form of ALA synthase is translationally regulated by the IRP regulatory system in a fashion analogous to ferritin regulation (Bhasker et al. 1993; Melefors et al. 1993). Other mRNAs that have IREs in the 5′end include mitochondrial aconitase (Dandekar et al. 1991; Kaptain et al. 1991) and succinate dehydrogenase of *Drosophila* (Au and Scheffler 1994; H.Y. Kim et al., unpubl.). In the latter two cases, it has not been shown that these IREs confer iron-dependent regulation in cells, although it is intriguing that both of these mRNAs encode Fe-S cluster proteins of the Krebs' cycle, a setting in which iron-regulated synthesis could be appropriate.

REGULATION OF FERRITIN TRANSLATION BY MEDIATORS OTHER THAN IRON

We now know of a number of Fe-S cluster proteins in which the cluster provides an environmentally sensitive switch that can regulate the activity of the enzyme. Such clusters are particularly sensitive to oxidizing conditions, including hyperbaric oxygen, superoxide, nitric oxide (NO), and possibly hydrogen peroxide. Superoxide will partially decompose the cluster of *Escherichia coli* aconitase, and it has been proposed that cluster degradation represents a regulatory response to oxidative stress (Gardner and Fridovich 1992). Recently, studies of the bacterial enzyme dihydroxyacid dehydratase (Flint et al. 1993) have demonstrated that the Fe-S cluster is reversibly assembled and disassembled by oxidizing conditions, such as hyperbaric oxygen. Studies of the bacterial transcription factor SoxR, a sensor of superoxide and NO, have demonstrated that the factor is a dimer containing a [4Fe-4S] cluster (Hidalgo and Demple 1994). The cluster is proposed to provide the environmental sensor to allow the bacteria to sense these oxidants. Additionally, Switzer and colleagues (Grandoni et al. 1989) have demonstrated that the rate-limiting enzyme for purine biosynthesis, PRPP amido transferase, contains an oxygen-sensitive Fe-S cluster that is disassembled by oxidation, resulting in the rapid degradation of the protein. Thus, the Fe-S cluster appears to be widely used in biology as an environmental signal transduction mechanism.

It has been known for at least 10 years that NO readily attacks iron-containing proteins and, in particular, proteins containing Fe-S clusters (Drapier and Hibbs 1986; Hibbs et al. 1987). For example, exposure to NO will result in the inactivation of mitochondrial aconitase (Drapier and Hibbs 1988). These observations led to the prediction that NO could result in the disassembly of the Fe-S cluster of IRP. Two groups demonstrated NO-mediated cluster disassembly (Drapier et al. 1993; Weiss et al. 1993) and showed that cytosolic aconitase activity was lost and, concomitantly, IRE-binding activity was acquired in response to the production or addition of NO. The Fe-S cluster of IRP1 appears to be destabilized by exposure to NO, possibly because the NO (Kennedy and Beinert 1994) or a NO byproduct forms a nitrosyl adduct with iron in the cluster (Castro et al. 1994; Hausladen and Fridovich 1994). Consistent with the destruction of the Fe-S cluster by NO and the increase in the IRE-binding activity of IRP1, ferritin synthesis was decreased by increases in NO (Drapier et al. 1993; Weiss et al. 1993; Pantopoulos et al. 1994). In the brain, the IRP is an immediate responder to the endogenous production of NO via brain *N*-methyl-D-aspatate (NMDA) receptors, which are coupled to the calcium-dependent NO synthase system in neurons (Jaffrey et al. 1994). There is an immediate and reversible activation of IRE-binding activity with a concomitant and quantitative loss of cytosolic aconitase activity that accompanies the physiologic stimulation of NMDA receptors in cerebellar brain slices. Furthermore, the distribution within the brain of IRP1 mRNA was evaluated by in situ hybridization. The pattern of distribution of IRP1 largely coincided with the distribution of NMDA receptors throughout the rodent brain. Interestingly, under conditions where there was a major loss of cytosolic aconitase activity, there was no accompanying loss of mitochondrial aconitase activity, suggesting either a different sensitivity to oxidants, compartment-specific protection, or the requirement for other compartment-specific factors that determine the fate of the aconitase Fe-S cluster.

Although little is known about the processes that facilitate Fe-S cluster assembly in cells, it is likely that assembly/dissassembly of the Fe-S cluster of IRP1 is an enzymatic process. In bacteria, two gene products have been identified that are required for cluster assembly, the NifS gene product, which functions as a cysteine desulfurase (Zheng et al. 1993), and the NifU gene product (Fu et al. 1994), which may bind iron for incorporation into Fe-S clusters. It is possible that Fe-S cluster disassembly is an ongoing process resulting from intermittent oxidative stresses in the cell, and similarly cluster assembly could be a process which is ongoing and which depends only on the availability of iron and sulfur in the correct redox form (Jaffrey et al. 1994).

OTHER POTENTIAL MODES OF REGULATION OF FERRITIN SYNTHESIS

In the absence of binding of IRP to IREs, IREs are thought to confer the capacity for high translational efficiency (Dix et al. 1992; Theil 1994). This is thought to be related to a possible increased efficiency of the IRE-containing ferritin mRNA in binding initiation factors.

Ferritin mRNA has been shown to be recruited to polyribosomes in response to interleukin-1β (IL-1β), which is released from macrophages during inflammation (Rogers et al. 1990). By preparing and transfecting chimeric constructs and assessing IL-1β response, Rogers et al. (1994) have mapped the region of ferritin H-chain mRNA responsible for IL-1β translational control to a GC-rich segment of the 5′UTR that is distinct from the IRE. Similar GC-rich sequence motifs exist in mRNAs encoding hepatic acute-phase proteins whose synthesis is increased during inflammation. Although it is clear that there is a transcriptional component to the acute-phase reaction, these data suggest that acute-phase genes and ferritin are also regulated translationally.

Treatment of HL-60 cells with phorbol esters stimulates phosphorylation of IRP1, and there is a concomitant increase in the high-affinity IRE-binding activity (Eisenstein et al. 1993). Given that protein kinase C has a role in the signal transduction pathways mediated by certain growth factors and hormones, these data suggest that the IRE/IRP system may mediate translational control of ferritin synthesis in response to agents other than iron. IRP1 was shown in this study to be a substrate for phosphorylation by protein kinase C, and two sites of phosphorylation were identified that correspond to serines S138 and S711 of IRP1. On the basis of comparison to mitochondrial aconitase, S711 is predicted to be near the entrance to the active-site cleft of the enzyme. In comparing IRP1 and IRP2, the residues surrounding the phosphorylation site are not conserved. Thus, it is possible that susceptibility to phosphorylation by protein kinase C is another way in which IRP1 and IRP2 differ. The functional implications of these phosphorylation events remain to be determined.

SUMMARY AND PERSPECTIVES

The regulation of ferritin translation by iron serves as a paradigm of eukaryotic translational control. Ferritin mRNAs are translated well when iron is abundant and poorly when iron is scarce. This regulation is mediated by binding of an IRP to an IRE near the 5′end of the transcript and attenuation of translation initiation through steric hindrance. Thus, a high-affinity RNA/protein-binding interaction can function as a simple repressor in a fashion analogous to the binding of bacterial transcriptional repressors.

In contrast to DNA-binding proteins which tend to be unifunctional, the IRP is an example of a nucleic-acid-binding protein that also func tions as a cytosolic enzyme with a defined activity. In the evolution of the bifunctional IRP, it is likely that the metabolic role predated the

RNA-binding role. Although bacteria contain and require aconitase, which is 60% identical to IRP1 in primary sequence (Prodromou et al. 1992), there is no evidence that the aconitase of bacteria or yeast can double as a regulatory RNA-binding protein. Thus, the RNA-binding function of IRP appears to be a later evolutionary development. Unlike the example of many RNA-binding proteins in which the RNA-binding motif is identifiable in the primary sequence as a linear motif, the RNA-binding site of the IRP depends on correct folding of the tertiary structure of the IRP along with participation of residues in domains 1–3 as well as in domain 4. Several other examples of bifunctional RNA-binding proteins that also function as enzymes include glyceraldehyde-3-phosphate dehydrogenase, which can also function as a specific tRNA-binding protein (Singh and Green 1993), and the mitochondrial tyrosyl tRNA synthase of *Neurospora*, in which the protein can function as either a synthase or a splicing protein (Guo and Lambowitz 1992). Perhaps the preexisting enzymes of simple organisms formed a pool of potential RNA-binding proteins that could be captured and used to perform a second function in an increasingly complex network of gene regulation. The fact that cytosolic aconitase already had the capacity for iron sensing based on its use of a labile Fe-S cluster may have provided an appropriate foundation on which to build an iron-dependent regulatory system. In this scenario, there would already be an entity present in the cytosol under conditions of iron starvation, namely, the apo-aconitase, in which residues of the former enzymatic active-site cleft could freely participate in the binding of appropriately shaped RNA structures.

In addition to the role of Fe-S clusters as iron sensors in the IRE-IRP regulatory network, the cluster may also permit the IRP to sense other stresses such as oxidative stress resulting from either nitric oxide production or the generation of free radicals. It is likely that enzymes participate in cluster assembly and perhaps disassembly, and the final status of the Fe-S cluster in IRP function integrates a number of environmental signals that are important to cells.

As our appreciation of the complexity of the IRP-IRE system expands, it is likely that we will recognize many more genes that are under the control of this regulatory system. The recognition that additional potential ligands can be bound, along with the observation that many diverse signaling pathways may participate, increases the potential importance of this regulatory network. As was discussed initially, requirements for iron are fundamental. In fact, it now seems possible that the function of the Krebs' cycle is itself modulated by this powerful regulatory system.

REFERENCES

Andrews, S.C., P. Arosio, W. Bottke, J.-F. Briat, M. von Darl, P.M. Harrison, J.-P. Laulhere, S. Levi, S. Lobreaux and S.J. Yewdall. 1992. Structure, function, and evolution of ferritins. *J. Inorg. Biochem.* **47:** 161–181.

Au, H.C. and I.E. Scheffler. 1994. Characterization of the gene encoding the iron-sulfur protein subunit of succinate dehydrogenase from *Drosophila melanogaster. Gene* **149:** 261–265.

Aziz, N. and H.N. Munro. 1987. Iron regulates ferritin mRNA translation through a segment of its 5′ untranslated regulated region. *Proc. Natl. Acad. Sci.* **84:** 8478–8482.

Barton, H.A., R.S. Eisenstein, A. Bomford, and H.N. Munro. 1990. Determinants of the interaction between the iron-responsive element-binding protein and its binding site in rat L-ferritin mRNA. *J. Biol. Chem.* **265:** 7000–7008.

Basilion, J.P., T.A. Rouault, C.M. Massinople, R.D. Klausner, and W.H. Burgess. 1994. The iron-responsive element-binding protein: Localization of the RNA-binding site to the aconitase active-site cleft. *Proc. Natl. Acad. Sci.* **91:** 574–578.

Baynes, R.D. 1994. Iron deficiency In *Iron metabolism in health and disease* (ed. J. Brock et al.), pp. 123–149. Saunders Scientific, London.

Beinert, H. and M.C. Kennedy. 1989. Engineering of protein bound iron-sulfur clusters. *Eur. J. Biochem.* **186:** 5–15.

———. 1993. Aconitase, a two-faced protein: Enzyme and iron-regulating factor. *FASEB J.* **7:** 1442–1449.

Bettany, A.J., R.S. Eisenstein, and H.N. Munro. 1992. Mutagenesis of the iron-regulatory element further defines a role for RNA secondary structure in the regulation of ferritin and transferrin receptor expression. *J. Biol. Chem.* **267:** 16531–16537.

Bhasker, C.R., G. Burgiel, B. Neupert, A. Emery-Goodman, L.C. Kühn, and B.K. May. 1993. The putative iron-responsive element in the human erythroid 5-aminolevulinate synthase mRNA mediates translational control. *J. Biol. Chem.* **268:** 12699–12705.

Binder, R., J.A. Horowitz, J.P. Basilion, D.M. Koeller, R.D. Klausner, and J.B. Harford. 1994. Evidence that the pathway of transferrin receptor mRNA degradation involves an endonucleolytic cleavage within the 3′UTR and does not involve poly(A) tail shortening. *EMBO J.* **13:** 1969–1980

Boyd, D., S.K. Jain, J. Crampton, K.J. Barrett, and J. Drysdale. 1984. Isolation and characterization of a cDNA clone for human ferritin heavy chain. *Proc. Natl. Acad. Sci.* **81:** 4751–4755.

Britton, R.S., A.S. Tavill, and B.R. Bacon. 1994. Mechanisms of iron toxicity In *Iron Metabolism in Health and Disease* (ed. J. Brock et al.), pp. 123–149. Saunders Scientific, London.

Cairo, G. and A. Pietrangelo. 1994. Transferrin receptor gene expression during rat liver regeneration. Evidence for post-transcriptional regulation by iron regulatory factor B, a second iron-responsive element-binding protein. *J. Biol. Chem.* **269:** 6405–6409.

Calnan, B.J., B. Tidor, S. Biancalana, D. Hudson, and A.D. Frankel. 1991. Arginine-mediated RNA recognition: The arginine fork. *Science* **252:** 1167–1171.

Caskey, J.H., C. Jones, Y.E. Miller, and P.A. Seligman. 1983. Human ferritin gene assigned to chromosome 19. *Proc. Natl. Acad. Sci.* **80:** 482–486.

Casey, J.L., M.W. Hentze, D.M. Koeller, S.W. Caughman, T.A. Rouault, R.D. Klausner, and J.B. Harford. 1988. Iron-responsive elements: Regulatory RNA sequences that control mRNA levels and translation. *Science* **240:** 924–928.

Castro, L., M. Rodriguez, and R. Radi. 1994. Aconitase is readily inactivated by peroxynitrite, but not by its precursor, nitric oxide. *J. Biol. Chem.* **269:** 29409–29415.

Cox, T.C., M.J. Bawden, A. Martin, and B.K. May. 1991. Human erthroid 5-aminolevulinate synthase: Promoter analysis and identification of an iron-responsive element in the mRNA. *EMBO J.* **10:** 1891–1902.

Coulson, R.M.R. and D.W. Cleveland. 1993. Ferritin synthesis is controlled by iron dependent translational derepression and by changes in synthesis/transport of nuclear ferritin mRNAs. *Proc. Natl. Acad. Sci.* **90:** 7613–7617.

Crichton, R.R. 1991. *Inorganic biochemistry of iron metabolism.* Ellis Horwood, New York.

Dandekar, T., R. Stripecke, N. Gray, B. Goossen, A. Constable, H.E. Johansson, and M.W. Hentze. 1991. Identification of a novel iron-responsive element in murine and human erythroid delta-aminolevulinic acid synthase mRNA. *EMBO J.* **10:** 1903–1909.

Dickey, L.F., Y.H. Wang, G.E. Shull, I.A.D. Wortman, and E.C. Theil. 1988. The importance of the 3′-untranslated region in the translational control of ferritin mRNA. *J. Biol. Chem.* **263:** 3071–3074.

Dierks, P. 1990. Molecular biology of eukaryotic 5-aminolevulinate synthase. In *Biosynthesis of heme and chlorophylls* (ed. H.A. Dailey), pp. 201–233. McGraw-Hill, New York.

Dix, D.J., P.N. Lin, Y. Kimata, and E.C. Theil. 1992. The iron regulatory region of ferritin mRNA is also a positive control element for iron-independent translation. *Biochemistry* **31:** 2818–2822.

Dix, D.J., P.N. Lin, A.R. McKenzie, W.E. Walden. and E.C. Theil. 1993. The influence of the base-paired flanking region on structure and function of the ferritin mRNA iron regulatory element. *J. Mol. Biol.* **231:** 230–240.

Drapier, J.C. and J.B. Hibbs. 1986. Murine cytotoxic activated macrophages inhibit aconitase in tumor cells. Inhibition involves the iron-sulfur prosthetic group and is reversible. *J. Clin. Invest.* **78:** 790–797.

———. 1988. Differentiation of murine macrophages to express nonspecific cytotoxicity for tumor cells results in L-arginine-dependent inhibition of mitochondrial iron-sulfur enzymes in the macrophage effector cells. *J. Immunol.* **140:** 2829–2838.

Drapier, J.C., H. Hirling, J. Wietzerbin, P. Kaldy, and L.C. Kühn. 1993. Biosynthesis of nitric oxide activates iron regulatory factor in macrophages. *EMBO J.* **12:** 3643–3649.

Eisenstein, R.S., P.T. Tuazon, K.L. Schalinske, S.A. Anderson, and J.A. Traugh. 1993. Iron responsive element-binding protein: Phosphorylation by protein kinase C. *J. Biol. Chem.* **268:** 27363–27370.

Emery-Goodman, A., H. Hirling, L. Scarpellino, B. Henderson, and L.C. Kühn. 1993. Iron regulatory factor expressed from recombinant baculovirus: Conversion between the RNA-binding apoprotein and Fe-S cluster containing aconitase. *Nucleic Acids Res.* **21:** 1457–1461.

Flint, D.H., M.H. Emptage, M.G. Finnegan, W. Fu, and M.K. Johnson. 1993. The role and properties of the iron-sulfur cluster in *Escherichia coli* dihydroxy-acid dehydratase. *J. Biol. Chem.* **268:** 14732–14742.

Fu, W., R.F. Jack, T.V. Morgan, D.R. Dean, and M.K. Johnson. 1994. NifU gene product from *Azotobacter vinelandii* is a homodimer that contains two identical [2Fe-2S] clusters. *Biochemistry* **33:** 13455–13463.

Gardner, P.R. and I. Fridovich. 1992. Superoxide sensitivity of the *Escherichia coli* aconitase. *J. Biol. Chem.* **266:** 19328–19333.

Goessling, L.S., S. Daniels-McQueen, M. Bhattacharyya-Pakrasi, J.-J. Lin, and R.E.

Thach. 1992. Enhanced degradation of ferritin repressor protein during induction of ferritin messenger RNA translation. *Science* **256:** 670–673.

Goessling, L.S., D.P. Mascotti, M. Bhattacharya-Pakrasi, H. Gang, and R.E. Thach. 1994. Irreversible steps in the ferritin synthesis induction pathway. *J. Biol. Chem.* **269:** 4343–438.

Goossen, B. and M.W. Hentze. 1992. Position is the critical determinant for function of iron-responsive elements as translational regulators. *Mol. Cell. Biol.* **12:** 1959–1966.

Goossen, B., S.W. Caughman, J.B. Harford, R.D. Klausner, and M.W. Hentze. 1990. Translational repression by a complex between the iron-responsive element of ferritin mRNA and its specific cytoplasmic binding protein is position dependent in vivo. *EMBO J.* **9:** 4127–4133.

Grandoni, J.A., R.L. Switzer, C.A. Makaroff, and H. Zalkin. 1989. Evidence that the iron-sulfur cluster of *Bacillus subtilis* glutamine phosphoribosylpyrophosphate amidotransferase determines stability of the enzyme to degradation in vivo. *J. Biol. Chem.* **264:** 6058–6064.

Gray, N.K. and M.W. Hentze. 1994. Iron regulatory protein prevents binding of the 43S translation pre-initiation complex to ferritin and eALAS mRNA. *EMBO J.* **13:** 3882–3891.

Gray, N.K., S. Quick, B. Goossen, A. Consable, H. Hirling, L.C. Kühn, and M.W. Hentze. 1993. Recombinant iron-regulatory factor functions as an iron-responsive-element-binding protein, a translational repressor and an aconitase. A functional assay for translational repression and direct demonstration of the iron switch. *Eur. J. Biochem.* **218:** 657–667.

Guo, B., Y. Yu, and E.A. Leibold. 1994. Iron regulates cytoplasmic levels of a novel iron-responsive element-binding protein without aconitase activity. *J. Biol. Chem.* **268:** 24252–24260.

Guo, Q. and A.M. Lambowitz. 1992. A tyrosyl-tRNA synthetase binds specifically to the group I intron catalytic core. *Genes Dev.* **6:** 1357–1372.

Gutteridge, J.M.C. 1989. Iron and oxygen: A biologically damaging mixture. *Acta Paediatr. Scand. Suppl.* **361:** 78–85.

Haile, D.J., M.W. Hentze, T.A. Rouault, J.B. Harford, and R.D. Klausner. 1989. Regulation of interaction of the iron-responsive element binding protein with iron-responsive RNA elements. *Mol. Cell. Biol.* **9:** 5055–5061.

Haile, D.J., T.A. Rouault, C.K. Tang, J. Chin, J.B. Harford, and R.D. Klausner. 1992a. Reciprocal control of RNA binding and aconitase activity in the regulation of the iron-responsive element binding protein: Role of the iron-sulfur cluster. *Proc. Natl. Acad. Sci.* **89:** 7536–7540.

Haile, D.J., T.A. Rouault, J.B. Harford, M.C. Kennedy, G.A. Blondin, H. Beinert, and R.D. Klausner. 1992b. Cellular regulation of the iron-responsive element binding protein: Disassembly of the cubane iron-sulfur cluster results in high affinity RNA binding. *Proc. Natl. Acad. Sci.* **89:** 11735–11739.

Harford, J.B. 1993. The regulated turnover of the transferrin receptor mRNA. In *Control of mRNA stability* (ed. J. Belasco and G. Brawerman), pp. 239–266. Academic Press, New York.

Harford, J.B., T.A. Rouault, and R.D Klausner. 1994. Iron-responsive elements and the control of cellular iron homeostasis. In *Iron metabolism in health and disease* (ed. J. Brock et al.), pp. 123–149. Saunders Scientific Ltd., London.

Harrell, C.M., A.R. McKenzie, M.M. Patino, W.E. Walden, and E.C. Theil. 1991. Ferritin mRNA: Interactions of iron regulatory element with translational regulator P-90 and the effect of base-paired flanking regions. *Proc. Natl. Acad. Sci.* **88:** 4166–4170.

Hausladen, A. and I. Fridovich. 1994. Superoxide and peroxynitrite inactivate aconitases, but nitric oxide does not. *J. Biol. Chem.* **269:** 29405–29408.

Henderson, B.R., C. Seiser, and L.C. Kühn. 1993. Characterization of a second RNA-binding protein in rodents with specificity for iron-responsive elements. *J. Biol. Chem.* **268:** 27327–27334.

Henderson, B.R., E. Menotti, C. Bonnard, and L.C. Kühn. 1994. Optimal sequence and structure of iron-responsive elements—selection of RNA stemloops with high-affinity for iron regulatory factor. *J. Biol. Chem.* **269:** 17481–17489.

Hentze, M.W. and P. Argos. 1991. Homology between IRE-BP, a regulatory RNA-binding protein, aconitase, and isopropylmalate isomerase. *Nucleic Acids Res.* **19:** 1739–1740.

Hentze, M.W., T.A. Rouault, J.B. Harford, and R.D. Klausner. 1989. Oxidation-reduction and the mechanism of a regulated interaction between RNA and protein. *Science* **244:** 357–359.

Hentze, M.W., T.A. Rouault, S.W. Caughman, A. Dancis, J.B. Harford, and R.D. Klausner. 1987. A *cis*-acting element is necessary and sufficient for translational regulation of human ferritin expression in response to iron. *Proc. Natl. Acad. Sci.* **84:** 6730–6734.

Hentze, M.W., S. Keim, P. Papadopoulos, S. O'Brien, W. Modi, J. Drysdale, W.J. Leonard, J.B. Harford, and R.D. Klausner. 1986. Cloning, characterization, expression, and chromosomal localization of a human ferritin heavy-chain gene. *Proc. Natl. Acad. Sci.* **83:** 7226–7230.

Hibbs, J.B., Jr., R.R. Taintor, and Z. Vavrin. 1987. Macrophage cytotoxicity: Role for L-arginine deiminase and imino nitrogen oxidation to nitrate. *Science* **235:** 473–476.

Hidalgo, E. and B. Demple. 1994. An iron-sulfur center essential for transcriptional activation by the redox-sensing SoxR protein. *EMBO J.* **13:** 138–146.

Hirling, H., B.R. Henderson, and L.C. Kühn. 1994. Mutational analysis of the [4Fe-4S]-cluster converting iron regulatory factor from its RNA-binding form to cytoplasmic aconitase. *EMBO J.* **13:** 453–461.

Hirling, H., A. Emery-Goodman, N. Thompson, B. Neupert, C. Seiser, and L.C. Kühn. 1992. Expression of active iron regulatory factor from a full-length human cDNA by in vitro transcription/translation. *Nucleic Acids Res.* **20:** 33–39.

Jain, S.K., K.J. Barrett, D. Boyd, M.F. Favreau, J. Crampton, and J. Drysdale. 1985. Ferritin H and L chains are derived from different multigene families. *J. Biol. Chem.* **260:** 11762–11768.

Jaffrey, S.R., D.J. Haile, R.D. Klausner, and J.B. Harford. 1993. The interaction between the iron-responsive element binding protein and its cognate RNA is highly dependent upon both RNA sequence and structure. *Nucleic Acids Res.* **21:** 4627–4631.

Jaffrey, S.R., N.A. Cohen, T.A. Rouault, R.D. Klausner, and S.H. Snyder. 1994. The iron responsive element binding protein: a target for synaptic actions of nitric oxide. *Proc. Natl. Acad. Sci.* **91:** 12994–12998.

Kaptain, S., W.E. Downey, C. Tang, C. Philpott, D. Haile, D.G. Orloff, J.B. Harford, T.A. Rouault, and R.D. Klausner. 1991. A regulated RNA binding protein also possesses aconitase activity. *Proc. Natl. Acad. Sci.* **88:** 10109–10113.

Kennedy, M.C. and H. Beinert. 1994. In vitro studies on the disassembly of the Fe-S cluster of cytosolic and mitochondreal aconitases on reaction with nitric oxide. *J. Inorg. Biochem.* **56:** L13.

Kennedy, M.C., L. Mende-Mueller, G.A. Blondin, and H. Beinert. 1992. Purification and characterization of cytosolic aconitase from beef liver and its relationship to the iron-responsive element binding protein (IRE-BP). *Proc. Natl. Acad. Sci.* **89:** 11730–11734.

Kim, H.Y., R.D. Klausner, and T.A. Rouault. 1995. Translational repressor activity is equivalent and is quantitatively predicted by in vitro RNA binding for two iron-responsive element binding proteins, IRP1 and IRP2. *J. Biol. Chem.* **270:** 4983–4986.

Klausner, R.D. and T.A. Rouault. 1993. A double life: Cytosolic aconitase as a regulatory RNA binding protein. *Mol. Biol. Cell* **4:** 1–5.

Klausner, R.D., T.A. Rouault, and J.B. Harford. 1993. Regulating the fate of mRNA: The control of cellular iron metabolism. *Cell* **72:** 19–28.

Kozak, M. 1986. Influences of mRNA secondary structure on initiation by eukaryotic ribosomes. *Proc. Natl. Acad. Sci.* **83:** 2850–2854.

Kühn, L.C. and M.W. Hentze. 1992. Coordination of cellular iron metabolism by post-transcriptional gene regulation. *J. Inorg. Biochem.* **47:** 183–195.

Lauble, H., M.C. Kennedy, H. Beinert, and C.D. Stout. 1992. Crystal structures of aconitase with isocitrate and nitroisocitrate bound. *Biochemistry* **31:** 2735–2748.

Leibold, E.A. and B. Guo. 1992. Iron-dependent regulation of ferritin and transferrin receptor expression by the iron-responsive element binding protein. *Annu. Rev. Nutr.* **12:** 345–368.

Leibold, E.A. and H.N. Munro. 1988. Cytoplasmic protein binds in vitro to a highly conserved sequence in the 5′ untranslated region of ferritin heavy-and light-subunit mRNAs. *Proc. Natl. Acad. Sci.* **85:** 2171–2175.

Leibold, E.A., A. Laudano, and Y. Yu. 1990. Structural requirements of iron-responsive-elements for binding of the protein involved in both transferrin receptor and ferritin mRNA post-transcriptional regulation. *Nucleic Acids Res.* **18:** 1819–1824.

Melefors, O., B. Goossen, H.E. Johansson, R. Stripecke, N.K. Gray, and M.W. Hentze. 1993. Translational control of 5-aminolevulinate synthase mRNA by iron-responsive elements in erythroid cells. *J. Biol. Chem.* **268:** 5974–5978.

Müllner, E.W., B. Neupert, and L.C. Kühn. 1989. A specific mRNA binding factor regulates the iron-dependent stability of cytoplasmic transferrin receptor mRNA. *Cell* **58:** 373–382.

Müllner, E.W., S. Rothenberger, A.M. Müller, and L.C. Kühn. 1992. In vivo and in vitro modulation of the mRNA-binding activity of iron regulatory factor: Tissue distribution and effects of cell proliferation, iron levels and redox state. *Eur. J. Biochem.* **208:** 597–605.

Munro, H.N. 1993. The ferritin genes: Their response to iron status. *Nutr. Rev.* **51:** 65–73.

Neupert, B., N.A. Thompson, C. Meyer, and L.C. Kühn. 1990. A high yield affinity purification method for specific RNA binding proteins: Isolation of the iron regulatory factor from human placenta. *Nucleic Acids Res.* **18:** 51–55.

Pantopoulos, K., N.K. Gray, and M.W. Hentze. 1995. Differential regulation of two related RNA-binding proteins, iron regulatory protein (IRP) and IRP_B. *RNA* **1:** 155–163.

Pantopoulos, K., G. Weiss, and M.W. Hentze. 1994. Nitric oxide and the posttranscriptional control of cellular iron traffic. *Trends Cell Biol.* **4:** 82–86.

Patino, M.M. and W.E. Walden. 1992. Cloning of a functional cDNA for the rabbit ferritin mRNA repressor protein: Demonstration of a tissue specific pattern of expression. *J. Biol. Chem.* **267:** 19011–19016.

Pelletier, J. and N. Sonenberg. 1985. Insertion mutagenesis to increase secondary structure within the 5′ noncoding region of a eukaryotic mRNA reduces translational efficiency. *Cell* **40:** 515–526.

Philpott, C.C., R.D. Klausner, and R.D. Rouault. 1994. The bifunctional iron-responsive element binding protein/cytosolic aconitase: The role of active site residues in ligand binding and regulation. *Proc. Natl. Acad. Sci.* **91:** 7321–7325.

Philpott, C.C., T.A. Rouault, and R.D. Klausner. 1991. Sequence and expression of the murine iron-responsive element binding protein. *Nucleic Acids Res.* **19:** 6333.

Prodromou, C., P.J. Artymiuk, and J.R. Guest. 1992. The aconitase of *E. coli. Eur. J. Biochem.* **204:** 599–609.

Puglisi, J.D., L. Chen, A.D. Frankel, and J.R. Williamson. 1993. Role of RNA structure in arginine recognition of TAR RNA. *Proc. Natl. Acad. Sci.* **90:** 3680–3684.

Puglisi, J.D., R. Tan, B.J. Calnan, A.D. Frankel, and J.R. Williamson. 1992. Conformation of the TAR RNA-arginine complex by NMR spectroscopy. *Science* **257:** 76–80

Rogers, J.T., K.R. Bridges, G.P. Durmowicz, J. Glass, P.E. Auron, and H.N. Munro. 1990. Translational control during acute phase response: Ferritin synthesis in response to interleukin-1. *J. Biol. Chem.* **265:** 14572–14578.

Rogers, J.T., J.L. Adriotakis, L. Lacroix, G.P. Durmowicz, K.D. Kasschau, and K.R. Bridges. 1994. Translational enhancement of H-ferritin mRNA by interleukin-1β acts through 5′ leader sequences distinct from the iron responsive element. *Nucleic Acids Res.* **22:** 2678–2686.

Rothenberger, S., E.W. Müllner, and L.C. Kühn. 1990. The mRNA-binding protein which controls ferritin and transferrin receptor expression is conserved during evolution. *Nucleic Acids Res.* **18:** 1175–1179.

Rouault, T.A., M.W. Hentze, S.W. Caughman, J.B. Harford, and R.D. Klausner. 1988. Binding of a cytosolic protein to the iron-responsive element of human ferritin messenger RNA. *Science* **241:** 1207–1210.

Rouault, T.A., M.W. Hentze, D.J. Haile, J.B. Harford, and R.D. Klausner. 1989. The iron-responsive element binding protein: A method for the affinity purification of a regulatory RNA-binding protein. *Proc. Natl. Acad. Sci.* **86:** 5768–5772.

Rouault, T.A., C.D. Stout, S. Kaptain, J.B.Harford, and R.D. Klausner. 1991. Structural relationship between an iron-regulated RNA-binding protein (IRE-BP) and aconitase: Functional implications *Cell* **64:** 881–883.

Rouault, T.A., M.W. Hentze, A. Dancis, S.W. Caughman, J.B. Harford, and R.D. Klausner. 1987. Influence of altered transcription on the translational control of human ferritin expression. *Proc. Natl. Acad. Sci.* **84:** 6335–6339.

Rouault, T.A., C.K. Tang, S. Kaptain, W.H. Burgess, D.J. Haile, F. Samaniego, W.O. McBride, J.B. Harford, and R.D. Klausner. 1990. Cloning of the cDNA encoding an RNA regulatory protein: The human iron responsive elements binding protein. *Proc. Natl. Acad. Sci.* **87:** 7958–7962.

Rouault, T.A., D.J. Haile, W.E. Downey, C.C. Philpott, C. Tang, F. Samaniego, J. Chin, I. Paul, D. Orloff, J.B. Harford, and R.D. Klausner. 1992. An iron-sulfur cluster plays a novel role in the iron-responsive element binding protein. *Biol. Metals* **5:** 131–140.

Samaniego, F., J. Chin, K. Iwai, T.A. Rouault, and R.D. Klausner. 1994. Molecular characterization of a second iron-responsive element binding protein, iron regulatory protein 2. *J. Biol. Chem.* **269:** 30904–30910.

Santoro, C., M. Marone, M. Ferrone, F. Costanzo, M. Columbo, C. Minganti, R. Cortese, and L. Silengo. 1986. Cloning of the gene coding for human L apoferritin. *Nucleic Acids Res.* **14:** 2863–2876.

Singh, R. and M.R. Green. 1993. Sequence-specific binding of transfer RNA by glyceraldehyde-3-phosphate dehydrogenase. *Science* **259:** 365–368.

Stripecke, R., C.C. Oliveira, J.E.G. McCarthy, and M.W. Hentze. 1994. Proteins binding to 5′ untranslated region sites: A general mechanism for translational regulation of mRNAs in human and yeast cells. *Mol. Cell. Biol.* **14:** 5898–5909.

Swenson, G.R. and W.E. Walden. 1994. Localization of an RNA binding element of the iron responsive element binding protein within a proteolytic fragment containing iron coordination ligands. *Nucleic Acids Res.* **22:** 2627–2633.

Tang, C.K., J. Chin, J.B. Harford, R.D. Klausner, and T.A. Rouault. 1992. The regulation of the iron reponsive element binding protein RNA binding activity occurs post-translationally. *J. Biol. Chem.* **267:** 24466–24470.

Theil, E.C. 1987. Ferritin: Structure, gene regulation, and cellular function in animals, plants, and microorganisms. *Annu. Rev. Biochem.* **56:** 289–315.

———. 1993. The IRE (iron regulatory element) family: Structures which regulate mRNA translation or stability. *BioFactors* **4:** 87–93.

———. 1994. Iron regulatory elements (IREs): A family of mRNA non-coding sequences. *Biochem. J.* **304:** 1–11.

Walden, W.E. 1993. Repressor-mediated translational control: The regulation of ferritin synthesis by iron. In *Translational expression of gene expression* (ed. J. Ilan), vol. 2, pp. 321–334. Plenum Press, New York.

Walden, W.E., M.M. Patino, and L. Gaffield. 1989. Purification of a specific repressor of ferritin mRNA translation from rabbit liver. *J. Biol. Chem.* **264:** 13765–13769.

Walden, W.E., S. Daniels-McQueen, P.H. Brown, L. Gaffield, D.A. Russell, D. Bielser, L.C. Bailey, and R.E. Thach. 1988. Translational repression in eukaryotes: Partial purification and characterization of a repressor of ferritin mRNA translation. *Proc. Natl. Acad. Sci.* **85:** 9503–9507.

Wang, Y.H., S.R. Sczekan, and E.C. Theil. 1990. Structure of the 5′untranslated regulatory region of ferritin mRNA studied in solution. *Nucleic Acids Res.* **18:** 4463–4468.

Weir, M.P., J.F. Gibson, and T.J. Peters. 1984. Biochemical studies on the isolation and characterization of human spleen haemosiderin. *Biochem. J.* **223:** 31–38.

Weiss, G., B. Goossen, W. Doppler, D. Fuchs, K. Pantopoulos, G. Werner-Felmayer, H. Wachter, and M.W. Hentze. 1993. Translational regulation via iron-responsive elements by the nitric oxide/NO-synthase pathway. *EMBO J.* **12:** 3651–3657.

Yu, Y., E. Radisky, and E.A. Leibold. 1992. The iron-responsive element binding protein: Purification, cloning, and regulation in rat liver. *J. Biol. Chem.* **267:** 19005–19010.

Zahringer, J., B.S. Baliga, and H.N. Munro. 1976. Novel mechanism for translational control in regulation of ferritin synthesis by iron. *Proc. Natl. Acad. Sci.* **73:** 857–861.

Zheng, L., M.C. Kennedy, H. Beinert, and H. Zalkin. 1992. Mutational analysis of active site residues in pig heart aconitase. *J. Biol. Chem.* **267:** 7895–7903.

Zheng, L., R.H. White, V.L. Cash, R.F. Jack, and D.R. Dean. 1993. Cysteine desulfurase activity indicates a role for NIFS in metallocluster biosynthesis. *Proc. Natl. Acad. Sci.* **90:** 2754–2758.

13

Translational Control of Ribosomal Protein mRNAs in Eukaryotes

Oded Meyuhas, Dror Avni, and Silvian Shama
Department of Developmental Biochemistry
Institute of Biochemistry, The Hebrew
University-Hadassah Medical School
Jerusalem 91120, Israel

Many complex processes that occur during normal growth, differentiation, development, or malignant transformation in eukaryotes involve regulation of ribosome biosynthesis. This organelle contains four species of ribosomal RNA and more than 70 different ribosomal protein molecules, all of which appear in equimolar amounts (for review, see Wool et al., this volume). Thus, to maintain the proper stoichiometry of the ribosomal components at varying cellular growth rates, there must be regulatory mechanisms to ensure both coordinate synthesis of all ribosomal components and modulated formation of ribosomes.

The equimolar accumulation of ribosomal proteins is maintained by coordinate regulation at various levels of gene expression (Meyuhas et al. 1987; Aloni et al. 1992 and references therein). It seems, however, that under many physiological conditions, control of ribosomal protein gene expression at the translational level is the most prevalent regulatory mechanism operating in both eukaryotes and prokaryotes. Interestingly, the synthesis of ribosomal proteins in yeast is generally not regulated at the translational level (Tsay et al. 1988), with the exception of a single documented case (Dabeva and Warner 1993).

This chapter focuses on the translational control of messenger RNAs encoding both ribosomal proteins and several related proteins during development, with fluctuations in growth rate and upon hormonal stimulation, as studied in various eukaryotic organisms and cell lines. Alterations in the translational efficiency of these mRNAs appear to be of a selective nature, to be coregulated with the accumulation of rRNA, and to involve the 5′-untranslated region (5′UTR) as the *cis*-regulatory element.

GROWTH-DEPENDENT TRANSLATIONAL CONTROL OF RIBOSOMAL PROTEIN mRNAS IN VERTEBRATES

The translation of vertebrate ribosomal protein mRNAs (rpmRNAs) is largely regulated in a growth-dependent manner, as illustrated by their selective shift from polysomes in growing cells into messenger ribonucleoprotein (mRNP) particles (subpolysomal fraction) in quiescent cells. This has been demonstrated during the transition of various cell types between growing and nongrowing states in response to a wide variety of physiological stimuli (Avni et al. 1994; Shama et al. 1995 and references therein). Similar specific translational fluctuations have been observed for rpmRNAs during the transition from the rapidly growing state in the fetal liver to the quiescent state in the adult and upon resumption of hepatocyte proliferation in the regenerating liver (Aloni et al. 1992).

The proportion of rpmRNAs actively engaged in protein synthesis, i.e., the proportion associated with polysomes, is significantly lower than that characteristic of other ubiquitous mRNAs, such as those encoding actin, nucleolin, various enzymes, and translation initiation factors (Meyuhas et al. 1987, 1990; Huang and Hershey 1989; Avni et al. 1994). On average, only 66% of rpmRNAs (ten different species) are engaged with ribosomes compared with 91% of the other housekeeping mRNAs (encoding nine different proteins). This selective translational repression of rpmRNAs becomes even more pronounced in cells that are growth-arrested, where only 26% of the rpmRNAs remain in polysomes compared with 84% of the non-rpmRNAs.

A detailed analysis of the kinetics of dissociation of rpmRNAs from, and recruitment into, polysomes upon growth arrest and stimulation, respectively, was performed with the *Xenopus* B 3.2 kidney cell line (Loreni and Amaldi 1992). The downshift of the polysomal association of rpmRNAs on serum starvation is very rapid, reaching its minimal value at about 40 minutes, with half-maximal change achieved after 10 minutes. The upshift upon serum refeeding is somewhat slower, reaching its maximal level within 60 minutes with half-maximal change at 25 minutes.

Treatment of mouse lymphosarcoma P1798 cells with dexamethasone leads to a selective translational repression of rpmRNAs. Translation of these mRNAs is restored within 3 hours after withdrawal of the hormone with half-maximal stimulation recorded at about 90 minutes (Meyuhas et al. 1987). The difference in kinetics of the up-regulation in these two cell lines might reflect a delayed response of the mouse cells because depletion of the intracellular hormone is not instantaneous.

DEVELOPMENTAL AND HORMONAL REGULATION OF THE TRANSLATION OF RIBOSOMAL PROTEIN mRNAs IN VARIOUS SPECIES

A selective translational repression of rpmRNAs has been observed during differentiation and development of a variety of organisms. Interestingly, this repression in some cases occurs despite rapid cell proliferation. Thus, the large stock of maternal rpmRNAs produced during *Drosophila* oogenesis to support rapid synthesis of ribosomes is selectively repressed in 0–5-hour-old embryos (Kay and Jacobs-Lorena 1985). At a later stage of embryogenesis, when the maternal ribosomes are diluted among the dividing cells, ribosome synthesis is reinitiated with a concomitant recruitment of rpmRNAs into polysomes (Al-Atia et al. 1985).

Similarly, the appearance of newly synthesized rpmRNAs during embryogenesis of *Xenopus laevis* seems to occur at gastrulation and precedes the onset of ribosomal protein synthesis, which occurs in tailbud embryos. This selective translational repression is evidenced by the fact that rpmRNAs initially appear in mRNP particles (stage 15) and start to be mobilized into polysomes around stage 26 (Pierandrei-Amaldi et al. 1982; Baum and Wormington 1985).

Dictyostelium discoideum responds to starvation by cessation of cell proliferation, formation of multicellular aggregates, and subsequent differentiation into two cell types, spore and stalk cells. This developmental stimulus leads to a very rapid and selective translational repression of rpmRNAs which precedes its effect on the abundance of these mRNAs (Steel and Jacobson 1987, 1991).

It has been shown for both resting chick embryo fibroblasts (DePhilip et al. 1980) and mouse myoblasts (Hammond and Bowman 1988) that the translational efficiency of rpmRNAs can be selectively induced by insulin treatment without a concomitant change in the growth rate. Similarly, copulation in *Drosophila melanogaster* leads to an abrupt elevation in the synthesis rate of the ribosomal proteins in the paragonial gland, which is not accompanied by increased mitogenic activity or the steady-state levels of the corresponding mRNAs. This observation suggests that translation of mRNAs is inefficient prior to copulation and is stimulated after copulation (Schmidt et al. 1985).

GENERAL FEATURES OF TRANSLATIONAL CONTROL OF rpmRNAs

The Bimodal Distribution of Ribosomal Protein mRNAs

Close inspection of the polysome profiles of rpmRNAs from growing or quiescent cells discloses a distinct bimodal distribution, suggesting two discrete populations of mRNA molecules: translationally active (polysome associated) and translationally inactive (masked in mRNP particles). Interestingly, the translationally active population, even in non-growing cells, is fully loaded with ribosomes, as indicated by the estimated spacing of about 31 codons per ribosome in the peak polysome fractions (Agrawal and Bowman 1987; Meyuhas et al. 1987, 1990; Loreni and Amaldi 1992). This spacing is considered to be close to the theoretical maximum (Kafatos 1972). Such a bimodal distribution, which has been observed in other types of mammalian cells (Yenofsky et al. 1983; Jefferies et al. 1994b) and in *Xenopus* embryos (Pierandrei-Amaldi et al. 1985a; Bagni et al. 1992), indicates that rpmRNAs alternate between repressed and active states and that when in the active state, they are translated at near maximum efficiency (an "all-or-none" phenomenon). This typical apportionment of rpmRNAs between mRNP and polysomes clearly indicates that translational repression results from a blockage at the translational initiation step.

The apparent underutilization of rpmRNAs, even during rapid proliferation, might reflect a waste of cellular resources due to a permanent overproduction of the respective mRNAs. Alternatively, the bimodal distribution of these mRNAs does not necessarily represent two distinct mRNA populations (translated and untranslated) within each individual cell, but rather cells at different phases of the cell cycle. It is therefore conceivable that in asynchronously growing cells, a constant proportion (approximately one third) is within the phase characterized by repressed translation. This notion is further supported by the observation that when growth of adult rat liver cells is stimulated by partial hepatectomy (Aloni et al. 1992) or when serum-starved NIH-3T3 cells are refed (S. Shama and O. Meyuhas, unpubl.), the proportion of rpmRNAs associated with polysomes increases to more than 90% within several hours following the stimulation. One plausible explanation for this dramatic increase in the translation efficiency is that it reflects synchronization of the cell population. Verification of this possibility will require a detailed analysis of the translational behavior of rpmRNAs through the cell cycle.

Storage and Recruitment of Ribosomal Protein mRNAs

An intriguing question is whether the sequestration of rpmRNAs in mRNP particles is associated with a modification that impedes their translatability. To address this issue, mRNA was extracted from subpolysomal or polysomal fractions and used to program protein synthesis in a cell-free system. Such analyses have suggested that rpmRNAs are sequestered in a translatable form (Baum and Wormington 1985; Kay and Jacobs-Lorena 1985; Thomas and Thomas 1986; Steel and Jacobson 1987; Hammond et al. 1991).

The recruitment of growth-dependent translationally controlled mRNAs into polysomes upon growth stimulation of quiescent vertebrate cells poses a fundamental question concerning the source of the mobilized mRNAs. Two lines of evidence suggest that these mRNAs are recruited from mRNP storage, rather than from a newly synthesized pool. First, the appearance (within 1 hour) of rpmRNAs in the polysomal fractions coincides with their disappearance from the subpolysomal fractions without a parallel alteration in the abundance or rate of accumulation of the respective mRNAs (Geyer et al. 1982; Meyuhas et al. 1987; Loreni and Amaldi 1992). Second, growth stimulation of resting Swiss-3T3 cells by serum induced the synthesis of several proteins, including elongation factor eEF1A and Q23, whose mRNAs are translationally coregulated with those of rpmRNAs (see details below). The increase in the synthesis rate of these proteins is not blocked by actinomycin D, suggesting that the respective mRNAs are recruited into polysomes from mRNP particles (Thomas et al. 1981; Thomas and Thomas 1986; Jefferies et al. 1994a).

Despite this circumstantial evidence, unambiguous proof that rpmRNAs are recruited into polysomes from mRNP particles will require further experimentation, such as (1) monitoring the redistribution of these mRNAs in the presence of an inhibitor of RNA polymerase II and (2) demonstrating a disproportionate appearance of unlabeled rpmRNAs in polysomes upon growth stimulation in the presence of labeled RNA precursor.

THE TRANSLATIONAL *CIS*-REGULATORY ELEMENT OF RIBOSOMAL PROTEIN mRNAS

The selective nature of the translational control of rpmRNAs suggests that they have a distinctive property that is recognized by the translational machinery and/or by proteins of the mRNP particles. A compilation of sequences in the vicinity of the cap site of all vertebrate ribosomal protein genes or mRNAs is presented in Table 1. Comparative analysis of these sequences indicates the following structural features: (1) rpmRNAs possess a short 5′UTR (~40 nucleotides on average) with a limited potential to form secondary structures, mostly of insignificant stability (see also Hammond et al. 1991); (2) the 5′UTR of rpmRNAs is essentially devoid of upstream AUGs; (3) the sequences spanning the AUG initiation codon in mammalian rpmRNAs conform to the general

Table 1 Sequences around the transcription start site of vertebrate ribosomal protein genes

Protein	Sequence (−30 … −20 … −10 … +1 … +10 … +20 … +30)	5′TOP (nucl.)	C/T ratio	5′UTR (nucl.)	Accession number
x rpL1	GCTTATAGCTGAGGAACGCGGCTCCTCCTTCCTTTTCTCTTCGTGGCCGCTGTGGAGAAG	12	5/7	52	X15678
x rpL14	CATAAGCCTGTTCGCCGCGCCTGTCTCTTT**C**CTTTCCTCCCCGAGAAGCCGCTGCTGCTA	12	8/4	41	X05025
x rpL22	**C**CTCTTTTCTCTTCCACCGCCGACATGCCT	15	7/8	24	X64207
x rpL32	**C**CTTTTCCTCCATCTTGGATACCAGTGCGG	11	6/5	62	V01440
x rpS1	ATATAGTCGCATCCGCCGGAAGTGGCCGCC**C**TTTCCTTCCAGCGGCGCTTAGCGAGATGG	10	5/5	26	Z34529
x rpS8	GAGATCTGGGTCGCTCTCGCGGGGTTTCGC**C**CTCTTCTAGGCCGTCACTGAGAAAGGACG	8	4/4	37	X71081
x rpS19	GATATGATTTGTCGAGGGCGGGCATTCCTT**C**CTTTCCTTCGTCACCGTGAGAGATAGCCG	10	5/5	35	M33517
x rpS22	**C**CTCTACACGTGGCTGTTCCTGACTCACCG	5	3/2	98	M34706
c rpS15	GCGCTCTCGCGAGATCTGCCGCTTCCGGCG**C**CCCTTTCCGCGCAGAGCAGGCCAAGATGG	9	6/3	26	D10167
c rpL37a	GCTCATACTGCGCCTGCGCACAGCAACTTC**C**TCTCTCTCTCTAGCGGCGCCGGGAAGGAA	12	6/6	31	D14167
c rpL7a	CATAGATATCCCCCAGCATGCCCTGTTCCG**C**CCTTTTTACTCTACTACCAAGATG	8	3/5	22	X62641
c rpL5	TAGGACCTGCGGCTGCGCAGTGGGAGCGGC**C**TTTTCCCCCCTGCAGTCGTTGCAGTCGTT	12	8/4	79	D10737
c rpL30	GGATAGGCGGTGCCCGCCGGGCGCTCGCTC**C**TTTCTGGCGGCCCGGCCATCTTGGTGGCG	6	2/4	69	D14521
m rpL7	CCTTTAAGGAGACAGCGATGCGCCACCTTT**C**TCTCTTCTTTTCCGGCTGGAACCATGGAG	14	6/8	24	M29015
m rpL7a	CCACTTACTTATCCCCCACAATTCCCTCTT**C**TTT**C**TTTCTCCAGCAGCCGAGCAAGATGC	12	5/7	26	X54067

m rpL13a	GGATAAGAAACCCTGCGAAAAGACCTCCTC**C**TTTCCCAGGCGGCTGCCGAAGATGGCGGA	7	3/4	22	X51528
m rpL30	TAGAAGAGCTTTGCATTGTGGGAGCTCCTT**C**CTTTCTCGCTCCCCGGCCGTCTTGGCGGC	8	4/4	38	K02928
m rpL32	TCATACCTTGCGCGCCGCCGCCGCCTCTTC**C**TTCTTCCTCGGCGCTGCCTACGAGGTGGC	10	5/5	52	K02060
m rpS4	AATTTCTTCGTTCGCCGGAAGAAGGCGAGT**C**TCTTTCCGTTCCTAGCGCAGCCATG	8	4/4	23	L24371
m rpS16	GAAAAATCGGCTGGGTTGGCCCCGCGCTT**CC**CTTTTCCGGTCGCGGCGCTGCGGTGTGGA	8	4/4	52	M11408
m rpP0	**C**TTCTCTCGCCAGGCGTCCTCGTTGGAGTG	8	4/4	77	X15267
r rpS15	CGATAACTGCGCAGCCGCTGACACCGCTTC**C**TTTTCCCAGCAGCCGCCAAGATGGCCGAA	8	4/4	21	D11388
r rpP2	TTTACCCCCGCCCACTGCGTCAGCATCTTC**C**TTTCGCCGCCGGACGCCGCCGAGGTCGCA	5	2/3	58	X55153
r rpL35a*	**C**TCTTTCTGCCATCTTGGCGTCTTTGGAGG	8	3/5	33	—
h rpL7a	ATACATATTACCCACAATTCCCTTTCCTTT**C**TCTCTCCTCCCGCCGCCCAAGATGCCGAA	12	8/4	22	X61923
h rpS4Y	GACTTTGGTTGCACCGGAAAAGAACAGATT**C**TCTTCCGTCGCAGAGTTTCGCCATGGCCC	7	4/3	23	L24370
h rpS4X	CAATTTCTACGCGCACCGGAAGACGGAGGT**C**CTCTTTCCTTGCCTAACGCAGCCATG	11	5/6	24	L24369
h rpS6	CTATGTGACCTGCGCTAAGCGGAAGTTGGC**C**CTCTTTTCCGTGGCGCCTCGGAGGCGTTC	10	5/5	42	X67309, M77232
h rpS8	TACAAACCGAACCGTGAATCTTTGCGGTTT**C**TCTTTCCAGCCAGCGCCGAGCGATGGGCA	8	4/4	23	X67247
h rpS14	GACCCCCGTCGCTCCGCCCTCTCCCACTCT**C**TCTTTCCGGTGTGGAGTCTGGAGACGACG	8	4/4	37	M13934
h rpS17	TTTAACAGGCTTCGCCTGTGCTTCCTGTTT**C**CTCTTTTACCAAGGACCCGCCAACATGGG	8	3/5	25	M18000

Sequences presented in this and the following tables represent only genes whose transcription start site has been rigorously determined thus far. (x) *Xenopus laevis*; (c) chicken; (m) mouse; (r) rat; (h) human. The nomenclature of amphibian ribosomal proteins differs from that of mammalian and avian ribosomal proteins.

*Kuzumaki et al. (1987).

consensus sequence (Perry and Meyuhas 1990); and (4) all vertebrate rpmRNAs have a C residue at the cap site, followed by an uninterrupted stretch of 4–13 pyrimidines (Table 1). The composition of this 5′-terminal oligopyrimidine tract (5′TOP), although varying among rpmRNAs even within a species, generally maintains a similar proportion of C and T residues. It should be emphasized that initiation of transcription at a C residue is rare among eukaryotic genes, which normally starts at an A residue (Bucher 1990), whereas the percentage of mammalian transcripts with a C residue at the cap site is only 17% (Schibler et al. 1977). The evolutionary conservation of the 5′TOP in all vertebrate rpmRNAs and the fact that some non-rpmRNAs which display the same translational control properties as rpmRNAs also have a 5′TOP (see below) suggest that this motif has a critical role within the translational regulatory element.

Initial evidence that the translational *cis*-regulatory element (TLRE) is located within the 5′UTR of an rpmRNA came from experiments in which *Xenopus* fertilized eggs were injected with chimeric genes. Thus, the 35-nucleotide-long 5′UTR of frog rpS19 mRNA was shown to be sufficient to confer translational repression on a reporter mRNA in a developmentally dependent manner (Mariottini and Amaldi 1990). Likewise, it has been shown that the first 29, 34, 27, or 32 nucleotides of mammalian mRNAs encoding for S16, L32, L13a, and L30 (Levy et al. 1991; Avni et al. 1994; S. Shama and O. Meyuhas, unpubl.) individually are not only necessary, but also sufficient to confer the typical behavior of rpmRNAs on human growth hormone mRNA. In addition, the activity of the mammalian TLRE appears to be position-dependent. Thus, TLRE fails to exert its effect when located internally at a distance of 69 or 41 nucleotides from the cap site or even when preceded by a single A residue (Hammond et al. 1991; Avni et al. 1994). Replacement of five out of the eight pyrimidines by purines, or just the C residue at the cap site by an A residue, completely abolished the translational regulatory properties of the TLRE of mouse rpS16 mRNA. These results indicate the critical role the 5′TOP has in the translational control mechanism (Levy et al. 1991).

The large majority of vertebrate rpmRNAs for which complete sequence information is available (27 of 32) have a 5′TOP of eight or more residues, whereas only two have a tract of only five pyrimidines (Table 1). Clearly, this is not the minimal structural requirement as translational control of rat rpP2 mRNA, which contains a five-pyrimidine-long 5′TOP, is indistinguishable from that of rpmRNAs with a 5′TOP of eight or more residues (Aloni et al. 1992; Avni et al. 1994). Shortening the TLRE of mouse mRNAs encoding rpL32 and rpS16 by removal of sequences downstream from the 5′TOP abolished the translational regulatory properties of these regions in mouse fibroblasts (Avni et al. 1994). These results suggest that full manifestation of the growth-dependent translational control requires, at least in fibroblasts, both the 5′TOP and sequences immediately downstream.

Heterologous transfection experiments revealed that the similar structural properties of the TLREs in mammalian and amphibian rpmRNAs reflect a complete evolutionary conservation of all the elements required for the growth-dependent translational control mechanism in these two distantly related vertebrate classes (Avni et al. 1994).

The human β_1-tubulin gene (M40), like the ribosomal protein genes, has a major transcription start site within an uninterrupted stretch of 24 pyrimidines (Lee et al. 1983). Nevertheless, the polysomal association of this mRNA resembles that of non-rpmRNAs, undergoing efficient translation under conditions where rpmRNAs are translationally repressed. Interestingly, the 5′UTR of human β_1-tubulin mRNA functions as a typical TLRE by conferring growth-dependent translational control on a reporter mRNA (D. Avni and O. Meyuhas, unpubl.). One plausible explanation for these results is that sequences downstream from the TLRE within the native β_1-tubulin mRNA neutralize the regulatory properties of the bona fide TLRE. Whatever the mechanism, these experiments clearly indicate that the presence of a 5′TOP, per se, cannot be used as an ultimate diagnostic tool to seek out mRNAs regulated at the translational level.

The translational *cis*-regulatory element of rpmRNAs of *Dictyostelium* (Steel and Jacobson 1991) and *Drosophila* (Patel and Jacobs-Lorena 1992), like that of their vertebrate counterparts, resides within its 5′UTR, as these regions are sufficient to confer translational repression on heterologous transcripts. Comparative analysis of sequences around the transcription start site of seven insect ribosomal protein genes suggests that most insect rpmRNAs initiate with a 5′TOP (Table 2). However, unlike higher eukaryotes, the 5′UTRs of *Dictyostelium* rpmRNAs lack an obvious common motif and are devoid of the 5′TOP (Steel et al. 1987; Singelton et al. 1989; Hoja et al. 1993; L.F. Steel, unpubl.).

mRNAs ENCODING ELONGATION FACTORS AND SEVERAL OTHER PROTEINS ARE TRANSLATIONALLY COREGULATED WITH RIBOSOMAL PROTEIN mRNAs THROUGH A COMMON MECHANISM

Transition of mammalian cells from resting to growing states elicits a rapid accumulation of poly(A)$^+$ mRNA in the cytoplasm (Johnson et al.

Table 2 Sequences around the transcription start site of insect ribosomal protein genes

Protein	Sequence	5′TOP (nucl.)	C/T ratio	5′UTR (nucl.)	Accession number
	−30	−20	−10	+1	+10	+20	+30					
d rpS6	GATCTGTCATCTCTACTTCCTCCTATCGGT**C**TTTTTCAACCTGCCGCTCGCAGCCAACAG	7	2/5	81	L01658, L02074
d rp49	CACTAATGGCTACACTTGTTGTGTCCTACC**A**GCTTCAAGATGACCATCCGCCCAGCATAC	—	—	9	X00848
d rpS17*	TTCACACAACCACCCTGGTCACACGGCATC**C**TTCTTTTTCTTTCGTTTCCGGCGAGCGGT	14	4/10	56	—
d rpS14a	TATAAATGTGGAAATTCCCTTGCTATTCCC**G**CAGTTGCTGATCGACAAACAAACCCAGAA	—	—	29	M21045
d rpS14b	GCCGGCAAATAAACATGGTGTTTTTTTCTC**T**TTTAGTTGCTGATCGACGACCAGACTCTG	4	0/4	35	M21045
d rpA1	CTCGACCGCGCGGCCACACTGACTTCCGCC**T**CTTTTTCGACGACAACTTGCAAAGGAAAG	8	2/6	89	X05016
mq rpL8	TTTCCTTGTTTTCGACATTTCGTCGCTTCG**C**TCTTTCTTTCCGGCTCGTACCAGAGTCTA	12	5/7	87	M99055

(d) *Drosophila melanogaster;* (mq) mosquito (*Aedes albopictus*).

*Maki et al. (1989).

1974), a concomitant twofold increase in the proportion of poly(A)$^+$ mRNA engaged in polysomes (Lee and Engelhardt 1978), and an increase in the rate of protein synthesis (Johnson et al. 1974; Rudland et al. 1975). Accordingly, the fraction of ribosomes engaged in translation increases by at least one half (Meyuhas et al. 1987; Kaspar et al. 1990). Since rpmRNAs collectively comprise about 8% of the cellular mRNA (Meyuhas and Perry 1980), it is conceivable that they cannot fully account for the recruitment of mRNAs and ribosomes into polysomes. An intriguing question is whether other mRNAs that are mobilized into polysomes in a growth-dependent manner are equipped with similar *cis*-regulatory elements.

Growth stimulation of resting Swiss-3T3 cells leads to an increase in the synthesis rate of a number of proteins without a concomitant increase in the abundance of the corresponding mRNAs, suggesting the involvement of a translational control mechanism (Thomas et al. 1981). One of these proteins was identified as elongation factor 1A (eEF1A) (Thomas and Thomas 1986). Subsequently, it was demonstrated that the corresponding mRNA shifted between mRNP particles and polysomes in parallel to the behavior of rpmRNAs in mammals (Rao and Slobin 1987; Avni et al. 1994; Jefferies et al. 1994a) or amphibia (Wormington 1989; Loreni et al. 1993). Sequence data available for the 5′ terminus of the human eEF1A mRNA indicate that this mRNA not only is translationally regulated in a growth-dependent manner, but also bears a typical 5′TOP (Table 3). *Xenopus laevis* eEF1A is encoded by three genes whose expression is developmentally regulated: 42Sp50, EF1AO, and EF1AS (Djé et al. 1990 and references therein). Interestingly, EF1AS mRNA, which is subject to translational control, includes a typical 5′TOP (Table 3). It is not known whether the *Xenopus* EF1AO gene or the two copies of the *D. melanogaster* eEF1A gene (F1 and F2) are also translationally regulated, but their mRNAs have a two-, two-, and five-nucleotide long 5′TOP, respectively (Table 3).

Treatment of mouse Swiss-3T3 cells with the immunosuppressant rapamycin leads to a selective repression of well-defined 5′TOP-containing mRNAs encoding rpS6 and eEF1A and of that encoding elongation factor-2 (eEF2) (Jefferies et al. 1994a; Terada et al. 1994). The 5′-terminal sequence of the mouse eEF2 mRNA is not currently available. Nevertheless, analysis of the transcription start site of the Chinese hamster gene encoding eEF2 has revealed that the resulting mRNA bears a typical 5′TOP (Table 3). Hence, it is conceivable that translational control of mouse eEF2 is mediated through this motif.

Two murine non-rpmRNAs encoding a 33-kD protein (originally

Table 3 Sequences around the transcription start site of translationally regulated non-ribosomal protein genes

Protein	Sequence (−30, −20, −10, +1, +10, +20, +30)	5′TOP (nucl.)	C/T ratio	5′UTR (nucl.)	Accession number
h eEF1A	TATATAAGTGCAGTAGTCGCCGTGAACGTT**C**TTTTTCGCAACGGGTTTGCCGCCAGAACA	7	2/5	63	J04617
x EF1AS	ATATAAAGGGTGGTTAAGGCCCGGTTCGCT**C**TCTTCCTCACCGGGTCTGCGGCGAGTTCT	9	5/4	60	M25697
x EF1AO	GGCCTTCCAGCAACTCCTCGCAGTCTGTAC**C**TGCTCTGCAATTGTCTAGTGGTCGCTGCC	2	1/1	33	M75873
d eEF1A F1	TCCTGTGCTGAGCTCTGCCGAACGCTCGTT**C**ACTTTGTTCGAATCCGTCGCCGCTTAGAC	1	1/0	80	X06869
d eEF1A F2	CGTTCACCTCAGCATCGTTGCTTTTTCGGT**C**TTTCCGTTTTGTGATTTCGAGGTAAGTGC	6	3/3	136	X06870
c.h. eEF2	TATAAAAGCGCGGGCGTTGACGTCAGCGGT**C**TCTTCCGCCGCAGCCGCCGCCATCGTCGG	7	4/3	83	J03200
m P21	**C**TTTTTTCCGCCCGCTCCCCCCTCCCCCCG	9	3/6	101	X06407
m P40	**C**TTTTTTCGCGCTACCCGGGGACGGGTCCA	8	2/6	79	J02870, X06406

(h) Human; (x) *Xenopus laevis;* (d) *Drosophila melanogaster;* (c.h.) Chinese hamster; (m) mouse.

referred to as P40) and a 21-kD protein (originally referred to as P21 and also known as P23 [Böhm et al. 1989] and Q23 [Jefferies et al. 1994a]), both of unknown function, are translationally repressed upon growth arrest (Yenofsky et al. 1982, 1983). The 33-kD protein that shares homology with both the laminin receptor (Rao et al. 1989; Auth and Brawerman 1992) and a yeast mitochondrial ribosomal protein (Davis et al. 1992) has been implicated as a component of the translation machinery (Auth and Brawerman 1992). These two translationally regulated mRNAs have been shown to possess a typical 5′TOP (see P21 and P40 in Table 3).

Ubiquitin is the most highly conserved protein known in eukaryotes and has been implicated in a wide variety of cellular processes (for review, see Finley and Chau 1991). Two members of the human ubiquitin gene family, UbB and UbC, encode three and nine direct repeats of mature ubiquitin, respectively (Baker and Board 1991 and references therein). The human UbA gene encodes a monomer of ubiquitin fused to a ribosomal protein of either 52 or 80 residues (UbA_{52} and UbA_{80}), providing a unique example of a naturally occurring ribosomal protein chimeric gene. The UbA_{52} mRNA possesses a 5′TOP that is separated by the 228-nucleotide-long ubiquitin-coding unit from the ribosomal protein-coding sequence, residing at the 3′end (Baker and Board 1991). In contrast, the mRNAs transcribed from the polyubiquitin genes lack a 5′TOP (Wiborg et al. 1985; Baker and Board 1987a,b). Analysis of the polysomal distribution of the various ubiquitin mRNAs in human cells or mouse lymphosarcoma cells under conditions where rpmRNAs are repressed demonstrates that the majority of UbA mRNA is stored in the mRNP particles while both UbB and UbC are efficiently translated (Avni et al. 1994).

Comparison of the function of proteins that are encoded by mRNAs which are translationally controlled in a growth-dependent manner suggests that the common denominator is their involvement in the translational apparatus. These include constituents of the cytoplasmic ribosome (ribosomal proteins and UbA), possibly a mitochondrial ribosomal protein (P40), and elongation factors (eEF1A and eEF2).

POSSIBLE INVOLVEMENT OF GENERAL COMPONENTS OF THE TRANSLATIONAL MACHINERY IN THE TRANSLATIONAL CONTROL OF RIBOSOMAL PROTEIN mRNAS

Besides a precise definition of the TLRE, it is also important to identify possible *trans*-acting factors that determine the activity of this regulatory element. The presence of such a factor is suggested by two lines of evidence: (1) Developmental or growth stimuli lead to a variable extent of recruitment into polysomes of individual rpmRNAs (Al-Atia et al. 1985; Meyuhas et al. 1987), suggesting that these mRNAs might differ in their affinity for the *trans*-acting factor(s), and (2) individual *Xenopus* embryos exhibit different developmental patterns of the recruitment of the same species of rpmRNA (Bagni et al. 1992), which might reflect diversity in the amount or activity of the *trans*-acting factor(s) among these individuals.

Translation Initiation Factor eIF4E and Translational Control of Ribosomal Protein mRNAs

A simple model to account for the selective growth-dependent translational control of rpmRNAs assumes the participation of a factor that is a component of the general protein synthesis machinery. If such a factor had a particularly low affinity for rpmRNAs due to their unique sequence at the 5′terminus, a decrease in its activity or amount might lead to a disproportionate discrimination against these mRNAs (Lodish 1974; Alton and Lodish 1977). According to the model of Lodish (1974), mRNAs with poor affinity for a limiting component of the initiation apparatus would be inefficiently initiated, as the limiting factor would be utilized primarily for the initiation of high-affinity mRNAs. One prediction of this model is that a partial block of elongation would impede the initiation of the efficient mRNAs, thereby making the factor more available for use by the low-affinity mRNAs. Indeed, treatment of eukaryotic cells with a low concentration of cycloheximide, which slows elongation of the nascent peptide, preferentially stimulates the synthesis of ribosomal proteins (Ignotz et al. 1981) and the polysomal association of rpmRNAs (Agrawal and Bowman 1987; Steel and Jacobson 1987; Pierandrei-Amaldi et al. 1991). These results are consistent with the assumption that rpmRNAs are sequestered in mRNP particles due to a low affinity for a limiting initiation component.

A prime candidate for such a factor is the limiting initiation factor eIF4E (for review, see Sonenberg; Merrick and Hershey; both this volume). Several lines of evidence support the notion that eIF4E could be involved in the selective translational control of rpmRNAs: (1) An increase in the translational efficiency of rpmRNAs during transition of Swiss-3T3 cells from nongrowing to growing states occurs simultaneously with enhanced phosphorylation of eIF4E (Kaspar et al. 1990), (2) the translation of mouse rpmRNAs in the rabbit reticulocyte lysate is relatively inefficient unless either eIF4F or eIF3 (possibly contaminated with eIF4F) are added (Hammond et al. 1991), and (3) the selective repression of rpmRNA translation during rat liver development and the stimulated translation of these mRNAs following partial hepatectomy are accompanied by parallel fluctuations in the abundance of the mRNAs encoding eIF4E and eIF4A (Aloni et al. 1992).

In addition, experiments directly aimed at examining the hypothesis that the translational efficiency of rpmRNAs is regulated by the amount or activity of eIF4E have revealed a good correlation between fluctuations in the utilization of rpmRNAs and the abundance of eIF4E in various cell lines. Nevertheless, two fundamental assumptions of this model were contradicted by demonstrating that (1) mRNAs containing or lacking a 5′TOP exhibit a similar sensitivity to the cap analog $m^7G(5')ppp(5')G$ when translated in a rabbit reticulocyte lysate (Shama et al. 1995) and (2) surplus amounts of phosphorylated eIF4E in an NIH-3T3-derived eIF4E-overexpressing cell line do not prevent the translational repression of rpmRNAs upon incubation of these cells in the presence of inhibitors of DNA synthesis (Shama et al. 1995). It therefore appears that the decrease in the level of eIF4E upon growth arrest does not have a key role in the repression of translation of rpmRNAs.

Currently, the possibility that the translational efficiency of rpmRNAs is determined by the activity of a single component or a combination of general components of the translational machinery cannot be formally excluded. However, two lines of circumstantial evidence argue against this notion: (1) A close inspection of the polysomal profiles of rpmRNAs in mouse cells (Agrawal and Bowman 1987; Meyuhas et al. 1987, 1990) and *Xenopus* kidney cells (Loreni and Amaldi 1992) does not reveal a gradual shift of these mRNAs into lighter polysomes upon growth arrest, as would be expected if the translational repression resulted from a diminished initiation rate. Instead, the majority of the rpmRNAs are totally excluded from polysomes in resting cells. (2) The translation of *Xenopus* rpmRNAs is repressed not only upon growth arrest, but also during early embryogenesis under conditions of extremely rapid proliferation and extensive protein synthesis (Amaldi and Pierandrei-Amaldi 1990). It is therefore highly unlikely that the selective translational repression of rpmRNAs in the developing *Xenopus* embryo can be ascribed simply to a temporary deficiency of one or more general translational factors.

Eukaryotic Ribosomal Protein mRNAs Are Not Autogenously Regulated at the Translational Level

The translation of bacterial rpmRNAs is regulated by an autogenous feedback mechanism (Nomura et al. 1984). Inferring from the bacterial precedent, a similar model might be suggested to account for the translational repression of rpmRNAs in eukaryotes. According to this model, the block in rRNA accumulation normally observed when vertebrate cells cease to proliferate (Aloni et al. 1992 and references therein) would lead to the accumulation of unassembled ribosomal proteins, which in turn would autogenously repress their own synthesis. Nevertheless, several lines of experimental and circumstantial evidence indicate that such a mechanism is not operative in the eukaryotes examined: (1) Synthesis of ribosomal proteins was shown to continue under conditions of repressed transcription of rDNA (Craig and Perry 1971; Warner 1977; Krauter et al. 1980). (2) Analysis of the synthesis of ribosomal proteins during embryogenesis of a *Xenopus* anucleolate mutant lacking the rRNA genes revealed that rpmRNAs were recruited normally into polysomes and translated at stage 30 even though rRNA was not synthesized (Pierandrei-Amaldi et al. 1985a). (3) Overexpression of rpmRNA did not change the translational efficiency of the respective mRNAs (Bowman 1987; Baum et al. 1988). (4) Direct microinjection of partially purified ribosomal proteins into *Xenopus* oocytes, or their addition to rabbit reticulocyte lysates engaged in translation of oocyte mRNA, had no repressive effect on the translation of rpmRNAs (Pierandrei-Amaldi et al. 1985b). Taken together, these experiments demonstrate that unlike the situation in *Escherichia coli*, eukaryotic rpmRNAs are efficiently translated even if the resulting proteins are in molar excess over those of free rRNA, thus disproving an autogenous feedback regulation of these mRNAs.

Do Free Ribosomes Exert Feedback Inhibition on the Translation of Ribosomal Protein mRNAs?

Free ribosomes have been suggested to mediate a feedback inhibition on the translation of rpmRNAs whenever the abundance of the former exceeds the cellular need (Amaldi and Pierandrei-Amaldi 1990). This model is supported by several observations: (1) The sequestration of rpmRNAs in mRNP particles during the early stages of *Xenopus* embryogenesis and their recruitment at later stages correlate with the proportion of ribosomes engaged in translation (Pierandrei-Amaldi et al. 1985a). (2) When quiescent *Xenopus* kidney cells commence growing, the mobilization of free ribosomes into polysomes precedes that of rpmRNAs (Loreni and Amaldi 1992). (3) The proportion of rpmRNAs associated with polysomes in anucleolate *Xenopus* embryos around stage 30 is considerably greater than that in normal embryos. This higher translation efficiency of rpmRNAs in anucleolate embryos accords with the apparent shortage of ribosomes in these embryos, due to the lack of rRNA genes (Pierandrei-Amaldi et al. 1985a). (4) Microinjection of partially purified ribosomes into fertilized *Xenopus* eggs, resulting in a 40% increase in ribosome content, selectively repressed the recruitment of rpmRNAs into polysomes as monitored at stage 32 (Pierandrei-Amaldi et al. 1991). (5) The aforementioned experiments, in which cycloheximide treatment was shown to increase the polysomal association of rpmRNAs, can be interpreted differently. Inhibition of elongation might lead to overloading of all mRNAs with ribosomes and consequently to a decrease in the free ribosome pool, which in turn could result in derepression of the translation of rpmRNAs (Pierandrei-Amaldi et al. 1991).

The ribosomal feedback model, although mechanistically attractive, is inconsistent with other observations: (1) rpmRNAs are efficiently translated during *Xenopus* oogenesis, even though free ribosomes are being accumulated at this developmental stage (Cardinali et al. 1987). (2) The polysomal association of rpmRNAs in adult rat liver is dramatically lower than that in fetal liver, yet the proportion of ribosomes not engaged in translation apparently remains constant during liver development (Aloni et al. 1992). For this model to be applicable for these two cases as well, one should be able to demonstrate that (1) most of the free ribosomes during *Xenopus* oogenesis are stored within a separate compartment and cannot interact with rpmRNAs and (2) rpmRNAs are localized in the adult rat liver within a subcellular compartment overloaded with free ribosomes.

The Phosphorylation State of rpS6 Tightly Correlates with the Translational Efficiency of Ribosomal Protein mRNAs

Numerous studies have shown that mitogenic stimulation of quiescent cells induces phosphorylation of rpS6 (for review, see Jefferies and Thomas, this volume). This phosphorylation has attracted attention due to its temporal correlation with the initiation of protein synthesis and the suggestion that ribosomes with the highest proportion of phosphorylated rpS6 have a selective advantage in mobilization into polysomes. Furthermore, rpS6 was localized within the mRNA-binding site in the ribosome (see Wool et al., this volume). These observations, together with the fact that eEF1A mRNA undergoes a partial shift into lighter polysomes upon growth arrest (Jefferies et al. 1994b), led Thomas and his colleagues to propose that rpS6 phosphorylation increases the affinity of ribosomes for

5′TOP-containing mRNAs and thus facilitates their initiation (Thomas and Thomas 1986; Jefferies et al. 1994b). This hypothesis is further supported by the fact that insulin treatment of quiescent cells does not elicit cell proliferation, yet it induces rpS6 phosphorylation (Thomas et al. 1982; Morley and Traugh 1993) and increased translational efficiency of rpmRNAs (DePhilip et al. 1980; Hammond and Bowman 1988). Moreover, treatment of cells with the immunosuppressant rapamycin, which selectively blocks the activation of the 70-kD S6 kinase and rpS6 phosphorylation (Chung et al. 1992; Price et al. 1992), also selectively represses the translation of 5′TOP-containing mRNAs (Jefferies et al. 1994b; Terada et al. 1994). It is therefore tempting to assume that S6 phosphorylation might be a determinant in the regulation of the translational efficiency of 5′TOP-containing mRNAs. Nevertheless, two observations disagree with this model: (1) S6 kinase activity is induced during meiotic maturation following progesterone treatment of stage VI *X. laevis* oocytes (Lane et al. 1992). However, not only does this treatment not increase the translational efficiency of rpmRNAs, but rather it also leads to its progressive cessation (Hyman and Wormington 1988). (2) If the extent of rpS6 phosphorylation correlates with the affinity of the ribosomes for 5′TOP-containing mRNAs, then alterations of its phosphorylation status should lead to parallel changes in the size of the respective polysomes, rather than the apparent "all-or-none" phenomenon reported for many rpmRNAs.

POSSIBLE INVOLVEMENT OF SPECIFIC *TRANS*-ACTING FACTORS IN THE TRANSLATIONAL CONTROL OF RIBOSOMAL PROTEIN mRNAs

One plausible explanation for the all-or-none translational behavior of the rpmRNAs is that the selective packaging of rpmRNAs into RNP particles renders them unavailable to the translational machinery as reviewed by A. Spirin (this volume). For the packaging to be selective, the rpmRNAs would have to interact directly or indirectly with a specific repressor as has been shown for mRNAs encoding ferritin (see Rouault et al., this volume) and 15-lipoxygenase (Ostareck-Lederer et al. 1994). Alternatively, recruitment of these mRNAs into polysomes might require a specific activator. In the context of these models, the selective translational repression of rpmRNA upon growth arrest might result from increasing amounts or activity of a specific repressor or from inactivation or depletion of a specific activator.

Establishing the role of the 5′TOP in the translational control of rpmRNAs and several other mRNAs has led to the hypothesis that this motif might bind a specific translational *trans*-acting factor. Indeed, a cytoplasmic protein of about 56 kD from mouse T lymphocytes was shown specifically to bind a sequence containing the 34-nucleotide-long TLRE of mouse rpL32 mRNA (Kaspar et al. 1992). This binding was dependent on the presence of an intact oligopyrimidine tract. Similarly, a sequence containing the first 52 nucleotides of *Xenopus* rpL1 mRNA was demonstrated to bind four proteins, of which two (57 and 47 kD) bound to the oligopyrimidine tract, whereas the binding sites of the other two (31 and 24 kD) were immediately downstream from the oligopyrimidine tract (Cardinali et al. 1993).

Another protein, polypyrimidine-tract-binding protein (PTB), with a similar binding specificity and size (57 kD) has been purified recently from the nuclei and cytosol of various mammalian cells (Hellen et al. 1993; Morris et al. 1993 and references therein). Several lines of evidence suggest, however, that the known pyrimidine-binding proteins are either irrelevant or insufficient for the selective translational control of 5′TOP-containing mRNAs: (1) The oligopyrimidine tract in rpmRNAs is strictly localized to the 5′terminus. In contrast, both PTB and the putative 5′TOP-binding proteins have been shown to bind to oligopyrimidine tracts in internal positions (Garcia-Blanco et al. 1989; Kaspar et al. 1992; Cardinali et al. 1993). (2) The binding of the *Xenopus* 57-kD protein was impaired but not abrogated by replacement of pyrimidines by purines within the 5′TOP sequence. (3) The binding activity of the 5′TOP-binding proteins, unlike the translational control of rpmRNAs, is not regulated in a growth-dependent manner in mouse cells (Kaspar et al. 1992) or developmentally regulated in *Xenopus* (Cardinali et al. 1993). It should be pointed out, however, that the binding activity does not have to directly or inversely correlate with the translation efficiency of the target mRNAs. A protein might constitutively bind to the recognition sequence and serve as an adapter for yet another protein that exerts its activity as a repressor or an activator. Alternatively, the binding protein might undergo a modification, like phosphorylation or dephosphorylation, that alters its repressor or activator potential in a growth- or developmental-dependent manner, without affecting its binding activity. An example of this latter possibility is the cyclic AMP response element binding protein (CREB), which is constitutively bound to its cognate sequence, yet it acquires its transcriptional *trans*-activation potential upon phosphorylation (Karin 1994).

CONCLUDING REMARKS

Significant progress has been made in recent years toward understanding the mechanism of the translational control of rpmRNAs in eukaryotes and particularly in vertebrates. Nevertheless, several fundamental questions remain unanswered, including the number and the nature of the *trans*-acting factors involved in the regulatory apparatus, the mechanism ensuring the coordination between the translation efficiency of rpmRNAs and the accumulation of the rRNA, and finally the pathway through which the growth signal is transduced to the translation machinery.

Currently, the nature of the *trans*-acting factor(s) involved in the translational control of rpmRNAs is enigmatic, as neither its mode of action (repressor or activator) nor its specificity (a general component of the translational apparatus or a specific factor) is known at present. Several general factors have been considered as possible candidates, but as discussed above, they either have been experimentally disproven or have failed to accord with all observations. Clues concerning putative specific *trans*-acting factors have been derived from RNA-protein-binding experiments (Kaspar et al. 1992; Cardinali et al. 1993). The relevance of the various oligopyrimidine-binding proteins has been questioned, however, due to the inconsistency between the binding activity and the translational behavior of rpmRNAs.

One plausible explanation, which might reconcile most of the contradictory observations, is based on the assumption that translational control of 5′TOP-containing mRNAs is carried out by an interaction between specific and general factors. According to this model, the translation of these mRNAs is completely repressed through the attachment of a specific protein to the 5′TOP. Translation can initiate, however, upon displacement of this repressor by a general component of the translational machinery. The binding activity of the repressor to its cognate sequence would remain constant regardless of the growth status, yet the general factor would acquire its displacement potential upon growth or hormonal stimulation. A prime candidate for the general factor is the small ribosomal subunit with the adjunct initiation factors, which could detach the repressor during the scanning of the 5′UTR. Conceivably, the displacement potential of the 40S subunit could be increased when it undergoes growth- or hormone-dependent phosphorylation on rpS6. The involvement of a putative specific repressor is consistent with the all-or-none phenomenon, whereas the presumptive engagement of the 40S subunit accords with the correlation between the phosphorylation status of rpS6 and the translation efficiency of 5′TOP-containing mRNAs.

The demonstration that sequences downstream from the 5′TOP are also essential for the translational control of rpmRNAs and the distribution of binding proteins along the 5′UTR of *Xenopus* rpL1 mRNA suggest that additional *trans*-acting factors might be involved in this regulatory mechanism. Whatever the number of binding proteins, establishing their function in the translational control of the target mRNAs cannot be based solely on their binding properties. Clearly, the role of a candidate protein as a repressor, activator, or adapter in this mode of regulation can only be demonstrated unambiguously by a functional assay or by genetic experiments.

ACKNOWLEDGMENTS

We are indebted to Francesco Amaldi, Robert P. Perry, and Paola Pierandrei-Amaldi for critically reviewing the manuscript and for helpful comments. We thank P. Pierandrei-Amaldi and L.F. Steel for sharing data prior to their publication. Research in the authors' laboratory is supported by grants to O.M. from The Council for Tobacco Research U.S.A., Basic Research Foundation administered by the Israel Academy of Sciences and Humanities, and the United States–Israel Binational Science Foundation (BSF-93-00032).

REFERENCES

Agrawal, A.G. and L.H. Bowman. 1987. Transcriptional and translational regulation of ribosomal protein formation during mouse myoblast differentiation. *J. Biol. Chem.* **262:** 4868–4875.

Al-Atia, G.R., P. Fruscoloni, and M. Jacobs-Lorena. 1985. Translational regulation of mRNAs for ribosomal proteins during early *Drosophila* development. *Biochemistry* **24:** 5798–5803.

Aloni, R., D. Peleg, and O. Meyuhas. 1992. Selective translational control and non-specific posttranscriptional regulation of ribosomal protein gene expression during development and regeneration of rat liver. *Mol. Cell. Biol.* **12:** 2203–2212.

Alton, T.H. and H.F. Lodish. 1977. Translational control of protein synthesis during early stages of differentiation of the slime mold *Dictyostelium discoideum*. *Cell* **12:** 301–310.

Amaldi, F. and P. Pierandrei-Amaldi. 1990. Translational regulation of the expression of ribosomal protein genes in *Xenopus laevis*. *Enzyme* **44:** 93–105.

Auth, D. and G. Brawerman. 1992. A 33-kDa polypeptide with homology to the laminin receptor: Component of translational machinery. *Proc. Natl. Acad. Sci.* **89:** 4368–4372.

Avni, D., S. Shama, F. Loreni, and O. Meyuhas. 1994. Vertebrate mRNAs with a 5′-terminal pyrimidine tract are candidates for translational repression in quiescent cells: Characterization of the translational *cis*-regulatory element. *Mol. Cell. Biol.* **14:** 3822–3833.

Baker, R.T. and P.G. Board. 1987a. The human ubiquitin gene family: Structure of a gene and pseudogenes from the Ub B subfamily. *Nucleic Acids Res.* **15:** 443–463.

———. 1987b. Nucleotide sequence of a human ubiquitin Ub B processed pseudogene. *Nucleic Acids Res.* **15:** 4352.

———. 1991. The human ubiquitin-52 amino acid fusion protein gene shares several structural features with mammalian ribosomal protein genes. *Nucleic Acids Res.* **19:**

1035–1040.

Bagni, C., P. Martiottini, L. Terrenato, and F. Amaldi. 1992. Individual variability in the translational regulation of ribosomal protein synthesis in *Xenopus laevis*. *Mol. Gen. Genet.* **234:** 60–64.

Baum, E.Z. and W.M. Wormington. 1985. Coordinate expression of ribosomal protein genes during *Xenopus* development. *Dev. Biol.* **111:** 488–489.

Baum, E.Z., L.E. Hyman, and W.M. Wormington. 1988. Post-translational control of ribosomal protein L1 accumulation in *Xenopus* oocytes. *Dev. Biol.* **126:** 141–149.

Böhm, H., R. Benndorf, M. Gaestel, B. Gross, P. Nürnberg, R. Kraft, A. Otto, and H. Bielka. 1989. The growth-related protein P23 of the Ehrlich ascites tumor: Translational control, cloning and primary structure. *Biochem. Int.* **19:** 277–286.

Bowman, L.H. 1987. The synthesis of ribosomal proteins S16 and L32 is not autogenously regulated during myoblast differentiation. *Mol. Cell. Biol.* **7:** 4464–4471.

Bucher, P. 1990. Weight matrix descriptions of four eukaryotic RNA polymerase II promoter elements derived from 502 unrelated promoter sequences. *J. Mol. Biol* **212:** 563–578.

Cardinali, B., N. Campioni, and P. Pierandrei-Amaldi. 1987. Ribosomal protein, histone and calmodulin mRNAs are differently regulated at the translational level during oogenesis of *Xenopus laevis*. *Exp. Cell Res.* **169:** 432–441.

Cardinali, B., M. Di Cristiana, and P. Pierandrei-Amaldi. 1993. Interaction of proteins with the mRNA for ribosomal protein L1 in *Xenopus*: Structural characterization of *in vivo* complexes and identifiction of proteins that bind *in vitro* to its 5′UTR. *Nucleic Acids Res.* **21:** 2301–2308.

Chung, J., C.J. Kuo, G.R. Crabtree, and J. Blenis. 1992. Rapamycin-FKBP specifically blocks growth-dependent activation of and signaling by the 70 kd S6 kinases. *Cell* **69:** 1227–1236.

Craig, N. and R.P. Perry. 1971. Persistent cytoplasmatic synthesis of ribosomal proteins during the selective inhibition of ribosome RNA synthesis. *Nat. New Biol.* **229:** 75–80.

Dabeva, M.D. and J.R. Warner. 1993. Ribosomal protein L32 of *Saccharomyces cerevisiae* regulates both splicing and translation of its own transcript. *J. Biol. Chem.* **268:** 19669-19674.

Davis, S.C., A. Tzagoloff, and S.R. Ellis. 1992. Characterization of a yeast mitochondrial protein structurally related to the mammalian 68-kDa high affinity laminin receptor. *J. Biol. Chem.* **267:** 5508–5514.

DePhilip, R.M., W.A. Rudert, and I. Lieberman. 1980. Preferential stimulation of ribosomal protein synthesis by insulin and in the absence of ribosomal and messenger ribonucleic acid formation. *Biochemistry* **19:** 1662–1669.

Djé, M.K., A. Mazabraud, A. Viel, M.I. Maire, H. Denis, E. Crawford, and D.D. Brown. 1990. Three genes under different developmental control encode elongation factor 1-α in *Xenopus laevis*. *Nucleic Acids Res.* **18:** 3489–3493.

Finley, D. and V. Chau. 1991. Ubiquitination. *Annu. Rev. Cell Biol.* **7:** 25–69.

Garcia-Blanco, M.A., S. Jamison, and P.A. Sharp. 1989. Identification and purification of a 62,000-dalton protein that binds specifically to the polypyrimidine tract of introns. *Genes Dev.* **3:** 1874–1886.

Geyer, P.K., O. Meyuhas, R.P. Perry, and L.F. Johnson. 1982. Regulation of ribosomal protein mRNA content and translation in growth-stimulated mouse fibroblasts. *Mol. Cell. Biol.* **2:** 685–693.

Hammond, M.L. and L.H. Bowman. 1988. Insulin stimulates the translation of ribosomal proteins and the transcription of rDNA in mouse myoblasts. *J. Biol. Chem.* **263:** 17785–17791.

Hammond, M.L., W. Merrick, and L.H. Bowman. 1991. Sequences mediating the translation of mouse S16 ribosomal protein mRNA during myoblast differentiation and *in vitro* and possible control points for the *in vitro* translation. *Genes Dev.* **5:** 1723–1736.

Hellen, C.U.T., G.W. Witherell, M. Schmid, S.H. Shin, T.V. Pestova, A. Gil, and E. Wimmer. 1993. A cytoplasmic 57-kDa protein that is required for translation of picornavirus RNA by internal ribosmal entry is identical to the nuclear pyrimidine tract-binding protein. *Proc. Natl. Acad. Sci.* **90:** 7642–7646.

Hoja, U., J. Hofmann, R. Marschalek, and T. Dingermann. 1993. Nucleotide sequence of a *Dictyostelium discoideum* gene encoding a protein homologous to the yeast ribosomal protein S31. *Biochem. Biophys. Res. Commun.* **190:** 134–139.

Huang, S. and J.W.B. Hershey. 1989. Translational initiation factor expression and ribosomal protein gene expression are repressed coordinately but by different mechanisms in murine lymphosarcoma cells treated with glucocorticoids. *Mol. Cell. Biol.* **9:** 3679–3684.

Hyman, L.E. and W.M. Wormington. 1988. Translational inactivation of ribosomal protein mRNAs during *Xenopus* oocyte maturation. *Genes Dev.* **2:** 598–605.

Ignotz, G.G., S. Hokari, R.M. DePhilip, K. Tsukada, and I. Lieberman. 1981. Lodish model and regulation of ribosomal protein synthesis by insulin-deficient chick embryo fibroblasts. *Biochemistry* **20:** 2550–2558.

Jefferies, H.B., G. Thomas, and G. Thomas. 1994a. Elongation factor-1α mRNA is selectively translated following mitogenic stimulation. *J. Biol. Chem.* **269:** 4367–4372.

Jefferies, H.B., C. Reinhard, S.C. Kozma, and G. Thomas. 1994b. Rapamycin selectively represses translation of the "polypyrimidine tract" mRNA family. *Proc. Natl. Acad. Sci.* **91:** 4441–4445.

Johnson, L.F., H.T. Abelson, H. Green, and S. Penman. 1974. Changes in RNA in relation to growth of the fibroblast. I. Amount of mRNA and tRNA in resting and growing cells. *Cell* **1:** 95–100.

Kafatos, F.C. 1972. The cocoonase zymogen cells of silk moths: A model of terminal cell differentiation for specific protein synthesis. *Curr. Top. Dev. Biol.* **7:** 125–191.

Karin, M. 1994. Signal transduction from the cell surface to the nucleus through the phosphorylation of transcription factors. *Curr. Opin. Cell Biol.* **6:** 415–424.

Kaspar, R.L., T. Kakegawa, H. Cranston, D.R. Morris, and M.W. White. 1992. A regulatory *cis* element and a specific binding factor involved in the mitogenic control of murine ribosomal protein L32 translation. *J. Biol. Chem.* **267:** 508–514.

Kaspar, R.L., W. Rychlik, M.W. White, R.E. Rhoads, and D.R. Morris. 1990. Simultaneous cytoplasmic redistribution of ribosomal protein L32 mRNA and phosphorylation of eukaryotic initiation factor 4E after mitogenic stimulation of Swiss 3T3 cells. *J. Biol. Chem.* **265:** 3619–3622.

Kay, M.A. and M. Jacobs-Lorena. 1985. Selective translational regulation of ribosomal protein gene expression during early development of *Drosophila melanogaster*. *Mol. Cell. Biol.* **5:** 3583–3592.

Krauter, K.S., R. Soeiro, and B. Nadal-Ginard. 1980. Uncoordinate regulation of ribosomal RNA and ribosomal protein synthesis during L_6E_9 myoblast differentiation. *J. Mol. Biol.* **142:** 145–159.

Kuzumaki, T., T. Tanaka, K. Ishikawa, and K. Ogata. 1987. Rat ribosomal protein L35a multigene family: Molecular structure and characterization of three L35a-related pseudogenes. *Biochim. Biophys. Acta* **909:** 99–106.

Lane, H.A., S.J. Morley, M. Dorée, S.C. Kozma, and G. Thomas. 1992. Identification and early activation of *Xenopus laevis* $p70^{s6k}$ following progesterone-induced meiotic maturation. *EMBO J.* **11:** 1743–1749.

Lee, G.T.-Y. and D.L. Engelhardt. 1978. Growth-related fluctuation in messenger RNA utilization in animal cells. *J. Cell Biol.* **79:** 85–96.

Lee, M.G.-S., S.A. Lewis, C.D. Wilde, and N.J. Cowan. 1983. Evolutionary history of a multigene family: An expressed human β-tubulin gene and three processed pseudogenes. *Cell* **33:** 477–487.

Levy, S., D. Avni, N. Hariharan, R.P. Perry, and O. Meyuhas. 1991. Oligopyrimidine tract at the 5′ end of mammalian ribosomal protein mRNAs is required for their translational control. *Proc. Natl. Acad. Sci.* **88:** 3319–3323.

Lodish, H.F. 1974. Model for the regulation of mRNA translation applied to haemoglobin synthesis. *Nature* **251:** 385–388.

Loreni, F. and F. Amaldi. 1992. Translational regulation of ribosomal protein synthesis in *Xenopus* cultured cells: mRNA relocation between polysomes and RNP during nutritional shifts. *Eur. J. Biochem.* **205:** 1027–1032.

Loreni, F., A. Francesconi, and F. Amaldi. 1993. Coordinate translational regulation in the synthesis of elongation factor 1α and ribosomal proteins in *Xenopus laevis*. *Nucleic Acids Res.* **21:** 4721–4725.

Maki, C., D.D. Rhoads, M.J. Stewart, B.V. Slyke, and D.J. Roufa. 1989. The *Drosophila melanogaster RPS17* gene encoding ribosomal protein S17. *Gene* **79:** 289–298.

Mariottini, P. and F. Amaldi. 1990. The 5′ untranslated region of mRNA for ribosomal protein S19 is involved in its translational regulation during *Xenopus* development. *Mol. Cell. Biol.* **10:** 816–822.

Meyuhas, O. and R.P. Perry. 1980. Construction and identification of cDNA clones for mouse ribosomal proteins: Application for the study of r-protein gene expression. *Gene* **10:** 113–127.

Meyuhas, O., A.E. Thompson, and R.P. Perry. 1987. Glucocorticoids selectively inhibit the translation of ribosomal protein mRNAs in P1798 lymphosarcoma cells. *Mol. Cell. Biol.* **7:** 2691–2699.

Meyuhas, O., V. Baldin, G. Bouche, and F. Amalric. 1990. Glucocorticoids repress ribosome biosynthesis in lymphosarcoma cells by affecting gene expression at the level of transcription, posttranscription and translation. *Biochim. Biophys. Acta* **1049:** 38–44.

Morley, S.J. and J.A. Traugh. 1993. Stimulation of translation in 3T3-L1 cells in response to insulin and phorbol ester is directly correlated with increased phosphate labelling of initiation factor (eIF-) 4F and ribosomal protein S6. *Biochimie* **75:** 985–989.

Morris, D.R., T. Kakegawa, R.L. Kaspar, and M.W. White. 1993. Polypyrimidine tracts and their binding proteins: Regulatory sites for posttranscriptional modulation of gene expression. *Biochemistry* **32:** 2931–2937.

Nomura, M., R. Gourse, and G. Baughman. 1984. Regulation of the synthesis of ribosome and ribosome components. *Annu. Rev. Biochem.* **53:** 75–117.

Ostareck-Lederer, A., D.H. Ostareck, N. Standart, and B.J. Thiele. 1994. Translation of 15-lipoxygenase mRNA is inhibited by a protein that binds to a repeated sequence in the 3′ untranslated region. *EMBO J.* **13:** 1476–1481.

Patel, R.C. and M. Jacobs-Lorena. 1992. *cis*-Acting sequences in the 5′ untranslated region of the ribosomal protein A1 mRNA mediate its translational regulation during early embryogenesis of *Drosophila*. *J. Biol. Chem.* **267:** 1159–1164.

Perry, R.P. and O. Meyuhas. 1990. Translational control of ribosomal protein production in mammalian cells. *Enzyme* **44:** 83–92.

Pierandrei-Amaldi, P., N. Campioni, and B. Cardinali. 1991. Experimental changes in the amount of maternally stored ribosomes affect the translation efficiency of ribosomal protein mRNA in *Xenopus* embryo. *Cell. Mol. Biol.* **37:** 227–238.

Pierandrei-Amaldi, P., E. Beccari, I. Bozzoni, and F. Amaldi. 1985a. Ribosomal protein production in normal and anucleate *Xenopus* embryos: Regulation at the posttranscriptional and translational level. *Cell* **42:** 317–323.

Pierandrei-Amaldi, P., N. Campioni, E. Beccari, I. Bozzoni, and F. Amaldi. 1982. Expression of ribosomal protein genes in *Xenopus laevis* development. *Cell* **30:** 163–171.

Pierandrei-Amaldi, P., N. Campioni, P. Gallinari, E. Beccari, I. Bozzoni, and F. Amaldi. 1985b. Ribosomal protein synthesis is not autogenously regulated at the translational level in *Xenopus laevis*. *Dev. Biol.* **126:** 141–146.

Price, D.J., J.R. Grove, V. Calvo, J. Avruch, and B.E. Bierer. 1992. Rapamycin-induced inhibition of the 70-kilodalton protein kinase. *Science* **257:** 973–977.

Rao, C.N., V. Castronovo, M.C. Schmitt, U.M. Wewer, A.P. Claysmith, L.A. Liotta, and M.E. Sobel. 1989. Evidence for a precursor of the high-affinity metastasis-associated murine laminin receptor. *Biochemistry* **28:** 7476–7486.

Rao, T.R. and L.I. Slobin. 1987. Regulation of the utilization of mRNA for eucaryotic elongation factor Tu in Friend erythroleukemia cells. *Mol. Cell. Biol.* **7:** 687–697.

Rudland, P.S., S. Weil, and A.R. Hunter. 1975. Changes in RNA metabolism and accumulation of presumptive messenger RNA during transition from growing to quiescent state of cultured mouse fibroblasts. *J. Mol. Biol.* **96:** 745–766.

Schibler, U., D.E. Kelley, and R.P. Perry. 1977. Comparison of methylated sequences in messenger RNA and heterogenous nuclear RNA from mouse L cells. *J. Mol. Biol.* **115:** 695–714.

Schmidt, T., P.S. Chen, and M. Pellegrini. 1985. The induction of ribosome biosynthesis in nonmitotic secretory tissue. *J. Biol. Chem.* **260:** 7645–7650.

Shama, S., D. Avni, R.M. Frederickson, N. Sonenberg, and O. Meyuhas. 1995. Overexpression of initiation factor eIF-4E does not relieve the translational repression of ribosomal protein mRNAs in quiescent cells. *Gene Expr.* **4:** 241–252

Singelton, C.K., S.S. Manning, and R. Ken. 1989. Primary structure and regulation of vegetative specific genes of *Dictyostelium discoideum*. *Nucleic Acids Res.* **17:** 9679–9692.

Steel, L.F. and A. Jacobson. 1987. Translational control of ribosomal protein synthesis during early *Dictyostelium discoideum* development. *Mol. Cell. Biol.* **7:** 965–972.

———. 1991. Sequence elements that affect mRNA translational activity in developing *Dictyostelium* cells. *Dev. Genet.* **12:** 98–103.

Steel, L.F., A. Smyth, and A. Jacobson. 1987. Nucleotide sequence and characterization of the transcript of a *Dictyostelium* ribosomal protein gene. *Nucleic Acids Res.* **15:** 10285–10298.

Terada, N., H.R. Patel, K. Takase, K. Kohno, A.C. Nairn, and E.W. Gelfand. 1994. Rapamycin selectively inhibits translation of mRNAs encoding elongation factors and ribosomal proteins. *Proc. Natl. Acad. Sci.* **91:** 11477–11481.

Thomas, G. and G. Thomas. 1986. Translational control of mRNA expression during the early mitogenic response in Swiss mouse 3T3 cells: Identification of specific proteins. *J. Cell Biol.* **103:** 2137–2144.

Thomas, G., G. Thomas, and H. Luther. 1981. Transcriptional and translational control of cytoplasmic proteins after serum stimulation of quiescent Swiss 3T3 cells. *Proc. Natl. Acad. Sci.* **78:** 5712–5716.

Thomas, G., J. Martin-Pérez, M. Siegmann, and A.M. Otto. 1982. The effect of serum, EGF, $PGF_{2\alpha}$ and insulin on S6 phosphorylation and the initiation of proten and DNA synthesis. *Cell* **30:** 235–242.

Tsay, Y.-F., J.R. Thompson, M.O. Rotenberg, J.C. Larkin, and J.L.J. Woolford. 1988. Ribosomal protein synthesis is not regulated at the translational level in *Saccharomyces cerevisiae*: Balanced accumulation of ribosomal proteins L16 and rp59 is mediated by turnover of excess protein. *Genes Dev.* **2:** 664–676.

Warner, J.R. 1977. In the absence of ribosomal RNA synthesis, the ribosomal protein of HeLa cells are synthesized normally and degraded rapidly. *J. Mol. Biol.* **115:** 315–333.

Wiborg, O., M.S. Pederson, A. Wind, L.E. Berglund, K.A. Marcker, and J. Vuust. 1985. The human ubiquitin multigene family: Some genes contain multiple directly repeated ubiquitin coding sequences. *EMBO J.* **4:** 755–759.

Wormington, W.M. 1989. Developmental expression and 5S rRNA-binding activity of *Xenopus laevis* ribosomal protein L5. *Mol. Cell. Biol.* **9:** 5281–5288.

Yenofsky, R., I. Bergman, and G. Brawerman. 1982. Messenger RNA species partially in a repressed state in mouse sarcoma ascites cells. *Proc. Natl. Acad. Sci.* **79:** 5876–5880.

Yenofsky, R., S. Careghini, A. Krowczynska, and G. Brawerman. 1983. Regulation of mRNA utilization in mouse erythroleukemia cells induced to differentiate by exposure to dimethyl sulfoxide. *Mol. Cell. Biol.* **3:** 1197–1203.

14

Ribosomal Protein S6 Phosphorylation and Signal Transduction

Harold B.J. Jefferies and George Thomas
Friedrich Miescher Institut
CH-4002 Basel, Switzerland

Growth factors, acting through single transmembrane tyrosine kinase receptors or G-protein-coupled serpentine receptors, induce cells in the animal and in culture to exit the G_0 state of the cell cycle, progress through G_1, synthesize DNA, and divide (Bourne et al. 1990; Cantley et al. 1991). This process is accomplished through the coordinate activation of a number of metabolic events (Pardee 1989). It has become evident during the last 5–10 years that the major intracellular mediator of these events is protein phosphorylation/dephosphorylation (Krebs 1994). The realization of the importance of this regulatory mechanism to cell growth has focused a great deal of attention on the identification of substrates and the signaling pathways that control cell cycle progression (Egan and Weinberg 1993; Downward 1994). One of the early obligatory metabolic events involved in the induction of cell growth is the activation of protein synthesis (Pardee 1989), a process that proceeds through a complex set of steps involving a large number of translational components (see Morley and Thomas 1991), including initiation factors, $tRNA^{Met}$, mRNA, and the 40S and 60S ribosomal subunits. The rate-limiting event in increased protein synthesis is thought to be the recognition and binding of mRNA by the 43S ribosomal preinitiation complex, a step that also requires a family of initiation factors termed 4A, 4B, and 4F (Sonenberg 1994; see Mathews et al.; Sonenberg; both this volume). In turn, this process is thought to be regulated by the concerted phosphorylation and dephosphorylation of key translational components (see Morley and Thomas 1991). One of these components is S6, a 40S ribosomal protein that is present in one copy per subunit and becomes multiply phosphorylated in response to growth factor stimulation. From a number of recent studies that use specific growth factors (Ballou et al. 1991), or agents that either block or induce the activation of specific kinases (Blenis et al.

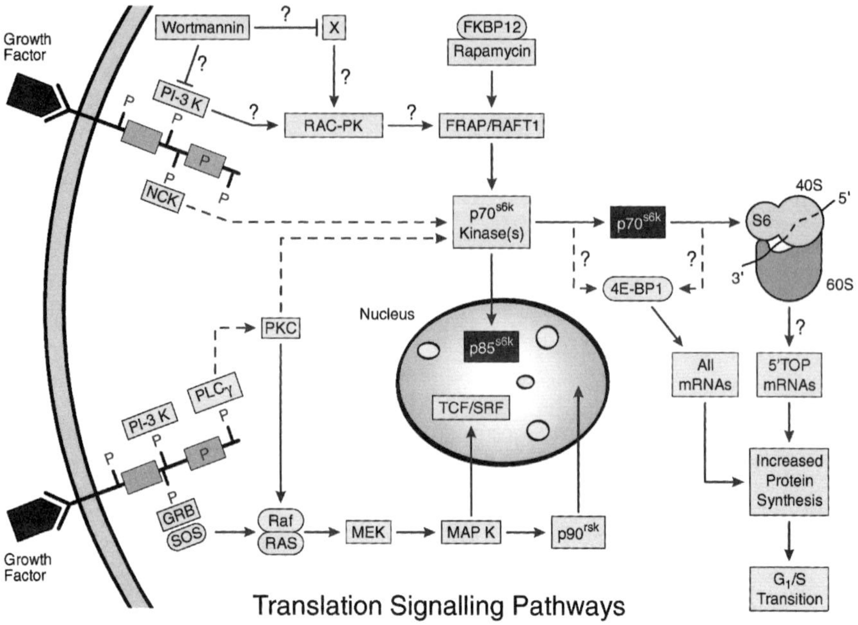

Figure 1 Schematic representation of cellular signaling pathways involved in controlling translation through ribosomal protein S6. (4E) Eukaryotic initiation factor-4E; (4E-BP1) eIF4E-binding protein 1; (5′TOP) 5′-terminal oligopyrimidine tract; (FKBP12) FK506-binding protein 12; (FRAP) FKBP-rapamycin associated protein; (GRB) growth factor receptor binding protein 2; (MAPK) mitogen activated protein kinase (p42/p44); (MEK) MAPK/ERK kinase; (NCK) adaptor molecule/growth factor receptor binding protein; (P) tyrosine autophosphorylation sites serving as specific signaling molecule docking sites; (PI-3K) phosphatidylinositol 3-OH kinase; (PKC) protein kinase C; (PLCγ) phospholipase Cγ; (SOS) mammalian homolog of *Drosophila son of sevenless*; (RAC-PK) related to A and C protein kinases; (Raf) cellular p74 serine/threonine kinase; (Ras) p21 GTPase.

1991; Kuo et al. 1992; Price et al. 1992), it has been argued that the enzyme responsible for controlling this process is $p70^{s6k}/p85^{s6k}$, two isoforms of the same kinase. This chapter focuses on our current knowledge of the role of S6 phosphorylation in protein synthesis, the mechanism by which $p70^{s6k}/p85^{s6k}$ kinase activation is brought about, and the signaling pathway that controls this response (see Fig. 1).

S6 PROTEIN

S6 is among 30 distinct proteins which with one molecule of 18S rRNA comprise a mature 40S ribosomal subunit (see Wool et al., this volume).

Phosphorylation of the 40S ribosomal subunit was first observed by Loeb and Blat (1970) in the livers of rats injected with $^{32}P_i$. By taking advantage of two-dimensional polyacrylamide gel electrophoresis (PAGE), Gressner and Wool (1974) demonstrated that partial hepatectomy (known to activate protein synthesis) induced the multiple phosphorylation of S6 in parallel. Indeed, the only protein in the 40S ribosomal subunit shown to undergo phosphorylation in vivo in response to growth factors and mitogens is S6. Phosphorylation of S6 occurs only on serine residues, with all of the sites residing in a 32-amino-acid carboxy-terminal cyanogen bromide fragment (Krieg et al. 1988; Bandi et al. 1993). Of the seven serine residues present in the fragment, the five phosphorylation sites were shown by direct Edman degradation to be Ser-235, Ser-236, Ser-240, Ser-244, and Ser-247 (Krieg et al. 1988; Bandi et al. 1993). Moreover, phosphate release studies have shown that phosphorylation proceeds in an ordered manner in vivo and in vitro: Ser-236→Ser-235 or Ser-240→Ser-244→Ser-247 (H.R. Bandi et al., unpubl.). The reason for ordered phosphorylation and the preference for one serine rather than another is not immediately apparent from examining the primary sequence (Bandi et al. 1993). Synthetic peptide studies have shown that the presence of a block of three arginines at positions 231, 232, and 233 of S6 is critical for the first phosphorylation event (Flotow and Thomas 1992). A model of the tertiary structure of the carboxyl terminus of S6, beginning from Lys-230, generated an α-helix in which the hydroxyl groups of the five phosphorylated serines face outward, in the same direction, whereas Ser-242 and Ser-246 lie on the opposite side of the helix (Ferrari et al. 1991). Thus, it might be envisaged that the kinase begins by phosphorylating first Ser-236 and then Ser-235, moving sequentially along the helix phosphorylating Ser-240, Ser-244, and finally Ser-247. Furthermore, from the studies described below, it may be speculated that the latter sites of phosphorylation (Ser-244 and Ser-247) contribute most to the activation of protein synthesis. Interestingly, *Saccharomyces cerevisiae* lack these last three sites of phosphorylation. Mutation to alanines of the serines in *S. cerevisiae* equivalent to Ser-236 and Ser-235 in mammalian S6 has little effect on protein synthesis or cell growth (Johnson and Warner 1987).

Franco and Rosenfeld (1990) have demonstrated that a nuclear pool of ribosomal protein S6 exists that is also phosphorylated in response to hormone stimulation. Ribosome biogenesis takes place in the nucleolus (Sommerville 1986; Nigg 1988). In a complex series of steps that are not well understood, a nascent 80S preribosomal particle, containing a single strand of 45S rRNA, is processed to 40S and 60S preribosomal particles containing either a single strand of 18S rRNA or a strand of both 28S rRNA and 5.8S rRNAs, respectively (Larson et al. 1991). The preribosomal particles are then exported to the cytoplasm, where the final stages of the assembly process are completed and mature ribosomal subunits are formed. S6 is one of the first proteins to be assembled into the 45S rRNA precursor (Todorov et al. 1983). Interestingly, the nuclear pool of phosphorylated S6 described by Franco and Rosenfeld (1990) can be separated into nucleoplasmic and nucleolar fractions. Immunofluorescence studies show that $p85^{s6k}$ is apparently excluded from the nucleolus (see $p70^{s6k}/p85^{s6k}$ below), suggesting that S6 is phosphorylated as a free nucleoplasmic protein before it is assembled into 80S preribosomal particles or that S6 may have a separate function in the nucleus (see Stewart and Thomas 1994). Both possibilities could be envisaged in models in which nuclear S6 phosphorylation is involved in regulating one or more specific steps of ribosome biogenesis. This concept is consistent with the findings that stimulation of ribosome biogenesis is concomitant with increased S6 phosphorylation (Thomas et al. 1979; Reinhard et al. 1994).

S6 has been localized within the mature 40S ribosomal subunit by chemical and UV cross-linking studies to a protrusion of the small head region of the 40S ribosomal subunit termed the beak or bill (Nygärd and Nilsson 1990). That S6 is also one of the few proteins of the 40S subunit to contact 28S rRNA directly indicates that it lies at the interface between the 40S and 60S subunits (Nygärd and Nika 1982). More importantly, S6 has been directly cross-linked to tRNA, initiation factors, and mRNA, suggesting that it resides at or very close to the mRNA-binding site (see Nygärd and Nilsson 1990; Stewart and Thomas 1994). The possibility that it may influence the binding of mRNA to the 40S subunit is supported by two additional findings. First, Terao and Ogata (1979a,b) have shown that S6 can be cross-linked to poly(U), a synthetic mRNA transcript, and can be protected from chemical modification in the presence of poly(U). Second, Fab fragments derived from antibodies generated against a carboxy-terminal peptide of S6 containing the phosphorylation sites block mRNA binding to 40S subunits (R.E.H. Wettenhall, pers. comm.).

In intact cells, phosphorylation of S6 is one of the earliest events detected after mitogenic stimulation of quiescent cells to reenter the cell cycle and is closely associated with increased rates of protein synthesis (see Stewart and Thomas 1994). Conversely, decreases in protein synthesis are paralleled by a decrease in S6 phosphorylation as is the case with serum deprivation (Thomas et al. 1979). Up-regulation of protein synthesis by mitogens is primarily controlled at the level of initiation and can be observed as the recruitment of inactive 80S ribosomes into actively translating polysomes (Thomas et al. 1979). Results from Duncan and McConkey (1982) and Thomas et al. (1982) showed that polysomes have a higher percentage of phosphorylated S6 than 80S ribosomes or free 40S subunits following mitogenic stimulation. It may be postulated that S6 acts to increase the rate of protein synthesis by increasing the affinity of 40S subunits for stored messenger ribonucleoprotein (mRNP) particles in a primary initiation event. Consistent with this hypothesis was the demonstration that following mitogenic stimulation, the translation of a number of mRNAs is up-regulated in parallel with increased S6 phosphorylation (Thomas et al. 1981; Thomas and Thomas 1986). More recently, a subfamily of these transcripts, one of which encodes elongation factor 1A (eEF1A), has been shown to be under selective translational control (Jefferies et al. 1994a). In quiescent cells, eEF1A transcripts are localized in mRNP particles and on monosomes and disomes (Jefferies et al. 1994a). Following mitogenic stimulation, both populations are recruited to very large polysomes. This observation is apparently at odds with the classic model of translational control based on mRNA competition (Lodish 1974; Godefroy-Colburn and Thach 1981). The model predicts that the affinity of the translational apparatus for specific mRNAs is constant. In this model, individual transcripts are present in either polysomes or stored mRNP particles as a function of the affinity of the translational apparatus for each mRNA (see Hershey 1991). Following mitogenic stimulation, the amount of polysomes increases, but the mean size remains constant, with most transcripts associated with the same size polysome. Thus, under these conditions, following mitogenic stimulation, eEF1A in mRNP particles would have been expected to shift to monosomes/disomes, and eEF1A mRNA already present in monosomes/disomes should have remained in this state. The data demonstrate a selective translational up-regulation of eEF1A transcripts. Notably, the transcript for eEF1A contains a 5′-terminal oligopyrimidine tract (5′TOP) at its transcriptional start site (Uetsuki et al. 1989), an element known to act as a translational regulator (Hammond et al. 1991; Levy et al. 1991). All mammalian ribosomal protein mRNAs in which the 5′ transcription start site has been mapped, as well as elongation factor 2 (eEF2) mRNA, contain this motif (see Meyuhas et al., this volume) and are apparently up-regulated in a manner similar to that of eEF1A (Jefferies et al. 1994b).

An early report of a direct link between the initiation of translation and S6 phosphorylation was provided by Palen and Traugh (1987) who employed an in vitro reconstituted protein synthesizing system to show functional changes in binding and translational activity of 40S ribosomal subunits containing phosphorylated S6. Using subunits phosphorylated by a protease activated kinase, termed PAK II, these authors observed up to a fourfold increase in translation of α- and β-globin mRNAs over that of nonphosphorylated 40S subunits. Surprisingly, the synthesis of β-globin was increased approximately 3.9-fold, whereas α-globin synthesis increased only 1.8-fold, suggesting that S6 phosphorylation could also selectively alter the ability of the 40S ribosomal subunit to interact with specific mRNAs. A criticism that might be leveled at this study is that the reconstituted system used was very low in total activity as compared to the unfractionated reticulocyte lysate. Indeed, what is lacking to date is corroborative experiments supporting these findings. With the recent production of active recombinant $p70^{s6k}$ (Kozma et al. 1993) and the development of reconstituted translation systems that exhibit high activity (Morley and Hershey 1990), it will be interesting to reexamine the direct role of S6 phosphorylation on general translation, as well as the selective up-regulation of specific transcripts.

One approach to examine the role of S6 phosphorylation would be to block the kinase responsible for controlling its level of phosphorylation. It has been shown recently that the immunosupressant rapamycin selectively inhibits the mitogen-induced phosphorylation and activation of $p70^{s6k}/p85^{s6k}$ as well as causing the rapid dephosphorylation and inactivation of the kinase within 10 minutes in mitogen-stimulated cells (see Fig. 1) (Kuo et al. 1992; Price et al. 1992). Rapamycin is a macrolide antibiotic produced by the soil bacterium *Streptomyces hygroscopicus*, which in association with an intracellular FK506- and rapamycin-binding protein (FKBP), either blocks or reduces the rate at which cells enter S phase and subsequently proliferate, depending on the cell type (Sehgal et al. 1994). The inhibition of $p70^{s6k}/p85^{s6k}$ by the FKBP12-rapamycin complex is not exerted directly on the kinase itself but on a component of its signaling pathway termed FRAP in humans (Brown et al. 1994) or RAFT1 in rats (Sabatini et al. 1994) which are the mammalian homologs of the products of two yeast genes *TOR1/DRR1* and *TOR2* (Kunz et al. 1993). FRAP and RAFT1 are 289-kD proteins that have homology with phosphatidylinositol (PtdIns) kinases. Recently, Brown et al. (1995) using FRAP variants that do not bind rapamycin-FKBP12 demonstrated that FRAP is a rapamycin-sensitive upstream regulator of $p70^{s6k}$. In parallel studies, Sabatini et al. (1995) have reported that immunoprecipitates of endogenous RAFT1 from rat brain or tissue culture cells contain PI4-kinase activity, although the rapamycin-FKBP12 complex has no effect on this activity. Indeed, the view that FRAP/RAFT1 is a PI4-kinase has been challenged (Brown et al. 1995; Hunter 1995). The effect of rapamycin on $p70^{s6k}/p85^{s6k}$ appears to be selective, as the activation of other kinases triggered in the early mitogenic response such as $p74^{raf}$, $p42^{mapk}$, or $p90^{rsk}$ is unaffected by the macrolide (Kuo et al. 1992; Price et al. 1992). As a consequence of rapamycin treatment, S6 phosphorylation is blocked, suggesting that this is a result of $p70^{s6k}/p85^{s6k}$ inactivation and that this kinase is the physiologically relevant S6 kinase. Since the major downstream target of rapamycin inhibition to date is $p70^{s6k}/p85^{s6k}$, it can be reasoned that the inhibitory effects on cell growth are through the inhibition of S6 phosphorylation and thus protein synthesis. Serum stimulation of quiescent Swiss-3T3 cells pretreated with rapamycin causes an approximately 10–15% inhibition of global protein synthesis (Jefferies et al. 1994b). This effect did not seem large enough to explain the greater effects observed on cell growth (Chung et al. 1992). Indeed, in contrast to the effect on global protein synthesis, the recruitment of eEF1A mRNA into polysomes following mitogenic stimulation in the presence of the macrolide was severely repressed (Jefferies et al. 1994b). In addition to suppressing the up-regulation of eEF1A transcripts, the macrolide also suppressed the translational up-regulation of three ribosomal protein mRNAs examined, as well as eEF2 mRNA (Jefferies et al. 1994b). No effect was observed on the distribution of transcripts that lacked the 5′TOP such as β-actin, eIF4A, and β-tubulin (Jefferies et al. 1994b). When serum-stimulated cells are subsequently treated with rapamycin for 1 hour, the approximate time taken for S6 to reach its basal phosphorylation state, 5′TOP mRNAs largely dissociate from polysomes and redistribute to monosomes/disomes and mRNP particles. These results argue that the translational repression exerted by rapamycin is selective for those mRNAs containing 5′TOPs. Indeed, recent studies using chimeric mRNAs have shown that an intact polypyrimidine tract is necessary for rapamycin to exert its inhibitory effect (C. Reinhard et al., unpubl.). Whether it is sufficient awaits further investigation. It should also be noted that the extent to which rapamycin treatment suppresses the translation of different 5′TOP mRNAs in 3T3 cells is not equal, with some more negatively affected than others. This may reflect a higher order of regulation based on distinct 5′TOP sequences (which are known to vary in length and composition from 5 to 14 pyrimidines; see Meyuhas et al., this volume) on specific proteins that bind to the 5′TOP (Kaspar et al. 1992; Cardinali et al. 1994) or on the potential influence of sequences downstream from the 5′TOP. The inhibitory effect of rapamycin on the translation of 5′TOP mRNAs in both pre- and post-serum-stimulated cells is not complete, consistent with its partial suppression of Swiss-3T3 cell growth (Chung et al. 1992), indicating the existence of at least one other mechanism in the translational up-regulation of 5′TOP mRNAs.

Accumulation of the data to date indicate that rapamycin exerts its effect on translation by inhibiting $p70^{s6k}/p85^{s6k}$, causing the dephosphorylation of S6. This implicates S6 phosphorylation in a selective translational control mechanism that up-regulates the translation of the 5′TOP mRNA family after mitogenic stimulation (see Fig. 1). Support for this hypothesis comes from two early observations. Terao and Ogata (1979a,b) demonstrated that S6 could bind to poly(U), and Gressner and van der Leur (1980) showed that the rate at which poly(U) dissociated from phosphorylated 40S ribosomes was threefold lower than that from nonphosphorylated subunits. If phosphorylated S6 is functioning in this manner, it must be established next whether phosphorylated S6 (1) directly recognizes the 5′TOP, (2) alters by conformational change the mRNA-binding site on the 40S subunit, or (3) interacts directly with proteins that bind to this motif (Kaspar et al. 1992; Cardinali et al. 1994).

$p70^{s6k}/p85^{s6k}$

At the time $p70^{s6k}$ was identified, a number of other kinases had been implicated in S6 phosphorylation (see Kozma et al. 1989). However, none of these kinases could fully phosphorylate S6 in vitro, and for most, the mechanism of activation was not consistent with a signaling pathway controlled by mitogens. The identification and purification of the $p70^{s6k}$ stemmed from earlier observations of Novak-Hofer and Thomas (1984), who employing phosphatase inhibitors during the preparation of cell extracts observed a 30-fold enhancement in the ability of serum-stimulated cell extracts to phosphorylate S6 in vitro versus extracts derived from quiescent cells. These findings provided the first evidence of a mitogen-activated serine/threonine kinase cascade involved in intracellular signaling and led to the eventual purification of the kinase (Jeno et al. 1988), the demonstration that its activity was directly controlled by phosphorylation (Ballou et al. 1988a), and the cloning of its cDNA (Banerjee et al. 1990; Harmann and Kilimann 1990; Kozma et al. 1990). Three laboratories identified cDNA clones encoding the $p70^{s6k}$. The first clone sequenced, Clone 1, encoded a protein of 502 amino acids (Kozma et al. 1990; Reinhard et al. 1992), whereas the second, termed Clone 2 (Reinhard et al. 1992), encoded a protein of 525 amino acids (Banerjee et al. 1990; Harmann and Kilimann 1990). The coding sequences of the two clones were identical, except that the product of Clone 2 contained a 23-

amino-acid extension at its amino terminus. Subsequently, it was demonstrated that both clones were derived from a common gene and that the 5′-untranslated region (5′UTR) and the region coding for the 23-amino-acid extension of Clone 2 were probably contiguous with the common coding sequence (Reinhard et al. 1992), a fact that has been confirmed in mice through the isolation of the first coding exon (Y. Chen et al., unpubl.).

Expression of Clone 1 and Clone 2 in vitro revealed that both encoded $p70^{s6k}$ but that Clone 2 also gave rise to a second product that migrated with an anomalously high molecular weight of 85,000 (Reinhard et al. 1992). The generation of two products by Clone 2 was consistent with two translational start sites, with the first giving rise to the M_r 85,000 form of the protein and the second to $p70^{s6k}$. This suspicion was confirmed by employing specific antibodies. In the early purification studies, Price et al. (1989) reported that a protein of M_r 85,000 copurified with $p70^{s6k}$, but they had initially discounted this protein as being related to the kinase as it did not react with the ATP analog, 8-azido ATP. However, it was soon confirmed that the M_r 85,000 kinase existed in vivo (Grove et al. 1991; Reinhard et al. 1992), that it was activated in parallel with $p70^{s6k}$ (Reinhard et al. 1992), and that it was activated by a common mechanism (Chung et al. 1992; Kozma et al. 1993). Thus, two isoforms of the kinase exist, the $p70^{s6k}$ and a second form termed the $p85^{s6k}$ (Reinhard et al. 1992).

Having established the existence of the $p85^{s6k}$, the enigma was: Why two forms of the kinase? The first clue came from the studies of Franco and Rosenfeld (1990) who, as mentioned earlier, had demonstrated the existence of a free form of S6 in the nucleoplasm, not associated with pre-40S ribosomal particles. This form of the protein also became rapidly phosphorylated in response to mitogenic stimulation. Closer examination of the 23-amino-acid extension of the $p85^{s6k}$ revealed that it contained all the hallmarks of a nuclear targeting sequence, suggesting that the $p85^{s6k}$ may be a resident nuclear protein. Use of affinity-purified antibodies specific for the 23-amino-acid extension confirmed this inference and also demonstrated that the kinase was excluded from the nucleolus (Reinhard et al. 1994). Furthermore, microinjection into cells of a mammalian expression vector encoding only the $p85^{s6k}$ isoform led to exclusive accumulation of the larger isoform in the nucleus. A similar approach was used to demonstrate that the putative nuclear targeting sequence of $p85^{s6k}$, when fused to the bacterial protein CAT, was sufficient to target the chimeric protein to the nucleus. Finally, inhibitory antibodies to the $p85^{s6k}$, when microinjected into the nucleus, but not when injected into the cytoplasm, serverly repressed G_1 progression (Reinhard et al. 1994).

In immunofluorescence studies employing an antibody that recognized both isoforms of the $p70^{s6k}/p85^{s6k}$, a strong punctate staining pattern was observed in the cytoplasm accompanied by much weaker nuclear staining (Reinhard et al. 1994). Since the $p70^{s6k}$ isoform is much more abundant than the $p85^{s6k}$ isoform in all cells examined to date (Ming et al. 1994; Reinhard et al. 1994), these results indicated that $p70^{s6k}$ is cytoplasmic. Microinjection of expression vectors encoding only $p70^{s6k}$ led to initial accumulation of the protein in the cytoplasm, but with time, overexpressed $p70^{s6k}$ could be detected in the nucleus (Reinhard et al. 1994). This last finding is not surprising, as overexpression of $p70^{s6k}$ (56,160 m.w.), which is not large enough to be excluded from the nucleus, may result in the localization of the protein to this compartment. That native $p70^{s6k}$ is not a nuclear protein was consistent with quantitative confocal microscopy studies. Furthermore, coinjection into nucleus of activated $p70^{s6k}$ together with $p85^{s6k}$ inhibitory antibodies rescues the G_1 block described above, whereas endogenous $p70^{s6k}$ fails to prevent the G_1 block (Reinhard et al. 1994). It has recently been claimed on the basis of cell fractionation studies that a portion of $p70^{s6k}$ is nuclear (de Groot et al. 1994). However, this approach is notorious for artifacts created by cross-contamination of intracellular compartments upon cell disruption and requires careful controls with marker proteins to guard against this problem (Gordon et al. 1981; Krek et al. 1992). Unfortunately, in this case, the control used was $p21^{ras}$, a membrane protein instead of, for example, a protein associated with the endoplasmic reticulum.

Excluding the nuclear targeting sequence, $p70^{s6k}/p85^{s6k}$ can be divided into three domains: (1) an amino-terminal stretch of 65 amino acids, rich in acidic residues, (2) the catalytic domain containing all of the 11 conserved motifs found in serine/threonine kinases, and (3) a carboxy-terminal domain that contains a putative autoinhibitory sequence (Banerjee et al. 1990), which has significant homology with the substrate S6. Support for the existence of an autoinhibitory region comes from two observations. First, synthetic peptides mimicking this region, amino acids 400–433, inhibit the activated $p70^{s6k}$ in the low micromolar range in vitro (Flotow and Thomas 1992; Mukhopadhayay et al. 1992), and second, four of the major sites of serum-induced phosphorylation associated with kinase activation (Ser-411, Ser-418, Thr-421, and Ser-424) are situated in this region (Ferrari et al. 1992). Indeed, more recent studies with synthetic peptides demonstrate that substitution of these four residues with acidic amino acids relieves peptide inhibition (H. Flotow, unpubl.). The results are consistent with a model in which the putative autoinhibitory region directly interacts with the catalytic domain, blocking its ability to bind the substrate. Phosphorylation of the four critical residues lying in the autoinhibitory domain would be hypothesized to participate in disrupting its interaction with the catalytic domain, allowing access to the substrate.

In support of the autoinhibitory model described above, transient expression of a mutant form of the kinase, in which the four sites of phosphorylation in the autoinhibitory domain had been substituted with acidic residues, led to expression of a kinase that was as active as the wild-type enzyme (Ferrari et al. 1993). Surprisingly, treatment of cells with rapamycin transiently expressing the mutant kinase led to the immediate dephosphorylation and inactivation of the enzyme. These observations eventually led to the finding that rapamycin causes the selective dephosphorylation of a novel set of phosphorylation sites distinct from the initial four identified following mitogenic stimulation. It is not known whether rapamycin exerts this inhibitory effect through inactivation of a kinase or activation of a phosphatase which selectively acts on these sites. The data to this point had also suggested that phosphate in these sites was turning over very slowly or not at all (Ferrari et al. 1993). However, recent studies demonstrate that these sites also become phosphorylated upon mitogenic stimulation (Han et al. 1995), and in the initial analysis were not detected due to the relative low abundance of the kinase and the protocol used in generating and analyzing tryptic phosphopeptides (Pearson et al. 1995). By employing a combination of techniques, three rapamycin-sensitive $p70^{s6k}$ phosphorylation sites have been identified as Thr-229, Thr-389, and Ser-404 (Pearson et al. 1995). Thr-229 resides at a conserved position in the T-loop of the catalytic domain, whose phosphorylation is involved in the activation of many mitogenically triggered kinases. However, the principal target of rapamycin-induced $p70^{s6k}$ inactivation is instead Thr-389, which lies in an unusual hydrophobic sequence outside the catalytic domain and which is conserved in many members of the second messenger family of protein kinases. Mutation of Thr-389 to an alanine abolishes kinase activity, whereas mutation to an acidic residue confers constitutive kinase activity and rapamycin resistance (Pearson et al. 1995). The identification of these three novel sites is valuable for determining the mechanism by which $p70^{s6k}$ is activated and also because they provide a unique possibility of elucidating the mechanism of rapamycin action and the potential for identifying $p70^{s6k}/p85^{s6k}$ kinase(s).

p70^{s6k}/p85^{s6k} SIGNALING PATHWAY

A number of observations initially supported a model in which the p42^{mapk}/p44^{mapk} linked p70^{s6k} activation to the receptor (see Fig. 1). These observations included (1) the kinetics of p42^{mapk}/p44^{mapk} activation, which precedes p70^{s6k} activation (Ahn and Krebs 1990; Ahn et al. 1990); (2) the sites of phosphorylation associated with p70^{s6k} activation contain p42^{mapk}/p44^{mapk} consensus phosphorylation sequences (Erickson et al. 1990; Alvarez et al. 1991); and (3) p42^{mapk}/p44^{mapk} could phosphorylate recombinant p70^{s6k} in vitro (Mukhopadhayay et al. 1992). Indeed, consistent with these findings, it was reported that p42^{mapk}/p44^{mapk} could reactivate phosphatase-treated p70^{s6k} in vitro (Gregory et al. 1989). However, in reexamining this last point, Ballou et al. (1991) found not only that p42^{mapk} would not reactivate phosphatase-treated p70^{s6k}, but also that it did not recognize the wild-type protein as a substrate in vitro. This last observation was also confirmed by Mukhopadhayay et al. (1992). More strikingly, it was found that some mitogens that potently activated p70^{s6k} in Swiss-3T3 cells did not activate p42^{mapk}/p44^{mapk} (Ballou et al. 1991). Consistent with this finding, cycloheximide activates p70^{s6k} but has no effect on p42^{mapk} activation (Blenis et al. 1991). Together, these findings led Ballou et al. (1991) to the hypothesis that the p70^{s6k} and p42^{mapk}/p44^{mapk} reside on distinct intracellular signaling pathways.

In the last few years, a major signaling pathway leading to the activation of the p42^{mapk}/p44^{mapk} has been elucidated (see Egan and Weinberg 1993). A principal component of this pathway is the GTPase p21^{ras}. Since Ha-Ras-transformed cells have elevated p70^{s6k} activity (Ballou et al. 1988b) and increased levels of S6 phosphorylation (Blenis and Erikson 1984), it was reasoned that bifurcation of the p70^{s6k} and p42^{mapk}/p44^{mapk} pathways would occur downstream from p21^{ras} (Ming et al. 1994). As p74^{raf} is the most proximal serine/threonine kinase to p21^{ras}, it was thought to be the most likely point of bifurcation. However, neither dominant-negative mutants of p74^{raf} nor p21^{ras} had any effect on p70^{s6k} activation in response to mitogens while ablating p42^{mapk} activation (Ming et al. 1994). Additionally, a deletion mutant of the platelet-derived growth factor receptor (PDGF-R) lacking the kinase insert domain, which is unaffected in its ability to activate p21^{ras} and p42^{mapk}/p44^{mapk} in response to PDGF (Burgering et al. 1994), fails to activate p70^{s6k}/p85^{s6k} (Ming et al. 1994). These findings argue that p21^{ras} is neither necessary nor sufficient to activate the p70^{s6k} and that the point of bifurcation is at the level of the receptor. Interestingly, phosphorylation of initiation factor 4E, which is thought to be a critical step in the initiation of protein synthesis, is p21^{ras}-dependent (Frederickson et al. 1992). Thus, the signaling pathways regulating the phosphorylation of eIF4E and S6 also appear to bifurcate at the level of the receptor, arguing that there are at least two distinct signaling pathways converging on translation.

To gain some insight into the identity of the signaling molecules that initiate the p70^{s6k}/p85^{s6k} signaling pathway, Chung et al. (1994) took advantage of point mutants of the PDGF-R and the inhibitor of PI3K activation, wortmannin (Chung et al. 1994). Their data showed that the critical docking site of the PDGF-R, which signals to p70^{s6k}, is formed by phosphorylated Tyr-740 and Tyr-751 and to a lesser extent phosphotyrosine 1021. They also found that wortmannin blocked p70^{s6k} activation in a mutant PDGF-R, in which five of the critical tyrosine phosphorylation sites mutated initially to phenylalanines had tyrosines reinserted into positions 740 and 751, but not when a tyrosine was reinserted at position 1021. As PI3K binds to Tyr-740 and Tyr-751 and phospholipase Cγ binds to Tyr-1021, it was concluded that PI3K and protein kinase C mediate the phosphorylation of the p70^{s6k} from these two sets of sites, respectively. However, Ming et al. (1994) found in analyzing the individual receptor point mutants 740 and 751 that only the Tyr-751 mutant was severely inhibited in its ability to induce p70^{s6k} activation in response to PDGF, whereas both mutants were blocked in their ability to bind PI3K. As the Tyr-740 mutant had little effect on p70^{s6k} activation in response to PDGF, it was concluded that PI3K binding to the receptor was not essential to p70^{s6k} activation (Ming et al. 1994). Furthermore, other investigators had shown that a large portion of PI3K activation lies downstream from p21^{ras} (Rodriguez-Viciana et al. 1994), leading Ming et al. (1994) to conclude that this source of PI3K is also not involved in signaling to p70^{s6k}. Instead, they favor a novel molecule that may bind to Tyr-751, such as Nck. Identification of the correct docking molecule will greatly facilitate the elucidation of this signaling pathway.

CONCLUDING REMARKS AND RECENT DEVELOPMENTS

The data reviewed here illustrate the complexity in delineating both the role of S6 phosphorylation in protein synthesis and the signal transduction pathway involved in modulating this response. Recent advances have provided a framework and opportunity to unravel some of these issues. In summary, translation of mRNAs containing polypyrimidine tracts is selectively up-regulated following mitogenic stimulation of cells to proliferate. Rapamycin apparently acts to block the activation of $p70^{s6k}/p85^{s6k}$ and consequently S6 phosphorylation. Recruitment of the 5′TOP mRNAs into polysomes is selectively repressed, indicating that S6 phosphorylation is central to the translational regulation of these transcripts. It is clear that to develop this model further demands the establishment of a direct causal relationship between $p70^{s6k}/p85^{s6k}$, S6 phosphorylation, and the translation of 5′TOP mRNA family. This could be studied in an in-vitro-reconstituted protein-synthesizing system employing recombinant $p70^{s6k}/p85^{s6k}$ and/or phosphorylated 40S ribosomes to analyze their effects on the translation of 5′TOP mRNAs or chimeric transcripts containing the 5′TOP. If it is the case that S6 phosphorylation directly regulates 5′TOP mRNA expression, it will be necessary to determine the precise mechanism involved, which will require both genetic and biochemical approaches. Early evidence may hint at S6 interacting directly with the tract itself. However, two laboratories have identified proteins of M_r 56,000–60,000 that bind specifically to the polypyrimidine tract, but whose binding activity remains unchanged regardless of the growth state of the cell (Kaspar et al. 1992; Cardinali et al. 1994). This suggests that although these proteins may act as translational modulators, their interaction with the mRNA is not the step being regulated. This step may in fact be one of protein-protein interaction involving phosphorylated S6 as one component. The differential regulation of individual 5′TOP transcripts could be due to the existence of specific binding proteins for each mRNA species and/or subtle 5′TOP sequence differences effecting the interaction with the mRNA-binding site. It also will be important to know the role of each of the five S6 phosphorylation sites and if S6 functionally interacts with other components of the translational apparatus, especially those involved in recognizing mRNA.

Along with S6 phosphorylation, recent studies have suggested that phosphorylation of 4E-BP1, a major repressor of eIF4E function (see Hu et al. 1994; Lin et al. 1994; Pause et al. 1994; Sonenberg, this volume), is also mediated by the FRAP/$p70^{s6k}$ signaling pathway and not $p42^{mapk}/p44^{mapk}$. This hypothesis is derived from studies demonstrating that rapamycin treatment of cells blocks 4E-BP1 phosphorylation (Beretta et al. 1995; Graves et al. 1995; Lin et al. 1995), and a number of agents known to induce 4E-BP1 phosphorylation did not activate $p42^{mapk}/p44^{mapk}$. Despite these findings, it was still maintained that $p42^{mapk}/p44^{mapk}$ has a critical role in regulating 4E-BP1 function (Graves et al. 1995; Lin et al. 1995). However, recent studies from our own group in collaboration with N. Sonenberg's group using specific inhibitors of the

$p70^{s6k}$ and the PDGF-R mutants described above clearly demonstrate that $p42^{mapk}/44^{mapk}$ is neither necessary nor sufficient in bringing about 4E-BP1 phosphorylation; instead, 4E-BP1 phosphorylation is mediated by the FRAP/$p70^{s6k}$ pathway (von Manteuffel et al. 1996). Overexpression of eIF4E has no effect on 5′TOP mRNA translation (Shama et al. 1995), suggesting that the effect of 4E-BP1 phosphorylation would be on general translation. Consistent with these findings, rapamycin strongly suppresses the initiation of translation in yeast (Barbet et al. 1996). On the basis of mutagenesis studies of the yeast equivalent of S6 (see above), this effect would not be argued to be mediated through the phosphorylation of this protein. In addition, yeast ribosomal protein mRNAs lack 5′TOPS, and protein expression appears to be largely controlled at the transcriptional level. Furthermore, the effect of rapamycin on yeast translation is much more dramatic than in mammalian cells, suggesting that rapamycin exerts its inhibitory effects through a different translational component, possibly an as yet unidentified homolog of 4E-BP1. Whether the elucidation of the translational component effected in yeast offers new insights into the mode of rapamycin action in mammalian translation awaits further experiments.

Although both the cytoplasmic $p70^{s6k}$ and the nuclear $p85^{s6k}$ isoforms share a common signaling pathway, the upstream kinases and mediators have not as yet been identified. It has been argued that this pathway may be largely controlled by PI3K (Chung et al. 1994); however, other studies, although not disproving this model, have called it into question (Ming et al. 1994). Furthermore, recent studies demonstrate that $p70^{s6k}/p85^{s6k}$ activation is controlled by the phosphorylation of two sets of sites, implying that at least two signaling pathways lead to its activation (Pearson et al. 1995). Identification of these sites will be critical for further identifying $p70^{s6k}/p85^{s6k}$ kinase and the regulatory pathways involved in controlling this response. Studies of receptor mutants where the tyrosine SH2 docking sites have been altered will facilitate this process by approaching the signaling pathway from the receptor.

ACKNOWLEDGMENTS

We thank Drs. P.B. Dennis, R.B Pearson, and C. Reinhard for critically reviewing the manuscript and D. Schofield for secretarial help. We are also indebted to M. Rothnie for assisting with drawing and design of the figure.

REFERENCES

Ahn, N.G. and E.G. Krebs. 1990. Evidence for an epidermal growth factor-stimulated protein kinase cascade in Swiss 3T3 cells. *J. Biol. Chem.* **265:** 11495–11501.

Ahn, N.G., J.E. Weiel, C.P. Chan, and E.G. Krebs. 1990. Identification of multiple epidermal growth factor-stimulated protein serine/threonine kinases from Swiss 3T3 cells. *J. Biol. Chem.* **265:** 11487–11494.

Alvarez, E., I.C. Northwood, F.A. Gonzalez, D.A. Latour, A. Seth, C. Abate, T. Curran, and R.J. Davis. 1991. Pro-Leu-Ser/Thr-Pro is a consensus primary sequence for substrate protein phosphorylation. *J. Biol. Chem.* **265:** 15277–15285.

Ballou, L.M., H. Luther, and G. Thomas. 1991. MAP2 kinase and 70k S6 kinase lie on distinct signalling pathways. *Nature* **349:** 348–350.

Ballou, L.M., M. Siegmann, and G. Thomas. 1988a. S6 kinase in quiescent Swiss 3T3 cells is activated by phosphorylation in response to serum treatment. *Proc. Natl. Acad. Sci.* **85:** 7154–7158.

Ballou, L.M., P. Jeno, R. Friis, and G. Thomas. 1988b. Regulation of S6 phosphorylation during the mitogenic response. In *Gene expression and regulation: The legacy of Luigu Gorini* (M.J. Bissell et al.), pp. 333–342. Elsevier, Amsterdam.

Bandi, H.R., S. Ferrari, J. Krieg, H.E. Meyer, and G. Thomas. 1993. Identification of 40 S ribosomal protein S6 phosphorylation sites in Swiss mouse 3T3 fibroblasts stimulated with serum. *J. Biol. Chem.* **268:** 4530–4533.

Banerjee, P., M.F. Ahamad, J.R. Grove, C. Kozlosky, D.J. Price, and J. Avruch. 1990. Molecular structure of a major insulin/mitogen-activated 70kDa S6 protein kinase. *Proc. Natl. Acad. Sci.* **87:** 8550–8554.

Barbet, N., U. Schneider, S.B. Helliwell, I. Standsfield, M. Tuite, and M.N. Hall. 1996. Tor controls translation initiation and early G1 progression in yeast. *Mol. Biol. Cell* (in press).

Beretta, L., A. Gingras, Y.V. Svitkin, M. Hall, and N. Sonenberg. 1995. Rapamycin blocks the phosphorylation of 4E-BP1 and inhibits cap-independent initiation of translation. *EMBO J.* (in press).

Blenis, J. and R.L. Erikson. 1984. Phosphorylation of the ribosomal protein S6 is elevated in cells transformed by a variety of tumor viruses. *J. Virol.* **50:** 966–969.

Blenis, J., J. Chung, E. Erikson, D.A. Alcorta, and R.L. Erikson. 1991. Distinct mechanisms for the activation of the RSK kinase/MAP2kinase/pp90^{rsk} and pp70-S6 kinase signaling systems are indicated by inhibition of protein synthesis. *Cell Growth Differ.* **2:** 279–285.

Brown, E.J., P.A. Beal, C.T. Keith, J. Chen, T.B. Shin, and S.L. Schreiber. 1995. Control of p70 S6 kinase by kinase activity of FRAP *in vivo. Nature* **377:** 441–446.

Brown, E.J., M.A. Albers, T.B. Shin, K. Ichikawa, C.T. Keith, W.S. Lane, and S.L. Schreiber. 1994. A mammalian protein targeted by G1-arresting rapamycin-receptor complex. *Nature* **369:** 756–758.

Bourne, H.R., D.A. Sanders, and F. McCormick. 1990. The GTPase superfamily: A conserved switch for diverse cell functions. *Nature* **348:** 125–132.

Burgering, B.M., E. Freed, L. van der Voorn, F. McCormick, and J.L. Bos. 1994. Platelet-derived growth factor-induced p21ras-mediated signaling is independent of platelet-derived growth factor receptor interaction with GTPase-activating protein or phosphatidylinositol-3-kinase. *Cell Growth Differ.* **5:** 341–347.

Cantley, L.C., K.R. Auger, C. Carpenter, B. Duckworth, A. Graziani, R. Kapeller, and S. Soltoff. 1991. Oncogenes and signal transduction. *Cell* **64:** 281–302.

Cardinali, B., M. Di Cristina, and P. Pierandrei-Amaldi. 1994. Interaction of proteins with the mRNA for ribosomal protein L1 in *Xenopus:* Structural charaterization of *in vivo* complexes and identifiation of proteins that bind *in vitro* to its 5′UTR. *Nucleic Acids Res.* **21:** 2301–2308.

Chung, J., C.J. Kuo, G.R. Crabtree, and J. Blenis. 1992. Rapamycin-FKBP specifically blocks growth-dependent activation of and signaling by the 70 kd S6 protein kinases. *Cell* **69:** 1227–1236.

Chung, J., T.C. Grammer, K.P. Lemon, A. Kazlauskas, and J. Blenis. 1994. PDGF- and insulin-dependent $pp70^{s6k}$ activation mediated by phosphatidylinositol-3-OH kinase. *Nature* **370:** 71–75.

de Groot, R.P., L.M. Ballou, and P. Sassone-Corsi. 1994. Postive regulation of the cAMP-responsive activator CREM by the p70 S6 kinase: An alternative route to mitogen-induced gene expression. *Cell* **79:** 81–91.

Downward, J. 1994. Regulating S6 kinase. *Nature* **371:** 378–379.

Duncan, R. and E.H. McConkey. 1982. Preferential utilization of phosphorylated 40-S ribosomal subunits during initiation complex formation. *Eur. J. Biochem.* **123:** 535–538.

Egan, S.E. and R.A. Weinberg. 1993. The pathway to signal achievement. *Nature* **365:** 781–782.

Erickson, A.K., D.M. Payne, P.A. Martino, A.J. Rossomando, J. Shabanowitz, M.J. Weber, D.F. Hunt, and T.W. Sturgill. 1990. Identification by mass spectrometry of threonine 97 in bovine myelin basic protein as a specific phosphorylation site for mitogen-activated protein kinase. *J. Biol. Chem.* **265:** 19728–19735.

Ferrari, S., H.R. Bandi, B.M Bussian, and G. Thomas. 1991. Mitogen-activated 70K S6 Kinase. *J. Biol. Chem.* **266:** 22770–22775.

Ferrari, S., W. Bannwarth, S.J. Morley, N.F. Totty, and G. Thomas. 1992. Activation of $p70^{s6k}$ is associated with phosphorylation of four clustered sites displaying Ser/Thr-Pro motifs. *Proc. Natl. Acad. Sci.* **89:** 7282–7285.

Ferrari, S., R.P. Pearson, M. Siegmann, S.C. Kozma, and G. Thomas. 1993. The immunosuppressant rapamycin induces inactivation of $p70^{s6k}$ through dephosphorylation of a novel set of sites. *J. Biol. Chem.* **268:** 16091–16094.

Flotow, H. and G. Thomas. 1992. Substrate recognition determinants of the mitogen-activated 70K S6 kinase from rat liver. *J. Biol. Chem.* **267:** 3074–3078.

Franco, R. and M.G. Rosenfeld. 1990. Hormonally inducible phosphorylation of a nuclear pool of ribosomal protein S6. *J. Biol. Chem.* **265:** 4321–4325.

Frederickson, R.M., W.E. Mushynski, and N. Sonenberg. 1992. Phosphorylation of translation initiation factor eIF-4E is induced in a *ras*-dependent manner during nerve growth factor-mediated PC12 cell differentiation. *Mol. Cell. Biol* **12:** 1239–1247.

Godefroy-Colburn, T. and R.E. Thach. 1981. The role of mRNA competition in regulating translation. IV. Kinetic model. *J. Biol. Chem.* **256:** 11762–11773.

Gordon, J.S., J. Bruno, and J.J. Lucas. 1981. Heterogenous binding of high mobility group chromosomal proteins to nuclei. *J. Cell Biol.* **88:** 373–379.

Graves, L.M., K.E. Bornfeld, G.M. Argast, E.G. Krebs, X. Kong, T.A. Lin, and J.C. Lawrence, Jr. 1995. cAMP- and rapamycin-sensitive regulation of the association of eukaryotic initiation factor 4E and the translational regulator PHAS-I in aortic smooth muscle cells. *Proc. Natl. Acad. Sci.* **92:** 7222–7226.

Gregory, J.S., T.G. Boulton, B.C. Sang, and M.H. Cobb. 1989. An insulin-stimulated ribosomal protein S6 kinase from rabbit liver. *J. Biol. Chem.* **264:** 18397–18401.

Gressner, A.M. and E. van de Leur. 1980. Interactions of synthetic polynucleotides with small rat liver ribosomal subunits possessing low and highly phosphorylated protein S6. *Biochim. Biophys. Acta* **608:** 459–468.

Gressner, A.M. and I.G. Wool. 1974. The phosphorylation of liver ribosomal proteins *in vivo*. Evidence that only a single small subunit (S6) is phosphorylated. *J. Biol. Chem.* **249:** 6917–6925.

Grove, J.S., P. Banerjee, A. Balasubramanyam, P.J. Coffer, D.J. Price, J. Avruch, and J.R. Woodgett. 1991. Regulation of an epitope-tagged recombinant Rsk-1 S6 kinase by phorbol ester and erk/Map kinase. *Mol. Cell. Biol.* **11:** 5541–5550.

Hammond, M.L., W. Merrick, and L.H. Bowman. 1991. Sequences mediating the translation of mouse S16 ribosomal protein mRNA during myoblast differentiation and in vitro and possible control points for the in vitro translation. *Genes Dev.* **5:** 1723–1736.

Han, J.-W., R.B. Pearson, P.B. Dennis, and G. Thomas. 1995. Rapamycin, wortmannin, and the Methylxanthine SQ20006 inactivate $p70^{s6k}$ by inducing dephosphorylation of the same subset of sites. *J. Biol. Chem.* **270:** 21396–21403.

Harmann, B. and M.W. Kilimann. 1990. cDNA encoding a 59kDa homolog of ribosomal S6 kinase from rabbit liver. *FEBS Lett.* **273:** 248–252.

Hershey, J.W.B. 1991. Translational control in mammalian cells. *Annu. Rev. Biochem.* **60:** 717–755.

Hu, C., S. Pang, X. Kong, M. Velleca, and J.C. Lawrence, Jr. 1994. Molecular cloning and tissue distribution of PHAS-I, an intracellular target for insulin and growth factors. *Proc. Natl. Acad. Sci.* **91:** 3730–3734.

Hunter, T. 1995. When is a lipid kinase not a lipid kinase? When it is a protein kinase. *Cell* **83:** 1–4.

Jefferies, H.B.J., G. Thomas, and G. Thomas. 1994a. Elongation factor-1α mRNA is selectively translated following mitogenic stimulation. *J. Biol. Chem.* **269:** 4367–4372.

Jefferies, H.B.J., C. Reinhard, S.C. Kozma, and G. Thomas. 1994b. Rapamycin selectively represses translation of the "polypyrimidine tract" mRNA family. *Proc. Natl. Acad. Sci.* **91:** 4441–4445.

Jeno, P., L.M. Ballou, I. Novak-Hofer, and G. Thomas. 1988. Identification and characterization of a mitogenic-activated S6 kinase. *Proc. Natl. Acad. Sci.* **85:** 406–410.

Johnson, S.P. and J.R. Warner. 1987. Phosphorylation of *Saccharomyces cerevisiae* equivalent of the ribosomal protein S6 has no detectable effect on growth. *Mol. Cell. Biol.* **7:** 1338–1345.

Kaspar, R.L., T. Kakagawa, H. Cranston, D.R. Morris, and M.W. White. 1992. A regulatory *cis* element and a specific binding factor involved in the mitogenic control of murine ribosomal protein L32 translation. *J. Biol. Chem.* **267:** 508–514.

Kozma, S.C., S. Ferrari, and G. Thomas. 1989. Unmasking a growth factor/oncogene-activated S6 phosphorylation cascade. *Cell. Signalling* **1:** 219–225.

Kozma, S.C., S. Ferrari, P. Bassand, M. Siegmann, N. Totty, and G. Thomas. 1990. Cloning of the mitogen-activated S6 kinase from rat liver reveals an enzyme of the second messenger subfamily. *Proc. Natl. Acad. Sci.* **87:** 7365–7369.

Kozma, S.C., E. McGlynn, M. Siegmann, C. Reinhard, S. Ferrari, and G. Thomas. 1993. Active Baculovirus recombinant $p70^{s6k}$ and $p85^{s6k}$ produced as a function of the infectious response. *J. Biol. Chem.* **268:** 7134–7138.

Krebs, E.G. 1994. The growth of research on protein phosphorylation. *Trends Biochem. Sci.* **19:** 439.

Krek, W., G. Maridor, and E.A. Nigg. 1992. Casein kinase II is a predominantly nuclear enzyme. *J. Cell Biol.* **116:** 43–55.

Krieg, J., J. Hofsteenge, and G. Thomas. 1988. Identification of the 40 S ribosomal protein S6 phosphorylation sites induced by cycloheximide. *J. Biol. Chem.* **263:** 11473–11477.

Kunz, J., R. Henriquez, U. Scheider, M. Deuter-Reinhard, N.R. Movva, and M.N. Hall. 1993. Target of rapamycin in yeast, TOR2, is an essential phosphatidylinositol kinase homolog required for G1 progression. *Cell* **73:** 585–596.

Kuo, C.J., J. Chung, D.F. Fiorentino, W.M. Flanagan, J. Blenis, and G.R. Crabtree. 1992. Rapamycin selectively inhibits interleukin-2 activation of p70 S6 kinase. *Nature* **358:** 70–73.

Larson, D.E., P. Zahradka, and B.H. Sells. 1991. Control points in eucaryotic ribosome biogenesis. *Biochem. Cell Biol.* **69:** 5–22.

Levy, S., D. Avni, N. Hariharan, R.P. Perry, and O. Meyuhas. 1991. Oligopyrimidine tract at the 5′ end of mammalian ribosomal protein mRNAs is required for their translational control. *Proc. Natl. Acad. Sci.* **88:** 3319–3323.

Lin, T.A., X. Kong, A.R. Saltiel, P.J. Blackshear, and J.C. Lawrence, Jr. 1995. Control of PHAS-I by insulin in 3T3-L1 adipocytes. *J. Biol. Chem.* **270:** 18531–18538.

Lin, T., X. Kong, T.A.J. Haystead, A. Pause, G. Belsham, N. Sonenberg, and J.C. Lawrence, Jr. 1994. PHAS-I as a link between mitogen-activated protein kinase and translation initiation. *Science* **266:** 653–656.

Lodish, H.F. 1974. Model for the regulation of mRNA translation applied to haemoglobin synthesis. *Nature* **251:** 385–388.

Loeb, J.E. and C. Blat. 1970. Phosphorylation of some rat liver ribosomal proteins and its activation by cyclic AMP. *FEBS Lett.* **10:** 105–108.

Ming, X.F., B.M.T. Burgerung, S. Wennström, L. Claesson-Welsh, C.H. Heldin, J.L. Bos, S.C. Kozma, and G. Thomas. 1994. Activation of p70/p85 S6 kinase by a pathway independent of $p21^{ras}$. *Nature* **371:** 426–429.

Morley, S.J. and J.W.B. Hershey. 1990. A fractionated reticulocyte lysate retains high efficiency for protein synthesis. *Biochimie* **72:** 259–264.

Morley, S.J. and G. Thomas. 1991. Intracellular messengers and the control of protein synthesis. *Pharmacol. Ther.* **50:** 291–319.

Mukhopadhayay, N.K., D.J. Price, J.M. Kyriakis, S. Pelech, J. Sanghera, and J. Avruch. 1992. An array of insulin-activated, proline-directed serine/threonine protein kinases phosphorylate the p70 S6 kinase. *J. Biol. Chem.* **267:** 3325–3335.

Nigg, E.A. 1988. Nuclear function and organization. *Int. Rev. Cytol.* **110:** 27–92.

Novak-Hofer, I. and G. Thomas. 1984. An activated S6 kinase in extracts from serum- and epidermal growth factor-stimulated Swiss 3T3 cells. *J. Biol. Chem.* **259:** 5995–6000.

Nygärd, O. and H. Nika. 1982. Identification by RNA-protein cross-linking of ribosomal proteins located at the interface between the small and the large subunits of mammalian ribosomes. *EMBO J.* **1:** 357–362.

Nygärd, O. and L. Nilsson. 1990. Translational dynamics. Interactions between the translational factors, tRNA and ribosomes during eukaryotic protein synthesis. *Eur. J. Biochem.* **191:** 1–17.

Palen, E. and J.A. Traugh. 1987. Phosphorylation of ribosomal protein S6 by cAMP-dependent protein kinase and mitogen-stimulated S6 kinase differentially alters translation of globin mRNA. *J. Biol. Chem.* **262:** 3518–3523.

Pardee, A.B. 1989. G_1 events and regulation of cell proliferation. *Science* **246:** 603–608.

Pause, A., G.J. Belsham, A.C. Gingras, O. Donzé, T.A. Lin, J.C. Lawrence, Jr., and N. Sonenberg. 1994. Insulin-dependent stimulation of protein synthesis by phosphorylation of a regulator of 5′-cap function. *Nature* **371:** 762–767.

Pearson, R.B., P.B. Dennis, J.-W. Han, N.A. Williamson, S.C. Kozma, R.E.H. Wettenhall, and G. Thomas. 1995. Rapamycin includes $p70^{s6k}$ inactivation by dephosphorylation of an essential site within a novel hydrophobic motif. *EMBO J.* **14:** 5279–5287.

Price, D.J., R.A. Nemenoff, and J. Avruch. 1989. Purification of a hepatic S6 kinase from cycloheximide-treated rats. *J. Biol. Chem.* **264:** 13825–13833.

Price, D.J., J.R. Grove, V. Calvo, J. Avruch, and B.E. Bierer. 1992. Rapamycin-induced inhibition of the 70-kilodalton S6 protein kinase. *Science* **257:** 973–977.

Reinhard, C., G. Thomas, and S.C. Kozma. 1992. A single gene encodes two isoforms of the p70 S6 kinase: Activation upon mitogenic stimulation. *Proc. Natl. Acad. Sci.* **89:** 4052–4056.

Reinhard, C., A. Fernandez, N.J.C. Lamb, and G. Thomas. 1994. Nuclear localization of $p85^{s6k}$: Functional requirement for entry into S phase. *EMBO J.* **13:** 1557–1565.

Rodriguez-Viciana, P., P.H. Warne, R. Dhand, B. Vanhaesebroeck, I. Gout, M.J. Fry, M.D. Waterfield, and J. Downward. 1994. Phosphatidylinositol-3-OH kinase as a direct target of Ras. *Nature* **370:** 527–532.

Sabatini, D.M., H. Erdjument-Bromage, M. Lui, P. Tempst, and S.H. Snyder. 1994. RAFT1: A mammalian protein that binds to FKBP12 in a rapamycin-dependent fashion and is homologous to yeast TORs. *Cell* **78:** 35–43.

Sabatini, D.M., B.A. Pierchala, R.K. Barrow, M.J. Schell, and S.H. Snyder. 1995. The rapamycin and FKBP12 target (RAFT) displays phosphatidylinositol 4-kinase activity. *J. Biol. Chem.* **270:** 20875–20878.

Sehgal, S.N., K. Molnar-Kimber, T.D. Ocain, and B.M. Wiechman. 1994. Rapamycin: A novel immunosupressive Macrolide. *Med. Res. Rev.* **14:** 1–22.

Shama, S., D. Avni, R.M. Frederickson, N. Sonenberg, and O. Meyuhas. 1995. Overexpression of initiation factor eIF-4E does not relieve the translational repression of ribosomal protein mRNAs in quiescent cells. *Gene Expr.* **4:** 241–252.

Sommerville, J. 1986. Nucleolar structure and ribosome biogenesis. *Trends Biochem. Sci.* **11:** 438–442.

Sonenberg, N. 1994. mRNA translation: Influence of the 5′ and 3′ untranslated region. *Curr. Opin. Genet. & Dev.* **4:** 310–315.

Stewart, M.J. and G. Thomas. 1994. Mitogenesis and protein synthesis: A role for ribosomal protein S6 phosphorylation? *BioEssays* **16:** 1–7.

Terao, K. and K. Ogata. 1979a. Proteins of small subunits of rat liver ribosomes. II. Cross-links between poly(U) and ribosomal proteins in 40S subunits induced by UV irradiation. *J. Biochem.* **86:** 605–617.

———. 1979b. Proteins of small subunits of rat liver ribosomes that interact with poly(U). I. Effects of preincubation of poly(U) with 40S subunits on the interactions of 40S subunit proteins with aurintricarboxylic acid and with N′N′-p-phenylenedimaleimide. *J. Biochem.* **86:** 597–603.

Thomas, G., M. Siegmann, and J. Gordon. 1979. Multiple phosphorylation of ribosomal protein S6 during transition of quiescent 3T3 cells into early G_1, and cellular compartmentalisation of the phosphate donor. *Proc. Natl. Acad. Sci.* **76:** 3952–3956.

Thomas, G., G. Thomas, and H. Luther. 1981. Transcriptional and translational control of cytoplasmic proteins after serum stimulation of quiescent Swiss 3T3 cells. *Proc. Natl. Acad. Sci.* **78:** 5712–5716.

Thomas, G., J. Martin-Pèrez, M. Siegmann, and A.M. Otto. 1982. The effect of serum EGF, $PGF_2\alpha$ and insulin on S6 phosphylation and the initiation of protein and DNA synthesis. *Cell* **30:** 235–242.

Thomas, G. and G. Thomas. 1986. Translational control of mRNA expression during the early mitogenic response in Swiss mouse 3T3 cells: Identification of specific proteins. *J. Cell Biol.* **103:** 2137–2144.

Todorov, I.T., F. Noll, and A.A. Hadjiolov. 1983. The sequential addition of ribosomal proteins during the formation of the small ribosomal subunit in Friend erythroleukemia cells. *Eur. J. Biochem.* **131:** 271–275.

Uetsuki, T., A. Naito, S. Nagata, and Y. Kaziro. 1989. Isolation and characterization of the human chromosomal gene for polypeptide chain elongation factor-1α. *J. Biol. Chem.* **264:** 5791–5798.

von Manteuffel, S., A.-C. Gingras, N. Sonenberg, and G. Thomas. 1996. 4E-BP1 phosphorylation is mediated by the FRAP-$p70^{s6k}$ pathway and is MAP kinase-independent. *Proc. Natl. Acad. Sci.* (in press).

15

Translational Control of Developmental Decisions

Marvin Wickens,[1] Judith Kimble,[1,2,3] and Sidney Strickland[4]
[1]Department of Biochemistry, University of Wisconsin
Madison, Wisconsin 53706
[2]Department of Medical Genetics and Laboratory of Molecular Biology
University of Wisconsin, Madison, Wisconsin 53706
[3]Howard Hughes Medical Institute
[4]Department of Pharmacology, University Medical Center
Stony Brook, New York 11794

At fertilization, the calm of oogenesis is broken, and the egg abruptly begins a flurry of activity. Many crucial steps—decisions concerning when and where to divide, specification of cell fates, and establishment of body axes—rely on materials the egg contains at that moment. In many animals, the first few hours of life proceed with little or no transcription. As a result, developmental regulation at these early stages is dependent on maternal cytoplasm, rather than the zygotic nucleus. The regulatory molecules accumulated during oogenesis might, in principle, be of any type, including RNA and protein. It is now clear that messenger RNAs present in the egg before fertilization (so-called maternal mRNAs) have a prominent role in early decisions. Viewed from this perspective, it is not surprising that oocytes and early embryos display an impressive array of posttranscriptional regulatory mechanisms, controlling mRNA stability, localization, and translation.

The mechanisms by which translation of specific maternal mRNAs are controlled and how those controls contribute to proper development are the main focus of this chapter. Translational regulation is vital throughout development in somatic and germ cells. The predominant mode of tissue-specific regulation in adult tissues is transcriptional; yet, several of the examples we discuss hint that the importance of translational control may be currently underestimated, perhaps dramatically so.

One conclusion emerges exceptionally clearly from studies of translational control during early development: The region between the termination codon and the poly(A) tail—the 3′-untranslated region (3′UTR)—is a key repository for the regulation of cytoplasmic mRNAs.

Other regions of the mRNA will no doubt be found to have critical roles in developmental regulation, but thus far, the 3′UTR is preeminent.

Translational control is defined broadly in this chapter. No mechanism is implied. Ideally, translational control is demonstrated by comparing the level of a specific mRNA to the rate of its translation. However, rates of translation can be difficult to measure directly in vivo. In several cases discussed in this chapter, only steady-state levels of the protein are known; however, translational control is inferred because the regulatory sequences responsible are located outside the protein-coding region.

In this chapter, we focus on translational controls that are vital for key developmental events. We first describe specific examples drawn from a broad range of biological contexts and organisms. Having presented the facts, we then turn our attention to what generalities may be drawn, to important puzzles that remain, and to speculation.

TRANSLATIONAL CONTROL OF DEVELOPMENTAL EVENTS: SPECIFIC EXAMPLES

Meiotic Maturation and the Early Embryonic Cell Cycle

A dramatic transition from cell cycle arrest to mitotic cleavage occurs upon fertilization. In some species, it is preceded by completion of the meiotic cell cycle, referred to as oocyte maturation. To control cell cycle transitions, eggs of many species contain mRNAs that encode cell cycle regulators, such as cyclin and cyclin-dependent kinases (CDKs). For the purposes of our discussion, it is necessary only to know that cyclin and CDK form a complex that is critical in governing the cell cycle.

Cyclin mRNAs

Frog oocytes contain mRNAs encoding several different cyclins. Translation of cyclin mRNAs appears to be important both for proper post-fertilization mitoses (see, e.g., Dagle et al. 1990) and perhaps for meiotic maturation. The analysis of cyclin function in the oocyte and embryo may be complicated by the presence of multiple cyclins with overlapping roles. Nevertheless, their translational regulation is striking and informative.

Cyclin A1, B1, and B2 mRNAs are activated at different times during maturation, and to different extents (Kobayashi et al. 1991). Each mRNA receives poly(A) concomitant with its translational stimulation (Sheets et al. 1994). To identify the signals involved in these controls, chimeric mRNAs were injected that contained each 3′UTR joined to a translational reporter. The different cyclin 3′ UTRs determined when, and how much, translation was stimulated during oocyte maturation. Invariably, translational stimulation required poly(A) addition (Sheets et al. 1994; for review, see Wormington 1994). Thus, 3′ UTRs, by controlling polyadenylation, can impose very different patterns of translation, stimulating translation at different times and to different extents.

Regulation of maternal cyclin mRNAs at the translational level may be common. In *Drosophila* embryos, for example, maternal cyclin B mRNA is localized to pole cells (the presumptive germ line) and is repressed until mitoses resume in the developing gonad, well after fertilization (Dalby and Glover 1993). The regulatory elements responsible for translational control and localization reside in its 3′ UTR (Dalby and Glover 1993). Similarly, in surf clams and sea urchins, certain cyclin mRNAs are repressed during oogenesis and then activated dramatically at fertilization, when they receive poly(A) (Rosenthal et al. 1980; Standart 1992). The common regulation of cyclin mRNAs presumably reflects their role after the cell cycle resumes at fertilization and the deleterious consequences of their premature expression. Other maternal mRNAs that participate in cell-cycle-related events, such as DNA replication and the synthesis of DNA precursors, are also subject to translational control (e.g., CDK2, histones, ribonucleotide reductase, and HGPRT; for review, see Standart 1992).

The oocyte and egg have served as model systems in the analysis of cell cycle control, with particular attention given to the regulation of preexisting cyclin and CDK proteins. Control of their synthesis likely also is critical in orchestrating the dramatic transition from quiescence to meiotic maturation and cell division.

c-mos *mRNA*

The c-*mos* proto-oncogene encodes a protein kinase that has been strongly implicated in the control of vertebrate meiosis and the early embryonic cell cycle (for review, see Yew et al. 1993; Vande Woude 1994). Consistent with these roles, c-*mos* mRNA is normally found only in the germ line. In frog oocytes, removal of c-*mos* mRNA prevents maturation, whereas its overexpression induces maturation (Sagata et al. 1988, 1990). Female mice lacking a functional c-*mos* gene display reduced fertility, as well as ovarian cysts and teratomas, consistent with a crucial role in oocyte growth (Colledge et al. 1994; Hashimoto et al. 1994).

In frogs, translation of c-*mos* mRNA apparently increases during oocyte maturation (Sagata et al. 1988). Fox et al. (1989) noted, by sequence inspection, that *Xenopus* c-*mos* mRNA contained signals that could cause cytoplasmic polyadenylation and proposed that cytoplasmic polyadenylation of c-*mos* mRNA might therefore be a critical control point in meiotic maturation, as depicted in Figure 1A. This hypothesis has since gained substantial support. c-*mos* mRNA receives poly(A) during maturation. Furthermore, the c-*mos* 3′UTR contains signals sufficient for cytoplasmic polyadenylation (Paris and Richter 1990; Sheets et

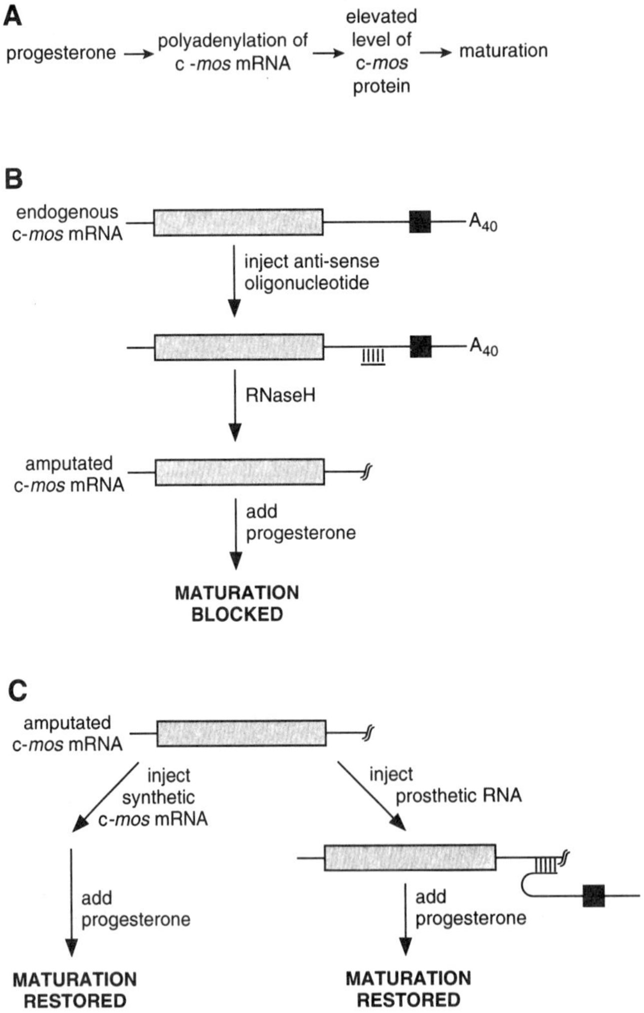

Figure 1 (*See facing page for legend.*)

al. 1994) and, when linked to a reporter, stimulates translation during maturation (Sheets et al. 1994). Removal of cytoplasmic polyadenylation signals from endogenous c-*mos* mRNA, achieved by targeted RNase H cleavage, prevents maturation (Fig. 1B) (Sheets et al. 1995). The amputated mRNA, lacking its polyadenylation signals, is stable. Maturation can be restored by injection of synthetic c-*mos* mRNA carrying polyadenylation signals (Fig. 1C, left) or of a prosthetic RNA that brings polyadenylation signals to the amputated endogenous mRNA by base pairing (Fig. 1C, right) (Sheets et al. 1995). These experiments strongly argue that polyadenylation is critical in the activation of c-*mos* mRNA. They do not, however, demonstrate that polyadenylation is the only process triggered by progesterone that is critical for its activation. For example, repressors might need to be removed from the mRNA as well.

Cytoplasmic polyadenylation of c-*mos* mRNA is also required for the maturation of mouse oocytes (Gebauer et al. 1994). In mouse oocytes, removal of the polyadenylation signals from c-*mos* mRNA does not block completion of the first meiosis as in frogs. Rather, these oocytes complete the first meiotic division but fail to progress normally to meiosis II. This phenotype mirrors that observed in oocytes derived from females homozygous for a disrupted c-*mos* gene, which undergo parthenogenetic activation after completing first meiosis (Colledge et al. 1994; Hashimoto et al. 1994). Together, these experiments demonstrate that

Figure 1 Polyadenylation of c-*mos* mRNA as a control point in the maturation of *Xenopus* oocytes. (*A*) The hypothesis (Fox et al. 1989) suggests that among the effects of progesterone addition to frog oocytes is polyadenylation of c-*mos* mRNA, which enhances its translation, thereby causing an increase in the level of c-Mos protein. The elevated level of c-Mos protein is required for the continuation of meiosis (see Yew et al. 1993). Control of c-*mos* proteolysis (Nishizawa et al. 1992), which may also contribute to the control of c-Mos protein levels, is not depicted. Proposed biochemical roles of the c-Mos protein kinase in frog oocyte maturation (e.g., as an activator of MAPK and MPF) are discussed elsewhere (Yew et al. 1993; Vande Woude 1994). (*B*) Removal of c-*mos* polyadenylation signals prevents maturation of frog oocytes in response to progesterone. (*Closed box*) Polyadenylation signals, including AAUAAA and a cytoplasmic polyadenylation element; (*gray box*) coding region; (*thin lines*) 5′ and 3′ UTRs. For details, see Sheets et al. (1995); for analogous experiments in mouse oocytes, see Gebauer et al. (1994). (*C*) Rescue of maturation by the injection of a synthetic form of c-*mos* mRNA or of a prosthetic RNA. Oocytes containing the amputated version of c-*mos* mRNA, generated as diagrammed in *B*, were injected with the RNAs indicated. For details, see Sheets et al. (1995).

lack of polyadenylation signals creates a functional null for c-*mos*-encoded protein in mouse oocytes and hence that polyadenylation is indispensable for normal meiosis and early development.

Specification of Cell Fate

As development unfolds, cells assume specific fates or paths of differentiation: One cell becomes a neuron, while another becomes a lymphocyte. In this section, we discuss evidence that cell fate can be regulated at the translational level and moreover that the spatial organization of cell fates within a tissue can rely on translational controls.

The Sperm/Oocyte Decision in the Nematode Caenorhabditis elegans

The germ line of a *C. elegans* hermaphrodite makes sperm first and then oocytes, beginning at the proximal end of the tubular gonad. The onset of spermatogenesis depends on translational regulation of one sex-determining gene, *tra-2* (Goodwin et al. 1993), whereas the switch from spermatogenesis to oogenesis depends on translational control of a second sex-determining gene, *fem-3* (Ahringer and Kimble 1991). The spatial organization of germ-line cell fates therefore depends on the execution of two translational controls.

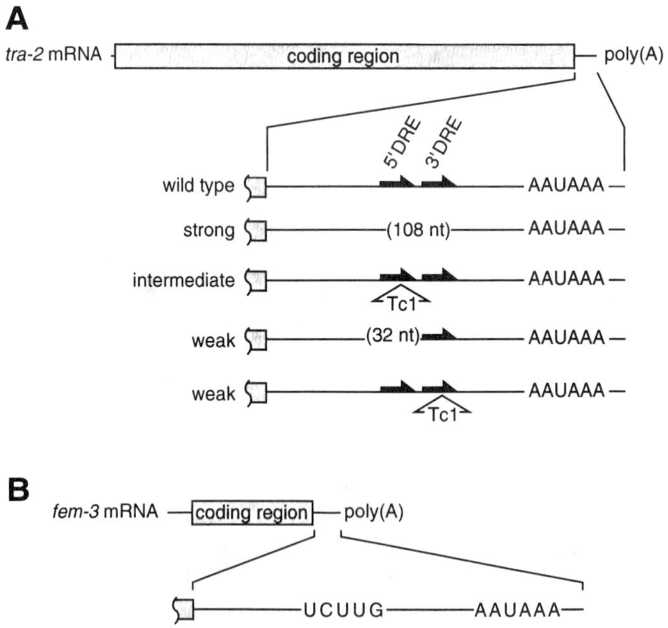

Figure 2 (*See facing page for legend.*)

tra-2

The *tra-2* gene normally directs female development (Hodgkin and Brenner 1977). Six regulatory mutations of *tra-2*, called *tra-2(gf)* (for *g*ain of-*f*unction), feminize the hermaphrodite germ line so that only oocytes are made (Doniach 1986; Schedl and Kimble 1988). Therefore, *tra-2* must normally be repressed to achieve the onset of hermaphrodite spermatogenesis. The *tra-2(gf)* mutants are defective in a *cis*-acting translational control element located in the *tra-2* 3′UTR (Fig. 2A) (Goodwin et al. 1993). This element consists of two tandem 28-nucleotide repeats and is called a direct repeat element (DRE). The DREs influence translational activity, as judged from Northern blots, polysome analyses, and chimeric reporter experiments (Goodwin et al. 1993). All the criteria thus far indicate that the DREs function as negative *cis*-acting regulatory elements to mediate translational repression of *tra-2* RNA.

The extent of *tra-2* translation varies with the number of DREs (Goodwin et al. 1993). Thus, the phenotype of a *tra-2(gf)* mutant with no

Figure 2 Regulatory elements in *tra-2* and *fem-3* mRNAs. (*A*) Schematic of *tra-2* 3′UTR regulatory elements. (*Gray box*) Coding region; (*thin line*) 5′and 3′UTRs. Each region of the mRNA is drawn to scale (5′UTR, 36 nucleotides; coding region, 4,425 nucleotides; 3′UTR, 218 nucleotides). At the bottom is an expanded view of wild-type and mutant *tra-2* 3′UTRs. Mutants are presented in order of the severity of their phenotype. The wild-type 3′UTR possesses two DREs (arrows) separated by four nucleotides; each DRE has the same 28-nucleotide sequence, UAGAUAUGAGUAGAUAAGAAAUUAAAUA; the four-nucleotide spacer is UGAG. The 5′DRE begins 88 nucleotides downstream from the termination codon. The strongest mutant lacks both DREs, an intermediate strength mutant harbors a transposable element (Tc1) in the more 5′DRE, and the weak mutants have one DRE left, due to a deletion or to a Tc1 insertion into the more 3′DRE. The Tc1 mutants have insertions at identical nucleotides in the 5′or 3′DRE. It is intriguing that disruption of the 5′DRE is more severe than disruption of the more 3′DRE; perhaps access to the DREs is limited by a process that moves from 5′to 3′, such as transcription or translation. For details, see text and Goodwin et al. (1993). (*B*) Schematic of *fem-3* 3′UTR regulatory element. (*Gray box*) Coding region; (*thin line*) 5′and 3′ UTRs. Each region of the mRNA is drawn to scale, the same scale as used in *A* (5′UTR, 248 nucleotides; coding region, 1164 nucleotides; 3′UTR, 268 nucleotides). At the bottom is an expanded view of wild-type and mutant *fem-3* 3′ UTR. Approximately midway in the wild-type *fem-3* 3′UTR (123 nucleotides from the UGA termination codon) is the sequence UCUUG, which is altered in *fem-3* gain-of-function mutants. A deletion of the region has the most severe phenotype. For details, see text and Ahringer and Kimble (1991).

DRE is more severely affected than a mutant with one DRE; furthermore, *tra-2(gf)* mRNA with no DRE is associated with larger polysomes than an mRNA with one DRE, which in turn is associated with larger polysomes than an mRNA with two DREs.

Strong candidates for the translational repressor of *tra-2* have been identified. Each DRE specifically binds a factor, called DRF, in crude extracts (Goodwin et al. 1993). Furthermore, a recently identified gene, *laf-1*, is predicted to either encode DRF or influence its activity (E. Goodwin and J. Kimble, unpubl.). Thus, a decrease in *laf-1* disrupts the translational repression of a chimeric reporter gene by the *tra-2* 3′UTR (E. Goodwin and J. Kimble, unpubl.).

fem-3

The *fem-3* gene normally directs male development (Hodgkin 1986; Barton et al. 1987). Nineteen *fem-3(gf)* mutations masculinize the hermaphrodite germ line so that it produces a vast excess of sperm and no oocytes (Barton et al. 1987). Therefore, *fem-3* must normally be repressed to achieve the switch in the hermaphrodite germ line from spermatogenesis to oogenesis. Each of 17 *fem-3(gf)* mutant genes carries a single nucleotide change in the middle of the *fem-3* 3′UTR; the remaining two *fem-3(gf)* mutations possess small deletions within the *fem-3* 3′UTR (Fig. 2B) (Ahringer 1991; Ahringer and Kimble 1991). Remarkably, all mutations either alter or remove nucleotides in a five-base-pair region, which presumably is part of the regulatory element (Fig. 2B). For simplicity, we refer to it as the point mutation element, or PME, although the regulatory element may comprise more than just these five nucleotides.

The PME appears to be a *cis*-acting translational control element. Thus, the *fem-3(gf)* mutations do not detectably affect transcription, splicing, or stability of *fem-3* RNA, and the *fem-3(gf)* mutant RNAs possess a longer poly(A) tail than their wild-type counterparts (Ahringer and Kimble 1991). Furthermore, in gel-retardation assays, the PME binds specifically to a factor in worm extracts, which we dub here PMF (Ahringer 1991). Finally, animals carrying a transgenic *fem-3* 3′UTR driven by the *fem-3* promoter exhibit germ-line masculinization (Ahringer and Kimble 1991). One interpretation of this effect is that RNA corresponding to the wild-type *fem-3* 3′UTR, which is synthesized from the transgene, titrates a negative regulator from endogenous *fem-3* RNA. The endogenous *fem-3* RNA would thereby be activated and direct spermatogenesis inappropriately.

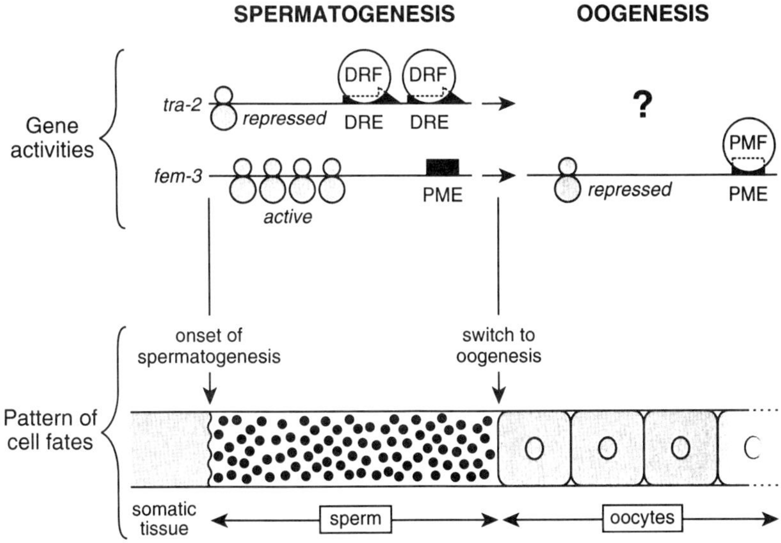

Figure 3 Speculative model for the role of the *tra-2* and *fem-3* 3′ UTR elements in controlling the pattern of cell fates in the hermaphrodite germ line of *C. elegans.* (*Top*) Translational activity of *tra-2* or *fem-3*; (*bottom*) pattern of cell fates. Pattern is generated as differentiation proceeds from proximal (*left*) to distal (*right*). First, repression of *tra-2* is required for the onset of spermatogenesis. DRF is proposed to bind DREs during larval development to repress *tra-2* translation and permit spermatogenesis; *fem-3*, which is translationally active at this time, directs spermatogenesis for a period of hours. Next, repression of *fem-3* is required for the switch from spermatogenesis to oogenesis. Thus, PMF is proposed to bind the PME to repress *fem-3* translation and permit oogenesis. The translational activity of *tra-2* during oogenesis is not known.

The strongest candidates for *trans*-acting gene products that act through the *fem-3* PME are encoded by six *mog* genes. Among these genes, *mog-1* is best characterized (Graham and Kimble 1993). The *mog-1* loss-of-function phenotype is a failure to switch from spermatogenesis to oogenesis, much like the *fem-3* gain-of-function phenotype. The phenotype of the other *mog* genes is similarly a masculinization of the hermaphrodite germ line (Graham et al. 1993). Intriguingly, all six *mog* genes not only are required for control of the sperm/oocyte switch, but also act maternally to influence embryogenesis. Thus, the *mog* genes may control not only *fem-3*, but other maternal mRNAs as well.

Translational Controls of tra-2 *and* fem-3 *in Germ Line and Soma*

Figure 3 depicts our current understanding of how translational controls contribute to the spatial pattern of cell fates in the hermaphrodite germ line. The temporal regulation of two translational controls appears to underlie the spatial pattern of differentiation in the gonad: proximal sperm and distal oocytes (Fig. 3). First, translational repression of *tra-2* permits the onset of spermatogenesis at one end of the gonad. Later, translational repression of *fem-3* mediates the switch from spermatogenesis to oogenesis in more distal cells. Thus, translational controls of *tra-2* and *fem-3* are central for determining the spatial distribution of those cell fates in the gonad.

As emphasized above, translational controls of *tra-2* and *fem-3* are essential for specification of germ-line cell fates. Importantly, translational controls of *tra-2* and *fem-3* also influence determination of sexual characteristics in somatic tissues (Doniach 1986; Schedl and Kimble 1988). The germ-line and somatic controls depend on the same 3′UTR regulatory elements and may depend on the same *trans*-acting regulatory machinery (E. Goodwin et al., unpubl.). For example, *laf-1* is required for translational repression via the DREs in both germ line and soma (E. Goodwin and J. Kimble, unpubl.).

Mesoderm Specification in Xenopus

In frogs, mesoderm arises through a process termed induction, in which a signal is secreted from endodermal cells at the bottom of the embryo to overlying cells, causing those cells to follow mesodermal fates (for review, see Melton 1994). Members of the fibroblast growth factor (FGF) and transforming growth factor β (TGF-β) families of secreted polypeptides are likely signals in this process, as are the cell-surface receptors to which they bind (for review, see Melton 1994).

Two forms of translational controls have been implicated in mesoderm induction. The first involves a maternal mRNA encoding an FGF receptor, FGFR-1 (Robbie et al. 1995). Expression in embryos of a dominant inhibitory form of the FGF receptor interferes with mesoderm induction in vivo, presumably by titrating wild-type receptors into inactive complexes (Amaya et al. 1991, 1993). These and other results strongly suggest that the FGF receptor and its ligand have a key role in mesoderm induction (for review, see Melton 1994). FGFR-1 mRNA is silent in oocytes, but activated during oocyte maturation, prior to fertilization (Musci et al. 1990). The repression is due to a negative regulatory element in the 3′UTR of FGFR-1 mRNA, in the 180 nucleotides immediately downstream from the termination codon. The temporal or spatial control of its derepression may be important in embryonic induction, although the existence of multiple receptors for FGF-related ligands may complicate the issue.

A second speculative role for translational control in mesoderm induction proposes that increased activity of eIF4E in the embryo specifically stimulates the translation of activin, a member of the TGF-β superfamily and a potent inducer of mesoderm (for review, see Melton 1994). This idea is based, in part, on the finding that overexpression of the general translation factor eIF4E in frog embryos induces mesodermal fates in cells that would otherwise form ectoderm (Klein and Melton 1994). Moreover, eIF4E overexpression specifically stimulates translation of injected activin mRNA, without affecting either total protein synthesis or other injected mRNAs (Klein and Melton 1994).

Activin and eIF4E may comprise a positive feedback loop. Mesoderm induction by eIF4E is blocked by coexpression of a dominant inhibitory form of the activin receptor (Klein and Melton 1994). Since mRNA injection experiments imply that activin translation may be stimulated by eIF4E, these data suggest a simple autocrine loop: Activin elevates eIF4E levels, which further enhances activin synthesis. A circuit of this type could both amplify the initial inducing signal and explain how one cell that has been induced to form mesoderm can induce mesoderm in an adjacent cell. This model predicts that the level of eIF4E activity is elevated during early development, at least in certain blastomeres, and that specific mRNAs involved in mesoderm induction should be stimulated as a result. Those mRNAs might encode activin or other mesoderm inducers.

Although the activin/eIF4E circuit is speculative, it closely parallels an apparent mechanism of neoplastic transformation of mammalian cells by overexpression of eIF4E (De Benedetti and Rhoads 1990; Lazaris-Karatzas et al. 1990; Sonenberg, this volume). In this case, as in mesoderm induction, elevation of the levels of a general translation factor has dramatic effects on cell fate.

Pattern Formation: Embryonic Axes

During oogenesis and early embryogenesis, asymmetries become evident that foreshadow the anterior-posterior, dorsal-ventral, and left-right axes of the mature organism. Translational controls are critical for establishing body axes (for review, see Wharton 1992; Curtis et al. 1995). In *Drosophila*, each of the maternal patterning systems (St Johnston and Nüsslein-Volhard 1992) requires the translational control of one or more mRNAs (for examples, see Table 1). Recent evidence from *C. elegans* and *Xenopus* suggests that one mechanism of controlling embryonic asymmetry—localized translational repression—may be a very primitive and universal means of laying down the anterior-posterior axis.

Table 1 Translational control in the four maternal patterning systems of *Drosophila:* Representative examples

Maternal system	Translationally controlled mRNA	Role of protein product
Anterior	*bicoid* (Driever and Nüsslein-Volhard 1988a)	anterior determinant, activates genes required for head and thorax formation (Frohnhoefer and Nüsslein-Volhard 1986; Driever and Nüsslein-Volhard 1988a,b); also required to repress translation of *caudal* mRNA (Struhl 1989), which encodes a homeobox protein (Mlodzik et al. 1985)
Posterior	*nanos* (Gavis and Lehmann 1994)	posterior determinant (Wang and Lehmann 1991; Wang et al. 1994); collaborates with *pumilio* to suppress translation of posterior maternal *hunchback* mRNA (Hulskamp et al. 1989; Irish et al. 1989; Struhl 1989a; Murata and Wharton 1995), which encodes a transcription factor (Hulskamp et al. 1990)
Terminal	*torso* (Casanova and Struhl 1989; Sprenger et al. 1989)	cell-surface receptor that responds to localized extracellular ligand to generate terminal structures (Stevens et al. 1990; Martin et al. 1994)
Dorsal-ventral	*toll* (Gay and Keith 1992)	cell-surface receptor that responds to localized extracellular ligand to generate ventral structures (Hashimoto et al. 1988; Stein et al. 1991; Morisato and Anderson 1994)

Coordinate Activation

The maternal transcripts of several axis-determining genes are translationally dormant in oocytes but are activated soon after fertilization. This coordinate activation may commonly require cytoplasmic polyadenylation. *bicoid*, *torso*, and *toll* mRNAs, encoding key regulatory proteins for the anterior, terminal, and dorsal-ventral patterning systems, respectively (Table 1), undergo polyadenylation concomitant with their activation. The role of polyadenylation in the activation of *bicoid* mRNA has been examined in some detail (Sallés et al. 1994). The poly(A) tail of endogenous *bicoid* mRNA increases from about 50 nucleotides in oocytes, where it is translationally silent, to 150 nucleotides in the early embryo, coincident with its activation. mRNA injection experiments indicate a critical role for cytoplasmic polyadenylation in activation: Injected *bicoid* mRNA possessing its wild-type 3′UTR receives poly(A) and rescues a *bicoid* mutant embryo, whereas a truncated transcript lacking part of the 3′UTR lacks both polyadenylation and rescuing activity. Importantly, the rescuing activity of the truncated mRNA is restored (although not to wild-type levels) by the addition, in vitro, of a 150-nucleotide poly(A) tail. A mere 50 nucleotides, as is present in the oocyte, do not suffice. These data demonstrate that polyadenylation is critical in the activation of *bicoid* mRNA (and by extrapolation, perhaps to *torso* and *toll* mRNAs as well) and indicate that signals required for activation lie, at least in part, in the 3′UTR.

Unlike *bicoid* mRNA, translational activation of mRNA encoding *nanos*, a posterior determinant, does not involve a detectable change in the length of its poly(A) tail upon fertilization (Sallés et al. 1994). Although the mechanism of *nanos* mRNA activation is not understood, the posterior location of the mRNA in the egg is critical (Gavis and Lehmann 1994). Normally, *nanos* mRNA is produced in nurse cells, imported into the oocyte at its anterior end, and then transported to its final destination at the posterior pole. *nanos* mRNA can be activated after fertilization if it resides at the posterior pole; placed in more anterior regions by manipulation of its localization signals, it remains repressed (Gavis and Lehmann 1994). The posterior localization of *nanos* mRNA may counteract an unlocalized *nanos* translational repressor. In wild-type embryos, although *nanos* mRNA is concentrated at the posterior pole, it is present at a low level throughout the embryo (Gavis and Lehmann 1994). Presumably, repression outside the posterior pole is required to avoid deleterious effects of Nanos protein in the anterior, namely, inhibition of head and thorax development (Wharton and Struhl 1989; Gavis and Lehmann 1992; Gavis and Lehmann 1994). The *nanos* 3′UTR contains elements that not only cause the localization of mRNA to the posterior, but prevent its translation when it is located elsewhere in the embryo (Gavis and Lehmann 1994).

Regulated Repression

The anterior-posterior axis is established by a set of maternally contributed mRNAs, including *bicoid, caudal, nanos, pumilio,* and *hunchback.* Establishing a gradient of Hunchback protein, with high Hunchback in the anterior and low Hunchback in the posterior, is essential. To achieve this gradient, Bicoid protein at the anterior pole activates zygotic transcription of *hunchback*, whereas Nanos and Pumilio proteins at the posterior pole repress translation of maternal *hunchback* mRNA.

Repression of maternal *hunchback* mRNA depends on regulatory elements in the *hunchback* 3′UTR, called Nanos Response Elements (NREs; Wharton and Struhl 1991). Both Pumilio and Nanos proteins are necessary for repression of maternal *hunchback* mRNA (Barker et al. 1992); neither protein alone is sufficient. Pumilio protein binds NRE-containing RNAs specifically, and this probably underlies its repressive activity (Murata and Wharton 1995). On the basis of the available evidence, the following model has been proposed to explain how these proteins might collaborate to repress *hunchback* translation: Uniformly distributed Pumilio binds directly to the NREs but is incapable alone of repressing translation. However, in the posterior of the embryo, Nanos protein is recruited to the NREs via protein-protein interactions, enabling suppression of the mRNA. In this elegant model, the asymmetric repression of *hunchback* mRNA function is provided by the asymmetric distribution of Nanos, which may not in fact interact directly with the RNA.

As if this complexity were not enough, Bicoid protein, in addition to its role as a transcriptional factor, is required for translational repression of *caudal* mRNA, another mRNA important in axis formation. In the ab-

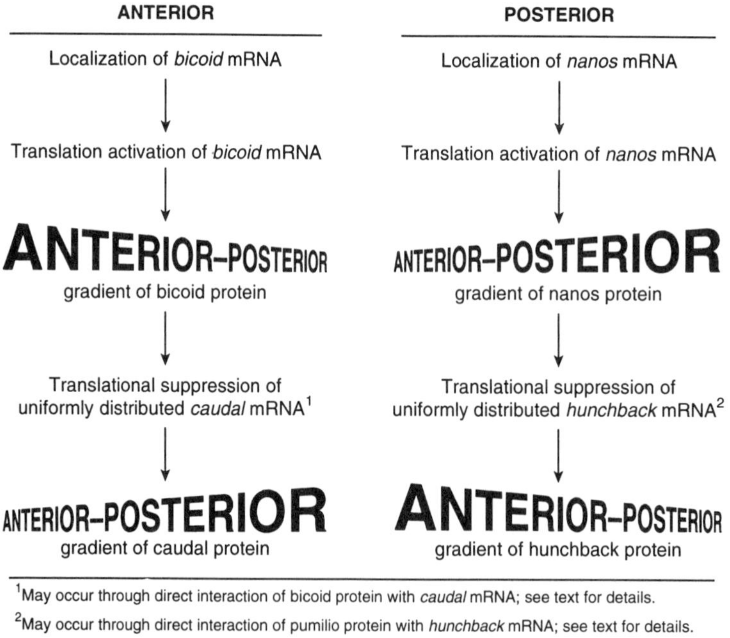

Figure 4 Anterior-posterior gradients generated by translational activation and repression. See text for details.

sence of *bicoid* activity, the normal gradient of Caudal protein (low in the anterior to high in the posterior) is disrupted, with high Caudal now found at the anterior as well (MacDonald and Struhl 1986; Mlodzik and Gehring 1987; Driever and Nüsslein-Volhard 1988b). Bicoid protein binds to the 3′UTR of *caudal* mRNA and this interaction appears to be essential for translational repression (Dubnau 1995; J. Dubnau and G. Struhl, in prep.). Remarkably, the homeodomain region of Bicoid protein is required to bind both to *caudal* mRNA and to DNA targets in its role as transcriptional activator (Dubnau 1995; J. Dubnau and G. Struhl, in prep.).

The regulation of axis-forming genes thus involves two parallel cascades of translational control at opposite ends of the embryo (Fig. 4). Many of the key players (*bicoid, nanos, caudal*, and *hunchback*) initially are translationally dormant and are activated only after fertilization. At the anterior, newly synthesized Bicoid protein represses *caudal* mRNA and ultimately activates zygotic transcription of *hunchback*; at the posterior, Nanos protein represses *hunchback* mRNA. The posteriorly localized *hunchback* mRNA is destroyed (Tautz and Pfeifle 1989). Thus, this web of interactions establishes opposing gradients of Hunchback and Caudal proteins.

A Potentially Primitive Mechanism for Establishing Asymmetry

Translational regulation is critical for polarity along the anterior-posterior axis in *C. elegans* as well as in *Drosophila* (Evans et al. 1994; Kimble 1994). This translational regulation effects the asymmetric expression of a membrane receptor, encoded by the *glp-1* gene. The Glp-1 protein is required for inductive cell interactions in the early nematode embryo (Priess et al. 1987; Hutter and Schnabel 1994; Mello et al. 1994; Moskowitz et al. 1994). When Glp-1 protein first appears, at the two- to four-cell stage, it is found in anterior but not posterior blastomeres; *glp-1* maternal mRNA, in contrast, is distributed uniformly (Evans et al. 1994). The anterior-posterior asymmetry of Glp-1 protein may be crucial for establishment of asymmetric patterns of cell-cell interactions, which in turn specify dorsal-ventral and left-right axes of the *C. elegans* embryo (Hutter and Schnabel 1994; Mello et al. 1994; Moskowitz et al. 1994).

glp-1 mRNA is subject to two distinct translational controls (Evans et al. 1994). One is temporal: *glp-1* is translationally silent in oocytes and one-cell embryos, but it becomes active by the two- to four-cell stage. The second is spatial: *glp-1* is translationally silent in posterior blastomeres but active in anterior blastomeres. The elements responsible for both controls reside in the 3′ UTR. Sequences responsible for spatial regulation lie in a 39-nucleotide region in the central region, whereas those required for temporal control lie at the 3′ end of the 3′ UTR (Evans et al. 1994; T. Evans and J. Kimble, unpubl.). Within the 39-nucleotide sequence is a bipartite sequence with striking similarity to an NRE from *Drosophila* (Fig. 5A). Therefore, asymmetry in early *C. elegans* and *Drosophila* embryos may be established by analogous mech-anisms: Translation of uniformly distributed *hunchback* and *glp-1* maternal transcripts is restricted to anterior regions of their embryos, through spatially controlled translational repression (Fig. 5B). The similarity in the sequence elements involved—the NREs of *Drosophila* (Wharton and Struhl 1991) and the NRE-like motif of *C. elegans* (Evans et al. 1994)—suggests that both the molecular components and the overall strategy may be conserved.

The molecular parallels between *hunchback* and *glp-1* regulation suggest the existence of an ancient mechanism for creating asymmetric patterns of gene expression in early embryos (Fig. 5B). A hint that this mechanism may function in vertebrates comes from the identification of

Figure 5 A potentially primitive mechanism for establishing asymmetry. (*A*) Similar regulatory elements. The 3′ UTRs of *Drosophila hunchback* and *C. elegans glp-1* contain sequence elements predicted to mediate translational control. The *Drosophila* NRE is a bipartite element: GUUGU separated by five nucleotides from AUUGUA. Two copies of this element reside in the *hunchback* 3′ UTR, and 1.5 copies reside in the *bicoid* 3′ UTR. The *C. elegans* NRE-like motif has a similar sequence, although the spacer region is larger. The significance of the NRE-like sequences of *C. elegans* remains to be shown by site-directed mutagenesis, although the coincidence of the motifs is striking. A 39-bp deletion that removes the 34-nucleotide sequence shown plus an additional 5 bp disrupts spatial control (T. Evans and J. Kimble, unpubl.). Each mRNA is drawn to the same scale. Two different *hunchback* mRNAs exist, differing in their 5′ UTRs (either 510 or 146 nucleotides); the coding region contains 2276 nucleotides, and the 3′ UTR contains 562 nucleotides. The first NRE begins 55 nucleotides downstream from the termination codon. For *glp-1* mRNA, the 5′ UTR contains ~90–100 nucleotides, the coding region 3885 nucleotides, and the 3′ UTR 365 nucleotides. The 5′-most AAUGA sequence, part of the *glp-1* NRE-like elements, lies 182 nucleotides downstream from the *glp-1* termination codon. (*B*) Proposed similarities in posterior translational repression in *Drosophila* and *C. elegans* embryos. *hunchback* and *glp-1* maternal mRNAs are uniformly distributed, whereas the Hunchback and Glp-1 proteins are expressed in the anterior. (*Black dots*) Polar granules in *Drosophila* and P granules in *C. elegans*. Maternal mRNA encoding nanos appears to be associated with polar granules; perhaps a translational repressor is associated with P granules in *C. elegans*.

a maternal *nanos*-like RNA, called Xcat-2, which is localized to the vegetal pole of *Xenopus* embryos (Mosquera et al. 1993). Although the function of Xcat-2 is unknown, its location suggests a role in early pattern formation. If similar molecular machinery regulates polarity in embryos as diverse as worms, flies, and frogs, it seems plausible that this mechanism participates in axis formation in all animal embryos, including mammals. "Molecular tinkering" (Jacob 1982) might then come into play to reinforce or modify this primitive strategy and derive other axes from it.

Temporal Control of Developmental Events

Translational controls are not restricted to maternal mRNAs and early embryos. Indeed, a particularly provocative form of translational control directs progression through the life cycle in the somatic tissues of the nematode *C. elegans*. Normally, *C. elegans* passes through four distinct larval stages, called L1, L2, L3, and L4, to reach maturity. This progression from L1 to adulthood depends on several "heterochronic" genes, including *lin-14* and *lin-4* (for review, see Ambros and Moss 1994).

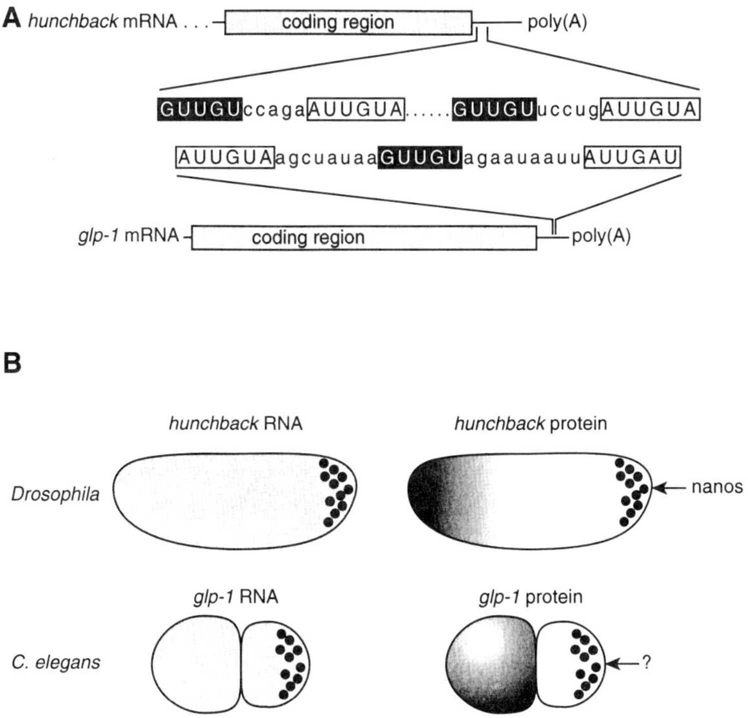

Figure 5 (*See facing page for legend.*)

Translational regulation of *lin-14* is essential for progression through the life cycle. The *lin-14* gene directs L1-specific events (Ambros and Horvitz 1984). Normally, L1 larvae possess abundant *lin-14*-encoded protein, whereas later stages in the life cycle possess little or none (Ruvkun and Giusto 1989). In contrast, *lin-14* mRNA is equally abundant throughout larval development (Wightman et al. 1993). Two *lin-14(gf)* mutants reiterate the L1 larval stage and possess Lin-14 protein throughout larval development (Ambros and Horvitz 1984; Ruvkun and Giusto 1989). The molecular defects in *lin-14(gf)* mutations reside in the DNA encoding the *lin-14* 3′UTR (Ruvkun et al. 1989; Wightman et al. 1991). Furthermore, the *lin-14* 3′UTR can confer upon a reporter gene a pattern of expression typical of Lin-14 protein (Wightman et al. 1993). Therefore, the *lin-14* 3′UTR is essential for the translational down-regulation of *lin-14* activity and progression into the L2 stage.

Both genetic and molecular analyses indicate that *lin-4* is required for the translational repression of *lin-14* (Ambros 1989; Arasu et al. 1991). Animals lacking *lin-4* activity reiterate L1-specific events (Chalfie et al. 1981), as do *lin-14(gf)* mutants. Furthermore, *lin-4* activity is essential for the translational repression of a chimeric reporter gene by the *lin-14* 3′UTR (Wightman et al. 1993). Remarkably, the *lin-4* gene encodes two short RNAs (22 and 61 nucleotides) with no apparent coding capacity for synthesis of a protein product. Instead, both RNAs have antisense complementarity to each of seven conserved elements present in the *lin-14* 3′UTR (Fig. 6) (Lee et al. 1993; Wightman et al. 1993). Therefore, the *lin-4* RNAs themselves are likely to be at least part of the *trans*-acting machinery that regulates *lin-14* translation. Proteins may also be critical, however. The secondary structures of each potential hybrid, and the sequence of the "looped-out" regions (Fig. 6B), are quite similar and could serve as protein-binding sites.

Figure 6A presents a simple binary switch, in which *lin-14* is translationally active during L1 and repressed in late L1 to effect the transition to L2. However, various lines of evidence suggest that the control may be more complex. Although the data are not yet definitive, a temporal gradient for *lin-14* activity has been suggested, with high, intermediate, and low levels of *lin-14* activity directing the L1, L2, and L3 stages of the life cycle, respectively (Austin and Kenyon 1994 and references therein). The presence of seven *cis*-acting elements in the *lin-14* 3′UTR provides a plausible molecular mechanism for generating such a temporal gradient: Increasing occupancy of *lin-14* sites by *lin-4* RNA might result in graded repression of translation.

The identification of the *lin-4* repressor is unambiguous, and its likely

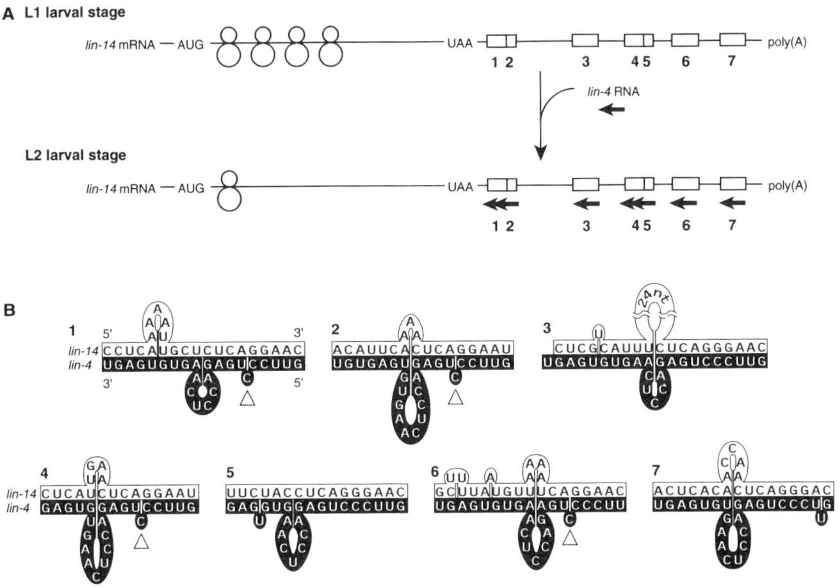

Figure 6 lin-14 mRNA and the *lin-4* repressor. (*Top*) Speculative model for the role of *lin-4* in translational repression of *lin-14* mRNA. (*Gray circles*) Ribosomes; (*thin lines*) *lin-14* mRNA; (*small open rectangles*) putative regulatory sites to which *lin-4* RNA may bind; (*thick black arrow*) *lin-4* RNA (the arrowhead is at the 3′end of *lin-4* RNA). The *lin-14* 3′UTR possesses seven conserved elements (1–7) that are likely to be translational regulatory elements. During the L1 larval stage, *lin-14* is translated; the translational repressor, *lin-4*, then associates with regulatory elements and *lin-14* becomes translationally repressed. The mRNA is not drawn to scale. (*Bottom*) Possible hybrids between *lin-14* mRNA and *lin-4* RNA. (*Open rectangles*) Elements in *lin-14* mRNA; (*closed rectangles*) *lin-4* RNA. The location of a point mutation in *lin-4* which reduces its activity is indicated by an open triangle in hybrids 1, 2, 4, and 6. Note that only a subset of these structures may be needed for repression and that a direct demonstration of the hybrids depicted is not yet available.

binding sites have been identified by sequence inspection. The biochemical mechanism of its action is unknown, however, and could involve collaboration with proteins. Nevertheless, the data hint that the *trans*-acting regulators of other mRNAs may, like the *lin-4* repressor, be RNA, not protein (see Wickens and Takayama 1994).

Terminal Differentiation

Certain genes are expressed late in differentiation, as cells take on their ultimate fates. In the first two examples of terminal differentiation described below—spermatogenesis and red blood cell differentiation—the nucleus is effectively silenced: The spermatid pronucleus is highly condensed and inactive, and in most mammals, red blood cells lose their nucleus entirely. In these cases as in the early embryo, the cell must exploit translational control to change the proteins it contains. In the third example, involving myogenesis, the nucleus remains active, and transcriptional regulation is in fact critical.

Spermatogenesis

The male gamete exhibits translational control of various mRNAs (for listing, see Braun et al. 1989). Mouse protamine-1 (mP1) mRNA is stored in a dormant form as a ribonucleoprotein particle for 1 week in round spermatids before it is activated in elongating spermatids (Balhorn et al. 1984; Kleene et al. 1984). Sequences in the mP1 3′UTR are necessary and sufficient for proper temporal regulation in transgenic mice (Braun et al. 1989).

Activation of mP1 mRNA is accompanied by shortening of its poly(A) tail (Kleene et al. 1984; Braun et al. 1989). This deadenylation may cause, or be a consequence of, translational activation. *Drosophila* spermatogenesis further complicates the issue, as translational activation of two mRNAs, Mst87D and Mst87F, in postmeiotic spermatids is correlated with an increase in poly(A) tail length (Schaefer et al. 1990; Kuhn et al. 1991).

15-Lipoxygenase and Red Blood Cell Differentiation

As mammalian reticulocytes differentiate into erythrocytes, their mitochrondria are destroyed. The enzyme 15-lipoxygenase (LOX) catalyzes deoxygenation of polyenoic fatty acids, even in intact membranes, and is thought to be critical for the destruction of mitochondria. Although LOX mRNA apparently is present even at early stages of erythropoiesis, it is not translated until reticulocytes mature into erythrocytes (Thiele et al. 1982).

The 3′UTR of rabbit LOX mRNA contains ten nearly perfect repeats of a 19-nucleotide sequence, whereas the mouse mRNA contains four similar repeats in a comparable location (Hunt 1989; Ostareck-Lederer et al. 1994). In recent elegant experiments, a 48-kD protein has been purified that binds to this sequence and causes translational repression specifically of LOX mRNA in vitro (Ostareck-Lederer et al. 1994). mRNAs containing as few as two repeats are fully repressed, whereas a single repeat yields only slight repression (B. Thiele, pers. comm.). Importantly, repression in vitro appears to be independent of any change in poly(A) length and of the 5′-terminal cap; we return to these points below.

Myogenesis: A Surprising Feedback Loop

Myoblasts fuse to form myotubes that express muscle-specific genes at high levels. Myogenic transcription factors, such as myoD, myogenin, and Myf5, are key players in the specification of muscle cell fate (for review, see Weintraub 1993; Olson and Klein 1994). However, a search for genes that influence muscle differentiation has revealed a remarkable regulatory circuit that may operate at the translational level. In cultured myoblasts, overexpression of the tropomyosin 3′UTR can cause transcriptional activation of muscle-specific genes, such as myosin, actin, and myoD (Rastinejad and Blau 1993; for review, see Wickens 1993). Thus, it appears that a positive feedback loop may reinforce the decision to become muscle. How does this work?

Myogenic differentiation and cell division appear to be mutually antagonistic (for review, see in Halevy et al. 1995; Parker et al. 1995). Remarkably, the tropomyosin 3′UTR inhibits cell growth in culture (Rastinejad and Blau 1993) and in transplanted tumor cells (Rastinejad et al. 1993). Thus, by inhibiting growth, the commitment to muscle differentiation might be reinforced and cause activation of muscle-specific genes. The mechanism by which the tropomyosin 3′UTR acts is unclear, but it may involve inhibition of the cell's translation apparatus. In in vitro translation assays, the tropomyosin 3′UTR inhibits translation of a reporter mRNA in *trans,* apparently by binding to and activating a cellular kinase, PKR (protein kinase, RNA-activated; Davis and Watson 1996). Active PKR phosphorylates eIF2α, blocking eIF2 function and thereby depressing overall protein synthesis (see Clemens, this volume). In vivo, inhibition of general protein synthesis might depress cell growth and thereby stimulate myogenesis. The translational inhibition presumably would be transient because muscle-specific mRNAs are ultimately translated efficiently. The finding that expression of inactive, dominantly interfering PKR mutants in NIH-3T3 cells elicits a transformed phenotype is consistent with a role for this kinase in growth control (Koromilas et al. 1992; Meurs et al. 1993).

An alternative explanation of the ability of tropomyosin 3′UTR to repress cell growth and stimulate transcription of muscle-specific genes involves effects on the recently discovered stimulation of CDK inhibitors by MyoD (Halevy et al. 1995; Parker et al. 1995; Skapek et al. 1995).

The tropomyosin 3′UTR could control a specific mRNA in that pathway in *trans*, perhaps by titrating a critical RNA-binding protein. With some of the key players in that circuit now identified, it may be possible to test this hypothesis directly.

A feedback circuit involving the tropomyosin 3′UTR has not yet been demonstrated in the intact mouse. Nevertheless, the work to date establishes the important principle that a single 3′UTR, acting in *trans*, can dramatically affect a cell's properties.

Masking

The masking hypothesis, initially proposed by Spirin (1966) nearly 30 years ago, suggests that specific mRNAs are repressed through the action of proteins that hide them from the translational apparatus. In response to a stimulus, such as fertilization, the masking proteins are removed, the mRNA is revealed, and its translation begins. The mRNA itself is unaltered. In its initial formulation, masking was proposed to explain the dramatic increase in protein synthesis observed in sea urchin eggs at fertilization, an effect that is in part due to stimulation of the translation machinery itself. As it is now understood (see, e.g., Standart 1992; Standart and Jackson 1994), masking during early development is operationally defined using extracts derived from oocytes and early embryos. In vivo, a specific mRNA is repressed in the oocyte; upon fertilization, it becomes active. The patterns of protein synthesis are maintained in extracts of oocytes and early embryos; in particular, in the form of mRNPs (mRNA-protein complexes), mRNAs that are repressed in vivo continue to be repressed when translated in vitro. Removal of the proteins activates (i.e., "unmasks") the mRNA in vitro. Protein removal can be accomplished crudely, for example, by extraction with organic solvents or by more subtle means, as described below (for review, see Standart 1992; Standart and Jackson 1994).

Clam ribonucleotide reductase mRNA provides a well-studied paradigm for masking (Standart et al. 1990). Unmasking of this mRNA in an extract of surf clam oocytes can be achieved by incubation in 0.5 M KCl and gel filtration, which presumably removes the "masking factor." Masking can be restored in the extract by removal of the salt prior to gel filtration, which presumably permits the factor to rebind. Remasking in this fashion requires sequences in the 3′UTR (Standart et al. 1990). An 82-kD protein that may bind to the critical sequences and participate in masking has been identified by UV cross-linking (Standart 1992).

Masked ribonucleotide reductase mRNA can be derepressed in oocyte extracts by severing the 3′UTR from the body of the mRNA, using targeted RNase H cleavage (Standart et al. 1990). The activation appears to be independent of polyadenylation, even though the mRNA receives poly(A) as it is activated in vivo. These data imply that removal of 3′UTR-bound factors is sufficient for derepression and that derepression in vitro can be uncoupled from poly(A) addition.

The approaches used to study masking of maternal mRNAs differ substantially from those in which regulatory elements have been identified either genetically or by mRNA injections. Masking of maternal mRNAs is classically defined as repression, which then is followed by activation, or unmasking. Only some of the mRNAs we have discussed in previous sections, *bicoid* and *glp-1,* for example, appear to behave in this way. In contrast, frog cyclin B1 mRNA is already expressed at a low level before oocyte maturation begins and so, at the least, may not be fully masked; *fem-3* and *lin-14* mRNAs are initially translationally active and then are shut off, the opposite of the situation in classical "masking." It is uncertain whether these different mRNAs are repressed by the "masking" mechanism exemplified by ribonucleotide reductase mRNA.

REGULATORY ELEMENTS: GENERAL ISSUES

Although the regulatory elements discussed here come from many different organisms and control a dramatic array of developmental decisions, they share certain unmistakable similarities.

3′UTRs

All of the elements we have described reside in the 3′UTR. It is remarkable that methods ranging from classical genetics to biochemistry should all converge on this region of the mRNA. Perhaps the strongest argument that regulatory elements commonly reside in 3′UTRs arises from genetic analysis: Screens for mutant animals that misregulate key genes have repeatedly yielded mutations in 3′UTRs, even though the selection schemes had no such bias.

Why the 3′UTR? The prevalence of regulatory elements in 3′UTRs may reflect the fact that 3′UTRs are unconstrained in evolution and so provide fertile ground for the derivation of new regulatory elements (see Wickens 1993). In contrast, the 5′UTR must be scanned prior to translation initiation, and alterations in its sequence, structure, or length can affect initiation. The coding region has even more obvious constraints.

Negative and Positive Control

By definition, negative elements repress translation, whereas positive elements either activate it or are required for translation per se. With the exception of the signals that control cytoplasmic polyadenylation, all of the other regulatory elements identified thus far in the 3′UTRs of mRNAs critical for development are negative. Negative control elements might work in either of two modes. Certain negative elements may repress translation as soon as the mRNA enters the cytoplasm, so that the mRNA begins life silently (e.g., *bicoid* and LOX mRNAs). Other negative elements may repress translation only after a period of translational activity (e.g., *fem-3* and *lin-14* mRNAs).

Sequences that control cytoplasmic polyadenylation may be both positive and negative. Most commonly, they are identified as positive elements required for polyadenylation and translational activation; mutant mRNAs lacking cytoplasmic polyadenylation elements do not receive poly(A) and their translation is not activated. However, the same elements may also be required to silence the mRNA as it emerges into the cytoplasm, through a process involving poly(A) shortening (Huarte et al. 1992).

Translational control elements in 3′UTRs may be either on-off switches or rheostats. Many regulatory elements in 3′UTRs are tandemly repeated. Indeed, the existence of repeated, conserved sequences in 3′UTRs is reasonable prima facie evidence of a control element. Elimination of some but not all of the regulatory sites in *tra-2* (Goodwin et al. 1993), *lin-14* (Wightman et al. 1993), and LOX mRNAs (B. Thiele et al., pers. comm.) yields an intermediate level of translation. Similarly, mRNAs containing a single NRE, rather than two, appear to be repressed less efficiently in vivo (Wharton and Struhl 1991). In wild-type mRNAs, partial occupancy of multiple sites may allow the level of translation to be modulated incrementally. Alternatively, multiple elements might facilitate cooperative binding of regulatory factors and facilitate concerted repression.

The genetic selections that have revealed negative translational control elements in 3′UTRs, as have been carried out with *tra-2* and *fem-3*, have been exceptionally powerful. As a result, even very rare alleles could be isolated. The frequency of their isolation was one to two orders of magnitude less than that of typical knockout mutations. Negative elements may well exist in many other genes but have escaped genetic detection.

Germ Line and Soma

Translational regulation is vital in both germ-line and somatic tissues. The importance of translational regulation in differentiation of somatic tissues is often disregarded, since the number of well-documented examples is far fewer in somatic cells than in eggs. Moreover, some of the examples that do exist actually undermine the notion that somatic regulation might be general: LOX and protamine mRNAs are translationally regulated in "peculiar" cells in which transcriptional control is impossible. Nevertheless, we suggest that translational control likely has a wider role than previously expected in somatic decisions. The case of *lin-14* is exemplary: It must be repressed in somatic cells at the L1 stage for cells to advance to their next developmental fate. Similarly, *fem-3* and *tra-2* mRNAs are translationally regulated in somatic tissues, as well as in the germ line, and that regulation is critical in determining cell fate. Oocytes and embryos, we suspect, do not exploit mechanisms peculiar unto themselves.

SPECULATIONS ON MECHANISM

The examples described earlier demonstrate that translational activity can be negatively controlled by sequences in the 3′UTR. Two central questions thus arise. First, how is translational repression exerted? Second, for those mRNAs that are first repressed and later activated, how is derepression accomplished? We discuss possible answers to these questions in turn. Both discussions are necessarily speculative as no conclusive answer to either of these problems is available in any system.

Translational Repression through the 3′ UTR

Translational repression could be achieved by interfering with initiation or elongation. In several instances, repressed mRNAs appear not to be associated with ribosomes or ribosomal subunits, suggesting that initiation is defective (for review, see Spirin, this volume). We therefore discuss five plausible mechanisms by which initiation may be prevented by elements in the 3′UTR. The proposed mechanisms of repression likely differ among mRNAs and are not mutually exclusive. For previous discussions of 3′UTR-mediated repression, presented with varying emphases, see Jackson and Standart (1990), Jackson (1993), Sonenberg (1994), Standart and Jackson (1994), and Curtis et al. (1995).

Nucleating Assembly of a Repressive Structure

In this model, mRNAs are repressed because they are assembled into a complex that effectively hides them from the translation apparatus. This complex might be an overall structure, which hides the mRNA in much the same way as chromatin condensation hides DNA from the transcription apparatus.

Y-box proteins, such as FRGY2 (also known as mRNP4), may be important in the formation of structures that cause repression (for review, see Wolffe et al. 1992; Wolffe 1994). FRGY2 is expressed in oocytes and not in somatic cells; homologs are present in somatic cells and may have comparable functions (for review, see Spirin, this volume). Y-box proteins, including FRGY2, are bona fide transcription factors (Tafuri and Wolffe 1990, 1992), yet are physically associated with many different maternal mRNAs (see, e.g., Darnbrough and Ford 1981; Dearsley et al. 1985; Murray et al. 1991; Tafuri and Wolffe 1993) and can inhibit their translation (Richter and Smith 1984; Kick et al. 1987; Ranjan et al. 1993; Bouvet and Wolffe 1994). These data suggest several provocative possibilities. For example, Y-box proteins might assemble with the mRNA to form a structure that effectively hides the mRNA. Dephosphorylation of Y-box proteins appears to enhance translational activation of the mRNA with which they are associated (Kick et al. 1987; Murray et al. 1991) yet may have little effect on the binding of Y-box proteins to RNA (Tafuri and Wolffe 1993; for a contrary view, see Kick et al. 1987). Thus, phosphorylation and dephosphorylation may influence Y-box protein activity, and hence translation, without modulating their association with RNA. More speculatively, dephosphorylation might conceivably "decondense" a complex structure and reveal the mRNA.

As yet, little sequence specificity has been demonstrated in either the RNA-binding or repressing activities of the Y-box proteins (Marello et al. 1992; Tafuri and Wolffe 1993). Thus, if the Y-box proteins do cause repression of some mRNAs but not others, some other factor must provide the sequence specificity.

Proteins bound to negative elements in the 3′ UTR could serve such a function, promoting the assembly of Y-box proteins into a repressive form or structure. Y-box proteins can be found associated with active mRNAs, arguing that their binding is insufficient for repression (Tafuri and Wolffe 1993). However, it may be instructive to bear in mind again the analogy to chromatin: Core histones are present on active and inactive genes, but their positions and higher-order structures differ and may have a critical role in regulating transcriptional activity.

Interfering with 5′ Cap Function

The 5′-terminal cap may be hidden from initiation factors through a specific RNA-protein interaction. In one form of this model (Standart and Jackson 1994), a protein bound to a regulatory site in the 3′UTRs prevents initiation by simultaneously binding the cap. 3′UTR-bound factors might interact directly with the cap or with proteins that do so. The finding that LOX mRNA without a cap can be repressed efficiently in vitro by LOX-BP (Ostareck-Lederer et al. 1994) argues against this model, at least in this case.

If the cap and poly(A) tail interact (discussed in Decker and Parker 1994; see Richter; Jacobson; both this volume), then repression could be caused by interfering with this interaction. 3′UTR-bound factors could interfere with the interaction by recognizing the cap, poly(A), or proteins associated with either element.

Keeping Poly(A) Tails Short

Although many studies suggest that polyadenylation is an integral part of translational activation, few address how poly(A) length is connected to repression. Do negative elements in the 3′UTR act by keeping the poly(A) tail too short? Do they repress through a mechanism that has nothing to do with having a short tail but which can be relieved by polyadenylation?

Two groups of mRNA injections support the hypothesis that the repression of a maternal mRNA is caused by the shortness of its poly(A) tail. In *Drosophila*, injected *bicoid* mRNAs with long poly(A) tails rescue *bicoid* mutant embryos, whereas the same mRNAs with shorter tails do not (Sallés et al. 1994); similarly, injected murine t-PA mRNAs are active with long, but not short, poly(A) tails, corresponding to their states before and after oocyte maturation (Huarte et al. 1992). The hypothesis that short tails cause repression leaves serious problems unresolved, however. For example, poly(A) tails as short as 25–50 nucleotides stimulate translation relative to an mRNA with no tail in several assay systems, both in vitro and in vivo (Jacobson; Richter; both this volume), yet repressed mRNAs often have tails longer than this. A further complication arises from the studies of ribonucleotide reductase mRNA discussed earlier. If its repression is due to its having "too short" a tail, then why does removal of its poly(A) tail (and 3′UTR) turn it on?

These considerations suggest that the repression of certain mRNAs may not be due to their having too short a tail but to a poly(A)-independent mechanism (such as the others discussed in this section). This does not preclude the possibility that poly(A) addition may have an important role in derepression.

Changing the Cellular Micro-Environment of the mRNA

Repressors bound to sites in the 3′UTR might move the mRNA into a cellular micro-environment that is translationally compromised. For example, the association of mRNAs or polyribosomes with the cytoskeleton may influence translation in vivo (for review, see Singer 1993a). Factors bound to the 3′UTR might prevent that association. One specific form of this hypothesis suggests that association with the cytoskeleton promotes interactions between translation factors and the mRNA, either because the translation factors themselves are associated with the cytoskeleton (Howe and Hershey 1984) or because association enhances ribosome recycling (Decker and Parker 1994).

Although mRNA sequences that cause cytoskeletal associations have not been identified, the elements that control mRNA "macrolocalization," such as movement to one end of a cell, often reside in the 3′UTR. These elements, sometimes called mRNA "zip codes," presumably are required for mRNA association with the cytoskeleton, movement and ultimate retention at its destination (for review, see Ding and Lipshitz 1993; Singer 1993b; Wilhelm and Vale 1993). Comparable elements may regulate more subtle movements as well.

Derepression

Changes in Regulatory Components

Negative *cis*-acting regulatory elements mediate repression; delete them, and the mRNA is activated aberrantly. If *trans*-acting factors are required for repression, which seems likely, then derepression presumably involves a change in 3′UTR-associated factors—dissociation from the mRNA, modification, or association with new components. In several systems, apparent regulatory factors are phosphorylated as repression is relieved (for review, see Standart and Jackson 1994). The temporal coincidence of phosphorylation and derepression suggests that phosphorylation may cause the change in translational activity, although no direct evidence yet demonstrates that this is so.

In the case of positive elements, such as those that cause polyadenylation, it appears that activation entails stimulation of a previously quiescent apparatus, probably through the activation of latent polyadenylation factors (Fox et al. 1992; Bilger et al. 1994; Richter, this volume). CPEB, a positive-acting polyadenylation factor (Hake and Richter 1994), is phosphorylated during maturation concomitant with polyadenylation of an mRNA to which it binds (Paris et al. 1991). Poly(A) polymerase also undergoes phosphorylation during maturation (Ballantyne et al. 1995). However, the role of phosphorylation in the activation of these or other polyadenylation factors, such as CPSF (see, e.g., Bilger et al. 1994) is uncertain.

Connections between Negative Elements, Derepression, and Polyadenylation

For many mRNAs, the transition from silence to activity is accompanied by an increase in poly(A) length. The connection between poly(A) and translation has been discussed elsewhere and so will not be recapitulated here (see Munroe and Jacobson 1990; Wickens 1992; Wormington 1993; Standart and Jackson 1994; Richter; Jacobson; both this volume). Instead, we focus explicitly on the connection between polyadenylation and negative translational control elements, so prevalent among critical maternal mRNAs. We consider three of many pathways that can explain how negative elements in the 3′UTR and polyadenylation might cause translational derepression. The pathways are depicted in Figure 7, in a simplified form intended to prompt consideration of other alternatives not considered here.

Pathway 1. Negative elements in the 3′UTR repress translation by keeping the tail short. The mRNA is derepressed by permitting polyadenylation. For example, a 3′UTR-bound factor that represses polyadenylation might dissociate, resulting in poly(A) lengthening, and, as a consequence, translational activation. This model accommodates the behavior of certain mRNAs very well, but it clearly does not account for those in which translational activation occurs without polyadenylation.

Pathway 2. Negative elements in the 3′UTR repress translation by preventing initiation: This repression is independent of poly(A) length. However, polyadenylation is required to inactivate the repressor. For example, repressors bound to the element might be removed or modified by

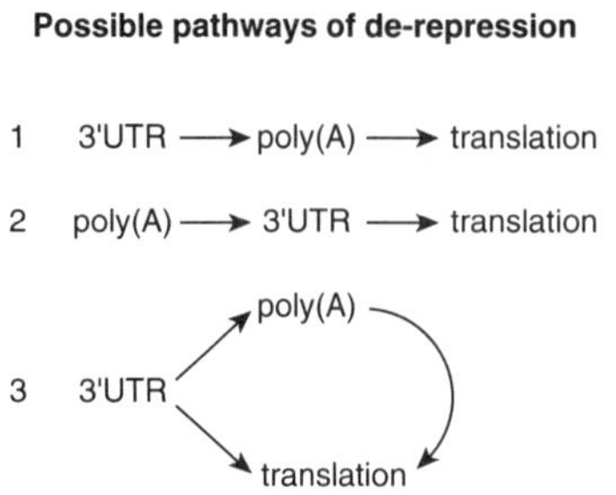

Figure 7 Possible pathways of derepression. "3′UTR" designates a negative regulatory element in the 3′UTR. For derepression, the activity of this element (or of the repressor to which it binds) must be shut off. "Poly(A)" indicates an increase in poly(A) length. "Translation" indicates derepression of translation. Thus, in the first pathway, shutting off the activity of the negative element (either by mutating the element or by regulating the repressor to which it binds) leads to an increase in poly(A) length, which in turn causes translational activation. Other possible pathways exist but are not depicted.

the binding of polyadenylation machinery. This pathway is suggested by experiments on G10 and Cl2 mRNAs in *Xenopus* in which the act of polyadenylation rather than the length of a poly(A) tail appears to be critical for activation (McGrew et al. 1989; Simon et al. 1992; Richter, this volume).

Pathway 3. Negative elements in the 3′UTR control polyadenylation and translation independently. For example, factors bound to the elements might repress by causing formation of an mRNP structure that hides the mRNA from both the translation apparatus and polyadenylation factors. Once the mRNA is exposed, both the translation and polyadenylation machineries act. Polyadenylation would then be required to maintain or enhance translational activity.

The third pathway accommodates most of the data. It posits that full derepression requires two experimentally separable steps: an initiation step that is independent of polyadenylation and a second step that is polyadenylation-dependent. Either process individually would yield incomplete, or improperly controlled, translation. The uncoupling of derepression and polyadenylation in vitro, as observed with LOX and ribonucleotide reductase mRNAs, would be due to execution of the initiation step alone; in vivo, polyadenylation would be required to complete or sustain the derepression. Conversely, the ability of polyadenylation to stimulate translation of an injected mRNA would reflect only the maintenance step; derepression of endogenous mRNAs in vivo would require a separate initiation step. Indeed, synthetic mRNAs injected into the cytoplasm, of the sort commonly used to assay the effect of polyadenylation on translation, may escape a form of repression to which endogenous mRNAs are subject (Bouvet and Wolffe 1994), and so may assay only part of the translational control pathway. Consistent with this view, mRNAs injected directly into the frog oocyte cytoplasm are more active than are the same mRNAs injected into the nucleus and transported to the cytoplasm subsequently (Bouvet and Wolffe 1994; Braddock et al. 1994). Although the mechanism by which "nuclear experience" causes lasting repression is unknown, antibody injection experiments suggest that it requires Y-box proteins, such as FRGY2 (Bouvet and Wolffe 1994; Braddock et al. 1994). Repression due to nuclear experience is relieved during oocyte maturation (Braddock et al. 1994), suggesting that such repression is reversible and may have a role in regulation in vivo.

NETWORKS, CASCADES, AND LOOPS

Networks of transcriptional control are commonplace and have crucial roles in development. A single transcription factor can activate and repress many genes, including those encoding other transcription factors; the intricate interactions of regulatory proteins at a promoter all provide inputs into the expression of a single gene. How similar might translational controls be, particularly during early development? Are there batteries of mRNAs that are interconnected through common factors?

Indications already exist of multiple translational control elements in a single 3′UTR: The *glp-1* 3′UTR contains separable spatial and temporal control elements; the *nanos* 3′UTR contains elements that move it to the posterior and permit its activation in the posterior but not elsewhere. In some respects, work on 3′UTRs may be at a stage equivalent to analysis in the early days of developmentally controlled promoters. We see separable elements, and hints of combinatorial regulation, but may not yet appreciate the depth of the complexity.

Cytoplasmic polyadenylation provides one obvious example of a potential network: Many mRNAs apparently require polyadenylation for translational activation or deadenylation for their inactivation. Thus, a change in polyadenylation activities or their regulators could coordinately control many mRNAs (for discussions, see Wickens 1992; Richter; Jacobson; both this volume).

In flies, the translational control circuits involved in formation of the anterior-posterior axis rely not only on the common activation of many mRNAs, but also on repression by one another's protein products. Do any of those repressors act on other, as yet unidentified, mRNAs? The observation that *pumilio* mutants, in addition to having effects on the anterior/posterior axis, have phenotypes during oogenesis and in somatic cells suggests that *pumilio* may regulate other mRNAs besides *hunchback.* Similarly, the *C. elegans mog* genes, which affect both the sperm/oocyte switch and embryogenesis more generally, may control the translation of numerous maternal mRNAs.

Regulatory components that act on more than one gene might be revealed by examining the phenotypes of animals overexpressing RNA regulatory signals. The feasibility of such an approach has been demonstrated by studies with the negative control element in the 3′UTR of *fem-3* mRNA; overexpression of this element, on its own, masculinizes the germ line (Ahringer and Kimble 1991). The simplest interpretation of this result is that the regulatory factor which binds to the element has been titrated out and can no longer repress the endogenous *fem-3* mRNA, leading to continuous spermatogenesis. Titration experiments of this type could, in principle, yield unexpected phenotypes that would strongly suggest new targets for the regulatory factor. Such was indeed the case with the tropomyosin 3′UTR and myogenesis.

Two positive feedback loops have been proposed that invoke effects on general translation factors: the "activin-eIF4E loop" in *Xenopus* mesoderm induction and the "tropomyosin 3′UTR-induced loop" in myogenesis. In both cases, a small incremental signal may be enhanced and reinforced through feedback to the general translation apparatus. Although neither loop has been demonstrated in an intact animal, the data raise the important possibility that other such loops might also exist to reinforce and commit cells to a particular fate.

WHY TRANSLATIONAL CONTROL?

It is striking that many key decisions in development rely on translational control. Clearly, during early embryogenesis, when pronuclei or zygotic nuclei are highly condensed and inactive, transcriptional control is not a major option. Why control maternal mRNA rather than maternal protein? Questions of this type are in one sense futile, as patterns of development evolve and so are restricted by contingency and history. However, within the constraints of a given developmental strategy, translational control can offer essential advantages. For example, activities involved in the earliest stages of pattern formation must be controlled in space and time. The Bicoid protein gradient cannot be established during oogenesis, because diffusion would collapse the gradient before it would have a chance to act in the early embryo. For regulatory proteins such as cyclin or Glp-1, premature translation might disrupt the timing of interaction with downstream factors or ligands. Moreover, in *Drosophila*, some translationally regulated proteins are themselves translational regulators and must be controlled in time and space. Testing these notions will require translation to be forced at the wrong time or in the wrong place as well as investigation of the biological consequences.

Factors that mediate translational control may be involved in transcription as well. For example, RNA-binding translational regulators, such as the FRGY2 and Bicoid proteins, are also DNA-binding transcription factors. This sort of dual function may be more common than we now realize, as it rarely would have been detected. The existence of translational regulatory factors in the embryo could provide molecular fodder for the subsequent evolution of transcriptional regulatory circuits, and vice versa. There is a sense in which the early embryo seems much like the primitive, prebiotic earth—an RNA world, free of the burden of DNA. Perhaps it is ontogeny's ultimate recapitulation of phylogeny.

ACKNOWLEDGMENTS

We thank Maria Gallegos, Scott Ballantyne, and Sunnie Thompson for comments on the manuscript and are grateful to Ann Helseley-Marchbanks, Carol Pfeffer, and Toni Daraio for help in its preparation. Laura VanderPloeg and Adam Steinberg of the Biochemistry Media Lab were very helpful preparing the figures. Work in the authors' laboratories is supported by National Institutes of Health research grants GM-31892 and GM-50942 to M.W., NIH grants GM-31816 and HD-24663 and National Science Foundation grant IBN9205215 to J.K, and NIH grants GM-51584, HD-25922, and HD-17875 to S.S.

REFERENCES

Ahringer, J. 1991. "Post-transcriptional regulation of *fem-3*, a sex-determining gene of *C. elegans*." Ph.D. thesis, University of Wisconsin, Madison.

Ahringer, J. and J. Kimble. 1991. Control of the sperm-oocyte switch in *Caenorhabditis elegans* hermaphrodites by the *fem-3* 3′ untranslated region. *Nature* **349:** 346–348.

Amaya, E., T. Musci, and M. Kirschner. 1991. Expression of a dominant negative mutant of the FGF receptor disrupts mesoderm formation in *Xenopus* embryos. *Cell* **66:** 257–270.

Amaya, E., P. Stein, T. Musci, and M. Kirschner. 1993. FGF signaling in the early specification of mesoderm in *Xenopus*. *Development* **118:** 177–187.

Ambros, V. 1989. A hierarchy of regulatory genes controls a larva-to-adult developmental switch in *C. elegans*. *Cell* **57:** 49–57.

Ambros, V. and H. Horvitz. 1984. Heterochronic mutants of the nematode *Caenorhabditis elegans*. *Science* **226:** 409–416.

Ambros, V. and E. Moss. 1994. Heterochronic genes and the temporal control *of C. elegans* development. *Trends Genet.* **10:** 123–127.

Arasu, P., B. Wightman, and G. Ruvkun. 1991. Temporal regulation of *lin-14* by the antagonistic action of two other heterochronic genes, *lin-4* and *lin-28*. *Genes Dev.* **5:** 1825–1833.

Austin, J. and C. Kenyon. 1994. Marking time with antisense. *Curr. Biol.* **4:** 366–369.

Balhorn, R., S. Weston, C. Thomas, and A. Wyrobek. 1984. DNA packaging in mouse spermatids: Synthesis of protamine variants and four transition proteins. *Exp. Cell Res.* **150:** 298–308.

Ballantyne, S., A. Bilger, J. Astrom, A. Virtanen, and M. Wickens. 1995. Poly(A) polymerases in the nucleus and cytoplasm of frog oocytes: Dynamic changes during oocyte maturation and early development. *RNA* **1:** 64–78.

Barker, D., C. Wang, J. Moore, L. Diskinson, and R. Lehmann. 1992. *Pumilio* is essential for function but for distribution of the *Drosophila* abdominal determinant nanos. *Genes Dev.* **6:** 2312–2326.

Barton, M., T. Schedl, and J. Kimble. 1987. Gain-of-function mutations of *fem-3*, a sex-determination gene in *Caenorhabditis elegans*. *Genetics* **115:** 107–119.

Bilger, A., C. Fox, E. Wahle, and M. Wickens. 1994. Nuclear polyadenylation factors recognize cytoplasmic polyadenylation elements. *Genes Dev.* **8:** 1106–1116.

Bouvet, P. and A. Wolffe. 1994. A role for transcription and FRGY2 in masking maternal mRNA within *Xenopus* oocytes. *Cell* **77:** 931–941.

Braddock, M., M. Muckenthaler, M. White, A. Thorburn, J. Sommerville, A. Kingsman, and S. Kingsman. 1994. Intron-less RNA injected into the nucleus of *Xenopus* oocytes accesses a regulated translation control pathway. *Nucleic Acids Res.* **22:** 5255–5264.

Braun, R., J. Peschon, R. Behringer, R. Brinster, and R. Palmiter. 1989. Protamine 3′-untranslated sequences regulate temporal translational control and subcellular localization of growth hormone in spermatids of transgenic mice. *Genes Dev.* **3:** 793–802.

Casanova, J. and G. Struhl. 1989. Localized surface activity of *torso*, a receptor tyrosine kinase, specifies terminal body pattern in *Drosophila*. *Genes Dev.* **3:** 2025– 2038.

Chalfie, M., H. Horvitz, and J. Sulston. 1981. Mutations that lead to reiterations in the cell lineages of *C. elegans*. *Cell* **24:** 59–69.

Colledge, W., M. Carlton, G. Udy, and M. Evans. 1994. Disruption of c-*mos* causes parthenogenetic development of unfertilized mouse eggs. *Nature* **370:** 65–68.

Curtis, D., R. Lehmann, and P. Zamore. 1995. Translational regulation and development. *Cell* **81:** 171–178.

Dagle, J., J. Walder, and D. Weeks. 1990. Targeted degradation of mRNA in *Xenopus* oocytes and embryos directed by modified oligonucleotides: Studies of An2 and cyclin in embryogenesis. *Nucleic Acids Res.* **18:** 4751–4757.

Dalby, B. and D. Glover. 1993. Discrete sequence elements control posterior pole accumulation and translational repression of maternal cyclin B RNA in *Drosophila*. *EMBO J.* **12:** 1219–1227.

Darnbrough, C. and P. Ford. 1981. Identification in *Xenopus laevis* of a class of oocyte-specific proteins bound to messenger RNA. *Eur. J. Biochem.* **113:** 415–424.

Davis, S. and J.C. Watson. 1996. *In vitro* activation of the interferon-induced, double-stranded RNA-dependent protein kinase PKR by RNA from the 3′ untranslated regions of human α-tropomyosin. *Proc. Natl. Acad. Sci.* (in press).

Dearsley, A., R. Johnson, P. Barrett, and J. Sommerville. 1985. Identification of a 60-kDa phosphoprotein that binds stored messenger RNA of *Xenopus* oocytes. *Eur. J. Biochem.* **150:** 95–103.

DeBenedetti, A. and R. Rhoads. 1990. Overexpression of eukaryotic protein synthesis initiation factor 4 in HeLa cells results in aberrant growth and morphology. *Proc. Natl. Acad. Sci.* **87:** 8212–8126.

Decker, C. and R. Parker. 1994. Mechanisms of mRNA degradation in eukaryotes. *Trends Biochem. Sci.* **19:** 336–340.

Ding, D. and H. Lipshitz. 1993. Localised RNAs and their functions. *BioEssays* **15:** 651–658.

Doniach, T. 1986. Activity of the sex-determining gene *tra-2* is modulated to allow spermatogenesis in the *C. elegans* hermaphrodite. *Genetics* **114:** 53–76.

Driever, W. and C. Nüsslein-Volhard. 1988a. A gradient of bicoid protein in *Drosophila* embryos. *Cell* **54:** 83–93.

———. 1988b. The bicoid protein determines position in the *Drosophila* embryo in a concentration-dependent manner. *Cell* **54:** 95–104.

Dubnau, J. 1995. "Homeodomain RNA regulation." Ph.D. thesis, Columbia University, New York, New York.

Evans, T., S. Crittenden, V. Kodoyianni, and J. Kimble. 1994. Translational control of maternal *glp-1* mRNA establishes an asymmetry in the *C. elegans* embryo. *Cell* **77:** 183–194.

Fox, C., M. Sheets, and M. Wickens. 1989. PolyA tail addition during maturation of frog oocytes: Distinct nuclear and cytoplasmic activities and regulation by the sequence UUUUUAU. *Genes Dev.* **3:** 2151–2162.

Fox, C., M. Sheets, E. Wahle, and M. Wickens. 1992. Polyadenylation of maternal mRNA during oocyte maturation: Poly(A) addition in vitro requires a regulated RNA binding activity and a poly(A) polymerase. *EMBO J.* **11:** 5021–5032.

Frohnhoefer, H. and C. Nüsslein-Volhard. 1986. Organization of anterior pattern in the *Drosophila* embryo by the maternal gene *bicoid*. *Nature* **324:** 120–125.

Gavis, E. and R. Lehmann. 1992. Localization of *nanos* RNA determines embryonic pathway. *Cell* **71:** 301–313.

———. 1994. Translational regulation of *nanos* by RNA localization. *Nature* **369:** 315–318.

Gay, N. and F. Keith. 1992. Regulation of translation and proteolysis during the development of embryonic dorso-ventral polarity in *Drosophila*. Homology of easter proteinase with *Limulus* proclotting enzyme and translational activation of *Toll* receptor synthesis. *Biochim. Biophys. Acta* **1132:** 290–296.

Gebauer, F., W. Xu, G. Cooper, and J. Richter. 1994. Translational control by cytoplasmic polyadenylation of c-*mos* mRNA is necessary for oocyte maturation in the mouse. *EMBO J.* **13:** 5712–5720.

Goodwin, E., P. Okkema, T. Evans, and J. Kimble. 1993. Translational regulation of *tra-2* by its 3′ untranslated region controls sexual identity in *C. elegans*. *Cell* **75:** 329–339.

Graham, P. and J. Kimble. 1993. The *mog-1* gene is required for the switch from spermatogenesis to oogenesis in *Caenorhabditis elegans*. *Genetics* **133:** 919–931.

Graham, P., T. Schedl, and J. Kimble. 1993. More *mog* genes that influence the switch from spermatogenesis to oogenesis in the hermaphrodite germ line of *Caenorhabditis elegans*. *Dev. Genet.* **14:** 471–484.

Hake, L. and J. Richter. 1994. CPEB is a specificity factor that mediates cytoplasmic polyadenylation during *Xenopus* oocyte maturation. *Cell* **79:** 617–627.

Halevy, O., B. Novitch, D. Spicer, S. Skapek, J. Rhee, G. Hannon, D. Beach, and A. Lassar. 1995. Correlation of terminal cell cycle arrest of skeletal muscle with induction of p21 by MyoD. *Science* **267:** 1018–1021.

Hashimoto, C., K. Hudson, and K. Anderson. 1988. The *Toll* gene of *Drosophila*, required for dorsal-ventral embryonic polarity, appears to encode a transmembrane receptor. *Cell* **52:** 269–279.

Hashimoto, N., N. Watanabe, Y. Furuta, H. Tamemoto, N. Sagata, M. Yokoyama, K. Okazaki, M. Nagayoshi, N. Takeda, Y. Ikawa, and S. Aizawa. 1994. Parthenogenetic activation of oocytes in c-*mos*-deficient mice. *Nature* **370:** 68–71.

Hodgkin, J. 1986. Sex determination in the nematode *C. elegans*: Analysis of *tra-3* suppressors and characterization of *fem* genes. *Genetics* **114:** 15–52.

Hodgkin, J. and S. Brenner. 1977. Mutations causing transformation of sexual phenotype in the nematode *Caenorhabditis elegans*. *Genetics* **86:** 275–287.

Howe, J.G. and J.W.B. Hershey. 1984. Translational initiation factor and ribosome-association with the cytoskeletal framework fraction from HeLa cells. *Cell* **37:** 85–93.

Huarte, J., A. Stutz, M. O'Connell, P. Gubler, D. Belin, A. Darrow, S. Strickland, and J.-D. Vassalli. 1992. Transient translational silencing by reversible mRNA deadenylation. *Cell* **69:** 1021–1030.

Hulskamp, M., C. Pfeifle, and D. Tautz. 1990. A morphogenetic gradient of hunchback protein organizes the expression of the gap genes *Kruppel* and *knirps* in the early *Drosophila* embryo. *Nature* **346:** 577–580.

Hulskamp, M., C. Schroeder, C. Pfeifle, H. Jaekle, and D. Tautz. 1989. Posterior segmentation of the *Drosophila* embryo in the absence of a maternal posterior organizer gene. *Nature* **338:** 629–632.

Hunt, T. 1989. On the translational control of suicide in red cell development. *Trends Biochem. Sci.* **14:** 393–394.

Hutter, H. and R. Schnabel. 1994. *glp-1* and inductions establishing embryonic axes in *C. elegans*. *Development* **120:** 2051–2064.

Irish, V., R. Lehmann, and M. Akam. 1989. The *Drosophila* posterior-group gene *nanos* functions by repressing *hunchback* activity. *Nature* **338:** 646–648.

Jackson, R. 1993. Cytoplasmic regulation of mRNA function: The importance of the 3′ untranslated region. *Cell* **74:** 9–14.

Jackson, R. and N. Standart. 1990. Do the poly(A) tail and 3′ untranslated region control mRNA translation? *Cell* **62:** 15–24.

Jacob, F. 1982. *The possible and the actual.* Pantheon, New York.

Kick, D., P. Barrett, A. Cummings, and J. Sommerville. 1987. Phosphorylation of a 60 kDa polypeptide from *Xenopus* oocytes blocks messenger RNA translation. *Nucleic Acids Res.* **15:** 4099–4109.

Kimble, J. 1994. An ancient molecular mechanism for establishing embryonic polarity? *Science* **266:** 577–578.

Kleene, K., R. Distel, and N. Hecht. 1984. Translational regulation and deadenylation of protamine mRNA during spermiogenesis in the mouse. *Dev. Biol.* **105:** 71–79.

Klein, P. and D. Melton. 1994. Induction of mesoderm in *Xenopus laevis* embryos by translation initiation factor 4E. *Science* **265:** 803–806.

Kobayashi, H., J. Minshull, C. Ford, R. Golsteyn, R. Poon, and T. Hunt. 1991. On the synthesis and destruction of A- and B-type cyclins during oogenesis and meiotic maturation in *Xenopus laevis*. *J. Cell Biol.* **114:** 755–765.

Koromilas, A.E., S. Roy, G. Barber, M. Katze, and N. Sonenberg. 1992. Malignant transformation by a mutant of the IFN-inducible dsRNA-dependent protein kinase. *Science* **257:** 1685–1689.

Kuhn, R., C. Kuhn, D. Borsch, K. Glatzer, U. Schaefer, and M. Schaefer. 1991. A cluster of four genes selectively expressed in the male germ line of *Drosophila melanogaster*. *Mech. Dev.* **35:** 143–151.

Lazaris-Karatzas, A., K. Montine, and N. Sonenberg. 1990. Malignant transformation by a eukaryotic initiation factor subunit that binds to mRNA 5′ cap. *Nature* **345:** 544–547.

Lee, R., R. Feinbaum, and V. Ambros. 1993. The *C. elegans* heterochronic gene *lin-4* encodes small RNAs with antisense complementarity to *lin-14*. *Cell* **75:** 843–854.

MacDonald, P. and G. Struhl. 1986. A molecular gradient in early *Drosophila* embryos and its role in specifying the body pattern. *Nature* **324:** 537–545.

Marello, K., J. LaRovere, and J. Sommerville. 1992. Binding of *Xenopus* oocyte masking proteins to mRNA sequences. *Nucleic Acids Res.* **20:** 5593–5600.

Martin, J., A. Raibaud, and R. Ollo. 1994. Terminal pattern elements in *Drosophila* embryo induced by the torso-like protein. *Nature* **367:** 741–745.

McGrew, L., E. Dworkin-Rastl, M. Dworkin, and J. Richter. 1989. Poly (A) elongation during *Xenopus* oocyte maturation is required for translational recruitment and is mediated by a short sequence element. *Genes Dev.* **3:** 803–815.

Mello, C., B. Draper, and J. Priess. 1994. The maternal genes *apx-1* and *glp-1* and establishment of dorsal-ventral polarity in the early *C. elegans* embryo. *Cell* **77:** 95–106.

Melton, D. 1994. Vertebrate embryonic induction: Mesodermal and neural patterning. *Science* **266:** 596–604.

Meurs, E.F., J. Galabru, G.N. Barber, M.G. Katze, and A.G. Hovanessian. 1993. Tumor suppressor function of the interferon-induced double-stranded RNA-activated protein kinase. *Proc. Natl. Acad. Sci.* **90:** 232–236.

Mlodzik, M. and S. Gehring. 1987. Hierarchy of the genetic interactions that specify the anteroposterior segmentation pattern of the *Drosophila* embryo as monitored by caudal protein expression. *Development* **101:** 421–435.

Mlodzik, M., A. Fjose, and W. Gehring. 1985. Isolation of *caudal*, a *Drosophila* homeobox-containing gene with maternal expression, whose transcripts form a concentration gradient at the pre-blastoderm stage. *EMBO J.* **4:** 2961–2969.

Morisato, D. and K. Anderson. 1994. The *spaetzle* gene encodes a component of the extracellular signaling pathway establishing the dorsal-ventral pattern of the *Drosophila* embryo. *Cell* **76:** 677–688.

Moskowitz, I., S. Gendreau, and J. Rothman. 1994. Combinatorial specification of blastomere identity by *glp-1*-dependent cellular interactions in the nematode *Caenorhabditis elegans*. *Development* **120:** 3325–3338.

Mosquera, L. C. Forristall, Y. Zhou, and M. King. 1993. A mRNA localized to the vegetal cortex of *Xenopus* oocytes encodes a protein with a *nanos*-like zinc finger domain. *Development* **117:** 377–386.

Munroe, D. and A. Jacobson. 1990. Tales of poly (A): A review. *Gene* **91:** 151–158.

Murata, Y. and R. Wharton. 1995. Binding of *pumilio* to maternal hunchback mRNA is required for posterior patterning in *Drosophila* embryos. *Cell* **80:** 7477–756.

Murray, M., G. Krohne, and W. Franke. 1991. Different forms of soluble mRNA binding proteins and particles in *Xenopus laevis* oocytes and embryos. *J. Cell Biol.* **112:** 1–11.

Musci, T., E. Amaya, and M. Kirschner. 1990. Regulation of fibroblast growth factor receptor in early *Xenopus* embryos. *Proc. Natl. Acad. Sci.* **87:** 8365–8369.

Nishizawa, M., K. Okazaki, N. Furuno, N. Watanabe, and N. Sagata. 1992. The "second-codon rule" and autophosphorylation govern the stability and activity of Mos during the meiotic cell cycle in *Xenopus* oocytes. *EMBO J.* **11:** 2433–2446.

Olson, E. and W. Klein. 1994. bHLH factors in muscle development: Dead lines and commitments, what to leave in and what to leave out. *Genes Dev.* **8:** 1–8.

Ostareck-Lederer, A., D. Ostareck, N. Standart, and B. Thiele. 1994. Translation of 15-lipoxygenase mRNA is controlled by a protein that binds to a repeated sequence in the 3′ untranslated region. *EMBO J.* **13:** 1476–1481.

Paris, J. and J. Richter. 1990. Maturation-specific polyadenylation and translational control: Diversity of cytoplasmic polyadenylation elements, influence of poly (A) tail size, and formation of stable polyadenylation complexes. *Mol. Cell. Biol.* **10:** 5634–5645.

Paris, J., K. Swenson, H. Piwnica-Worms, and J. Richter. 1991. Maturation-specific polyadenylation: In vitro activation by $p34^{cdc2}$ kinase and phosphorylation of a 58 kD CPE-binding protein. *Genes Dev.* **5:** 1697–1708.

Parker, S., G. Eichele, P. Zhang, A. Rawls, A. Sands, A. Bradley, E. Olson, J. Harper, and S. Elledge. 1995. p53-independent expression of p21Cip1 in muscle and other terminally differentiating cells. *Science* **267:** 1024–1027.

Priess, J., H. Schnabel, and R. Schnabel. 1987. The *glp-1* locus and cellular interactions in early *C. elegans* embryos. *Cell* **51:** 601–611.

Ranjan, M., S. Tafuri, and A. Wolffe. 1993. Masking mRNA from translation in somatic cells. *Genes Dev.* **7:** 1725–1736.

Rastinejad, F. and H. Blau. 1993. Genetic complementation reveals a novel regulatory role for 3′ untranslated regions in growth and development. *Cell* **72:** 903–917.

Rastinejad, F., M. Conboy, T. Rando, and H. Blau. 1993. Tumor suppression by RNA from the 3′ untranslated region of α-tropomyosin. *Cell.* **75:** 1107–1117.

Richter, J. and L. Smith. 1984. Reversible inhibition of translation by *Xenopus* oocyte-specific proteins. *Nature* **309:** 378–380.

Robbie, E., M. Peterson, E. Amaya, and T. Musci. 1995. Temporal regulation of the *Xenopus* FGF receptor in development: A translation inhibitory element in the 3′ untranslated region. *Development* **121:** 1775–1785.

Rosenthal, E., T. Hunt, and J. Ruderman. 1980. Selective translation of mRNA controls the pattern of protein synthesis during early development of the surf clam, *Spisula solidissima*. *Cell* **20:** 487–496.

Ruvkun, G. and J. Giusto. 1989. The *Caenorhabditis elegans* heterochronic gene *lin-14* encodes a nuclear protein that forms a temporal developmental switch. *Nature* **338:** 313–319.

Ruvkun, G., V. Ambros, A. Coulson, R. Waterston, J. Sulston, and H. Horvitz. 1989. Molecular genetics of the *Caenorhabditis elegans* heterochronic gene *lin-14*. *Genetics* **121:** 501–516.

Sagata, N., I. Daar, M. Oskarsson, S. Showalter, and G. Vande Woude. 1990. The product of the c-*mos* proto-oncogene as a candidate "initiator" for oocyte maturation. *Science* **245:** 643–646.

Sagata, N., M. Oskarsson, T. Copeland, J. Brumbaugh, and G. Vande Woude. 1988. Function of c-*mos* proto-oncogene product in meiotic maturation in *Xenopus* oocytes. *Nature* **335:** 519–525.

Sallés, F., M. Lieberfarb, C. Wreden, J. Gergen, and S. Strickland. 1994. Coordinate initiation of *Drosophila* development by regulated polyadenylation of maternal messenger RNAs. *Science* **266:** 1996–1999.

Schaefer, M., R. Kuhn, F. Bosse, and U. Schaefer. 1990. A conserved element in the leader mediates post-meiotic translation as well as cytoplasmic polyadenylation of a *Drosophila* spermatocyte mRNA. *EMBO J.* **9:** 4519–4525.

Schedl, T. and J. Kimble. 1988. *fog-2*, a germ-line-specific sex determination gene required for hermaphrodite spermatogenesis in *Caenorhabditis elegans. Genetics* **123:** 755–7679.

Sheets, M., M. Wu, and M. Wickens. 1995. Polyadenylation of c-*mos* mRNA as a control point in *Xenopus* meiotic maturation. *Nature* **374:** 511–516.

Sheets, M., C. Fox, T. Hunt, G. Vande Woude, and M. Wickens. 1994. The 3′-untranslated regions of c-*mos* and cyclin mRNAs stimulate translation by regulating cytoplasmic polyadenylation. *Genes Dev.* **8:** 926–938.

Simon, R., J. Tassen, and J. Richter. 1992. Translational control by poly(A) elongation during *Xenopus* development: Differential repression and enhancement by a novel cytoplasmic polyadenylation element. *Genes Dev.* **6:** 2580–2591.

Singer, R. 1993a. The cytoskeleton and mRNA localization. *Curr. Opin. Cell Biol.* **4:** 15–19.

———. 1993b. RNA zipcodes for cytoplasmic addresses. *Curr. Biol.* **3:** 719–721.

Skapek, S., J. Rhee, D. Spicer, and A. Lassar. 1995. Inhibition of myogenic differentiation in proliferating myoblasts by a cyclin D1-dependent kinase. *Science* **267:** 1022–1024.

Sonenberg, N. 1994. mRNA translation: Influence of the 5′and 3′untranslated regions. *Curr. Opin. Genet. Dev.* **4:** 310–315.

Spirin, A. 1966. On "masked" forms of messenger RNA in early embryogenesis and in other differentiating systems. *Curr. Top. Dev. Biol.* **1:** 1–63.

Sprenger, F., L. Stevens, and C. Nüsslein-Volhard. 1989. The *Drosophila* gene *torso* encodes a putative receptor tyrosine kinase. *Nature* **338:** 478–483.

Standart, N. 1992. Masking and unmasking of maternal mRNA. *Semin. Dev. Biol.* **3:** 367–379.

Standart, N. and R. Jackson. 1994. Regulation of translation by specific protein/mRNA interactions. *Biochimie* **76:** 867–879.

Standart, N., M. Dale, E. Stewart, and T. Hunt. 1990. Maternal mRNA from clam oocytes can be specifically unmasked in vitro by antisense RNA complementary to the 3′-untranslated region. *Genes Dev.* **4:** 2157–2168.

Stein, D., S. Roth, E. Vogelsang, and C. Nüsslein-Volhard. 1991. The polarity of the dorsoventral axis in the *Drosophila* embryo is defined by an extracellular signal. *Cell* **65:** 725–735.

Stevens, L., H. Frohnhoefer, M. Klinger, and C. Nüsslein-Volhard. 1990. Localized requirement for *torso*-like expression in follicle cells for development of terminal anlagen of the *Drosophila* embryo. *Nature* **346:** 660–663.

St Johnston, D. and C. Nüsslein-Volhard. 1992. The origin of pattern and polarity in the *Drosophila* embryo. *Cell* **68:** 201–219.

Struhl, G. 1989a. Differing strategies for organizing the anterior and posterior body pattern in *Drosophila* embryos. *Nature* **338:** 741–744.

———. 1989b. Morphogen gradients and the control of body pattern in insect embryos. *Ciba Found. Symp.* **144:** 65–91.

Tafuri, S. and A. Wolffe. 1990. *Xenopus* Y-box transcription factors: Molecular cloning, functional analysis and developmental regulation. *Proc. Natl. Acad. Sci.* **87:**

9028–9032.

———. 1992. DNA binding, multimerization and transcription stimulation by the *Xenopus* Y box proteins in vitro. *New Biol.* **4:** 349–359.

———. 1993. Selective recruitment of masked maternal mRNA from messenger ribonucleoprotein particles containing FRGY2 (mRNP4). *J. Biol. Chem.* **268:** 24255–24261.

Tautz, D. and C. Pfeifle. 1989. A non-radioactive *in situ* hybridization method for the localization of specific RNAs in *Drosophila* embryos reveals translational control of the segmentation gene hunchback. *Chromosoma* **98:** 81–85.

Thiele, B., H. Andree, M. Hohne, and S. Rapoport. 1982. Lipoxygenase mRNA in rabbit reticulocytes. Its isolation, characterization, and translational repression. *Eur. J. Biochem.* **129:** 133–141.

Vande Woude, G. 1994. On the loss of *Mos*. *Nature* **370:** 20–21.

Wang, C. and R. Lehmann. 1991. *nanos* is the localized posterior determinant in *Drosophila*. *Cell* **66:** 637–648.

Wang, C., L. Dickinson, and R. Lehmann. 1994. Genetics of *nanos* localization in *Drosophila*. *Dev. Dyn.* **199:** 103–115.

Weintraub, H. 1993. The MyoD family and myogenesis: Redundancy, networks, and thresholds. *Cell* **75:** 1241–1244.

Wharton, R. 1992. Regulated expression from maternal mRNAs in *Drosophila*. *Semin. Dev. Biol.* **3:** 391–397.

Wharton, R. and G. Struhl. 1989. Structure of the *Drosophila* Bicaudal D protein and its role in localizing the posterior determinant nanos. *Cell* **59:** 881–892.

———. 1991. RNA regulatory elements mediate control of *Drosophila* body pattern by the posterior morphogen *nanos*. *Cell* **67:** 955–967.

Wickens, M. 1992. Forward, backward, how much, when: Mechanisms of poly(A) addition and removal and their role in early development. *Semin. Dev. Biol.* **3:** 399–412.

———. 1993. Springtime in the desert. *Nature* **363:** 305–306.

Wickens, M. and K. Takayama. 1994. Deviants—Or emissaries. *Nature* **367:** 17–18.

Wightman, B., I. Ha, and G. Ruvkun. 1993. Post-transcriptional regulation of the heterochronic gene *lin-14* by *lin-4* mediates temporal pattern formation in *C. elegans*. *Cell* **75:** 855–862.

Wightman, B., T. Bürglin, J. Gatto, P. Arasu, and G. Ruvkun. 1991. Negative regulatory sequences in the *lin-14* 3′-untranslated region are necessary to generate a temporal switch during *Caenorhabditis elegans* development. *Genes Dev.* **5:** 1813–1824.

Wilhelm, J. and R. Vale. 1993. RNA on the move: The mRNA localisation pathway. *J. Cell Biol.* **123:** 269–274.

Wolffe, A. 1994. Structural and functional properties of the evolutionary ancient Y-box family of nucleic acid binding proteins. *BioEssays* **16:** 245–251.

Wolffe, A., S. Tafuri, M. Ranjan, and M. Familari. 1992. The Y-box factors: A family of nucleic acid binding proteins conserved from *Escherichia coli* to man. *New Biol.* **4:** 290–298.

Wormington, M. 1993. Poly(A) and translation: Development control. *Curr. Opin. Cell Biol.* **5:** 950–954.

———. 1994. Unmasking the role of the 3′UTR in the cytoplasmic polyadenylation and translational regulation of maternal mRNAs. *BioEssays* **16:** 533–535.

Yew, N., M. Strobel, and G. Vande Woude. 1993. *Mos* and the cell cycle: The molecular basis of the transformed phenotype. *Curr. Opin. Genet. Dev.* **3:** 19–25.

16

Poly(A) Metabolism and Translation: The Closed-loop Model*

Allan Jacobson
Department of Molecular Genetics and Microbiology
University of Massachusetts Medical School
Worcester, Massachusetts 01655-0122

More than two decades ago, experiments on the ribonuclease susceptibility of different RNAs led to the realization that uninterrupted tracts of polyadenylic acid (poly[A]) are present within eukaryotic messenger RNAs (Lim and Canellakis 1970; Darnell et al. 1971a,b; Edmonds et al. 1971; Kates 1971; Lee et al. 1971). It is now well recognized that almost all mRNAs whose biosynthesis originates within nuclei contain a 3′ poly(A) tail. Poly(A) sequences are not encoded within genes (Philipson et al. 1971; Birnboim et al. 1973; Jacobson et al. 1974) but are added to nascent pre-mRNAs in a processing reaction that involves site-specific cleavage and subsequent polyadenylation. The site of cleavage is determined by a highly conserved AAUAAA sequence usually 5–30 nucleotides 5′ to the site and by other less well-conserved sequences 3′, and sometimes 5′, to the site. Cleavage and polyadenylation are dependent on poly(A) polymerase and several other factors that impart specificity and processivity to the reaction. For reviews on the sequence and factor requirements of polyadenylation, see Wickens (1990), Proudfoot (1991), Wahle and Keller (1992), and Manley and Proudfoot (1994).

Pulse-chase experiments demonstrated that a large fraction of newly synthesized nuclear poly(A) was conserved in cytoplasmic mRNA (Puckett et al. 1975). These experiments provided early support for the notion that heterogeneous nuclear RNA (hnRNA) was the precursor to mRNA and also forced the conclusion that at least one function of poly(A) must be cytoplasmic. Evidence that the latter conclusion had merit is presented in this chapter, where I consider the role of the poly(A) tail, and changes in its length, in the translatability of individual mRNAs. Translatability is taken broadly here and is meant to encompass both the traditional concept of translational efficiency and the issue of mRNA availability, i.e., relative mRNA abundance. As will be proposed, the two

*This chapter is dedicated to the memory of David H. Gillespie, my mentor and friend.

are intimately linked: A poly(A) tail of appropriate length stimulates translation, possibly by promoting interactions between factors associated with opposite ends of the mRNA, and the loss or shortening of the poly(A) tail can disrupt this functional complex and promote rapid mRNA decay.

mRNA POLY(A) TAIL LENGTHS ARE NEITHER UNIFORM NOR STATIC

The Apparent Homogeneity of Newly Synthesized Poly(A) Tails

On the basis of the narrow distribution of lengths of the poly(A) tracts detected in pulse-labeled nuclear RNA (Brawerman and Diez 1975; Sheiness et al. 1975; Palatnik et al. 1979), specific pre-mRNA substrates polyadenylated in vitro in nuclear extracts (Sheets and Wickens 1989; Butler et al. 1990; Bienroth et al. 1992), or mRNAs accumulating in mutants incapable of poly(A) metabolism (Piper and Aamand 1989; Sachs and Davis 1989), it has been inferred that the newly synthesized poly(A) tails of different transcripts are relatively homogeneous in length. Such *initial* poly(A) tail lengths (see Fig. 1, poly$(A)_i$) range from 200 to 250 adenylate residues in mammals and from 60 to 80 adenylate residues in yeast; in organisms of intermediate phylogenetic ranking, poly(A) tail lengths fall between these two extremes (Adams and Jeffery 1978; Palatnik et al. 1979). At least in mammalian systems, the size limitation on poly(A) tail length appears to be attributable to a loss of processivity by the polyadenylation complex after a tract of approximately 200 As has been synthesized (Wahle 1991; Sachs and Wahle 1993). The apparent uniformity of the tract lengths of newly synthesized poly(A) tails (in a given species) may, however, be a function of the limited number of nuclear RNAs analyzed. Precedent for nonuniformity of poly(A) tail lengths comes from cytoplasmic polyadenylation in developing systems. The extent of such polyadenylation differs for different mRNAs (McGrew et al. 1989; Paris and Richter 1990; Salles et al. 1992; Sheets et al. 1994) yet depends largely on factors, and at least one *cis*-acting sequence element (AAUAAA), important in nuclear polyadenylation (Bilger et al. 1994). Indeed, some reports have identified newly synthesized mRNAs with atypical poly(A) lengths (Carrazana et al. 1988; Zingg et al. 1988).

In the Cytoplasm, Poly(A) Tails Shorten with mRNA Age

Early disagreements on the poly(A) lengths of specific mRNAs arose as a consequence of the failure to recognize that poly(A) length is dynamic, not static. Following transport of mRNA to the cytoplasm, poly(A) tracts

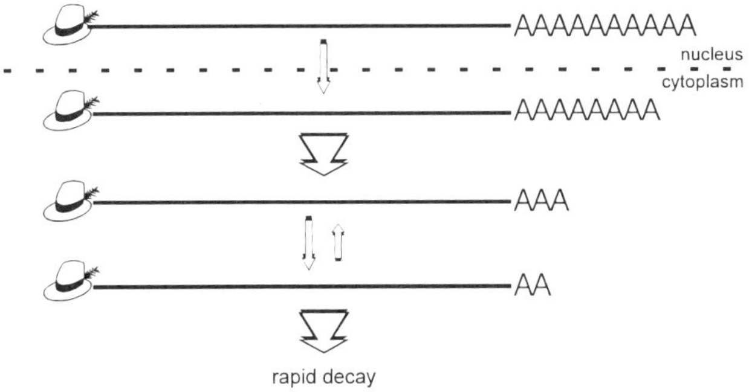

Figure 1 Poly(A) tails shorten with mRNA age. A long poly(A) tail is added to pre-mRNAs in a nuclear cleavage and polyadenylation reaction. Such initial poly(A) tails, poly(A)$_i$, are then shortened in the cytoplasm, a step that precedes decay of many mRNAs. The size of the arrows is proportional to the relative speed of the respective steps. The single arrow in reverse orientation indicates the existence of a pathway for modest readenylation.

are gradually shortened to lengths of approximately 10–60 adenylate residues (depending on the mRNA and the organism) and, in some instances, may be completely removed (Fig. 1) (Brawerman 1981; Baker 1993). It is thus incorrect to think in terms of the size of the poly(A) tail on specific mRNAs, unless it is made clear that such size refers to a specific point in the "life cycle" of that mRNA. Most measurements of poly(A) length have been made on steady-state RNA samples. Since the ratio of "aged" versus newly synthesized mRNAs is likely to be higher in steady state for stable mRNAs than for unstable mRNAs, the average poly(A) lengths of steady-state mRNAs are often inversely related to mRNA half-life (Palatnik et al. 1979, 1980).

Changes in mRNA poly(A) tail length have been detected by several independent procedures, including (1) pulse-chase analysis of total poly(A)$^+$ RNA (Sheiness and Darnell 1973; Sheiness et al. 1975; Palatnik et al. 1979; Hsu and Stevens 1993), (2) fractionation of total poly(A)$^+$ RNA and individual mRNAs by thermal elution from poly(U)-Sepharose (Palatnik et al. 1979, 1980, 1981, 1984; Saunders et al. 1980; Colot and Rosbash 1982), (3) Northern blotting of individual mRNAs with or without prior deadenylation or liberation of a 3′fragment of mRNA by RNase H digestion (Mercer and Wake 1985; Restifo and Guild 1986; Carrazana et al. 1988; Shyu et al. 1991; Decker and Parker 1993), (4) 3′-end labeling and RNase digestion of individual mRNAs and mRNA populations (Ahlquist and Kaesberg 1979; Steel and Jacobson 1987; Sachs and Davis 1989; Minvielle-Sebastia et al. 1991; Proweller and Butler 1994), and (5) analysis of the relative binding of individual mRNAs to immobilized oligo(dT) (Storti et al. 1980; Rosenthal et al. 1983; Herrick et al. 1990; Hsu and Stevens 1993). The most successful approaches to the analysis of changes in poly(A) tail length have been those in which transcription of a given mRNA occurs in a limited burst and the resultant population of mRNAs proceeds through the shortening pathway in a relatively synchronous manner (Saunders et al. 1980; Palatnik et al. 1981; Mercer and Wake 1985; Shyu et al. 1991; Decker and Parker 1993; Muhlrad et al. 1994, 1995). Experiments of this sort have demonstrated that different mRNAs shorten poly(A) tails at different overall rates and that shortening of a specific poly(A) tail occurs in at least two kinetically distinct steps (Shyu et al. 1991; Decker and Parker 1993). The initial, relatively slow rate of shortening is followed by a rapid phase at the end of which the reduction in poly(A) size usually reaches a plateau before complete deadenylation (Fig. 1). The plateau may represent the legitimate limit to the shortening reaction or a third phase of the shortening reaction in which modest poly(A) readdition may compensate to some extent for additional shortening occurring at a reduced rate (Brawerman and Diez 1975; Mercer and Wake 1985; Shyu et al. 1991), or it may simply be an apparent plateau occurring as a consequence of rapid mRNA turnover that ensues once the poly(A) length has been reduced below a critical point (Decker and Parker 1994; see below).

The overall rates of poly(A) shortening are slower for stable mRNAs than for unstable mRNAs. For example, in yeast, poly(A) tails of stable mRNAs are shortened at a rate of approximately 2–8 adenylate residues per minute, whereas the poly(A) tails of unstable mRNAs shorten approximately two- to tenfold faster (Herrick et al. 1990; Decker and Parker 1993). In mammalian cells, poly(A) shortening rates are on the order of 0.2–1.0 and 10–20 adenylate residues per minute for stable and unstable mRNAs, respectively (Mercer and Wake 1985; Shyu et al. 1991). These differences in poly(A) shortening rates in stable and unstable mRNAs are attributable to specific sequences; unstable mRNAs appear to have *cis*-acting destabilizing elements that also accelerate the rate of poly(A) shortening. These sequences include the well-known AU-rich elements (AREs) located in the 3′-untranslated regions (3′UTRs) of mammalian proto-oncogene, cytokine, and lymphokine mRNAs, as well as other, less-well-characterized sequences in mRNA-coding regions and 3′UTRs (Peltz et al. 1991; Sachs 1993). Our present understanding of the role of these sequence elements in the shortening reaction is limited to the recognition that they bind specific proteins and may require ongoing translational elongation in *cis* to function (Peltz and Jacobson 1992; Sachs 1993). Possible mechanisms for the shortening reaction are discussed below.

Poly(A) Tail Lengths Can Be Regulated

Selective adenylation or deadenylation of specific mRNAs has been observed in cells responding to developmental programs or physiological stimuli. The best-characterized examples of regulated adenylation occur in the cytoplasm of oocytes and early embryos where posttranscriptional changes in poly(A) length generally correlate with changes in translational activity (McGrew et al. 1989; Vassalli et al. 1989; Fox and Wickens 1990; Paris and Richter 1990; Huarte et al. 1992; Salles et al. 1992, 1994). In these cells, adenylation of specific mRNAs has been shown to precede their translational recruitment, whereas deadenylation of other mRNAs accompanies their loss of translatability. These events and the importance of specific sequences and factors in determining the rate, timing, and extent of polyadenylation and deadenylation are discussed at length elsewhere in this volume (see Richter; Wickens et al.).

Abrupt changes in the poly(A) tail lengths of specific mRNAs have also been detected in somatic cells responding to extracellular stimuli of growth or differentiation. Poly(A) tails of the rat vasopressin mRNA appear to lengthen during osmotic stress, whereas in the suprachiasmatic nucleus, the same mRNA undergoes rhythmic, biological clock-controlled changes in poly(A) size (Carrazana et al. 1988; Robinson et al. 1988; Zingg et al. 1988; Carter and Murphy 1989, 1991; Murphy and Carter 1990). Likewise, thyroid hormone depletion or glucocorticoid addition promotes an increase in the length of the poly(A) tail associated with growth hormone mRNA (Paek and Axel 1987; Jones et al. 1990; Murphy et al. 1992), and growth of insulinoma cells in serum-free, hormonally defined media leads to an increase in abundance and poly(A) tail length of the insulin mRNA (Muschel et al. 1986). There is no evidence, however, that any of these changes in poly(A) length is attributable to posttranscriptional mechanisms, i.e., to the specific shortening or lengthening of the poly(A) tails of preexisting mRNAs. Indeed, recent experiments demonstrate that the apparent lengthening of the vasopressin mRNA poly(A) tail can be explained by increased transcription of the vasopressin gene (Murphy and Carter 1990; Carter and Murphy 1991) and hence accumulation of an mRNA population that normally has long poly(A) tails.

Poly(A) shortening rates for some mRNAs can be reduced or enhanced in response to decreases in translation rates. Insertion of strong secondary structures into the 5′ UTRs of several yeast mRNAs markedly inhibits their translation (Beelman and Parker 1994; Vrecken and Raue 1992); in the yeast *PGK1* mRNA, the same structures also increase the rate of poly(A) shortening (Muhlrad et al. 1995). Inhibition of translational elongation by cycloheximide inhibits shortening of the poly(A) tails on mammalian c-*fos* and c-*myc* mRNAs (Wilson and Treisman 1988; Laird-Offringa et al. 1990), whereas poly(A) tails of *Chlamydomonas* α-tubulin mRNAs are rapidly shortened in the presence of this inhibitor (Baker et al. 1989). These opposing effects may indicate that coding region instability sequences present in the c-*fos* and c-*myc* mRNAs (Shyu et al. 1991; Wisdom and Lee 1991) promote poly(A) shortening in a ribosome translocation-dependent manner and that such sequences are absent from the α-tubulin mRNAs. The effects observed with all three mRNAs in cycloheximide-treated cells are selective since poly(A) tail lengths of other mRNAs in the same cells were unaltered. Global effects of translational inhibition on poly(A) shortening rates may account for the dramatic accumulation of poly(A)-deficient mRNAs in heat-shocked *Drosophila* cells or early developing (starved) *Dictyostelium* cells (Storti et al. 1980; Palatnik et al. 1984). Both phenomena are accompanied by substantial reductions in cellular translation rates.

Factors and Possible Mechanisms Involved in Poly(A) Shortening

Insight into the mechanism of poly(A) shortening has followed from an analysis of the structure and function of the ubiquitous poly(A)-binding protein (PABP). By far the most abundant of the numerous mRNP proteins, PABPs associate with the poly(A) tracts of essentially all polyadenylated mRNAs, spaced approximately 25 nucleotides apart and requiring a minimal binding site of 12 adenosines (Baer and Kornberg 1980, 1983; Sachs et al. 1986, 1987). Cloning of the yeast PABP gene (*PAB1*; Adam et al. 1986; Sachs et al. 1986) led to the demonstration that it is essential but that its absence could be suppressed by mutations in at least seven different genes, all of which appear to affect either the structure or maturation of 60S ribosomal subunits (see below; Sachs et al. 1987; Sachs and Davis 1989). Studies of the latter suppressors, and other conditional *pab1* mutants, showed that strains deficient in PABP fail to shorten mRNA poly(A) tails (Sachs and Davis 1989). This observation prompted Sachs et al. to purify a poly(A) nuclease (PAN) from yeast that is dependent on PABP for activity (Lowell et al. 1992; Sachs and Deardorff 1992). In vitro, PAN is a 3′ to 5′ exonuclease that shortens poly(A) tracts to a length of 15–25 nucleotides, liberating 5′mononucleotides in the process (Lowell et al. 1992). Substrate recognition depends not only on PABP, but also on the proximity of the poly(A) tract to the 3′end of the mRNA (Lowell et al. 1992). Yeast poly(A) nuclease is a multimeric enzyme, and a gene thought to encode a possible subunit (*PAN1*) has been cloned and sequenced (Sachs and Deardorff 1992). However, recent experiments demonstrate that its 161-kD polypeptide product is not required for poly(A) shortening activity in vitro or in vivo (A. Sachs, pers. comm.). Exonuclease activities specific for poly(A) have also been identified in metazoans, but their relationship to the structure and function of the yeast enzyme remains to be determined (Astrom et al. 1991, 1992; Varnum et al. 1992).

The PABP dependence of the yeast poly(A) nuclease activity is consistent with several models for poly(A) recognition and shortening (Bernstein and Ross 1989; Lowell et al. 1992; Sachs and Deardorff 1992; Baker 1993). The simplest interpretation is that PAN recognizes and degrades "exposed" adenosine residues 3′to bound PABP. Dissociation of a PABP monomer would create a substrate of approximately 25 adenosine residues, provided at least one upstream molecule of PABP remained. As such, PABP could be considered both to protect poly(A) from degradation and to simultaneously be required for its degradation. A shortening limit to degradation would follow from the requirement for at least one molecule of bound PABP. Consistent with this model is the observation of a 25-nucleotide periodicity in poly(A) tail lengths (Baer and Kornberg 1980; Baker et al. 1989) and the limit often seen to the extent of poly(A) shortening (see above). Both the regulation of shortening rates and the extent of shortening could thus depend on the rate of dissociation of PABP and may in turn be influenced by other sites or factors with which PABP might preferentially interact. A particularly interesting class of such sites and factors may be the 3′UTR instability elements, such as the mammalian ARE class of elements and their binding proteins (Bernstein and Ross 1989).

A ROLE FOR POLY(A) IN TRANSLATION

The occurrence of mRNAs with and without poly(A) tails and the routine and often substantial changes in poly(A) tract length during the cytoplasmic lifetime of an mRNA imply that poly(A) content and poly(A) size must be relevant to mRNA function. In recent years, a variety of different experimental approaches have substantiated this conclusion, demonstrating that poly(A) status can be a determinant of both mRNA translational efficiency and the time of onset of mRNA decay. The former is discussed in this section and the latter in the next section.

Poly(A) Tails Stimulate mRNA Translation In Vitro

Early studies comparing the translational capacity of adenylated and deadenylated forms of various mRNAs (Bard et al. 1974; Sippel et al. 1974; Soreq et al. 1974; Williamson et al. 1974; Spector et al. 1975) or the effects of obstructing the poly(A) tail by annealing with poly(U) (Munoz and Darnell 1974) indicated that poly(A) was neither a requirement for nor a stimulus to in vitro translational activity of mRNA. It was soon recognized that these conclusions were based on translation systems that reinitiated poorly; in more active reticulocyte extracts, poly(A)$^+$ mRNA had a translational advantage over poly(A)$^-$ mRNA that was attributable to differences in the respective efficiencies of initiation (Doel and Carey 1976). Translational preference for poly(A)$^+$ mRNAs was subsequently observed in other cell-free extracts, including those prepared from Ehrlich ascites tumor cells (Hruby and Roberts 1977) and yeast (Iizuka et al. 1994).

The functional difference between poly(A)$^+$ and poly(A)$^-$ mRNAs in vitro was established in experiments analyzing the efficiency with which two sets of synthetic mRNAs (rabbit β-globin and vesicular stomatitis virus N mRNAs), differing only in their respective poly(A) tail lengths or cap structures, were assembled into mRNPs, recruited into polysomes, translated, or degraded in a reticulocyte cell-free translation system (Munroe and Jacobson 1990a). These experiments confirmed that poly(A)$^-$ mRNAs had approximately one-half to one-third the translational capacity of poly(A)$^+$ mRNAs in reticulocyte extracts, a difference reflected in reduced recruitment into polysomes and association with fewer ribosomes than poly(A)$^-$ mRNAs. The differences in translational activity could not be attributed to differences in mRNA stabilities, extents of capping, or translational elongation rates, but they could be ascribed to the reduced ability of poly(A)$^-$ mRNAs to join 60S ribosomal subunits to 48S preinitiation complexes during the later stages of translational initiation (Munroe and Jacobson 1990a). These experiments also showed that the effect of poly(A) on translation in vitro is directly related to its length. Maximal recruitment of mRNA into polysomes required a poly(A) tail of more than five adenylate residues. Beyond this size, the efficiency of polysome formation increased as a function of poly(A) tail length, with the most pronounced effect occurring between poly(A) lengths of 5–32 adenylate residues. Thus, in this in vitro system, the presence of a poly(A) tail that exceeded a minimal length stimulated the formation of 80S translational initiation complexes by enhancing the rate of 60S subunit joining. A role for poly(A) in the 60S joining step is also suggested by observations that in the same in vitro system, poly(A)$^-$ histone mRNA formed a significantly higher percentage of "half-mers" (polysomes containing bound 40S subunits, but lacking 60S subunits) than poly(A)$^+$ globin mRNA (Nelson and Winkler 1987).

The stimulatory effects of poly(A) observed in vitro may also occur in *trans*. Addition of exogenous poly(A) to reticulocyte extracts inhibits translation of poly(A)$^+$ mRNAs (see Poly(A) Effects and the Poly(A)-binding Protein below) but stimulates the translation of unadenylated mRNAs, provided the latter are capped (Jacobson and Favreau 1983; Munroe and Jacobson 1990a). The selective augmentation of the translation of capped mRNAs implies that events in translational initiation involve the recognition of both 5′cap and 3′poly(A) structures (together with their associated proteins), a conclusion supported by other experiments discussed below.

mRNA Poly(A) Status and Translation In Vivo

Numerous experimental systems manifest a correlation between mRNA adenylation status and translational activity in vivo. Prototypical, and most illustrative, of these is early development in frogs, mice, and flies, in which translational activation or inactivation of many maternal mRNAs is paralleled by the respective lengthening or shortening of their poly(A) tails (Bachvarova 1992; Wickens 1992; Wormington 1993). Microinjection experiments with synthetic mRNAs, and other approaches, have been exploited to show that (1) specific 3′UTR sequences govern the timing and extent of polyadenylation, (2) deadenylation is a default reaction that occurs in the absence of polyadenylation, and (3) the presence of a long poly(A) tract is both necessary and sufficient for the translational recruitment of at least three mRNAs, the *Xenopus* B4 mRNA (Paris and Richter 1990), the mouse t-PA mRNA (Vassalli et al. 1989), and the *Drosophila bicoid* mRNA (Salles et al. 1994). Further discussion of the relationship between translational discrimination and polyadenylation status in early development and details of the mechanisms of adenylation, deadenylation, and translational activation in these systems are presented elsewhere in this volume (see Richter; Wickens et al.).

Additional evidence that poly(A) has a translational role in vivo has been obtained in studies of the expression of heterologous mRNAs microinjected into *Xenopus* oocytes or electroporated into plant, animal, or yeast cells. Translation in oocytes of mRNAs encoding $\alpha_{2\mu}$-globulin, lysozyme, preprochymosin, and zein was shown to be up to 20-fold more effective with mRNAs containing long poly(A) tails (Deshpande et al. 1979; Drummond et al. 1985; Galili et al. 1988). In one study, adenylated zein and β-globin mRNAs were found on oocyte polysomes with a larger average number of ribosomes than their unadenylated counterparts, indicating that the latter mRNAs were deficient in translational initiation (Galili et al. 1988). In electroporated tobacco, carrot, maize, rice, Chinese hamster ovary, and yeast cells, expression of synthetic α-glucuronidase or luciferase mRNAs with a 50-nucleotide poly(A) tail was up to two orders of magnitude greater than that from the same mRNAs lacking a poly(A) tail (Gallie et al. 1989; Gallie 1991). This stimulation was most pronounced with capped mRNAs, suggesting a synergistic effect of the presence of both a cap and a poly(A) tail (Gallie 1991). The extent of poly(A)-mediated stimulation is so much more substantial in electroporated cells than in reticulocyte extracts (see above; Munroe and Jacobson 1990a) that the possibility of effects on steps other than translation must be considered in the former situation. Although attempts were made to exclude the possibility that the stimulation observed in electroporated cells was due to enhanced mRNA stability (Gallie 1991), this remains a legitimate prospect because nothing is known about the fraction of mRNA in electroporated cells that actually associates with polysomes.

Although these examples illustrate the stimulatory role of a poly(A) tail, discrimination against poly(A)-deficient mRNAs is not a routine observation in vivo, and the presence of a poly(A) tail is not always sufficient for translational activity. Poly(A)$^-$ mRNAs are known to be translated in vivo (Adesnik and Darnell 1972; Greenberg 1979), and there are examples of translationally inactive polyadenylated mRNAs (Raff 1980; Rosenthal et al. 1983) as well as examples of mRNAs that appear to lose their poly(A) tracts as they become translationally active (Hruby and Roberts 1977; Iatrou and Dixon 1977; Kleene 1989). Perhaps the most striking inconsistency with the correlation between adenylation status and translation has been observed in a yeast mutant with a temperature-sensitive mutation (*pap1-1*) in poly(A) polymerase. After a shift to the nonpermissive temperature, cells of this mutant rapidly accumulate mRNAs lacking poly(A) tails, but overall cellular translation rates and polysome association of individual mRNAs are virtually unaffected for up to 2 hours (Patel and Butler 1992; Proweller and Butler 1994). These apparent contradictions may simply reflect three themes. First, it is likely that not all adenylated mRNAs are affected comparably by the presence of a poly(A) tail. For example, an mRNA with a strong ribosome-binding site may be less poly(A)-dependent than an mRNA with a weak ribosome-binding site. Second, the mechanisms that ensure the efficient translational initiation of poly(A)$^+$ and poly(A)$^-$ mRNAs may have subtle differences. Third, the stimulatory effects of poly(A) may be a function of competition among mRNAs for components of the translational machinery. A corollary of the last point is that the poly(A) tail may not be essential in the absence of such competition. This would explain the observations in the *pap1-1* mutant, since a secondary consequence of the mutation in this strain is a twofold reduction in the total amount of mRNA (Proweller and Butler 1994), and would also explain why translational discrimination is detected in starved, but not in logarithmically growing, *Dictyostelium* cells (Palatnik et al. 1984; Steel and Jacobson 1987; Shapiro et al. 1988). Relative to cells in balanced growth, starved *Dictyostelium* amoebae have markedly reduced rates of translational initiation (Alton and Lodish 1977; Cardelli and Dimond 1981) and preferentially translate mRNAs with long poly(A) tails (Palatnik et al. 1984; Steel and Jacobson 1987; Shapiro et al. 1988).

Insight into a possible mechanism of translational discrimination against poly(A)$^-$ mRNAs in vivo has been obtained in studies of the yeast cytoplasmic double-stranded RNA viruses L-A, M, and L-BC. mRNAs (+ strands) of these viruses are uncapped and unadenylated, and wild-type yeast cells encode a set of genes, the *SKI* genes, that repress their expression (Wickner 1992; Mathews, this volume). In a recent study, it was shown that the *SKI2* protein was a negative regulator of translation not only of viral mRNAs, but also of the *lacZ* mRNA if expressed in an uncapped, poly(A)$^-$ form (Widner and Wickner 1993). *SKI2* regulation of the latter was eliminated if the mRNA was both capped and polyadenylated (Widner and Wickner 1993). These results raised the possibility that a normal component of the protein synthesis apparatus represses translation unless antagonized by the presence of a poly(A) tail (or a cap) on an mRNA. Further studies from the same laboratory have demonstrated that the effect is specific for the poly(A) tail and dependent on the products of the *SKI3* and *SKI8* genes as well as the *SKI2* gene product (Masison et al. 1995). Expression of poly(A)$^-$ viral mRNAs can also be reduced by mutations in *MAK* genes, most of which appear to reduce the levels of 60S ribosomal subunits (Ohtake and Wickner 1995). The simultaneous reduction of both 60S subunits and the translation of poly(A)$^-$ mRNA implies, as did the translation results in vitro (see above), that the poly(A) tail stimulates 60S joining.

Poly(A) Effects and the Poly(A)-binding Protein

The cytoplasmic PABP is the most likely mediator of the translational effects of poly(A). This conclusion follows principally from experiments on the genetics of the yeast *PAB1* gene but also from less direct experiments, including those in which exogenous poly(A) is used as a competitor of PABP activity, proteins cross-linked to specific mRNAs are compared, or the stability of the PABP is analyzed.

The yeast *PAB1* gene is essential for viability (Adam et al. 1986; Sachs et al. 1986, 1987), and the nature of its indispensable function has been probed by controlling its expression with a regulatable promoter or by direct mutation of its encoded protein. Depletion of PABP by promoter repression and inactivation of its function with a temperature-sensitive (ts) lesion both lead to reductions in cellular translation rates (Sachs and Davis 1989). The latter are paralleled by decreases in the level of polysomes and relative increases in the amounts of 80S monosomes but not by changes in the size of polysomes (Sachs and Davis 1989). Changes in polysome size would have been expected if depletion of PABP simply reduced the rate of translational initiation on all mRNAs. A screen for extragenic suppressors of the temperature-sensitive mutation identified seven complementation groups of *spb* mutations, each of which was also able to suppress a *PAB1* deletion (Sachs and Davis 1989). All of the *spb* mutants showed marked decreases in their relative levels of 60S ribosomal subunits and two showed more direct connections to ribosome structure. *spb2-1* has been localized to the gene for the 60S ribosomal protein L46 and *spb4-1* has been localized to a gene encoding a putative rRNA helicase involved in the maturation of 25S rRNA (Sachs and Davis 1989, 1990). These data indicate that translation in the absence of functional PABP may require an alteration in the structure of the 60S ribosomal subunit, a conclusion that is consistent with the function ascribed to poly(A) in reticulocyte extracts and in yeast viral systems (see above).

A set of indirect experiments also implicating a role for the PABP in translation demonstrated that, under conditions of translational discrimination between poly(A)$^+$ and poly(A)$^-$ mRNAs in vitro, the mRNP proteins cross-linked to otherwise identical representatives of these two classes of mRNAs are the same, except for proteins that associate with poly(A) (Munroe 1989; Munroe and Jacobson 1990a). These experiments identified PABP monomers, dimers, and trimers bound to the same molecule of poly(A); the appearance of PABP multimers was dependent on poly(A) tail length. Increased binding of PABP correlated with increases in poly(A) length that in turn correlated with increases in translational efficiency. Other indirect experiments showed that exogenous poly(A) is a potent and specific inhibitor of the translation of capped poly(A)$^+$, but not poly(A)$^-$, mRNAs in rabbit reticulocyte, wheat-germ, L-cell, and pea seed extracts (Jacobson and Favreau 1983; Bablanian and Banerjee 1986; Lemay and Millward 1986; Sieliwanowicz 1987; Grossi de Sa et al. 1988; Munroe and Jacobson 1990a) as well as in *Xenopus* oocytes (Drummond et al. 1985). Comparable inhibition is not observed with other ribopolymers, and inhibition activity is inversely related to poly(A) size, with the smallest effective size being similar to that protected by a monomer of PABP (Jacobson and Favreau 1983). Inhibition does not affect the average size of the polypeptides synthesized, suggesting that elongation is not the target, and the inhibitory effect can be overcome by increased mRNA concentrations or by translating mRNPs instead of mRNA (Jacobson and Favreau 1983). These results have been interpreted to indicate that exogenous poly(A) inhibits translation by limiting the availability of unbound PABP, a conclusion supported by the observation that addition of purified PABP to reticulocyte lysates overcomes the inhibitory activity (Grossi de Sa et al. 1988). These experiments do not, however, exclude the possibility that exogenous poly(A) has a high affinity for another component(s) of the translation apparatus and that addition of purified PABP simply competes for this interaction. Indeed, this conclusion was drawn by Gallie and Tanguay (1994) to explain the observation that uncapped mRNAs are much more sensitive to poly(A) inhibition than capped mRNAs, regardless of their adenylation status (Munroe and Jacobson 1990a; Gallie and Tanguay 1994). Their experiments demonstrate that poly(A) inhibition of translation can be reversed by simultaneous addition of translation initiation factors eIF4F, eIF4B, and eIF4A (Gallie and Tanguay 1994).

Three other sets of experiments support a translational function for the PABP: the demonstration that the PABP of germinated pea embryo axes stimulates translation from poly(A)$^+$ mRNAs in vitro (Sieliwanowicz 1987), the correlation between the abundance and stability of PABPs and the rate of translational initiation in developing or heat-shocked *Dictyostelium* amoebae (Manrow and Jacobson 1986, 1987), and the recent observation that HIV-1 mRNAs that are translationally incapacitated in the absence of the Rev protein are also devoid of bound PABP (Campbell et al. 1994).

Given the plethora of direct and indirect evidence supporting the contention that the poly(A)-binding protein facilitates poly(A) function, it is thus surprising that Zelus et al. (1989) find that *Xenopus* oocytes contain very little of this protein. This raises several possibilities. *Xenopus* oocytes may contain as yet unrecognized PABPs, or alternative mechanisms and factors may promote the poly(A)-dependent regulation of translation occurring in early development. The chapter by Richter (this volume) includes a proposal for such an alternative mechanism.

Stimulation of Translational Initiation by Poly(A) and the Poly(A)-binding Protein: The Closed-loop Model

The translational effects of the poly(A) tail summarized above can be accounted for by a model, henceforth dubbed the closed-loop model, whose principal tenet is that factors associated with the 5′and 3′ends of an mRNA interact to facilitate translational initiation (Fig. 2A). This model, originally proposed more than a decade ago (Jacobson and Favreau 1983; Palatnik et al. 1984) and upgraded more recently (Munroe and Jacobson 1989, 1990a,b; Sachs and Davis 1989; Brawerman 1993; Muhlrad et al. 1994), implies that optimal function of an mRNP depends on the existence of at least a transient pseudo-circular structure. Promoting the closed-loop from the 3′side is postulated to be the role of the PABP and, possibly, proteins bound to it or other 3′UTR-binding proteins. The interacting component(s) on the 5′side is at present undefined, but the end result of its interaction with the 3′factor(s) is thought to be a stimulation of the 60S subunit joining reaction. As such, the 5′factor(s) could be an initiation factor (Gallie and Tanguay 1994), a large subunit ribosomal protein, or, as suggested by Widner and Wickner (1993), a repressor of 60S joining that is antagonized by the 3′factor. The interaction of the mRNP ends is not considered to be essential for translation; rather, it is postulated to improve translational efficiency for many but not all mRNAs, particularly when some component of the translational apparatus is limiting. Poly(A) and possibly other 3′UTR sequences are thus considered to be the formal equivalent of transcriptional enhancers, and the mRNAs that respond to such stimuli may be those with 5′UTR sequences that are suboptimal for initiation.

If existence of the closed-loop depends on bound PABP, then this in turn would depend on the availability of a poly(A) tail of minimal length. Poly(A) shortening is thus a possible mechanism to open the loop (and reduce translational efficiency), and poly(A) lengthening is a mechanism to ensure loop formation and maximize translation. The kinetics of loop

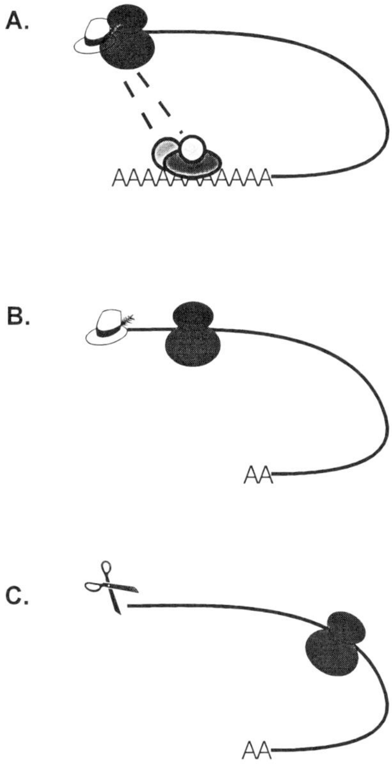

Figure 2 The closed-loop model. (*A*) Poly(A)-binding protein (PABP; *gray oval*), bound to a poly(A) tail, is shown stimulating the 60S joining step. The stimulus may be a direct or indirect effect. The latter possibility is favored and the figure depicts proteins (*circles*) bound to the PABP that are involved in such indirect stimulation. The 5′ cap is rendered inaccessible to decapping enzymes during this step (see text). (*B*) After sufficient shortening of the poly(A) tail, PABP is no longer capable of binding the mRNA. In the absence of PABP promotion of the 60S joining step, the rate of removal of the 5′ cap can increase markedly. (*C*) The 5′ cap has been removed and exonucleolytic decay has commenced.

formation, or the extent of 5′ stimulation, should be directly dependent on poly(A) length and would thus vary with the age of the mRNA in the cytoplasm. Under conditions of significant competition, newly synthesized mRNA would be expected to have a translational advantage over preexisting mRNA, an event for which there is experimental precedent (Palatnik et al. 1984; Shapiro et al. 1988).

From a strictly translational perspective, the location of poly(A) sequences on the mRNA may be unimportant, i.e., the important event is postulated to be the interaction of a factor(s) normally associated with mRNA 3′ends with translational components at the mRNA 5′end. Several mRNAs contain 5′poly(A) sequences that should be capable of binding PABP, including poxvirus mRNAs (Bertholet et al. 1987; Patel and Pickup 1987; Schwer et al. 1987; see Schneider, this volume) and PABP mRNAs (Adam et al. 1986; Sachs et al. 1986; Grange et al. 1987), and stimulation of translation by poly(A) in *trans* has been observed in vitro (Jacobson and Favreau 1983; Munroe and Jacobson 1990a). Proximity considerations (Howard et al. 1970) make poly(A) enhancement in *trans* an unlikely mechanism in vivo, but the conservation of the 5′A-rich domain between yeast and human PABP mRNAs and among the mRNAs of the members of the poxvirus family suggests that it provides some advantage. One possible advantage would be the enhancement of translational initiation even on relatively "old" mRNAs with shorter 3′poly(A) tracts. A requirement for 3′*cis*-linkage of the poly(A) tail does, however, provide the cell with a mechanism for linking mRNA translation and turnover. Deadenylation, endonucleolytic cleavage of an mRNA, and other decay events that occur independent of deadenylation will, at a minimum, disrupt the closed-loop state. These events, and the connection between the closed loop and mRNA decay, are discussed below in the next section.

Stimulation of translational initiation by the 3′end of an mRNA could certainly be explained by other mechanisms. For example, the PABP could be involved in tethering the mRNA to a particular subcellular site whose microenvironment is more favorable to initiation. Although ample evidence exists for a variety of 3′UTR-dependent posttranscriptional stimuli (Jackson and Standart 1990; Kislauskis and Singer 1992; Decker and Parker 1995), interaction between 5′and 3′ends is the favored model because of the 60S effects observed in vitro, in *spb* mutants, and in *mak* mutants, and because electron micrographs of polysomes with interacting 5′and 3′ends have been observed (Warner et al. 1962; Dubochet et al. 1973; Hsu and Coca-Prodos 1979; Ladhoff et al. 1981; Christensen et al. 1987).

Kinetic preference for a closed-loop state may have consequences or causative mechanisms other than those proposed here. For example, proximity of mRNA ends may facilitate recycling of ribosomes on the same mRNA. Evidence for ribosome recycling has been reported (Adamson et al. 1970; Baglioni et al. 1970; Howard et al. 1970) as has its dependence on a poly(A) tail (Galili et al. 1988) and other 3′UTR elements (L.E. Hann and L. Gehrke, in prep.). An effect specific for ribosome recycling would imply the existence of two modes of initiation:

de novo initiation by "new" ribosomes on a "new" mRNA and reinitiation by "old" ribosomes on an "old" mRNA. As suggested by Galili et al. (1988), only the latter would respond to 3′-end stimuli. Finally, the possibility remains that 5′/3′ interactions can occur by a variety of different mechanisms in different cells or with different mRNAs. For example, Richter (this volume) hypothesizes that the complex responsible for polyadenylation of mRNAs in early development may also be capable of stimulating translation by promoting methylation of the 5′ cap.

mRNA STABILITY AND MAINTENANCE OF THE CLOSED LOOP

Poly(A) Shortening Often Precedes mRNA Decay

A relationship between poly(A) metabolism and mRNA stability was originally suggested by observations that decay of individual mRNAs is often preceded by poly(A) shortening (see, e.g., Wilson et al. 1978; Mercer and Wake 1985; Brewer and Ross 1988; Wilson and Treisman 1988; Laird-Offringa et al. 1990) and by experiments demonstrating that deadenylated mRNAs could be preferred substrates for decay in oocytes or in crude extracts (Huez et al. 1975, 1978; Nudel et al. 1976; Peltz et al. 1987; Brewer and Ross 1988; Bernstein et al. 1989). Skepticism about the significance of these observations, arising primarily from the recognition of an equivalent number of examples to the contrary (Jacobson and Favreau 1983; Herrick et al. 1990; Peltz et al. 1991; see below for consideration of the "counter examples"), has been somewhat muted recently in light of experiments that provide a high-resolution analysis of the changes in individual mRNA poly(A) lengths and the relationship of those changes to the timing of mRNA decay. These studies synchronized poly(A) metabolism by transcriptional pulsing and showed that shortening of the poly(A) tract to limit lengths (see Fig. 1) precedes decay of mRNAs in both yeast and mammals (Shyu et al. 1991; Decker and Parker 1993; Muhlrad et al. 1995). The relationship between deadenylation and decay was shown to hold for mRNAs with a broad range of decay rates (Decker and Parker 1993; Muhlrad et al. 1995) and to be dependent on specific sequence elements. In unstable mRNAs, these elements frequently promote both rapid deadenylation and rapid onset of decay, and mutations that block one of these functions generally block the other (Shyu et al. 1991; Muhlrad and Parker 1992; Decker and Parker 1993; Chen et al. 1994). Further evidence that deadenylation preceded decay followed from the unraveling of a decay pathway in yeast in which the final step was 5′→3′ exonucleolytic decay (see below). Decay intermediates that accumulated when the exonuclease was inhibited all had shortened poly(A) tails (Hsu and Stevens 1993; Muhlrad et al. 1994).

Poly(A) as a Negative Regulator of Decapping

A potential role for poly(A) shortening in mRNA decay became apparent when Hsu and Stevens (1993) discovered that yeast cells with a deletion of the 5′→3′ exonuclease encoded by the *XRN1* gene accumulated uncapped mRNAs with shortened poly(A) tails. Use of this mutant and physical blocks to decay identified a pathway that proceeded from poly(A) shortening, to decapping (or 5′-proximal cleavage), to very rapid 5′→3′ exonucleolytic decay (Hsu and Stevens 1993; Decker and Parker 1994; Muhlrad et al. 1994). Decapping and decay appear to commence when poly(A) tails have been shortened to 15 nucleotides or less, a size approaching the lower limit for recognition and binding by the yeast PABP (Sachs et al. 1987). Poly(A) (and PABP) could thus be seen as a negative regulator of decapping and PABP as a rate-limiting step for commencement of exonucleolytic digestion (Decker and Parker 1994; Muhlrad et al. 1994; Beelman and Parker 1995).

These observations provide a missing link that explains the concurrent dependence of mRNA translation and turnover on poly(A) lengths. If stimulation of translational initiation by poly(A) and PABP markedly reduces the rate of decapping, then the closed-loop model can accommodate the interdependence of the two phenomena. mRNAs with poly(A) tails shortened beyond a critical minimum length would be unable to bind PABP efficiently, stimulate initiation efficiently, or protect the cap efficiently (Fig. 2B,C). Although such mRNAs should, in general, be substrates for rapid decay, this fate might be avoided if translation were enhanced by other, non-poly(A)-dependent mechanisms (e.g., the *spb* effects on 60S subunits; see above) or if decapping and exonucleolytic digestion were also regulated (Decker and Parker 1994). Such alternative mechanisms of translation enhancement may account for at least some of the mRNAs that fail to be destabilized by substantial poly(A) shortening (see above).

The universality of this hypothesis depends on a further understanding of the relationship between poly(A) shortening and PABP-binding sites in mammalian cells. Present data suggest that decay of some mammalian mRNAs ensues when poly(A) tails have been shortened to lengths that certainly could accommodate a monomer of yeast PABP (see above). This discrepancy could be explained if the mammalian PABPs bound cooperatively or if they had a larger binding-site size than their yeast counterparts.

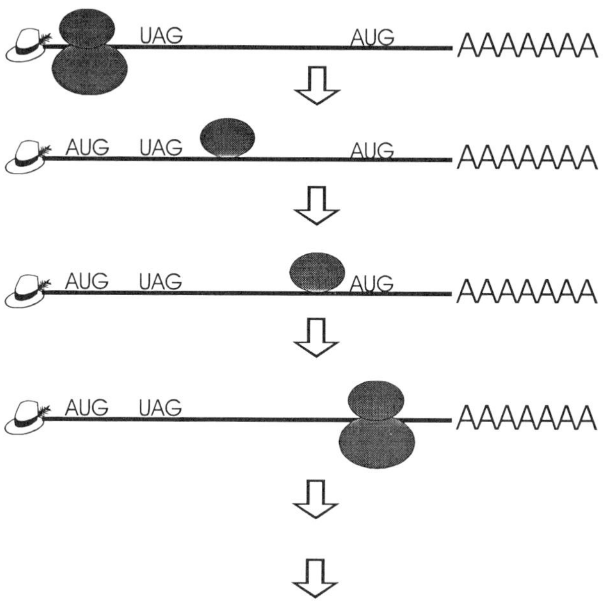

Figure 3 Nonsense-mediated mRNA decay. Rapid decay of nonsense-containing mRNAs in the cytoplasm requires a premature translational termination codon and a downstream sequence that appears to be a site of translational reinitiation. The latter event is thought to trigger mRNA decay.

mRNA Decay without Prior Poly(A) Shortening

Deadenylation is not a prerequisite for decay of all mRNAs, and this raises the question of whether the closed-loop model only applies to a particular class of mRNPs. At least two deadenylation-independent mechanisms of decay are still compatible with the model. In one, degradation is initiated by endonucleolytic cleavage of polyadenylated mRNAs (see, e.g., Stoeckle 1992; Binder et al. 1994; Christiansen et al. 1994). Here, cleavage can be considered an alternate means to disrupt the closed loop and generate substrates for subsequent exonucleolytic mechanisms. The connection to the closed loop for the second mechanism, nonsense-mediated mRNA decay (Peltz et al. 1994; see Theodorakis and Cleveland, this volume), is less straightforward. Substrates of this decay pathway are mRNAs containing premature translational termination signals, and decay depends on both an early nonsense codon and downstream sequences (Peltz et al. 1993, 1994). Current evidence suggests that the latter are sites of translational reinitiation used by ribosomes that had terminated upstream. A fraction of these ribosomes

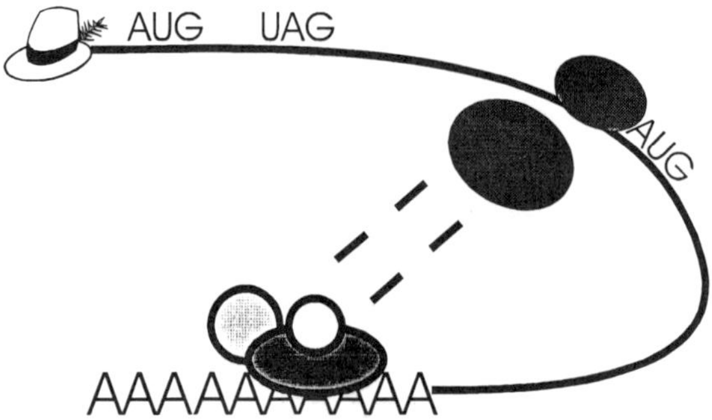

Figure 4 Disruption of the closed loop by nonsense-stimulated reinitiation. Translation on a nonsense-containing mRNA has terminated at the premature UAG, and a 40S subunit that resumed scanning at that position is about to reinitiate at a downstream AUG. The 3′poly(A)/PABP complex is shown stimulating the 60S joining step at the reinitiation site, thereby leaving the 5′cap unprotected.

presumably resume scanning, and the act of reinitiation appears to trigger decay (Fig. 3) (Peltz et al. 1993, 1994; R. Ganesan and A. Jacobson; E. Welch and A. Jacobson; both in prep.). A hypothetical reconciliation of these observations with the closed loop is described in Figure 4. Here, internal reinitiation is depicted as an event that competes with normal initiation for the PABP or other 3′UTR-specific binding proteins. Such competition in turn reduces the effectiveness with which these proteins normally protect the 5′cap. The cap's rate of removal is thus accelerated, and 5′→3′digestion ensues. This hypothesis predicts that nonsense-containing mRNAs should be decapped without deadenylation and should be substrates, at least in yeast, for the XRN1 protein exonuclease. That both predictions have been substantiated (Muhlrad and Parker 1994; Hagan et al. 1995; R. Ganesan and A. Jacobson, in prep.) lends some support to this interpretation. In mammalian cells, available evidence indicates that premature translational termination enhances cytoplasmic decay rates of some mRNAs and may alter nuclear metabolism of others (for review, see Maquat 1995; Theodorakis and Cleveland, this volume). The relationship of the latter to the mechanism postulated here is, at present, unclear.

CONCLUSIONS AND PERSPECTIVES

The discovery of poly(A) ushered in the modern era of eukaryotic molecular biology, as it permitted for the first time routine identification and purification of mRNA and ultimately cloning of individual cDNAs. Most investigators simply exploited the technical aspects of this discovery, whereas others pondered the significance and function of this mRNA appendage, leading to more than two decades of controversy as to whether poly(A) specifically affects mRNA turnover or translation. Recent experiments support earlier predictions that mRNA stability is in part a consequence of changes in translational efficiency attributable to changes in poly(A) lengths. Poly(A)-mediated promotion of 5′/3′ interactions appears to be an event that optimizes translation of the mRNP while limiting its availability to factors that trigger the onset of rapid decay. Ironically, after all these years of research and debate, the favored model bears remarkable resemblance to a hypothesis that predated essentially all of the studies discussed here. The ticketing hypothesis postulated that an mRNA has a finite lifetime dictated by the systematic removal, with each round of translation, of a segment of a repetitive RNA tract essential for mRNA activity (Sussman 1970); in so doing, it presaged poly(A) and poly(A) shortening and their roles in translational control.

ACKNOWLEDGMENTS

Work in my laboratory is supported by grants from the National Institutes of Health (GM-27757) and the American Cancer Society (NP-843). I thank Andrew Bond, Feng He, David Mangus, Ellen Welch, and Dorit Zuk for helpful comments on drafts of this chapter.

REFERENCES

Adam, S.A., T. Nakagawa, M.S. Swanson, T.K. Woodruff, and G. Dreyfuss. 1986. mRNA polyadenylate-binding protein: Gene isolation and sequencing and identification of a ribonucleoprotein consensus sequence. *Mol. Cell. Biol.* **6:** 2932–2943.

Adams, D.S. and W.R. Jeffery. 1978. Poly(adenylic acid) degradation by two distinct processes in the cytoplasmic RNA of *Physarum polycephalum. Biochemistry* **17:** 4519–4524.

Adamson, S.D., G.A. Howard, and E. Herbert. 1970. The ribosome cycle in a reconstituted cell-free system from reticulocytes. *Cold Spring Harbor Symp. Quant. Biol.* **34:** 547–554.

Adesnik, M. and J.E. Darnell. 1972. Biogenesis and characterization of histone messenger RNA in HeLa cells. *J. Mol. Biol.* **67:** 397–406.

Ahlquist, P. and P. Kaesberg. 1979. Determination of the length distribution of poly(A) at the 3′ terminus of the virion RNAs of EMC virus, poliovirus, rhinovirus, RAV-61, and CPMV and of mouse globin mRNA. *Nucleic Acids Res.* **7:** 1195–1204.

Alton, T.H. and H.F. Lodish. 1977. Translational control of protein synthesis during early stages of differentiation of the slime mold *Dictyostelium discoideum*. *Cell* **12:** 301–310.

Astrom, J., A. Astrom, and A. Virtanen. 1991. In vitro deadenylation of mammalian mRNA by a HeLa cell 3′ exonuclease. *EMBO J.* **10:** 3067–3071.

———. 1992. Properties of a HeLa cell 3′ exonuclease specific for degrading poly(A) tails of mammalian mRNA. *J. Biol. Chem.* **267:** 18154–18159.

Bablanian, R. and A. Banerjee 1986. Poly(riboadenylic acid) preferentially inhibits in vitro translation of cellular mRNAs compared with vaccinia virus mRNAs: Possible role in vaccinia virus cytopathology. *Proc. Natl. Acad. Sci.* **83:** 1290–1294.

Bachvarova, R.F. 1992. A maternal tail of poly(A): The long and the short of it. *Cell* **69:** 895–897.

Baer, B. and R.D. Kornberg. 1980. Repeating structure of cytoplasmic poly(A) ribonucleoprotein. *Proc. Natl. Acad. Sci.* **77:** 1890–1892.

———. 1983. The protein responsible for the repeating structure of cytoplasmic poly(A)-ribonucleoprotein. *J. Cell. Biol.* **96:** 717–721.

Baglioni, C., C. Vesco, and M. Jacobs-Lorena. 1970. The role of ribosomal subunits in mammalian cells. *Cold Spring Harbor Symp. Quant. Biol.* **34:** 555–565.

Baker, E.J. 1993. Control of poly(A) length. In *Control of messenger RNA Stability* (ed. J.G. Belasco and G. Brawerman), pp. 367–415. Academic Press, San Diego.

Baker, E.J., D.R. Diener, and J.L. Rosenbaum. 1989. Accelerated poly(A) loss on α-tubulin mRNAs during protein synthesis inhibition in *Chlamydomonas*. *J. Mol. Biol.* **207:** 771–781.

Bard, E., D. Efron, A. Marcus, and R.P. Perry. 1974. Translational capacity of deadenylated messenger RNA. *Cell* **1:** 101–106.

Beelman, C.A. and R. Parker. 1994. Differential effects of translational inhibition in *cis* and *trans* on the decay of the unstable yeast *MFA2* mRNA. *J. Biol. Chem.* **269:** 9687–9692.

———. 1995. Degradation of mRNA in eukaryotes. *Cell* **81:** 179–183.

Bernstein, P. and J. Ross. 1989. Poly(A), poly(A) binding protein and the regulation of mRNA stability. *Trends Biochem. Sci.* **14:** 373–377.

Bernstein, P., S.W. Peltz, and J. Ross. 1989. The poly(A)-poly(A)-binding protein complex is a major determinant of mRNA stability in vitro. *Mol. Cell. Biol.* **9:** 659–670.

Bertholet, C., E. Van Meir, B. ten Heggeler-Bordier, and R. Wittek 1987. Vaccinia virus produces late mRNAs by discontinuous synthesis. *Cell* **50:** 153–162.

Bienroth, S., W. Keller, and E. Wahle. 1992. Assembly of a processive messenger RNA polyadenylation complex. *EMBO J.* **12:** 585–594.

Bilger, A., C.A. Fox, E. Wahle, and M. Wickens. 1994. Nuclear polyadenylation factors recognize cytoplasmic polyadenylation elements. *Genes Dev.* **8:** 1106–1116.

Binder, R., J.A. Horowitz, J.P. Basilion, D.M. Koeller, R.D. Klausner, and J.B. Harford. 1994. Evidence that the pathway of transferrin receptor mRNA degradation involves an endonucleolytic cleavage within the 3′ UTR and does not involve poly(A) shortening. *EMBO J.* **13:** 1969–1980.

Birnboim, H.C., R.E.J. Mitchel, and N.A. Straus. 1973. Analysis of long pyrimidine polynucleotides in HeLa cell nuclear DNA: Absence of polydeoxythymidylate. *Proc. Natl. Acad. Sci.* **70:** 2189–2192.

Brawerman, G. 1981. The role of the poly(A) sequence in mammalian messenger RNA. *Crit. Rev. Biochem.* **10:** 1–38.

———. 1993. mRNA degradation in eukaryotic cells: an overview. In *Control of messenger RNA stability* (ed. J.G. Belasco and G. Brawerman), pp. 149–159. Academic

Press, San Diego.
Brawerman, G. and J. Diez. 1975. Metabolism of the polyadenylate sequence of nuclear RNA and messenger RNA in mammalian cells. *Cell* **5:** 271–280.
Brewer, G. and J. Ross. 1988. Poly(A) shortening and degradation of the 3′ A+U-rich sequences of human c-*myc* mRNA in a cell-free system. *Mol. Cell. Biol.* **8:** 1697–1708.
Butler, J.S., P.P. Sadhale, and T. Platt. 1990. RNA processing in vitro produces mature 3′ ends of a variety of *Saccharomyces cerevisiae* mRNAs. *Mol. Cell. Biol.* **10:** 2599–2605.
Campbell, L.H., K.T. Borg, J.K. Haines, R.T. Moon, D.R. Schoenberg, and S.J. Arrigo. 1994. Human immunodeficiency virus type 1 rev is required in vivo for binding of poly(A)-binding protein to rev-dependent RNAs. *J. Virol.* **68:** 5433–5438.
Cardelli, J.A. and R.L. Dimond. 1981. Regulation of protein synthesis in *Dictyostelium discoideum:* Effects of starvation and anoxia on initiation. *Biochemistry* **20:** 7391–7398.
Carrazana, E., K. Pasieka, and J. Majzoub. 1988. The vasopressin mRNA poly(A) tract is unusually long and increases during stimulation of vasopressin gene expression in vivo. *Mol. Cell. Biol.* **8:** 2267–2274.
Carter, D.A. and D. Murphy. 1989. Independent regulation of neuropeptide mRNA level and poly(A) tail length. *J. Biol. Chem.* **264:** 6601–6603.
———. 1991. Nuclear mechanisms mediate rhythmic changes in vasopressin mRNA expression in the rat suprachiasmatic nucleus. *Mol. Brain Res.* **12:** 315–321.
Chen, C.-Y.A., T.M. Chen, and A.B. Shyu. 1994. Interplay of two functionally and structurally distinct domains of the c-*fos* AU-rich element specifies its mRNA-destabilizing function. *Mol. Cell. Biol.* **14:** 416–426.
Christensen, A.K., L.E. Kahn, and C.M. Bourne. 1987. Circular polysomes predominate on the rough endoplasmic reticulum of somatotropes and mammotropes in the rat anterior pituitary. *Am. J. Anat.* **178:** 1–10.
Christiansen, J., M. Kofod, and F.C. Nielsen. 1994. A guanosine quadruplex and two stable hairpins flank a major cleavage site in insulin-like growth factor II mRNA. *Nucleic Acids Res.* **22:** 5709–5716.
Colot, H.V. and M. Rosbash. 1982. Behavior of individual maternal pA^+ RNAs during embryogenesis of *Xenopus laevis. Dev. Biol.* **94:** 79–86.
Darnell, J.E., R. Wall, and R.J. Tushinski. 1971a. An adenylic acid-rich sequence in messenger RNA of HeLa cells and its possible relationship to reiterated sites in DNA. *Proc. Natl. Acad. Sci.* **68:** 1321–1325.
Darnell, J.E., L. Philipson, R. Wall, and M. Adesnik. 1971b. Polyadenylic acid sequences: Role in conversion of nuclear RNA into messenger RNA. *Science* **174:** 507–510.
Decker, C.J. and R. Parker. 1993. A turnover pathway for both stable and unstable mRNAs in yeast: Evidence for a requirement for deadenylation. *Genes Dev.* **7:** 1632–1643.
———. 1994. Mechanisms of mRNA degradation in eukaryotes. *Trends Biochem. Sci.* **19:** 336–340.
———. 1995. Diversity of cytoplasmic functions for the 3′ untranslated region of eukaryotic transcripts. *Curr. Opin. Cell Biol.* **7:** 386–392.
Deshpande, A., B. Chatterjee, and A. Roy. 1979. Translation and stability of rat liver messenger RNA for $\alpha_{2\mu}$-globulin in *Xenopus* oocyte. *J. Biol. Chem.* **254:** 8937–8942.
Doel, M.T. and N.H. Carey. 1976. The translational capacity of deadenylated ovalbumin

messenger RNA. *Cell* **8:** 51–58.

Drummond, D.R., J. Armstrong, and A. Colman. 1985. The effect of capping and polyadenylation on the stability, movement, and translation of synthetic messenger RNAs in *Xenopus* oocytes. *Nucleic Acids Res.* **13:** 7375–7394.

Dubochet, J., C. Morel, B. Lebleu, and M. Herzberg. 1973. Structure of globin mRNA and mRNA-protein particles: Use of dark-field electron microscopy. *Eur. J. Biochem.* **36:** 465–472.

Edmonds, M., M.H. Vaughn, Jr., and H. Nakazoto. 1971. Polyadenylic acid sequences in the heterogenous nuclear RNA and rapidly-labelled polyribosomal RNA of HeLa cells: Possible evidence for a precursor-product relationship. *Proc. Natl. Acad. Sci.* **68:** 1336–1340.

Fox, C.A. and M.P. Wickens. 1990. Poly (A) removal during oocyte maturation: A default reaction selectively prevented by specific sequences in the 3′-UTR of certain maternal mRNAs. *Genes Dev.* **4:** 2287–2298.

Galili, G., E. Kawata, L.D. Smith, and B.A. Larkins. 1988. Role of the 3′-poly(A) sequence in translational regulation of mRNAs in *Xenopus laevis* oocytes. *J. Biol. Chem.* **263:** 5764–5770.

Gallie, D.R. 1991. The cap and poly(A) tail function synergistically to regulate mRNA translational efficiency. *Genes Dev.* **5:** 2108–2116.

Gallie, D.R. and R. Tanguay. 1994. Poly(A) binds to initiation factors and increases cap-dependent translation in vitro. *J. Biol. Chem.* **269:** 17166–17173.

Gallie, D.R., W. Lucas, and V. Walbot. 1989. Visualizing mRNA expression in plant protoplasts: factors influencing efficient mRNA uptake and translation. *Plant Cell* **1:** 301–311.

Grange, T., C. Martins de Sa, J. Oddos, and R. Pictet. 1987. Human mRNA polyadenylate binding protein: Evolutionary conservation of a nucleic acid binding motif. *Nucleic Acids Res.* **15:** 4771–4787.

Greenberg, J.R. 1979. Ultraviolet light-induced crosslinking of messenger RNA to proteins. *Nucleic Acids Res.* **6:** 715–732.

Grossi de Sa, M., N. Standart, C. Martins de Sa, O. Akhayat, M. Huesca, and K. Scherrer. 1988. The poly(A)-binding protein facilitates in vitro translation of poly(A)-rich mRNA. *Eur. J. Biochem.* **176:** 521–526.

Hagan, K.W., M.J. Ruiz-Echevarria, Y. Quan, and S.W. Peltz. 1995. Characterization of *cis*-acting sequences and decay intermediates involved in nonsense-mediated mRNA turnover. *Mol. Cell. Biol.* **15:** 809–823.

Herrick, D., R. Parker, and A. Jacobson. 1990. Identification and comparison of stable and unstable mRNAs in the yeast *Saccharomyces cerevisiae. Mol. Cell. Biol.* **10:** 2269–2284.

Howard, G., S. Adamson, and E. Herbert. 1970. Subunit recycling during translation in a reticulocyte system. *J. Biol. Chem.* **245:** 6237–6239.

Hruby, D.E. and W.K. Roberts. 1977. Encephalomyocarditis virus RNA. II. Polyadenylic acid requirement for efficient translation. *J. Virol.* **23:** 338–344.

Hsu, C.L. and A. Stevens. 1993. Yeast cells lacking 5′→3′exoribonuclease 1 contain mRNA species that are poly(A) deficient and partially lack the 5′cap structure. *Mol. Cell. Biol.* **13:** 4826–4835.

Hsu, M.T. and M. Coca-Prodos. 1979. Electron microscopic evidence for the circular form of RNA in the cytoplasm of eukaryotic cells. *Nature* **280:** 339–340.

Huarte, J., A. Stutz, M.L. O'Connell, P. Gubler, D. Belin, A.L. Darrow, S. Strickland,

and J.D. Vassalli. 1992. Transient translational silencing by reversible mRNA deadenylation. *Cell* **69:** 1021–1030.

Huez, G., G. Marbaix, D. Gallwitz, E. Weinberg, R. Devos, E. Hubert, and Y. Cleuter. 1978. Functional stabilization of HeLa cell histone messenger RNAs injected into *Xenopus* oocytes by 3′-OH polyadenylation. *Nature* **271:** 572–573.

Huez, G., G. Marbaix, E. Hubert, Y. Cleuter, M. Leclercq, H. Chantrenne, R. Devos, H. Soreq, U. Nudel, and U.Z. Littauer. 1975. Readenylation of polyadenylate-free globin messenger RNA restores its stability in vivo. *Eur. J. Biochem.* **59:** 589–592.

Iatrou, K. and G.H. Dixon. 1977. The distribution of poly(A)$^+$ and poly(A)- protamine messenger RNA sequences in the developing trout testis. *Cell* **10:** 433–441.

Iizuka, N., L. Najita, A. Franzusoff, and P. Sarnow. 1994. Cap-dependent and cap-independent translation by internal initiation of mRNAs in cell extracts prepared from *Saccharomyces cerevisiae. Mol. Cell. Biol.* **14:** 7322–7330.

Jackson, R.J. and N. Standart. 1990. Do the poly(A) tail and 3′untranslated region control mRNA translation? *Cell* **62:** 15–24.

Jacobson, A. and M. Favreau. 1983. Possible involvement of poly(A) in protein synthesis. *Nucleic Acids Res.* **11:** 6353–6368.

Jacobson, A., R.A. Firtel, and H.F. Lodish. 1974. Transcription of poly (dT) sequences in the genome of the cellular slime mold, *Dictyostelium discoideum. Proc. Natl. Acad. Sci.* **71:** 1607–1611.

Jones, P.M., J.M. Burrin, M.A. Ghatel, D.J. O'Halloran, S. Legon, and S.R. Bloom. 1990. The influence of thyroid hormone status on the hypothalamo-hypophyseal growth hormone axis. *Endocrinology* **126:** 1374–1379.

Kates, J. 1971. Transcription of the vaccinia virus genome and the occurrence of polyriboadenylic acid sequences in messenger RNA. *Cold Spring Harbor Symp. Quant. Biol.* **35:** 743–752.

Kislauskis, E.H. and R.H. Singer. 1992. Determinants of mRNA localization. *Curr. Opin. Cell Biol.* **4:** 975–978.

Kleene, K. 1989. Poly(A) shortening accompanies the activation of translation of five mRNAs during spermiogenesis in the mouse. *Development* **106:** 367–373.

Ladhoff, A.M., I. Uerlings, and S. Rosenthal. 1981. Electron microscopic evidence of circular molecules of 9-S globin mRNA from rabbit reticulocytes. *Mol. Biol. Rep.* **7:** 101–106.

Laird-Offringa, I.A., C.L. De Wit, P. Elfferich, and A.J. van der Eb. 1990. Poly(A) tail shortening is the translation-dependent step in c-*myc* mRNA degradation. *Mol. Cell. Biol.* **10:** 6132–6140.

Lee, S.Y., J. Mendecki, and G. Brawerman. 1971. A polynucleotide segment rich in adenylic acid in the rapidly-labelled polysomal RNA component of mouse sarcoma 180 ascites cells. *Proc. Natl. Acad. Sci.* **68:** 1331–1335.

Lemay, G. and S. Millward. 1986. Inhibition of translation in L-cell lysates by free polyadenylic acid: Differences in sensitivity among different RNAs and possible involvement of an initiation factor. *Arch. Biochem. Biophys.* **249:** 191–198.

Lim, L. and K.S. Canellakis. 1970. Adenine-rich polymer associated with rabbit reticulocyte messenger RNA. *Nature* **227:** 710–712.

Lowell, J.E., D.Z. Rudner, and A.B. Sachs. 1992. 3′-UTR-dependent deadenylation by the yeast poly(A) nuclease. *Genes Dev.* **6:** 2088–2099.

Manley, J.L. and N.J. Proudfoot. 1994. RNA 3′ends: Formation and function-meeting review. *Genes Dev.* **8:** 259–264.

Manrow, R.E. and A. Jacobson. 1986. Identification and characterization of developmentally regulated mRNP proteins of *Dictyostelium discoideum. Dev. Biol.* **116:** 213–227.

———. 1987. Increased rates of decay and reduced levels of accumulation of the major poly(A)-associated proteins of *Dictyostelium* during heat shock and development. *Proc. Natl. Acad. Sci.* **84:** 1858–1862.

Maquat, L.E. 1995. When cells stop making sense: Effects of nonsense codons on RNA metabolism in vertebrate cells. *RNA* **1:** 453–465.

Masison, D.C., A. Blanc, J.C. Ribas, K. Carroll, N. Sonenberg, and R.B. Wickner. 1995. Decoying the cap$^-$ mRNA degradation system by a dsRNA virus and poly(A)$^-$ mRNA surveillance by a yeast antiviral system. *Mol. Cell. Biol.* **15:** 2763–2771.

McGrew, L., E. Dworkin-Rastl, M.B. Dworkin, and J.D. Richter. 1989. Poly(A) elongation during *Xenopus* oocyte maturation is required for translational recruitment and is mediated by a short sequence element. *Genes Dev.* **3:** 803–815.

Mercer, J.F.B. and S.A. Wake. 1985. An analysis of the rate of metallothionein mRNA poly(A)-shortening using RNA blot hybridization. *Nucleic Acids Res.* **13:** 7929–7943.

Minvielle-Sebastia, L., B. Winsor, N. Bonneaud, and F. Lacroute. 1991. Mutations in the yeast *RNA14* and and *RNA15* genes result in an abnormal mRNA decay rate: Sequence analysis reveals an RNA-binding domain in the RNA15 protein. *Mol. Cell. Biol.* **11:** 3075–3087.

Muhlrad, D. and R. Parker. 1992. Mutations affecting stability and deadenylation of the yeast MFA2 transcript. *Genes Dev.* **6:** 2100–2111.

———. 1994. Premature translational termination triggers mRNA decapping. *Nature* **340:** 578–581.

Muhlrad, D., C.J. Decker, and R. Parker. 1994. Deadenylation of the unstable mRNA encoded by the yeast *MFA2* gene leads to decapping followed by 5′→3′ digestion of the transcript. *Genes Dev.* **8:** 855–866.

———. 1995. Turnover mechanisms of the stable yeast *PGK1* mRNA. *Mol. Cell. Biol.* **15:** 2145–2156.

Munoz, R.F. and J. Darnell. 1974. Poly(A) in mRNA does not contribute to secondary structure neccessary for protein synthesis. *Cell* **2:** 247–252.

Munroe, D. 1989. "mRNA poly(A) tail: A 3′ enhancer of translational initiation." Ph.D. thesis, University of Massachusetts Medical School, Worcester.

Munroe, D. and A. Jacobson, 1989. Poly(A) is a 3′ enhancer of translational initiation. In *Structure, function, and evolution of ribosomes* (ed. W. Hill), pp. 299–305. American Society for Microbiology, Washington, D.C.

———. 1990a. mRNA poly(A) tail: A 3′ enhancer of translational initiation. *Mol. Cell. Biol.* **10:** 3441–3455.

———. 1990b. Tales of poly(A)—A review. *Gene* **91:** 151–158.

Murphy, D. and D. Carter. 1990. Vasopressin gene expression in the rodent hypothalamus: Transcriptional and posttranscriptional responses to physiological stimulation. *Mol. Endocrinol.* **4:** 1051–1059.

Murphy, D., K. Pardy, V. Seah, and D. Carter. 1992. Posttranscriptional regulation of rat growth hormone gene expression: Increased message stability and nuclear polyadenylation accompany thyroid hormone depletion. *Mol. Cell. Biol.* **12:** 2624–2632.

Muschel, R., G. Khoury, and L.M. Reid. 1986. Regulation of insulin mRNA abundance and adenylation: Dependence on hormones and matrix substrata. *Mol. Cell. Biol.* **6:** 337–341.

Nelson, E.M. and M.M. Winkler. 1987. Regulation of mRNA entry into polysomes: Parameters affecting polysome size and the fraction of mRNA in polysomes. *J. Biol. Chem.* **262:** 11501–11506.

Nudel, U., H. Soreq, U.Z. Littauer, G. Marbaix, G. Huez, M. Leclercq, E. Hubert, and H. Chantrenne. 1976. Globin mRNA species containing poly(A) segments of different lengths: their functional stability in *Xenopus* oocytes. *Eur. J. Biochem.* **64:** 115–121.

Ohtake, Y. and R.B. Wickner. 1995. Yeast virus propagation depends critically on free 60S ribosomal subunit concentration. *Mol. Cell. Biol.* **15:** 2772–2781.

Paek, I. and R. Axel. 1987. Glucocorticoids enhance stability of human growth hormone mRNA. *Mol. Cell. Biol.* **7:** 1496–1507.

Palatnik, C.M., R.V. Storti, and A. Jacobson. 1979. Fractionation and functional analysis of newly synthesized and decaying messenger RNAs from vegetative cells of *Dictyostelium discoideum. J. Mol. Biol.* **128:** 371–397.

———. 1981. Partial purification of a developmentally regulated mRNA from *Dictyostelium discoideum* by thermal elution from poly(U)-Sepharose. *J. Mol. Biol.* **150:** 389–398.

Palatnik, C.M., C. Wilkins, and A. Jacobson. 1984. Translational control during early *Dictyostelium* development: possible involvement of poly(A) sequences. *Cell* **36:** 1017–1025.

Palatnik, C.M., A.K. Capone, R.V. Storti, and A. Jacobson. 1980. Messenger RNA stability in *Dictyostelium discoideum:* Does poly (A) have a regulatory role? *J. Mol. Biol.* **141:** 99–118.

Paris, J. and J.D. Richter. 1990. Maturation-specific polyadenylation and translational control: Diversity of cytoplasmic polyadenylation elements, influence of poly(A) tail size, and formation of stable polyadenylation complexes. *Mol. Cell Biol.* **10:** 5634–5645.

Patel, D. and J.S. Butler. 1992. Conditional defect in mRNA 3′ end processing caused by a mutation in the gene for poly(A) polymerase. *Mol. Cell. Biol.* **12:** 3297–3304.

Patel, D.D. and D.J. Pickup. 1987. Messenger RNAs of a strongly-expressed late gene of cowpox virus contain 5′-terminal poly(A) sequences. *EMBO J.* **6:** 3787–3794.

Peltz, S.W. and A. Jacobson. 1992. mRNA stability: In *trans*-it. *Curr. Opin. Cell Biol.* **4:** 979–983.

Peltz, S.W., A.H. Brown, and A. Jacobson. 1993. mRNA destabilization triggered by premature translational termination depends on at least three *cis*-acting sequence elements and one *trans*-acting factor. *Genes Dev.* **7:** 1737–1754.

Peltz, S.W, G. Brewer, G. Kobs, and J. Ross. 1987. Substrate specificity of the exonuclease activity that degrades H4 histone mRNA. *J. Biol. Chem.* **262:** 9382–

Peltz, S.W., F. He, E. Welch, and A. Jacobson. 1994. Nonsense-mediated mRNA decay in yeast. *Prog. Nucleic Acid Res. Mol. Biol.* **47:** 271–298.

Peltz, S.W., G. Brewer, P. Bernstein, P.A. Hart, and J. Ross. 1991. Regulation of mRNA turnover in eukaryotic cells. *Crit. Rev. Eukaryotic Gene Expr.* **1:** 99–126.

Philipson, L., R. Wall, G. Glickman, and J.E. Darnell. 1971. Addition of polyadenylate sequences to virus-specific RNA during adenovirus replication. *Proc. Natl. Acad. Sci.* **68:** 2806–2809.

Piper, P.W. and J.L. Aamand. 1989. Yeast mutation thought to arrest mRNA transport markedly increases the length of the 3′poly(A) on polyadenylated mRNA. *J. Mol. Biol.* **208:** 697–700.

Proudfoot, N. 1991. Poly(A) signals. *Cell* **64:** 671–674.

Proweller, A. and S. Butler. 1994. Efficient translation of poly(A)-deficient mRNAs in

Saccharomyces cerevisiae. *Genes Dev.* **8:** 2629–2640.

Puckett, L., S. Chambers, and J.E. Darnell. 1975. Short-lived messenger RNA in HeLa cells and its impact on the kinetics of accumulation of cytoplasmic polyadenylate. *Proc. Natl. Acad. Sci.* **72:** 389–393.

Raff, R.A. 1980. Masked messenger RNA and the regulation of protein synthesis in eggs and embryos. In *Cell biology: A comprehensive treatise,* vol. 4; *Gene expression: Translation and the behavior of proteins* (ed. D.M. Prescott and L. Goldstein), pp. 107–136. Academic Press, New York.

Restifo, L.L. and G.M. Guild. 1986. Poly(A) shortening of coregulated transcripts in *Drosophila*. *Dev. Biol.* **115:** 507–510.

Robinson B.G., D.M. Frim, W.J. Schwartz, and J.A. Majzoub. 1988. Vasopressin mRNA in the suprachiasmatic nuclei: Daily regulation of polyadenylate tail length. *Science* **241:** 342–344.

Rosenthal, E.T., T.R. Tansey, and J.V. Ruderman. 1983. Sequence-specific adenylations and deadenylations accompany changes in the translation of maternal messenger RNA after fertilization of *Spisula* oocytes. *J. Mol. Biol.* **166:** 309–327.

Sachs, A.B. 1993. Messenger RNA degradation in eukaryotes. *Cell* **74:** 413–421.

Sachs, A.B. and R.W. Davis. 1989. The poly(A)-binding protein is required for poly(A) shortening and 60S ribosomal subunit dependent translation initiation. *Cell* **58:** 857–867.

———. 1990. Translation initiation and ribosomal biogenesis: Involvement of a putative rRNA helicase and RPL46. *Science* **247:** 1077–1079.

Sachs, A.B. and J.A. Deardorff. 1992. Translation initiation requires the PAB-dependent poly(A) ribonuclease in yeast. *Cell* **70:** 961–973.

Sachs, A. and E. Wahle. 1993. Poly(A) tail metabolism and function in eucaryotes. *J. Biol. Chem.* **268:** 22955–22958.

Sachs, A.B., M.W. Bond, and R.D. Kornberg. 1986. A single gene from yeast for both nuclear and cytoplasmic polyadenylate-binding protein: Domain structure and expression. *Cell* **45:** 827–835.

Sachs, A.B., R.W. Davis, and R.D. Kornberg. 1987. A single domain of yeast poly(A)-binding protein is neccessary and sufficient for RNA binding and cell viability. *Mol. Cell. Biol.* **7:** 3268–3276.

Salles, F.J., A.L. Darrow, M.L. O'Connell, and S. Strickland. 1992. Isolation of novel murine maternal mRNAs regulated by cytoplasmic polyadenylation. *Genes Dev.* **6:** 1202–1212.

Salles, F.J., M.E. Lieberfarb, C. Wreden, J.P. Gergen, and S. Strickland. 1994. Coordinate initiation of *Drosophila* development by regulated polyadenylation of maternal messenger RNAs. *Science* **266:** 1996–1999.

Saunders, C.A., K.A. Bostian, and H.O. Halvorson. 1980. Post-transcriptional modification of the poly(A) length of galactose-1-phosphate uridyl transferase mRNA in *Saccharomyces cerevisiae*. *Nucleic Acids Res.* **8:** 3841–3849.

Schwer, B., P. Visca, J.C. Vos, H.G. Stunnenberg 1987. Discontinuous transcription or RNA processing of vaccinia virus late messengers results in a 5′poly(A) leader. *Cell* **50:** 163–169.

Shapiro, R.A., D. Herrick, R.E. Manrow, D. Blinder, and A. Jacobson. 1988. Determinants of mRNA stability in *Dictyostelium discoideum* amoebae: Differences in poly(A) tail length, ribosome loading, and mRNA size cannot account for the heterogeneity of mRNA decay rates. *Mol. Cell. Biol.* **8:** 1957–1969.

Sheets, M.D. and M. Wickens. 1989. Two phases in the addition of a poly(A) tail. *Genes Dev.* **3:** 1401–1412.

Sheets, M.D., C.A. Fox, T. Hunt, G. Vande Woude, and M. Wickens. 1994. The 3′ UTRs of c-*mos* and cyclin mRNAs control translation by regulation cytoplasmic polyadenylation. *Genes Dev.* **8:** 926–938.

Sheiness, D. and J.E. Darnell. 1973. Polyadenylic acid segment in mRNA becomes shorter with age. *Nat. New Biol.* **241:** 265–268.

Sheiness, D., L. Puckett, and J.E. Darnell. 1975. Possible relationship of poly(A) shortening to mRNA turnover. *Proc. Natl. Acad. Sci.* **72:** 1077–1081.

Shyu, A.-B., J.G. Belasco, and M.E. Greenberg. 1991. Two distinct destabilizing elements in the c-*fos* message trigger deadenylation as a first step in rapid mRNA decay. *Genes Dev.* **5:** 221–231.

Sieliwanowicz, B. 1987. The influence of poly(A)-binding proteins on translation of poly(A)$^+$ RNA in a cell-free system from embryo axes of dry pea seeds. *Biochim. Biophy. Acta* **908:** 54–59.

Sippel, A.E., J.G. Stavrianopoulos, G. Schutz, and P. Feigelson. 1974. Translational properties of rabbit globin mRNA after specific removal of poly(A) with ribonuclease H. *Proc. Natl. Acad. Sci.* **71:** 4635–4639.

Soreq, H., U. Nudel, R. Salomon, M. Revel, and U.Z. Littauer. 1974. In vitro translation of polyadenylic acid-free rabbit globin messenger RNA. *J. Mol. Biol.* **88:** 233–245.

Spector, D.H., L. Villa-Komaroff, and D. Baltimore. 1975. Studies on the function of polyadenylic acid on poliovirus RNA. *Cell* **6:** 41–44.

Steel, L.F. and A. Jacobson. 1987. Translational control of ribosomal protein synthesis during early *Dictyostelium discoideum* development. *Mol. Cell. Biol.* **7:** 965–972.

Stoeckle, M.Y. 1992. Removal of a 3′ non-coding sequence is an initial step on degradation of *groα mRNA and is regulated by interleukin-1. Nucleic Acids Res.* **20:** 1123–1127.

Storti, R.V., M.P. Scott, A. Rich, and M.L. Pardue. 1980. Translational control of protein synthesis in response to heat shock in *D. melanogaster* cells. *Cell* **22:** 825–834.

Sussman, M. 1970. Model for quantitative and qualitative control of mRNA translation in eukaryotes. *Nature* **225:** 1245–1246.

Varnum, S.M., C.A. Hurney, and W.M. Wormington. 1992. Maturation-specific deadenylation in *Xenopus* oocytes requires nuclear and cytoplasmic factors. *Dev. Biol.* **153:** 283–290.

Vassalli, J.-D., J. Huarte, D. Belin, P. Gubler, A. Vassalli, M.L. O'Connell, L.A. Parton, R.J. Rickles, and S. Strickland. 1989. Regulated polyadenylation controls mRNA translation during meiotic maturation of mouse oocytes. *Genes Dev.* **3:** 2163–2171.

Vreken, P. and H.A. Raue. 1992. The rate-limiting step in yeast *PGK1* mRNA degradation is an endonucleolytic cleavage in the 3′-terminal part of the coding region. *Mol. Cell. Biol.* **12:** 2986–2996.

Wahle, E. 1991. A novel poly(A)-binding protein acts as a specificity factor in the second phase of messenger RNA polyadenylation. *Cell* **66:** 759–768.

Wahle, E. and W. Keller. 1992. The biochemistry of 3′-end cleavage and polyadenylation of messenger RNA precursors. *Annu. Rev. Biochem.* **61:** 419–440.

Warner, J.R., A. Rich, and C.E. Hall. 1962. Electron microscope studies of ribosomal clusters synthesizing hemoglobin. *Science* **138:** 1399–1403.

Wickens, M.P. 1990. How the messenger got its tail: Addition of poly(A) in the nucleus. *Trends Biochem. Sci.* **15:** 277–280.

———. 1992. Forward, backward, how much, when: Mechanisms of poly (A) addition and removal and their role in early development. *Semin. Dev. Biol.* **3:** 399–412.

Wickner, R.B. 1992. Double-stranded and single-stranded RNA viruses of *Saccharomyces cerevisiae. Annu. Rev. Microbiol.* **46:** 347–375.

Widner, W.R. and R.B. Wickner. 1993. Evidence that the SKI antiviral system of *Saccharomyces cerevisiae* acts by blocking expression of viral mRNA. *Mol. Cell. Biol.* **13:** 4331–4341.

Williamson, R., J. Crossley, and S. Humphries. 1974. Translation of mouse globin messenger ribonucleic acid from which the poly(adenylic acid) sequence has been removed. *Biochemistry* **13:** 703–707.

Wilson, M.C., S.G. Sawicki, P.A. White, and J.E. Darnell. 1978. A correlation between the rate of poly(A) shortening and half-life of messenger RNA in adenovirus transformed cells. *J. Mol. Biol.* **126:** 23–36.

Wilson, T. and R. Treisman. 1988. Removal of poly(A) and consequent degradation of c-*fos* mRNA facilitated by 3′ AU-rich sequences. *Nature* **336:** 396–399.

Wisdom, R. and W. Lee. 1991. The protein coding region of c-*myc* RNA contains a sequence that specifies rapid mRNA turnover and induction by protein synthesis inhibitors. *Genes Dev.* **5:** 232–243.

Wormington, M. 1993. Poly(A) and translation: Development control. *Curr. Opin. Cell Biol.* **5:** 950–954.

Zelus, B.D., D.H. Giebelhaus, D.W. Eib, K.A. Kenner, and R.T. Moon. 1989. Expression of the poly(A)-binding protein during development of *Xenopus laevis. Mol. Cell. Biol.* **9:** 2756–2760.

Zingg, H.H., D.L. Lefebvre, and G. Almazan. 1988. Regulation of poly(A) tail size of vasopressin mRNA. *J. Biol. Chem.* **263:** 11041–11043.

17

Dynamics of Poly(A) Addition and Removal during Development

Joel D. Richter
Worcester Foundation for Experimental Biology
Shrewsbury, Massachusetts 01545

Nuclear polyadenylation of pre-mRNA generally occurs without regard to RNA sequence or cell type. Cytoplasmic polyadenylation of messenger RNA, on the other hand, is both mRNA sequence-specific and cell-type-specific. With few exceptions, the consequence of cytoplasmic polyadenylation, which occurs in the eggs and embryos of many species, is a highly ordered activation of translationally dormant mRNAs that is essential for normal development. Antithetically, the silencing of certain translating mRNAs is also under developmental control and, perhaps not surprisingly, is the result of message-specific deadenylation. This chapter reviews the current state of our knowledge of poly(A) addition and removal during development. For related aspects of polyadenylation and translational control, see Wickens et al. and Jacobson (both this volume).

To place cytoplasmic polyadenylation in context, it is worthwhile to consider its historical antecedent. In an interpretive review of late 19th and early 20th century embryology, Wilson (1925) noted that "the behavior of nuclei in eggs is determined by the cytoplasm in which they lie." Although Wilson was speaking at that time specifically of the continuity of germ plasm, we now know that cytoplasmic components inherited by the egg at the time of fertilization program early development, probably in all metazoans. These components, for the most part mRNAs, are generally quiescent in oocytes, but they are translationally activated following the resumption of meiosis (oocyte maturation) or after fertilization. How these "masked" or "maternal" mRNAs are translationally regulated has been the subject of much speculation since they were first discovered in the 1960s. Although hindsight has shown that no single mechanism is responsible for this regulation, the recently discovered poly(A) tract on the 3′ ends of mRNAs suggested to Slater et al. (1972, 1973) and Wilt (1973) that this tract might influence translational recruitment in fertilized sea urchin eggs. Indeed, these investigators were the first to show that cytoplasmic polyadenylation of maternal mRNAs does occur and that this is coincident with their assembly into polysomes. Although attractive, the hypothesis that polyadenylation is responsible for translational activation after fertilization remained just that, due largely to the low-resolution techniques available at the time that limited analysis to bulk populations of RNA instead of individual species.

CYTOPLASMIC POLYADENYLATION AND TRANSLATIONAL ACTIVATION: CORRELATIVE STUDIES

The results of Slater and Wilt prompted investigations of cytoplasmic polyadenylation in several organisms, most notably in eggs of marine invertebrates, mouse, and *Xenopus*. Because of their ready availability, sea urchin eggs were used again by Wilt (1977) and Dolecki et al. (1977) to demonstrate not only that fertilization stimulated an approximately twofold increase in poly(A) content of maternal mRNA, but also that this same poly(A) was lost by the two-cell stage. Thus, poly(A) appeared to be under a dynamic control, which offered the appealing possibility that poly(A) elongation promotes translation, whereas deadenylation inhibits it. However, this idea could not immediately be reconciled with the earlier observation of Mescher and Humphreys (1974), who had asked whether polyadenylation *might* control translation. These investigators incubated sea urchin eggs and embryos in 3′-deoxyadenosine (cordycepin), an agent that upon incorporation into a growing poly(A) tail results in its immediate termination. Although cordycepin did indeed inhibit polyadenylation, it had no effect on the overall recruitment of maternal mRNA into polysomes. Thus, Mescher and Humphreys (1974) concluded that poly(A) either was a superfluous by-product of fertilization or had some function unrelated to translational activation, such as enhancing mRNA stability. Bear in mind, however, that this study, as well as many others of this era, suffered from the fact that populations of mRNAs were examined instead of individual species. As a consequence, the regulation of an individual mRNA might be hidden by the behavior of the message population as a whole.

Although a number of investigators subsequently showed that cytoplasmic polyadenylation and deadenylation are widespread in the animal kingdom (Cabada et al. 1977; Slater et al. 1978; Darnbrough and Ford 1979; Egrie and Wilt 1979; Sagata et al. 1980; Colot and Rosbash 1982; Clegg and Piko 1983), for the most part they were still examining overall changes in poly(A), which naturally limited their ability to present defined cases where these changes were linked to translation. It was not until the work of Rosenthal et al. (1983) that very compelling, albeit still correlative, data were presented indicating that the cytoplasmic polyadenylation of specific mRNAs controls translation. These investigators probed Northern blots of polysomal RNA prepared from oocytes and fertilized eggs of the surf clam *Spisula* and showed that four specific RNAs that have short poly(A) tails in the oocyte are recruited into polysomes in the egg when their tails are considerably elongated. Moreover, they noted that tubulin mRNA, which contains a long poly(A) tail and is translated in oocytes, undergoes deadenylation and is coincidently released from polysomes. These results, which were extended in a further study (Rosenthal and Ruderman 1987), rejuvenated the idea that cytoplasmic polyadenylation could be a major regulator of maternal mRNA translation and spawned similar analyses in other systems.

In *Xenopus*, for example, Dworkin and colleagues showed that a number of oocyte mRNAs receive poly(A) during meiotic maturation (Dworkin and Dworkin-Rastl 1985; Smith et al. 1988a,b) and enter polysomes after fertilization (Dworkin et al. 1985). In mouse oocytes, specific mRNAs such as those encoding hypoxanthine phosphoribosyl transferase (HPRT) and Mos also receive poly(A) during oocyte maturation (Goldman et al. 1988; Paynton et al. 1988; Paynton and Bachvarova 1994). At least in the case of c-*mos* mRNA, this correlates with translational activation (O'Keefe et al. 1989; cf. Gebauer et al. 1994). In addition, at least one mRNA, encoding actin, loses its poly(A) tail during the same time period and becomes translationally inactivated (Bachvarova et al. 1985). Finally, in *Urechis caupo*, a marine worm, several mRNAs are polyadenylated upon fertilization and simultaneously shift into polysomes from a nontranslating ribonucleoprotein compartment (Rosenthal and Wilt 1986).

In the aggregate, the aforementioned studies present a strong argument that polyadenylation is coincident with, and may be responsible for, the translational activation of a number of mRNAs in early development. However, one must also consider apparently contradictory results from *Xenopus* oocytes, where histone H4 mRNA, which contains a poly(A) tail in oocytes, undergoes deadenylation during maturation (Ballantine and Woodland 1985) when it is translationally activated (Woodland 1980). In addition, it is important to note that the translational activation of several mRNAs during mouse spermatogenesis correlates with poly(A) shortening, rather than with poly(A) lengthening (Kleene et al. 1984; Kleene 1989, 1993). Thus, these results are either odd exceptions to a general rule or the correlation between polyadenylation and translation is just that, a correlation without causation.

CYTOPLASMIC POLYADENYLATION: ANALYTICAL STUDIES

Cis Elements Required for Polyadenylation during Oocyte Maturation

In 1985, Huarte et al. noted that the tissue-type plasminogen activator (t-PA), a serum protein involved in the coagulation process, is present in secondary (mature) mouse oocytes but not in primary germinal vesicle stage oocytes. t-PA mRNA, however, is present in both primary and mature oocytes, suggesting that t-PA production is controlled at the translational level. Subsequently, Huarte et al. (1987) demonstrated that oocyte maturation stimulated not only elongation of the poly(A) tail of t-PA mRNA, but its translational activation as well. In the first attempt to disrupt the cytoplasmic polyadenylation of a specific mRNA, Strickland et al. (1988) injected antisense oligonucleotides corresponding to various portions of the 5′- and 3′-untranslated regions (UTRs), as well as the coding region, of t-PA mRNA into mouses oocytes. As a result of an endogenous RNase-H-like activity that destroys RNA-DNA hybrids, these authors were able to demonstrate that the most distal 103 nucleotides of the 3′UTR were necessary for both polyadenylation and translational activation (Strickland et al. 1988). Although a further definition of the precise *cis*-acting element(s) within this 103-base region was not attempted, it does contain the hexanucleotide AAUAAA, which is notable for its role in nuclear pre-mRNA cleavage and polyadenylation. Thus, it seemed reasonable to suggest that this hexanucleotide might also be involved in cytoplasmic polyadenylation as well.

In a further series of important experiments, Vassalli et al. (1989) injected the 3′UTR of t-PA mRNA into mouse oocytes and showed that this sequence, which contained the 103-base region noted above, is sufficient to promote polyadenylation during oocyte maturation. Additional injection experiments demonstrated that although the hexanucleotide AAUAAA is a necessary element for polyadenylation, at least one additional sequence in the 3′UTR is required for this process.

While the Strickland and Vassalli laboratories were examining maturation-specific polyadenylation of t-PA mRNA in mouse oocytes, parallel studies were being conducted using *Xenopus* mRNAs that undergo polyadenylation and translational activation during oocyte maturation (Dworkin and Dworkin-Rastl 1985). G10 and D7 RNAs (the functions of the encoded proteins are unknown) were synthesized in vitro and injected into oocytes that were induced to mature with progesterone. Both of these RNAs were polyadenylated during maturation, and one, G10 mRNA, was translated as well (Fox et al. 1989; McGrew et al. 1989). A mutational analysis of the RNAs showed that the hexa-

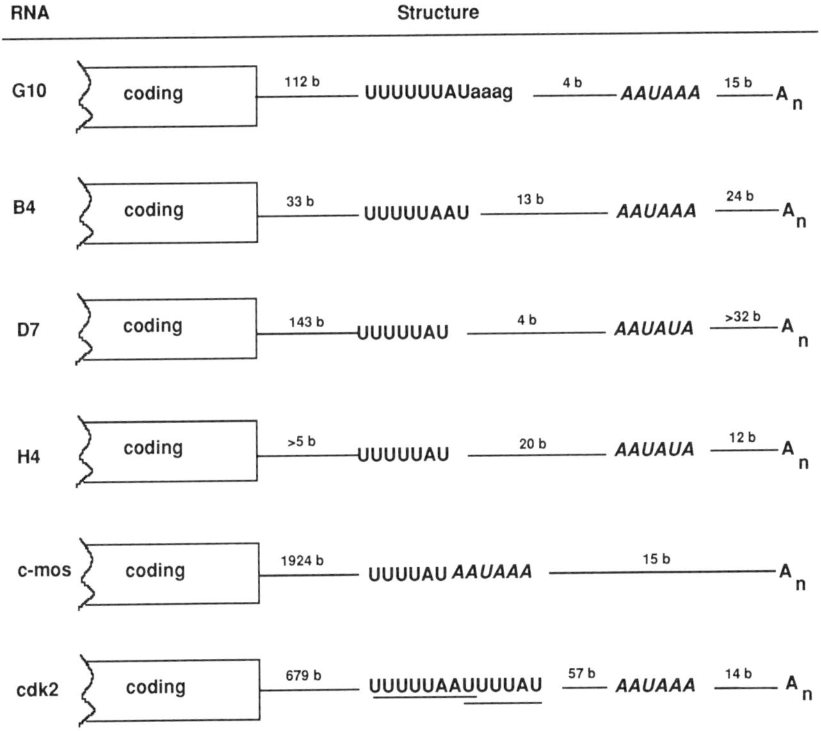

Figure 1 Maturation-specific cytoplasmic polyadenylation signals from six *Xenopus* RNAs. The CPEs and polyadenylation hexanucleotides (italicized), and their positions within the 3′UTRs, are shown. The lowercase letters in the G10 RNA refer to nucleotides that, when removed, inhibit polyadenylation. These bases therefore may be a part of the CPE. cdk2 RNA contains two CPEs (underlined). (G10) McGrew et al. 1989; McGrew and Richter 1990; (B4) Paris and Richter 1990; (D7) Fox et al. 1989; (H4) Fox et al. 1989; (c-*mos*) Paris and Richter 1990; (cdk2) Stebbins-Boaz and Richter 1994.

nucleotide AAUAAA is an essential sequence for polyadenylation and in this respect resembled t-PA mRNA (Fox et al. 1989; McGrew et al. 1989). However, because the hexanucleotide is present in many RNAs that do not undergo polyadenylation, another controlling element seemed likely. This other "specificity element," also present in the 3′UTR, has the general structure of UUUUUUAU and is called the cytoplasmic polyadenylation element (CPE). The CPE, as well as AAUAAA, has been shown to be necessary for the cytoplasmic polyadenylation of all other mRNAs tested during oocyte maturation in *Xenopus* and mouse. However, the CPE is not identical in all RNAs, nor is its position constant relative to AAUAAA. Figure 1 shows the sequences of different CPEs that have been identified in a number of *Xenopus* RNAs and their positions relative to AAUAAA. It appears that the smallest sequence that can stimulate polyadenylation, which would be considered the minimal CPE, is UUUUAU (Paris and Richter 1990).

Cis-acting Elements Required for Polyadenylation during Embryogenesis

Paris et al. (1988) and Paris and Philippe (1990) identified two sets of mRNAs from *Xenopus* that are differentially polyadenylated and translated during maturation or early embryogenesis. These observations raised the interesting possibility that perhaps there are two types of CPE, one specific for maturation and a second specific for embryogenesis. Indeed, Simon et al. (1992) and Simon and Richter (1994) identified a second "embryonic-type" CPE that consists of oligo(U), with 12 bases being the minimum number necessary to promote efficient polyadenylation. This CPE directs polyadenylation up to, and possibly beyond, the 4000-cell mid blastula stage.

One surprise that came out of the investigation of Simon et al. (1992) was that the embryonic-type CPE could function during oocyte maturation if other sequences in the 3′UTR were deleted (cf. Simon and Richter 1994). Such additional sequences comprise a large 400-base-plus "masking" element that prevents precocious polyadenylation during maturation. Because the full masking effect requires not only the 400 bases, but also the CPE itself, Simon et al. (1992) speculated that some secondary structure within this region prevents the binding of an essential polyadenylation factor. Figure 2 illustrates the organization of the regulatory elements that control the polyadenylation of Cl2 RNA.

Cis-acting Elements That Control Polyadenylation in Invertebrate Development

In mouse and *Xenopus*, the only vertebrates so far examined, the U-rich CPEs are essential control elements for cytoplasmic polyadenylation. In invertebrates, however, this may not be the case. For example, in *Spisula*, the polyadenylation of ribonucleotide reductase mRNA in fertilized egg extracts apparently requires neither a CPE nor AAUAAA (Standart and Dale 1993). In *Drosophila* embryos, where at least three mRNAs, *bicoid*, *Toll*, and *torso*, undergo cytoplasmic polyadenylation, the necessary sequences have not been defined. However, if a CPE and AAUAAA are required for cytoplasmic polyadenylation, they probably are not sufficient (Salles et al. 1994).

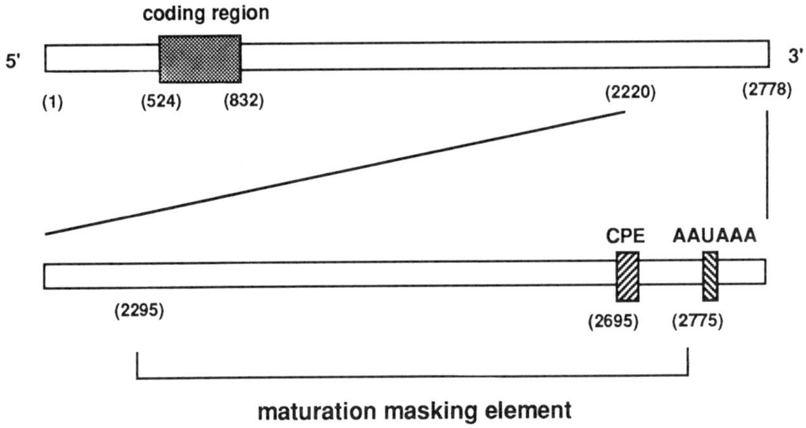

Figure 2 Organization of Cl2 mRNA polyadenylation regulatory elements. Cl2 RNA, which is polyadenylated in *Xenopus* embryos, contains three regulatory elements in its unusually long 3′UTR. The elements that promote polyadenylation in embryos are an oligo(U) CPE and AAUAAA. A large masking element prevents precocious polyadenylation during oocyte maturation. The positions of relevant nucleotide sequence numbers are noted. (Modified from Simon et al. 1992.)

Of all the cases described thus far, the *cis*-acting elements that direct cytoplasmic polyadenylation reside in the 3′UTR, irrespective of whether they include the CPE or AAUAAA. However, there are instances where *cis*-acting elements directing cytoplasmic polyadenylation reside in the 5′UTR. For example, the transcript encoded by the *Drosophila Mst87F* gene is synthesized in primary spermatocytes, stored in a translationally dormant form for 3 days during spermiogenesis, and then undergoes cytoplasmic poly(A) elongation and translational activation. Schafer and colleagues (Kuhn et al. 1988, 1991; Schafer et al. 1990; Kempe et al. 1993) have identified a regulatory sequence for these events called the translational control element (TCE), which consists of the 12-base sequence ACAUCAAAAUUU. Interestingly, the TCE, which always begins at base +28 relative to the start of transcription, is conserved in four additional RNAs that have similar poly(A) elongation patterns in *Drosophila* spermatogenesis (Kuhn et al. 1991).

Trans-acting Factors That Control Cytoplasmic Polyadenylation

Two approaches have generally been taken in the analysis of factors involved in cytoplasmic polyadenylation. The first is based on the identification and isolation of proteins that interact with the CPE, which would specify those mRNAs that are to be polyadenylated. The second approach is to ask whether factors known to be important for nuclear polyadenylation might also be involved in cytoplasmic polyadenylation. As will become evident below, both approaches have been successful.

Because of their ready availability and well-known facility as biochemical systems, *Xenopus* oocytes and eggs were used to prepare extracts that mimic the in vivo developmental control of cytoplasmic polyadenylation. In such extracts, CPE-binding activities were detected by RNA gel-shift analyses (Paris and Richter 1990; Fox et al. 1992) and were further resolved by UV cross-linking. For example, the CPE of G10 RNA, defined by nucleotide-substitution mutagenesis as UUUUUUAU AAAG, is cross-linked to an 82-kD protein found in polyadenylation-proficient egg extracts but not in polyadenylation-deficient oocyte extracts (McGrew and Richter 1990). In contrast, the CPE of B4 RNA, UUUUUAAU, is bound by a 62-kD protein in both oocyte and egg extracts (Paris et al. 1991). Interestingly, the 62-kD protein appears to be phosphorylated by the kinase $p34^{cdc2}$ in extracts at a time coincident with the induction of polyadenylation (Paris et al. 1991), suggesting that the phosphorylation of this protein is a key event for the cytoplasmic polyadenylation of this RNA.

Using single-step RNA affinity chromatography, Hake and Richter (1994) isolated the 62-kD protein, obtained amino acid sequence information, and cloned and sequenced a cDNA corresponding to the entire coding region. This protein, referred to as the CPE-binding protein (CPEB), contains two RNA recognition motifs (RRMs) and has striking homology with one particular protein in the data bases. This protein, Orb, is a *Drosophila* oocyte RNA-binding protein that is 77% identical to CPEB in the RRM region (Hake and Richter 1994). At the morphological level, Orb is necessary for egg chamber formation and the establishment of embryonic polarity; at the molecular level, it is essential for the correct localization of such developmentally important mRNAs as *K10*, *bicaudal-D* and *gurkin* (Lantz et al. 1992, 1994; Christersen and McKearin 1994). Although the putative Orb-binding site on these mRNAs has not yet been defined, recent evidence indicates that in *orb* mutant flies, the cytoplasmic polyadenylation of all three mRNAs is abolished. In addition, this lack of polyadenylation correlates with lower levels of protein synthesis directed by the three RNAs (P. Schedl, pers. comm.). Thus, as will become evident, CPEB and Orb are probably homologs.

The main question concerning CPEB is whether it is necessary for cytoplasmic polyadenylation. Using an antibody raised against a glutathione-*S*-transferase (GST)-CPEB fusion protein synthesized in bacteria, Hake and Richter (1994) immunodepleted an egg extract of CPEB, which resulted in the complete loss of polyadenylation activity of exogenous B4 RNA. When CPEB mRNA was translated in a rabbit reticulocyte lysate and the lysate added to the depleted extract, polyadenylation activity was restored. On the basis of these results, they concluded that CPEB is necessary for cytoplasmic polyadenylation.

Using a different approach, Fox et al. (1992) fractionated egg extracts into two activities, a CPE-binding activity and a poly(A) polymerase activity. Further fractionation of the CPE-binding activity suggested that it could be similar to cleavage-polyadenylation specificity factor (CPSF), a heterotrimeric (or heterotetrameric) protein complex that is indispensable for nuclear polyadenylation (Wahle and Keller 1992). To assess whether CPSF could also be important for cytoplasmic polyadenylation, Bilger et al. (1994) combined bovine CPSF with bovine poly(A) polymerase and showed that this mixture catalyzes CPE-dependent polyadenylation in vitro. This result was somewhat unexpected because CPSF is only known to interact with AAUAAA (Keller et al. 1991; Jenny et al. 1994). Bilger et al. (1994) suggested, however, that CPSF recognizes both AAUAAA and the CPE. Moreover, they proposed that cytoplasmic CPSF (or a CPSF-like molecule) is activated or synthesized during maturation to effect polyadenylation.

Although the data of Bilger et al. (1994) certainly implicate CPSF in cytoplasmic polyadenylation, it is unclear whether this complex plus poly(A) polymerase is all that is required for the cytoplasmic polyadenylation of different mRNAs that occurs at specific times during maturation (Sheets et al. 1994) and embryogenesis (Paris and Philippe 1990; Simon et al. 1992; Simon and Richter 1994). Moreover, these CPSF results would appear to be incompatible with the CPEB data described above.

As suggested by both Bilger et al. (1994) and Hake and Richter (1994), the results of these two studies need not be mutually exclusive. For example, CPEB and CPSF may be components of a larger complex that promotes cytoplasmic polyadenylation. Perhaps CPEB, when phosphorylated, recruits, stabilizes, or activates CPSF in an mRNA-specific and/or development-specific manner. Indeed, perhaps CPEB is analogous to cleavage stimulatory factor (CstF), a complex that stabilizes CPSF binding to the AAUAAA of pre-mRNA in the nucleus. A model for the activation of polyadenylation by CPEB, and its possible interaction with CPSF, is illustrated in Figure 3.

A third factor involved in cytoplasmic polyadenylation is poly(A)

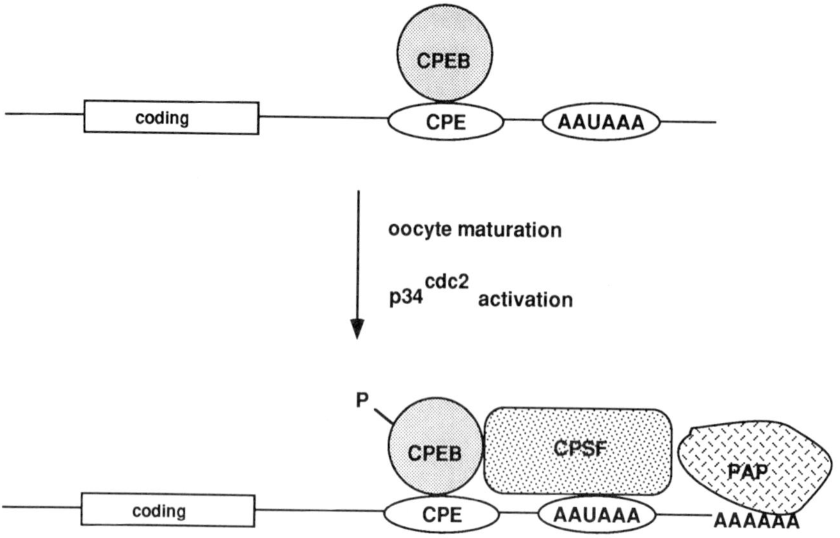

Figure 3 Model for CPEB activation of polyadenylation during *Xenopus* oocyte maturation. In immature oocytes, the CPE is bound by CPEB (Paris et al. 1991; Hake and Richter 1994); it is not known whether any protein interacts with AAUAAA at this time. During oocyte maturation, the activation of several kinases, including $p34^{cdc2}$, results in the phosphorylation of CPEB (Paris et al. 1991). This could cause the recruitment of CPSF (Bilger et al. 1994), which in turn recruits poly(A) polymerase (PAP).

polymerase. As noted above, Fox et al. (1992) obtained two fractions from egg extracts, one that contained an RNA-binding activity and one that contained a poly(A) polymerase activity. They subsequently showed that substitution of the poly(A) polymerase-containing compartment with bovine poly(A) polymerase, which catalyzes nuclear polyadenylation, could catalyze CPE-dependent polyadenylation as well. On the basis of this result, these investigators as well as Ballantyne et al. (1995) suggested that the same poly(A) polymerase probably catalyzes both nuclear and cytoplasmic polyadenylation.

Recent evidence, however, indicates that oocytes might have a unique cytoplasmic poly(A) polymerase. Gebauer and Richter (1995) cloned a cDNA for poly(A) polymerase that, when expressed as a fusion protein in bacteria, catalyzes polyadenylation in vitro. In addition, antibody raised against this same fusion protein neutralizes polyadenylation when added to egg extracts. (It should be noted, however, that several poly(A) polymerases are likely to be recognized by this antibody due to sequence conservation [cf. Ballantyne et al. 1995].) Finally, the poly(A) polymerase identified by Gebauer and Richter (1995), which lacks an obvious nuclear localization signal, concentrates in the cytoplasm when injected into oocytes. These results imply that this poly(A) polymerase might be one of several that participate in cytoplasmic polyadenylation and that the nuclear and cytoplasmic poly(A) polymerases are not necessarily the same.

Although polyadenylation in *Xenopus* embryos is likely to involve most of the same factors as polyadenylation during maturation, at least some may be different. For example, the oligo(U) CPEs of RNAs that are polyadenylated during *Xenopus* embryogenesis are not bound by CPEB, but rather by two proteins of 36 and 45 kD (Simon and Richter 1994). Such proteins may be functionally homologous to CPEB and could perhaps recruit or stabilize CPSF during embryogenesis.

Importance of Cytoplasmic Polyadenylation for Development

Perhaps one of the most important issues regarding cytoplasmic polyadenylation is its relationship to early development; i.e., although t-PA, G10, and B4 RNAs are all polyadenylated and translated during mouse or *Xenopus* oocyte maturation, it was unclear whether the proteins they encode are necessary for early development. Indeed, this appears not to be the case for D7 protein (Smith et al. 1991). However, three recent studies have addressed this issue directly and have shown that this process is essential for normal development.

In the mouse, Mos is necessary for full meiotic progression in maturing oocytes (O'Keefe et al. 1989; Paules et al. 1989). In addition, c-*mos* mRNA is polyadenylated during maturation (Goldman et al. 1988). To determine whether c-*mos* mRNA polyadenylation is required for meiosis, Gebauer et al. (1994) injected oocytes with an antisense oligonucleotide complementary to the coding region of this message. As shown initially by O'Keefe et al. (1989), this resulted not only in the destruction of c-*mos* mRNA, presumably by an RNase-H-like enzyme, but also in the failure of the oocytes to progress from meiosis I to meiosis II. In an attempt to "rescue" these oocytes, Gebauer et al. (1994) constructed a c-*mos* mRNA that encoded the wild-type Mos protein but contained alternate codons in the region complementary to the antisense oligonucleotide. This mRNA, when injected into oocytes, would not anneal with the antisense oligonucleotide and thus would escape destruction. Indeed, antisense oligonucleotide-injected oocytes additionally injected with this altered mRNA encoding the wild-type Mos protein (i.e., full coding plus 3′UTR) were rescued and progressed to meiosis II. However, when antisense oligonucleotide-injected oocytes were injected with the same altered c-*mos* mRNA noted above, but which additionally lacked functional CPEs, rescue was substantially reduced. Therefore, cytoplasmic polyadenylation of c-*mos* mRNA is necessary for normal development in the mouse.

Similar experiments have also been performed with *Xenopus* oocytes by Sheets et al. (1995). These investigators truncated c-*mos* mRNA after the coding region by injecting an oligonucleotide complementary to its 3′UTR. These oocytes, when incubated with progesterone, the natural inducer of maturation in *Xenopus*, failed to re-enter the meiotic divisions. Such oocytes were rescued (i.e., they underwent progesterone-induced oocyte maturation), however, when they were additionally injected with a "prosthetic" RNA containing the cytoplasmic polyadenylation signals (for experimental details, see Wickens et al., this volume). Thus, as was observed for the mouse, cytoplasmic polyadenylation is critical for oocyte maturation in *Xenopus*.

In *Drosophila*, several maternally inherited transcripts are responsible for setting up dorsal-ventral polarity and the differentiation of terminal structures in the early embryo (St Johnston and Nüsslein-Vollhard 1992). Of these, *bicoid*, *Toll*, and *torso* mRNAs undergo cytoplasmic polyadenylation at a time coincident with their translational activation (Salles et al. 1994). To assess whether polyadenylation is necessary for subsequent development, two sets of experiments were performed (Salles et al. 1994). First, wild-type *bicoid* RNA was injected into mutant fly embryos that do not produce Bicoid protein; complete rescue of the mutant *bicoid* phenotype was observed, which indicates that Bicoid protein was produced from the injected RNA. The next set of experiments was to inject an altered *bicoid* mRNA that lacked the 3′UTR and hence the signals necessary for cytoplasmic polyadenylation. This RNA was unable to rescue the *bicoid* phenotype, suggesting that the injected RNA was not translated. However, this same RNA when furnished with a long poly(A) tail was able to rescue the *bicoid* phenotype, albeit not as well as wild-type *bicoid* mRNA. Thus, these experiments demonstrate that in both flies and mice, cytoplasmic polyadenylation is a critical regulator of early development.

mRNA DEADENYLATION IN DEVELOPMENT

Oocyte Maturation in *Xenopus*

In addition to poly(A) elongation, poly(A) removal from maternal mRNA is a characteristic of mature *Xenopus* oocytes. As a part of their studies on ribosomal protein mRNA (rpmRNA) translation, Hyman and

Wormington (1988) noted that L1 rpmRNA undergoes deadenylation and translational inactivation during maturation. Subsequently, Varnum and Wormington (1990) and Fox and Wickens (1990) showed that certain RNAs that already contain poly(A) tails, including L1, Xfin, and actin mRNAs, were deadenylated when injected into oocytes after the oocytes were induced to mature. Moreover, at least in the case of L1 RNA, deadenylation was coincident with translational inactivation (Varnum and Wormington 1990).

Further analysis of these and additional RNAs failed to reveal a specific *cis*-acting element responsible for maturation-specific deadenylation. Indeed, the homopolymer poly(C), or a nonspecific vector sequence, followed by poly(A), was deadenylated when injected into maturing oocytes. However, when the CPE and hexanucleotide were included in the sequence, deadenylation was prevented (Fox and Wickens 1990; Varnum and Wormington 1990). These investigators concluded that deadenylation during maturation occurs by default; i.e., an RNA either undergoes CPE-mediated polyadenylation or, if it does not contain a CPE, is "automatically" deadenylated.

Because deadenylation takes place after dissolution of the nuclear envelope (germinal vesicle breakdown or GVBD), there was no immediate clue as to whether the factors responsible were nuclear or cytoplasmic. However, if oocytes are first enucleated and then induced to mature, deadenylation does not occur (Varnum et al. 1992), suggesting that nuclear components are essential. Further studies showed not only that both nuclear and cytoplasmic compartments are necessary for deadenylation (Varnum et al. 1992), but also that the nucleus is the repository of the "deadenylase," whereas an activator molecule is stored in the cytoplasm (M. Wormington, pers. comm.).

Embryogenesis in *Xenopus*

In contrast to mRNA deadenylation during oocyte maturation, deadenylation of mRNAs in embryos requires specific *cis* elements. This was first shown by Legagneaux et al. (1992), who injected two-cell *Xenopus* embryos with polyadenylated chimeric RNAs composed of the CAT-coding sequence fused to various portions of the 3′UTR of Eg2 RNA. The 410 most distal nucleotides of Eg2 contained the information necessary for embryonic deadenylation, and the signal was later refined to as few as 17 nucleotides (Bouvet et al. 1994). Similar experiments were also performed with cdk2 RNA, although in this case, two sequence elements, one 3′of the hexanucleotide and the other 5′of the CPE, are

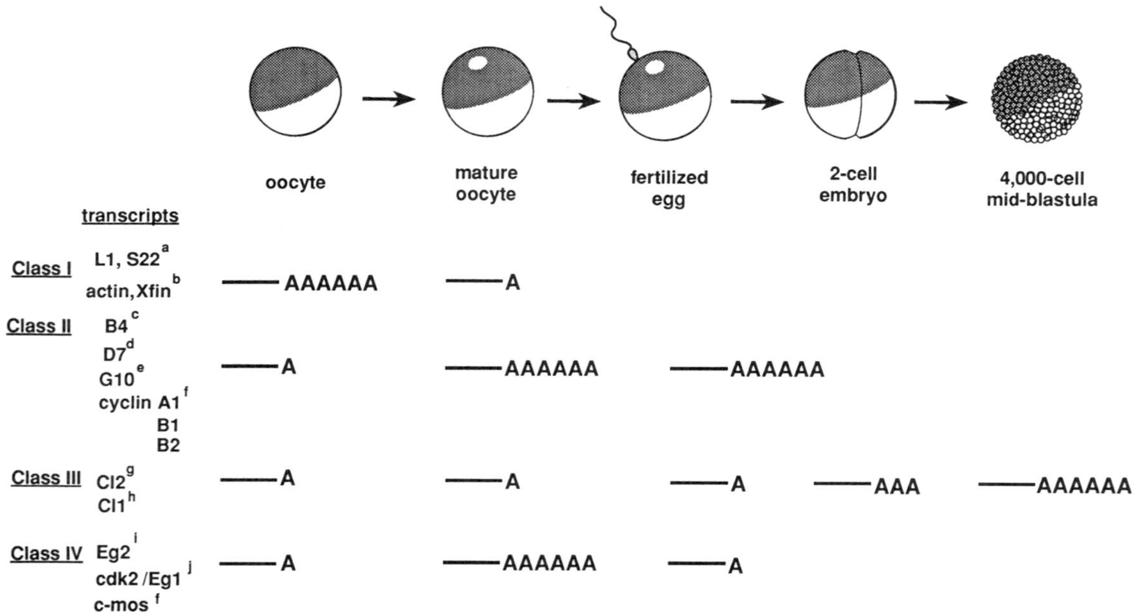

Figure 4 Poly(A) is under a dynamic control in *Xenopus* development. mRNAs are categorized according to the metabolism of their poly(A) tails. Class I RNAs have long poly(A) tails in oocytes, which are lost during maturation. Class II RNAs have short tails initially but are elongated during maturation. These long tails are maintained at least for several cleavage divisions in the embryo. Class III RNAs have short tails through maturation but are elongated only after fertilization. Class IV RNAs acquire long poly(A) tails during maturation only to lose them soon after fertilization. (a) Varnum and Wormington 1990; (b) Fox and Wickens 1990; (c) Paris et al. 1991; (d) Fox et al. 1989; (e) McGrew et al. 1989; (f) Sheets et al. 1994; (g) Simon et al. 1992; (h) Simon and Richter 1994; (i) Legagneaux et al. 1992; (j) Stebbins-Boaz and Richter 1994.

essential for embryonic deadenylation (Stebbins-Boaz and Richter 1994). There is no apparent similarity between the deadenylation signals of Eg2 and cdk2 mRNAs.

Figure 4 summarizes the dynamic changes in poly(A) tail length that occur in early *Xenopus* development. RNAs belonging to class I, such as those encoding ribosomal proteins L1 and S22, contain long poly(A) tails and are almost exclusively polysomal in oocytes. During maturation, these messengers are deadenylated by default and are commensurately released from polysomes. RNAs included in class II (e.g., the cyclins) have short poly(A) tails in oocytes and are translationally dormant. During oocyte maturation, they acquire long poly(A) tails and as a result are translationally activated. These mRNAs generally keep their long poly(A) tails for some time after fertilization but are probably eventually deadenylated (cf., e.g., McGrew et al. 1989). Class III mRNAs, of which Cl1 and Cl2 are the best examples (for other RNAs that may fall in this class, see Simon and Richter 1994), only undergo cytoplasmic polyadenylation and translational activation during early embryogenesis. As mentioned previously, these mRNAs have inhibitor elements that prevent polyadenylation during maturation and contain unique CPEs that drive polyadenylation in embryos. Finally, mRNAs belonging to class IV (e.g., cdk2) undergo poly(A) elongation and translational activation during maturation but are deadenylated and translationally repressed soon after fertilization. It should be emphasized that this fertilization-stimulated deadenylation does not occur by default but instead requires specific *cis*-acting sequences.

Translational Repression in Mouse Oocytes by Specific Deadenylation

Although data from several laboratories make it clear that a short poly(A) tail appended to mRNAs is responsible for translational dormancy, they do not address the question of why specific messages have a short tail in the first place. Huarte et al. (1987) report that in somatic cells, t-PA mRNA, like most other mRNAs, has a long poly(A) tail of 200–300 bases. Yet in mouse oocytes, it is one of a small group of RNAs that contain tails of only 30–40 bases (Huarte et al. 1987). These observations led Huarte et al. (1992) to propose two alternatives: In contrast to most other RNAs, t-PA mRNA acquires a short poly(A) tail in the nucleus of oocytes, which is maintained until maturation, or t-PA mRNA acquires the usual long poly(A) tail in the nucleus, but it is specifically shortened once it reaches the cytoplasm.

Using an intron-specific primer together with oligo(dT), these investigators amplified, by polymerase chain reaction (PCR), unspliced but polyadenylated sequences from growing and fully grown primary oocytes and determined the sizes of the products. (Because the oligo[dT] primer could anneal anywhere along the length of a poly[A] tail, amplified products derived from transcripts with short tails would be small and of a discrete size, whereas those derived from transcripts with long tails would be larger and heterogeneous in size.) They showed that unspliced t-PA RNA receives a long poly(A) tail (up to 400 bases) in both growing and primary oocytes, and thereby concluded that t-PA pre-mRNA, like almost all other pre-mRNAs, receives the normal long poly(A) tail in the nucleus. Therefore, the alternative proposal, that t-PA mRNA is specifically deadenylated once it reaches the cytoplasm, was investigated. t-PA RNA that was polyadenylated in vitro and then injected into growing oocytes was indeed deadenylated. Moreover, a mutational analysis of the t-PA RNA sequences conferring the specific deadenylation showed that the CPE was a necessary component. Consequently, this is not a default deadenylation because it requires a specific *cis*-acting sequence. Thus, the CPE has a dual function: It directs RNA deadenylation in growing mouse oocytes and polyadenylation in maturing mouse oocytes (Salles et al. 1992). For this reason, Huarte et al. (1992) have used the term adenylation control element (ACE) to describe the CPE.

TRANSLATIONAL CONTROL BY CYTOPLASMIC POLYADENYLATION

A discussion of how polyadenylation induces translation must take into account that it appears to do so in two ways. The first was shown by McGrew et al. (1989), who demonstrated that RNAs injected into oocytes could be prevented from being translated during maturation by deleting either the CPE or the AAUAAA or by blocking the 3′ terminus with cordycepin. Although this indicated that polyadenylation is required for translation, the addition of a poly(A) tail prior to injection was insufficient to stimulate translation. These authors thus concluded that the process of polyadenylation, but not the poly(A) tail per se, is necessary for translational recruitment. A similar conclusion was reached by Simon et al. (1992) in their study of polyadenylation during *Xenopus* embryogenesis.

The second way polyadenylation induces translation was demonstrated by Vassalli et al. (1989) and Paris and Richter (1990). Performing fundamentally the same experiments as McGrew et al. (1989) but with different RNAs, they showed that a long poly(A) tail was all that was necessary for translational activation in immature mouse or *Xenopus* oocytes. In this context, several studies have shown that, in general, mRNAs containing poly(A) translate with greater efficiency than those lacking poly(A) (for review, see Munroe and Jacobson 1990a; Jacobson; Wickens et al.; both this volume). Munroe and Jacobson (1990a,b) have argued that this may be due to the interaction between poly(A) and poly(A)-binding protein (PABP), which somehow facilitates ribosome entry at the 5′end of the mRNA. However, at least in *Xenopus* oocytes, PABP may not have an important role in mediating translation due its relative paucity (Zelus et al. 1989). If this is the case, how, in mechanistic terms, does polyadenylation stimulate translation in early development?

Consider the following observations: (1) 5′cap methylation correlates with translational activation in fertilized sea urchin eggs (Caldwell and Emerson 1985), although this has been disputed (Showman et al. 1987); (2) cap ribose-methylated RNAs translate with greater efficiency in vitro than their nonmethylated counterparts, albeit only at saturating concentrations (Muthukrishnan et al. 1978); (3) the vaccinia virus genome encodes both a poly(A) polymerase and a cap ribose methyltransferase that form a heterodimer and hence may act in concert (Gershon et al. 1991; Schnierle et al. 1992); and (4) the vaccinia virus cap ribose methyltransferase is required for virus replication in tissue culture cells, thereby demonstrating its essential nature (Schnierle et al. 1992). On the basis of these results, one could hypothesize that polyadenylation during maturation leads to cap ribose methylation, which in turn facilitates translational activation. Moreover, perhaps this is brought about by the interaction of a poly(A) polymerase and a methyltransferase, as shown for vaccinia virus, which in effect leads to circularity of the mRNA (for a review of this subject, see Jacobson, this volume).

This hypothesis has been tested in recent experiments from the author's laboratory (Kuge and Richter 1995) based on the resistance that ribose methylation confers against RNase hydrolysis. Injected RNA that is polyadenylated during oocyte maturation underwent cap ribose methylation, whereas inhibition of poly(A) elongation, by mutating the necessary *cis*-acting signals or by treatment with cordycepin, prevented cap ribose methylation. Interestingly, when an in-vitro-polyadenylated RNA was injected into oocytes, no cap ribose methylation was detected either before or after maturation. We therefore propose that cap ribose methylation requires ongoing polyadenylation, or, more specifically, poly(A) polymerase. Possibly, poly(A) polymerase forms a heterodimer with a methyltransferase, which in turn catalyzes cap ribose methylation. Finally, to assess the effects of cap ribose methylation on translation, progesterone-treated oocytes were incubated in medium containing *S*-isobutyladenosine, a specific inhibitor of methyltransferase reactions. Although this agent had little effect on polyadenylation or general protein synthesis, it significantly reduced the translational activation of a reporter mRNA that contained the cytoplasmic polyadenylation signals in its 3′UTR. Thus, it may be that translational control by poly(A) elongation, at least during *Xenopus* oocyte maturation, is mediated by cap ribose methylation.

SUMMARY

Although cytoplasmic polyadenylation has been studied for more than two decades, only in the past half dozen years or so has it been recognized as a major translational regulatory mechanism. Indeed, in many ways it typifies the recognition that 3′UTRs have generally received as the residence of translational, degradative, and localization signals (Wickens 1993; Wickens et al., this volume). Two of the main tasks before us are clearly the identification and characterization of the factors that control cytoplasmic polyadenylation. If nuclear polyadenylation is a fair comparison, many more proteins are likely to be involved than what one might have thought. A second area of investigation is the molecular mechanisms by which poly(A) controls translation. Cap ribose methylation has been mentioned as one possibility, but there may be several others. A third related topic is how deadenylation seems to induce translational inactivation so abruptly. Perhaps this leads to some other as yet unknown change in the mRNA. These are just some of the topics with which we must grapple over the next several years.

ACKNOWLEDGMENTS

I thank P. Schedl, S. Strickland, and M. Wormington for communicating results prior to publication and F. Gebauer, H. Kuge, C. de Moor, L. Lorenz, and R. Simon for comments on the manuscript. The work in my laboratory has been supported by grants from the National Institutes of Health, the National Science Foundation, and the March of Dimes Birth Defects Foundation.

REFERENCES

Bachvarova, R., V. De Leon, A. Johnson, G. Kaplan, and B.V. Paynton. 1985. Changes in total RNA, polyadenylated RNA, and actin mRNA during meiotic maturation of mouse oocytes. *Dev. Biol.* **108:** 325–331.

Ballantine, J.E.M. and H.R. Woodland. 1985. Polyadenylation of histone mRNA in *Xenopus* oocytes and embryos. *FEBS Lett.* **180:** 224–228.

Ballantyne, S., A. Bilger, J. Astrom, A. Virtanen, and M. Wickens. 1995. Poly(A) polymerases in the nucleus and cytoplasm of frog oocytes: Dynamic changes during oocyte maturation and early development. *RNA* **1:** 64–78.

Bilger, A., C.A. Fox, E. Wahle, and M. Wickens. 1994. Nuclear polyadenylation factors recognize cytoplasmic polyadenylation elements. *Genes Dev.* **8:** 1106–1116.

Bouvet, P., F. Omilli, Y. Arlot-Bonnemains, V. Legagneaux, C. Roghi, T. Bassez, and H.B. Osborne. 1994. The deadenylation conferred by the 3′ untranslated region of a developmentally controlled mRNA in *Xenopus* embryos is switched to polyadenylation by deletion of a short sequence element. *Mol. Cell. Biol.* **14:** 1893–1900.

Cabada, M.O., C. Darnbrough, P.J. Ford, and P.C. Turner. 1977. Differential accumulation of two size classes of poly(A) associated with messenger RNA during oogenesis in *Xenopus laevis*. *Dev. Biol.* **57:** 427–439.

Caldwell, D.C. and C.P. Emerson, Jr. 1985. The role of cap methylation in the translational activation of stored maternal histone mRNA in sea urchin embryos. *Cell* **42:** 691–700.

Clegg, K.B. and L. Piko. 1983. Poly(A) length, cytoplasmic adenylation and synthesis of poly(A)+ RNA in early mouse embryos. *Dev. Biol.* **95:** 331–341.

Christersen, L.B. and D.M. McKearin. 1994. *orb* is required for anteroposterior and dorsoventral patterning during *Drosophila* oogenesis. *Genes Dev.* **8:** 614–628.

Colot, H.V. and M. Rosbash. 1982. Behavior of individual maternal pA+ mRNAs during embryogenesis of *Xenopus laevis*. *Dev. Biol.* **94:** 79–86.

Darnbrough, C. and P.J. Ford. 1979. Turnover and processing of poly(A) in full-grown oocytes and during progesterone-induced oocyte maturation in *Xenopus laevis*. *Dev. Biol.* **71:** 323–340.

Dolecki, G.J., R.F. Duncan, and T. Humphreys. 1977. Complete turnover of poly(A) on maternal mRNA of sea urchin embryos. *Cell* **11:** 339–344.

Dworkin, M.B. and E. Dworkin-Rastl. 1985. Changes in RNA titers and polyadenylation during oogenesis and oocyte maturation in *Xenopus laevis*. *Dev. Biol.* **112:** 451–457.

Dworkin, M.B., A. Shrutkowsi, and E. Dworkin-Rastl. 1985. Mobilization of specific maternal mRNA species into polysomes after fertilization in *Xenopus laevis*. *Proc. Natl. Acad. Sci.* **82:** 7636–7640.

Egrie, J.C. and F.H. Wilt. 1979. Changes in poly(adenylic acid) polymerase activity during sea urchin embryogenesis. *Biochemistry* **18:** 269–274.

Fox, C.A. and M. Wickens. 1990. Poly(A) removal during oocyte maturation: A default reaction selectively prevented by specific sequences in the 3′ UTR of certain maternal mRNAs. *Genes Dev.* **4:** 2287–2298.

Fox, C.A., M.D. Sheets, and M.P. Wickens. 1989. Poly(A) addition during maturation of frog oocytes: Distinct nuclear and cytoplasmic activities and regulation by the sequence UUUUUAU. *Genes Dev.* **3:** 2151–2162.

Fox, C.A., M.D. Sheets, E. Wahle, and M. Wickens. 1992. Polyadenylation of maternal mRNA during oocyte maturation: Poly(A) addition *in vitro* requires a regulated RNA-binding activity and a poly(A) polymerase. *EMBO J.* **11:** 5021–5032.

Gebauer, F. and J.D. Richter. 1995. Cloning and characterization of a *Xenopus* poly(A) polymerase. *Mol. Cell. Biol.* **15:** 1422–1430.

Gebauer, F., W. Xu, G.M. Cooper, and J.D. Richter. 1994. Translational control by cytoplasmic polyadenylation of c-*mos* mRNA is necessary for oocyte maturation in the mouse. *EMBO J.* **13:** 5712–5720.

Gershon, P.D., B.-Y.Ahn, M. Garfield, and B. Moss. 1991. Poly(A) polymerase and a dissociable polyadenylation stimulatory factor encoded by vaccinia virus. *Cell* **66:** 1269–1278.

Goldman, D.S., A.A. Kiessling, and G.M. Cooper. 1988. Post-transcriptional processing suggests that c-*mos* functions as a maternal message in mouse eggs. *Oncogene* **3:** 159–162.

Hake, L.E. and J.D. Richter. 1994. CPEB is a specificity factor that mediates cytoplasmic polyadenylation during *Xenopus* oocyte maturation. *Cell* **79:** 617–628.

Huarte, J., D. Belin, and J.-D. Vassalli. 1985. Plasminogen activator in mouse and rat oocytes: Induction during meiotic maturation. *Cell* **43:** 551–558.

Huarte, J., D. Belin, A. Vassalli, and S. Strickland. 1987. Meiotic maturation of mouse oocytes triggers the translation and polyadenylation of dormant tissue-type plasminogen activator mRNA. *Genes Dev.* **1:** 1201–1211.

Huarte, J., A. Stutz, M.L. O'Connell, P. Gubler, D. Belin, A.L. Darrow, S. Strickland, and J.-D. Vassalli. 1992. Transient translational silencing by reversible mRNA deadenylation. *Cell* **69:** 1021–1030.

Hyman, L.E. and M.W. Wormington. 1988. Translational inactivation of ribosomal protein mRNAs during *Xenopus* oocyte maturation. *Genes Dev.* **2:** 598–605

Jenny, A., A.-P. Hauri, and W. Keller. 1994. Characterization of cleavage and polyadenylation specifity factor and cloning of its 100-kilodalton subunit. *Mol. Cell. Biol.* **14:** 8183–8190.

Keller, W., K.M. Bienroth, K.M. Lang, and G. Cristofori. 1991. Cleavage and polyadenylation factor CPF specifically interacts with the pre-mRNA 3′processing signal AAUAAA. *EMBO J.* **10:** 4241–4249.

Kempe, E., B. Muhs, and M. Schafer. 1993. Gene regulation in *Drosophila* spermatogenesis: Analysis of protein binding at the translational control element TCE. *Dev. Genet.* **14:** 449–459.

Kleene, K.C. 1989. Poly(A) shortening accompanies the activation of translation of five mRNAs during spermiogenesis in the mouse. *Development* **106:** 367–373.

———. 1993. Multiple controls over the efficiency of translation of the mRNAs encoding transition proteins, protamines, and the mitochondiral capsule selenoprotein in late spermatids in mice. *Dev. Biol.* **159:** 720–731.

Kleene, K.C., R.J. Distel, and N.B. Hecht. 1984. Translational regulation and deadenylation of a protamine mRNA during spermatogenesis in the mouse. *Dev. Biol.* **105:** 71–79

Kuge, H. and J.D. Richter. 1995. Cytoplasmic 3′poly(A) addition induces 5′cap ribose methylation: Implications for translational control of maternal mRNA. *EMBO J.* (in press).

Kuhn, R., U., Schafer, and M. Schafer. 1988. *Cis*-acting regions sufficient for spermatocyte-specific transcriptional control and spermatid-specific translational control of the *Drosophila melanogaster* gene *mst(3)gl-9*. *EMBO J.* **7:** 447–454.

Kuhn, R., C. Kuhn, D. Borsch, K.H. Glatzer, U. Schafer, and M. Schafer. 1991. A cluster of four genes selectively expressed in the male germ line of *Drosophila melanogaster*.

Mech. Dev. **35:** 143–151.

Lantz, V., L. Ambrosio, and P. Schedl. 1992. The *Drosophila orb* gene is predicted to encode sex-specific germline RNA-binding protein and has localized transcripts in ovaries and early embryos. *Development* **115:** 75–88.

Lantz, V., J.S. Chang, J.I. Horabin, D. Bopp, and P. Schedl. 1994. The *Drosophila* Orb RNA-binding protein is required for the formation of the egg chamber and the establishment of polarity. *Genes Dev.* **8:** 598–613.

Legagneaux, V., P. Bouvet, F. Omilli, S. Chevalir, and H.B. Osborne. 1992. Identification of RNA-binding proteins specific to *Xenopus* Eg mRNAs: Association with the portion of Eg mRNA that promotes deadenylation in embryos. *Development* **116:** 1193–1202.

McGrew, L.L. and J.D. Richter. 1990. Translational control by cytoplasmic polyadenylation during *Xenopus* oocyte maturation: Characterization of *cis* and *trans* elements and regulation by cyclin/MPF. *EMBO J.* **9:** 3743–3751.

McGrew, L.L., E. Dworkin-Rastl, M.B. Dworkin, and J.D. Richter. 1989. Poly(A) elongation during *Xenopus* oocyte maturation is required for translational recruitment and is mediated by a short sequence element. *Genes Dev.* **3:** 803–815.

Mescher, A. and T. Humphreys. 1974. Activation of maternal mRNA in the absence of poly(A) formation in fertilised sea urchin eggs. *Nature* **249:** 138–139

Munroe, D. and A. Jacobson. 1990a. Tales of poly(A): A review. *Gene* **91:** 151–158.

———. 1990b. mRNA poly(A) tail, a 3′enhancer of translational intitiation. *Mol. Cell. Biol.* **10:** 3441–3455.

Muthukrishnan, S., B. Moss, J.A. Cooper, and E.S. Maxwell. 1978. Influence of 5′-terminal cap structure on the initiation of translation of vaccinia virus mRNA. *J. Biol. Chem.* **253:** 1710–1715.

O'Keefe, S.J., H. Wolfes, A.A. Kiessling, and G.M. Cooper. 1989. Microinjection of antisense c-*mos* oligonucleotides prevents meiosis II in the maturing mouse egg. *Proc. Natl. Acad. Sci.* **86:** 7038–7042.

Paris, J. and M. Philippe. 1990. Poly(A) metabolism and polysomal recruitment of maternal mRNAs during early *Xenopus* development. *Dev. Biol.* **140:** 221–224.

Paris, J. and J.D. Richter. 1990. Maturation-specific polyadenylation and translational control: Diversity of cytoplasmic polyadenylation elements, influence of poly(A) tail size, and formation of stable polyadenylation complexes. *Mol. Cell. Biol.* **10:** 5634–5645.

Paris, J., K. Swenson, H. Piwnica-Worms, and J.D. Richter. 1991. Maturation-specific polyadenylation: *in vitro* activation by $p34^{cdc2}$ and phosphorylation of a 58-kDa CPE-binding protein. *Genes Dev.* **5:** 16970–1708.

Paris, J., H.B. Osborne, A. Couturier, A. LeGuellec, and M. Philippe. 1988. Changes in the polyadenylation of specific stable RNA during early the development of *Xenopus laevis*. *Gene* **72:** 169–176.

Paules, R., R. Buccione, R. Moschel, and G.F. Vande Woude. 1989. Mouse *mos* protooncogene product is present and functions during oogenesis. *Proc. Natl. Acad. Sci.* **86:** 5395–5399.

Paynton, B.V. and R. Bachvarova. 1994. Polyadenylation and deadenylation of maternal mRNAs during oocyte growth and maturation in the mouse. *Mol. Reprod. Dev.* **37:** 172–180.

Paynton, B.V., R. Rempel, and R. Bachvarova. 1988. Changes in state of adenylation and time course of degradation of maternal mRNAs during oocyte maturation and early embryonic development in the mouse. *Dev. Biol.* **129:** 304–314.

Rosenthal, E. and J.V. Ruderman. 1987. Widespread changes in the translation and adenylation of natural messenger RNAs following fertilization of *Spisula* oocytes. *Dev. Biol.* **121:** 237–246.

Rosenthal, E.T. and F.H. Wilt. 1986. Patterns of maternal messenger RNA accumulation and adenylation during oogenesis in *Urechis caupo*. *Dev. Biol.* **117:** 55–63.

Rosenthal, E.T., T.R. Tansey, and J.V. Ruderman. 1983. Sequence-specific adenylations and deadenylation accompany changes in the translation of maternal messenger RNA after fertilization of *Spisula* oocytes. *J. Mol. Biol.* **166:** 309–327.

Sagata, N., K. Shiokawa, and K. Yamana. 1980. A study on the steady-state population of poly(A)+ RNA during the early development of *Xenopus laevis*. *Dev. Biol.* **77:** 431–448.

Salles, F.J., A.L. Darrow, M.L. O'Connell, and S. Strickland. 1992. Isolation of novel murine maternal mRNAs regulated by cytoplasmic polyadenylation. *Genes Dev.* **6:** 1202–1212.

Salles, F.J., M.E. Lieberfarb, C. Wreden, J.P. Gergen, and S. Strickland. 1994. Regulated polyadenylation of maternal mRNAs allows coordinate initiation of *Drosophila* development. *Science* **266:** 1996–1998.

Schafer, M., R. Kuhn, F. Bosse, and U. Schafer. 1990. A conserved element in the leader mediates post-meiotic translation as well as cytoplasmic polyadenylation of a *Drosophila* spermatocyte mRNA. *EMBO J.* **9:** 4519–4525.

Schnierle, B.S., P.D. Gershon, and B. Moss. 1992. Cap-specific mRNA (nucleoside-$O^{2'}$-)-methyltransferase and poly(A) polymerase stimulatory activities of vaccinia virus mediated by a single protein. *Proc. Natl. Acad. Sci.* **89:** 2897–2901.

Sheets, M.D., M. Wu, and M. Wickens. 1995. Polyadenylation of c-*mos* mRNA as a control point in *Xenopus* meiotic maturation. *Nature* **374:** 511–516.

Sheets, M.D., C.A. Fox, T. Hunt, G. Vande Woude, and M. Wickens. 1994. The 3′ UTRs of c-*mos* and cyclin mRNAs control translation by regulating cytoplasmic polyadenylation. *Genes Dev.* **8:** 926–938.

Showman, R.M., D.S. Leaf, J.A. Anstrom, and R.A. Raff. 1987. Translation of maternal histone mRNAs in sea urchin embryos: A test of control by 5′ cap methylation. *Dev. Biol.* **121:** 284–287.

Simon, R. and J.D. Richter. 1994. Further analysis of cytoplasmic polyadenylation in *Xenopus* embryos and identification of embryonic cytoplasmic polyadenylation element-binding proteins. *Mol. Cell. Biol.* **14:** 7867–7875.

Simon, R., J.-P. Tassan, and J.D. Richter. 1992. Translational control by poly(A) elongation during *Xenopus* development: Differential repression and enhancement by a novel cytoplasmic polyadenylation element. *Genes Dev.* **6:** 2580–2591.

Slater, D.W., I. Slater, and F.J. Bollum. 1978. Cytoplasmic poly(A) polymerase from sea urchin eggs, merogones, and embryos. *Dev. Biol.* **63:** 94–110.

Slater, D.W., I. Slater, and D. Gillespie. 1972. Post-fertilization synthesis of polyadenylic acid in sea urchin embryos. *Nature* **240:** 333–337.

Slater, I., D. Gillespie, and D.W. Slater. 1973. Cytoplasmic adenylation and processing of maternal mRNA. *Proc. Natl. Acad. Sci.* **70:** 406–411.

Smith, R.C., M.B. Dworkin, and E. Dworkin-Rastl. 1988a. Destruction of a translationally controlled mRNA in *Xenopus* oocytes delays progesterone-induced maturation. *Genes Dev.* **2:** 1296–1306.

———. 1991. The maternal gene product D7 is not required for early *Xenopus* development. *Mech. Develop.* **35:** 213–225.

Smith, R.C., E. Dworkin-Rastl, and M.B. Dworkin. 1988b. Expression of a histone H1-like protein is restricted to early *Xenopus* development. *Genes Dev.* **1:** 1284–1295.

Standart, N. and M. Dale. 1993. Regulated polyadenylation of clam maternal mRNAs *in vitro*. *Dev. Genet.* **14:** 492–499.

Stebbins-Boaz, B. and J.D. Richter. 1994. Multiple sequence elements and a maternal mRNA product control cdk2 RNA polyadenylation and translation during early *Xenopus* development. *Mol. Cell. Biol.* **14:** 5870–5880.

St Johnston, D. and C. Nüsslein-Volhard. 1992. The origin of pattern and polarity in the *Drosophila* embryo. *Cell* **68:** 201–219.

Strickland, S., J. Huarte, D. Belin, A. Vassalli, R.J. Rickles, and J.-D. Vassalli. 1988. Antisense RNA directed against the the 3′ noncoding region prevents dormant mRNA activation in mouse oocytes. *Science* **241:** 680–684.

Varnum, S.M. and W.M. Wormington. 1990. Deadenylation of maternal mRNAs during *Xenopus* oocyte maturation does not require specific cis-sequences: A default mechanism for translational control. *Genes Dev.* **4:** 2278–2286.

Varnum, S.M., C.A. Hurney, and W.M. Wormington. 1992. Maturation-specific deadenylation in *Xenopus* oocytes requires nuclear and cytoplasmic factors. *Dev. Biol.* **153:** 283–290.

Vassalli, J.-D., J. Huarte, D. Belin, P. Gubler, A. Vassalli,, M.L. O'Connell, L.A. Parton, R.J. Rickles, amd S. Strickland. 1989. Regulated polyadenylation controls mRNA translation during meiotic maturation of mouse oocytes. *Genes Dev.* **3:** 2161–2171.

Wahle, E. and W. Keller. 1992. The biochemistry of 3′-end cleavage and polyadenylation of messenger RNA precursors. *Annu. Rev. Biochem.* **61:** 419–440.

Wickens, M. 1993. Messenger RNA; Springtime in the desert. *Nature* **363:** 305–306.

Wilson, E.B. 1925. The cell in development and heredity. Macmillan Press, New York.

Wilt, F.H. 1973. Polyadenylation of maternal RNA of sea urchin eggs after fertilization. *Proc. Natl. Acad. Sci.* **70:** 2345–2349.

———. 1977. The dynamics of maternal poly(A)-containing mRNA in fertilized sea urchin eggs. *Cell* **11:** 673–681.

Woodland, H.R. 1980. Histone synthesis during the early development of *Xenopus laevis*. *FEBS Lett.* **121:** 1–7.

Zelus, B.D., D.H. Giebelhaus, D.W. Eib, K.A. Kenner, and R.T. Moon. 1989. Expression of the poly(A) binding protein during development of *Xenopus laevis*. *Mol. Cell. Biol.* **9:** 2756–2760.

18

Interactions between Viruses and the Cellular Machinery for Protein Synthesis

Michael B. Mathews
Cold Spring Harbor Laboratory
Cold Spring Harbor, New York 11724

Viruses are obligate intracellular parasites or symbionts. They are incapable of independent existence because they lack the enzymes and associated apparatus for conducting most metabolic and biosynthetic reactions. Instead, they rely on the cells that they infect to supply the energy, chemicals, and machinery required for virus replication. Nowhere is this dependence more apparent than in protein synthesis. Many viruses encode enzymes for nucleic acid biosynthesis, but—with the exception of some tRNAs—none of them is known to encode any part of the translational apparatus. They are therefore obligated to make use of the cellular translational apparatus for the synthesis of one of their chief components. As a consequence, and because they can be manipulated with some ease, viral systems have provided many insights into the workings of the cellular protein synthetic machinery. For example, analysis of mutations in the coat protein of tobacco mosaic virus and the *r*II cistron of bacteriophage T4 contributed to breaking the genetic code (Barnett et al. 1967; Wittmann and Wittmann-Liebold 1967), and biochemical evidence for the existence of messenger RNA (mRNA) came first from phage-infected bacteria (Volkin and Astrachan 1956; Brenner et al. 1961; Gros et al. 1961). Viral RNA was the natural template of choice for studies of messenger-directed protein synthesis in both prokaryotes and eukaryotes, leading to the development of faithfully initiating cell-free translation systems and thence to the identification of the initiator tRNAs and characterization of ribosome-binding sites (Nathans et al. 1962; Adams and Capecchi 1966; Kerr et al. 1966; Webster et al. 1966; Hindley and Staples 1969; Steitz 1969; Mathews and Korner 1970; Smith and Marcker 1970; Smith et al. 1970; Dasgupta et al. 1975; Lazarowitz and Robertson 1977). In eukaryotes, the 7-methyl guanosine cap structure was discovered at the 5′end of viral mRNAs (Furuichi et al. 1975; Wei and Moss 1975), and the scanning model for initiation site selection was founded largely on data from viruses (Kozak 1978).

In their interactions with the host translation system, viruses do not simply coexist with the cellular machinery, however. They have adopted regulatory mechanisms from their hosts, and perhaps invented some of their own. Moreover, viruses have evolved to take advantage of the cellular translation system in ways that are less orthodox, and sometimes less benign. Some of the earliest instances of translational control were noted in viral systems (see Mathews et al., this volume). More recently, the existence of seemingly bizarre and idiosyncratic mechanisms such as internal ribosome entry, shunting, and hopping has emerged from viral studies. Several of these are now known to be used by cellular mRNAs as well, but they seem to play an especially prominent part in viral protein synthesis. In addition, viruses may impose sophisticated and sweeping changes upon the cellular machinery, modifying the system so as to favor the synthesis of their own proteins at the cells' expense. Furthermore, viruses ingeniously contrive to appropriate components of the translation system for entirely different purposes (e.g., in nucleic acid replication).

For their part, cells confronted with a virus do not simply await their fate with stoic resignation. Infection triggers defensive measures—in prokaryotic as well as eukaryotic cells—and these are met by viral countermeasures. Cellular defenses responding to the threat include products acting at the translational level, designed to inhibit protein synthesis and limit virus multiplication. In their turn, these defenses are countered by viral mechanisms that aim to neutralize the cell response and sustain virus multiplication. Such interactions contribute to the virulence of a viral infection, which can range from mild to acute depending on numerous factors, both cellular and viral.

This chapter presents an overview of the extensive interplay between viruses and the translation system of the cell. Most of the discussion centers on viruses that infect mammals, especially humans, but examples are also drawn from bacteriophages and from viruses that infect other eukaryotes. Throughout the chapter, groups of viruses are usually referred to by their common names or by the name of the best-studied prototype. Formal nomenclature, as well as more comprehensive background information, can be found in standard virology texts such as Fields et al. (1990). For detailed reviews of translational control in specific viruses, the reader is referred to chapters in this monograph by Ehrenfeld (picornaviruses), Katze (influenza virus and reovirus), and Schneider (adenovirus and vaccinia virus). Pertinent facets of the translation system are treated in the chapters by Clemens; Jackson; Merrick and Hershey; Sonenberg; and Trachsel (all this volume).

ASPECTS OF VIRAL INFECTION

Viruses as a group are exceptionally heterogeneous and almost certainly polyphyletic in origin. Their strategies for infecting cells and replicating within them are diverse; accordingly, their interactions with the host protein synthesis machinery are rich and varied. To set these in perspective, it is useful to consider features of viral structure and virus-cell interactions that impact the translation system.

Viral Structure and Complexity

All viruses consist of a nucleic acid genome surrounded by a protective shell of viral protein(s) which form the capsid. In enveloped viruses (e.g., retroviruses, herpesviruses, and influenza virus), the virion acquires an additional membranous covering, containing viral proteins (often glycoproteins) together with cellular lipids, as it "buds" through cell membranes. During this process, some viruses also incorporate cytosolic constituents such as tRNA, which plays a notable part in the life cycle of retroviruses as described below.

Viral particles vary in size over a wide range. At one extreme, the virions of picornaviruses and hepatitis delta virus are comparable to a ribosome (~30 nm diameter), whereas at the other extreme, the largest virions approach the dimensions of a mitochondrion (vaccinia particles are a few hundred nanometers in diameter). Viral genomes vary correspondingly: The simplest contain 3–4 kb of nucleic acid (e.g., RNA phages and parvoviruses)—even less (1.7 kb) in the helper-dependent hepatitis delta virus—whereas the most complex have about 200 kb (herpesvirus and vaccinia virus). The genetic material, which is either DNA or RNA, may be single- or double-stranded (or partially both) and, if single-stranded, of positive or negative polarity (i.e., equivalent to mRNA or to its complement). Moreover, the genome may be circular or linear, and unipartite or segmented into several "chromosomes." This listing does not exhaust the catalogue of variations, but it covers the main virus types discussed here.

The reproductive strategy of a virus is determined in large measure by its genome structure and genetic complexity and by the battery of gene products that it can muster. The gene products number from as few as two or three to several hundred proteins and may include noncoding

RNAs (in adeno-, Epstein-Barr, vaccinia, and human immunodeficiency viruses). Viral genomes must contain all the *information* needed for their own multiplication. This does not mean that they necessarily encode all the *enzymes* needed for the replication of their genetic material, however. Many viruses utilize host-cell DNA and RNA polymerases, and helper-dependent viruses (such as adeno-associated virus and hepatitis delta virus) also require functions provided by larger viruses (adenovirus or herpesvirus and hepatitis B virus, respectively, in these instances). Even if they are autonomous for such activities, the nucleic acid genomes of some viruses are not infectious in the absence of virus-coded enzymes that are packaged into the viral particle; thus, retroviruses need reverse transcriptase to convert their RNA genomes to DNA, some other RNA viruses need RNA-dependent RNA polymerases, and vaccinia virus encodes numerous enzymes including RNA polymerase. The possession of such enzymes has far-reaching consequences, with implications for viral interactions with biochemical pathways and for the site of virus replication within the cell (discussed below).

Apart from genes for replication functions, viral genomes typically encode one or more structural proteins together with a variable number of regulatory products. These function through interactions with both viral and host components, including components of the translation apparatus. As a general rule, the larger and more complex viruses encode more gene products that interact with the translational machinery, but even the simplest viruses engage in sophisticated regulatory interactions.

Life Cycles and Switches

Despite many variations, virus life cycles follow the general pattern illustrated in Figure 1A. Infection takes place after the virus has adsorbed to receptors on the cell surface. Susceptible cells carry suitable receptor molecules for the virus in question; other cells are resistant, although they may be infected by means such as microinjection or transfection. After penetrating the cell, the virus is uncoated, thereby releasing the genome as naked nucleic acid or nucleoprotein, or exposing it in a more limited way as nucleocapsid. Infectious virus then becomes undetectable (the "eclipse" phase) until progeny virions are elaborated.

Many viral genomes, especially those of DNA viruses, are programmed to generate products in an orderly fashion and temporal sequence. This is determined largely through transcriptional controls, but also by regulation at the translational level (as in the RNA phages). In productive infections with most DNA viruses, the infectious cycle is

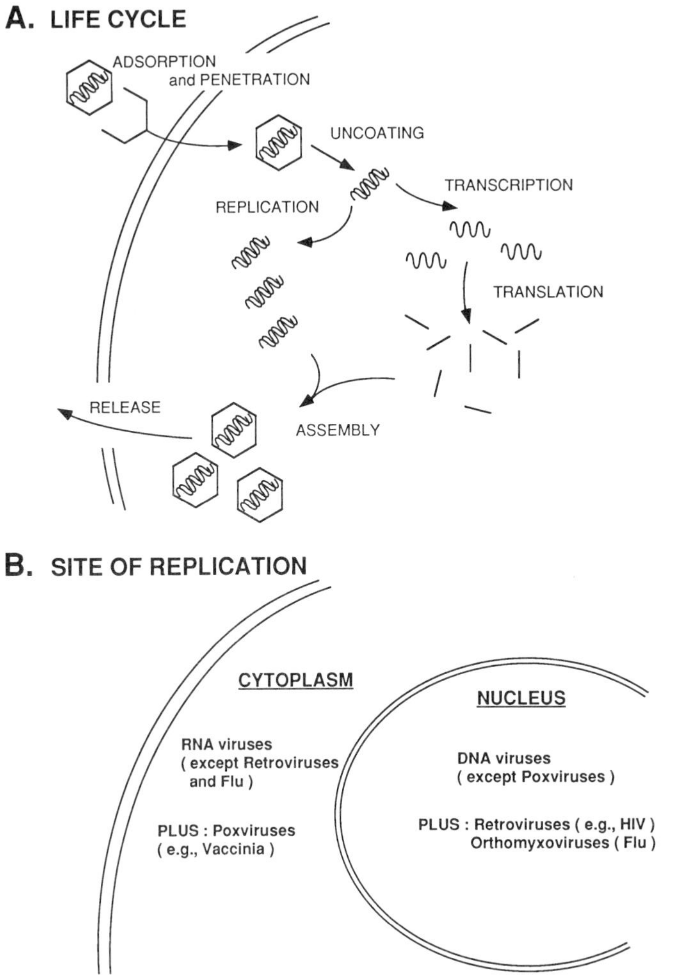

Figure 1 Virus replication. (*A*) A generalized virus life cycle. (*B*) Sites of replication for different groups of viruses.

divided into two phases, early and late, demarcated by the onset of viral DNA replication. During the early phase, cell metabolism is little disturbed and a subset of the viral genes is expressed, generally at a modest level. Early viral products include replication enzymes and regulatory products that set the scene for a more extensive redirection of the host-cell activities. These regulatory proteins exert a multitude of functions: Some are transcriptional activators (e.g., adenovirus E1A and SV40 T antigen) that activate cellular genes needed for virus replication, whereas others (such as some adenovirus E1B and E3 products) suppress host defenses by interceding in apoptosis or antigen presentation. In the more complex DNA viruses, such as herpesviruses, the early phase is subdivided into immediate early and intermediate stages, depending on whether the genes are expressed autonomously or require viral proteins for their expression. During the late phase, template number and transcriptional activity both increase, resulting in the abundant production of viral mRNA. The synthesis of early proteins generally declines, and a new class of late proteins accumulates. This is composed of the coat protein(s) and other virion components, as well as proteins required for morphogenesis and related functions.

Infections with RNA viruses are not characterized by well-defined phases of this sort, although their infectious cycles are not wholly undifferentiated. Temporal regulation may be accomplished by mechanisms operating at the level of transcription, mRNA splicing/transport, or translation. Influenza and reovirus operate switches at several levels, causing the differential transcription and translation of their RNA genomes. The mechanisms involved appear to include preferential translational initiation upon mRNAs whose products are required earlier or in larger amounts for virus assembly segments (Schiff and Fields 1990; Kingsbury 1990; Katze, this volume). In cells infected with complex retroviruses such as the human immunodeficiency virus type 1 (HIV-1), another kind of early-to-late switch operates at the level of RNA splicing or nuclear-cytoplasmic transport. The HIV-1 Rev protein is made at a relatively late stage of infection and permits the accumulation of "underspliced" RNAs containing fewer than the maximum possible number of splice junctions. Such RNAs include the mRNAs encoding structural proteins of the virion, such as Gag and Env, as well as the virion RNA itself (Feinberg et al. 1986; Kim et al. 1989; Cullen 1991).

The clearest examples of temporal switches operating at the translational level are afforded by RNA viruses whose genomes serve directly as mRNAs (see Viral Regulatory Strategies, below). It has long been recognized that synthesis of the replicase protein of the single-stranded RNA phages (f2, R17, MS2, Qβ, etc.) begins shortly after infection but is soon curtailed, well before the synthesis of the other cistrons has peaked (van Duin 1988). The translation of replicase is first activated by translation of the coat protein cistron and then switched off by the coat protein itself. Protein synthesis in poliovirus-infected cells exemplifies a different kind of early/late transition. A rapid decline in amino acid incorporation is followed by an upswing due to translation of the increasing number of viral templates. The initial inhibition of host-protein synthesis is attributable to proteolysis of eukaryotic initiation factor eIF4G (previously p220), catalyzed by a virus-coded enzyme whose primary role is to cleave the viral polyprotein (for details, see Ehrenfeld, this volume; see also Modifications of the Translational Apparatus and Their Relationships to Host Shutoff, below).

Sites of Virus Replication

Critical to virus-host interactions is the site of viral genome replication within the cell. Many viruses replicate and are assembled in specific structures termed inclusion bodies, replication compartments, or viral factories. These comprise both viral and cellular components, including cytoskeletal elements and the viral core in the case of reovirus, and DNA replication enzymes in the case of adenovirus and herpesvirus. Whether replication takes place in the nuclear or cytoplasmic compartment of the cell is a fundamental characteristic of each individual virus family that appears to be determined by its strategy for mRNA production (see Fig. 1B). Viruses that depend on cellular transcription enzymes replicate in the nucleus. This group includes all the DNA viruses except for vaccinia, together with two groups of RNA viruses: the retroviruses (whose genomes go through a chromosomally integrated DNA phase) and influenza virus (which pirates the capped 5′ end of nuclear mRNA precursors as primers for viral transcription). All other viruses replicate in the cytoplasm, which they are equipped to do because they encode their own replicases or import them in the virion. Vaccinia virions contain viral RNA polymerase as well as enzymes for capping, methylation, and polyadenylation of the products and are therefore self-sufficient for transcription. Among the RNA viruses (retroviruses and influenza viruses excepted), those whose genomes are plus strands (e.g., picornaviruses) generate the requisite enzymes directly by translation; those whose genomes are double-stranded (e.g., reovirus) or of negative polarity (e.g., vesicular stomatitis virus [VSV]) must package RNA-dependent RNA polymerases in their virions to permit them to generate plus strands and mRNA. Evidently, a cytoplasmic location is preferred, because no virus that is equipped for replication in this compartment has opted to billet itself in the nucleus; although other rationales can be entertained, a likely interpretation is that ready access to the protein synthesis machinery (or its products) confers a decisive advantage.

Pathogenicity

Infection does not lead inexorably to virus multiplication and cell death, although this is the kind of interaction that is most readily studied in the laboratory. Other possible outcomes include abortive, persistent, and latent infections and oncogenesis. Whether, and to what extent, a virus multiplies depends on the virulence of the virus as well as the permissivity of the cell, both of which are multifactorial properties operating at many levels. Correspondingly, there exists a spectrum of pathogenicity ranging from acute to chronic effects.

In a cell that contains all the factors required by the virus, productive infection occurs and viral progeny are assembled. If the virus is endowed with lytic functions, the cell breaks open and virus is released; otherwise, viruses emerge by budding through the cell membrane or are released passively after cell death. Translational control is most often and more easily studied in productive infections because the virus comes to dominate all aspects of cell macromolecular synthesis. Many cells are nonpermissive or only partially permissive, however. At one extreme, the virus may disappear altogether, or it may integrate some or all of its genome into the cell genome, resulting in cellular transformation and possibly in malignancy. Alternatively, there is slow or intermittent production of infectious virus. Persistently infected cultures continue to produce virus at low levels, either because few cells are productive at any one time or because the infected cells produce virus at a slow rate and survive undamaged. Latently infected cells (such as nerve cells infected with herpesvirus) contain the viral genome in a quiescent state, but virus production is undetectable until triggered by imperfectly understood mechanisms. These more subtle interactions of virus and cell are common but more difficult to study from the perspective of translational control. In such infections, it is assumed that one or more permissivity factors are missing or limiting. The nature of these factors is generally not known, but in some cases at least they appear to include factors operating at the translational level—for example, in VSV infections (Schmidt et al. 1995). This should be a fertile field for investigation in the future.

Host-cell Shutoff

During the later stages of a productive infection, many viruses interfere with the production, maturation, or stability of cellular DNA, RNA, and proteins. The inhibition of host-cell macromolecular synthesis is referred to as the "host-cell shutoff phenomenon." Together with the increasing rate of synthesis of viral products, host shutoff contributes to the viral domination of cellular biosynthetic pathways that is seen in many infections. The rationale normally given—but hardly established—is that shutoff allows viruses to usurp the cellular machinery by alleviating competition for precursors. This would be expected to accelerate replication and perhaps enhance virus yield.

Things might not be so simple, however. Although shutoff is widespread, it is not ubiquitous and it is certainly not a single phenomenon with a unique mechanism. Thus, herpesvirus inhibits ribosomal RNA synthesis and destabilizes host mRNA, whereas adenovirus inhibits ribosomal RNA processing (rather than synthesis) and cellular mRNA turnover is not accelerated, although its transport to the cytoplasm is interrupted. Related viruses may employ different mechanisms to accomplish the same end, or even do without shutoff altogether, as illustrated by members of the picornavirus family (see Modifications of the Translational Apparatus and Their Relationships to Host Shutoff, below). Shutoff usually requires viral protein synthesis but may be independent of it because the factor responsible is introduced by the infecting virions (e.g., vaccinia virus and herpesvirus). Moreover, shutoff is not tightly linked to the abundant synthesis of viral proteins: In poliovirus-infected cells, shutoff precedes the appearance of high levels of viral proteins, whereas the reverse is seen with adenovirus. Finally, it has been pointed out that virus production is not necessarily more efficient when shutoff occurs than in comparable infections (with another cell line or virus strain) where it does not (see Kozak 1986, 1992). Indeed, shutoff during adenovirus infection has less to do with the accumulation of viral proteins than with the release of completed virions from degenerating cells (Zhang and Schneider 1994).

The existence and severity of shutoff depend on many factors including cell type and multiplicity of infection. This probably explains why shutoff, despite extensive research, remains a controversial area with discordant reports on both phenomenology and mechanism in the literature. Remarkably, the preferential translation of viral mRNAs over host mRNAs can often be reproduced and examined in cell-free systems; however, such studies suffer from the drawback that it is difficult to be sure whether the natural mechanisms are in operation. One possible resolution of this quandary might come through analysis of cell or viral mutants. Mutations that affect protein synthesis shutoff have been recorded in a number of viruses, including herpesvirus, poliovirus, and VSV (Stanners et al. 1977; Read and Frenkel 1983; Bernstein et al. 1985), but the underlying mechanisms are not yet fully established.

In a few cases, the selective translation of viral mRNA has been traced to modifications of host-cell translational components (for discussion, see below; and Ehrenfeld; Schneider; both this volume). In most cases, however, the picture is more nebulous, and some form of mRNA competition has generally been held responsible for the phenomenon. For example, the evidence reviewed by Katze and by Schneider (this volume) suggests that influenza virus produces intrinsically "strong" mRNAs, which are translated with relatively high efficiency, whereas vaccinia virus may prevail by the sheer abundancy of its transcripts. Arguments for both of these mechanisms have been adduced in the case of reovirus (Katze, this volume). In cells infected with VSV, the accumulation of overwhelming concentrations of viral mRNA appears to have a major role (Lodish and Porter 1980, 1981), but other factors including changes in the ionic milieu favoring viral mRNA translation have also been implicated (Nuss et al. 1975). Dramatic effects of monovalent cation concentration have also been observed with other viruses, such as encephalomyocarditis virus (EMCV) (Alonso and Carrasco 1981; Lacal and Carrasco 1982). Like other cardioviruses, this picornavirus does not cause eIF4G cleavage so shutoff must be attributed to a different cause (see Modifications of the Translational Apparatus, below).

Regardless of the advantages that might accrue from monopolizing cellular resources, viruses need to keep their hosts alive and functioning to the extent that the virus life cycle can be completed. Accordingly, many viruses produce proteins that promote cell functions, including translation (see Cellular Defenses and Viral Countermeasures, below), and shutoff must be a delicately balanced affair.

EXPLOITATION OF THE HOST PROTEIN SYNTHETIC APPARATUS BY VIRUSES

Although they depend unconditionally on the cellular apparatus for translating their mRNAs, individual viruses have evolved the means to exploit the host apparatus in two distinct ways. First, they appropriate parts of the apparatus for wholly other purposes, and second, they take advantage of an extensive repertoire of unconventional mechanisms.

Redeployment of the Translational Apparatus

Components of the translational machinery have been coopted by viruses and pressed into service in other functions. Since the protein synthesis system is of great antiquity, the four examples that follow are undoubtedly the thin end of a long wedge.

RNA Phages

The enzyme that replicates the genome of the plus-strand RNA bacteriophage Qβ (and its relatives R17, f2, etc.) is composed of four subunits, of which one is the product of the viral replicase gene and three are host-encoded (Blumenthal and Carmichael 1979; van Duin 1988). The three proteins contributed by the bacterial cell are components of the translational system—namely, elongation factors EF1A and EF1B (formerly EF-Tu and EF-Ts) and the ribosomal protein S1. Involvement of a fifth protein, HF (for host factor) has been reported; this protein is also ribosome-associated. The workings of the replicase, and its relationship to the functions of the cellular proteins in protein synthesis, have been studied intensively but are not yet fully understood. In RNA phage replication, the replicase host factors are needed for the initiation reaction but not for elongation. During protein synthesis, EF1A forms a ternary complex with GTP and aminoacyl-tRNA: Since replication initiates with GTP and phage RNA has a tRNA-like structure at its 3′end, it seems likely that the elongation factor involvement is concerned with template and nucleotide binding. The involvement of S1 may have implications for protein synthesis. This protein was previously known as translational control factor, i, which interferes with the synthesis of phage coat protein while favoring that of replicase (Groner et al. 1972). It appears that S1 facilitates the binding of natural mRNAs to the 30S ribosomal subunit and that it (and the holo-replicase) binds to the Qβ coat protein translation initiation site (van Duin 1988). The holo-enzyme also binds to sites in Qβ RNA that are essential for replication, suggesting that competition between ribosomes and replicase for S1 prevents collisions between them on the same RNA template. Compatible with this idea, S1 is not needed for copying the minus strand (which does not function as mRNA). Thus, abduction of translational proteins may serve a dual purpose: It provides the replicase with necessary biochemical attributes and also couples replication to translation.

Plant Viruses

Many plant viral RNAs, including those of tobacco mosaic virus and cucumber mosaic virus, carry distinctly tRNA-like sequences at their 3′ ends. These can interact with tRNA-specific enzymes, including nucleotidyl- and methyltransferases, elongation factors, and aminoacyl-tRNA synthetases (Haenni et al. 1982). Indeed, the capability of turnip yellow mosaic virus RNA to accept valine has been known for more than 25 years. The structure at the 3′end of brome mosaic virus RNA is required for minus-strand RNA replication as well as for charging with tyrosine and nucleotidyl transfer (Dreher and Hall 1988a,b), raising the possibility that cellular enzymes of protein synthesis and tRNA metabolism are coopted into the replicase complexes of these plant viruses as well.

Epstein-Barr Virus

Epstein-Barr virus is an oncogenic herpesvirus. The most abundant viral transcript found in human B lymphocytes transformed by this virus is a short noncoding RNA, EBER-1, about 170 nucleotides long. EBER-1 is synthesized by RNA polymerase III and is found in association with two cellular proteins, the La antigen (Lerner et al. 1981) and ribosomal protein L22 (Toczyski et al. 1994), normally a component of the large ribosomal subunit (see Wool et al., this volume). As much as half of the cells' L22 is present in these ribonucleoprotein particles, which seem to reside in the nucleoplasmic compartment. The implications for transformation, protein synthesis, and ribosome structure are uncertain at present, but two connections have been made. First, EBER-1 can bind to and inactivate the protein kinase PKR, whose best known substrate is the initiation factor eIF2 (Clarke et al. 1990a,b; see Modifications, below); dominant negative mutants of PKR can transform cells and render them tumorigenic (Koromilas et al. 1992; Meurs et al. 1993; Clemens, this volume). Second, the gene for L22 is the target of a chromosomal translocation in some leukemia patients (Nucifora et al. 1993). Whether these phenomena are mechanistically related remains to be seen.

Retroviruses

The final example of redeployment involves RNA rather than protein as the cellular component. Host-cell tRNA plays an essential part in the conversion of retroviral genomes into DNA by the viral enzyme reverse transcriptase (Coffin 1990). Both the enzyme and the tRNA are associated with the nucleocapsid in the viral particle. The 3′ terminus of the tRNA is base-paired with the viral RNA, at a point approximately 100–500 nucleotides from the 5′ end of the genome. The 3′ end of the tRNA serves as primer for reverse transcription, generating a chimeric polynucleotide that contains a repeated DNA sequence, R, at its 3′ end. The complement of the R sequence is present at both the 5′ end and the 3′ end of the viral RNA. The RNA-DNA chimera (called "strong-stop" DNA) then moves to the 3′ copy of the R segment and primes DNA synthesis from this end of the genome, allowing reverse transcriptase to complete transcribing the entire genomic RNA (plus strand) into minus-strand DNA. A consequence of this two-step priming mechanism is the assembly of viral sequences that comprise the long terminal repeat (LTR), which performs numerous essential functions during the virus life cycle. Each virus (or viral subgroup) commandeers a specific tRNA for this purpose (e.g., the primer for HIV-1 and other lentiviruses is a lysine acceptor); at least five different tRNA species are thus exploited for retrovirus replication.

Unorthodox Translational Tactics

Viruses provide a rich seam of exceptions to some of the most firmly entrenched concepts in the field of translation. This may seem paradoxical because so many of the fundamental mechanisms in protein synthesis were first established using viral systems (see Introduction). Indeed, several of the unconventional gambits first recognized in viral systems have also been found to operate in cellular systems, so one might question how unorthodox they really are. Nevertheless, it is difficult to avoid the impression that viruses resort to "tricks" at a higher frequency than do cells themselves, often as part of a regulatory strategy but also as a way to expand their coding capacity within a confined genome.

These unorthodox tactics, depicted in Figure 2, represent departures from the orderly and sequential readout of an mRNA by conventional scanning and decoding from the 5′end. Signals that are ordinarily respected are bypassed or ignored, either in whole or in part; such behavior is dictated by overriding or additional signals that specify deviations from the standard mechanism. Many of these events are treated in detail elsewhere in this volume (see Atkins and Gesteland; Geballe; Jackson). The purpose of the present discussion is to illustrate the translational virtuosity of viral systems, principally with eukaryotes in mind, and to note the relevance of the unconventional tactics to translational control.

Internal Ribosome Entry

The entry of ribosomes to mRNAs at internal sites constitutes a radical departure from scanning (Kozak 1989). It occurs in response to a signal known as the internal ribosome entry site (IRES), which has been found in all picornaviruses as well as a growing number of other viruses (Jang et al. 1988; Pelletier and Sonenberg 1988; Chen and Sarnow 1995; Ehrenfeld; Jackson; both this volume). An IRES typically extends over 400 or 500 nucleotides in the 5′-untranslated region (5′UTR) of the

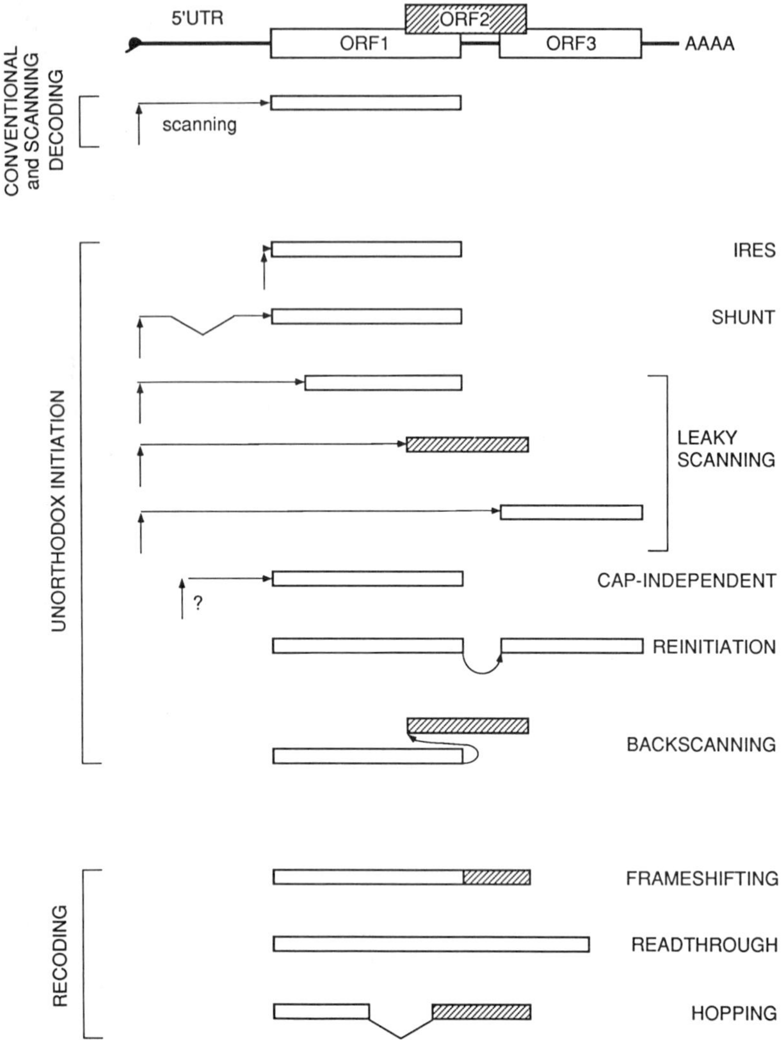

Figure 2 Unconventional translational strategies. The top line represents a capped and polyadenylated mRNA with a 5′-untranslated region (5′UTR) and three open reading frames (ORFs 1–3). Various strategies for initiating translation of this mRNA and for decoding its information are depicted on the lines below. Not that, in some cases, IRES-dependent initiation entails limited scanning from the 3′ border of the IRES. The ribsome entry site for uncapped mRNAs that lack an IRES has not been defined.

mRNA. The signal is not fully defined, but it appears to involve higher-order structure in the RNA. Cellular proteins recognize features of the IRES and are apparently sufficient for its function, although the poliovirus protease 2A has a stimulatory role (Hambidge and Sarnow 1992). Once it has become associated with the mRNA, the ribosome may initiate immediately after the IRES, or it may scan briefly to a nearby downstream initiation codon (Fig. 2).

IRES-mediated initiation confers three potential advantages upon a viral mRNA. First, ribosomes gain the ability to ignore upstream sequences that are irrelevant to translation but have other roles in the virus life cycle—for example, in packaging or replication (Borman et al. 1994). Thus, sequences replete with AUG triplets and secondary structure that would inhibit initiation via the scanning mechanism can be tolerated in the 5′end of these viral mRNAs. Second, the presence of a 5′cap structure is irrelevant for IRES-mediated initiation. This relieves the virus from the need to replicate in the nucleus (where the cellular capping enzyme resides) or to provide its own capping enzyme (as most cytoplasmically replicating viruses do). Third, although the same complement of initiation factors is required, the IRES element frees the mRNA that contains it from dependence on the integrity of initiation factor eIF4G. This feature provides the basis for one of the clearest and best understood examples of host-cell protein synthesis shutoff (see Modifications of the Translational Apparatus, below).

Ribosome Shunting

Another mechanism for avoiding primary and secondary structure in the 5′UTR is the ribosome shunt (see Jackson, this volume). This mechanism was first characterized in cauliflower mosaic virus (CaMV) 35S RNA (Fütterer et al. 1993) and has recently been shown to operate on adenovirus RNAs containing the tripartite leader (see Schneider, this volume); it may also operate in other viruses such as hepatitis B. Shunting allows ribosomes to scan for a short distance from the 5′terminus of the mRNA, then skip across several hundred nucleotides to a site near the 3′border of the UTR without scanning through the intervening sequences (Fig. 2). The determinants of the signal are not yet established, but in CaMV, the 5′UTR has a considerable degree of secondary structure and the efficiency of shunting seems to depend on host factors. Although no viral protein is essential, the product of the viral open reading frame (ORF) VI gene, TAV, is stimulatory (Bonneville et al. 1989). Like the IRES, the shunt permits ribosomes to avoid 5′UTR sequences and structure—even large ORFs—which are incompatible with conventional scanning. Unlike the IRES, however, the shunt would not be expected to alleviate the 5′ cap and eIF4G requirements. On the other hand, the shunt boasts the interesting feature that its takeoff and landing sites can be on separate molecules. Its relationship to translational control mechanisms remains to be determined.

Leaky Scanning

The leaky scanning hypothesis accommodates mRNAs in which a downstream AUG is used for initiation in addition to, or in preference to, the first (i.e., 5′ proximal) AUG. Up to 10% of eukaryotic mRNAs are in this class, and many of these encode two proteins (for review, see Geballe; Jackson; both this volume). The proteins may be in the same or different reading frames (Fig. 2), and dicistronic mRNAs of this type are found in a large number of viruses. One of the principal features defining the initiation site on an mRNA is the context in which the initiation codon is set. The most favorable context, deduced by sequence comparisons and experimentation, is **A**CC<u>AUG</u>**G**, where the A at position –3 and the G at position +4 (relative to the A of the initiator AUG triplet, assigned +1) are the chief determinants of a strong initiation site (Kozak 1989). In dicistronic mRNAs, the first AUG is often in a suboptimal context, suggesting that it functions as a weak initiator. According to the scanning model, this site would be bypassed at a substantial frequency allowing for initiation at a subsequent site downstream.

Regardless of the detailed mechanism, leaky scanning is widely used in viruses, where it presumably helps to economize on coding space and signals for transcription and RNA processing. In HIV-1, for example, the essential envelope protein (Env) is translated from mRNAs that contain an upstream ORF encoding the accessory protein Vpu in a different reading frame. To permit Env synthesis, the *vpu* initiation site is required to be weak (Schwartz et al. 1992). When the two ORFs are in-frame with each other, the result is a nested pair of proteins with overlapping carboxy-terminal sequences and related functions (e.g., SV40 coat proteins VP2 and VP3). On the other hand, when the two ORFs are in different reading frames, the resultant proteins need not be functionally related (e.g., HIV-1 Vpu and Env); nevertheless, they sometimes are (e.g., adenovirus-5 E1B 19 kD and 55 kD proteins), perhaps as a reflection of their evolutionary origins. In either case, regulatory possibilities abound. For example, there might be interactions between the two ORFs. This has been suggested for the overlapping but out-of-frame ORFs in the reovirus S1 transcript, where ribosomes translating the σ1 ORF may impede those translating the downstream p14 ORF (Fajardo and Shatkin

1990; Belli and Samuel 1993). Furthermore, subtle adjustments in the translation apparatus might lead to changes in the ratio of the proteins produced by leaky scanning; however, it is not known whether this potential is realized in the infected cell and, if so, under what circumstances.

Translation of Uncapped RNA

The RNAs of some plant viruses, such as satellite tobacco necrosis virus (Wimmer et al. 1968), and of the yeast virus L-A (Nemeroff and Bruenn 1987; Masison et al. 1995) are exceptional in lacking 5′cap and 3′poly (A) structures. Unlike picornavirus RNA, which is also not capped, they function without the benefit of an IRES element. In the case of L-A, this depends on interactions with products of the cellular *SKI* genes (see Cellular Defenses and Viral Countermeasures, below). Judging from experiments with an artificial uncapped RNA encoding HIV-1 Tat, initiation involves a scanning mechanism despite the absence of a cap (Gunnery and Mathews 1995).

Reinitiation

Since scanning and 5′caps are specific to eukaryotes, the foregoing events apply exclusively to viruses that infect eukaryotic cells. Reinitiation provides another means whereby two proteins can be made from a single mRNA. Although rare in eukaryotes, it is common in prokaryotes where internal ribosome entry, rather than scanning, is standard. Prokaryotic cellular and viral mRNAs frequently contain non-overlapping ORFs in tandem, and their ribosomes can generally gain access to all of the initiation sites within a prokaryotic polycistronic mRNA, allowing for several proteins to be translated independently. In phage ϕX174, for example, two proteins are translated from the same sequence using separate ribosome-binding sites that give access to different reading frames (Ravetch et al. 1977). The individual sites are not necessarily utilized with equal efficiencies, however, and in several cases, the translation of a downstream cistron depends on the translation of an upstream cistron. Such translational coupling is discussed below (see Viral Regulatory Strategies).

Evidence that reinitiation can be used by eukaryotes to gain access to downstream ORFs in a dicistronic mRNA has been derived from work with artificially constructed gene fusions (Peabody et al. 1986). In such mRNAs, it appears that the ribosomes can even "backscan" across a small interval to engage a suitable initiation site (see Fig. 2). In the natural eukaryotic cases that have been studied most intensively, however, the upstream ORFs are generally short and often appear to be regulatory in nature. Typically, they exert a negative influence over the translation of the downstream ORF, although the effects may be complex (for review, see Geballe, this volume). Thus, yeast GCN4 mRNA, which constitutes the canonical example, contains four short upstream ORFs that are all intrinsically negative in character; nonetheless, these ORFs confer inducibility upon the long downstream ORF (see Hinnebusch, this volume). In the GCN4 case, the short encoded peptides appear to play little part in the phenomenon; in other cases, however, the effects are sensitive to changes in the upstream coding sequence, implying that the peptide product itself mediates the inhibition. For example, the 22-residue peptide produced by translating an upstream ORF in the cytomegalovirus gp48 mRNA is thought to act by blocking the scanning process (Cao and Geballe 1995). On the other hand, an upstream ORF in CaMV 35S RNA exerts a positive effect on the translation of downstream ORFs. This stimulatory action requires the presence of the viral *trans*-activator protein TAV; furthermore, the ORF must encode a peptide of at least 30 amino acids, although its composition does not seem to be critical (Fütterer and Hohn 1992). Evidently, the reinitiation mechanism is rarely used in eukaryotic viruses but when it is, it is subject to elaborate controls.

Frameshifting

During the decoding of some mRNAs, the advancing ribosome slips forward or back by one nucleotide, resulting in a programmed +1 or –1 change of reading frame (Jacks and Varmus 1985; Atkins and Gesteland, this volume). Such frameshifting events are common in retroviruses, for many of which a –1 shift is an essential event in reverse transcriptase synthesis. The shift takes place at a "slippery" site, where tRNA can move along the template by one base and reestablish codon:anticodon pairing. Generally, the efficiency of frameshifting is enhanced by a second element, either a pseudoknot or a hairpin positioned downstream, which presumably acts by causing ribosomes to pause at the shift site. The proportion of ribosomes that change reading frames is characteristic of each site and is usually rather low. Therefore, the translation products consist of two proteins, a majority of the conventionally decoded polypeptide and a minority of the "recoded" form. In retroviruses, the Gag-Pol shifted product is usually about 5% of the unshifted Gag product. Most retroviruses translate *pol* this way, and similar events occur in coronaviruses and yeast L-A virus. There is little evidence that the proportion of frameshifting at a particular site is modulated, however, so the mechanism seems to be a device for increasing the coding capacity of the viral genome and for producing the two products in a fixed ratio. The latter may be of critical importance since a small increase or decrease in the ratio of Gag to Gag-Pol is detrimental to virus assembly and proliferation (Dinman and Wickner 1992; Karacostas et al. 1993).

Read-through

The suppression or read-through of stop codons is also used by many viruses to generate a carboxy-terminally extended protein at a fixed ratio to the conventionally translated product (see Fig. 2). Examples are found in viruses infecting plants, bacteria, and mammals (e.g., the Moloney murine leukemia virus Gag-Pol protein) but rarely in cellular mRNAs (see Atkins and Gesteland, this volume). Again, there is no firm evidence as yet that the process is controlled during infection.

Hopping

A final instance of recoding is the bypass of 50 nucleotides which occurs during the translation of T4 gene *60* mRNA. This extraordinary event is mediated by the nascent peptide and requires duplications flanking the bypassed sequence, as well as a stop codon (Weiss et al. 1990). Its regulatory significance is unknown.

VIRAL REGULATORY STRATEGIES

Two principal types of mechanisms, serving contrasting purposes, have been implicated in the regulation of protein synthesis by viruses. The first controls ribosome access to translational initiation sites via RNA:RNA and RNA:protein interactions. This mechanism is well known in bacteriophage systems, where it subserves autoregulatory roles. The second type of mechanism controls the activity of cellular translational components and can influence the translation of both viral and cellular mRNAs. Such modifications, which are more prominent with viruses that infect eukaryotes, have important roles in the shutoff of host-cell protein synthesis.

Initiation Site Availability and Viral Autoregulation

A major determinant of translational efficiency in bacteria is mRNA secondary structure, particularly in the region of the ribosome-binding site (de Smit and van Duin 1990; Voorma, this volume). The degree of secondary structure is subject to regulation by ribosomes (serving as activators of translation) and by *trans*-acting proteins (which generally function as translational repressors). Proteins binding in the vicinity of the initiation site can also influence ribosome binding directly or indirectly (Gold 1988; McCarthy and Gualerzi 1990). Both of these mechanisms are practiced by phages to regulate the translation of their own genes.

Translational Coupling

The initiation site is frequently engaged in higher-order structure that wholly or partly obscures some essential feature, such as the AUG or the Shine-Dalgarno sequence, thereby restricting initiation. In some viral and cellular mRNAs, the cistrons are arranged such that the restriction is lifted by ribosomes traversing a different cistron (usually, but not necessarily, upstream). This mechanism is termed translational coupling; in essence, translating ribosomes are acting as derepressors. The RNA phages and phage T4 have provided a rich supply of examples and have been intensively studied, but well-characterized examples occur in many other phages.

What is probably the best-known case governs the synthesis of the replicase protein of the RNA phages such as f2, MS2, and Qβ. The synthesis of replicase depends on translation of the coat protein cistron which lies upstream, as first evidenced by the observation that an amber mutation early in the coat protein gene exerts a polar effect on replicase synthesis, whereas an amber mutation later in the gene does not (Lodish and Zinder 1966; Lodish 1975). Similarly, other mutations that prevent the passage of ribosomes across the coat protein cistron also down-regulate replicase synthesis. The effect is due to long-range base pairing between a sequence in the coat protein cistron and nucleotides immediately upstream of the replicase AUG. This interaction hinders access to the replicase initiation site. The limitation is relieved by elongating ribosomes that disrupt the base pairing. Consistent with this model, mutations or chemical treatments that weaken or destroy the RNA secondary structure result in constitutive expression and up-regulation of replicase translation.

More often, the interactions occur over a short range and involve terminating (rather than elongating) ribosomes. For example, the ORF encoding the lysis peptide of the RNA phages f2 and MS2 overlaps the 3′end of the coat protein cistron. Removal of the coat protein initiation signal or insertion of amber mutations into the coat protein gene eliminates lysis protein production, consistent with translational coupling. Moving the stop codon downstream has the same effect, however, indicating that the mechanism has something to do with the termination event rather than with elongation. Further mutagenesis has led to the proposal that lysis protein initiation makes use of ribosomes that have just finished synthesizing coat protein. Here, the stability of the hairpin containing the lysis protein AUG codon has an important role. A somewhat different mechanism operates in phage f1, however, where gene VII is translationally coupled to gene V. The gene VII initiation site is intrinsically defective and incapable of directly binding a 30S ribosomal subunit. In this case, it appears that coupling results from the need for the gene VII initiation site to be supplied with such subunits by the upstream gene V (Ivey-Hoyle and Steege 1989, 1992). Finally, in postulating a model for regulation of the assembly protein of phage f2, Robertson (1975) has emphasized that RNA structure is dynamic. Since translation takes place principally on nascent mRNAs in bacteria (and potentially in animal cells infected with viruses that replicate in the cytoplasm), it is necessary to consider temporal effects on protein synthesis due to the growth and folding of the RNA strand.

Why are such mechanisms not known in eukaryotic viruses? The answer is that suspected cases do exist but have not yet been subjected to such rigorous analysis. A possible occurrence in the reovirus σ3 transcript was mentioned above; another is in parainfluenza virus P/C mRNA (Boeck et al. 1992), and further candidates are discussed by Geballe (this volume).

Repression and Activation

Translation of the coat protein cistron of the RNA phages is responsible for up-regulating replicase synthesis, as described above, but the coat protein itself subsequently down-regulates replicase synthesis while its own production is amplified. It appears that a declining level of replicase synthesis late in infection is advantageous, whereas a sustained high level of coat protein synthesis is plainly required for virion production. The mechanism rests on an RNA-protein interaction that is particularly well-understood (Witherell et al. 1991). Coat protein binds to a stem-loop structure that encompasses the replicase AUG initiation site, thereby excluding ribosomes: It apparently blocks initiation by obscuring both the Shine-Dalgarno sequence and the AUG codon. The protein-binding site is a segment of about 21 nucleotides containing a 7-base-pair stem interrupted by an unpaired A residue: Features of both the loop and the stem (including the bulged base) are necessary for its function. Interestingly, coat protein attached at this site then facilitates virion assembly.

Similar principles apply to several translational repressors found in phage T4: Although different in detail, they all interfere with the formation of initiation complexes on their mRNA targets (Gold 1988). Thus, T4 Reg A protein represses the translation of many phage early mRNAs by occluding the AUG codon but not the Shine-Dalgarno sequence. T4 DNA polymerase, on the other hand, represses its own synthesis by binding to a region of the initiation site that includes the Shine-Dalgarno sequence but not the AUG itself. Similarly, autogenous regulation of T4 gene *32* protein is due to the occlusion of its Shine-Dalgarno sequence. In this case, several molecules of the single-stranded DNA-binding protein assemble on the mRNA, beginning at an upstream pseudoknot and extending through cooperative binding interactions until the translation initiation site is covered.

Whereas most such mRNA:protein interactions known to date are negative, phage Mu furnishes a precedent for a positive translational effector. Translation of the Mom protein, a DNA modification enzyme, depends on the Com protein encoded by the same operon. The *mom* initiation site is structured such that the AUG and part of the Shine-Dalgarno sequence are sequestered in a stem. The Com protein binds specifically to a site upstream of the cryptic *mom* initiation site, causing a conformational change that renders it accessible to ribosomes (Hattman et al. 1991; Wulczyn and Kahmann 1991).

No such translational activators or repressors have been reported in eukaryotic viruses to date, but it would be premature to exclude their existence a priori. One candidate is the adenovirus 100K protein that accumulates to high levels in the cytoplasm at late times of infection. This protein is an RNA-binding protein, although no specificity has yet been demonstrated. A mutation in the protein leads to a defect in the translation of late viral mRNAs without deterring the translation of early viral mRNAs or the shutoff of host-protein synthesis in the late phase (Hayes et al. 1990). Thus, adenovirus-2 100K has many of the hallmarks of a translational activator. Positive translational functions have also been ascribed to two RNA-binding proteins of HIV-1, Tat (Cullen 1986; SenGupta et al. 1990) and—more compellingly—Rev (Arrigo and Chen 1991; D'Agostino et al. 1992; Campbell et al. 1994), as well as to CaMV TAV (see Ribosome Shunting and Reinitiation, above): These activities merit further investigation.

Modifications of the Translational Apparatus and Their Relationships to Host Shutoff

Shutoff at the translational level generally entails two events. The virus places itself at an advantage by imposing a limitation upon the translation system and concomitantly creating a bypass so that it escapes the limitation. Four kinds of virus-induced covalent modifications of translational components are presently known to take place during infection. Three are found in eukaryotes and one in prokaryotes. The modifications and their implications for host-cell shutoff are discussed below.

tRNA Cleavage

The T-even phages enjoy the double distinction that they alone are recorded to encode components of the translational machinery—namely tRNAs—and that they engender a large number of alterations in the bacterial protein synthetic apparatus (for review, see Mosig and Eiserling 1988). Of the modifications that take place in T4-infected *Escherichia coli*, tRNA cleavage reactions are the most conspicuous. The destruction of host tRNAs accentuates the dependence of protein synthesis on T4-derived mRNA. For example, the $tRNA^{Leu}$ species that recognizes the codon CUG, rare in T4 mRNAs, is cleaved soon after infection. Phage T4 encodes eight tRNAs (for Arg, Ile, Thr, Ser, Pro, Gly, Leu, and Gln), which are nonessential under standard laboratory conditions; they may be advantageous in a natural setting, however, either because they serve codons that are frequent in T4 mRNAs but rare in host mRNAs, or because they circumvent a virus-induced lesion. Thus, in some *E. coli* strains, the cellular $tRNA^{Ile}$ is cleaved in its anticodon loop and is functionally replaced by the phage-encoded isoaccepting species. Such changes are part of a widespread alteration of the cellular machinery that takes place after T4 infection to the benefit of the phage: In contrast, the cleavage of $tRNA^{Lys}$, which is not covered by a phage gene, is part of a cellular defense mechanism (the *prr* exclusion system, discussed below).

eIF2 Phosphorylation

Eukaryotic initiation factor 2, eIF2, is composed of three nonidentical subunits. In the form of a ternary complex (eIF2•GTP•Met-$tRNA_i$), it serves to transport the initiator tRNA to the 40S ribosomal particle; it also appears to assist in mRNA binding and initiation site selection. The activity of eIF2 is modulated by phosphorylation on Ser-51 of its α-subunit (see Clemens; Hinnebusch; Merrick and Hershey; Trachsel; all this volume). Eukaryotes possess three kinases that can inhibit initiation by phosphorylating eIF2. GCN2 is present in yeast, the heme-controlled repressor (HCR or HRI) is principally found in red cells and their precursors, and the double-stranded RNA-activated inhibitor (referred to as PKR, DAI, dsI, or P1) is widespread in higher cells.

PKR is intimately linked with the host response to viral infection. Phosphorylation of eIF2 limits its function by trapping the GTP exchange factor (eIF2B) which is required for eIF2 to recycle, resulting in ternary complex depletion. Since eIF2B is less abundant than eIF2, phosphorylation of a fraction of the eIF2 (~30%) can sequester all of the recycling factor and lead to a complete block to protein synthesis. This outcome may contribute to the interferon-induced antiviral response (see Cellular Defenses and Viral Countermeasures, below). A lesser degree of eIF2 phosphorylation is observed in cells infected with many different viruses (for references, see Kozak 1986, 1992), and it has been speculated that this might lead to host shutoff. Support for this idea has been drawn from the observation that host-cell shutoff does not take place when PKR-deficient cells are infected with adenovirus (O'Malley et al. 1989; Huang and Schneider 1990). Although the basis for the implied translational selectivity is not clear, two possibilities have been entertained. The first suggests that viral mRNAs are intrinsically more efficient or abundant, and hence less sensitive to a reduction in the effective concentration of initiation factor; the second supposes that the inhibitory effect of eIF2 phosphorylation is compartmentalized in the cell, so that the host mRNAs are preferentially inhibited while the viral mRNAs are spared (see Schneider, this volume).

eIF4E Dephosphorylation

The entry of mRNA into the initiation pathway is mediated by factors in the eIF4 group (see Merrick and Hershey; Sonenberg; both this volume). In many circumstances, the cap-binding protein, eIF4E, is the rate-limiting initiation factor. It binds to the cap structure found at the 5′end of viral and cellular mRNAs. It also complexes with two other factors, eIF4A and eIF4G, to form the cap-binding complex, eIF4F. This complex, together with eIF4B, possesses helicase activity and facilitates unwinding of secondary structure in the 5′end of the mRNA. The eIF4F complex then catalyzes the binding of the mRNA to the 40S ribosomal subunit. Phosphorylation of eIF4E at Ser-209 correlates with its increased activity in the initiation pathway, possibly because of an elevated affinity for the 5′cap and for other components of the eIF4F complex (Sonenberg, this volume). In cells infected with a number of viruses, including adenovirus and influenza virus, the extent of eIF4E phosphorylation falls (Huang and Schneider 1991; Feigenblum and Schneider 1993). Such dephosphorylation is believed to contribute to host-cell shutoff by placing weak cellular mRNAs at a disadvantage when they are competing against strong viral mRNAs. The strong adenovirus and influenza virus mRNAs presumably have relatively little secondary structure in their 5′ends and a correspondingly low requirement for eIF4 activity (see Schneider; Katze; both this volume).

A similar effect is apparently brought about by other means in cells infected with the cardiovirus EMCV. The activity of initiation factor eIF4E can be modulated by a protein called 4E-BP1; in its underphosphorylated form, 4E-BP is a repressor of eIF4E function (Sonenberg, this volume). Recent work has demonstrated that the phosphorylation of 4E-BP1 decreases concomitantly with the inhibition of host-protein synthesis in EMCV-infected cells, suggesting that translation of the uncapped viral RNA is favored by diminished cap-binding activity (A. Gingras et al., unpubl.).

eIF4G Cleavage

The largest subunit of eIF4F, namely eIF4G (previously p220), is cleaved into two fragments in cells infected with certain picornaviruses (see Ehrenfeld; Jackson; both this volume). Cleavage is mediated by some of the proteolytic enzymes that process the primary viral translation product (a polyprotein chain composed of ten or more linked polypeptides) to yield mature viral proteins. The enzymes responsible for eIF4F cleavage are protease 2A in the enteroviruses (including poliovirus and coxsackievirus) and rhinoviruses (the common cold virus), and protease L (in foot-and-mouth disease virus)—although earlier indications had led to the suggestion that the poliovirus enzyme cuts eIF4G indirectly, via a cellular protease. Cleavage of eIF4G effectively separates it into two domains; an amino-terminal part that interacts with eIF4E, and a carboxy-terminal part that interacts with eIF4A and eIF3 (Lamphear et al. 1995; Mader et al. 1995). As a result, cap-dependent initiation is severely inhibited. Since the IRES-dependent initiation mechanism is unaffected or even facilitated (Ohlmann et al. 1995), eIF4G cleavage sets these picornavirus mRNAs at an advantage. The proteases of other picornavirus groups (the cardioviruses and hepatoviruses) do not cleave eIF4G, but the cardioviruses at least have contrived a different mechanism to achieve protein synthesis shutoff as outlined above.

CELLULAR DEFENSES AND VIRAL COUNTERMEASURES

In addition to the immune system, higher organisms are furnished with enzymes that act at the cellular level as first-line defenses against viral infection. These include the eIF2 kinase PKR, introduced above (see Modifications of the Translational Apparatus), and a ribonuclease known as RNase L, both of which affect protein synthesis. They are expressed constitutively at a low level, from which they can moderate viral infection, and they have key roles in the antiviral state that is established by interferon. In response, a plethora of viral products serve to neutralize or attenuate the impact of these defensive enzymes on translation. Although interferon is restricted to vertebrates, comparable systems must be widespread in nature since cellular defense mechanisms are emplaced in organisms as distant as *E. coli* and yeast; moreover, as described below, their viruses are also armed with appropriate countermeasures.

Translational Inhibition in Higher Cells

The defense system composed of PKR and RNase L, together with 2′,5′oligoadenylate synthetase (2-5A synthetase, a regulator of RNase L), is present in untreated, uninfected cells but is neither fully mobilized in the absence of interferon induction nor fully primed without viral infection. These enzymes contribute an important element to the intensively studied antiviral state that is elicited by this family of cytokines, and they have also been implicated more broadly in the control of cell growth, differentiation, and transformation (Lengyel 1993).

Interferon Induction

Interferon is not a single protein species, but a multigene family of secreted proteins with wide-ranging effects on cell growth and differentiation as well as susceptibility to viruses (for review, see Pestka et al. 1987; Staeheli 1990; Samuel 1991). Although individual types and species of interferon have distinct properties, for present purposes they can be treated as a single entity. The synthesis of interferon is induced by viral infection, especially by RNA viruses, and by the exposure of cells to other stimuli such as double-stranded RNA (Gilmour and Reich 1995). Once released from the primary infected cell, interferon diffuses to adjacent cells in the culture, tissue, or organism, binds to specific surface receptors, and triggers a signal transduction pathway that results in the transcriptional activation of more than 30 genes. Their products establish an antiviral state in the recipient cells, during which virus replication may be blocked at a number of levels. In many cases, infection follows a normal course up to and including the synthesis of viral mRNA, but this mRNA does not become stably associated with polysomes as a result of the actions of PKR and RNase L. Both of these enzymes are found at relatively low levels in many uninduced cells, and their synthesis is induced by interferon at the transcriptional level, albeit to differing degrees—about five- to tenfold and two- to fourfold, respectively (Laurent et al. 1985; Zhou et al. 1993). The 2-5A synthetases, of which there are four isozymes differing in size and subcellular distribution, are also inducible; their levels are greatly elevated by interferon treatment (Kerr et al. 1977).

Activation of PKR and RNase L

Both of these enzymes are present in uninfected cells mainly in an inactive (latent) form, and both are activated by double-stranded RNA (dsRNA) although via different mechanisms, as depicted in Figure 3. PKR activation occurs as a direct response to dsRNA and is accompanied by autophosphorylation. The enzyme has two copies of a dsRNA-binding motif (dsRBM); the binding of dsRNA is thought to cause a conformational change in the enzyme which unmasks its kinase activity. Autophosphorylation then might allow PKR to phosphorylate the α-subunit of eIF2, resulting in the inhibition of initiation (Manche et al. 1992; Green et al. 1995; Clemens, this volume; also see Modifications of the Translational Apparatus, above). The activation of RNase L occurs by a more indirect route. Although it does not appear to contain a dsRBM (Patel and Sen 1992), the 2-5A synthetases are activated by dsRNA, producing a series of short, 2′–5′-linked oligoadenylates of the form pppA(pA)$_n$, where *n* is commonly 2 or 3. These oligonucleotides, known collectively as 2-5A, specifically activate RNase L. In turn, this nuclease degrades RNA, chiefly by cutting at the 3′side of UpUp and UpAp sequences.

What is the physiological activator of this system? Conventional wisdom holds that it is viral dsRNA which is generated during the course of virus replication or transcription, or is the viral genome itself in the case of dsRNA-containing viruses such as reovirus. This remains a strong likelihood despite doubts as to the existence, free in the cytoplasm, of dsRNA genomes or replicative intermediates. Such reservations are compounded by the presence in cells of nucleases and modifying/unwinding enzymes which seem designed to dispose of potentially toxic, exposed dsRNA; on the other hand, it is possible that these protective mecha-

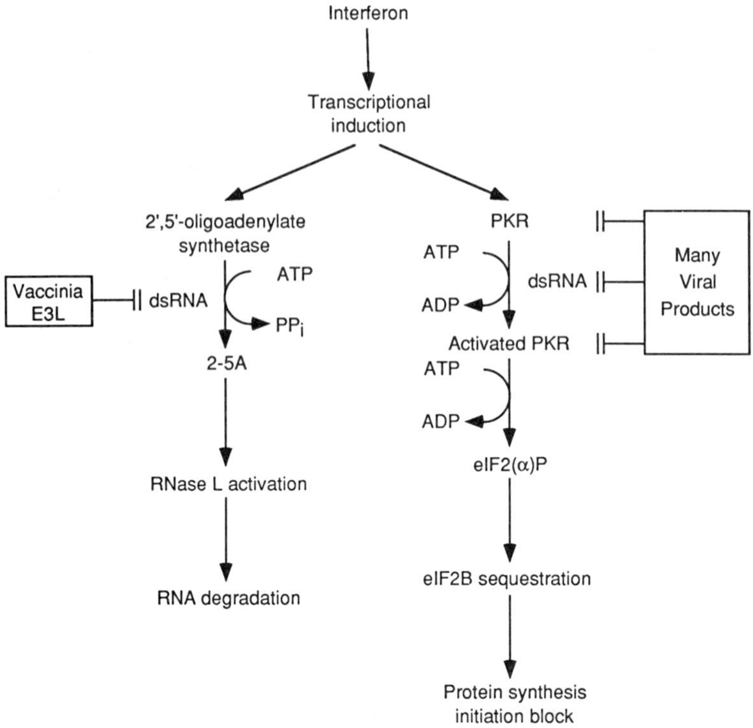

Figure 3 Translational inhibition by interferon-induced pathways. The flowchart summarizes the pathways for induction and activation of PKR and RNase L, and the results of their activation. Also indicated are positions in the pathways where viral products can intercede to overcome translational inhibition.

nisms are swamped during viral infection. Definitive evidence about the activator is scarce and hard to obtain, but dsRNA has been isolated from adenovirus-infected cells (Maran and Mathews 1988) and EMCV RNA has been found associated with 2-5A synthetase in immunoprecipitates from infected cells (Gribaudo et al. 1991). These findings argue that viral dsRNA is indeed the activator. However, it should be borne in mind that PKR can be activated in vitro by ostensibly single-stranded RNAs such as σ3 reovirus mRNA or hepatitis delta virus RNA (Thomis and Samuel 1993; Robertson et al. 1996), and even by unrelated polyanions such as heparin (Hovanessian and Galabru 1987).

The Antiviral State

The antiviral actions of 2-5A synthetase are principally directed against picornaviruses and possibly vaccinia virus, whereas PKR affects a broader range of viruses (Samuel 1988; Staeheli 1990). These generalizations are based on indirect and correlative evidence, as well as data from overexpression of the proteins. Thus, constitutive expression of the 40-kD form of 2-5A synthetase confers resistance to mengovirus but not to VSV or herpesvirus type 2 (Chebath et al. 1987; Coccia et al. 1990); overexpression of PKR, on the other hand, inhibits the replication of vaccinia virus and EMCV, although not of VSV (Lee and Esteban 1993; Meurs et al. 1992). Persuasive support for the antiviral roles of these enzymes can be inferred from the study of viral countermeasures (described below). For example, the absence of VA RNA_I, an adenoviral PKR antagonist, is more deleterious to virus multiplication in interferon-treated cells than in untreated cells (Kitajewski et al. 1986). Therefore, the uninduced levels of these enzymes confer partial resistance to viral infection, and interferon treatment raises cells to a state of heightened readiness from which they can mount a more effective response.

The roles of these enzymes and of interferon in confronting viral infection are firmly established, yet some unanswered questions persist. An enigma of long-standing stems from the fact that neither PKR nor RNase L is intrinsically specific for viral mRNA. So why are they part of an antiviral strategy? One possible answer is that they act in a localized fashion so as to affect viral mRNAs while sparing cellular mRNA. This could be achieved if the dsRNA activator were localized near to, or actually associated with, the viral mRNA. Although this situation has been modeled in various ways and has received indirect support (Nilsen and Baglioni 1979; De Benedetti and Baglioni 1984; De Benedetti et al. 1985; Kaufman and Murtha 1987; Maran et al. 1994), it raises again the question of the nature of the activator. In addition, since the effective inhibitor is not the dsRNA-dependent enzyme itself, but is removed from it by one or two biochemical steps involving presumably diffusible intermediates, it is not obvious how the inhibitor is tethered to the dsRNA activator.

Another explanation to be considered posits that the selective inhibition of viral translation is more apparent than real. Possibly the organism is prepared to sacrifice infected cells in the interest of restricting virus spread. Alternatively, a temporary hiatus in protein synthesis might damage the virus more than the cell—either because the viral replicative program is disrupted or because the cell mobilizes additional defenses in the interim. In this connection, it is worth noting that the two defense pathways have built-in reversibility: 2-5A is labile because of an active 2′,5′phosphodiesterase, and cellular phosphatases can remove phosphate groups from both PKR and eIF2.

Viral Countermeasures against PKR

Despite considerable variability in the response of cells and viruses to interferon, most viruses induce interferon and many of them produce dsRNA that can activate PKR and the 2-5A synthetase/RNase L system. PKR in particular poses a serious threat to virus multiplication, judging from the number of mechanisms that viruses have elaborated to counteract its effects on protein synthesis. These countermeasures are diverse in kind, involving viral RNAs and proteins as well as host proteins; PKR function is inhibited at different levels, as illustrated in Figure 4. Since this topic is the subject of several recent reviews (Sonenberg 1990; Mathews and Shenk 1991; Mathews 1993; Katze 1995 and this volume; Schneider, this volume), the following account represents a synopsis of the best-understood mechanisms. No doubt other mechanisms are awaiting discovery.

Sequestration of dsRNA

Both vaccinia virus and reovirus produce proteins that bind dsRNA, rendering it unavailable to activate PKR and 2-5A synthetase (Katze; Schneider; both this volume). The vaccinia virus protein, E3L, is identical with a specific kinase inhibitory factor (SKIF) detected in infected cell extracts. Its dsRNA-binding capability is due to a single copy of the same dsRBM that is found as a tandem repeat in PKR. Vaccinia virus mutants lacking E3L exhibit increased sensitivity to interferon, and cells infected with them display high activity of PKR and the 2-5A synthetase/RNase L system. The product of the reovirus S4 gene, σ3, also binds dsRNA tightly, although it lacks a dsRBM. The σ3 protein inhibits PKR activation and can partially substitute for vaccinia E3L or adenovirus VA RNA_I; it also can counter the effects of interferon (Beattie et al. 1995).

Inhibition by dsRNA Analogs

The transcription of adenovirus by RNA polymerase III produces VA RNA_I, a small (160-nucleotide) RNA that accumulates to high concentrations in the cytoplasm at late times of infection. It imitates dsRNA in that it interacts with the dsRNA-binding region of PKR, but instead of causing activation of the kinase, the binding of VA RNA *prevents* PKR activation. Critical features of the RNA molecule are a stem that ensures efficient binding to PKR, and a region of complex tertiary structure including a pseudoknot that blocks activation (Y. Ma and M.B. Mathews, in prep.). VA RNA_I enhances virus multiplication, confers interferon resistance, and may be involved in host shutoff (Mathews and Shenk

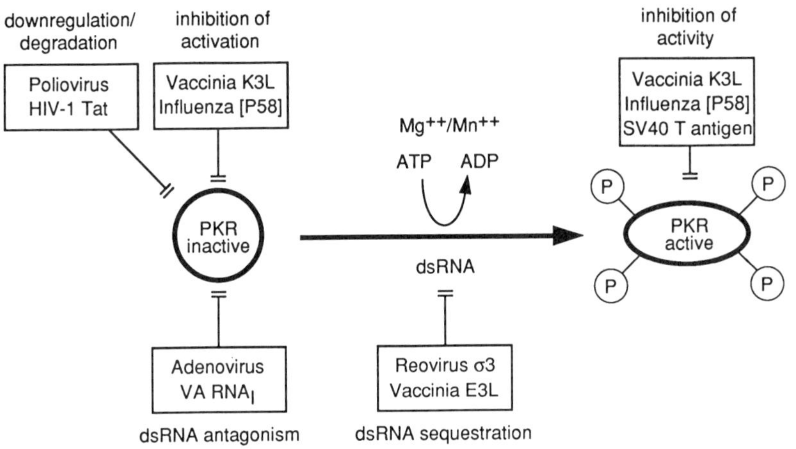

Figure 4 Activation and inhibition of PKR. The activation of PKR by dsRNA-mediated autophosphorylation is shown schematically. Mechanisms whereby viruses interfere with PKR are also indicated, together with the viral product if it is known. (Note that P58 is a cellular protein.)

1991; Schneider, this volume). Other viruses also produce small RNAs that have the capacity to inhibit PKR, but the significance of the findings for viral infection is problematic. The EBERs of Epstein-Barr virus resemble the VA RNAs in many respects and can inhibit PKR activation in vitro as noted above (see Redeployment of the Translational Apparatus). Evidence that the EBERs subserve this function in vivo, however, remains sparse (Bhat and Thimmappaya 1985; Swaminathan et al. 1991, 1992). TAR RNA, an HIV-1 transcript of approximately 60 nucleotides, is generally agreed to bind to PKR, but controversy surrounds the result of this interaction. Some work (most recently that of Maitra et al. 1994) suggests that TAR RNA activates the kinase; however, other studies have attributed this effect to dsRNA contaminants and have demonstrated the ability of TAR RNA to block PKR activation both in vivo and in vitro (Gunnery et al. 1992). At present, no data are available linking either of these observations to events occurring in HIV-1 infection.

Inhibition of Kinase Function

In addition to E3L, described above, vaccinia virus encodes a second protein that interacts with PKR. This protein, known as K3L, is homologous to the α-subunit of eIF2 in the vicinity of its phosphorylation domain. K3L lacks an equivalent of Ser-51, however, and appears to act as a pseudosubstrate for the kinase. It blocks the activation of PKR (and of HRI) as well as the activity of the kinase once activated. The inhibitor has been shown to counteract the effects of PKR and interferon in vivo (see Schneider, this volume). Influenza-virus-infected cells also contain an inhibitor of PKR activation and activity called P58 (see Katze, this volume). P58 is a cellular protein, a member of the tetratricopeptide family, whose mode of action is under investigation. Its capacity to block PKR is masked in uninfected cells by means that are presently obscure. Both K3L and P58 appear to inhibit eIF2 phosphorylation more efficiently than autophosphorylation. Another viral protein that antagonizes PKR activity, although not its activation, is SV40 T antigen. Accordingly, T antigen can functionally substitute for VA RNA deficiency in an adenovirus infection (Rajan et al. 1995). HIV-1 Tat inhibits PKR activation and also serves as a substrate for the enzyme (McMillan et al. 1995; S.R. Brand and M.B. Mathews, in prep.), but again the physiological import of the observations remains to be elucidated.

Down-regulation of PKR Levels

Two viruses contrive to reduce the concentration of the kinase. Poliovirus seems to destabilize PKR (Black et al. 1989), as well as eIF4G (see above). Cells expressing the *tat* gene or infected with HIV-1 contain reduced amounts of PKR, although the mRNA for 2-5A synthetase is unaffected (Roy et al. 1990). In neither case is the mechanism yet known.

Host-Virus Interactions in Yeast and Bacterial Cells

It would be surprising if the kind of defensive and evasive systems described above, which are so ubiquitous in virus-infected animal cells, were not to be found elsewhere. Indeed, systems with interesting parallels to those induced by interferon have recently been uncovered in yeast and bacteria, as diagrammed in Figure 5, although it must be a coincidence that one is strictly translational and the other concerns mRNA degradation.

Bacteriophage T4 and the prr *Exclusion System*

Some *E. coli* strains carry a nonessential operon, *prr*, that confers resistance to phage T4 infection by blocking protein synthesis at the elongation stage (Fig. 5A). The *prr* locus contains four ORFs, one of which (*prrC*) encodes a specific endonuclease called anticodon nuclease.

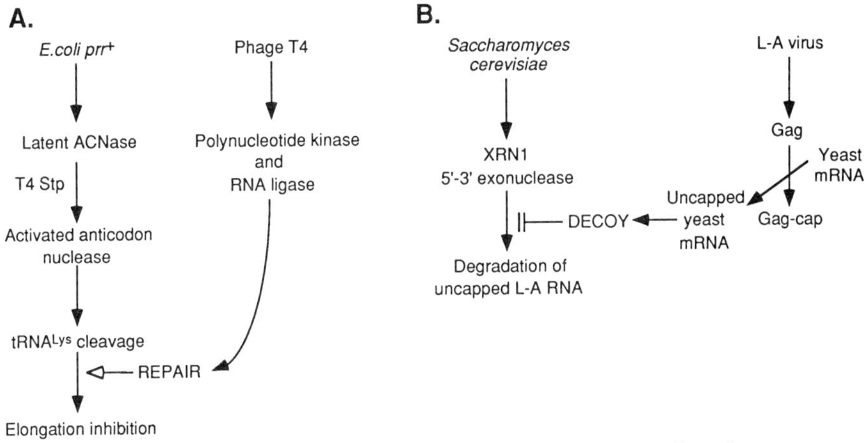

Figure 5 Host/virus interactions in bacteria and yeast. The flowcharts summarize pathways for translational inhibition in virus-infected bacterial (*A*) and yeast (*B*) cells, as well as the viral mechanisms that restore translational activity.

This enzyme inactivates the cellular $tRNA^{Lys}$ by cleaving it at the 5′border of its anticodon (Penner et al. 1995; Shterman et al. 1995). (Parenthetically, the nuclease also cleaves the mammalian $tRNA^{Lys,3}$ isoacceptor which primes HIV-1 reverse transcriptase.) In uninfected bacteria, the nuclease is present in a latent state, masked by the products of the other three genes of the *prr* locus (*prr A*, *B*, and *D*). Remarkably, unmasking is elicited by a short polypeptide, only 18 amino acids long, that is produced by the phage T4 *stp* gene. Thus, in a scenario that echoes the interferon-induced defense system, the phage Stp polypeptide brings about activation of the cellular anticodon nuclease, resulting in tRNA cleavage, inhibition of late T4 protein synthesis, and aborting the infection.

The bacteriophage is equipped with countermeasures to meet this cellular response to infection. Phage T4 encodes two products that overcome the translational inhibition. These are the familiar enzymes polynucleotide kinase and RNA ligase, to which no biological functions had previously been ascribed. Acting consecutively, they repair the severed tRNA and restore the cell to translational competence (Fig. 5A). In formal terms, the *prr* system is analogous to the PKR system: In both cases, a viral product activates a latent cellular enzyme, resulting in the modification of a component of the translation apparatus. Moreover, the analogy extends further, in that viruses have elaborated products to neutralize the threats posed by both systems.

Yeast L-A Virus and the SKI *System*

The *SKI1* gene is one of a group of genes that regulates the "superkiller" phenotype observed in yeast carrying the L-A virus. As noted above, the mRNA of this virus is uncapped and nonpolyadenylated. Mutations in the *SKI* genes allow enhanced replication of L-A and of its satellite virus, M_1, which produces a secreted toxin (killer toxin). The functions of the *SKI* genes are incompletely understood, but they are exerted at least partly at the level of protein synthesis: For example, mutations in several of them (*ski2*, *ski3*, and *ski8*) allow elevated translation of uncapped, nonpolyadenylated RNA (Masison et al. 1995).

Whereas these three *SKI* genes are dispensable for normal yeast growth, *SKI1* is an essential gene that was recently identified with *XRN1:* The product of the *SKI1/XRN1* gene is a 5′–3′ exoribonuclease specific for uncapped RNA; mutants in this gene permit the accumulation of elevated concentrations of uncapped L-A virus mRNA (as well as uncapped cellular mRNA fragments). Thus, the wild-type SKI1 nuclease is an antiviral protein, as well as an enzyme in the normal RNA degradative pathway (Masison et al. 1995). Although RNase L does not seem to have such a major role in cell metabolism, in the context of the present discussion, both nucleases may be viewed as parts of cellular defense mechanisms.

In the case of L-A virus, the countermeasure consists of an unusual enzymatic activity. The coat protein of L-A virus, Gag, has the ability to detach the cap structure from mRNA, leaving an uncapped mRNA while the cap remains covalently associated with the Gag protein (Blanc et al. 1992, 1994). The uncapped RNA can then serve as a substrate or "decoy" for the exonuclease, diverting it away from the L-A mRNA which is therefore spared destruction (Fig. 5B). This scheme is reminiscent of the 2-5A synthetase/RNase L system, which is blocked by the vaccinia K3L protein albeit by a different mechanism.

PROSPECTS

Although viruses do not encode a translational system, they engage the cellular system in many ways. If it is "an extremely difficult matter to present current ideas about protein synthesis in a stimulating form," as averred by Crick (1958) in an admittedly different context, then it is not because viruses haven't been doing their bit. Viral systems have provided insights into the general mechanism of protein synthesis, and they continue to afford unrivaled opportunities to probe translational control mechanisms. By the same token, appreciation of virus life cycles and of the interactions between viruses and their host cells has been deepened by increasing knowledge of the mechanism and control of the translation process. Yet some broad issues remain to be addressed.

Given that many viruses encode their own enzymes for DNA and RNA synthesis, it is curious that they do not also encode proteins in the translation system, as far as we know. Why should this be? One plausible explanation takes note of the great complexity of the translation system, which numbers well over 200 gene products, vis-à-vis the small size of many viruses. Although the compactness of most viral genomes argues that their coding capacity is limited, the enormous diversity of viral products and strategies leaves no doubt that if there were an advantage to specifying a translational protein, some viruses would have done so. That they have not seized the opportunity suggests that the translation system of the cell is either so efficient and accommodating toward viral mRNAs or else so intricate and finely tuned that no advantage would accrue. An alternative view considers viral origins, which are, of course, unknown, although numerous theories are debated. If viruses are degenerate cells or escaped cellular replicons, they might be so well adapted to the preexisting translational machinery that changes would not bring any improvement. On the other hand, if they evolved from earlier self-replicating molecules (perhaps denizens of the RNA world), they might be encumbered with constraints that prohibit undue tampering with the translation machinery; some of the cases of "redeployment," described above, are potential examples of such constraints. The correct explanation, if it is ever learnt, might well be some combination of these factors.

The foregoing is not to say that viruses do not exhibit ingenious and versatile mechanisms to regulate, modify, and exploit the host-protein synthetic machinery, and to defend themselves against changes in it that compromise their replication. In so doing, viruses apparently seek to capitalize on the full potential of the translational system, or perhaps strain to increase its range. This raises some further questions: Are there any truly virus-specific mechanisms, or do viruses merely leverage small differences for their own profit? And is it conceivable that a pathway used by a virus might be untrodden in the uninfected host cell? Even though the viral area has been rather thoroughly explored, at present we can point to few viral mechanisms for which cellular counterparts are wholly lacking. (Ribosome shunting in the 5′ UTR and eIF4G cleavage may fall into this category.) Answers to these questions are of more than academic importance because virus/host-cell differences are potential points of therapeutic intervention. The seeming ease with which viruses subvert host-cell processes and elude cellular controls gives reason to believe that such distinctions exist at the molecular level and that discrimination is within the realm of the possible.

ACKNOWLEDGMENTS

I thank Shobha Gunnery, Huey-Jane Liao, Tsafrira Pe'ery, and Hugh Robertson for comments on the manuscript, Tsafrira Pe'ery for contributions to its genesis, colleagues in the Weizmann Institute for hospitality during the initial stages of writing, and members of my laboratory present and past for their efforts and inspirations. The author is supported by grants from the National Cancer Institute and National Institute of Allergy and Infectious Diseases, National Institutes of Health.

REFERENCES

Adams, J.M. and M. Capecchi. 1966. *N*-formylmethionyl-sRNA as the initiator of protein synthesis. *Proc. Natl. Acad. Sci.* **55:** 147–155.

Alonso, M.A. and L. Carrasco. 1981. Reversion by hypotonic medium of the shutoff of protein synthesis induced by encephalomyocarditis virus. *J. Virol.* **37:** 535–540.

Arrigo, S.J. and I.S.Y. Chen. 1991. Rev is necessary for translation but not cytoplasmic accumulation of HIV-1 *vif*, *vpr*, and *env*/*vpu* 2 RNAs. *Genes Dev.* **5:** 808–819.

Barnett, L., S. Brenner, F.H.C. Crick, R.G. Schulman, and R.J. Watts-Tobin. 1967. Phase shift and other mutants in the first part of the *rIIB* cistron of bacteriophage T4. *Philos. Trans. R. Soc. Lon. B Biol. Sci.* **252:** 487–560.

Beattie, E., K.L. Denzler, J. Tartaglia, M.E. Perkus, E. Paoletti, and B.L. Jacobs. 1995. Reversal of the interferon-sensitive phenotype of a vaccinia virus lacking E3L by expression of the reovirus S4 gene. *J. Virol.* **69:** 499–505.

Belli, B.A. and C.E. Samuel. 1993. Biosynthesis of reovirus-specified polypeptides: Identification of regions of the bicistronic reovirus S1 mRNA that affect the efficiency of translation in animal cells. *Virology* **193:** 16–27.

Bernstein, H.D., N. Sonenberg, and D. Baltimore. 1985. Poliovirus mutant that does not selectively inhibit host cell protein synthesis. *Mol. Cell. Biol.* **5:** 2913–2923.

Bhat, R.A. and B. Thimmappaya. 1985. Construction and analysis of additional adenovirus substitution mutants confirm the complementation of VAI RNA function by two small RNAs encoded by Epstein-Barr virus. *J. Virol.* **56:** 750–756.

Black, T.L., B. Safer, A. Hovanessian, and M.G. Katze. 1989. The cellular 68,000-M_r protein kinase is highly autophosphorylated and activated yet significantly degraded during poliovirus infection: Implications for translational regulation. *J. Virol.* **63:** 2244–2251.

Blanc, A., C. Goyer, and N. Sonenberg. 1992. The coat protein of the yeast double-stranded RNA virus L-A attaches covalently to the cap structure of eukaryotic mRNA. *Mol. Cell. Biol.* **12:** 3390–3398.

Blanc, A., J.C. Ribas, R.B. Wickner, and N. Sonenberg. 1994. His-154 is involved in the linkage of the *Saccharomyces cerevisiae* L-A double-stranded RNA virus Gag protein to the cap structure of mRNAs and is essential for M_1 satellite virus expression. *Mol. Cell. Biol.* **14:** 2664–2674.

Blumenthal, T. and G.C. Carmichael. 1979. RNA replication: Function and structure of Qβ-replicase. *Annu. Rev. Biochem.* **48:** 525–548.

Boeck, R., J. Curran, Y. Matsuoka, R. Compans, and D. Kolakofsky. 1992. The parainfluenza virus type 1 P/C gene uses a very efficient GUG codon to start its C′ protein. *J. Virol.* **66:** 1765–1768.

Bonneville, J.M., H. Sanfaçon, J. Fütterer, and T. Hohn. 1989. Postranscriptional *trans*-activation in cauliflower mosaic virus. *Cell* **59:** 1135–1143.

Borman, A.M., F.G. Deliat, and K.M. Kean. 1994. Sequences within the poliovirus internal ribosome entry segment control viral RNA synthesis. *EMBO J.* **13:** 3149–3157.

Brenner, S., F. Jacob, and M. Meselson. 1961. An unstable intermediate carrying information from genes to ribosomes for protein synthesis. *Nature* **190:** 576–581.

Campbell, L.H., K.T. Borg, J.K. Haines, R.T. Moon, D.R. Schoenberg, and S.J. Arrigo. 1994. Human immunodeficiency virus type 1 rev is required in vivo for binding of poly(A)-binding protein to rev-dependent RNAs. *J. Virol.* **68:** 5433–5438.

Cao, J. and A.P. Geballe. 1995. Translational inhibition by a human cytomegalovirus upstream open reading frame despite inefficient utilization of its AUG codon. *J. Virol.* **69:** 1030–1036.

Chebath, J., P. Benech, M. Revel, and M. Vigneron. 1987. Constitutive expression of (2′-5′) oligo A synthetase confers resistance to picornavirus infection. *Nature* **330:** 587–588.

Chen, C. and P. Sarnow. 1995. Initiation of protein synthesis by the eukaryotic translational apparatus on circular RNAs. *Science* **268:** 415–417.

Clarke, P.A., N.A. Sharp, and M.J. Clemens. 1990a. Translational control by the Epstein-Barr virus small RNA EBER-1. Reversal of the double-stranded RNA-induced inhibition of protein synthesis in reticulocyte lysates. *Eur. J. Biochem.* **193:** 635–641.

Clarke, P.A., M. Schwemmle, J. Schickinger, K. Hilse, and M.J. Clemens. 1990b. Binding of Epstein-Barr virus small RNA EBER-1 to the double-stranded RNA-activated protein kinase DAI. *Nucleic Acids Res.* **19:** 243–248.

Coccia, E.M., G. Romeo, A. Nissim, G. Marziali, R. Albertini, E. Affabris, A. Battistini, G. Fiorucci, R. Orsatti, G.B. Rossi, and J. Chebath. 1990. A full-length murine 2-5A synthetase cDNA transfected in NIH-3T3 cells impairs EMCV but not VSV replication. *Virology* **179:** 228–233.

Coffin, J.M. 1990. Retroviridae and their replication. In *Fields virology* (ed. B.N. Fields et al.), pp. 1437–1500. Raven Press, New York.

Crick, F.H.C. 1958. On protein synthesis. *Symp. Soc. Exp. Biol.* **12:** 138–163.

Cullen, B.R. 1986. *Trans*-activation of human immunodeficiency virus occurs via a bimodal mechanism. *Cell* **46:** 973–982.

———. 1991. Human immunodeficiency virus as a prototypic complex retrovirus. *J. Virol.* **65:** 1053–1056.

D'Agostino, D.M., B.K. Felber, J.E. Harrison, and G.N. Pavlakis. 1992. The rev protein of human immunodeficiency virus type 1 promotes polysomal association and translation of *gag/pol* and *vpu/env* mRNAs. *Mol. Cell. Biol.* **12:** 1375–1386.

Dasgupta, R., D.S. Shih, C. Saris, and P. Kaesberg. 1975. Nucleotide sequence of a viral RNA fragment that binds to eukaryotic ribosomes. *Nature* **256:** 624–628.

De Benedetti, A. and C. Baglioni. 1984. Inhibition of mRNA binding to ribosomes by localized activation of dsRNA-dependent protein kinase. *Nature* **311:** 79–81.

De Benedetti, A., G.J. Williams, L. Comeau, and C. Baglioni. 1985. Inhibition of viral mRNA translation in interferon-treated L cells infected with reovirus. *J. Virol.* **55:**

588–593.
de Smit, M.H. and J. van Duin. 1990. Control of prokaryotic translational initiation by mRNA secondary structure. *Prog. Nucleic Acid Res. Mol. Biol.* **38:** 1–35.
Dinman, J.D. and R.B. Wickner. 1992. Ribosomal frameshifting efficiency and *gag/gag-pol* ratio are critical for yeast M_1 double-stranded RNA virus propagation. *J. Virol.* **66:** 3669–3676.
Dreher, T.W. and T.C. Hall. 1988a. Mutational analysis of the sequence and structural requirements in brome mosaic virus RNA for minus strand promoter activity. *J. Mol. Biol.* **201:** 31–40.
———. 1988b. Mutational analysis of the tRNA mimicry of brome mosaic virus RNA. *J. Mol. Biol.* **201:** 41–55.
Fajardo, J.E. and A.J. Shatkin. 1990. Translation of bicistronic viral mRNA in transfected cells: Regulation at the level of elongation. *Proc. Natl. Acad. Sci.* **87:** 328–332.
Feigenblum, D. and R.J. Schneider. 1993. Modification of eukaryotic initiation factor 4F during infection by influenza virus. *J. Virol.* **67:** 3027–3035.
Feinberg, M.B., R.F. Jarrett, A. Aldovini, R.C. Gallo, and F. Wong-Staal. 1986. HTLV-III expression and production involve complex regulation at the levels of splicing and translation of viral RNA. *Cell* **46:** 807–817.
Fields, B.N., D.M. Knipe, R.M. Chanock, M.S. Hirsch, J.L. Melnick, T.P. Monath, and B. Roizman, eds. 1990. *Fields virology*. Raven Press, New York.
Furuichi, Y., M. Morgan, S. Muthukrishnan, and A.J. Shatkin. 1975. Reovirus messenger RNA contains a methylated, blocked 5′-terminal structure: $m^7G(5')ppp(5')G^mpCp$-. *Proc. Natl. Acad. Sci.* **72:** 362–366.
Fütterer, J. and T. Hohn. 1992. Role of an upstream open reading frame in the translation of polycistronic mRNAs in plant cells. *Nucleic Acids Res.* **20:** 3851–3857.
Fütterer, J., Z. Kiss-László, and T. Hohn. 1993. Nonlinear ribosome migration on cauliflower mosaic virus 35S RNA. *Cell* **73:** 789–802.
Gilmour, K.C. and N.C. Reich. 1995. Signal transduction and activation of gene transcription by interferons. *Gene Expr.* **5:** 1–18.
Gold, L. 1988. Posttranscriptional regulatory mechanisms in *Escherichia coli*. *Annu. Rev. Biochem.* **57:** 199–233.
Green, S.R., L. Manche, and M.B. Mathews. 1995. Two functionally distinct RNA-binding motifs in the regulatory domain of the protein kinase DAI. *Mol. Cell. Biol.* **15:** 358–364.
Gribaudo, G., D. Lembo, G. Cavallo, S. Landolfo, and P. Lengyel. 1991. Interferon action: Binding of viral RNA to the 40-kilodalton 2′–5′-oligoadenylate synthetase in interferon-treated HeLa cells infected with encephalomyocarditis virus. *J. Virol.* **65:** 1748–1757.
Groner, Y., R. Scheps, R. Kamen, D. Kolakofsky, and M. Revel. 1972. Host subunit of Qβ replicase is translation control factor i. *Nat. New Biol.* **239:** 19–20.
Gros, F., H. Hiatt, W. Gilbert, G. Kurland, R.W. Risebrough, and J.D. Watson. 1961. Unstable ribonucleic acid revealed by pulse labelling of *Escherichia coli*. *Nature* **190:** 581–585.
Gunnery, S. and M.B. Mathews. 1995. Functional mRNA can be generated by RNA polymerase III. *Mol. Cell. Biol.* **15:** 3597–3607.
Gunnery, S., S.R. Green, and M.B. Mathews. 1992. Tat-responsive region RNA of human immunodeficiency virus type 1 stimulates protein synthesis *in vivo* and *in vitro*: Relationship between structure and function. *Proc. Natl. Acad. Sci.* **89:** 11557–11561.

Haenni, A.-L., S. Joshi, and F. Chapeville. 1982. tRNA-like structures in the genomes of RNA viruses. *Prog. Nucleic Acid Res. Mol. Biol.* **27:** 85–104.

Hambidge, S. and P. Sarnow. 1992. Translational enhancement of the poliovirus 5′ non-coding region mediated by virus-encoded polypeptide 2A. *Proc. Natl. Acad. Sci.* **89:** 10272–10276.

Hattman, S., L. Newman, H.M. Krishna Murthy, and V. Nagaraja. 1991. Com, the phage Mu *mom* translational activator, is a zinc-binding protein that binds specifically to its cognate mRNA. *Proc. Natl. Acad. Sci.* **88:** 10027–10031.

Hayes, B.W., G.C. Telling, M.M. Myat, J.F. Williams, and S.J. Flint. 1990. The adenovirus L4 100-kilodalton protein is necessary for efficient translation of viral late mRNA species. *J. Virol.* **64:** 2732–2742.

Hindley, J. and D.H. Staples. 1969. Sequence of a ribosome binding site in bacteriophage Q-β-RNA. *Nature* **224:** 964–967.

Hovanessian, A.G. and J. Galabru. 1987. The double-stranded RNA-dependent protein kinase is also activated by heparin. *Eur. J. Biochem.* **167:** 467–473.

Huang, J. and R.J. Schneider. 1990. Adenovirus inhibition of cellular protein synthesis is prevented by the drug 2-aminopurine. *Proc. Natl. Acad. Sci.* **87:** 7115–7119.

———. 1991. Adenovirus inhibition of cellular protein synthesis involves inactivation of cap binding protein. *Cell* **65:** 271–280.

Ivey-Hoyle, M. and D.A. Steege. 1989. Translation of phage f1 gene VII occurs from an inherently defective initiation site made functional by coupling. *J. Mol. Biol.* **208:** 233–244.

———. 1992. Mutational analysis of an inherently defective translation initiation site. *J. Mol. Biol.* **224:** 1039–1054.

Jacks, T. and H.E. Varmus. 1985. Expression of the Rous sarcoma virus *pol* gene by ribosomal frameshifting. *Science* **230:** 1237–1242.

Jang, S.K., H.-G. Krausslich, M.J.H. Nicklin, G.M. Duke, A.C. Palmenberg, and E. Wimmer. 1988. A segment of the 5′nontranslated region of encephalomyocarditis virus RNA directs internal entry of ribosomes during in vitro translation. *J. Virol.* **62:** 2636–2643.

Karacostas, V., E.J. Wolffe, K. Nagashima, M.A. Gonda, and B. Moss. 1993. Overexpression of the HIV-1 Gag-Pol polyprotein results in intracellular activation of HIV-1 protease and inhibition of assembly and budding of virus-like particles. *Virology* **193:** 661–671.

Katze, M.G. 1995. Regulation of the interferon-induced PKR: Can viruses cope? *Trends Microbiol.* **3:** 75–78.

Kaufman, R.J. and P. Murtha. 1987. Translational control mediated by eucaryotic initiation factor-2 is restricted to specific mRNAs in transfected cells. *Mol. Cell. Biol.* **7:** 1568–1571.

Kerr, I.M., R.E. Brown, and A.G. Hovanessian. 1977. Nature of inhibitor of cell-free protein synthesis formed in response to interferon and double-stranded RNA. *Nature* **268:** 540–542.

Kerr, I.M., N. Cohen, and T.S. Work. 1966. Factors controlling amino acid incorporation by ribosomes from Krebs II mouse ascites-tumor cells. *Biochem. J.* **98:** 826–835.

Kim, S., R. Byrn, J. Groopman, and D. Baltimore. 1989. Temporal aspects of DNA and RNA synthesis during human immunodeficiency virus infection: Evidence for differential gene expression. *J. Virol.* **63:** 3708–3713.

Kingsbury, D.W. 1990. Orthomyxoviridae and their replication. In *Fields virology* (ed.

B.N. Fields et al.), pp. 1075–1089. Raven Press, New York.

Kitajewski, J., R.J. Schneider, B. Safer, S.M. Munemitsu, C.E. Samuel, B. Thimmappaya, and T. Shenk. 1986. Adenovirus VAI RNA antagonizes the antiviral action of interferon by preventing activation of the interferon-induced eIF-2α kinase. *Cell* **45:** 195–200.

Koromilas, A.E., S. Roy, G.N. Barber, M.G. Katze, and N. Sonenberg. 1992. Malignant transformation by a mutant of the IFN-inducible dsRNA-dependent protein kinase. *Science* **257:** 1685–1689.

Kozak, M. 1978. How do eucaryotic ribosomes select initiation regions in messenger RNA? *Cell* **15:** 1109–1123.

———. 1986. Regulation of protein synthesis in virus-infected animal cells. *Adv. Virus Res.* **31:** 229–292.

———. 1989. The scanning model for translation: An update. *J. Cell Biol.* **108:** 229–241.

———. 1992. Regulation of translation in eukaryotic systems. *Annu. Rev. Cell Biol.* **8:** 197–225.

Lacal, J.C. and L. Carrasco. 1982. Relationship between membrane integrity and the inhibition of host translation in virus-infected mammalian cells. Comparative studies between encephalomyocarditis virus and poliovirus. *Eur. J. Biochem.* **127:** 359–366.

Lamphear, B.J., R. Kirchweger, T. Skern, and R.E. Rhoads. 1995. Mapping of functional domains in eukaryotic protein synthesis factor 4G (eIF4G) with picornaviral proteases. Implications for cap-dependent and cap-independent translation initiation. *J. Biol. Chem.* **270:** 21975–21983.

Laurent, A.G., B. Krust, J. Galabru, J. Svab, and A.G. Hovanessian. 1985. Monoclonal antibodies to an interferon-induced M_r 68,000 protein and their use for the detection of double-stranded RNA-dependent protein kinase in human cells. *Proc. Natl. Acad. Sci.* **82:** 4341–4345.

Lazarowitz, S.G. and H.D. Robertson. 1977. Initiator regions from the small size class of reovirus mRNA protected by rabbit reticulocyte ribosomes. *J. Biol. Chem.* **252:** 7842–7849.

Lee, S.B. and M. Esteban. 1993. The interferon-induced double-stranded RNA-activated human p68 protein kinase inhibits the replication of vaccinia virus. *Virology* **193:** 1037–1041.

Lengyel, P. 1993. Tumor-suppressor genes: News about the interferon connection. *Proc. Natl. Acad. Sci.* **90:** 5893–5895.

Lerner, M.R., N.C. Andrews, G. Miller, and J.A. Steitz. 1981. Two small RNAs encoded by Epstein-Barr virus and complexed with protein are precipitated by antibodies from patients with systemic lupus erythematosus. *Proc. Natl. Acad. Sci.* **78:** 805–809.

Lodish, H.F. 1975. Regulation of in vitro protein synthesis by bacteriophage RNA by RNA tertiary structure. In *RNA phages* (ed. N.D. Zinder), pp. 301–318. Cold Spring Harbor Laboratory, Cold Spring Harbor, New York.

Lodish, H.F. and M. Porter. 1980. Translational control of protein synthesis after infection by vesicular stomatitis virus. *J. Virol.* **36:** 719–733.

———. 1981. Vesicular stomatitis virus mRNA and inhibition of translation of cellular mRNA—Is there a P function in vesicular stomatitis virus? *J. Virol.* **38:** 504–517.

Lodish, H.F. and N.D. Zinder. 1966. Mutants of the bacteriophage f2. VIII, control mechanisms for phage-specific syntheses. *J. Mol. Biol.* **19:** 333–348.

Mader, S., H. Lee, A. Pause, and N. Sonenberg. 1995. The translation initiation factor eIF-4E binds to a common motif shared by the translation factor eIF-4γ and the translational repressors 4E-binding proteins. *Mol. Cell. Biol.* **15:** 4990–4997.

Maitra, R.K., N.A.J. McMillan, S. Desai, J. McSwiggen, A.G. Hovanessian, G. Sen, B.R.G. Williams, and R.H. Silverman. 1994. HIV-1 TAR RNA has an intrinsic ability to activate interferon-inducible enzymes. *Virology* **204:** 823–827.

Manche, L., S.R. Green, C. Schmedt, and M.B. Mathews. 1992. Interactions between double-stranded RNA regulators and the protein kinase DAI. *Mol. Cell. Biol.* **12:** 5238–5248.

Maran, A. and M.B. Mathews. 1988. Characterization of the double-stranded RNA implicated in the inhibition of protein synthesis in cells infected with a mutant adenovirus defective for VA RNA_1. *Virology* **164:** 106–113.

Maran, A., R.K. Maitra, A. Kumar, B. Dong, W. Xiao, G. Li, B.R.G. Williams, P.F. Torrence, and R.H. Silverman. 1994. Blockage of NF-κB signaling by selective ablation of an mRNA target by 2-5A antisense chimeras. *Science* **265:** 789–792.

Masison, D.C., A. Blanc, J.C. Ribas, K. Carroll, N. Sonenberg, and R.B. Wickner. 1995. Decoying the cap^- mRNA degradation system by a double-stranded RNA virus and $poly(A)^-$ mRNA surveillance by a yeast antiviral system. *Mol. Cell. Biol.* **15:** 2763–2771.

Mathews, M.B. 1993. Viral evasion of cellular defense mechanisms: Regulation of the protein kinase DAI by RNA effectors. *Semin. Virol.* **4:** 247–257.

Mathews, M.B. and A. Korner. 1970. Mammalian cell-free protein synthesis directed by viral ribonucleic acid. *Eur. J. Biochem.* **17:** 328–338.

Mathews, M.B. and T. Shenk. 1991. Adenovirus virus-associated RNA and translational control. *J. Virol.* **65:** 5657–5662.

McCarthy, J.E.G. and C. Gualerzi. 1990. Translational control of prokaryotic gene expression. *Trends Genet.* **6:** 78–85.

McMillan, N.A.J., R.F. Chun, D.P. Siderovski, J. Galabru, W.M. Toone, C.E. Samuel, T.W. Mak, A.G. Hovanessian, K.-T. Jeang, and B.R.G. Williams. 1995. HIV-1 Tat directly interacts with the interferon-induced, double-stranded RNA-dependent kinase, PKR. *Virology* **213:** 413–424.

Meurs, E.F., J. Galabru, G.N. Barber, M.G. Katze, and A.G. Hovanessian. 1993. Tumor suppressor function of the interferon-induced double-stranded RNA-activated protein kinase. *Proc. Natl. Acad. Sci.* **90:** 232–236.

Meurs, E.F., Y. Watanabe, S. Kadereit, G.N. Barber, M.G. Katze, K. Chong, B.R.G. Williams, and A.G. Hovanessian. 1992. Constitutive expression of human double-stranded RNA-activated p68 kinase in murine cells mediates phosphorylation of eukaryotic initiation factor 2 and partial resistance to encephalomyocarditis virus growth. *J. Virol.* **66:** 5805–5814.

Mosig, G. and F. Eiserling. 1988. Phage T4 structure and metabolism. In *The bacteriophages* (ed. R. Calendar), vol. 2, pp. 521–606. Plenum Press, New York.

Nathans, D., G. Notani, J.H. Schwartz, and N.D. Zinder. 1962. Biosynthesis of the coat protein of coliphage f2 by *E. coli* extracts. *Proc. Natl. Acad. Sci.* **48:** 1424–1431.

Nemeroff, M.E. and J.A. Bruenn. 1987. Initiation by the yeast viral transcriptase *in vitro*. *J. Biol. Chem.* **262:** 6785–6787.

Nilsen, T.W. and C. Baglioni. 1979. Mechanism for discrimination between viral and host mRNA in interferon-treated cells. *Proc. Natl. Acad. Sci.* **76:** 2600–2604.

Nucifora, G., C.R. Begy, P. Erickson, H.A. Drabkin, and J.D. Rowley. 1993. The 3;21 translocation in myelodysplasia results in a fusion transcript between the *AML1* gene and the gene for EAP, a highly conserved protein associated with the Epstein-Barr virus small RNA EBER 1. *Proc. Natl. Acad. Sci.* **90:** 7784–7788.

Nuss, D.L., H. Oppermann, and G. Koch. 1975. Selective blockage of initiation of host protein synthesis in RNA virus-infected cells. *Proc. Natl. Acad. Sci.* **72:** 1258–1262.

O'Malley, R.P., R.F. Duncan, J.W.B. Hershey, and M.B. Mathews. 1989. Modification of protein synthesis initiation factors and shut-off of host protein synthesis in adenovirus-infected cells. *Virology* **168:** 112–118.

Ohlmann, T., M. Rau, S.J. Morley, and V.M. Pain. 1995. Proteolytic cleavage of initiation factor eIF-4γ in the reticulocyte lysate inhibits translation of capped mRNAs but enhances that of uncapped mRNAs. *Nucleic Acids Res.* **23:** 334–340.

Patel, R.C. and G.C. Sen. 1992. Identification of the double-stranded RNA-binding domain of the human interferon-inducible protein kinase. *J. Biol. Chem.* **267:** 7671–7676.

Peabody, D.S., S. Subramani, and P. Berg. 1986. Effect of upstream reading frames on translational efficiency in simian virus 40 recombinants. *Mol. Cell. Biol.* **6:** 2704–2711.

Pelletier, J. and N. Sonenberg. 1988. Internal initiation of translation of eukaryotic mRNA directed by a sequence derived from poliovirus RNA. *Nature* **334:** 320–325.

Penner, M., I. Morad, L. Snyder, and G. Kaufmann. 1995. Phage T4-coded Stp: Double-edged effector of coupled DNA and tRNA-restriction systems. *J. Mol. Biol.* **249:** 857–868.

Pestka, S., J.A. Langer, K.C. Zoon, and C.E. Samuel. 1987. Interferons and their actions. *Annu. Rev. Biochem.* **56:** 727–777.

Rajan, P., S. Swaminathan, J. Zhu, C.N. Cole, G. Barber, M.J. Tevethia, and B. Thimmapaya. 1995. A novel translation regulation function for the simian virus 40 large-T antigen gene. *J. Virol.* **69:** 785–795.

Ravetch, J.V., P. Model, and H.D. Robertson. 1977. Isolation and characterisation of ϕX174 ribosome binding sites. *Nature* **265:** 698–702.

Read, G.S. and N. Frenkel. 1983. Herpes simplex virus mutants defective in the virion-associated shutoff of host polypeptide synthesis and exhibiting abnormal synthesis of α (immediate early) polypeptides. *J. Virol.* **46:** 498–512.

Robertson, H.D. 1975. Functions of replicating RNA in cells infected by RNA bacteriophages. In *RNA phages* (ed. N.D. Zinder), pp. 113–145. Cold Spring Harbor Laboratory, Cold Spring Harbor, New York.

Robertson, H.D., L. Manche, and M.B. Mathews. 1996. Paradoxical interactions between human delta hepatitis agent RNA and the cellular protein kinase PKR. *J. Virol.* (in press).

Roy, S., M.G. Katze, N.T. Parkin, I. Edery, A.G. Hovanessian, and N. Sonenberg. 1990. Control of the interferon-induced 68-kilodalton protein kinase by the HIV-1 *tat* gene product. *Science* **247:** 1216–1219.

Samuel, C.E. 1988. Mechanisms of the antiviral action of interferons. *Prog. Nucleic Acid Res. Mol. Biol.* **35:** 27–72.

———. 1991. Antiviral actions of interferon. Interferon-regulated cellular proteins and their surprisingly selective antiviral activities. *Virology* **183:** 1–11.

Schiff, L.A. and B.N. Fields. 1990. Reoviruses and their replication. In *Fields virology* (ed. B.N. Fields et al.), pp. 1275–1306. Raven Press, New York.

Schmidt, M.R., K.A. Gravel, and R.T. Woodland. 1995. Progression of a vesicular stomatitis virus infection in primary lymphocytes is restricted at multiple levels during B cell activation. *J. Immunol.* **155:** 2533–2544.

Schwartz, S., B.K. Felber, and G.N. Pavlakis. 1992. Mechanism of translation of monocistronic and multicistronic human immunodeficiency virus type 1 mRNAs. *Mol. Cell. Biol.* **12:** 207–219.

SenGupta, D.N., B. Berkhout, A. Gatignol, A. Zhou, and R.H. Silverman. 1990. Direct evidence for translational regulation by leader RNA and Tat protein of human immunodeficiency virus type 1. *Proc. Natl. Acad. Sci.* **87:** 7492–7496.

Shterman, N., O. Elroy-Stein, I. Morad, M. Amitsur, and G. Kaufmann. 1995. Cleavage of the HIV replication primer $tRNA^{Lys,3}$ in human cells expressing bacterial anticodon nuclease. *Nucleic Acids Res.* **23:** 1744–1749.

Smith, A.E. and K.A. Marcker. 1970. Cytoplasmic methionine transfer RNAs from eukaryotes. *Nature* **226:** 607–610.

Smith, A.E., K.A. Marcker, and M.B. Mathews. 1970. Translation of RNA from encephalomyocarditis virus in a mammalian cell-free system. *Nature* **225:** 184–187.

Sonenberg, N. 1990. Measures and countermeasures in the modulation of initiation factor activities by viruses. *New Biol.* **2:** 402–409.

Staeheli, P. 1990. Interferon-induced proteins and the antiviral state. *Adv. Virus Res.* **38:** 147–200.

Stanners, C.P., A.M. Francoeur, and T. Lam. 1977. Analysis of VSV mutant with attenuated cytopathogenicity: Mutation in viral function, P, for inhibition of protein synthesis. *Cell* **11:** 273–281.

Steitz, J.A. 1969. Polypeptide chain initiation: Nucleotide sequences of the three ribosomal binding sites in bacteriophage R17 RNA. *Nature* **224:** 957–964.

Swaminathan, S., B. Tomkinson, and E. Kieff. 1991. Recombinant Epstein-Barr virus with small RNA (EBER) genes deleted transforms lymphocytes and replicates *in vitro*. *Proc. Natl. Acad. Sci.* **88:** 1546–1550.

Swaminathan, S., B.S. Huneycutt, C.S. Reiss, and E. Kieff. 1992. Epstein-Barr virus-encoded small RNAs (EBERs) do not modulate interferon effects in infected lymphocytes. *J. Virol.* **66:** 5133–5136.

Thomis, D.C. and C.E. Samuel. 1993. Mechanism of interferon action: Evidence of intermolecular autophosphorylation and autoactivation of the interferon-induced, RNA-dependent protein kinase PKR. *J. Virol.* **67:** 7695–7700.

Toczyski, D.P., A.G. Matera, D.C. Ward, and J.A. Steitz. 1994. The Epstein-Barr virus (EBV) small RNA EBER1 binds and relocalizes ribosomal protein L22 in EBV-infected human B lymphocytes. *Proc. Natl. Acad. Sci.* **91:** 3463–3467.

van Duin, J. 1988. Single-stranded RNA bacteriophages. In *The bacteriophages* (ed. R. Calendar), vol. 1, pp. 117–167. Plenum Press, New York.

Volkin, E. and L. Astrachan. 1956. Phosphorus incorporation in *Escherichia coli* ribonucleic acid after infection with bacteriophage T2. *Virology* **2:** 146–161.

Webster, R.E., D.L. Engelhardt, and N.D. Zinder. 1966. *In vitro* protein synthesis: Chain initiation. *Proc. Natl. Acad. Sci.* **55:** 155–161.

Wei, C.M. and B. Moss. 1975. Methylated nucleotides block 5′-terminus of vaccinia virus messenger RNA. *Proc. Natl. Acad. Sci.* **72:** 318–322.

Weiss, R.B., W.M. Huang, and D.M. Dunn. 1990. A nascent peptide is required for ribosomal bypass of the coding gap in bacteriophage T4 gene *60*. *Cell* **62:** 117–126.

Wimmer, E., A.Y. Chang, J.M. Clark, Jr., and M.E. Reichmann. 1968. Sequence studies of satellite tobacco necrosis virus RNA. Isolation and characterization of a 5′-terminal trinucleotide. *J. Mol. Biol.* **38:** 59–73.

Witherell, G.W., J.M. Gott, and O.C. Uhlenbeck. 1991. Specific interaction between

RNA phage coat proteins and RNA. *Prog. Nucleic Acid Res. Mol. Biol.* **40:** 185–220.

Wittmann, H.G. and B. Wittmann-Liebold. 1967. Protein chemical studies of two RNA viruses and their mutants. *Cold Spring Harbor Symp. Quant. Biol.* **31:** 163–172.

Wulczyn, F.G. and R. Kahmann. 1991. Translational stimulation: RNA sequence and structure requirements for binding of Com protein. *Cell* **65:** 259–269.

Zhang, Y. and R.J. Schneider. 1994. Adenovirus inhibition of cell translation facilitates release of virus particles and enhances degradation of the cytokeratin network. *J. Virol.* **68:** 2544–2555.

Zhou, A., B.A. Hassel, and R.H. Silverman. 1993. Expression cloning of 2-5A-dependent RNAase: A uniquely regulated mediator of interferon action. *Cell* **72:** 753–765.

19

Initiation of Translation by Picornavirus RNAs

Ellie Ehrenfeld
Department of Molecular Biology and Biochemistry
University of California
Irvine, California 92717-3900

About 15 years ago, soon after the initial development of methods to apply recombinant DNA and powerful sequencing technology to RNA molecules, the complete sequence of poliovirus RNA was determined (Kitamura et al. 1981; Racaniello and Baltimore 1981). It was immediately apparent that the primary structure of the 5′end of the viral RNA was inconsistent with the cap-dependent scanning model for translation initiation that was so successfully being developed and presented for most cellular and viral messenger RNAs (Kozak 1978, 1989). There was a striking absence of an m^7G cap on the 5′terminus of poliovirus RNA, and the initiating AUG codon was located at a position 743 nucleotides downstream from the 5′end. The resulting long 5′-untranslated region (5′UTR) contained numerous unused AUG triplets as well as a complex predicted secondary structure that would present significant obstacles to a scanning ribosome. Evidence for an alternate ribosome-binding and translation initiation mechanism accumulated slowly, until poliovirus and another related viral RNA, encephalomyocarditis virus (EMCV) RNA, became the first demonstrated examples of cap-independent internal translation initiation (Jang et al. 1988; Pelletier and Sonenberg 1988). The sequences and structures within the viral 5′UTR required for translation initiation are known as the internal ribosome entry site (IRES).

The rationale for evolving a unique structure and mechanism of ribosome binding in the case of these rapidly growing, lytic viruses might appear to be to enable the viruses to interfere with cellular translation, thereby eliminating competition for essential cellular components, and to more efficiently subvert the cell to synthesis of viral macromolecules. Indeed, poliovirus appears to accomplish just such a takeover of cellular translation activity, by eliminating cap-dependent ribosome binding to cellular mRNAs. In this chapter, a view of the cap-independent, internal translation initiation mechanism utilized by members of

Translational Control © 1996 Cold Spring Harbor Laboratory Press 0-87969-458-0/96 $5 + .00

Table 1 Classification of members of the picornavirus family

Picornavirus group	Representative members
Enteroviruses	poliovirus, coxsackievirus, echovirus
Rhinoviruses	human common cold virus
Aphthoviruses	foot-and-mouth disease virus
Cardioviruses	EMCV, mengo, Theiler's virus
Hepatoviruses	hepatitis A
Unnamed	echoviruses 22 and 23

the picornavirus family is described, followed by a discussion of the specific interference with cellular cap-dependent mRNA translation induced by poliovirus and many of its relatives.

THE PICORNAVIRUS FAMILY

The Picornaviridae consist of a group of small icosahedral viruses with positive-sense, single-stranded RNA genomes. The physical structure, genome size and organization, and general replication strategy of all members of the family are similar; family members differ in host target specificity, pathogenic potential, and various details of structure and gene expression. The various members of the picornavirus family have been classified into five genera based on physicochemical properties and molecular sequence comparisons. Recent data point to the segregation of a possible sixth genus consisting of echoviruses 22 and 23 (Hyypiä et al. 1992; Stanway et al. 1994). Table 1 indicates the current classification scheme and some of the better-known members of each genus.

Picornavirus genomes are approximately 7500 nucleotides in length. They are polyadenylated at their 3′ends and covalently attached to a small protein, VPg, at their 5′ends. After entry into a permissive host cell, the encapsidated RNAs are released and translated by cellular components to generate viral proteins that engage in RNA replication and amplification of viral macromolecules. mRNA isolated from polyribosomes in infected cells has the same polarity and sequence as RNA extracted from virion particles, but it lacks the 5′-linked protein and bears a 5′-terminal nucleoside (uridine) monophosphate. Translation of the viral mRNA occurs from a single open reading frame that begins with an AUG codon located hundreds of nucleotides downstream from the 5′ end. The uninterrupted coding sequence that follows the initiating AUG generates a single approximately 250-kD polyprotein that is proteolytically processed both during and subsequent to translation by viral

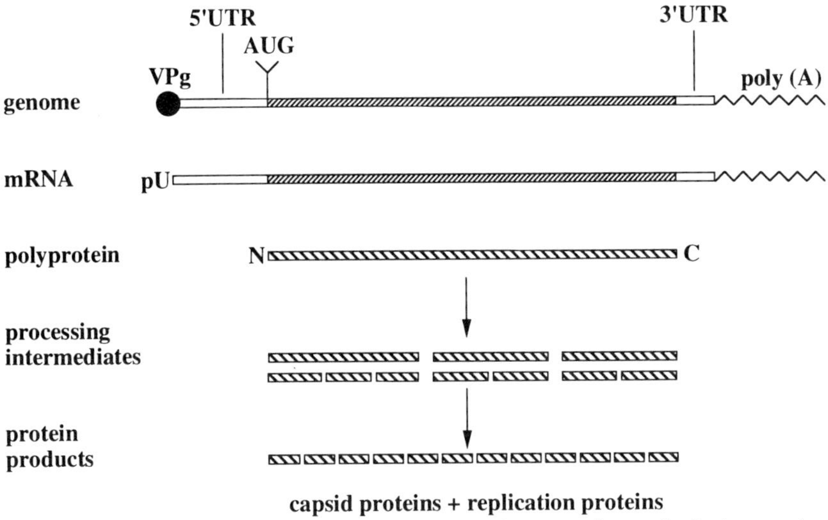

Figure 1 Schematic representation of the genetic map of a typical picornavirus. The regions of the genome and protein products are not drawn to scale.

proteases acting both in *cis* and in *trans*. The final products of translation and processing include the viral capsid proteins and a number of proteins involved in virus replication, not all of whose functions have been identified (for review, see Wimmer et al. 1993). Figure 1 shows a simplified scheme of the gene expression map of a picornavirus.

THE INTERNAL RIBOSOME ENTRY SEGMENT

IRES Structures in Different Picornavirus RNAs

The 5′UTRs of the members of the various virus genera are from 600 to more than 1200 nucleotides in length, and all contain IRES structures extending over approximately 450 nucleotides. Although these regions perform the same function for each virus, the different groups contain IRES elements of quite different sequence and predicted structure (Jackson et al. 1990). The entero- and rhinoviruses share considerable conservation of IRES structure, as do the cardio- and aphthoviruses; however, almost no similarity exists in the pattern of motifs between these two groups (see Fig. 2A,B). The hepatoviruses contain an IRES element that appears to be related to the cardio/aphtho element (Brown et al. 1994). For each of the basic IRES structures, evidence for internal entry of ribosomes was demonstrated by translation studies in vitro and in cultured cells in which dicistronic mRNAs with downstream cistrons containing a picornavirus 5′UTR were translated even under conditions where the

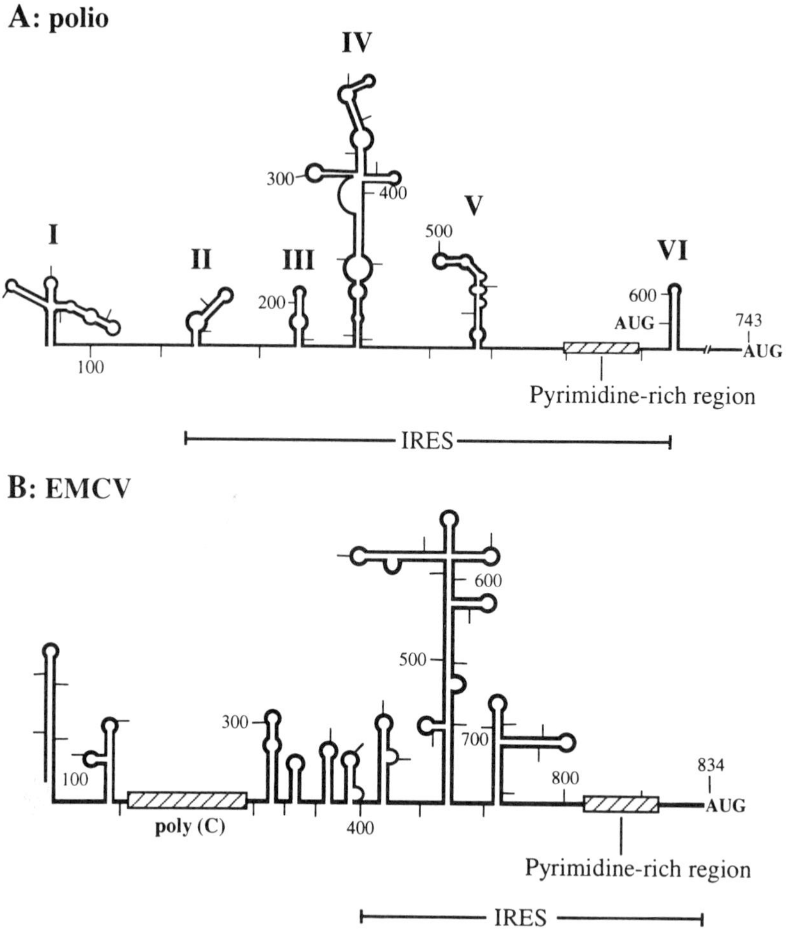

Figure 2 Schematic representation of the predicted stem-loop structures in the poliovirus (*A*) and EMCV (*B*) 5′-untranslated region. The nomenclature of the stem-loop domains, I to VI, is from Wimmer et al. (1993).

upstream cistron was prevented from being translated (Jang et al. 1988; Pelletier and Sonenberg 1988; Belsham and Brangwyn 1990; Borman et al. 1993; Glass et al. 1993). More recently, circular RNAs containing the EMCV IRES were constructed that bound to ribosomes in the presence of the elongation inhibitor, sparsomycin, and that directed translation of the predicted protein product (Chen and Sarnow 1995). These studies provided compelling evidence that ribosomes bound internally to the 5′UTR of picornavirus RNAs without the usual mode of scanning from a free 5′terminus.

Poliovirus: The Prototype for the Entero/Rhinovirus IRES

Intensive efforts have been directed toward defining the features of the poliovirus 5′UTR responsible for its unique, cap-independent binding to ribosomal subunits. Secondary structure maps have been predicted by several independent folding procedures (Rivera et al. 1988; Pilipenko et al. 1989b; Skinner et al. 1989; Le and Zucker 1990), and all predicted similar overall structures, with some differences in the details of particular stem-loop domains. A drawing of a consensus structure is presented in Figure 2A. Biochemical probings for nucleotides involved in base pairing were consistent with many of the structural predictions (Pilipenko et al. 1989b; Skinner et al. 1989). In addition, an analysis of sequence variation found in independent poliovirus isolates supported the existence of most of the proposed stem-loop structures by revealing extensive structure-conserving substitutions within stems, a high degree of sequence conservation in loops, and "hot spots" of sequence divergence in apparent spacer regions between domains (Poyry et al. 1992).

To dissect which regions of the poliovirus 5′UTR control translation initiation, mutations were introduced into cDNAs representing the 5′UTR to generate RNA transcripts whose translation could be evaluated both in vitro and in vivo. Deletion of contiguous nucleotides from the 3′end of the UTR placed the downstream border of the translation element around nucleotide 600 (Kuge and Nomoto 1987; Pelletier et al. 1988; Pilipenko et al. 1992). Similar approaches to define the 5′border showed that removal of the first 79 nucleotides had no effect on translation in vitro (Pelletier and Sonenberg 1988; Pelletier et al. 1988), whereas removal of 139 nucleotides slightly reduced translational efficiency. The upstream 5′-terminal portion of the RNA contains signals for RNA replication and possibly other functions, although one mutant virus was described that exhibits a fivefold decrease in viral translation caused by a region in domain I (Simoes and Sarnow 1991). In general, the segment of RNA that includes sequences required for cap-independent, internal ribosome binding and initiation lies within the boundaries of nucleotides ~130 and ~600, a region that embraces secondary structure domains II–V and part of the sequences in domain VI (Fig. 2A). Within this region, most deletions and other mutations eliminate or reduce IRES function (Trono et al. 1988; Nicholson et al. 1991; Percy et al. 1992; for review, see Ehrenfeld and Semler 1995), although deletion of the entire domain III is tolerated (Dildine and Semler 1989; Nicholson et al. 1991; Percy et al. 1992). Often spacing, rather than sequence, between stem-loops is important, and usually maintenance of base pairing within a stem is required rather than specific nucleotide sequence (Poyry et al. 1992). The importance of secondary structure is supported by the fact that phenotypic revertants of linker substitution and point mutations include second-site suppressor mutations that restore the base pairing found in the wild-type sequence (Kuge and Nomoto 1987; Haller and Semler 1992).

Upstream of the base of the stem-loop in domain VI, there is a 21-nucleotide pyrimidine-rich region that is highly sensitive to mutation (Kuge and Nomoto 1987; Iizuka et al. 1989; Meerovitch et al. 1991; Nicholson et al. 1991; Pestova et al. 1991), which precedes an essential AUG triplet in the hairpin stem of domain VI by a critical spacing of 20–25 nucleotides (Pilipenko et al. 1992). Mutation of this AUG (nucleotides 586–588 in poliovirus type 1) reduced translation efficiency in vitro from the normal downstream initiation site (nucleotide 743) and compromised the growth of virus harboring the mutation (Pelletier et al. 1988). It is thought that this motif (pyrimidine-rich region/spacer/AUG) may constitute the actual site of ribosome binding (Meerovitch and Sonenberg 1993; Jackson et al. 1994). The AUG triplet at the terminus of this element marks the 3′boundary of the IRES. It is normally not used to initiate translation, but, if its context is improved by mutation of surrounding nucleotides, it can function to initiate translation (Pestova et al. 1994). In the wild-type sequence, an approximately 150-nucleotide unstructured spacer of nonessential sequence separates the end of the IRES from the start of the coding sequence (Agol 1991). Ribosome movement from the entry site to the initiating AUG may occur by conventional scanning or some other form of shunt mechanism (see Jackson, this volume).

EMCV: The Prototype for the Cardio/Aphthovirus IRES

Parallel studies with the 5′UTR of EMCV showed that this sequence contains a genetic element functionally related to that of poliovirus, but which forms a series of stem-loop structures of quite different pattern (for review, see Hellen and Wimmer 1995). The availability of a large base of nucleotide sequence data has facilitated determination of potential secondary structures using computational and phylogenetic approaches (Pilipenko et al. 1989a; Duke et al. 1992), which have been experimentally supported by mutational analyses (Jang and Wimmer 1990) and by chemical or enzymatic probings for nucleotides involved in base pairing (Pilipenko et al. 1989b; Evstafieva et al. 1991). A schematic drawing for the EMCV 5′UTR is shown in Figure 2B. Little or no similarity in sequence or structure between the EMCV IRES and the poliovirus IRES is apparent. Although the various stem-loop domains within each of the basic IRES structures undoubtedly interact to form higher-order structures, it is not apparent that the different secondary structures converge to form similar three-dimensional presentations.

The 3′boundary of the EMCV IRES is marked by a pyrimidine-rich tract/spacer/AUG motif that represents the only universally conserved element in all the picornavirus IRES structures. For EMCV, however, the AUG at the end of this motif is the initiating codon for translation. Although there is no obvious reason a priori for IRES sequences not to extend into the coding region of the viral RNA, and there exist some anecdotal claims that reporter sequences fused directly to the EMCV IRES translate less well than if those sequences were fused a short distance downstream from the AUG, the only reported direct test indicated no significant effect on translation efficiency by sequences within the coding region of the EMCV genome (Hunt et al. 1993).

A View of IRES Structure-Function

The division of the long IRES segment into a series of stem-loop domains, as indicated in Figure 2, provides an important experimental approach for dissecting its function. There is some evidence that artificial RNA transcripts representing individual domains can assume structures that are similar to functional elements within the intact 5′UTR. For example, transcripts representing only domain VI and parts of the upstream linker region (including the pyrimidine-rich tract/spacer/AUG motif) of poliovirus RNA can efficiently bind a 52-kD protein that also binds and affects initiation of translation from the intact 5′UTR (Meerovitch et al. 1989, 1993); transcripts containing only domain IV sequences of poliovirus RNA compete efficiently with the intact 5′UTR for both binding of specific proteins (Blyn et al. 1995) and translation of poliovirus RNA (L. Blyn et al., unpubl.). At the same time, this division into secondary structure-determined domains poses some constraints on our thinking, since these domains must interact with one another in ways that preclude their functioning as independent units. Some efforts to elucidate such higher-order interactions have been made (Muzychenko et al. 1991; Le et al. 1992, 1993), although experimental determination of the spatial organization of the entire IRES element is currently difficult with the available technology for RNA structure analysis.

The evolution of the picornavirus IRES likely occurred by the gradual addition of domains and elements that improved the utilization of the IRES in specific host-cell environments. The resulting structures are composed of several hundred contiguous nucleotides (although some have insertions that can be deleted without disrupting the essential structure) upstream of the conserved pyrimidine-rich tract/spacer/AUG element which appears to serve as the site of ribosome attachment. The role of the approximately 450 nucleotides that guide the ribosome to its internal entry site may be envisioned as forming a structural base from which a few important sequences are presented, as though on a platform or stage with positions marked for important actors. These interacting sequences are located in the RNA loops and bulges extended in space from their stems or stalks. The important sequences are thus fixed to contact proteins or other RNA segments in a spatial context about which we have very little information. When ribosome attachment at the 3′end of the IRES is completed, mediated by proteins interacting with the exposed RNA loop sequences, AUG selection for translation initiation occurs. For some picornaviruses (e.g., EMCV and perhaps hepatitis A virus [HAV]), the AUG at the end of the IRES is used directly; for others (e.g., poliovirus and rhinovirus), the ribosome scans or translocates to the next AUG codon, which may be located a variable distance from the IRES; in some cases (e.g., foot-and-mouth disease virus [FMDV]), both options are used, resulting in two alternative initiation events.

INTERACTIONS OF PROTEINS WITH THE IRES

Cellular Proteins

Recognition and utilization of the IRES structural domains are mediated by an unknown number of proteins. At least the majority of these proteins must be of cellular origin, since the infecting viral RNA is translated prior to the production of any viral proteins. It is generally believed that internal initiation directed by a picornavirus IRES requires all or most of the same set of initiation factors that are utilized by typical capped cellular mRNAs (Staehelin et al. 1975; Scheper et al. 1992). A first step in recognition of capped mRNAs is binding of the 5′end by eIF4F (see Merrick and Hershey, this volume). This protein appears to act as a complex of three polypeptides, including eIF4E, which contains the cap-binding activity, eIF4A, an RNA-dependent ATPase and helicase, and eIF4G, a large polypeptide of unknown biochemical activity. Although the cap-binding role of eIF4F predictably is not required for translation of the uncapped picornavirus RNAs, the RNA helicase activity manifested by this protein may still have an essential or important role in internal initiation (Rozen et al. 1990). The large subunit of eIF4F becomes structurally modified and at least partially inactivated following infection with entero-, rhino-, and aphthoviruses (see below), and thus if it is required for IRES-driven initiation events, it must function in a structurally modified form.

In addition to the standard set of initiation factors, other *trans*-acting protein factors mediate internal initiation by picornavirus IRESs. Indeed, the availability and functional forms of the required spectrum of proteins in a given cell type may contribute to the growth and potential pathogenesis of a specific virus in that host. For example, translation of poliovirus, rhinovirus, or HAV RNA does not occur in wheat-germ extracts, and it occurs extremely inefficiently and usually incorrectly in rabbit reticulocyte lysates (RRL) (see, e.g., Brown and Ehrenfeld 1979; Jia et al. 1991; Borman et al. 1993). Translation of RNAs with the EMCV or FMDV IRES (the cardio/aphtho-type IRES structure), however, occurs accurately and with high efficiency in RRL (see, e.g., Shih et al. 1978), although not in wheat-germ extracts. The poliovirus IRES is utilized equally as well as the EMCV IRES in HeLa cell extracts (see, e.g., Pelletier and Sonenberg 1989). Its poor translation in wheat germ and RRL is markedly improved by addition of protein factors from HeLa cells (Brown and Ehrenfeld 1979; Dorner et al. 1984; Phillips and Emmert 1986; Pelletier et al. 1988; Svitkin et al. 1988). HAV RNA translation remains poor in HeLa cell extracts or in RRL supplemented with HeLa cell protein factors; its translation may require some proteins expressed specifically in mammalian liver cells (Glass and Summers 1993).

The existence of cell-specific factors that mediate IRES utilization prompted a search for cellular proteins that could interact specifically with picornavirus IRES structures. Most commonly, the IRES element is used as a radioactively labeled probe for cross-linking with UV light to proteins in cell extracts. In some instances, subdomains of the IRES are prepared and examined for mobility shifts in gels following incubation with cellular proteins. In this way, a 52-kD protein, subsequently identified as the La autoantigen, has been shown to bind to nucleotides 559–624 of poliovirus RNA, a region that includes the pyrimidine-rich tract/spacer/AUG motif described above (Meerovitch et al. 1989). Addition of recombinant La to RRL specifically stimulated poliovirus translation and corrected the inefficient and aberrant translation usually seen in RRL, although very high concentrations were required, in excess of the levels normally present in HeLa cell extracts (Meerovitch et al. 1993; Svitkin et al. 1994). In uninfected cells, La protein is reportedly involved in RNA polymerase III transcription termination (Gottlieb and Steitz 1989). Recently, it has been reported to bind and unwind double-stranded RNA (Xiao et al. 1994). A biochemical role for La in poliovirus RNA translation has not been identified.

Another protein of 57 kD binds to the 5′UTR of poliovirus RNA at multiple binding sites (Hellen et al. 1994). This same protein binds with even higher affinity to the cardio/aphtho IRES (Borman et al. 1993), recognizing a stem-loop toward the 5′end (Borovjagin et al. 1990; Jang and Wimmer 1990; Luz and Beck 1990). This protein has been identified as polypyrimidine-tract-binding (PTB) protein (Borman et al. 1993; Hellen et al. 1993). Depletion of PTB from HeLa cell extracts selectively prevented the translation of IRES-driven mRNAs (Hellen et al. 1993), and addition of physiological concentrations of recombinant PTB completely restored that capacity to the depleted extracts (A. Kaminski et al., in prep.). Additional proof of an essential role for PTB in EMCV translation was obtained by analysis of mutations introduced into the region of the PTB-binding site. Disruption of base pairing in a stem near the binding site (by changing two G residues at nucleotide positions 415–416 [see Fig. 2B] to C residues) abrogated both protein binding and internal initiation of translation, whereas both properties were restored by compensating mutations on the other side of the stem that restored base pairing (Jang and Wimmer 1990). The PTB protein appears to function in conjunction with another protein, PSF, in nuclear pre-mRNA metabolism (Patton et al. 1993). Whether it has additional roles in cellular protein synthesis is not known.

Numerous other proteins have been shown to bind specifically to various regions of the IRES structure (delAngel et al. 1989; Pestova et al. 1991; Dildine and Semler 1992; Gebhard and Ehrenfeld 1992; Haller et al. 1993), but the identities of these proteins and their possible functions in translation have not been determined. One additional protein of some interest is a 97-kD protein that was purified biochemically as an activity which specifically stimulates rhinovirus IRES-driven translation in RRL, rather than as a protein that simply binds to the IRES RNA (Borman et al. 1993). Its characterization and biochemical function await identification.

Specific Activation of Entero/Rhino IRES Translation after Viral Infection

Although initial translation of the infecting viral RNA must rely on the proteins and structures provided by the host cell, the bulk of subsequent viral protein synthesis occurs after inhibition of cellular protein synthesis and after modification of eIF4F by entero-, rhino-, and aphthoviruses (see below). Hambidge and Sarnow (1992) reported that mRNAs whose translation was initiated from the poliovirus IRES were translated at enhanced rates compared to other mRNAs in infected cells. These authors suggested that the virus produced or induced a factor that *trans*-activated utilization of the poliovirus IRES. Genetic analyses identified the viral protein, 2A, as the *trans*-activating factor. This protein is a protease that catalyzes a primary cleavage event during processing of the viral polypeptide (Toyoda et al. 1986), and recent evidence suggests that it may also have a role in viral RNA replication (Molla et al. 1993). Specific enhancement of IRES-driven translation has been demonstrated in vitro as well, in HeLa cell extracts preincubated with purified 2A protein (Liebig et al. 1993). The involvement of 2A in translation of poliovirus RNA is also suggested by the observation that destabilizing mutations in the poliovirus 5′UTR that affect translation efficiency can be suppressed by second-site mutations in the 2A gene (Macadam et al. 1994). Although 2A has been clearly implicated in the virus-induced inhibition of cellular protein synthesis (see below), the enhancement of viral translation appears to be independent of the inhibition of cellular protein synthesis (Hambidge and Sarnow 1992; Macadam et al. 1994; E. Ehrenfeld, unpubl.), and thus to result not merely from elimination of competition for general translation components. Rather, a specific effect of 2A or its reaction products on poliovirus RNA translation appears likely. Indeed, it is known that 2A is responsible for eIF4F modification, and early experiments had suggested that the structurally modified eIF4F present in poliovirus-infected cells specifically stimulated the translation of poliovirus RNA, but not capped mRNAs, in vitro, although the stimulation was only about twofold (Buckley and Ehrenfeld 1987). No evidence for direct interaction of 2A with the poliovirus 5′UTR has been obtained; it is possible that the *trans*-activation is a function of the structural modification to eIF4F induced by 2A. A mechanism for the specific enhancement has not been formulated.

Neurovirulence Determinants in IRES of Poliovirus RNA

The Sabin vaccine strains of poliovirus were developed almost 50 years ago by blind passage of wild-type isolates in cultured cells and selection for temperature-sensitive growth, followed by empirical screening for reduced neurovirulence. The recent ability to apply molecular sequence analyses and to construct chimeric viruses and site-specific mutant viruses has allowed a dissection of the mutations that contribute to the attenuated phenotypes of the three Sabin strains (Minor 1992). A major determinant in all three strains is a single base change in nearby positions in the stem of domain V (see Fig. 2A) in the 5′UTR. These mutations cause reduced translational efficiency compared to wild-type neurovirulent strains. This reduced efficiency of translation is manifested specifically in neuronal cells (Agol et al. 1989; La Monica and Racaniello 1989) but not in HeLa cells in culture or, presumably, in epithelial cells lining the intestine of an infected individual, where the attenuated viruses grow and induce protective immunity. RNAs isolated from Sabin strains of poliovirus translate in vitro with reduced efficiency as well (Svitkin et al. 1985), and the translational deficiency was shown conclusively to be the result of the single transition mutation at nucleotide 472 in domain V (see Fig. 2A) of the 5′UTR of poliovirus type 3. It is believed that limiting levels of an essential factor in neuronal cells prevents productive binding to the structure formed by a weakened base pair in the IRES. Thus, in this case, the RNA-protein interactions that control translation initiation contribute to the pathogenesis and virulence manifestations of the viral infection.

INACTIVATION OF THE CAP RECOGNITION SYSTEM BY SOME PICORNAVIRUSES

The existence of a cap-independent, internal initiation mechanism as has evolved in picornavirus RNAs provides a biochemical basis upon which viral and cellular protein synthesis can be distinguished and thereby selectively inhibited. Three of the five classified genera of picornaviruses appear to have taken advantage of this opportunity, since entero-, rhino-, and aphthoviruses actively and selectively inhibit host-cell translation after infection, whereas the cardioviruses may take over the translation machinery by more passive means; the hepatoviruses, which establish a slow, persistent infection in cultured cells, manifest no effects on host macromolecular synthesis. It is not known whether the virus-induced inhibition of host-cell protein synthesis is an essential viral function in those cases where it occurs, since viral mutants that fail to inhibit protein synthesis without affecting other replication activities have not been isolated (Yu et al. 1995).

Cleavage of eIF4G

One of the hallmarks of poliovirus infection of cultured cells is the marked and rapid inhibition of cellular protein synthesis, which is followed by the exclusive production of viral proteins (for review, see Sonenberg 1990; Wyckoff 1993). Figure 3 illustrates the rate of protein synthesis in a suspension culture of HeLa cells infected at high multi-

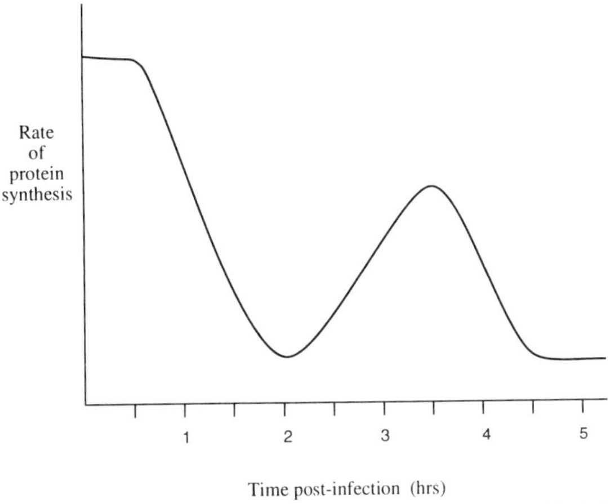

Figure 3 Rates of protein synthesis in HeLa cells infected with poliovirus.

plicity with poliovirus. By 2 hours postinfection, virtually all polyribosomes have disaggregated, and the synthesis of almost all cellular proteins has ceased. A new burst of protein synthesis follows, resulting exclusively from translation of viral mRNA. By 3.5–4 hours postinfection, viral protein synthesis has peaked, and the infection cycle declines.

Extracts prepared from infected cells are unable to translate capped mRNAs, but they remain fully functional for the translation of poliovirus RNA (Helentjaris and Ehrenfeld 1978). Extensive biochemical analyses of the various components of the protein synthesis pathway demonstrated that infected cells were defective in the activity of only one translation initiation factor, eIF4F (Wyckoff 1993), the factor that recognizes and binds to the m^7G cap group of capped mRNAs and thereby identifies the 5′end of the mRNA for subsequent unwinding and ribosome binding (see Merrick and Hershey, this volume). The inactivation of this factor during poliovirus infection is caused by the cleavage of its eIF4G subunit (formerly called p220; Etchison et al. 1982) to a set of amino-terminal products with apparent molecular weights estimated from SDS-PAGE of approximately 130,000, and a carboxy-terminal product with an apparent molecular weight of 100,000. Similar cleavage products of eIF4G have been detected in cells infected with various members of the entero-, rhino-, and aphthovirus groups (Etchison and Fout 1985; Lloyd et al. 1988; Kirchweger et al. 1994); no cleavage is detected during infection with cardiovirus or hepatoviruses.

The biochemical function of the 4G subunit of eIF4F has not been defined. It has a calculated molecular mass of 154 kD (from the cloned cDNA; Yan et al. 1992), although its previous designation as p220 indicates an aberrantly retarded mobility on SDS-PAGE. Neither the aberrant mobility nor the heterogeneity of bands exhibited by p220 or its amino-terminal cleavage products is understood, although both are a function of the amino-terminal third of the protein (D. Etchison and E. Ehrenfeld, unpubl.). It can be envisioned that eIF4G may serve as a scaffold: Its association with eIF4E, the cap-binding protein, brings it in association with capped mRNAs; its affinity for eIF3, which is present on 40S ribosomal subunits, localizes it to the ribosome; and its binding to eIF4A, the RNA helicase, brings together the whole complex, allowing activation of helicase activity at the 5′end of the mRNA. The result of cleavage of eIF4G is therefore release of the amino-terminal domain, which is bound to eIF4E, from the complex. Cap-dependent recruitment of mRNA to the ribosome is thereby prevented (B.J. Lamphear et al., pers. comm.). The carboxy-terminal portion of eIFG, bound to eIF4A and the ribosomal subunit via eIF3 interactions, would remain available for translation of RNAs that were not dependent on capped-end initiation. This eIF4G cleavage product, or some complex of this fragment with other polypeptides, is likely responsible for the specific activation of poliovirus RNA translation described earlier.

Some data indicate that the complete inhibition of cellular protein synthesis observed after poliovirus infection requires the contribution of other mechanisms, in addition to the cleavage of the 4G subunit of eIF4F. For example, under certain conditions of poliovirus infection, all detectable eIF4G was cleaved with only a modest decrease in cellular mRNA translation (Bonneau and Sonenberg 1987; Perez and Carrasco 1992). Another group reported that recombinant 2A protein expressed in COS-1 cells inhibited cellular RNA polymerase II transcription as well as causing eIF4G cleavage and suggested that the latter effect itself may be insufficient for the inhibition of translation (Davies et al. 1991). Since inhibition of cellular protein synthesis appears to be a common theme during many viral infections (see Mathews; Schneider; Katze; all this volume), it is possible that more than one mechanism is operative in poliovirus-infected cells.

The Mediator of eIF4G Cleavage

The synthesis of viral proteins is required for the cleavage of eIF4G and the inhibition of host-cell protein synthesis. Infected cells contain an activity that cleaves eIF4G (Etchison et al. 1982; Lloyd et al. 1987), and it was initially suspected that the cleavage was catalyzed by one of the proteases that process the viral polyprotein encoded by the viral genome. A number of experiments implicated the 2A protease as the viral gene product required for the shutoff reaction. Poliovirus with mutations in the 2A gene neither specifically inhibits protein synthesis nor cleaves eIF4G (Bernstein et al. 1985; O'Neill and Racaniello 1989). When RNAs transcribed from various portions of the poliovirus genome were translated in vitro, only transcripts coding for an intact 2A protein could induce eIF4G cleavage (Kräusslich et al. 1987; Lloyd et al. 1988). In addition, incubation of 2A expressed from a recombinant plasmid in *Escherichia coli* with an extract from uninfected HeLa cells causes cleavage of the HeLa cell eIF4G (Alvey et al. 1991). Finally, transfection of cells with plasmids in which 2A was the only poliovirus-encoded gene product inhibited the expression of genes cotransfected with 2A and induced cleavage of eIF4G (Sun and Baltimore 1989). Transfection of the 2A gene in a recombinant vaccinia virus vector caused parallel and virtually complete eIF4G cleavage and shutoff of cell protein synthesis (R. Chen and E. Ehrenfeld, unpubl.). These data collectively demonstrated that 2A is the only viral protein required for eIF4G cleavage in vitro and in vivo. The 2A protease is a small protein, encoded in the middle of the poliovirus genome, and is thought to be related to the class of small serine proteases, although a cysteine residue replaces the serine in the catalytic triad (Argos et al. 1984; Gorbalenya et al. 1986; Bazan and Fletterick 1988; Yu and Lloyd 1992).

Although the requirement for 2A to effect eIF4G cleavage is firmly established, several lines of evidence indicated that it did so indirectly and was not itself the enzyme that catalyzed the cleavage reaction. This conclusion was originally based on both biochemical and immunological data which indicated that $2A^{pro}$ and the eIF4G cleavage activity in infected cells did not copurify and that eIF4G cleavage was not inhibited by anti-$2A^{pro}$ serum that did inhibit 2A-mediated polyprotein cleavage (Lloyd et al. 1986; Kräusslich et al. 1987; Wyckoff et al. 1990). A model was therefore proposed that the 2A protease activated a latent cellular protease that in turn cleaved eIF4G (Kräusslich et al. 1987; Wyckoff et al. 1990, 1992). This model was strengthened by the demonstration that partially purified preparations of eIF4G were not cleaved by bacterial extracts containing active recombinant 2A, unless an additional factor was added. The required factor was identified as eIF3 (Wyckoff et al. 1990, 1992).

Recently, however, the 2A proteases of human rhinovirus 2 and coxsackievirus B4 were expressed and purified from *E. coli* (Sommergruber et al. 1992; Liebig et al. 1993). These purified 2A proteases were able to cleave the eIF4G subunit in purified eIF4F directly to generate the characteristic cleavage products found in infected cells (Lamphear et al. 1993; Sommergruber et al. 1994). The cleavage site selected in eIF4G (Arg-486/Gly-487) shows similar specificity to the 2A cleavage site(s) in the viral polyprotein, and both proteases can cleave a small peptide containing the cleavage sequence. Although it is highly unlikely that the mechanism of polio's 2A-induced cleavage of eIF4G is fundamentally different from that of rhinovirus or coxsackievirus, it is disappointing that the experiment has not been performed directly with the poliovirus enzyme. Expression of poliovirus 2A in *E. coli* or other recombinant systems has been accomplished (Alvey et al. 1991), but the enzyme aggregates and manifests heterogeneous properties during standard biochemical purification protocols (R. Lloyd and E. Ehrenfeld, unpubl.). Thus, the issue of whether the poliovirus 2A protease induces cleavage of eIF4G directly or via a cellular intermediate awaits final resolution. One possible explanation for the conflicting data may lie in the utilization of an eIF4G substrate for cleavage by the rhinovirus and coxsackievirus 2A proteases that had been purified as an intact complex of subunits by affinity chromatography on m^7G-agarose columns to select for the eIF4E, cap-binding subunit, with which eIF4G is associated. The studies with the poliovirus 2A protease, on the other hand, utilized an eIF4G substrate that had been purified by phosphocellulose column chromatography followed by calmodulin-Sepharose affinity chromatography, in a form deliberately dissociated from the other eIF4F polypeptides (Wyckoff et al. 1992). A possible explanation is that when eIF4G is complexed with the other polypeptide components of functional eIF4F protein, it assumes a conformation that is a substrate for 2A. The eIF4G polypeptide alone, on the other hand, may adopt a conformation that is resistant to 2A cleavage. Since eIF3 is known to interact with eIF4G, the addition of eIF3 to this preparation may alter its conformation to make the cleavage site available or susceptible to cleavage. Thus, eIF3 may act as a kind of chaperonin to assist in the proper folding of eIF4G, and simultaneously, to permit presentation of the cleavage site.

Other Picornavirus Groups

The FMDV members of the aphthovirus group of picornaviruses manifest the same pattern of inhibition of host-cell protein synthesis and eIF4G cleavage as the entero- and rhinoviruses (Lloyd et al. 1988); however, they have no homologous 2A gene. Cleavage of eIF4G in FMDV-infected cells is mediated by the viral leader (L) protease (Devaney et al.

1988), which is encoded in a location different from that of 2A on the viral genome and which appears to be related to a papain-like class of thiol proteases (Gorbalenya et al. 1991). Entero- and rhinoviruses contain no L-protease-coding sequences. The cleavage site in eIF4G recognized by the FMDV L protease is seven amino acids upstream of the bond hydrolyzed by the rhino- or coxsackievirus 2A protease, between Gly-479 and Arg-480 (Kirchweger et al. 1994). The proximity of the two cleavage sites implies the presence of an exposed domain which is susceptible to proteolysis, perhaps linking separate functional domains of eIF4G.

The cardioviruses (e.g., EMCV) show no specific cleavage of eIF4G in infected cells (Mosenkis et al. 1985), despite an equally efficient replication capacity. Consistent with the absence of eIF4G cleavage in cardiovirus-infected cells, neither the L protein nor the 2A protein of these viruses contains protease consensus motifs or demonstrable proteolytic activities. Recently, a protein known as 4E-BP1 (eIF4E-binding protein 1) was described that binds to the 4E-subunit of eIF4F and represses its activity (Pause et al. 1994). Binding to eIF4E occurs when 4E-BP1 is underphosphorylated. Activation of 4E-BP1 by dephosphorylation occurs in cells infected with EMCV, as well as with poliovirus (N. Sonenberg, pers. comm.). Thus, suppression of cap-dependent protein synthesis occurs in cardiovirus-infected cells by an alternate mechanism, whereas other members of the picornavirus family may have evolved a redundant mechanism to ensure complete inhibition of cellular translation. Since the hepatoviruses manifest a slow-growth phenotype with no associated cytopathology and establish a persistent infection in cultured cells with no apparent effect on cellular protein synthesis, it is not surprising that cleavage of eIF4G was not observed (E. Wyckoff and E. Ehrenfeld, unpubl.). No proteolytic activity of its 2A gene product has been detected (Jia et al. 1991).

CONCLUDING REMARKS

The observation that poliovirus infection induced a specific inhibition of host-cell protein synthesis was first made more than 30 years ago. Progress in understanding the involvement of the inactivator of the cap recognition system awaited discovery of the capped structure on cellular mRNAs and the identification of eIF4F. The description by Etchison et al. (1982) of the cleavage of the eIF4G subunit of eIF4F provided the first major insight into a biochemical basis for the cessation of cellular translation. The fact that viral translation occurred under conditions when cellular protein synthesis was inhibited implied that there was some fundamental difference in the mechanism of the two reactions. It was not until the late 1980s, however, that the concept of a large structure which formed an internal ribosome entry segment in the 5′UTR was developed and its implications were understood.

In the last few years, good progress has been made in understanding both the biochemistry of the host-cell shutoff reaction by some picornaviruses and the ribosome-binding mechanism used by viral mRNAs that allows them to bypass the inhibition imposed on the host. A full understanding of the complex reaction of IRES-directed initiation of translation will ultimately require identification and characterization of all of the *trans*-acting factors involved, as well as a complete picture of the *cis*-acting elements and their spatial arrangement in the RNA. It is likely that the complete spectrum of mRNA-specific, cell-specific, and general translation initiation factors utilized by these IRES-containing viral RNAs will be quite complex.

Studies of the translation events in picornavirus-infected cells provided the impetus for many studies leading to important breakthroughs in understanding the pathway of normal translation initiation, as well as a stimulus for examining other viruses and some cellular mRNAs for other examples of internal initiation. Picornavirus mRNAs have become the paradigm for internal initiation reactions. The methods developed for their analyses have become the standard for criteria to define new examples of internal initiation of translation (Jackson, this volume). It remains to be seen how common these examples are, and how variations on this theme have evolved.

ACKNOWLEDGMENT

The author thanks past and present members of the laboratory for their intellectual input and informative data.

REFERENCES

Agol, V. 1991. The 5′untranslated region of picornaviral genomes. *Adv. Virus Res.* **40:** 103–180.

Agol, V.I., S.G. Drozdov, T.A. Ivannikova, M.S. Kolesnikova, M.B. Korolev, and E.A. Tolskaya. 1989. Restricted growth of attenuated poliovirus strains in cultured cells of a human neuroblastoma. *J. Virol.* **63:** 4034–4038.

Alvey, J.C., E.E. Wyckoff, S.F. Yu, R. Lloyd, and E. Ehrenfeld. 1991. *Cis* and *trans* activities of poliovirus 2A protease expressed in *Escherichia coli. J. Virol.* **65:** 6077–6083.

Argos, P., G. Kamer, M.J.H. Nicklin, and E. Wimmer. 1984. Similarity in gene organisation and homology between proteins of animal picornaviruses suggest common ancestry of these virus families. *Nucleic Acids Res.* **12:** 7251–7267.

Bazan, J.F. and R.J. Fletterick. 1988. Viral cysteine proteases are homologous to the trypsin-like family of serine proteases: Structural and functional implications. *Proc. Natl. Acad. Sci.* **85:** 7872–7876.

Belsham, G.J. and J.K. Brangwyn. 1990. A region of the 5′noncoding region of foot-and-mouth disease virus RNA directs efficient internal initiation of protein synthesis within cells: Involvement with the role of L protease in translational control. *J. Virol.* **64:** 5389–5395.

Bernstein, H.D., N. Sonenberg, and D. Baltimore. 1985. Poliovirus mutant that does not selectively inhibit host cell protein synthesis. *Mol. Cell. Biol.* **5:** 2913–2923.

Blyn, L., R. Chen, B.L. Semler, and E. Ehrenfeld. 1995. HeLa cell proteins binding to domain IV of the 5′noncoding region of poliovirus RNA. *J. Virol.* **69:** 4381–4389.

Bonneau, A.-M. and N. Sonenberg. 1987. Proteolysis of the p220 component of the cap-binding protein complex is not sufficient for complete inhibition of host cell protein synthesis after poliovirus infection. *J. Virol.* **61:** 986–991.

Borman, A., M.T. Howell, J.G. Patton, and R.J. Jackson. 1993. The involvement of a spliceosome component in internal initiation of human rhinovirus RNA translation. *J. Gen. Virol.* **74:** 1775–1788.

Borovjagin, A.V., A.G. Evstafieva, T.Y. Ugarova, and I.N. Shatsky. 1990. A factor that specifically binds to the 5′-untranslated region of EMCV RNA. *FEBS Lett.* **261:** 237–240.

Brown, B.A. and E. Ehrenfeld. 1979. Translation of poliovirus RNA *in vitro*: Changes in cleavage pattern and initiation sites by ribosomal salt wash. *Virology* **97:** 396–405.

Brown, E., A. Zajac, and S.M. Lemon. 1994. In vitro characterization of an internal ribosomal entry site (IRES) present within the 5′nontranslated region of hepatitis A virus RNA: Comparison with the IRES of encephalomyocarditis virus. *J. Virol.* **68:** 1066–1074.

Buckley, B. and E. Ehrenfeld. 1987. The cap-binding protein complex in uninfected and poliovirus-infected HeLa cells. *J. Biol. Chem.* **262:** 13599–13606.

Chen, C. and P. Sarnow. 1995. Initiation of protein synthesis by the eukaryotic apparatus on circular RNAs. *Science* **268:** 415–417.

Davies, M., J. Pelletier, K. Meerovitch, N. Sonenberg, and R.J. Kaufman. 1991. The effect of poliovirus protease $2A^{pro}$ expression on cellular metabolism. *J. Biol. Chem.* **266:** 14714–14720.

delAngel, R.M., A.G. Papavassiliou, C. Fernandez-Tomas, S.J. Silverstein, and V. Racaniello. 1989. Cell proteins bind to multiple sites within the 5′untranslated region of poliovirus RNA. *Proc. Natl. Acad. Sci.* **86:** 8299–8303.

Devaney, M.A., V.N. Vakharia, R.E. Lloyd, E. Ehrenfeld, and M.J. Grubman. 1988. leader protein of foot-and-mouth disease virus is required for cleavage of the p220 component of the cap-binding protein complex. *J. Virol.* **62:** 4407–4409.

Dildine, S.L. and B.L. Semler. 1989. The deletion of 41 proximal nucleotides reverts a poliovirus mutant containing a temperature-sensitive lesion in the 5′noncoding region of genomic RNA. *J. Virol.* **63:** 847–862.

———. 1992. Conservation of RNA-protein interactions among picornaviruses. *J. Virol.* **66:** 4364–4376.

Dorner, A.J., B.L. Semler, R.J. Jackson, R. Hanecak, E. Duprey, and E. Wimmer. 1984.

In vitro translation of poliovirus RNA: Utilization of internal initiation sites in reticulocyte lysate. *J. Virol.* **50:** 507–514.

Duke, G.M., M.A. Hoffman, and A.C. Palmenberg. 1992. Sequence and structural elements that contribute to efficient encephalomyocarditis virus RNA translation. *J. Virol.* **66:** 1602–1609.

Ehrenfeld, E. and B. Semler. 1995. Anatomy of the poliovirus internal ribosome entry site. *Curr. Top. Microbiol. Immunol.* **203:** 65–83.

Etchison, D. and S. Fout. 1985. Human rhinovirus 14 infection of HeLa cells results in the protolytic cleavage of the p220 cap-binding complex subunit and inactivates globin mRNA translation in vitro. *J. Virol.* **54:** 634–638.

Etchison, D., S.C. Milburn, I. Edery, N. Sonenberg, and J.W.B. Hershey. 1982. Inhibition of HeLa cell protein synthesis following poliovirus infection correlates with the proteolysis of a 220,000 Da polypeptide associated with eukaryotic initiation factor 3 and a cap binding protein complex. *J. Biol. Chem.* **257:** 14806–14810.

Evstafieva, A.G., T.Y. Ugarova, B.K. Chernov, and I.N. Shatsky. 1991. A complex RNA sequence determines the internal initiation of encephalomyocarditis virus RNA translation. *Nucleic Acids Res.* **19:** 665–671.

Gebhard, J.R. and E. Ehrenfeld. 1992. Specific interactions of HeLa cell proteins with proposed translation domains of the poliovirus 5′ noncoding region. *J. Virol.* **66:** 3101–3109.

Glass, M.J. and D.F. Summers. 1993. Identification of a *trans*-acting activity from liver that stimulates HAV translation in vitro. *Virology* **193:** 1047–1050.

Glass, M.J., X.Y. Jia, and D.F. Summers. 1993. Identification of the hepatitis A virus internal ribosome entry site: In vivo and in vitro analysis of bicistronic RNAs containing the HAV 5′ noncoding region. *Virology* **193:** 842–852.

Gorbalenya, A.E., V.M. Blinov, and A.M. Donchenko. 1986. Poliovirus-encoded proteinase 3C: A possible evolutionary link between cellular and serine and cysteine proteinases. *FEBS Lett.* **194:** 253–257.

Gorbalenya, A.E., E.V. Koonin, and M.M.-C. Lai. 1991. Putative papain-related thiol proteases of positive-strand RNA viruses. *FEBS Lett.* **288:** 201–205.

Gottlieb, E. and J.A. Steitz. 1989. Function of the mammalian La protein: Evidence for its action in transcription termination by RNA polymerase III. *EMBO J.* **8:** 851–861.

Haller, A.A. and B.L. Semler. 1992. Linker scanning mutagenesis of the internal ribosome entry site of poliovirus RNA. *J. Virol.* **66:** 5075–5086.

Haller, A.A., J.H.C. Nguyen, and B.L. Semler. 1993. Minimum internal ribosome entry site required for poliovirus infectivity. *J. Virol.* **67:** 7461–7471.

Hambidge, S. and P. Sarnow. 1992. Translational enhancement of the poliovirus 5′ noncoding region mediated by virus-encoded polypeptide 2A. *Proc. Natl. Acad. Sci.* **89:** 10272–10276.

Helentjaris, T. and E. Ehrenfeld. 1978. Control of protein synthesis in poliovirus-infected cells. I. mRNA discrimination by crude initiation factors. *J. Virol.* **26:** 510–521.

Hellen, C.U.T. and E. Wimmer. 1995. Translation of EMCV RNA by internal ribosomal entry. *Curr. Top. Microbiol. Immunol.* **203:** 31–63.

Hellen, C.U.T., T.V. Pestova, M. Litterst, and E. Wimmer. 1994. The cellular polypeptide p57 pyrimidine tract-binding protein binds to multiple sites in the poliovirus 5′ nontranslated region. *J. Virol.* **68:** 941–950.

Hellen, C.U.T., G.W. Witherell, M. Schmid, S.H. Shin, T.V. Pestova, A. Gil, and E. Wimmer. 1993. The cellular polypeptide p57 that is required for translation of picornavirus RNA by internal ribosome entry is identical to the nuclear pyrimidine-tract binding protein. *Proc. Natl. Acad. Sci.* **90:** 7642–7646.

Hunt, S.L., A. Kaminski, and R.J. Jackson. 1993. The influence of viral coding sequences on the efficiency of internal initiation of translation of cardiovirus RNAs. *Virology* **197:** 801–807.

Hyypiä, T., C. Horsnell, M. Maaronen, M. Khan, N. Kalkkinen, P. Auvinen, L. Kinnunen, and G. Stanway. 1992. A distinct picornavirus group identified by sequence analysis. *Proc. Natl. Acad. Sci.* **89:** 8847–8851.

Iizuka, N., M. Kohara, K. Hagino-Yamagishi, S. Abe, T. Komatsu, K. Tago, M. Arita, and A. Nomoto. 1989. Construction of less neurovirulent polioviruses by introducing deletions into the 5′ noncoding sequence of the genome. *J. Virol.* **63:** 5354–5363.

Jackson, R.J., M.T. Howell, and A. Kaminski. 1990. The novel mechanism of initiation of picornavirus RNA translation. *Trends. Biochem. Sci.* **15:** 477–483.

Jackson, R., S. Hunt, C. Gibbs, and A. Kaminski. 1994. Internal initiation of translation of picornavirus RNAs. *Mol. Biol. Rep.* **19:** 147–159.

Jang, S.K. and E. Wimmer. 1990. Cap-independent translation of encephalomyocarditis virus RNA: Structural elements of the internal ribosomal entry site and involvement of a cellular 57-kD RNA-binding protein. *Genes Dev.* **4:** 1560–1572.

Jang S.K., H.-G. Kräusslich, M.J.H. Nicklin, G.M. Duke, A.C. Palmenberg, and E. Wimmer. 1988. A segment of the 5′ nontranslated region of encephalomyocarditis virus RNA directs internal entry of ribosomes during in vitro translation. *J. Virol.* **62:** 2636–2643.

Jia, X.-Y., G. Scheper, D. Brown, W. Updike, S. Harmon, D. Richards, D. Summers, and E. Ehrenfeld. 1991. Translation of hepatitis A virus RNA in vitro: Aberrant internal initiations influenced by 5′ noncoding region. *Virology* **182:** 712–722.

Kirchweger, R., E. Ziegler, B. Lamphear, D. Waters, H. Liebig, W. Sommergruber, F. Sobrino, C. Hohenadl, D. Blaas, R. Rhoads, and T. Skern. 1994. Foot-and-mouth disease virus leader proteinase: Purification of the Lb form and determination of its cleavage site on eIF-4γ. *J. Virol.* **68:** 5677–5683.

Kitamura, N., B.L. Semler, P.G. Rothberg, G.R. Larsen, C.J. Adler, A.J. Dorner, E.A. Emini, R. Hanecak, J.J. Lee, S. van der Werf, C.W. Anderson, and E. Wimmer. 1981. Primary structure, gene organization and polypeptide expression of poliovirus RNA. *Nature* **291:** 547–553.

Kozak, M. 1978. How do eukaryotic ribosomes select initiation regions in messenger RNA? *Cell* **15:** 1109–1123.

———. 1989. The scanning model for translation: An update. *J. Cell Biol.* **108:** 229–241.

Kräusslich, H.G., M.J. Nicklin, H. Toyoda, D. Etchison, and E. Wimmer. 1987. Poliovirus protease 2A induces cleavage of eukaryotic initiation factor 4F polypeptide p220. *J. Virol.* **61:** 2711–2718.

Kuge, S. and A. Nomoto. 1987. Construction of viable deletion and insertion mutants of the Sabin strain of type 1 poliovirus: Function of the 5′ noncoding sequence in viral replication. *J. Virol.* **61:** 1478–1487.

La Monica, W. and V.R. Racaniello. 1989. Differences in replication of attenuated and neurovirulent polioviruses in human neuroblastoma cell line SH-SY5Y. *J. Virol.* **63:** 2357–2360.

Lamphear, B.J., R. Yan, F. Yang, D. Waters, H.-D. Liebig, E. Klump, E. Kuechler, T. Skern, and R.E. Rhoads. 1993. Mapping the cleavage site in protein synthesis initiation factor eIF-4γ of the 2A proteases from human coxsackievirus and rhinovirus. *J. Biol. Chem.* **268:** 19200–19203.

Le, S.-Y. and M. Zucker. 1990. Common structures of the 5′noncoding RNA in enteroviruses and rhinoviruses: Thermodynamic stability and statistical significance. *J. Mol. Biol.* **216:** 729–741.

Le, S.-Y., J.H. Chen, N. Sonenberg, and J. Maizel. 1992. Conserved tertiary structural elements in the 5′untranslated region of human enteroviruses and rhinoviruses. *Virology* **191:** 858–866.

———. 1993. Conserved tertiary structural elements in the 5′nontranslated region of cardiovirus, aphthovirus and hepatitis A virus RNAs. *Nucleic Acids Res.* **21:** 2445–2451.

Liebig, H., E. Ziegler, R. Yan, K. Hartmuth, H. Klump, H. Kowalski, D. Blaas, W. Sommergruber, L. Frosel, B. Lamphear, R. Rhoads, E. Kuechler, and T. Skern. 1993. Purification of two picornaviral 2A proteinases: Interaction with eIF-4γ and influence on translation. *Biochemistry* **32:** 7581–7588.

Lloyd, R.E., M. Grubman, and E. Ehrenfeld. 1988. Relationship of p220 cleavage during picornavirus infection to 2A protease sequences. *J. Virol.* **62:** 4216–4223.

Lloyd, R.E, H.G. Jense, and E. Ehrenfeld. 1987. Restriction of translation of capped mRNA in vitro as a model for poliovirus-induced inhibition of host cell protein synthesis: Relationship to p220 cleavage. *J. Virol.* **61:** 2480–2488.

Lloyd, R.E., H. Toyoda, D. Etchison, E. Wimmer, and E. Ehrenfeld. 1986. Cleavage of the cap binding protein complex polypeptide p220 is not effected by the second poliovirus protease 2A. *Virology* **150:** 299–303.

Luz, N. and E. Beck. 1990. A cellular 57 kDa protein binds to two regions of the internal translation initiation site of foot-and-mouth disease virus. *FEBS Lett.* **269:** 311–314.

Macadam, A.J., G. Ferguson, T. Fleming, D.M. Stone, J.W. Almond, and P.D. Minor. 1994. Role for poliovirus protease 2A in cap independent translation. *EMBO J.* **13:** 924–927.

Meerovitch, K. and N. Sonenberg. 1993. Internal initiation of picornavirus RNA translation. *Semin. Virol.* **4:** 217–227.

Meerovitch, K., R. Nicholson, and N. Sonenberg. 1991. In vitro mutational analysis of *cis*-acting RNA translational elements within the poliovirus type 2 5′untranslated region. *J. Virol.* **65:** 5895–5901.

Meerovitch, K., J. Pelletier, and N. Sonenberg. 1989. A cellular protein that binds to the 5′noncoding region of poliovirus RNA: Implications for internal translation initiation. *Genes Dev.* **3:** 1026–1034.

Meerovitch, K., Y.V. Svitkin, H.S. Lee, F. Lejbkowicz, D.J. Kenan, E.K.L. Chan, V.I. Agol, J.D. Keene, and N. Sonenberg. 1993. La autoantigen enhances and corrects aberrant translation of poliovirus RNA in reticulocyte lysate. *J. Virol.* **67:** 3798–3807.

Minor, P.D. 1992. The molecular biology of poliovaccines. *J. Gen. Virol.* **73:** 3065–3077.

Molla, A., A.V. Paul, M. Schmid, S.K. Jang, and E. Wimmer. 1993. Studies on dicistronic poliovirus implicate viral proteinase $2A^{pro}$ in RNA replication. *Virology* **196:** 739–742.

Mosenkis, J.S., S. Daniels-McQueen, S. Janovec, R. Duncan, J.W.B. Hershey, J.A. Grifo, W.C. Merrick, and R.E. Thach. 1985. Shutoff of host translation by encephalomyocarditis virus infection does not involve cleavage of the eucaryotic initiation factor 4F polypeptide that accompanies poliovirus infection. *J. Virol.* **54:** 643–645.

Muzychenko, A.R., G.Y. Lipskaya, S.V. Maslova, Y.V. Svitkin, E.V. Pilipenko, B.K. Nottay, O.M. Kew, and V.I. Agol. 1991. Coupled mutations in the 5′-untranslated region of the Sabin poliovirus strains during in vivo passages: Structural and functional implications. *Virus Res.* **21:** 111–122.

Nicholson, R., J. Pelletier, S.Y. Le, and N. Sonenberg. 1991. Structural and functional analysis of the ribosome landing pad of poliovirus type 2: In vivo translation studies. *J. Virol.* **65:** 5886–5894.

O'Neill, R.E. and V.R. Racaniello. 1989. Inhibition of translation in cells infected with a poliovirus $2A^{pro}$ mutant correlates with phosphorylation of the alpha subunit of eukaryotic initiation factor 2. *J. Virol.* **63:** 5069–5075.

Patton, J.G., E.B. Porro, J. Galceran, P. Tempst, and B. Nadal-Girard. 1993. Cloning and characterization of PSF, a novel pre-mRNA splicing factor. *Genes Dev.* **7:** 393–406.

Pause, A., G.J. Belsham, A.C. Gingras, D. Donze, T.A. Lin, J. Lawrence, Jr., and N. Sonenberg. 1994. Insulin-dependent stimulation of protein synthesis by phosphorylation of a reglator of 5′-cap function. *Nature* **371:** 762–767.

Pelletier, J. and N. Sonenberg. 1988. Internal initiation of translation of eukaryotic mRNA directed by a sequence derived from poliovirus RNA. *Nature* **334:** 320–325.

———. 1989. Internal binding of eucaryotic ribosomes on poliovirus RNA: Translation in HeLa cell extracts. *J. Virol.* **63:** 441–444.

Pelletier, J., G. Kaplan, V.R. Racaniello, and N. Sonenberg. 1988. Cap-independent translation of poliovirus mRNA is conferred by sequence elements within the 5′ noncoding region. *Mol. Cell. Biol.* **8:** 1103–1112.

Percy, N., G.J. Belsham, J.K. Brangwyn, M. Sullivan, D.M. Stone, and J.W. Almond. 1992. Intracellular modifications induced by poliovirus reduce the requirement for structural motifs in the 5′ noncoding region of the genome involved in internal initiation of protein synthesis. *J. Virol.* **66:** 1695–1701.

Perez, L. and L. Carrasco. 1992. Lack of correlation between p220 cleavage and the shut-off of host translation after poliovirus infection. *Virology* **189:** 178–186.

Pestova, T.V., C.U.T. Hellen, and E. Wimmer. 1991. Translation of poliovirus RNA: Role of an essential *cis*-acting oligopyrimidine element within the 5′ nontranslated region and involvement of a cellular 57-kilodalton protein. *J. Virol.* **65:** 6194–6204.

———. 1994. A conserved AUG triplet in the 5′ noncoding region of poliovirus RNA can function as an initiation codon in vitro and in vivo. *Virology* **204:** 729–738.

Phillips, B.A. and A. Emmert. 1986. Modulation of the expression of poliovirus proteins in reticulocyte lysates. *Virology* **148:** 255–267.

Pilipenko, E.V., V.M. Blinov, B.K. Chernov, T.M. Dimitrieva, and V.I. Agol. 1989a. Conservation of the secondary structure elements of the 5′ untranslated region of cardio- and aphthovirus RNAs. *Nucleic Acids Res.* **17:** 5701–5711.

Pilipenko, E.V., V.M. Blinov, L.I. Romanova, A.N. Sinyakov, S.V. Maslova, and V.I. Agol. 1989b. Conserved structural domains in the 5′ untranslated region of picornaviral genomes: An analysis of the segment controlling translation and neurovirulence. *Virology* **168:** 201–209.

Pilipenko, E.V., A.P. Gmyl, S.V. Maslova, Y.V. Svitkin, A.N. Sinyakov, and V.I. Agol. 1992. Prokaryotic-like *cis* elements in the cap-independent internal initiation of translation on picornavirus RNA. *Cell* **68:** 119–131.

Poyry, T., L. Kinnunen, and T. Hovi. 1992. Genetic variation in vivo and proposed functional domains of the 5′ noncoding region of poliovirus RNA. *J. Virol.* **66:** 5313–5319.

Racaniello, V. and D. Baltimore. 1981. Cloned poliovirus complementary DNA is in-

fectious in mammalian cells. *Science* **214:** 916–919.
Rivera, V.M., J.D. Welsh, and J.V. Maizel. 1988. Comparative sequence analysis of the 5′ noncoding region of the enteroviruses and rhinoviruses. *Virology* **165:** 42–50.
Rozen, G., I. Edery, K. Meerovitch, T.E. Dever, W.C. Merrick, and N. Sonenberg. 1990. Bidirectional RNA helicase activity of eucaryotic initiation factors 4A and 4F. *Mol. Cell. Biol.* **10:** 1134–1144.
Scheper, G.C., H.O. Voorma, and A.A.M. Thomas. 1992. Eukaryotic initiation factors-4E and -4F stimulate 5′ cap-dependent as well as internal initiation of protein synthesis. *J. Biol. Chem.* **267:** 7269–7274.
Shih, D.S., C.T. Shih, O. Kew, M. Pallansch, R. Rueckert, and P. Kaesberg. 1978. Cell-free synthesis and processing of the proteins of poliovirus. *Proc. Natl. Acad. Sci.* **75:** 5807–5811.
Simoes, E.A. and P. Sarnow. 1991. An RNA hairpin at the extreme 5′end of the poliovirus RNA genome modulates viral translation in human cells. *J. Virol.* **65:** 913–921.
Skinner, M.A., V.R. Racaniello, G. Dunn, J. Cooper, P.D. Minor, and J.W. Almond. 1989. New model for the secondary structure of the 5′ noncoding RNA of poliovirus is supported by biochemical and genetic data that also show that RNA secondary structure is important in neurovirulence. *J. Mol. Biol.* **207:** 379–392.
Sommergruber, W., H. Ahorn, A. Zoephal, I. Maurer-Fogy, F. Fessl, G. Schnorrenberg, H. Liebig, D. Blaas, E. Kuechler, and T. Skern. 1992. Cleavage specificity on synthetic peptide substrates of human rhinoviruses proteinases 2A. *J. Biol. Chem.* **267:** 22639–22644.
Sommergruber, W., H. Ahorn, H. Klump, J. Seipelt, A. Zoephel, F. Fessl, E. Krystek, D. Blaas, E. Kuechler, H.-D. Liebig, and T. Skern. 1994. 2A proteinases of coxsackie- and rhinovirus cleave peptides derived from eIF-4γ via a common recognition motif. *Virology* **198:** 741–745.
Sonenberg, N. 1990. Poliovirus translation. *Curr. Top. Microbiol. Immunol.* **161:** 23–47.
Staehelin, T., H. Trachsel, B. Erni, A. Boschetti, and M.H. Schreier. 1975. The mechanism of initiation of mammalian protein synthesis. *FEBS Symp.* **39:** 309–323.
Stanway, G., N. Kalkkinen, M. Roivainen, F. Ghozi, M. Khan, M. Smyth, O. Meurman, and T. Hyypia. 1994. Molecular and biological characteristics of echovirus 22, a representative of a new picornavirus group. *J. Virol.* **68:** 8232–8238.
Sun, X.-H. and D. Baltimore. 1989. Human immunodeficiency virus tat-activated expression of poliovirus protein 2A inhibits mRNA translation. *Proc. Natl. Acad. Sci.* **86:** 2143–2146.
Svitkin, Y.V., S.V. Maslova, and V.I. Agol. 1985. The genomes of attenuated and virulent poliovirus strains differ in their in vitro translation efficiencies. *Virology* **147:** 243–252.
Svitkin, Y.V., T.V. Pestova, S.V. Maslova, and V.I. Agol. 1988. Point mutations modify the response of poliovirus RNA to a translation initiation factor: A comparison of neurovirulent and attenuation strains. *Virology* **166:** 394–404.
Svitkin, Y.V., K. Meerovitch, H.S. Lee, J.N. Dholakia, D.J. Kenan, V.I. Agol, and N. Sonenberg. 1994. Internal translation initiation on poliovirus RNA: Further characterization of La function in poliovirus translated in vitro. *J. Virol.* **68:** 1544–1550.
Toyoda, H., M.J.H. Nicklin, M.G. Murray, C.W. Anderson, J.J. Dunn, F.W. Studier, and E. Wimmer. 1986. A second virus-encoded protinase involved in proteolytic processing of poliovirus polyprotein. *Cell* **45:** 761–770.

Trono, D., R. Andino, and D. Baltimore. 1988. An RNA sequence of hundreds of nucleotides at the 5′end of poliovirus RNA is involved in allowing viral protein synthesis. *J. Virol.* **22:** 2291–2299.

Wimmer, E., C. Hellen, and X. Cao. 1993. Genetics of poliovirus. *Annu. Rev. Genet.* **27:** 353–436.

Wyckoff, E. 1993. Inhibition of host cell protein synthesis in poliovirus-infected cells. *Semin. Virol.* **4:** 209–215.

Wyckoff, E., J.W.B. Hershey, and E. Ehrenfeld. 1990. Eukaryotic initiation factor 3 is required for poliovirus 2A protease-induced cleavage of the p220 component of eukaryotic initiation factor 4F. *Proc. Natl. Acad. Sci.* **87:** 9529–9533.

Wyckoff, E.E., R.E. Lloyd, and E. Ehrenfeld. 1992. The relationship of eukaryotic initiation factor 3 to the poliovirus-induced p220 cleavage activity. *J. Virol.* **66:** 2943–2951.

Xiao, Q., T.V. Sharp, I.W. Jeffrey, M.C. James, G.J. Pruijin, W.J. van Venrooij, and M.J. Clemens. 1994. The La antigen inhibits the activation of the interferon-inducible protein kinase PKR by sequestering and unwinding double-stranded RNA. *Nucleic Acids Res.* **22:** 2512–2518.

Yan, R., W. Rychlik, D. Etchison, and R. Rhoads. 1992. Amino acid sequence of the human protein synthesis initiation factor eIF-4γ. *J. Biol. Chem.* **267:** 23226–23231.

Yu, S.F. and R.E. Lloyd. 1992. Characterization of the roles of conserved cysteine and histidine residues in poliovirus 2A proteases. *Virology* **186:** 725–735.

Yu, S.F., P. Benton, M. Bovee, J. Sessions, and R.E. Lloyd. 1995. Defective RNA replication by poliovirus mutants deficient in 2A protease cleavage activity. *J. Virol.* **69:** 247–252.

20

Adenovirus and Vaccinia Virus Translational Control

Robert J. Schneider
Department of Biochemistry and Kaplan Cancer Center
New York University School of Medicine
New York, New York 10016

Adenovirus (Ad) and vaccinia virus (VV) provoke a similar set of cellular antiviral responses during infection that are designed to block viral translation. Both viruses effectively prevent these antiviral responses, using strategies that are mechanistically quite distinct. Ad and VV also dominate host-cell protein synthesis, implementing sophisticated approaches that assure the exclusive translation of viral messenger RNAs while simultaneously suppressing those of the cell. Consequently, virus–host cell interactions at the level of translation can be especially complex, particularly if the virus employs different mechanisms to dominate protein synthesis in different types of cells. In the case of Ad and VV, certain of these mechanisms are better understood than others, and it will be evident from this review that only an incomplete understanding currently exists.

This chapter first summarizes the evidence suggesting that translational regulation of viral gene expression occurs during infection by both viruses, including the early observations that Ad and VV block the host-cell antiviral response that takes place at the level of protein synthesis. The current understanding of translational control in Ad- and VV-infected cells is then discussed, with a particular emphasis on establishing likely mechanisms for translational dominance of the host cell by each virus. (For a more detailed discussion of specific features of infection by Ad and VV and of host cell–virus interactions, see Bablanian 1984; Schneider and Shenk 1987; Moss 1990; Traktman 1990b; Mathews and Shenk 1991; Schneider and Zhang 1993; Mathews, this volume.)

INFECTION AND GENE EXPRESSION BY ADENOVIRUS AND VACCINIA VIRUS

Overview

Ad and the poxviruses, of which VV is a member, are large viruses that infect a variety of tissues in humans and animals. Both contain double-stranded linear DNA genomes. The Ad genome is typically 36 kb, whereas genomes in poxviruses are larger, averaging 186 kb for VV (for review, see Moss 1990). Like most DNA viruses, Ad and VV gene expression is organized into early and late phases, corresponding to expression of genes prior to, or commensurate with, viral DNA replication, respectively. Gene products synthesized during the early phase of infection are associated with functions required for viral DNA replication and for activation of late viral genes. Late gene products generally encode structural and nonstructural polypeptides that are required in large amounts for assembly of viral capsids (for review, see Moss 1990; Schneider and Zhang 1993).

Adenovirus Infection

Early viral gene expression takes place in the nucleus using cellular RNA polymerases and can be detected within an hour of infection (Ginsberg 1984). Early Ad transcription initiates from six different transcription units distributed throughout the genome and located on both DNA strands. Viral early gene expression gives rise to only a small fraction of the total pool of mRNAs and proteins found in the cell, and there is no evidence for the dramatic inhibition of host-cell macromolecular processes during this period.

Late Ad gene expression commences with viral DNA replication, which takes place in the nucleus, typically between 10 and 16 hours after infection and utilizes several virus-encoded replication factors, including a DNA polymerase (for review, see Ginsberg 1984). Most, but not all, late Ad transcription is initiated from a single major late promoter (MLP), which synthesizes one long transcript corresponding to 80% of the Ad genome. Five families of late Ad transcripts (L1–5) are derived from the MLP primary late transcript by differential splicing and polyadenylation. Abundant amounts of mRNAs that encode pIX and pIVa2 proteins are also transcribed during late phase, but from promoters independent of the MLP. Because many of the early transcription units are in opposition to the late transcription unit, double-stranded RNA (dsRNA) is apparently formed after Ad enters late phase by the pairing of early and late mRNAs (Maran and Mathews 1988).

All mRNAs derived from the MLP possess a common 5′-noncoding region of 200 nucleotides in length called the tripartite leader, produced by the splicing of three small exons (Berget et al. 1977). As described below, the tripartite leader is essential for the exclusive translation of late Ad mRNAs during late phase, and it has a central role in the mechanism of shutoff of cell translation. The pIX and IVa2 mRNAs do not contain the tripartite leader 5′-noncoding region common to MLP mRNAs. Ad also encodes two small RNA polymerase III transcribed products referred to as virion-associated or VA RNAs I and II (for review, see Mathews and Shenk 1991). The VA RNAs (see below) are a vital part of a successful strategy evolved by Ad to block the intracellular antiviral response. The VA RNAs might also participate in selective translation of late Ad mRNAs. The exclusive translation of late viral mRNAs and shutoff of cell protein synthesis occur only after Ad has entered the late phase of infection. The infectious cycle lasts from 30 to 60 hours and is associated with dramatic cytopathic effects (CPE), including loss of cellular structural integrity and release of viral particles from nuclei during cell lysis (for a detailed review, see Ginsberg 1984). The late phase of replication is therefore marked by inhibition of host-cell protein synthesis and production of large amounts of late viral polypeptides.

Vaccinia Virus Infection

Viral particles undergo a two-step process of uncoating in the cytoplasm, for which several early viral gene functions are required (Moss 1990; Traktman 1990a). A number of enzymes are contained in partially uncoated VV cores, including those sufficient to carry out transcription. Early genes are transcribed within the intact enveloped cores, and then mRNAs are transported to the cytoplasm where they are translated (Kates and McAuslan 1967). The entire life cycle of VV takes place in the cytoplasm, unlike that of Ad. Products of early gene expression include proteins required to complete the uncoating process, which occurs several hours later; enzymes involved in viral DNA synthesis; and transcription factors that activate late gene expression. Early transcripts are encoded from a variety of sites, mapping approximately to 25–50% of the viral genome (Boone and Moss 1978). The proportion of total, cytoplasmic mRNAs corresponding to early transcripts is considerably higher in VV-infected cells (~15–25%) (Boone and Moss 1978) than the very small fraction present during early Ad infection (Ginsberg 1984). With the onset of viral DNA replication, several intermediate mRNAs are synthesized that include activators of the late gene group, after which late mRNAs are produced (Wright and Moss 1989; Keck et al. 1990). In this chapter, the intermediate stage of infection and this class of mRNAs are included in the late class since they are structurally related (Baldick and Moss 1993).

The kinetics of VV infection are greatly influenced by the multiplicity of infection, reflected by the wide variation observed for early and late phases. The early phase commences immediately following infection and typically lasts from 2 to 4 hours (Moss 1968). The late phase typically begins between 2 and 6 hours and is marked by uncoating of viral cores, replication of viral genomic DNA, and activation of late gene expression. VV replication generally ceases by 10 hours postinfection (for review, see Traktman 1990b). The late VV mRNAs, like those of Ad, possess a common 5′-noncoding region. The common 5′-noncoding region found in late VV mRNAs, however, consists of a 5′poly(A) tract adjacent to the initiating AUG, ranging in size from 8 to 100 nucleotides (Bertholet et al. 1987; Patel and Pickup 1987; Schwer et al. 1987; Ahn and Moss 1989). All VV mRNAs, like those of Ad, are also 5′-capped (Boone and Moss 1977; Thimmappaya et al. 1982). The capped 5′A-tract arises by treadmilling of the viral RNA polymerase during initiation at conserved sequences in the noncoding DNA strand (de Magistris and Stunnenberg 1988; Davison and Moss 1989). A complex of VV enzymes caps, methylates, and 3′ polyadenylates its mRNAs. Several early mRNAs also possess capped 5′A-tracts, but most do not and instead contain very short 5′-noncoding regions (Moss 1990). Since synthesis of late VV mRNAs encompass approximately 80% of the viral genome, and suppression of early gene expression is not complete (Cooper and Moss 1979; Moss 1990; McDonald et al. 1992), viral dsRNA is apparently generated during VV infection (Boone et al. 1979). Finally, assembly of viral particles, biosynthesis, and packaging of viral genomic DNAs take place in circumscribed cytoplasmic "factories" prior to transport to the cell periphery where they are enveloped.

ADENOVIRUS AND VACCINIA VIRUS ANTAGONIZE CELLULAR AND INTERFERON-INDUCED ANTIVIRAL RESPONSES THAT ACT AT THE TRANSLATIONAL LEVEL

Cellular and Interferon α/β Antiviral Responses at the Translational Level

Infection of animal cells with many different viruses induces intracellular antiviral activities that prevent virus growth at the level of protein synthesis. In addition, treatment of animal cells with interferons impairs infection of many, but not all, viruses by inducing related if not identical responses (for review, see Stewart 1979; Samuel 1991). Type I interferons consist of α (leukocyte) and β (fibroblast) types, which are induced by viral infection. Type II interferon refers to γ-interferon, which is induced by T cells in response to mitogens and antigens. In addition to their effects on cell growth and immune responsiveness, interferons can also induce antiviral states in animal cells at a variety of different levels. Some of these include inhibition of viral adsorption, uncoating, replication, transcription, and translation (Samuel 1991). The best-studied antiviral activities mediated by interferons are manifested at the translational level and coincide with two well-established intracellular antiviral responses.

One antiviral translational response induced by interferon comprises the 2′-5′-oligoadenylate (2–5A) synthetase/nuclease system (for review, see Pestka et al. 1987). This group of interferon-inducible enzymes, which are activated by dsRNA, synthesizes a class of related oligonucleotides with the structure ppp(A2′p5′)nA, which in turn activates endoribonuclease (RNase) L. Once activated, the RNase then degrades both cellular and viral RNAs. Careful examination of interferon-treated cells has shown that the 2–5A synthetase system is an effective antiviral response only to picornavirus infection (Chebath et al. 1987; Samuel 1988; Coccia et al. 1990). A second interferon-inducible antiviral system consists of the 68-kD protein kinase known variously as DAI, dsI, eIF2 kinase, and PKR, herein called PKR. Like 2–5A synthetase, PKR is activated by dsRNA. It phosphorylates the α-subunit of translation factor eIF2, preventing its participation in initiation (see Clemens; Mathews; Trachsel; all this volume). PKR is believed to mediate antiviral activities against many types of viruses (Samuel 1991).

Adenovirus and Vaccinia Virus Prevent Host Cell and Interferon-mediated Antiviral Responses That Act at the Level of Protein Synthesis

A number of early studies showed that interferon treatment of responsive cell lines could impair infection by some viruses and not others at the level of viral protein synthesis (for review, see Stewart 1979). In some cases, coinfection with resistant viruses could also rescue the translation of interferon-sensitive viruses. For instance, type I interferon treatment had little effect on productive infection by VV or Ad (Munoz and Carrasco 1984; Paez and Esteban 1984; Kitajewski et al. 1986a), and VV or Ad could rescue translation of interferon-sensitive viruses (Whitaker-Dowling and Youngner 1983; Munoz and Carrasco 1984; Paez and Esteban 1984). Other studies demonstrated that VV and Ad prevent the activation of the 2–5A synthetase-nuclease pathway during infection (Schneider et al. 1984; Rice et al. 1985). In the case of VV, cells accumulated 2–5A synthetase during infection, but activation of RNase L was slowed (Rice et al. 1985). In Ad-infected cells, the block has not been as well-studied, and a mechanistic understanding is lacking. Described below are studies that elucidated many of the molecular details by which Ad and VV counter the host-cell antiviral response and the antiviral state induced by type I interferons, both of which are implemented at the level of protein synthesis.

Adenovirus Inhibition of Cellular Antiviral Response

A large body of work has elucidated a sophisticated mechanism by which Ad prevents activation of cellular antiviral activities mediated by PKR. Because this work is covered elsewhere in this volume (see Mathews; Clemens) and in several recent reviews (Mathews 1990; Mathews and Shenk 1991; Thimmappaya et al. 1993), only the main features of these studies are described here. All Ads express one or two genes (depending on the serotype) transcribed by RNA polymerase III that give rise to a small, highly structured RNA with no coding capacity, known as VA RNA (Mathews and Shenk 1991). The two VA RNAs (I and II) of group C Ad2 and Ad5 are the best studied. VA RNA_I, the major species, accumulates rapidly in infected cells during late infection. VA RNA_I is found in remarkably large amounts, approaching 10^8 copies per cell, whereas VA RNA_{II} levels are generally 10–40-fold lower (Söderlund et al. 1976). The VA RNAs are found mostly in the cytoplasm, some bound to ribosomes, possibly RNAs, and several proteins including PKR (Mathews 1980; Schneider et al. 1984; Katze et al. 1987; Galabru et al. 1989; Kostura and Mathews 1989; Mellits et al. 1990; Ghadge et al. 1991).

A genetic approach was central to developing an understanding of the role of VA RNA_I in protein synthesis. An Ad mutant deleted of the VA RNA_{II} gene was phenotypically wild type, whereas deletion of the VA RNA_I gene (Ad *dl*331) significantly impaired virus growth (Thimmappaya et al., 1982). Impaired growth of Ad *dl*331 was found to result from global inhibition of viral and cellular protein synthesis (Thimmappaya et al. 1982). The translation defect was traced to an early step in initiation (Schneider et al. 1984; Reichel et al. 1985), corresponding to elevated phosphorylation of the α-subunit of eIF2, and inactivation of the initiation factor through sequestration with eIF2B (Schneider et al. 1985; Reichel et al. 1985; O'Malley et al. 1986a,b). The role of phosphorylation in eIF2 activity, the mechanism for GTP to GDP exchange on eIF2 by eIF2B, and the ability of PKR to inhibit GTP recycling on eIF2 through phosphorylation of eIF2α are reviewed in this volume (see Trachsel; Merrick and Hershey; Clemens; Mathews).

A considerable amount of evidence has established that a primary function of VA RNA_I during natural viral infection is to antagonize PKR activation: (1) Addition of eIF2 or eIF2B can restore translation in extracts prepared from Ad *dl*331-infected cells (Schneider et al. 1984; Reichel et al. 1985). In addition, unsupplemented extracts contain impaired eIF2B activity (Siekierka et al. 1985). (2) The kinetics for activation of PKR and phosphorylation of eIF2α in Ad *dl*331-infected cells parallels the global shutoff of protein synthesis (O'Malley et al. 1986b). (3) Ad generates dsRNA capable of activating PKR as it enters the late phase of infection (O'Malley et al. 1986b; Maran and Mathews 1988). (4) Translation of Ad *dl*331 is normal in cell lines that cannot activate PKR and phosphorylate eIF2α (Schneider et al. 1985; O'Malley et al. 1986b). In addition, overexpression of an eIF2α mutant that cannot be phosphorylated by PKR rescues protein synthesis during Ad *dl*331 infection (Davies et al. 1989). (5) The translation defect demonstrated by VA RNA_I (–) mutant Ad *dl*331 and a double VA $RNA_{I/II}$ (–) mutant is significantly increased in cells pretreated with type I interferon (Kitajewski et al. 1986a,b; Davies et al. 1989). (6) Ad does not inhibit an eIF2α phosphatase activity, excluding other mechanisms for control of PKR activity (Siekierka et al. 1985; Kitajewski et al. 1986a).

Our understanding of the mechanism by which Ad VA RNA_I prevents activation of PKR is still evolving, but many parts are known in considerable detail (Mathews 1990; Mathews and Shenk 1991; Thimmappaya et al. 1993). Activation of PKR by dsRNA can be described by a model of autophosphorylation in which two inactive PKR proteins bind a dsRNA molecule, which triggers mutual phosphorylation, activation, and release from the dsRNA (Kostura and Mathews 1989). VA RNA_I directly binds PKR and blocks interaction with authentic dsRNA (Kostura and Mathews 1989). The dsRNA-binding site of PKR consists of two copies of a dsRNA-binding motif (dsRBM), often found in RNA-binding proteins (Green and Mathews 1992; St Johnston et al. 1992). Secondary structure mapping and coordinated mutation analysis of VA RNA_I support a model in which a short stem-loop region of the RNA, known as the central domain, interacts with the dsRBM sites and blocks activation of PKR by dsRNA (Mellits and Mathews 1988; Furtado et al. 1989; Mellits et al. 1990; Ghadge et al. 1991; Clarke et al. 1994; Clarke and Mathews 1995).

A number of important issues still remain to be resolved regarding the role of the VA RNAs in antagonizing antiviral responses during Ad infection. First, the function of VA RNA_{II} is not known. Although it is true that a double VA $RNA_{I/II}$ (–) Ad mutant is more sensitive to inhibition by interferon, it is also apparent that VA RNA_{II} is a much poorer inhibitor of PKR activation than is VA RNA_{I} (Kitajewski et al. 1986b; Davies et al. 1989; Ma and Mathews 1993). It would therefore make sense to consider whether VA RNA_{II} has been evolutionarily engineered for a different purpose. It is not known how Ad prevents activation of the 2-5A pathway during infection, although an argument has been advanced for a role of VA RNA_{II} (Ma and Mathews 1993). There is also some evidence that the VA RNAs may perform other functions in addition to preventing dsRNA activation of PKR. VA RNA_{I} expression was reported to stabilize reporter mRNAs coexpressed in a transient transfection system (Bhat et al. 1989; Strijker et al. 1989). No similar role for VA RNA_{I} in stabilization of Ad mRNAs was found, however. Finally, the VA RNAs seem to be associated with ribosomes in the infected cell (Schneider et al. 1984) and possibly with late Ad mRNAs (Mathews 1980). The large number of VA RNAs produced during infection approximately matches the number of ribosomes. Given these associations (see below), VA RNA_{I} may also have a role in selective translation of late Ad mRNAs (Mathews 1980).

Vaccinia Virus Inhibition of Cellular Antiviral Response

VV resistance to the effects of type I interferon (Paez and Esteban 1984; Rice and Kerr 1984) and its ability to protect translation of interferon-sensitive viruses during coinfection and to block activation of PKR and 2-5A systems have been well documented (Rice and Roberts 1983; Whitaker-Dowling and Youngner 1983, 1984, 1986; Paez and Esteban 1984; Rice et al. 1984; Akkaraju et al. 1989). A number of observations support the conclusion that the function of two VV genes is to block antiviral translational effects: (1) Two VV genes responsible for resistance to type I interferon were genetically mapped on the genome (Paez and Esteban 1985). Both are early VV gene products that prevent phosphorylation of eIF2α by PKR in vivo and in vitro (Whitaker-Dowling and Youngner 1983, 1985; Paez and Esteban 1984; Rice and Kerr 1984). (2) Biochemical purification demonstrated that one of these proteins, originally called SKIF for specific kinase inhibitory factor, could prevent PKR activation by dsRNA in vitro (Whitaker-Dowling and

Youngner 1984; Akkaraju et al. 1989; Watson et al. 1991). SKIF corresponds to the products of the VV *E3L* gene, which are a full-length 25-kD protein and an amino-truncated 20-kD protein that possess dsRNA-binding activity and a dsRBM (Chang et al. 1992). Additionally, expression of the *E3L* gene during transfection of tissue culture cells prevents dsRNA activation of PKR and enhances translation (Davies et al. 1993). (3) Growth of a VV mutant specifically deleted of the *E3L* gene (vP1080) is severely inhibited by type I interferon (Beattie et al. 1995). Cells infected with the E3L(–) VV strongly activate the 2-5A/RNase L system and degrade rRNA. (4) Specific deletion of a second VV early gene called *K3L* also impairs growth of VV mutant vP872 in interferon-treated cells (Beattie et al. 1991). The *K3L* gene encodes a small polypeptide, a portion of which is homologous to the amino terminus of eIF2α, including the PKR phosphorylation site (Beattie et al. 1991), but lacks a serine residue corresponding to the eIF2 phosphorylation site (Davies et al. 1992). Thus, as originally suggested (Beattie et al. 1991), the K3L product appears to act as a competitive binding-inhibitor of the PKR kinase. In further support of this suggestion, addition of the K3L gene product to in vitro extracts prevents phosphorylation of eIF2α by active PKR (Carroll et al. 1993; Davies et al. 1993) and prevents activation of PKR itself (Carroll et al. 1993).

The functioning of E3L and K3L gene products during VV infection needs to be characterized in molecular detail. It is not known how the *E3L* and *K3L* genes block activation of the 2-5A synthetase/RNase L and PKR systems during VV infection. The *E3L* gene produces a potent scavenger of dsRNA, which presumably removes dsRNA before it can activate the 2-5A and PKR pathways. With respect to *E3L* activity, studies need to determine during natural infection how the p20 and p25 products encoded by the *E3L* gene (Watson et al. 1991; Chang et al. 1992; Yuwen et al. 1993) actually block activation of the PKR and 2-5A systems, whether the p20 product displays activities distinct from the much more abundant p25 protein, and whether scavenging and unmasking of dsRNA represents an authentic in vivo function of these proteins. The p20/p25 E3L proteins were found to be distributed in both the nucleus and cytoplasm of infected cells (Yuwen et al. 1993). The role of a nuclear location is not understood and may indicate other functions for the E3L proteins (Yuwen et al. 1993), particularly since VV transcription occurs in cytoplasmic factories, not in the nucleus. The E3L(–) VV also displays a conditionally defective host-range phenotype in certain cells (Beattie et al. 1995), again indicating that perhaps the E3L product possesses additional functions unrelated to sequestration of dsRNA.

The K3L product, like E3L, is expressed during the early phase of VV infection, but apparently with somewhat different functional kinetics (Jagus and Gray 1994; E. Beattie, in prep.). In a transient transfection assay, the E3L product was reported to be more effective than the K3L protein at preventing PKR activation and loss of translation activity (Davies et al. 1993). In contrast, in vitro studies using equimolar amounts of K3L and E3L products demonstrate similar activities (Jagus and Gray 1994). The levels of dsRNA, K3L, and E3L products are not known in the transient transfection system, probably accounting for the different observations. Nevertheless, the different mechanisms of PKR inactivation and the different kinetics of expression likely indicate that *E3L* and *K3L* genes are expressed by VV for somewhat different purposes during infection, rather than simply representing overlapping strategies for inhibition of dsRNA-mediated antiviral activities. The known functions may represent only a partial understanding of the parts played by E3L and K3L during natural VV infection.

SELECTIVE VIRAL TRANSLATION IN ADENOVIRUS-INFECTED AND VACCINIA-VIRUS-INFECTED CELLS

Overview

Ad and VV both inhibit cellular protein synthesis and preferentially accumulate late polypeptides. Both viruses also induce a series of complex metabolic changes in the host cell during infection, several of which could potentially promote accumulation of late viral proteins without a need for translational control. This section therefore first reviews the historical findings which concluded that Ad and VV exercise control over cellular and viral translation and then explores how they facilitate exclusive translation of viral mRNAs while suppressing those of the cell.

Evidence for Translational Control in Late Adenovirus-infected Cells

As Ad enters the late phase of infection, it induces the inhibition of cell division (for review, see Ginsberg 1984) and greatly impairs the cytoplasmic accumulation of cellular rRNAs and mRNAs (Beltz and Flint 1979). The block in cytoplasmic accumulation of cellular mRNAs results from inhibition of RNA transport from the nucleus to the cytoplasm, rather than impaired transcription of cellular genes (Babich et al. 1983; Castiglia and Flint 1983; Flint et al. 1983). Ad mRNAs overide the block and are preferentially transported to the cytoplasm using the Ad E1B-55K/E4-34K protein complex (Babiss and Ginsberg 1984; Halbert et al. 1984; Pilder et al. 1986), although the mechanism by which this is achieved is not understood. The majority of newly synthesized mRNAs that accumulate in the cytoplasm during late Ad infection are therefore viral in origin (Beltz and Flint 1979; Castiglia et al. 1983). Ad does not induce the degradation of cellular mRNAs, which remain functionally intact (Petterson and Philipson 1975). Cellular mRNAs remain stable, capped, and polyadenylated and can be efficiently translated when extracted and translated in vitro (Anderson et al. 1974; Thimmappaya et al. 1982). Therefore, exclusive translation of late Ad mRNAs does not result simply from a decreased abundance of host mRNAs. This point was clearly established by analysis of individual mRNAs (Babich et al. 1983; Flint et al. 1983, 1984; Yoder et al. 1983). Two studies provide additional evidence that Ad specifically inhibits host-cell translation through a mechanism that is not directly linked to the block in host mRNA transport. First, certain mRNAs, such as β-tubulin, escape the viral block to mRNA transport but are still excluded from translation (Moore et al. 1987). Second, Ad shutoff of cell translation can be prevented with inhibitors, such as 2-aminopurine, without eliminating the block in transport of cellular mRNAs (Huang and Schneider 1990).

During the late phase of Ad infection, viral genomic replication expands the number of templates for late transcription so that the majority of new mRNAs entering the cytoplasmic pool are late transcripts derived from the MLP, as well as pIX and IVa2 promoters (Bhaduri et al. 1972; Lindberg and Sunquist 1974). Ad late mRNAs also progressively dominate the protein synthetic machinery of the cell (Beltz and Flint 1979), displacing cellular mRNAs from polyribosomes and eventually constituting approximately 95% of those undergoing translation (Ginsberg et al. 1967). This raises the possibility that preferential accumulation of late Ad polypeptides merely reflects the cytoplasmic abundance of mRNAs, rather than a specific mechanism for exclusive translation of late mRNAs. Several studies, however, have shown that late Ad mRNAs generally constitute only about 20% of the cytoplasmic pool during late times in infection, at a time when synthesis of late polypeptides is maximal and cell protein synthesis is suppressed (Price and Penman 1972; Lindberg and Sunquist 1974; Tal et al. 1975). In addition, in vitro translation in reticulocyte lysates of mRNAs extracted from uninfected and late Ad-infected cells gives rise to identical levels of cellular polypeptides, excluding the possibility that cellular mRNAs are simply diluted. Thus, although late Ad mRNAs constitute the majority of those translated during late infection, their exclusive translation cannot be attributed to dilution, degradation, or inactivation of cellular mRNAs.

Several Mechanisms Likely Contribute to Suppression of Cellular Translation by Adenovirus

Several mechanisms have been described that may contribute to the exclusive translation of late Ad mRNAs in infected cells. It is not yet known whether these different mechanisms represent an integrated strategy by which Ad dominates cell protein synthesis or whether they act individually. Since Ad naturally infects a wide variety of cell types, it is possible that the virus has evolved multiple strategies to achieve translational control of the cell. A strategy of redundancy would assure translational dominance under a variety of conditions and in many different types of tissues.

Adenovirus Dephosphorylation of eIF4E and the Viral Late 5′- noncoding Region Comprise a System for Translational Discrimination between late Adenovirus and Cellular mRNAs

Mutational studies demonstrated that the tripartite leader found on late Ad mRNAs transcribed from the MLP promotes their translation at late, but not early, times after viral infection (Logan and Shenk 1984; Berkner and Sharp 1985). Several studies showed that the tripartite leader provides mRNAs with the ability to translate in cells that are superinfected by poliovirus (Castrillo and Carrasco 1987; Dolph et al. 1988). The finding that poliovirus inactivates cap-dependent translation by proteolytic degradation of the p220 component of eIF4F, impairing its ability to promote cap-dependent initiation, suggested that the tripartite leader reduces the dependence of late Ad mRNAs on eIF4F activity. Through studies conducted during the past few years, the mechanism by which the tripartite leader promotes translation is beginning to emerge. Viral gene products are not essential for the reduced dependence of the tripartite leader on eIF4F (Dolph et al. 1988, 1990). The leader also does not function as an internal ribosome entry site (IRES), at least as described for picornavirus mRNAs, but actually requires a free 5′end to recruit ribosomes (Dolph et al. 1990). Secondary structure mapping and mutagenic analyses showed that the tripartite leader possesses a relaxed 5′-proximal secondary structure, which decreases the dependence for eIF4F (Zhang et al. 1989; Dolph et al. 1990). Reduced 5′secondary structure is thought to correlate with either a reduced requirement for eIF4F or an increased ability to recruit limiting amounts of eIF4F to the mRNA (Lawson et al. 1986; Fletcher et al. 1990). In vitro translation studies with the tripartite leader confirm that it does not eliminate, but probably reduces, a requirement for eIF4F (Thomas et al. 1992). It has recently been demonstrated that a novel form of ribosome loading referred to as "shunting" (Fütterer et al. 1993; see Jackson; Mathews, both this volume) is used by the tripartite leader to recruit ribosomes (A. Yueh and R.J. Schneider, in prep.). Ribosome shunting, as it occurs in the tripartite leader, loads 40S ribosome subunits onto the 5′end of the mRNA via scanning but then directly translocates the subunits to the downstream AUG. The element that promotes shunting was roughly mapped to the last half of the tripartite leader. The role of ribosome shunting in late Ad translation is not known, but it might facilitate translation when eIF4F is limiting.

The fact that late Ad mRNAs translate with a reduced requirement for eIF4F suggested that inactivation of this factor might provide a way for Ad to discriminate late viral mRNAs from all others in the cell, possibly providing for their selective translation. In late Ad-infected cells, eIF4F is not inactivated by proteolysis of the p220 component, as in poliovirus-infected cells (Dolph et al. 1988). Instead, Ad causes the quantitative dephosphorylation (>95%) of the eIF4E component of eIF4F (Huang and Schneider 1991). eIF4F is thought to be regulated by the phosphorylation state of its eIF4E component (see Sonenberg, this volume). In short, phosphorylation of eIF4E correlates with increased translation activity and dephosphorylated eIF4E with decreased translation. Several lines of evidence indicate that dephosphorylation of eIF4E constitutes an important part of the mechanism for inhibition of host-cell protein synthesis during Ad infection: (1) The 10–20-fold reduction in eIF4E phosphorylation occurs with kinetics that correlate with suppression of cellular protein synthesis in all cell lines tested (Huang and Schneider 1991; Zhang et al. 1994). (2) In cell lines resistant to Ad shutoff of translation, eIF4E dephosphorylation does not occur (Huang and Schneider 1991; Zhang and Schneider 1994; Zhang et al. 1994). (3) The protein kinase inhibitor 2-aminopurine prevents Ad shutoff of cell protein synthesis and its ability to mediate dephosphorylation of eIF4E (Huang and Schneider 1991). The mechanism for prevention of translation shutoff by 2-aminopurine is under study. (4) Inhibition of cell protein synthesis and eIF4E dephosphorylation are linked and coordinately induced by a late Ad gene function (Zhang et al. 1994). This gene function was shown biochemically to block eIF4E phosphorylation, rather than to increase its dephosphorylation, and to correlate genetically with activation of the viral major late promoter, thereby corresponding either to production of viral dsRNA or to synthesis of one or more late viral polypeptides, or to both (Zhang et al. 1994). (5) At late times in infection, Ad suppresses (or reduces) the translation of all mRNAs that lack the tripartite leader, in-

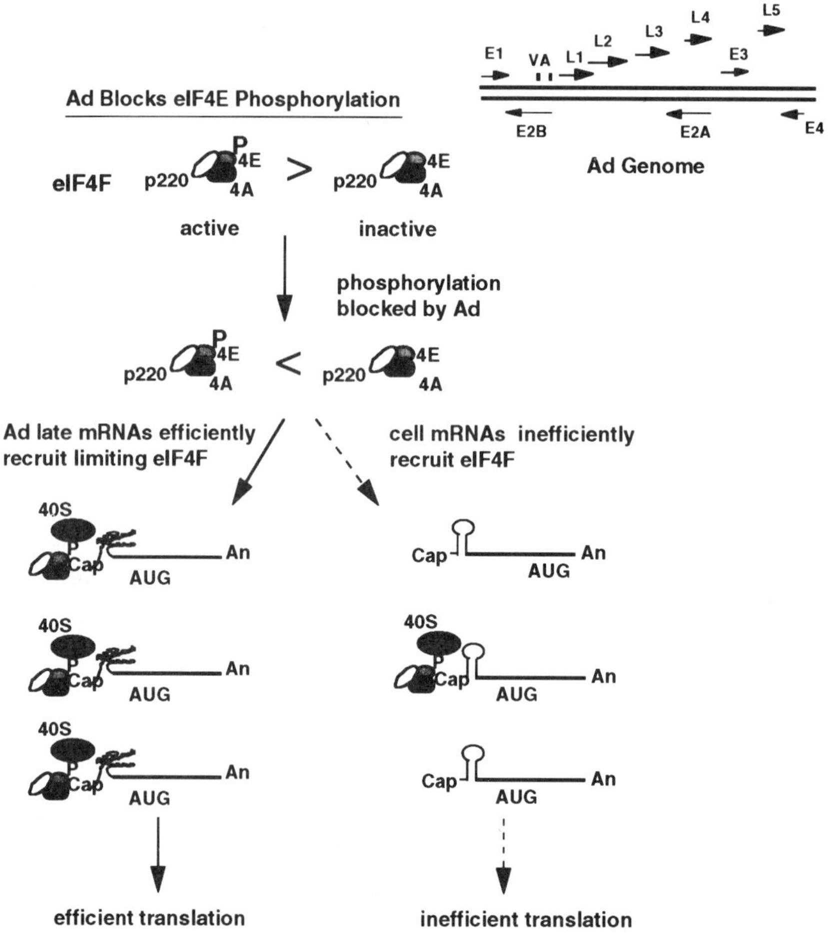

Figure 1 Model for Ad translational control during the late phase of infection. Inhibition of PKR activation by VA RNA_I is shown on the right side of the diagram. Activation of PKR by dsRNA occurs by autophosphorylation, probably by bridging a preformed PKR dimer, or by bringing two PKR proteins together. Activated PKR in turn phosphorylates the α-subunit of eIF2. eIF2B normally exchanges GDP for GTP on eIF2, permitting it to participate in another round of initiation. Phosphorylated eIF2α tightly binds eIF2B, however, sequestering it in an inactive complex unable to exchange GDP for GTP, effectively blocking initiation. VA RNA_I competitively binds PKR, blocking activation by dsRNA. Inhibition of cell protein synthesis and preferential translation of

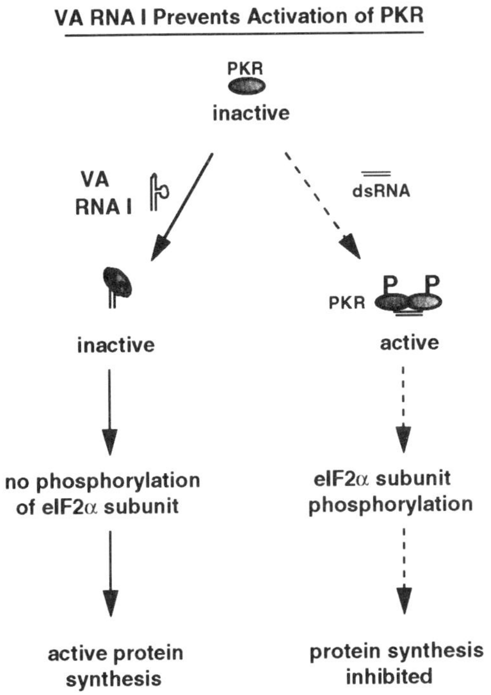

late Ad mRNAs are shown on the left side of the diagram. A late Ad gene function blocks the phosphorylation of the majority of eIF4E (cap-binding protein), a component of eIF4F, so that the pool of inactive eIF4F (dephosphorylated eIF4E) is about 20-fold greater than the active (phosphorylated eIF4E) fraction. Phosphorylation of eIF4E is thought to promote eIF4F cap-dependent initiation, whereas dephosphorylation is thought to severely reduce or alter the activity of the pool of eIF4F. The tripartite leader 5′-noncoding region on late Ad mRNAs transcribed from the MLP permits translation when levels of phosphorylated eIF4E are low, probably by efficiently recruiting the small amounts of active (eIF4E phosphorylated) eIF4F. The translation of other mRNAs (cell, early, and late Ad mRNAs without a tripartite leader) is therefore suppressed, and translation of most late Ad mRNAs is enhanced. Potential roles also exist for the late viral 100K protein and for the slightly elevated levels of active PKR in shutoff of cell translation (see text), which are not as clearly defined and have not been included.

cluding its own early mRNAs, late mRNAs that encode proteins IVa2 and pIX, and those of the cell (Huang and Schneider 1990, 1991; Zhang and Schneider 1994; Zhang et al. 1994). The Ad-mediated block in eIF4E phosphorylation, together with the ability of late viral mRNAs to translate with little dependence on eIF4F, therefore comprises a system that establishes inhibition of host-cell protein synthesis and exclusive translation of late Ad tripartite-leader-bearing mRNAs. A summary diagram of the mechanism described above can be found in Figure 1.

Adenovirus 100K Protein: A Nonstructural Polypeptide Involved in Particle Assembly That Is also Involved in Late Viral Translation

The 100K protein is a late Ad polypeptide encoded by the L4 transcription unit that is synthesized early during the late phase of viral infection in very large amounts (Oosterom-Dragon and Ginsberg 1980). The 100K protein is required for morphogenesis of the Ad hexon capsid protein (Oosterom-Dragon and Ginsberg 1980) and is therefore essential for production of viral particles (Oosterom-Dragon and Ginsberg 1981; Cepko and Sharp 1983). 100K protein is also very strongly associated with mRNAs and can be removed only with high concentrations of salt (Adam and Dreyfuss 1987). Thus, the 100K protein possesses the biochemical characteristics of an mRNA-binding protein, suggesting a second potential role in the regulation of translation (Adam and Dreyfuss 1987). The 100K protein has been directly linked to a role in late Ad translation (Hayes et al. 1990; Riley and Flint 1993). An Ad with a specific temperature-sensitive (ts) mutation in the 100K protein (Ad2*ts*1) near its RNA-binding domain displays a severe defect in late viral protein synthesis at the restrictive temperature. Early and late viral mRNA transcription is unchanged during infection with the 100K mutant at nonpermissive temperature, but translation of viral mRNAs at late times is severely impaired. The translation defect is not restricted to tripartite leader viral mRNAs, but it includes protein IX and IVa2 mRNAs that are normally translated to some extent during the late phase. 100K protein is therefore involved in efficient translation initiation of cellular and viral mRNAs during late phase. The 100K mutant virus was found to efficiently prevent cellular protein synthesis at the restrictive temperature during late infection (Hayes et al. 1990), supporting the contention that inhibition of cell translation and preferential translation of late Ad mRNAs are not linked. 100K protein does not specifically associate with any particular class of mRNA (viral or cellular) (Riley and

Flint 1993), which has made it difficult to provide a mechanistic understanding for the translation function of 100K protein during late infection.

VA RNA_I Might Participate in Selective Translation of Late Adenovirus mRNAs

The function of Ad VA RNA_I in countering the antiviral activation of PKR has already been described. It has been proposed, in addition, that VA RNA_I and PKR might cooperate in the preferential translation of late Ad mRNAs and the shutoff of host-cell protein synthesis (O'Malley et al. 1989). There are several reasons for proposing such a model: (1) Ad does not inhibit host translation in cells deficient in PKR activity (O'Malley et al. 1989; Huang and Schneider 1990). (2) Severalfold activation of PKR and a similar increase in phosphorylation of eIF2α are frequently observed during late Ad infection (O'Malley et al. 1989; Duncan 1990; Huang and Schneider 1990). (3) The level of eIF2α phosphorylation can influence the specificity with which certain mRNAs are translated (Kaufman and Murtha 1987; Kaufman et al. 1989; Huang and Schneider 1990; Dever et al. 1991), and several cellular and viral mRNAs have been shown to translate despite what would normally be considered inhibitory levels of active PKR and phosphorylated eIF2α (Jacobsen et al. 1983; Black et al. 1989; Kaufman et al. 1989; O'Malley et al. 1989; Huang and Schneider 1990; Dever et al. 1991; Choi et al. 1992). (4) Like many translation factors, PKR (Zilberstein et al. 1976) and VA RNA_I (Schneider et al. 1984) are associated with ribosomes. These components are therefore in a location consistent with the ability to exercise specific translational regulation of mRNAs.

Given the compelling group of observations described above, a regulatory scheme has been envisioned in which preferential translation of late Ad mRNAs might take place via segregation of viral mRNAs from cellular mRNAs into functionally or physically distinct compartments (O'Malley et al. 1989; Mathews and Shenk 1991). The model proposes that cellular mRNA translation might be specifically inhibited, whereas viral mRNA translation might be enhanced by activation of PKR and a slight increase in phosphorylated eIF2α levels. Late Ad mRNAs, by virtue of its reported ability to bind VA RNA_I (Mathews 1980), would then block activation of ribosome-associated PKR at the level of the mRNA. However, to circumvent global inhibition of translation that should occur through release of phosphorylated eIF2α (eIF2α[P]) into the free pool, it was further proposed that Ad might mediate compartmentalization of the cellular translational machinery either physically or functionally (O'Malley et al. 1989; Mathews and Shenk 1991). A functionally active compartment would then include VA RNA_I, late viral mRNA, and a pool of active (nonphosphorylated) eIF2 and eIF2B. The functionally inactive compartment would contain cellular mRNAs and phosphorylated eIF2α and would probably be depleted of nonphosphorylated eIF2 and active eIF2B. It is known that Ad does alter the structure of cellular intermediate filaments during infection (for review, see Ginsberg 1984). In addition, some studies suggest that eIF2B and active eIF2 may be segregated from phosphorylated (inactive) eIF2 in cells, possibly through an interaction with 60S ribosomal subunits (Ginsberg 1984; Gross et al. 1985; Thomas et al. 1985; Chakrabarti and Maitra 1992). Studies now need to directly test various aspects of this model.

Evidence for Translational Control in Vaccina-virus-infected Cells

VV infection results in a series of striking biochemical alterations to the host cell. The onset and severity of these changes are linked to the multiplicity of infection, occurring more rapidly and to a greater extent with the use of high levels of infectious virus. Biochemical alterations induced by VV occur at many different levels of gene regulation. VV inhibits host DNA synthesis several hours after infection (Jungwirth and Launer 1968; Vassef et al. 1982), probably through the activity of virus-encoded DNases (Olgaiti et al. 1976). VV inhibits transcription of host genes by RNA polymerases I and II, as well as processing and transport of some RNAs to the cytoplasm (for review, see Moss 1990). Inhibition of host-cell DNA and RNA synthesis are generally evident by about 2–3 hours after infection (for review, see Bablanian 1993); their role in shutoff of cell protein synthesis has not been established. Therefore, these effects likely contribute to, but are not the basis for, inhibition of cell translation. In VV-infected cells, host-cell and early viral mRNAs are destabilized (Becker and Joklik 1964; Salzman et al. 1964; Bablanian 1993), probably by reduction in their cytoplasmic half-lives (Rice and Roberts 1983). The cytoplasmic level of host mRNAs is reduced by no more than half, however, at a time when shutoff of translation is largely complete (Rice and Roberts 1983). This suggests that degradation of cellular mRNAs per se is not the main mechanism for inhibition of host translation, unless it also includes the loss of an unknown protein product involved in promoting cellular mRNA translation. VV infection is also associated with an early and transient increase in cell membrane permeability (Carrasco and Esteban 1982), which given its short-lived effect is unlikely to contribute significantly to shutoff of cell protein synthesis.

There is some evidence that VV establishes a system for translational regulation in infected cells, independent of the block in host mRNA synthesis, transport, or stability. First, VV suppresses translation of its own early mRNAs after entry into late phase (McAuslan 1963; Vassef et al. 1982; Hruby 1985), recently shown to be unrelated to the reduced half-life of early mRNAs (McDonald et al. 1992). During late VV infection, abundant amounts of early viral and cellular mRNAs are in the cytoplasm, and these are well translated in vitro despite poor translation during the late phase (Moss 1990). On the other hand, VV late mRNAs accumulate to very high levels in late infected cells and account for a major fraction of the cytoplasmic pool of messengers (Boone and Moss 1978; Cooper and Moss 1979). The sheer abundance of late VV mRNAs likely contributes to the exclusive synthesis of late VV proteins.

Evidence for translational regulation during late VV infection (other than mRNA abundance) is derived from several different sources. First, suppression of host-cell protein synthesis during VV infection is processive, initiating shortly after infection (within an hour or so), advancing significantly while the virus is still in early phase and culminating during late infection with the marked inhibition of both host-cell and early VV mRNA translation (Moss 1968, 1990; Moss and Salzman 1968; Cooper and Moss 1979; Bablanian 1993). Significant shutoff of host-cell protein synthesis therefore occurs before late mRNAs accumulate to very high levels. Second, early VV mRNA translation is rapidly suppressed during the transition from early to late phase (see, e.g., Christen et al. 1992). Although early VV mRNAs are unstable, enhancing the stability of early transcripts does not promote their translation during late phase (McDonald et al. 1992 and references therein). Most early VV mRNAs, but not all, therefore appear to undergo rapid translational suppression as VV enters late phase, which occurs in addition to the shutoff of host-cell mRNA translation.

Translational Control in Vaccinia-virus-infected Cells

There is not yet a clear understanding or a model that can adequately describe the translational events that occur during VV infection. More extensive biochemical studies on VV translation are essential. For instance, some studies have concluded that viral gene expression is not necessary for inhibition of host-protein synthesis (Moss 1968), and some have implicated either a small VV core protein in shutoff (Ben-Hamida and Beaud 1978; Cooper and Moss 1978) or the viral surface tubule protein in the process (Mbuy et al. 1982). These VV proteins were found to suppress host mRNAs disproportionately when cotranslated with VV mRNAs in vitro. In contrast, other studies indicate that viral gene expression is essential to initiate shutoff of host-protein synthesis, which is not complete until viral DNA replication commences (Bablanian et al. 1981). Some studies have also found no selective translation effect by VV proteins (Pelham et al. 1978; Ben-Hamida et al. 1983). It is clear that a genetic approach to this problem is needed both to test biochemical findings and to explore new models for VV translational control. Consequently, this next section is mostly descriptive, and molecular models are suggested in rough outline only.

Vaccina Virus Does Not Detectably Modify Translation Initiation Factors

Several studies have investigated whether changes in translation initiation factors accompany VV infection. No significant or even minor changes were observed in eIF2α phosphorylation as occurs during Ad infection, nor were there changes in factors eIF4A, eIF4B, eIF2β, or eIF4Fγ (p220) (Duncan 1990). Activation of PKR is therefore efficiently inhibited by VV E3L and K3L gene products, probably excluding a potential role for enhanced eIF2α phosphorylation in selective translation of VV mRNAs. Two studies also demonstrated that VV does not cause underphosphorylation of eIF4E (Gierman et al. 1992; Schnierle and Moss 1992), in contrast to events described in late Ad- and influenza-virus-infected cells (Huang and Schneider 1991; Feigenblum and Schneider 1993; see Katze, this volume), ruling out control of eIF4F by virus-mediated covalent modifications. VV infection was reported to be associated with phosphorylation of several 40S ribosomal proteins but not S6 (Buendia et al. 1987). It has not been addressed whether the changes described in ribosomal protein phosphorylation are linked to detectable effects on cellular or VV mRNA translation.

Potential Role of 5′ Poly(A) in Late Vaccina Virus Translation

Most late and several early VV mRNAs contain capped 5′oligo(A) sequences generally 8–30 residues in length (Moss 1990). A role for the 5′ poly(A) sequences has not been firmly established, but it might be expected to influence translation of the mRNA, as discussed for 3′poly(A) tails by Jacobson (this volume). A tract of A residues might be less structured and thereby facilitate translation by reducing the requirement for eIF4F during initiation. However, when eIF4F activity was blocked with cap analogs during in vitro translation, early and late VV mRNAs were only slightly (approximately twofold) more resistant to inhibition than cellular mRNAs (Bablanian et al. 1991). Alternatively, it has been suggested that because late infection produces large amounts of complementary RNAs, some of which could block mRNA translation by duplex formation, 5′poly(A) tracts might reduce pairing and permit normal ribosome binding (Moss 1990). The potential role of 5′poly(A) sequences in VV translation, whether to reduce the requirement for eIF4F or to simply permit 40S ribosome binding, remains unresolved.

Potential Role for Nontranslated Small Polyadenylated RNAs in Vaccina Virus Translational Control

Early studies that examined the role of VV genes in translation shutoff used moderate concentrations of actinomycin D or UV-irradiation to prevent VV gene expression (Rosemond-Hornbeak and Moss 1975; Gershowitz and Moss 1979). Small poly(A)-containing RNA molecules were abundantly produced in these cases, and it was speculated that they might have a role in translation shutoff under these conditions. High doses of actinomycin D or UV-irradiation were later found to prevent formation of the aberrant RNAs and the shutoff of cell protein synthesis (Bablanian et al. 1978, 1981). Their addition to in vitro translation extracts at low levels also prevented translation of cellular but not VV mRNAs (Bablanian and Banerjee 1986; Bablanian et al. 1986, 1987). More recently, these small polyadenylated RNAs (called POLADS) were found in VV cores and shown to require the 3′poly(A) sequence to preferentially inhibit cellular mRNA translation in vitro (Su and Bablanian 1990). It was also reported that POLADS preferentially inhibit cellular mRNAs, followed by early and then late VV mRNAs as POLAD concentrations are increased in vitro (Bablanian et al. 1991). It is not yet known, however, whether POLADS actually accumulate to effective inhibitory levels during natural VV infection, although one study suggests that they might (Bablanian et al. 1993).

The mechanism for possible POLAD action is unclear, but it may include sequestration of poly(A)-binding protein (PABP). Suppression of cellular mRNA translation in vitro by POLADS is prevented by addition of PABP and eIF4A (Bablanian et al. 1991), raising the suggestion that they function as free poly(A) and sequester PABP (for review, see Bablanian 1993). A variety of studies implicate poly(A) in facilitating the initiation of translation (see Jacobson, this volume). It was proposed that VV mRNAs, particularly late messages, may be less sensitive to inhibition due to 5′poly(A) residues, although a mechanistic explanation is not in hand. Models now need to be developed and tested to resolve how PABP, eIF4A, and 5′poly(A) sequences might participate in selective translation of VV mRNAs.

CONCLUDING REMARKS

It is evident for both Ad and VV that future research must describe in much more complete terms the process by which translation of cellular mRNAs is inhibited and viral mRNAs is promoted. The identities of viral gene products that induce or mediate shutoff of host-cell translation are not known for either Ad or VV, and studies for both viruses need to address these issues genetically, in addition to current biochemical approaches.

For Ad, studies need to clarify the roles of 100K protein and VA RNA_I/PKR in late viral translation. Studies also need to determine whether the different possible mechanisms form an integrated viral strategy for dominance of cell translation. For VV, studies need to determine more clearly the function of 5′poly(A) sequences in late mRNA translation, as well as the roles played by late mRNA abundance and by POLADS. It is also clear that additional regulatory paths are emerging that very likely impact on both Ad and VV translational control. The discovery of eIF4E sequestering protein termed PHAS-I or 4E-BP1 (Pause et al. 1994), and its regulation by MAP kinase signaling cascades (Lin et al. 1994), raises the possibility that it may have a role in both Ad and VV infection. In VV-infected cells, studies have shown a decrease in cap-dependent eIF4F activity (R. Jagus, pers. comm.), suggesting the possibility of PHAS-I involvement. Ongoing studies in our laboratory are exploring the role of PHAS-I protein in the control of host and Ad translation during infection.

Finally, the shutoff of host-protein synthesis during viral infection is likely involved in functions apart from maximizing the translation of viral mRNAs. In Ad-infected cells, inhibition of cell protein synthesis was found to enhance late viral mRNA translation by five- to tenfold but is also essential for cell lysis and release of viral particles (Zhang and Schneider 1994). Ad was shown to degrade cytokeratins K7 and K18 using the late viral L3-23-kD protease (Chen et al. 1993). Shutoff of host protein synthesis was found to promote destruction of the cytokeratin intermediate filament network, which provides structural integrity of the cell, by preventing synthesis-repair of the degraded filaments (Zhang and Schneider 1994). Ad shutoff of cell protein synthesis, coupled to degradation of the cytokeratin network, therefore causes a severe loss of cell structural integrity, promoting lysis and release of Ad particles. Thus, the exclusive translation of late Ad mRNAs promotes the efficient production of viral particles and is also partly responsible for viral pathogenesis.

ACKNOWLEDGMENTS

I thank B. Moss, R. Babalanian, R. Jagus, E. Beattie, and J. Flint for taking the time to share unpublished work and insightful ideas. The author's work described in this review was supported by a grant from the National Institutes of Health (CA-42357).

REFERENCES

Adam, S.A. and G. Dreyfuss. 1987. Adenovirus proteins associated with mRNA and hnRNA in infected HeLa cells. *J. Virol.* **61:** 3276–3283.

Ahn, B.-Y. and B. Moss. 1989. Capped poly(A) leaders of variable lengths at the 5′ ends of vaccinia virus late mRNAs. *J. Virol.* **63:** 226–232.

Akkaraju, G.R., P. Whitaker-Dowling, J.S. Youngner, and R. Jagus. 1989. Vaccinia specific kinase inhibitory factor prevents translational inhibition by double-stranded RNA in rabbit reticulocyte lysate. *J. Biol. Chem.* **264:** 10321–10325.

Anderson, C.W., J.B. Lewis, J.F. Atkins, and R.F. Gesteland. 1974. Cell-free synthesis of adenovirus 2 proteins programmed by fractionated mRNA. *Proc. Natl. Acad. Sci.* **71:** 2756–2760.

Babich, A., C.T. Feldman, J.R. Nevins, J.E. Darnell, and C. Weinberger. 1983. Effect of adenovirus on metabolism of specific host mRNAs: Transport control and specific translational discrimination. *Mol. Cell. Biol.* **3:** 1212–1221.

Babiss, L.E. and H.S. Ginsberg. 1984. Adenovirus type 5 early region 1b gene product is required for efficient shutoff of host protein synthesis. *J. Virol.* **50:** 202–212.

Bablanian, R. 1984. Poxvirus cytopathogenicity: Effects on cellular macromolecular synthesis. *Compr. Virol.* **19:** 391–429.

———. 1993. Translational regulation by vaccinia virus. In *Translational regulation of gene expression* (ed. J. Ilan), vol. 2, pp. 187–202. Plenum Press, New York.

Bablanian, R. and A.K. Banerjee. 1986. Poly(riboadenylic acid) preferentially inhibits in vitro translation of cellular mRNA compared with vaccinia virus mRNA: Possible role in vaccinia virus cytopathology. *Proc. Natl. Acad. Sci.* **83:** 1290–1294.

Bablanian, R., S. Scribani, and M. Esteban. 1993. Amplification of polyadenylated non-translated small RNA sequences (POLADS) during superinfection correlates with the inhibition of viral and cellular protein synthesis. *Cell. Mol. Biol. Res.* **39:** 243–255.

Bablanian, R., G. Coppola, P. Masters, and A.K. Banerjee. 1986. Characterization of vaccinia virus transcripts involved in selective inhibition of host protein synthesis. *Virology* **148:** 375–380.

Bablanian, R., G. Coppola, S. Scribani, and M. Esteban. 1981. Inhibition of protein synthesis by vaccina virus. IV. The role of low molecular weight virus RNA in the inhibition of protein synthesis. *Virology* **112:** 13–24.

Bablanian, R., M. Esteban, B. Baxt, and J.A. Sonnabend. 1978. Studies on the mechanism of vaccinia virus cytopathic effects. Inhibition of protein synthesis in infected cells is associated with virus induced RNA synthesis. *J. Gen. Virol.* **39:** 391–402.

Bablanian, R., S.K. Goswami, M. Esteban, and A.K. Banerjee. 1987. Selective inhibition of protein synthesis by authentic and vaccinia virus-core synthesized poly-(riboadenylic acids). *Virology* **161:** 366–373.

Bablanian, R., S.K. Goswami, M. Esteban, A.K. Banerjee, and W. C. Merrick. 1991. Mechanism of selective translation of vaccinia virus mRNAs: Differential role of poly(A) and initiation factors in the translation of viral and cellular mRNAs. *J. Virol.* **65:** 4449–4460.

Baldick, C.J. and B. Moss. 1993. Characterization and temporal regulation of mRNAs encoded by vaccinia virus intermediate stage genes. *J. Virol.* **67:** 3515–3527.

Beattie, E., J. Tartaglia, and E. Paoletti. 1991. Vaccinia virus-encoded eIF2α homolog abrogates the antiviral effect of interferon. *Virology* **183:** 419–422.

Beattie, E., K.L. Denzler, J. Tartaglia, M.E. Perkus, E. Paoletti, and B.L. Jacobs. 1995. Reversal of the interferon-sensitive phenotype of a vaccinia virus lacking E3L by expression of the reovirus S4 gene. *J. Virol.* **69:** 499–505.

Becker, Y. and W.K. Joklik. 1964. Messenger RNA in cells infected with vaccinia virus. *Proc. Natl. Acad. Sci.* **51:** 577–585.

Beltz, G.A. and S.J. Flint. 1979. Inhibition of HeLa cell protein synthesis during adenovirus infection. *J. Mol. Biol.* **131:** 353–373.

Ben-Hamida, F. and G. Beaud. 1978. In vitro inhibition of protein synthesis by purified cores from vaccinia virus. *Proc. Natl. Acad. Sci.* **75:** 175–179.

Ben-Hamida, F., A. Person, and G. Beaud. 1983. Solubilization of a protein synthesis inhibitor from vaccinia virions. *J. Virol.* **45:** 452–455.

Berget, S.M., C. Moore, and P. Sharp. 1977. Spliced segments at the 5′ terminus of Ad2 late mRNA. *Proc. Natl. Acad. Sci.* **74:** 3171–3175.

Berkner, K.E. and P.A. Sharp. 1985. Effect of tripartite leader on synthesis of a non-viral protein in an adenovirus 5′ recombinant. *Nucleic Acids Res.* **13:** 841–857.

Bertholet, C., E. Van Meir, B. ten Heggeler-Bordier, and R. Wittek. 1987. Vaccinia virus produces late mRNAs by discontinuous synthesis. *Cell* **50:** 153–162.

Bhaduri, S., H.J. Raskas, and M. Green. 1972. Procedure for the preparation of milligram quantities of adenovirus messenger ribonucleic acid. *J. Virol.* **10:** 1126–1129.

Bhat, R.A., M.R. Furtado, and B. Thimmappaya. 1989. Efficient expression of small polymerase III genes from a novel simian virus 40 vector and their effect on viral gene expression. *Nucleic Acids Res.* **17:** 1159–1176.

Black, T.L., B. Safer, A. Hovanessian, and H.G. Katze. 1989. The cellular 68,000 Mr protein kinase is highly phsophorylated and activated yet significantly degraded during poliovirus infection: Implications for translational regulation. *J. Virol.* **63:** 2244–2251.

Boone, R.F. and B. Moss. 1977. Methylated 5′ terminal sequences of vaccinia virus mRNA species made in vivo at early and late times after infection. *Virology* **79:** 67–80.

———. 1978. Sequence complexity and relative abundance of vaccinia virus mRNAs synthesized in vivo and in vitro. *J. Virol.* **26:** 554–569.

Boone, R.F., R.P. Parr, and B. Moss. 1979. Intermolecular duplexes formed from polyadenylated vaccinia virus RNA. *J. Virol.* **30:** 365–374.

Buendia, B., A. Person-Fernandez, G. Beaud, and J.-J. Madjar. 1987. Ribosomal protein phosphorylation in vivo and in vitro by vaccinia virus. *Eur. J. Biochem.* **162:** 95–103.

Carrasco, L. and M. Esteban. 1982. Modification of membrane permeability in vaccinia virus infected cells. *Virology* **117:** 62–69.

Carroll, K., O. Elroy-Stein, B. Moss, and R. Jagus. 1993. Recombinant vaccinia virus K3L gene product prevents activation of double-stranded RNA-dependent, initiation factor 2 alpha-specific protein kinase. *J. Biol. Chem.* **268:** 12837–12842.

Castiglia, C.L. and S.J. Flint. 1983. Effects of adenovirus infection on RNA synthesis and maturation in HeLa cells. *Mol. Cell. Biol.* **3:** 662–671.

Castrillo, J.L. and L. Carrasco. 1987. Adenovirus late protein synthesis is resistant to the inhibition of translation induced by poliovirus. *J. Biol. Chem.* **262:** 7328–7334.

Cepko, C.L. and P.A. Sharp. 1983. Assembly of major adenovirus capsid protein is mediated by a nonvirion protein. *Virology* **129:** 137–154.

Chakrabarti, A. and U. Maitra. 1992. Release and recycling of eukaryotic initiation factor 2 in the formation of an 80S ribosomal polypeptide chain initiation complex. *J. Biol. Chem.* **267:** 12964–12972.

Chang, H.-W., J.C. Watson, and B.L. Jacobs. 1992. The *E3L* gene of vaccinia virus encodes an inhibitor of the interferon-induced, double-stranded RNA-dependent protein kinase. *Proc. Natl. Acad. Sci.* **89:** 4825–4829.

Chebath, J., P. Benech, M. Revel, and M. Vigneron. 1987. Constitutive expression of (2′-5′) oligo A synthetase confers resistance to picornavirus infection. *Nature* **330:** 587–588.

Chen, P.H., D.A. Ornelles, and T. Shenk. 1993. The adenovirus L3 23-kilodalton proteinase cleaves the amino-terminal head domain from cytokeratin 18 and disrupts the cytokeratin network of HeLa cells. *J. Virol.* **67:** 3507–3514.

Choi, S.Y., B.J. Scherer, J. Schnier, M.V. Davies, R.J. Kaufman, and J.W.B. Hershey. 1992. Stimulation of protein synthesis in COS cells transfected with variants of the alpha-subunit of initiation factor eIF2. *J. Biol. Chem.* **267:** 286–293.

Christen, L., M.A. Higman, and E.G. Niles. 1992. Phenotypic characterization of three temperature-sensitive mutations in the vaccinia virus early gene transcription initiation factor. *J. Gen. Virol.* **73:** 3155–3167.

Clarke, P.A. and M.B. Mathews. 1995. Interactions between the double-stranded RNA binding motif and RNA: Definition of the binding site for the interferon-induced protein kinase (PKR) on adenovirus VA RNA. *RNA* (in press).

Clarke, P.A., T. Pe'ery, Y. Ma, and M.B. Mathews. 1994. Structural features of adenovirus 2 virus-associated RNA required for binding to the protein kinase DAI. *Nucleic Acids Res.* **22:** 4364–4374.

Coccia, E.M., G. Romeo, A. Nissim, G. Marziali, R. Albertini, E. Affabris, A. Battistini, G. Fiorucci, R. Orsatti, G.B. Rossi, and J. Chebath. 1990. A full-length murine 2-5Å synthetase cDNA transfected in NIH-3T3 cells impairs EMCV but not VSV replication. *Virology* **179:** 228–233.

Cooper, J.A. and B. Moss. 1978. Transcription of vaccinia virus mRNA coupled to translation in vitro. *Virology* **88:** 149–165.

———. 1979. In vitro translation of immediate early, early and late classes of RNA from vaccinia virus-infected cells. *Virology* **96:** 368–380.

Davies, M.V., H.W. Chang, B.L. Jacobs, and R.J. Kaufman. 1993. The E3L and K3L vaccina virus gene products stimulate translation through inhibition of the double-stranded RNA-dependent protein kinase by different mechanisms. *J. Virol.* **67:** 1688–1692.

Davies, M.V., O. Elroy-Stein, R. Jagus, B. Moss, and R.J. Kaufman. 1992. The vaccinia virus K3L gene product potentiates translation by inhibiting double-stranded RNA- activated protein kinase and phosphorylation of the alpha subunit of eukaryotic initiation factor 2. *J. Virol.* **66:** 1943–1950.

Davies, M.V., M. Furtado, J.W.B. Hershey, B. Thimmappaya, and R.J. Kaufman 1989. Complementation of adenovirus-associated RNA I gene deletion by expression of a mutant eukaryotic translation initiation factor. *Proc. Natl. Acad. Sci.* **86:** 9163–9167.

Davison, A.J. and B. Moss. 1989. The structure of vaccinia virus late promoters. *J. Mol. Biol.* **210:** 771–784.

de Magistris, L. and H.G. Stunnenberg. 1988. Cis-acting sequences affecting the length of the poly(A) head of vaccinia virus late transcripts. *Nucleic Acids Res.* **16:** 3141–3156.

Dever, T.E., L. Feng, R.C. Wek, A.M. Cigan, T.F. Donahue, and A.G. Hinnebusch. 1991. Phosphorylation of initiation factor 2-alpha by protein kinase GCN2 mediates gene-specific translational control of GCN4 in yeast. *Cell* **68:** 585–596.

Dolph, P.J., J. Huang, and R.J. Schneider. 1990. Translation by the adenovirus tripartite leader: Elements which determine independence from cap-binding protein complex. *J. Virol.* **64:** 2669–2677.

Dolph, P.J., V. Racaniello, A. Villamarin, F. Palladino, and R.J. Schneider. 1988. The adenovirus tripartite leader eliminates the requirement for cap binding protein during translation initiation. *J. Virol.* **62:** 2059–2066.

Duncan, R.F. 1990. Protein synthesis initation factor modifications during viral infections: Implications for translational control. *Electrophoresis* **11:** 219–227.

Feigenblum, D. and R.J. Schneider. 1993. Modification of eukaryotic initiation factor 4F during infection by influenza virus. *J. Virol.* **67:** 3027–3035.

Fletcher, L., S.D. Corbin, K.G. Browning, and J.M. Ravel. 1990. The absence of a m7G cap on beta-globin mRNA and alfalfa mosaic virus 4 increases the amounts of initiation factor 4F required for translation. *J. Biol. Chem.* **265:** 19582–19587.

Flint, S.J., G.A. Beltz, and D.I.H. Linzer. 1983. Synthesis and processing of simian virus 40 specific RNA in adenovirus infected, simian virus 40 transformed human cells. *J. Mol. Biol.* **167:** 335–359.

Flint, S.J., M.A. Plumb, U.C. Yang, S. Stein, and J.L. Stein. 1984. Effect of adenovirus infection on expression of human histone genes. *Mol. Cell. Biol.* **4:** 1363–1371.

Furtado, M.R., S. Subramanian, R.A. Bhat, D.M. Wowlkes, B. Safer, and B. Thimmappaya. 1989. Functional dissection of adenovirus VA1 RNA. *J. Virol.* **63:** 3423–3434.

Fütterer, J., Z. Kiss-Laszlo, and T. Hohn. 1993. Nonlinear ribosome migration on cauliflower mosaic virus 35S RNA. *Cell* **73:** 789–802.

Galabru, J., M.G. Katze, N. Robert, and A.G. Hovanessian. 1989. The binding of double-stranded RNA and adenoviruse VAI RNA to the interferon-induced protein kinase. *Eur. J. Biochem.* **178:** 581–589.

Gershowitz, A. and B. Moss. 1979. Abortive transcription products of vaccinia virus are guanylated, methylated and polyadenylated. *J. Virol.* **31:** 849–853.

Ghadge, G.D., S. Swaminathan, M.G. Katze, and B. Thimmappaya. 1991. Binding of the adenovirus VAI RNA to the interferon induced 68 -kD protein kinase correlates with function. *Proc. Natl. Acad. Sci.* **88:** 7140–7144.

Gierman, T.M., R.M. Frederickson, N. Sonenberg, and D.J. Pickup. 1992. The eukaryotic translation initiation factor 4E is not modified during the course of vaccinia virus replication. *Virology* **188:** 934–937.

Ginsberg, H.S. 1984. *The adenoviruses.* Plenum Press, New York.

Ginsberg, H.S., L.S. Bello, and A.J. Levine. 1967. The molecular biology of adenoviruses. In *The molecular biology of viruses* (ed. J.S. Cotter and W. Paranchych), pp. 547–572. Academic Press, New York.

Green, S.R. and M.B. Mathews. 1992 Two RNA-binding motifs in the double-stranded RNA-activated protein kinase, DAI. *Genes Dev.* **6:** 2478–2490.

Gross, M., R. Redman, and D.A. Kaplansky. 1985. Evidence that the primary effect of phosphorylation of eukaryotic initiation factor 2 alpha in rabbit reticulocyte lysate is inhibition of the release of eukaryotic initiation factor 2-GDP from 60S ribosome subunits. *J. Biol. Chem.* **260:** 9491–9500.

Halbert, D.N., J.R. Cutt, and T. Shenk. 1984. Adenovirus early region 4 encodes functions required for efficient DNA replication, late gene expression and host cell shutoff. *J. Virol.* **56:** 250–257.

Hayes, B.W., G.C. Telling, M.M. Myat, J.F. Williams, and S.J. Flint. 1990. The adenovirus L4 100 kilodalton protein is necessary for efficient translation of viral late mRNA species. *J. Virol* **64:** 2732–2742.

Hruby, D.E. 1985. Inhibition of vaccinia virus thymidine kinase by the distal products of its own metabolic pathway. *Virus Res.* **2:** 151–156.

Huang, J. and R.J. Schneider. 1990. Adenovirus inhibition of cellular protein synthesis is prevented by the drug 2-aminopurine. *Proc. Natl. Acad. Sci.* **87:** 7115–7119.

———. 1991. Adenovirus inhibition of cellular protein synthesis involves inactivation of cap binding protein. *Cell* **65:** 271–280.

Jacobsen, H., D.A. Epstein, R.M. Friedman, B. Safer, and P.F. Torrenie. 1983. Double-stranded RNA-dependent phosphorylation of protein P1 and eukaryotic initiation factor 2 alpha does not correlate with protein synthesis inhibition in a cell-free system from interferon-treated mouse L cells. *Proc. Natl. Acad. Sci.* **80:** 41–45.

Jagus, R. and M.M. Gray. 1994. Proteins that interact with PKR. *Biochimie* **76:** 779–791.

Jungwirth, C. and J. Launer. 1968. Effect of poxvirus infection on host cell DNA synthesis. *J. Virol.* **2:** 401–408.

Kates, J.R. and B. McAuslan. 1967. Poxvirus DNA-dependent RNA polymerase. *Proc. Natl. Acad. Sci.* **58:** 134–141.

Katze, M.G., D. DeCorato, B. Safer, J. Galabru, and A.G. Hovanessian. 1987. Adenovirus VA1 RNA complexes with the 68,000 M_r protein kinase to regulate its autophosphorylation and activity. *EMBO J.* **6:** 689–697.

Kaufman, R.J. and P. Murtha. 1987. Translational control mediated by eucaryotic initiation factor 2 is restricted to specific mRNAs in transfected cells. *Mol. Cell. Biol.* **7:** 1568–1571.

Kaufman, R.J., M.V. Davies, V.K. Pathak, and J.W.B. Hershey. 1989. The phosphorylation state of eukaryotic initiation factor 2 alters translational efficiency of specific mRNAs. *Mol. Cell. Biol.* **9:** 946–958.

Keck, J.G., C.J. Baldick, and B. Moss. 1990. Role of DNA replication in vaccinia virus gene expression: A naked template is required for transcription of three late *trans*-activator genes. *Cell* **61:** 801–809.

Kitajewski, J., R.J. Schneider, B. Safer, and T. Shenk. 1986a. An adenovirus mutant unable to express VA1 RNA displays different growth responses and sensitivity to interferon in various host cell lines. *Mol. Cell. Biol.* **6:** 4493–4498.

Kitajewski, J., R.J. Schneider, B. Safer, S.M. Munemitsu, C.E. Samuel, and T. Shenk. 1986b. Adenovirus VA1 RNA antagonizes the antiviral action of interferon by preventing activation of the interferon induced eIF2 alpha kinase. *Cell* **45:** 195–200.

Kostura, M. and M.B. Mathews. 1989. Purification and activation of the double-stranded RNA dependent eIF2 kinase DAI. *Mol. Cell. Biol.* **9:** 1576–1586.

Lawson, T.G., B.K. Ray, J.T. Dodds, J.A. Grifo, R.D. Abramson, W.C. Merrick, D.F. Betsch, H.L. Weith, and R.E. Thach. 1986. Influence of 5′ proximal secondary structure on the translational efficiency of eukaryotic mRNAs and on their interaction with initiation factors. *J. Biol. Chem.* **261:** 13979–13989.

Lin, T.-A., X. Kong, T.A.J. Haystead, A. Pause, G. Belsham, N. Sonenberg, and J.C. Lawrence. 1994. PHAS-1 as a link between mitaogen-activated protein kinase and translation initiation. *Science* **266:** 653–656.

Lindberg, U. and B. Sunquist. 1974. Isolation of messenger ribonucleoproteins from mammalian cells. *J. Mol. Biol.* **86:** 451–468.

Logan, J. and T. Shenk. 1984. Adenovirus tripartite leader sequence enhances translation of mRNAs late after infection. *Proc. Natl. Acad. Sci.* **81:** 3655–3659.

Ma, Y. and M.B. Mathews. 1993. Comparative analysis of the structure and function of adenovirus virus-associated RNAs. *J. Virol.* **67:** 6605–6617.

Maran, A. and M.B. Mathews. 1988. Characterization of the double-stranded RNA implicated in inhibition of protein synthesis in cells infected with a mutant adenovirus defective for VA RNA I. *Virology* **164:** 106–113.

Mathews, M.B. 1980. Binding of adenovirus VA RNA to mRNA: A possible role in splicing. *Nature* **285:** 575–577.

———. 1990. Control of translation in adenovirus-infected cells. *Enzyme* **44:** 250–264.

Mathews, M.B. and T. Shenk. 1991. Adenovirus virus-associated RNA and translational control. *J. Virol.* **65:** 5657–5662.

Mbuy, G.N., R.E. Morris, and H.C. Bubel. 1982. Inhibition of cellular protein synthesis by vaccinia virus surface tubules. *Virology* **116:** 137–147.

McAuslan, B.R. 1963. Control of induced thymidine kinase activity in the poxvirus infected cells. *Virology* **20:** 162–168.

McDonald, W.F., V. Crozel-Goudot, and P. Traktman. 1992. Transient expression of the vaccinia virus DNA polymerase is an intrinsic feature of the early phase of infection and is unlinked to DNA replication and late gene expression. *J. Virol.* **66:** 534–547.

Mellits, K.H. and M.B. Mathews. 1988. Effects of mutations in stem and loop regions on the structure and function of adenovirus VA RNA_I. *EMBO J.* **7:** 2849–2859.

Mellits, K.H., M. Kostura, and M.B. Mathews. 1990. Interaction of adenovirus VA RNA I with the protein kinase DAI: Nonequivalence of binding and function. *Cell* **61:** 843–852.

Moore, M., J. Schaack, S.B. Baim, R.I. Morimoto, and T. Shenk. 1987. Induced heat shock mRNAs escape the nucleocytoplasmic transport block in adenovirus infected HeLa cells. *Mol. Cell. Biol.* **7:** 4505–4512.

Moss, B. 1968. Inhibition of HeLa cell protein synthesis by the vaccinia virion. *J. Virol.* **2:** 1028–1037.

———. 1990. Poxviridae and their replication. In *Virology*, 2nd edition (ed. B.N. Fields and D.M. Knipe), pp. 2079–2111. Raven Press, New York.

Moss, B. and N.P. Salzman. 1968. Sequential protein synthesis following vaccinia virus infection. *J. Virol.* **2:** 1016–1027.

Munoz, A. and L. Carrasco. 1984. Action of human lymphoblastaoid interferon on HeLa cells infected with RNA containing animal viruses. *J. Gen. Virol.* **65:** 377–390.

Olgaiti, D., B.G.T. Pogo, and S. Dales. 1976. Biogenesis of vaccinia: Specific inhibition of rapidly labeled host DNA in vaccinia inoculated cells. *Virology* **71:** 325–335.

O'Malley, R.P., R.F. Duncan, J.W.B. Hershey, and M.B. Mathews. 1989. Modification of protein synthesis initiation factors and the shut-off of host protein synthesis in adenovirus infected cells. *Virology* **168:** 112–118.

O'Malley, R.P., T.M. Mariano, J. Siekierka, and M.B. Mathews. 1986a. A mechanism for the control of protein synthesis by adenovirus VA RNA_I. *Cell* **44:** 391–400.

O'Malley, R.P., T.M. Mariano, J. Siekierka, W.C. Merrick, P.A. Reichel, and M.B. Mathews. 1986b. The control of protein synthesis by adenovirus VA RNA. *Cancer Cells* **4:** 291–301.

Oosterom-Dragon, E.A. and H.S. Ginsberg. 1980. Purification and preliminary immunological characterization of the type 5 adenovirus nonstructural 100,000 dalton protein. *J. Virol.* **33:** 1203–1207.

———. 1981. Characterization of two temperature sensitive mutants of type 5 adenovirus with mutations in the 100,000k dalton protein gene. *J. Virol.* **40:** 491–500.

Paez, E. and M. Esteban. 1984. Resistance to vaccinia virus to interferon is related to an interference phenomenon between the virus and the interferon system. *Virology* **134:** 12–28.

———. 1985. Interferon prevents the generation of spontaneous deletions at the left terminus of vaccinia virus DNA. *J. Virol.* **56:** 75–84.

Patel, D.D. and D.J. Pickup. 1987. Messenger RNAs of a strongly-expressed late gene of cowpox virus contain 5′-terminal poly(A) sequences. *EMBO J.* **6:** 3787–3794.

Pause, A., G.J. Belsham, A.-C. Gingras, O. Donze, T.-A. Lin, J.C. Lawrence, and N. Sonenberg. 1994. Insulin-dependent stimulation of protein synthesis by phosphorylation of a regulator of 5′-cap function. *Nature* **371:** 762–767.

Pelham, H.B., J.M.M. Sykes, and T. Hunt. 1978. Characteristics of a coupled cell-free transcription translation system directed by vaccinia virus cores. *Eur. J. Biochem.* **82:** 199–209.

Pestka, S., J.A. Langer, K.C. Zoon, and C.E. Samuel. 1987. Interferons and their actions. *Annu. Rev. Biochem.* **56:** 727–777.

Petterson, V. and L. Philipson. 1975. Location of sequences on the adenovirus genome coding for the 5.5S RNA. *Cell* **6:** 1–4.

Pilder, S., J. Logan, and T. Shenk. 1986. The adenovirus E1B-55k transforming polypeptide modulates transport or cytoplasmic stablization of viral and host cell mRNAs. *Mol. Cell. Biol.* **6:** 470–476.

Price, R. and S. Penman. 1972. Transcription of the adenovirus genome by an alpha amanitin sensitive ribonucleic acid polymerase. *J. Virol.* **9:** 621–626.

Reichel, P.A., W.C. Merrick, J. Siekierka, and M.B. Mathews. 1985. Regulation of a protein initiation factor by adenovirus associated RNA_I. *Nature* **313:** 196–200.

Rice, A.P. and I.M. Kerr. 1984. Interferon mediated, double-stranded RNA-dependent protein kinase is inhibited in extracts from vaccinia virus-infected cells. *J. Virol.* **50:** 229–236.

Rice, A.P. and B.E. Roberts. 1983. Vaccinia virus induces cellular mRNA degradation. *J. Virol.* **47:** 529–539.

Rice, A.P., W.K. Roberts, and I.M. Kerr. 1984. 2-5Å accumulates to high levels in interferon-treated, vaccinia virus infected cells in the absence of any inhibition of virus replication. *J. Virol.* **50:** 220–228.

Rice, A.P., S.M. Kerr, W.K. Roberts, R.E. Brown, and I.M. Kerr. 1985. Novel 2′,5′-oligoadenylates synthesized in interferon-treated vaccinia virus-infected cells. *J. Virol.* **56:** 1041–1044.

Riley, D. and S.J. Flint. 1993. RNA-binding properties of a translational activator, the adenovirus L4 100-kilodalton protein. *J. Virol.* **67:** 3586–3595.

Rosemond-Hornbeak, H. and B. Moss. 1975. Inhibition of host protein synthesis by vaccinia virus: Fate of cell mRNA and synthesis of small poly(A)-rich polyribonucleotides in the presence of actinomycin D. *J. Virol.* **16:** 34–42.

Salzman, N.P., A.J. Shatkin, and E.D. Sebring. 1964. The synthesis of a DNA-like RNA in the cytoplasm of HeLa cells infected with vaccinia virus. *J. Mol. Biol.* **8:** 405–416.

Samuel, C.E. 1988. Mechanisms of the antiviral action of interferon. *Prog. Nucleic Acid Res. Mol. Biol.* **35:** 27–72.

———. 1991. Antiviral actions of interferon: Interferon-regulated cellular proteins and their surprisingly selective antiviral activities. *Virology* **183:** 1–11.

Schneider, R.J. and T. Shenk. 1987. Impact of virus infection on host cell protein synthesis. *Annu. Rev. Biochem.* **56:** 317–332.

Schneider, R.J. and Y. Zhang. 1993. Translational regulation in adenovirus infected cells. In *Translational regulation of gene expression* (ed. J. Ilan), vol. 2, pp. 227–250. Plenum Press, New York.

Schneider, R.J., C. Weinberger, and T. Shenk. 1984. Adenovirus VA1 RNA facilitates the initiation of translation in virus infected cells. *Cell* **37:** 291–298.

Schneider, R.J., B. Safer, S. Munemitsu, C.E. Samuel, and T. Shenk. 1985. Adenovirus VA1 RNA prevents phosphorylation of the eukaryotic initiation factor 2 alpha subunit subsequent to infection. *Proc. Natl. Acad. Sci.* **82:** 4321–4325.

Schnierle, B.S. and B. Moss. 1992. Vaccinia virus-mediated inhibition of host protein synthesis involves neither degradation nor underphosphorylation of components of the cap binding eukaryotic translation initiation factor eIF4F. *Virology* **188:** 931–933.

Schwer, B., P. Visca, J.C. Vos, and H.G. Stunnenberg. 1987. Discontinuous transcription or RNA processing of vaccinia virus late messengers results in a 5′ poly(A) leader. *Cell* **50:** 163–169.

Siekierka, J., T.M. Marino, P.A. Reichel, and M.B. Mathews. 1985. Translational control by adenovirus: Lack of virus-associated RNA1 during adenovirus infection results in phosphorylation of initiation factor eIF2 and inhibition of protein synthesis. *Proc. Natl. Acad. Sci.* **82:** 1959–1963.

Söderlund, H., U. Petterson, B. Vennström, L. Philipson, and M.B. Mathews. 1976. A new species of virus coded low molecular weight RNA from cells infected with adenovirus type 2. *Cell* **7:** 585–593.

Stewart, W.E. 1979. *The interferon system.* Springer Verlag, New York.

St Johnston, D., N.H. Brown, J.G. Gall, and M. Jantsch. 1992. A conserved double-stranded RNA-binding domain. *Proc. Natl. Acad. Sci.* **89:** 10979–10983.

Strijker, R., D.T. Fritz, and A.D. Levinson. 1989. Adenovirus VAI-RNA regulates gene expression by controlling stability of ribosome bound RNAs. *EMBO J.* **8:** 2669–2675.

Su, M.J. and R. Bablanian. 1990. Polyadenylated RNA sequences from vaccnia virus- infected cells selectively inhibit translation in a cell-free system. *Virology* **179:** 679–693.

Tal, J.T., E.A. Craig, and H.J. Raskas. 1975. Sequence relationships between adenovirus 2 early RNA and viral RNA size classes synthesized at 18 hours after infection. *J. Virol.* **15:** 137–144.

Thimmappaya, B., G.D. Ghadge, P. Rajan, and S. Swaminathan. 1993. Translational control by adenovirus-associated RNA I. In *Translational regulation of gene expression* (ed. J. Ilan), vol. 3, pp. 203–226. Plenum Press, New York.

Thimmappaya, B., C. Weinberger, R.J. Schneider, and T. Shenk. 1982. Adenovirus VA1 RNA is required for efficient translation of viral mRNA at late times after infection. *Cell* **31:** 543–551.

Thomas, A.M., G.C. Scheper, M. Kleijn, M. DeBoer, and H.O. Voorma. 1992. Dependence of the adenovirus tripartite leader on the p220 subunit of eukaryotic initiation factor 4F during in vitro translation. *Eur. J. Biochem.* **207:** 471–477.

Thomas, N.S.B., R.L. Matts, D.H. Levin, and I.M. London. 1985. The 60S ribosomal subunit as a carrier of eucaryotic initiation factor 2 and the site of reversing factor activity during protein synthesis. *J. Biol. Chem.* **260:** 9860–9866.

Traktman, P. 1990a. The enzymology of poxvirus DNA replication. *Curr. Top. Microbiol. Immunol.* **163:** 93–123.

———. 1990b. Poxviruses: An emerging portrait of biological strategy. *Cell* **62:** 621–626.

Vassef, A., F. Ben-Hamida, A. Dru, and G. Beaud. 1982. Translational control of early protein synthesis at the late stage of vaccinia virus infection. *Virology* **118:** 45–53.

Watson, J.C., H.-W. Chang, and B.L. Jacobs. 1991. Characterization of a vaccinia virus-encoded double-stranded RNA binding protein that may be involved in inhibition of the double-stranded RNA-dependent protein kinase. *Virology* **185:** 206–216.

Whitaker-Dowling, P. and J.S. Youngner. 1983. Vaccinia rescue of VSV from interferon-induced resistance: reversal of translation block and inhibition of protein kinase acivity. *Virology* **131:** 128–136.

———. 1984. Characterization of a specific kinase inhibitory factor produced by vaccinia virus which inhibits the interferon-induced protein kinase. *Virology* **137:** 171-181.

———. 1986. Vaccinia-mediated rescue of encephalomyocarditis virus from the inhibitory effects of interferon. *Virology* **152:** 50–57.

Wright, C.F. and B. Moss. 1989. Identification of factors specific for transcription of the late class of vaccinia virus genes. *J. Virol.* **63:** 4224–4233.

Yoder, S.S., B.L. Robberson, E.J. Leys, A.G. Hook, M. Al-Ubaidi, C.Y. Yeung, R.E. Kellems, and S.M. Berget . 1983. Control of cellular gene expression during adenovirus infection: Induction and shut off of dihydrofolate reductase gene expression by adenovirus type 2. *Mol. Cell. Biol.* **3:** 819–828.

Yuwen, H., J.H. Cox, J.W. Yewdell, J.R. Bennink, and B. Moss. 1993. Nuclear localization of a double-stranded RNA binding protein encoded by the vaccinia virus E3l gene. *Virology* **195:** 732–744.

Zhang, Y. and R.J. Schneider. 1994. Adenovirus inhibition of cell translation facilitates release of virus particles and enhances degradation of the cytokeratin network. *J. Virol.* **68:** 2544–2555.

Zhang, Y., P.J. Dolph, and R.J. Schneider. 1989. Secondary structure analysis of adenovirus tripartite leader. *J. Biol. Chem.* **264:** 10679–10684.

Zhang, Y., D. Feigenblum, and R.J. Schneider. 1994. A late adenovirus factor blocks eIF 4E phosphorylation and induces inhibition of cell protein synthesis. *J. Virol.* **68:** 7040–7050.

Zilberstein, A., P. Federmann, L. Schulman, and M. Revel. 1976. Specific phosphorylation in vitro of a protein associated with ribosomes of interferon treated mouse L cells. *FEBS Lett.* **68:** 119–124.

21

Translational Control in Cells Infected with Influenza Virus and Reovirus

Michael G. Katze
Department of Microbiology (SC-42)
School of Medicine, University of Washington
Seattle, Washington 98195

The two RNA viruses, influenza virus and reovirus, utilize dramatically distinct tactics to optimize the translation of viral messenger RNAs. Presented here is a comprehensive review of the strategies used by these two virus groups to ensure both the selective and efficient synthesis of viral proteins. Among other topics are a discussion of the virus-induced "host shutoff" and documentation of the mechanisms by which influenza virus and reovirus down-regulate the double-stranded RNA (dsRNA)-activated, interferon-induced protein kinase, PKR.

INFLUENZA VIRUS

Influenza virus is a enveloped, negative-stranded segmented RNA virus that causes a disease which is responsible for up to 70,000 deaths a year in the United States and in the worst epidemic years, has killed 20,000,000 worldwide. It is a highly cytopathic virus that interferes dramatically with host macromolecular synthesis. Unlike most non-oncogenic RNA viruses, the influenza replicative cycle includes both a nuclear and cytoplasmic phase (Krug et al. 1989), and this has a definite influence on the translational strategies of the virus. In the early 1970s, it was recognized that a shutoff of host-cell protein synthesis occurs at times when viral proteins are maximally made (Lazarowitz et al. 1971; Skehel 1972). What was not appreciated at the time were the numerous and complex strategies required by influenza virus to accomplish these goals (Garfinkel and Katze 1993b, 1994). In terms of translational control, influenza virus has two major objectives, neither of which is unique to this virus (see Mathews, this volume): (1) to selectively translate viral mRNAs and stop the translation of cellular mRNAs and (2) to keep the infected cells translationally competent such that the steps involved in

Table 1 Influenza virus translational regulatory mechanisms

Strategy	References
Host shutoff of cellular protein synthesis	Lazarowitz et al. (1971); Skehel (1972); Katze and Krug (1984)
Inhibition of cellular mRNA transport/ degradation of mRNAs in nucleus	Katze and Krug (1984)
Inhibition of the interferon-induced, double-stranded RNA-activated protein kinase, PKR	Katze et al. (1984, 1986a, 1988); Lee et al. (1990, 1992); Feigenblum and Schneider (1993)
Recruitment of P58, a TPR protein and possible oncogene	Lee et al. (1994); Barber et al. (1994)
Cap-dependent, selective translation of influenza virus mRNAs	Katze et al. (1984, 1986b); Alonso-Caplen et al. (1988); Garfinkel and Katze (1992, 1993)
Inhibition of cellular mRNA translation at initiation and elongation stages	Katze et al. (1986b, 1989); Garfinkel and Katze (1992)
Degradation of cellular mRNAs in the cytoplasm	Inglis (1982); Beloso et al. (1992)
Structure of influenza virus mRNAs	Alonso-Caplen et al. (1988); Garfinkel and Katze (1992, 1993)
Dephosphorylation of eukaryotic initiation factor 4E	Feigenblum and Schneider (1993)
Temporal control of influenza virus protein synthesis	Yamanaka et al. (1988, 1991); Enami et al. (1994)

the protein synthetic pathway are not compromised. Both of these tactics are discussed. The major stategies are summarized in Table 1.

Selective Translation of Influenza Viral mRNAs, "the Host Shutoff"

The Fate of Nuclear and Cytoplasmic Cellular mRNAs

What is known about the down-regulation of cellular gene expression and protein synthesis in influenza-virus-infected cells? There is a modest decrease in cellular mRNA transcription rates (about twofold) after viral infection (Katze and Krug 1984, 1990) and there are reports of cellular mRNA degradation in an influenza-virus-infected cell, particularly late after infection (Inglis 1982; Beloso et al. 1992). It appears, however, that more complex controls are exerted by the virus to ensure maximal protein synthesis. For example, newly synthesized cellular mRNAs never reach the cytoplasm of influenza-virus-infected cells (Katze and Krug 1984). Labeling experiments with [^{3}H]uridine showed that the failure of newly synthesized cellular mRNAs to reach the cytoplasm after infection is not due to a lack of cellular mRNA transport as occurs in adenovirus-infected cells (Babich et al. 1983). Rather, it is due to the degradation of nuclear RNA early after infection. It is reasonable to assume that influenza-virus-induced cellular nuclear RNA degradation is initiated by the cleavage of the 5′ends of cellular RNA polymerase II transcripts by the viral cap-dependent endonuclease. The decapped RNAs would likely be more susceptible to degradation by cellular nucleases as it is well established that the 5′cap structure stabilizes mRNAs (Krug et al. 1989).

For a complete cessation of cellular protein synthesis, it is not sufficient for influenza virus to block the transport of newly made cellular RNAs. One must account for the cellular mRNAs that were in the cytoplasm prior to viral infection. In contrast to the nuclear RNAs, these cellular mRNAs are stable (Katze and Krug 1984). Despite the earlier cited reports of cellular mRNA degradation during influenza virus infection (Inglis 1982; Beloso et al. 1992), only minor decreases in cellular mRNA levels at times of maximal host shutoff were found (Garfinkel and Katze 1994). Moreover, these cellular mRNAs were functional when tested for translatability in reticulocyte extracts. Thus, the shutoff of cellular protein synthesis is not *primarily* due to either the degradation or modification of cellular mRNAs. Cellular mRNA translation is blocked at both the initiation and elongation steps of protein synthesis (Katze et al. 1986a). Polysome analysis revealed that although there is a significant decrease in the association of cellular mRNAs with large polyribosomes, many cellular mRNAs remain polysome-associated even though these mRNAs were not translated inside the infected cell. It was next determined whether the polysomes containing the nontranslated cellular mRNAs were associated with the cytoskeleton (Katze et al. 1989), since suggestions in the literature indicated that actively translated mRNAs were associated with a pool of polyribosomes that were cytoskeleton-bound (Lenk and Penman 1979). It transpires, however, that both cellular and viral polysome-associated mRNAs are associated with the cytoskeleton framework. Parenthetically, this was not found to be the case in poliovirus-infected cells in which cellular mRNAs are dissociated from the polysomes and the cellular cytoskeleton (Lenk and Penman 1979; Katze et al. 1989). There was therefore still not a clear understanding of the mechanisms underlying preferential influenza virus protein synthesis.

Contribution of Influenza Virus mRNA Structure

It was conceivable that translational selectivity, at least at the level of initiation, could be due to competition between cellular and viral mRNAs for limiting components of the translational machinery as shown in other systems including the reoviruses (see below) (Lodish and Porter 1980; Walden et al. 1981; Ray et al. 1983). Since viral mRNAs do not have an advantage merely due to sheer mass (Garfinkel and Katze 1994), it was likely that the influenza virus mRNAs are intrinsically better initiators of translation due to certain unknown structural qualities. Such structural features could trick the cellular protein synthesizing machinery into making only viral proteins (Garfinkel and Katze 1993b, 1994). The first convincing evidence that influenza virus mRNAs are intrinsically efficient initiators of translation was provided by studies in which cells were doubly infected with influenza virus and adenovirus (Katze et al. 1984; 1986b). These early experiments also demonstrated that influenza virus has a strategy to maintain overall high levels of protein synthesis. As described in much more detail below, this had less to do with selective translation and more to do with the down-regulation of the interferon-induced protein kinase, PKR.

Studies of these adenovirus/influenza virus doubly infected cells showed that influenza virus proteins were made at essentially the same levels as in cells infected by influenza virus alone (Katze et al. 1984). The data showed that influenza virus is able to overcome the blocks on host-cell mRNA transport and translation exerted by adenovirus (Babich et al. 1983; Schneider, this volume) and furthermore that influenza virus establishes its own translational and transport regulatory mechanisms. Finally, influenza virus mRNAs were shown to be more efficiently translated than late adenovirus mRNAs, believed to be strong mRNAs due to the presence of the tripartite leader sequences (Logan and Shenk 1984). However, there was still no *direct* evidence that the structure of influenza virus mRNAs has a key role in their selective translation. This was provided by the analysis of a recombinant adenovirus expressing the gene for the nucleocapsid protein (NP) (Alonso-Caplen et al. 1988). The translation rate of NP mRNA encoded by the recombinant adenovirus NP was as efficient as the native NP expressed by influenza virus itself. The recombinant NP mRNA translation rates were high in the absence of all other influenza virus gene products. Thus, these experiments showed that the sequence and/or structure alone of an influenza virus RNA can confer enhanced translatability. These studies provided the foundation for the later detailed structural studies described below.

Because these studies were done with a "genetics unfriendly" segmented, negative-stranded virus, alternative strategies were developed to produce more decisive data on the contribution of mRNA structure to selective translation. A transfection/infection assay was devised in which representative viral or cellular cDNAs were transfected into COS-1 cells, which were then infected with influenza virus (Garfinkel and Katze 1992). mRNA translation directed by cellular transfected genes such as interleukin-2 (IL-2) or secreted embryonic alkaline phosphatase (SEAP) was markedly shut off after viral infection similar to endogenous cellular mRNAs. In contrast, a shortened version of the influenza virus NP gene, when transfected, was not subjected to the host shutoff: This mRNA was translated with equal efficiency in influenza-virus-infected cells and uninfected cells. Northern blot analysis demonstrated that SEAP and IL-2 mRNA levels were not diminished, and polysome analysis revealed that the block to their translation was at the level of initiation and elongation (since cellular mRNAs were still associated with polyribosomes and a proportion were released from the polysomes). Thus, this assay reproduced the natural shutoff of endogenous cellular mRNAs in influenza-virus-infected cells. More critically, the selectivity of viral mRNA translation in this assay directly demonstated that mRNA structure by itself dictates whether an mRNA will be translated in influenza-virus-infected cells.

The Importance of the Influenza Virus mRNA 5′-untranslated Region

It was then determined which exact sequences and/or structure direct selective viral mRNA translation. Attention was focused on the very 5′ end and the 5′-untranslated region (5′UTR) on the basis of extensive literature demonstrating the critical nature of this region in the regulation of protein synthesis (Kozak 1989, 1991; see Sonenberg; Hinnebusch; Jackson; Geballe; all this volume). The cap dependence of the translation of influenza virus mRNAs was examined first since it was known that both adenovirus and poliovirus mRNA translation are cap-independent (see Schneider; Ehrenfeld; both this volume). If a virus can direct this type of translation for its own mRNAs, it would provide a straightforward way to shut off host mRNA translation since virtually all cellular mRNA translation is cap-dependent. It was found, however, that influenza mRNA translation is very likely cap-dependent as influenza virus protein synthesis was reduced in cells doubly infected by both influenza and poliovirus (Garfinkel and Katze 1992). The possible contribution of the viral 5′UTR to selective translation was examined next by constructing chimeras between viral and cellular genes and using them in a transfection/infection assay (Garfinkel and Katze 1993a). Cellular mRNAs that contained a viral 5′UTR, in place of a cellular 5′UTR, were now capable of being translated in an influenza-virus-infected cell. Conversely, viral mRNAs that had their 5′UTR replaced by a cellular 5′UTR were no longer translated during viral infection. Thus, the host translational apparatus distinguishes between viral and cellular mRNAs based on the sequence and/or structure of the 5′UTR. Defining the exact sequences present within this region and identifying possible *trans*-acting factors that interact with these sequences comprise current goals. Relevant to this story are observations by Schneider and colleagues (Feigenblum and Schneider 1993), who found a modest dephosphorylation of eukaryotic initiation factor eIF4E in influenza-virus-infected cells. Dephosphorylation of eIF4E leads to a functional limitation in eIF4F, one activity of which is to unwind secondary structure in the 5′UTRs of mRNAs (Hershey 1991; Lazaris-Karatzas et al. 1990; Merrick 1992; see relevant chapters in this volume). One can speculate that influenza virus mRNA UTRs may require less functional eIF4E than cellular mRNAs due to reduced higher-order structure and that this may contribute to mRNA selective translation.

Temporal Regulation of Viral mRNA Translation

In addition to regulating cellular mRNA translation, influenza virus carries out temporal regulation of its own viral gene expression at the level of translation. For example, Yamanaka et al. (1988, 1991) transfected HeLa cells with a CAT reporter gene appended to the 5′UTR of each of the viral mRNAs separately and superinfected these cells with influenza virus. They measured CAT activity at early and late times after infection. When the CAT construct contained the 5′UTR of an mRNA encoding an early protein such as NS, CAT activity was up-regulated early. Conversely, when this was done with a late viral protein such as neuraminidase, CAT activity was higher late after infection. This avenue of investigation has not been pursued further and exact sequences or *trans*-acting factors responsible for this regulation have not been identified. More recently, however, it was found that the influenza virus NS1 protein can stimulate the synthesis of the viral M1 protein (Enami et al. 1994). Site-directed mutagenesis studies showed that specific sequences contained within the M1 mRNA 5′UTR are required for this stimulation. These data, taken together with earlier cited results, suggest a key role for the 5′UTR of influenza virus mRNAs in dictating temporal and selective translation in the infected cell.

Regulation of the Interferon-induced, dsRNA-activated Protein Kinase, PKR, by Influenza Virus

A History of Influenza Virus and PKR

Not only must influenza virus devise mechanisms to ensure the preferential translation of its viral mRNAs, but it also must guarantee that the cell remains maximally translationally competent during infection (Katze 1992, 1993, 1995). Without both of these major strategies, virus replication might be compromised, a scenario unacceptable to an actively replicating cytopathic virus. The translational competence of the infected cell is assured because influenza virus has developed an intricate strategy to down-regulate the protein kinase referred to as PKR (for *p*rotein *k*inase, *R*NA-dependent). PKR, which is discussed thoroughly within this volume (see Clemens; Schneider; Mathews; all this volume), is an interferon-induced kinase that is activated by dsRNA and other polyanions (Katze et al. 1991). Since the kinase is described in detail elsewhere, I reiterate only briefly here the reason influenza must regulate this important enzyme. As a result of activation by dsRNA (and many viral specific RNAs can activate PKR [see Maran and Mathews 1988; Black et al. 1989; Roy et al. 1991; Clemens; Mathews; both this volume]), PKR phosphorylates the α-subunit of eukaryotic initiation factor 2 (eIF2) (Hovanessian 1989; Samuel 1993). This leads to limitations in functional eIF2 (see Hershey 1991; Merrick 1992; Rhoads 1993; Merrick and Hershey; Clemens; both this volume) and, as a result, an inhibition of protein synthesis initiation, a situation unfavorable to viruses that must make their proteins to replicate. As discussed elsewhere and below for reovirus, many viruses, in addition to influenza virus, have developed elegant ways to make sure that PKR is not activated during infection.

It was first determined that influenza virus down-regulated PKR by analyzing cells doubly infected with influenza virus and the adenovirus VA RNA_I-negative mutant *dl*331 (Katze et al. 1984, 1986b). In cells infected by *dl*331 alone, there was a dramatic reduction in the levels of both viral and cellular protein synthesis (Thimmappaya et al. 1982). This was due to excessive phosphorylation of the eIF2 α-subunit by an active PKR that cannot be down-regulated because of the absence of the virus-encoded VA RNA_I (for review, see Mathews and Shenk 1991). In contrast, when *dl*331-infected cells were superinfected with influenza virus, a reduction of the protein kinase activity normally detected during *dl*331 infection was observed (Katze et al. 1984, 1986b). These data provided the first evidence that influenza virus encodes or activates a gene product that, analogous to VA RNA_I, inhibits PKR and prevents any resultant inhibition of protein synthesis initiation. It was subsequently shown that the suppression of PKR activity also occurs in cells infected by influenza virus alone (Katze et al. 1988).

P58 Is a Cellular Protein and the PKR Inhibitor

The next objective was to identify this PKR inhibitor. In an in vitro assay developed to measure PKR inhibition quantitatively, the inhibitor was purified to homogeneity (Lee et al. 1990). The final product, which has an apparent molecular weight of 58,000 (thus referred to as P58), inhibited both the autophosphorylation of PKR and phosphorylation of the eIF2 α-subunit. P58 does not function as a ribonuclease, phosphatase, protease, or ATPase. Furthermore, P58 does not act by binding or sequestering dsRNA as do other viral inhibitors of PKR. Preliminary evidence now shows that P58 likely inhibits PKR function through a direct interaction with the protein kinase (S. Polyak et al., in prep.).

Perhaps the most suprising finding was that P58 is not an influenza-virus-encoded protein but rather a cellular protein (Lee et al. 1990, 1992; Lee and Katze 1993). This was first suggested by immunoblot analysis and later confirmed by the molecular cloning of the P58 gene (Lee et al. 1994). The cloning and subsequent sequence analysis also revealed that P58 is a member of the TPR (tetratricopeptide) family of proteins. These proteins contain internal 34-amino-acid repeats, motifs that are thought to form helix-turn structures, each with a knob and hole, acting as helix-associating domains. The TPR family includes a diverse group of fungal and nonfungal proteins that function in wide areas of gene expression, including (but not limited to) mitosis, transcription, protein import, RNA splicing, and stress (for reviews, see Goebl and Yanagida 1991; Sikorski et al. 1991). In addition to these homologies, a region near the P58 carboxyl terminus (amino acids 392–463) shares significant homology with the conserved J region domain of the DnaJ heat shock family of proteins (Silver and Way 1993). Finally, limited similarity also exists between P58 and eIF2α, the natural substrate of PKR. Mutagenesis of the P58 gene is under way in an effort to determine which portions of P58 are critical for function.

Another major goal is to determine how influenza virus activates this P58/PKR pathway during infection. It is not known whether a specific viral gene product is responsible. Interestingly, kinase inhibitory activity is undetectable in crude extracts prepared from uninfected cells; however, after extracts are subjected to ammonium sulfate fractionation, kinase inhibitory activity appears (Lee et al. 1992). On the basis of biochemical analysis and in vitro mixing experiments, it has been pro-

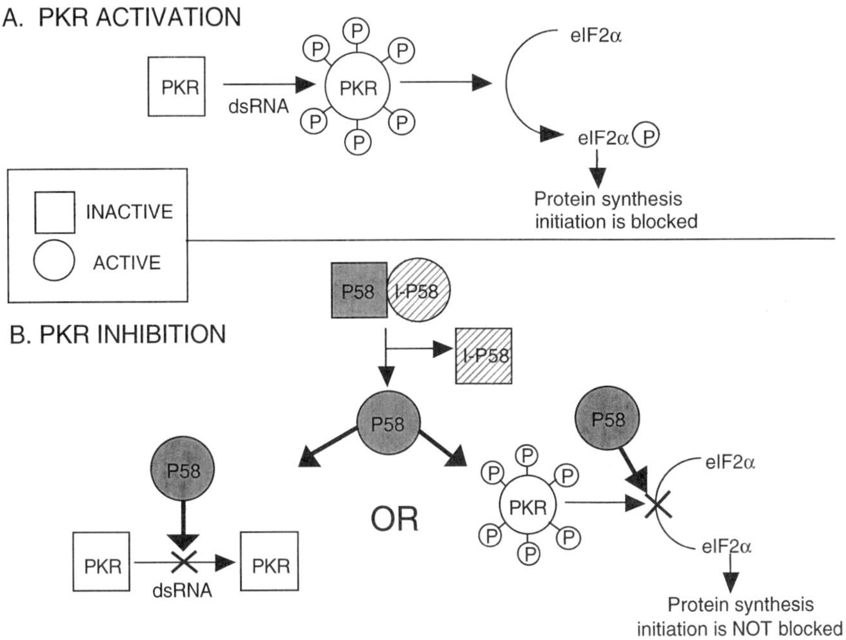

Figure 1. A model for PKR regulation in influenza-virus-infected cells (see text for details).

posed that a factor (referred to as an anti-inhibitor or I-P58 for inhibitor of P58) is dissociated from P58 during high-salt treatment (and also during viral infection). This anti-inhibitory activity has been identified and our laboratory is in the process of purifying I-P58 based on in vitro assays.

A Model

Figure 1 illustrates a model for PKR regulation in influenza-virus-infected cells. Figure 1A depicts the scenario in the absence of any regulation: The kinase is activated by dsRNA (possibly viral dsRNA) and protein synthesis is inhibited. Figure 1B depicts PKR inhibition in an influenza-virus-infected cell. After viral infection and synthesis of virus-specific RNAs, P58, which has been in an inactive complex with its own inhibitor (I-P58), becomes dissociated from I-P58 and can then block the autophosphorylation and/or activity of the kinase most likely through a direct interaction with PKR. The critical end result is that viral protein synthetic rates are not diminished and thus virus replication can occur maximally. The model also might be applicable even in the absence of viral infection: It is possible that P58 would become dissociated from I-P58 (or even associated with a positive acting regulator) in response to a specific environmental signal or stimulus (e.g., stress).

PKR, P58, and Malignant Transformation

Work on the cellular P58 inhibitor and its regulation should provide insight into PKR control not only in virus-infected cells, but also in uninfected cells. This may now be particularly important as accumulating evidence indicates that PKR has a key role in the regulation of cellular gene expression (and cell proliferation) in the absence of interferon treatment and viral infection (Lengyel 1993; Clemens, this volume). It is therefore reasonable to assume that an inhibitor of PKR may also have alternative cell regulatory roles. For example, if PKR is confirmed to regulate signal transduction and transcriptional regulatory pathways (Kumar et al. 1994; Maran et al. 1994), then P58 (or I-P58) may have a role in these types of control. Perhaps the most compelling evidence for a role of PKR in growth regulation comes from recent data suggesting that PKR may be a tumor suppressor gene. Introduction of catalytically inactive PKR mutants into NIH-3T3 cells induced malignant transformation possibly as a result of transdominant inhibition of the endogenous mouse PKR (Koromilas et al. 1992; Meurs et al. 1993). Furthermore, data now exist showing that overexpression of the cellular PKR inhibitor, P58, in NIH-3T3 cells induces their malignant transformation, also potentially through down-regulation of the murine protein kinase (Barber et al. 1994). The demonstration of P58's oncogenic potential provides another important clue to the biological function of the kinase inhibitor and an additional critical link between the PKR regulatory pathway and tumorigenesis. The role of PKR and other translationally important proteins involved in the regulation of malignant transformation is addressed in considerable detail elsewhere in this volume (see Clemens; Sonenberg).

REOVIRUSES

Reoviruses, although they are RNA viruses like influenza viruses, have a completely different molecular structure and, more importantly, totally different strategies to ensure the selective and efficient synthesis of viral proteins. These viruses first came to the attention of virologists as "viruses in search of a disease" and were therefore called reoviruses for *r*espiratory, *e*nteric, *o*rphan viruses (Joklik 1974, 1984). Six genera of reoviruses now exist: three that infect animals (reovirus, orbivirus, and rotavirus) and three that infect only plants and insects (cypovirus, phytoreovirus, and figivirus). There are three serotypes of mammalian reoviruses: type 1, type 2, and type 3. Reoviruses are nonenveloped viruses that contain ten segments of dsRNA surrounded by a double capsid shell of proteins. The outer capsid consists of three polypeptides, σ1, σ3, and u1C (Smith et al. 1969), which are encoded by the S1, S4, and M2 dsRNA segments, respectively. In addition to its role as a structural protein, σ3 has a critical role in the regulation of translation in a reovirus-infected cell as discussed in more detail below.

Reovirus and Translational Control: A Brief Overview and History

In a seminal paper published in 1982 on the inhibition of protein (and RNA) synthesis in reovirus-infected cells, Sharpe and Fields showed that type-2 reovirus inhibited host-protein synthesis in mouse L cells more rapidly and efficiently than type 3. By using recombinant reoviruses containing various combinations of dsRNA segments derived from reovirus types 2 and 3, these investigators determined that the S4 dsRNA segment (that encoding the σ3 major capsid polypeptide) is responsible for the ability of type-2 reovirus to shut off host protein synthesis. Although the molecular mechanisms underlying this shutoff are still not known, this study was one of the first to look carefully at the problem and reveal the importance of σ3 in these regulatory mechanisms. Very early on, the Thach laboratory also made important contributions by demonstrating that selective reovirus mRNA translation may be due to mRNA competition for limited concentrations of select protein synthesis discriminatory initiation factors (Walden et al. 1981; Ray et al. 1983). These investigators suggested that the huge abundance of reovirus mRNAs in infected cells may allow the viral mRNAs to outcompete cellular mRNAs, thus contributing to the shutoff of host-protein synthesis. The possible role of viral mRNA abundance and competition in the regulation of reoviral protein synthesis has been revisited more recently by a group that utilized actinomycin D in a study on an avian reovirus (Mallo et al. 1991). These researchers found that drug treatment stimulates the growth of the avian reovirus in L cells and that this enhancement correlates with increased viral protein synthesis (and decreased levels of cellular protein synthesis presumably due to actinomycin D inhibition of cellular RNA transcription). Although mRNA competition may indeed have a role in selective translation, it will become clear that other, more subtle mechanisms also must be at work.

Some other factors contributing to translational control may relate to the unusual structural characteristics of reovirus mRNAs. For example, reovirus mRNAs do not contain a poly(A) tract at their 3′termini (Lemay 1988). It was demonstrated that reovirus mRNA translation is unaffected compared to cellular mRNAs when free poly(A) is added to in vitro translation extracts (Lemay and Millward 1986). More work focusing on the biological significance of these observations and how they relate to the in vivo regulation of protein synthesis is necessary. Perhaps more interesting and certainly provocative is the extended debate over whether late reoviral mRNAs are capped and if not, whether the lack of a cap contributes to their selective translation in virus-infected cells (Lemay 1988). One group of investigators reported that, unlike early viral and cellular mRNAs, late viral mRNAs are not capped presumably due, at least in part, to the lack of appropriate enzymatic activities (Skup and Millward 1980; Lemay 1988). In contrast, others reported that late mRNAs are capped partly based on studies utilizing cap analogs (Detjen et al. 1982).

Highlighted for the remainder of this chapter are the more recent advances that have furthered our understanding of reovirus translational control. Of particular focus is the multifaceted σ3 protein, a major player in this regulation despite it being a viral structural protein. Much of the discussion focuses on PKR regulation in vitro and in vivo as was the case for influenza virus, although also stressed are the novel aspects of the reovirus system. Most of the studies concentrate on work done on the mammalian reoviruses but also included is a description of a very recent report on a related member of the family, a porcine rotavirus. For a summary of reovirus translational control strategies, see Table 2.

σ3, a Reovirus Protein

σ3 Is a dsRNA-binding Protein

The reovirus σ3 protein is a major outer capsid protein of 41 kD encoded by dsRNA segment S4 (McCrae and Joklik 1978). σ3 can bind both synthetic and natural forms of dsRNA efficiently (Imani and Jacobs 1988), although this RNA-binding protein does not appear to have a role in the packaging of reovirus dsRNA genomes (Miller and Samuel 1992). σ3 does not bind DNA or single-stranded RNA nor does it appear to recognize specific sequences in the dsRNA (Huismans and Joklik 1976). Analysis of proteolytic fragments of virion-derived σ3 protein revealed that the dsRNA-binding region is localized in a 16-kD carboxy-terminal fragment that is distinct from the zinc finger motif which is localized in

Table 2 Reovirus translational regulatory mechanisms

Strategy	References
Host shutoff of cellular protein synthesis	Joklik (1974, 1984); Sharpe and Fields (1982); Lemay (1988)
Role of mRNA competition	Walden et al. (1981); Ray et al. (1983); Mallo et al. (1991)
Capped vs. uncapped late reoviral mRNAs	Skup and Millward (1980); Detjen et al. (1982); Lemay (1988)
σ3 as a dsRNA-binding protein	Huismans and Joklik (1976); Miller and Samuel (1992); Denzler and Jacobs (1994)
σ3 as a PKR inhibitor	Imani and Jacobs (1988); Giantini and Shatkin (1989); Seliger et al. (1992); Lloyd and Shatkin (1992); Beattie et al. (1995)
σ3 and preferential virus mRNA translation	Lemieux et al. (1987)
Regulation of reovirus mRNA differential translation	Munemitsu and Samuel (1988); Roner et al. (1989); Bishoff and Samuel (1989); Samuel and Brody (1990); Fajardo and Shatkin (1990); Doohan and Samuel (1992, 1993)
Porcine rotavirus NSP 3 protein is a dsRNA-binding protein and PKR inhibitor	Mattion et al. (1992); Langland et al. (1994)

the amino-terminal portion of the protein (Miller and Samuel 1992). It is important to mention that σ3 protein is not a member of the newly emerging family of dsRNA-binding proteins, which includes PKR and other viral proteins (see below), and therefore does not possess the well-characterized dsRNA-binding motifs present in these other proteins (Green and Mathews 1992; St Johnston et al. 1992).

Miller and Samuel (1992) performed a structural analysis of recombinant σ3 and proposed that an 85-amino-acid region present in the carboxy-terminal portion is responsible for dsRNA binding. This region contains a repeated basic amino acid motif that is conserved in all three serotypes of reovirus. Curiously, this 85-amino-acid region shares homology with a region of the regulatory subunit of PP2A, an eIF2α phosphatase (Miller and Samuel 1992). There is no direct in vivo or in vitro evidence that σ3 possesses phosphatase activity or can regulate the activity of any protein phosphatase. It is quite possible, however, that this issue has not been addressed. In a related paper that is in disagreement with this earlier study, Jacobs and colleagues (Denzler and Jacobs 1994) found that deletion of as few as ten amino acids from either the amino- or carboxy-terminal part of the σ3 protein abrogates dsRNA binding. Even the exact amino-terminal deletion mutant (containing amino acids 221–365), shown by Miller and Samuel to bind dsRNA, failed to bind dsRNA when tested by Denzler and Jacobs in their assay. This discrepancy may be explained by differences between dsRNA-binding assays employed. Whereas Miller and Samuel utilized a Northwestern assay, Denzler and Jacobs tested binding in a soluble assay. As proposed (Denzler and Jacobs 1994), it may well be that the soluble assay is more physiologically relevant as σ3 mutants that failed to bind dsRNA (in the soluble assay) also failed to complement vaccinia virus mutants lacking the *E3L* gene (see below) (Beattie et al. 1995). Further experimentation is clearly needed to resolve these conflicting data and to identify the σ3 dsRNA-binding regions unequivocally.

σ3 Can Inhibit the Interferon-induced PKR

The dsRNA-binding properties of σ3 make it a candidate as a down-regulator of PKR. Such control is important as reovirus dsRNAs are efficient activators of PKR. Moreover, reovirus mRNAs, like many other viral mRNAs, also can activate PKR. However, not all these mRNAs activate PKR to the same extent: S1 mRNA is a much better activator of PKR than S4 mRNA (which encodes the σ3 protein!) (Bischoff and Samuel 1989). A discussion of the relative translational efficiency of reovirus mRNAs, including how the translation may be related to PKR activity, is addressed below. At any rate, it is imperative that reovirus encode a mechanism to down-regulate PKR so that viral protein synthesis is not compromised. Different reovirus types markedly differ in their sensitivity to interferon, with type 3 being vastly more sensitive than type 1 to its antiviral effects (Jacobs and Ferguson 1991). With other virus groups, there often is a correlation between a virus's ability to down-regulate PKR and to avoid interferon's negative effects on virus replication (Katze 1992, 1993, 1995; see Mathews, this volume). Whether this is the case with the reoviruses remains to be shown definitively.

Imani and Jacobs (1988) provided the first direct evidence that the reovirus σ3 protein may be involved in PKR regulation, most likely by sequestering dsRNA activators. Mixing experiments demonstrated that extracts prepared from reovirus-infected cells blocked the activation of PKR present in extracts from interferon-treated cells. The majority of the inhibitory activity could be removed by prereacting extracts with σ3 antibody or immobilized dsRNA, strongly suggesting that σ3 was involved. In a final definitive experiment, purified σ3 protein was directly shown to inhibit PKR function in vitro and this inhibition could be reversed by addition of excess dsRNA. These experiments were carried out with the reovirus type-1 σ3 protein, the serotype less sensitive to interferon.

A series of experiments was subsequently reported, mainly from the Shatkin laboratory, on the ability of σ3 to regulate PKR activity and mRNA translation in vivo. Giantini and Shatkin (1989) showed that cotransfection of COS cells with plasmids containing the serotype-3 S4 gene (encoding σ3) with a CAT reporter gene resulted in enhanced CAT activity mainly at the level of CAT mRNA translation. In a follow-up study to determine whether enhancement of CAT mRNA translation was a general property of σ3, the effect of cotransfecting the S4 genes from all three serotypes was analyzed (Seliger et al. 1992). Although all three serotypes stimulated CAT expression, the type-2 and type-3 S4 genes were more stimulatory than type 1. The significance of these observations is yet to be determined. Interestingly, coexpression of the M2 and S4 genes prevented stimulatory effects of σ3 on CAT expression presumably because of the inactivation of σ3 as a result of complex formation with the M2 gene product μ1C (Tillotson and Shatkin 1992). Thus, at least indirectly, μ1C also may be considered a possible modulator of PKR. Finally, the best evidence that cotransfected S4 DNA encoding σ3 actually was working in vivo through an inactivation of PKR was provided by studies comparing σ3 and the adenovirus-encoded PKR inhibitor VA RNA_I (Lloyd and Shatkin 1992). S4 gene transfections partially restored protein synthetic capacity of 293 cells infected by the adenovirus VA_I RNA^- mutant *dl*331, even though σ3 and VA RNA_I function by completely different mechanisms (Thimmappaya et al. 1982). Importantly, this effect directly correlated with decreases in eIF2α phosphorylation levels.

A recent study provided the most compelling evidence thus far that σ3 can regulate PKR activity and even affect interferon sensitivity (Beattie et al. 1995). The S4 gene was inserted into a vaccinia mutant lacking E3L. E3L is itself a dsRNA-binding protein involved in PKR regulation, and viruses lacking this gene are more sensitive to the antiviral effects of interferon (Beattie et al. 1991). The resultant S4/vaccinia virus recombinant is now more resistant to the antiviral effects of interferon compared to the parental E3L⁻ mutant. In addition, the S4-containing virus is able to rescue the interferon-sensitive vesicular stomatitis virus similarly to the wild-type vaccinia virus. Experiments like these help provide a genetic approach not previously possible with reoviruses. This potentially allows one to perform a detailed dissection of the reovirus proteins involved in PKR regulation and resistance to interferon as well as those regulating selective translation.

Is σ3 Involved with the Host Shutoff?

An important question that arises is whether these σ3/PKR regulatory controls have anything to do with the reovirus-induced host-protein synthesis shutoff. The S4 gene and σ3 protein have been implicated by genetic studies in the shutoff of protein synthesis (Sharpe and Fields 1982). Furthermore, the σ3 protein has been shown to regulate the translation of late reoviral mRNAs in experiments conducted in vitro (Lemieux et al. 1987). Imani and Jacobs suggested that the amount of σ3 in cells infected with different serotypes of reovirus may be type 1>type 3>type 2, i.e., inversely related to the strength of the host shutoff (Imani and Jacobs 1988; Giantini and Shatkin 1989). Thus, the more σ3, the more PKR inhibition and the less inhibition of cellular mRNA translation. Despite these observations, there is still no *direct* evidence that PKR can modulate mRNA selective translation in this way, at least not in mammalian cells. The evidence is only correlative at this point: When PKR activity is effectively suppressed, cellular protein synthesis occurs unabated such as might occur in cells infected with reovirus type 1 (possibly because there will not be limitations in functional eIF2).

Regulation of Reovirus mRNA Translational Efficiency

Not all reovirus translational control studies have focused directly on PKR down-regulation or the host-protein synthesis shutoff. There is a differential ability of various reovirus mRNAs to be translated. It has been known for some time that the monocistronic S4 mRNA (encoding the σ3 protein) is translated five- to tenfold more efficiently than the polycistronic S1 mRNA (encoding σ1 and the nonstructural σ1 NS). This observation spurred a number of studies seeking to explain the differences in translation of these two reoviral mRNAs. One contributing factor could include the sequence and/or structure of the 5′ UTR, particularly nucleotides surrounding the AUG initiation codon (Munemitsu and Samuel 1988; Roner et al. 1989). Another possible explanation for differential reovirus mRNA translation brings us back to the PKR pathway. Bischoff and Samuel (1989) found that the poorly translated S1 mRNA was a potent activator of PKR, whereas the better S4 mRNA was a poor activator, perhaps because its gene product is itself a PKR inhibitor. Moreover, a follow-up study showed that treatment of cells with 2-aminopurine (a drug known to block PKR function among its other inhibitory activities) increased the translational efficiency of the reovirus S1 mRNA but had little effect on S4 mRNA translation (Samuel and Brody 1990). Taken together, these studies indirectly suggest that PKR might be involved in differential viral mRNA translation. However, these data also must be interpreted with caution as must any studies using drug treatment. Furthermore, many of these studies examined mRNA encoded by transiently transfected genes and did not look at bona fide viral mRNA translation in a reovirus-infected cell.

Fortunately, other non-PKR-related approaches were undertaken to look at differential reovirus mRNA translation. Doohan and Samuel (1992, 1993) embarked on a path that examined the relative distribution of translating ribosomes on the polycistronic S1 mRNA compared to the monocistronic S4 mRNA using a sensitive modified ribosome protection assay. The first experiments employed a cell-free protein synthesizing system. Briefly, they found that ribosomes paused at several positions on the S1 mRNA relative to the S4 mRNA, suggesting that translating ribosomes were less evenly distributed along the coding region of the inefficiently translated S1 mRNA compared to the efficiently translated S4 mRNA (Doohan and Samuel 1992). In a follow-up study, Doohan and Samuel (1993) analyzed the dynamics of ribosome-mRNA interaction in cultured cells. The distribution of translating ribosomes on polysome-bound reovirus S1 and S4 mRNAs was examined in virus-infected cells where all reovirus mRNAs were present. The pattern of ribosome pausing was nevertheless similar to that of the in vitro system, again with ribosomes less evenly distributed along the poorly translated polycistronic S1 mRNA. Although still inconclusive, these complex studies have opened the door to an improved molecular understanding of reovirus mRNA translation. Related studies using alternative approaches have led Shatkin and colleagues to propose their own model to explain inefficient S1 mRNA translation: Ribosomes engaged in translation of one S1 reading frame interfere with the movement of ribosomes in the other S1 frame due to differences in codon usage (Fajardo and Shatkin 1990). This complex area remains fertile for future investigation.

Porcine Rotaviruses and PKR Regulation

Before closing the section on reoviruses, I would like to mention a recent report analyzing PKR regulation by a different member of the reovirus family, a porcine rotavirus. Rotaviruses cause severe disease such as viral gastroenteritis in humans and animals (Joklik 1984). There are six distinct groups of rotaviruses (referred to as A through F). Group A viruses cause severe diarrheal disease in young and elderly humans. Of main interest to this current review are the group C rotaviruses that cause diarrhea in pigs, cows, and humans. One particular group C virus gene product is relevant: the NSP3 45-kD protein (formerly referred to as NS34). This rotavirus protein contains significant homology with the dsRNA-binding family of proteins, of which the human and murine PKR are members (St Johnston et al. 1992). Besides these interesting sequence homologies, a clue to the possible function of NSP3 was first revealed by experiments showing that the protein binds to viral genomic dsRNA (Mattion et al. 1992). Jacobs and colleagues went on to provide direct evidence that NSP3 indeed functions as a bona fide PKR inhibitor (Langland et al. 1994). Furthermore, like the reovirus σ3 protein, NSP3 was able to rescue replication of the interferon-sensitive E3L–vaccinia virus mentioned above. Given the results with σ3, it was expected that other members of the reovirus family would encode a specific mechanism to down-regulate PKR. What is surprising, however, is that σ3 is not a member of the family of PKR-related dsRNA-binding proteins, whereas NSP3 does belong to this group. It will thus be of interest to examine different reovirus types to identify other PKR inhibitors.

CONCLUDING REMARKS

Progress in both the influenza virus and reovirus virus fields (in and out of translational control regulation studies) has been hampered by the lack of genetics so readily available to investigators studying DNA viruses or even positive-strand RNA viruses. Although not yet widely applicable, this situation with influenza has improved somewhat with the reverse genetics technology, originally pioneered by Palese and colleagues (Luytjes et al. 1989). No such progress has been made with the more intractable dsRNA-containing reoviruses. Despite the lack of genetics, remarkable progress has been made in translation studies of both virus groups, although certainly more work is needed. Future research should answer key unanswered questions concerning mechanisms of selective translation. What exactly are the sequences present in influenza and reovirus mRNAs that dictate preferential translation? Are all these sequences to be found in the 5′UTR or are there critical *cis*-acting sequences present in the coding regions or 3′UTRs as well? There are certainly precedents in the literature suggesting that this might be possible. Furthermore, what are the *trans*-acting factors involved in this selectivity? Will viral and cellular proteins direct the preferential synthesis of viral proteins? As reported elsewhere is this book, perhaps the most progress has been made with the picornaviruses (Sonenberg 1991; Ehrenfeld, this volume) in this respect. However, the story is far from clear and a consensus has not been reached in the picornavirus field as to what cellular proteins are involved in selective translation. Any concluding thoughts must include a discussion of where the experiments on PKR regulation are leading. At least in terms of influenza virus, attempts to dissect the molecular pathways involving PKR will very likely be more complex than previously imagined. It must be determined how influenza virus activates the P58/PKR regulatory pathway. Before this is accomplished, it might be necessary to identify I-P58 and other potential regulatory players.

Despite the progress made in examining the role of σ3 in regulation of PKR activity and reovirus mRNA translation, more work in this area is needed. Although much closer than ever (due largely to the work with the vaccinia recombinant viruses), it is still not known how σ3 acts at the molecular level inside a reovirus-infected cell to inhibit PKR and contribute to the ability of the reovirus to avoid the effects of interferon. New genetic technology will likely permit such in vivo experiments in the near future. Certainly, the role of σ3 in the preferential translation of reovirus mRNAs needs to clarified. Can this important protein accomplish all that is attributed to it? One of the real mysteries is how this structural protein performs so many nonstructural jobs. We should be grateful that there are enough unanswered questions to keep today's influenza and reovirus graduate students and postdoctoral fellows busy into the next millennium.

ACKNOWLEDGMENTS

I thank all members of my laboratory past and present who have carried out much of the work included in the influenza virus section. I am also grateful to my many outside collaborators. The work presented from the author's laboratory was supported by a grant from RiboGene Inc. of Hayward California, and by National Institutes of Health grants AI-22646 and RR-00166.

REFERENCES

Alonso-Caplen, F.V., M.G. Katze, and R.M. Krug. 1988. Efficient transcription, not translation, is dependent on adenovirus tripartite leader sequences at late times of infection. *J. Virol.* **62:** 1606–1616.

Babich, A., L. Feldman, J. Nevins, J.E. Darnell, and C. Weinberger. 1983. Effect of adenovirus on metabolism of specific host mRNAs: Transport control and specific translation discrimination. *Mol. Cell. Biol.* **3:** 1212–1221.

Barber, G.N., S. Thompson, T.G. Lee, T. Strom, R. Jagus, A. Darveau, and M.G. Katze. 1994. The 58-kilodalton inhibitor of the interferon-induced double-stranded RNA-activated protein kinase is a tetratricopeptide repeat protein with oncogenic properties. *Proc. Natl. Acad. Sci.* **91:** 4278–4282.

Beattie, E., J. Tartaglia, and E. Paoletti. 1991. Vaccinia virus-encoded eIF-2 alpha homolog abrogates the antiviral effect of interferon. *Virology* **183:** 419–422.

Beattie, E., K.L. Denzler, J. Tartaglia, M.E. Perkus, E. Paoletti, and B.L. Jacobs. 1995. Reversal of the interferon-sensitive phenotype of an E3L-minus vaccinia virus by expression of the reovirus S4 gene. *J. Virol.* (in press).

Beloso, A., C. Martinez, J. Valcarcel, J.F. Santaren, and J. Ortin. 1992. Degradation of cellular mRNA during influenza virus infection: Its possible role in protein synthesis shutoff. *J. Gen. Virol.* **73:** 575–581.

Bischoff, J.R. and C.E. Samuel. 1989. Mechanism of interferon action: Activation of the human PI/eIF-2 protein kinase by individual reovirus-class mRNAs: S1 mRNA is a potent activator relative to S4 mRNA. *Virology* **172:** 106–115.

Black, T., B. Safer, A.G. Hovanessian, and M.G. Katze. 1989. The cellular 68,000 M_r protein kinase is highly autophosphorylated and activated yet significantly degraded during poliovirus infection: Implications for translational regulation. *J. Virol.* **63:** 2244–2252.

Denzler, K.L. and B.L. Jacobs. 1994. Site-directed mutagenic analysis of reovirus sigma 3 protein binding to dsRNA. *Virology* **204:** 190–199.

Detjen, B.M., W.E. Walden, and R.E. Thach. 1982. Translational specificity in reovirus-infected mouse fibroblasts. *J. Biol. Chem.* **257:** 9855–9860.

Doohan, J.P. and C.E. Samuel. 1992. Biosynthesis of reovirus-specified polypeptides: Ribosome pausing during the translation of reovirus S1 mRNA. *Virology* **186:** 409–425.

———. 1993. Biosynthesis of reovirus specified polypeptides. *J. Biol. Chem.* **268:** 18313–18320.

Enami, K., T.A. Sato, S. Nakada, and M. Enami. 1994. Influenza virus NS1 protein stimulates translation of the M1 protein. *J. Virol.* **68:** 1432–1437.

Fajardo, J.E. and A.J. Shatkin. 1990. Translation of bicistronic viral mRNA in transfected cells: Regulation at the level of elongation. *Proc. Natl. Acad. Sci.* **87:** 328–332.

Feigenblum, D. and R.J. Schneider. 1993. Modification of eukaryotic initiation factor 4F during infection by influenza virus. *J. Virol.* **67:** 3027–3035.

Garfinkel, M.S. and M.G. Katze. 1992. Translational control by influenza virus: Selective and cap dependent translation of viral mRNAs in infected cells. *J. Biol. Chem.* **267:** 9383–9390.

———. 1993a. Translational control by influenza virus. Selective translation is mediated by sequences within the viral mRNA 5′-untranslated region. *J. Biol. Chem.* **268:** 22223–22226.

———. 1993b. How does influenza virus regulate gene expression at the level of transla-

tion: Let us count the ways. *Gene Expression* **3:** 109–118.
———. 1994. Influenza virus control of protein synthesis. *Sci. Am. Sci. Med.* **1:** 64–73.
Giantini, M. and A. Shatkin. 1989. Stimulation of chloramphenicol acetyl transferase mRNA translation by reovirus capsid polypeptide sigma 3 in cotransfected Cos cells. *J. Virol.* **63:** 2415–2421.
Goebl, M. and M. Yanagida. 1991. The TPR snap helix: A novel protein repeat motif from mitosis to transcription. *Trends Biochem. Sci.* **16:** 173–177.
Green, S.R. and M.B. Mathews. 1992. Two RNA-binding motifs in the double-stranded RNA-activated protein kinase, DAI. *Genes Dev.* **6:** 2478–2490.
Hershey, J.W.B. 1991. Translational control in mammalian cells. *Annu. Rev. Biochem.* **60:** 717–755.
Hovanessian, A.G. 1989. The double stranded RNA-activated protein kinase induced by interferon: dsRNA-PK. *J. Interferon Res.* **9:** 641–647.
Huismans, H. and W.K. Joklik. 1976. Reovirus-coded polypeptides in infected cells: Isolation of two native monomeric polypeptides with affinity for single-stranded and double-stranded RNA, respectively. *Virology* **70:** 411–424.
Imani, F. and B.L. Jacobs. 1988. Inhibitory activity for the interferon-induced protein kinase is associated with the reovirus serotype 1 sigma 3 protein. *Proc. Natl. Acad. Sci.* **85:** 7887–7891.
Inglis, S.C. 1982. Inhibition of host protein synthesis and degradation of cellular mRNAs during infection by influenza and herpes simplex virus. *Mol. Cell. Biol.* **2:** 1644–1648.
Jacobs, B.L. and R.E. Ferguson. 1991. The Lang strain of reovirus serotype 1 and the Dearing strain of reovirus serotype 3 differ in their sensitivities to beta interferon. *J. Virol.* **65:** 5102–5104.
Joklik, W.K. 1974. *Comprehensive virology* (ed. H. Fraenkel-Conrat and R. Wagner), pp. 297–334. Plenum Press, New York.
———. 1984. *The reoviridae.* Plenum Press, New York.
Katze, M.G. 1992. The war against the interferon-induced dsRNA activated protein kinase: Can viruses win? *J. Interferon Res.* **12:** 241–248.
———. 1993. Games viruses play: A strategic initiative against the interferon induced dsRNA activated 68,000 M_r protein kinase. *Semin. Virol.* **4:** 259–268.
———. 1995. Regulation of the interferon induced-PKR: Can viruses cope? *Trends Microbiol.* **3:** 75–78.
Katze, M.G. and R.M. Krug. 1984. Metabolism and expression of RNA polymerase II transcripts in influenza virus infected cells. *Mol. Cell. Biol.* **4:** 2198–2206.
———. 1990. Translational control in influenza virus-infected cells. *Enzyme* **44:** 265–277.
Katze, M.G., Y.T. Chen, and R.M. Krug. 1984. Nuclear-cytoplasmic transport and VAI RNA-independent translation of influenza viral messenger RNAs in late adenovirus-infected cells. *Cell* **37:** 483–490.
Katze, M.G., D. DeCorato, and R.M. Krug. 1986a. Cellular mRNA translation is blocked at both the initiation and elongation following infection by influenza virus or adenovirus. *J. Virol.* **60:** 1027–1039.
Katze, M.G., J. Lara, and M. Wambach. 1989. Nontranslated cellular mRNAs are associated with cytoskeletal framework in influenza virus or adenovirus infected cells. *Virology* **169:** 312–322.
Katze, M.G., B.M. Detjen, B. Safer, and R.M. Krug. 1986b. Translational control by influenza virus: Suppression of the kinase that phosphorylates the alpha subunit of initia-

tion factor eIF-2 and selective translation of influenza viral mRNAs. *Mol. Cell. Biol.* **6:** 1741–1750.

Katze, M.G., J. Tomita, T. Black, R.M. Krug, B. Safer, and A.G. Hovanessian. 1988. Influenza virus regulates protein synthesis during infection by repressing the autophosphorylation and activity of the cellular 68,000 M_r protein kinase. *J. Virol.* **62:** 3710–3717.

Katze, M.G., M. Wambach, M.-L. Wong, M.S. Garfinkel, E. Meurs, K.L. Chong, B.R.G. Williams, A.G. Hovanessian, and G.N. Barber. 1991. Functional expression and RNA binding analysis of the interferon-induced, dsRNA activated 68,000 M_r protein kinase in a cell-free system. *Mol. Cell. Biol.* **11:** 5497–5505.

Koromilas, A.E., S. Roy, G.N. Barber, M.G. Katze, and N. Sonenberg. 1992. Malignant transformation of the IFN-inducible dsRNA dependent protein kinase. *Science* **257:** 1685–1689.

Kozak, M. 1989. The scanning model for translation: An update. *J. Cell Biol.* **108:** 229–241.

———. 1991. An analysis of vertebrate mRNA sequences: Intimations of translational control. *J. Cell Biol.* **115:** 887–903.

Krug, R.M., F. Alonso-Caplen, I. Julkunen, and M.G. Katze. 1989. Expression and replication of the influenza virus genome. In *The influenza viruses* (ed. R.M. Krug), pp. 89–152. Plenum Press, New York.

Kumar, A., J. Haque, J. Lacoste, J. Hiscott, and B.R.G. Williams. 1994. The dsRNA-dependent protein kinase, PKR, activates transcription factor NFkB by phosphorylating IkB. *Proc. Natl. Acad. Sci.* **91:** 6288–6292.

Langland, J.O., S. Pettiford, B. Jiang, and B.L. Jacobs. 1994. Products of the porcine group C rotavirus NSP3 gene bind specifically to double-stranded RNA and inhibit activation of the interferon-induced protein kinase PKR. *J. Virol.* **68:** 3821–3829.

Lazaris-Karatzas, A., K.S. Mantine, and N. Sonenberg. 1990. Malignant transformation by a eukaryotic initiation factor subunit that binds to mRNA 5′ cap. *Nature* **345:** 544–547.

Lazarowitz, S.G., R.W. Compans, and P.W. Choppin. 1971. Influenza virus structural and nonstructural proteins in infected cells and their plasma membrane. *Virology* **46:** 830–843.

Lee, T.-G. and M.G. Katze. 1993. Cellular inhibitors of the interferon-induced dsRNA activated protein kinase. *Prog. Mol. Subcell. Biol.* **14:** 48–65.

Lee, T.-G., J. Tomita, A.G. Hovanessian, and M.G. Katze. 1990. Purification and partial characterization of a cellular inhibitor of the interferon-induced 68,000 M_r protein kinase from influenza virus-infected cells. *Proc. Natl. Acad. Sci.* **87:** 6208–6212.

———. 1992. Characterization and regulation of the 58,000 dalton cellular inhibitor of the interferon-induced, dsRNA activated protein kinase. *J. Biol. Chem.* **267:** 14238–14243.

Lee, T.G., N. Tang, S. Thompson, J. Miller, and M.G. Katze. 1994. The 58,000-dalton cellular inhibitor of the interferon-induced double-stranded RNA-activated protein kinase (PKR) is a member of the tetratricopeptide repeat family of proteins. *Mol. Cell. Biol.* **14:** 2331–2342.

Lemay, G. 1988. Transcriptional and translational events during reovirus infection. *Biochem. Cell Biol.* **66:** 803–812.

Lemay, G. and S. Millward. 1986. Inhibition of translation in L-cell lysates by free polyadenylic acid: Differences in sensitivity among different mRNAs and possible in-

volvement of an initiation factor. *Arch. Biochem. Biophys.* **249:** 191–198.
Lemieux, R., G. Lemay, and S. Millward. 1987. The viral protein sigma 3 participates in translation of late viral mRNA in reovirus-infected L cells. *J. Virol.* **61:** 2472–2479.
Lengyel, P. 1993. Tumor-suppressor genes: News about the interferon connection. *Proc. Natl. Acad. Sci.* **90:** 55893–55895.
Lenk, R. and S. Penman. 1979. The cytoskeleton framework and poliovirus metabolism. *Cell* **16:** 289–301.
Lloyd, R.M. and A.J. Shatkin. 1992. Translational stimulation by reovirus polypeptide 3: Substitution for VAI RNA and inhibition of phosphorylation of the a subunit of eukaryotic initiation factor 2. *J. Virol.* **66:** 6878–6884.
Lodish, H.F. and M. Porter. 1980. Translational control of protein synthesis after infection by vesicular stomatitis virus. *J. Virol.* **36:** 719–733.
Logan, J. and T. Shenk. 1984. Adenovirus tripartite leader sequence enhances translation of mRNAs late after infection. *Proc. Natl. Acad. Sci.* **81:** 3655–3659.
Luytjes, W., M. Krystal, M. Enami, J.D. Parvin, and P. Palese. 1989. Amplification, expression, and packaging of a foreign gene by influenza virus. *Cell* **59:** 1107–1113.
Mallo, M., J. Martinez-Costas, and J. Benavente. 1991. The stimulatory effect of actinomycin D on avian reovirus replication in L cells suggests that translational competition dictates the fate of the infection. *J. Virol.* **65:** 5506–5512.
Maran, A. and M.B. Mathews. 1988. Characterization of the dsRNA implicated in the inhibition of protein synthesis in cells infected with a mutant adenovirus defective for VA RNA 1. *Virology* **164:** 106–113.
Maran, A., R.K. Maitra, A. Kumar, B. Dong, W. Xiao, G. Li, B.R.G. Williams, P.F. Torrence, and R.H. Silverman. 1994. Blockage of NF-κB Signaling by selective ablation of an mRNA Target by 2-5A antisense chimeras. *Science* **265:** 789–792.
Mathews, M.B. and T. Shenk. 1991. Adenovirus virus-associated RNA and translation control. *J. Virol.* **65:** 5657–5662.
Mattion, N.M., J. Cohen, C. Aponte, and M. Estes. 1992. Characterization of an oligomerization domain and RNA-binding properties on rotavirus nonstructural protein NS34. *Virology* **190:** 68–83.
McCrae, M.A. and W.K. Joklik. 1978. The nature of the polypeptide encoded by each of the 10 double-stranded RNA segments of reovirus type 3. *Virology* **89:** 578–593.
Merrick, W.C. 1992. Mechanism and regulation of eukaryotic protein synthesis. *Microbiol. Rev.* **56:** 291–315.
Meurs, E., J. Galabru, G.N. Barber, M.G. Katze, and A.G. Hovanessian. 1993. Tumour suppressor function of the interferon-induced double-stranded RNA-activated 68,000-M_r protein kinase. *Proc. Natl. Acad. Sci.* **90:** 232–236.
Miller, J.E. and C.E. Samuel. 1992. Proteolytic cleavage of the reovirus sigma 3 protein results in enhanced double-stranded RNA-binding activity: Identification of a repeated basic amino acid motif within the C-terminal binding region. *J. Virol.* **66:** 5347–5356.
Munemitsu, S.M. and C.E. Samuel. 1988. Biosynthesis of reovirus-specified polypeptides: Effect of point mutation of the sequences flanking the 5′-proximal AUG initiator codons of the reovirus S1 and S4 genes on the efficiency of mRNA translation. *Virology* **163:** 643–646.
Ray, B.K., T.G. Brendler, S. Adya, S. Daniels-McQueen, J.K. Miller, J.W.B. Hershey, J.A. Grifo, W.C. Merrick, and R.E. Thach. 1983. Role of mRNA competition in regulating translation: Further characterization of mRNA discriminatory initiation factors. *Proc. Natl. Acad. Sci.* **80:** 663–667.

Rhoads, R.E. 1993. Regulation of eukaryotic protein synthesis by initiation factors. *J. Biol. Chem.* **268:** 3017–3020.

Roner, M.R., R.K. Gaillard, Jr., and W.K. Joklik. 1989. Control of reovirus messenger RNA translation efficiency by the regions upstream of initiation codons. *Virology* **168:** 292–301.

Roy, S., M.B. Agy, A.G. Hovanessian, N. Sonenberg, and M.G. Katze. 1991. The integrity of the stem structure of human immunodeficiency virus type 1 Tat-responsive sequence RNA is required for interaction with the interferon-induced 68,000-M_r protein kinase. *J. Virol.* **65:** 632–640.

Samuel, C.E. 1993. The eIF-2 alpha protein kinases, regulators of translation in eukaryotes from yeasts to humans. *J. Biol. Chem.* **268:** 7603–7606.

Samuel, C.E. and M.S. Brody. 1990. Biosynthesis of reovirus-specified polypeptides. 2-aminopurine increases the efficiency of translation of reovirus s1 mRNA but not s4 mRNA in transfected cells. *Virology* **176:** 106–113.

Seliger, L.S., M. Giantini, and A.J. Shatkin. 1992. Translational effects and sequence comparisons of the three serotypes of the reovirus S4 gene. *Virology* **187:** 202–210.

Sharpe, A.H. and B.N. Fields. 1982. Reovirus inhibition of cellular RNA and protein synthesis: Role of the S4 gene. *Virology* **122:** 381–391.

Sikorski, R.S., W.A. Michaud, J.C. Wootton, M.S. Boguski, C. Connelly, and P. Hieter. 1991. TPR proteins as essential components of the yeast cell cycle. *Cold Spring Harbor Symp. Quant. Biol.* **56:** 663–673.

Silver, P.A. and J.C. Way. 1993. Eukaryotic DnaJ homologs and the specificity of Hsp70 activity. *Cell* **74:** 5–6.

Skehel, J.J. 1972. Polypeptide synthesis in influenza virus-infected cells. *Virology* **49:** 23–36.

Skup, D. and S. Millward. 1980. mRNA capping enzymes are masked in reovirus progeny subviral particles. *J. Virol.* **34:** 490–496.

Smith, R.E., H.J. Zweerink, and W.K. Joklik. 1969. Polypeptide components of virions, top component, and cores of reovirus type 3. *Virology* **39:** 791–798.

Sonenberg, N. 1991. Picornavirus RNA translation continues to surprise. *Trends. Genet.* **7:** 105–106.

St Johnston, D., N.H. Brown, J.G. Gall, and M. Jantsch. 1992. A conserved double stranded RNA-binding domain. Proc. *Natl. Acad. Sci.* **89:** 10979–10983.

Thimmappaya, B., C. Weinberger, R.J. Schneider, and T. Shenk. 1982. Adenovirus VA1 RNA is required for efficient translation of viral mRNAs at late times after infection. *Cell* **31:** 543–551.

Tillotson, L. and A.J. Shatkin. 1992. Reovirus polypeptide sigma 3 and N-terminal myristoylation of polypeptide u1 are required for site-specific cleavage to u1C in transfected cells. *J. Virol.* **66:** 2180–2186.

Walden, W.E., T. Godefroy-Colburn, and R.E. Thach. 1981. The role of mRNA competition in regulating translation. I. Demonstration of competition *in vivo*. *J. Biol. Chem.* **256:** 11739–11746.

Yamanaka, K., A. Ishihama, and K. Nagata. 1988. Translational regulation of influenza virus mRNAs. *Virus Genes* **1:** 19–30.

Yamanaka, K., K. Nagata, and A. Ishihama. 1991. Temporal control for translation of influenza virus mRNAs. *Arch. Virol.* **120:** 33–42.

22

Translationally Coupled Degradation of mRNA in Eukaryotes

Nicholas G. Theodorakis
Johns Hopkins University Medical School
Division of Pediatric Surgery
Baltimore, Maryland 21205

Don W. Cleveland
Ludwig Institute for Cancer Research
University of California, San Diego
La Jolla, California 92093

The abundance of cytoplasmic messenger RNA in a cell is determined by both the rate of synthesis of the mRNA (i.e., its transcription rate) and the rate at which the mRNA is degraded. Thus, both processes contribute equally to gene expression. If the cell is to change the level of an mRNA rapidly, then it would be advantageous for that mRNA to be short-lived, so that it could respond to rapid changes in transcription. Similarly, a long half-life would be useful if an mRNA was to encode a protein that is needed in great abundance. Therefore, it would seem likely that different mRNAs would have different stabilities, depending on their cellular role.

It is becoming increasingly clear that protein synthesis may have a role in mRNA degradation. There are many examples in which the inhibition of protein synthesis stabilizes mRNA (see below). In principle, this increase in mRNA stability could be a result of the decay of a labile factor, or that the translation machinery itself is involved in the degradation of mRNA. We have limited this discussion to mechanisms of mRNA degradation in eukaryotes; for a recent review of the role of translation in prokaryotic mRNA degradation, see Peterson (1993). In the first part of this chapter, we focus on the degradation of specific mRNAs that carry message instability elements; the second part deals with a global mechanism for degradation of mRNA.

EXAMPLES OF SPECIFIC mRNAS IN WHICH TRANSLATION IS COUPLED TO DEGRADATION

β-Tubulin

One of the best studied examples in which the degradation of a specific mRNA is coupled to translation is that of the autoregulated degradation of β-tubulin mRNA (for more comprehensive reviews of tubulin mRNA degradation, see Theodorakis and Cleveland 1993; Cleveland and Theodorakis 1994). When the level of tubulin subunits rises in the cell (e.g., when microtubules are depolymerized or when excess tubulin subunits are microinjected into the cell), the major changes in gene expression that occur are the repression of the synthesis of α- and β-tubulin, which is the consequence of the reduction in the level of tubulin mRNAs (Ben Ze'ev et al. 1979; Cleveland et al. 1981, 1983). A variety of experiments have shown that this response is dictated at the posttranscriptional level. First of all, no changes in the rate of transcription of tubulin genes could be detected when tubulin mRNA levels were lowered in response to microtubule-depolymerizing drugs (Cleveland and Havercroft 1983). Second, hybrid gene constructs between tubulin and thymidine kinase produced mRNAs that were still responsive to changes in tubulin subunit levels even when no tubulin promoter sequences were present, as long as sequences from the tubulin-coding region were present (Gay et al. 1987; Yen et al. 1988b; see below). Finally, cells that were enucleated could still respond to increases in tubulin subunit levels by decreasing the synthesis of tubulin (Caron et al. 1985; Pittenger and Cleveland 1985).

Detailed molecular dissection of the β-tubulin gene revealed that a 13-nucleotide region that encoded the first four translated codons (MREI) was necessary and sufficient for conferring onto a heterologous gene the ability to respond to changes in the level of tubulin subunits (Yen et al. 1988a). These results suggested two models for the recognition of tubulin mRNA by the cell degradation machinery. One possibility is that some cellular factor could recognize the tubulin mRNA directly; alternatively, since the sequence that confers responsiveness to subunit levels resides in a translated portion of the mRNA, the nascent peptide itself, rather than the mRNA per se, could be the autoregulatory sequence. A variety of observations suggest that the latter model is the correct one. First of all, only tubulin mRNAs that are associated with polysomes are degraded in response to microtubule-depolymerizing drugs; furthermore, protein synthesis inhibitors that disrupt polysomes such as puromycin or pactamycin inhibit the autoregulatory process, whereas a low concentration of cycloheximide (10 μg/ml), which maintains polysomes, actually stimulates autoregulated degradation (Pachter et al. 1987; Bachurski et

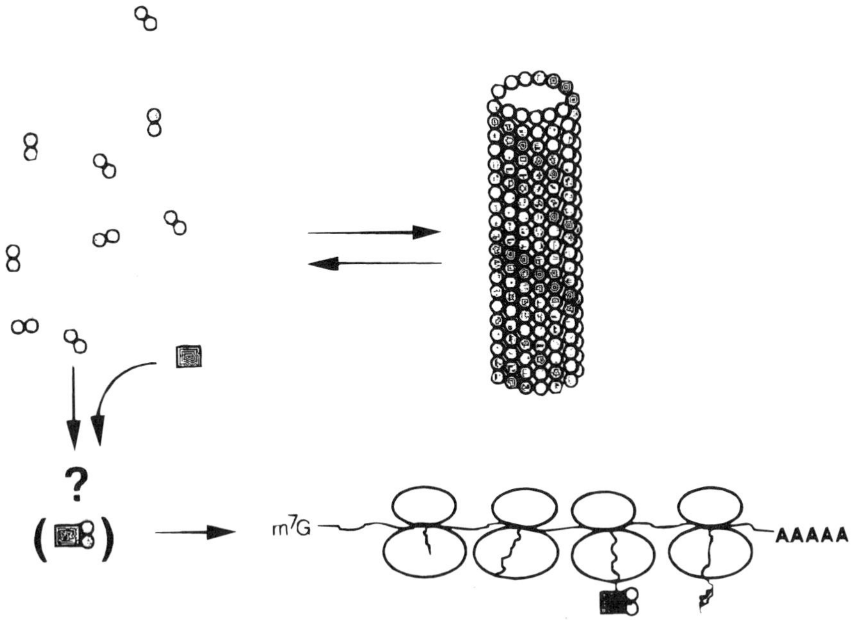

Figure 1 A model for autoregulated β-tubulin mRNA instability. Dynamic instability of the microtubule (large cylinder) leads to a pool of free tubulin subunits (*small circles*). Tubulin subunits, perhaps in association with an unknown cellular factor (*small square*), bind to the β-tubulin nascent peptide as it emerges from the ribosome (*large ovals*). This binding event activates a nuclease that degrades the polysome-bound tubulin mRNA.

al. 1994). Second, β-tubulin mRNAs or β-tubulin/thymidine kinase hybrid mRNAs that terminate translation prior to codon 42 are unresponsive to changes in tubulin subunit levels (Pachter et al. 1987; Yen et al. 1988b), suggesting that the nascent peptide needs to be exposed. Third, β-tubulin mRNAs that had mutations in the autoregulatory sequence could still respond to changes in tubulin subunit levels only if the protein-coding capacity of the regulatory sequence MREI (or alternatively, MRDI) was maintained (Yen et al. 1988a; Bachurski et al. 1994). Fourth, the autoregulatory sequence must be translated in its correct reading frame to function (Yen et al. 1988a). Finally, antibodies that bind to the nascent peptide disrupt the autoregulatory process when introduced into cells (Theodorakis and Cleveland 1992). Together, these data indicate that the binding of some cellular factor (as yet unknown, but presumably tubulin subunits) to the β-tubulin nascent peptide is the trigger for the degradation of β-tubulin mRNA (see Fig. 1).

The translation process may have another role in the degradation of tubulin mRNA beyond its role in the synthesis of the autoregulatory sequence. Studies with translation inhibitors indicate that translational elongation is required for autoregulated degradation of β-tubulin mRNA (Pachter et al. 1987; Gay et al. 1989). Although low concentrations of cycloheximide or anisomycin (translation elongation inhibitors) enhance the autoregulated degradation of tubulin mRNA, high concentrations of either drug used alone, or a combination of low concentrations of both drugs, block the autoregulated degradation of tubulin mRNA. Similar results are also obtained with emetine, another translation elongation inhibitor that freezes ribosomes on mRNA (Gong and Brandhorst 1988; Gay et al. 1989). One interpretation for these results is that the process of translation elongation itself is necessary for degrading tubulin mRNA.

Histone

The synthesis of histones is tightly coupled to DNA synthesis; when cells leave the synthetic phase (S phase) of the cell cycle or when DNA synthesis is inhibited by drugs such as hydroxyurea, the mRNAs encoding the "replication-dependent" histones are rapidly degraded (for review, see Marzluff and Hanson 1993). In animal cells, the replication-dependent histones are unique in that their 3′ ends are not formed by a polyadenylation signal. Rather, a stem-loop structure in the mRNA is necessary for the formation of the 3′ end and for the regulated degradation of histone mRNA (Pandey and Marzluff 1987). In contrast, the so-called "replacement" variants of histone mRNAs lack this characteristic stem-loop structure, are polyadenylated, and are not rapidly degraded in the absence of DNA synthesis (Sittman et al. 1983).

Two lines of evidence suggest that histone mRNA degradation is coupled to translation. The first is that for histones, as is the case for many cellular mRNAs, inhibition of protein synthesis stabilizes the mRNA (Stimac et al. 1984). The second is that histone mRNA constructs, in which the 3′ stem-loop structure is more than 300 nucleotides from the stop codon, are not regulated in response to changes in DNA synthesis (Graves et al. 1987). A model that explains these data is that the ribosome carries a nuclease or nuclease-activating protein that recognizes the histone mRNA 3′ stem-loop region; if the ribosome falls off the mRNA before reaching this vicinity, then the mRNA is not degraded (Marzluff and Hanson 1993). Support for this model comes from the observation that nucleases that degrade histone mRNA in vitro can be purified from the polysome fraction of cells (Ross et al. 1986). However, another explanation for the requirement that ribosomes must approach the 3′stem-loop for histone mRNA degradation is that long stretches of untranslated sequence upstream of the stem-loop might allow the formation of alternative secondary structures involving base pairing with the sequence at the 3′end, thus destroying the secondary structure of the stem-loop. In this context, the requirement for ribosome translocation to close proximity of the stem-loop is that the ribosomes might unwind other competing secondary structures. This hypothesis is supported by the observation that the stem-loop must be at the extreme 3′terminus of the mRNA for its ability to function as a destabilizing element; sequences 3′to the stem-loop might be able to disrupt its secondary structure (Graves et al. 1987).

AU-rich Element-containing mRNAs

A wide variety of mRNAs that encode cytokines, proto-oncogenes, and serum-induced mRNAs are unstable and, in addition, contain in their 3′-untranslated regions (3′UTRs) a sequence that is very rich in adenylate and uridylate residues. In several cases, it has been shown that this AU-rich element (ARE) can confer on a test gene low mRNA levels that are associated with rapid degradation (Shaw and Kamen 1986; Wilson and Treisman 1988; Shyu et al. 1989; Aharon and Schneider 1993). Although the exact sequence of the ARE has not been rigorously defined, sequence inspection of several ARE-containing mRNAs suggests that a consensus may be multiple copies of the sequence AUUUA (Shaw and Kamen 1986), although experimental evidence indicates that the ARE is likely not to be this simple (Chen and Shyu 1994; Lagnado et al. 1994; Zubiaga et al. 1995). It should also be noted that it is not clear whether all AREs are regulated in the same manner (see, e.g., Schuler and Cole 1988). As is the case for many unstable mRNAs, ARE-containing mRNAs are stabilized in the presence of protein synthesis inhibitors (Shaw and Kamen 1986; Wilson and Treisman 1998). However, whether translation of a labile *trans*-acting factor or whether translation of the mRNA in *cis* is required is a complex issue. In the case of the granulocyte-macrophage–colony-stimulating factor (GM-CSF) ARE, the data indicate that translation of the mRNA in *cis* is required for destabilization. In one example, a β-globin mRNA containing the GM-CSF ARE, which is normally expressed at low levels, was rendered untranslatable by mutating its initiating codon to a stop codon (Savant-Bhonsale and Cleveland 1992). The level of the resulting mRNA was increased greater than tenfold such that its level approximated that of a similar mRNA with no functional ARE. In another example, a stable stem-loop structure was introduced into the 5′UTR of a hybrid mRNA containing the hepatitis B surface antigen coding region with the GM-CSF ARE in its 3′ UTR (Aharon and Schneider 1993). This structure reduced its translational efficiency by greater than 20-fold and increased the stability of its mRNA to a similar degree. The possibility that the 5′stable stem-loop protected the mRNA from a 5′-3′exonuclease was eliminated by inserting an internal ribosome entry site (IRES) 3′to the stem-loop structure. The IRES both restored its translational efficiency and decreased its stability.

In the case of c-*fos*, there are at least two domains that can specify rapid mRNA degradation. One domain is in the translated portion of the mRNA; the other is a 75-nucleotide AU-rich sequence in the 3′UTR that contains several copies of the AUUUA motif (Shyu et al. 1991). Either domain could specify rapid degradation on a test mRNA; however, the removal of only one of these elements from the c-*fos* gene had no effect on the stability of its mRNA due to the presence of the remaining destabilizing element. Curiously, the two elements may act through different pathways. Whereas the destabilizing element in the coding region can function in the absence of continued transcription, the ARE requires continuing transcription to function. Either the ARE or the domain in the protein-coding region may function by stimulating the rapid removal of the poly(A) tail, which precedes degradation of the mRNA (Wilson and Treisman 1988; Schiavi et al. 1994).

The effects of translation on the degradation of c-*fos* mRNA are complex. Like many mRNAs, inhibition of protein synthesis stabilizes c-*fos* mRNA as well as inhibiting poly(A) removal (Wilson and Treisman 1988), indicating either the requirement for translation of the mRNA in *cis* or the continued synthesis of a labile *trans*-acting factor. The stabilization of c-*fos* mRNA occurs almost immediately after the addition of the protein synthesis inhibitors, suggesting that the translation of the mRNA per se was required for degradation (Wilson and Treisman 1988). Experiments using mutational analysis of the c-*fos* gene to affect its translatability gave conflicting results, possibly because the elements are regulated differently. In one experiment designed to regulate the translation of a hybrid c-*fos* mRNA, Koeller et al. (1991) inserted the c-*fos* ARE into an mRNA that contains an iron regulatory element (IRE) in its 5′UTR, allowing the translation of the hybrid mRNA to be regulated by cellular iron levels (see Rouault et al., this volume). These authors observed that the stability of this mRNA was unaffected by changes in the translation state of the hybrid mRNA. However, unlike the c-*fos* ARE, the instability element in the c-Fos protein-coding region requires translation of the mRNA in *cis* in order to function (Schiavi et al. 1994). Although this result is reminiscent of the instability determinant in β-tubulin mRNA, which is actually the nascent peptide (see above), the c-Fos protein-coding region instability determinant is recognized as RNA, not as protein (Wellington et al. 1993).

A wide variety of other unstable mRNAs also contain AREs in their 3′UTRs, including, but not limited to, those encoding *groα* and 9E3 (members of the platelet factor 4 family), tumor necrosis factor-α (TNF-α), β-interferon, and monocyte-derived neutrophil-activating factor (Raj and Pitha 1981; Caput et al. 1986; Shaw and Kamen 1986; Kowalski and Denhardt 1989; Stoeckle and Hanafusa 1989; Stoeckle 1991, 1992). It is not known whether the AREs in those mRNAs function via the same pathway as those in GM-CSF or c-*fos*. Intriguingly, these mRNAs may not always be degraded by a pathway involving poly(A) shortening, since in some cases, presumptive decay intermediates have been found that appear as if the mRNA was endonucleolytically cleaved without the prior loss of poly(A) (Stoeckle 1992).

c-*myc*

There are many superficial similarities between the regulation of c-*myc* gene expression and that of c-*fos*. As is the case for c-*fos*, c-*myc* mRNA is induced by serum stimulation (albeit at a slightly later time than is the case for c-*fos*; Lau and Nathans 1987), is very unstable, with a half-life of about 15 minutes (Dani et al. 1984), and is stabilized by the inhibition of protein synthesis (Linial et al. 1985). Moreover, like the c-*fos* mRNA, the removal of the poly(A) tail precedes the degradation of the mRNA, and continued protein synthesis is required for the rapid removal of poly(A) and mRNA degradation (Laird-Offringa et al. 1990). Furthermore, the c-*myc* mRNA contains in its 3′-untranslated sequence a region rich in adenylate and uridylate residues with one or two copies of the AUUUA motif that has been implicated in the instability of GM-CSF and c-*fos* mRNAs (Shaw and Kamen 1986; Jones and Cole 1987). Nevertheless, multiple regions of the c-*myc* mRNA appear to be involved in its stability (Jones and Cole 1987; Pei and Calame 1988; Laird-Offringa et al. 1991; Wisdom and Lee 1991; Laird-Offringa 1992). Although early studies implicated the 5′UTR in the instability of c-*myc* mRNA (Pei and Calame 1988), attention soon shifted to the 3′UTR containing the AU-rich sequences (Jones and Cole 1987; Laird-Offringa et al. 1991). However, unlike the cases of c-*fos* and GM-CSF mRNAs, it became clear that the AUUUA pentanucleotide sequences in the 3′UTR of c-*myc* were neither necessary nor sufficient for mRNA instability (Bonnieu et al. 1990; Cole and Mango 1990; Laird-Offringa et al. 1991; Wisdom and Lee 1991). Nevertheless, sequences in the 3′UTR other than the AUUUA motif contain an instability element (Laird-Offringa 1992).

In addition to these sequences, there is also an instability determinant in the protein-coding region of the c-*myc* mRNA (Wisdom and Lee 1991), which is also reminiscent of c-*fos* mRNA. Translation through this segment is apparently not required for the rapid degradation of c-*myc* mRNA (Laird-Offringa et al. 1991), although in some cases using hybrid mRNAs, there is an approximately twofold stabilization of c-*myc* mRNA by terminating translation prior to this sequence (Herrick and Ross 1994).

Although translation of the entire body of the c-*myc* mRNA is not necessary to maintain at least a portion of its instability, the requirement of protein synthesis for mRNA degradation does appear to be in *cis*, rather than through a *trans*-acting labile factor. Mutation of the initiating AUG codon to a stop codon to render the mRNA untranslatable stabilizes c-*myc* mRNA (Cole and Mango 1990).

Summary

What is the translational requirement for degradation of mRNAs containing instability elements? Most, if not all, short-lived messages examined are stabilized when protein synthesis is inhibited. Of the two possible explanations for this phenomenon—either that an extremely short-lived protein is required for message degradation or that translation of the mRNA in *cis* is required for its degradation—the latter possibility seems more plausible and has greater experimental support.

What could be the mechanism by which ribosome transit couples translation to degradation? In the case of β-tubulin, translation is required in order to synthesize the peptide recognition sequence; this mechanism has also been suggested for the c-Myc protein-coding RNA instability determinant (Laird-Offringa 1992). However, this is clearly not the case for other unstable mRNAs such as those encoding c-*fos* or GM-CSF. Instead, other possibilities include the following models (which are not necessarily exclusive): (1) The ribosomes or other parts of the translational machinery are associated with factors that recognize the instability determinants or with nucleases that target mRNA for degradation. (2) Translation alters the secondary structure of the mRNA, thereby making the instability determinant accessible to its recognition factors or making the mRNA more accessible to nucleases. (3) Translation removes protective proteins from the mRNA that might be protecting the mRNA from degradation. (4) Translation might localize the RNA to a subcellular region that is the location of the degradation machinery (see Cole and Mango 1990).

DEGRADATION OF mRNAS THAT CONTAIN PREMATURE TRANSLATION TERMINATORS

Perhaps the best example in which mRNA degradation is intimately linked with the process of translation is in the accelerated degradation of some mRNAs that have premature translation terminators, a process that is sometimes referred to as nonsense-mediated mRNA degradation (Peltz and Jacobson 1993). In *Saccharomyces*, it has long been observed that many mRNAs that harbor nonsense mutations have shorter half-lives than their normal counterparts (Losson and Lacroute 1979). That this altered half-life is due to premature translation termination and not due to altered mRNA structure is supported by the fact that an amber suppressor tRNA can restore the proper half-life to an mRNA that contains a premature amber mutation (Losson and Lacroute 1979). In vertebrate cells, the reduction in mRNA levels caused by premature translation termination has been observed in a wide variety of mRNAs, including those encoding globins (Lim et al. 1992; Lim and Maquat 1992), Rous sarcoma virus (RSV) *gag* (Barker and Beemon 1991), dihydrofolate reductase (Urlaub et al. 1989), and triosephosphate isomerase (Daar and Maquat 1988). In mammalian cells, it has been difficult to show that this lowering of mRNA levels by premature translation termination is due to a reduction in cytoplasmic mRNA stability; these sorts of complications necessitate that premature termination in mammalian cells be discussed separately.

The main conceptual problem for the case of nonsense-mediated mRNA decay is the following question: How does a cell recognize that an mRNA is prematurely terminated? One old idea was that translating polysomes somehow "protect" an mRNA from degradation, so that premature translation termination allows access of the remaining portions of the mRNA to nuclease attack. This is consistent with the observation that there is often a position effect of premature translation terminators on mRNA degradation; 5′-proximal mutations are usually more destabilizing than 3′-proximal mutations. However, several lines of evidence render this hypothesis untenable.

First, inhibiting translation by using drugs that freeze ribosomes on mRNA or that remove ribosomes from mRNA is usually seen to stabilize, not destabilize, mRNA (see above). Second, it seems likely that the reason for position effects of premature terminators on mRNA stability is due to sequence context (see below). Finally, a long region of untranslated mRNA 3′to the terminator is not sufficient to render an mRNA unstable. This can be clearly seen in the case of the *gag* gene of RSV. Although *gag* is translated from an unspliced mRNA that is nearly 10 kb long, *gag* normally terminates translation at position 2485, leaving a 3′UTR of greater than 7 kb. However, whereas premature termination at position 2319 allows normal levels of *gag* mRNA, premature termination at position 1924 (and positions 5′to this) reduces its stability severalfold (Barker and Beemon 1991). Another example in which the length of the 3′UTR has no effect on the stability of the mRNA is the demonstration that lengthening the 3′UTR of the yeast *PGK1* gene (which produces an mRNA that is subject to nonsense-mediated mRNA decay; see below) does not change its stability (Peltz et al. 1993).

Nonsense-mediated mRNA Decay in Yeast

The relative ease by which *Saccharomyces* can be genetically manipulated promises to reveal the *cis* and *trans* mediators of nonsense-mediated mRNA decay. Moreover, temperature-sensitive mutants in yeast RNA polymerase II provide a useful tool to examine the effects on the half-lives of mRNAs when factors that mediate mRNA decay are modified (for review, see Peltz and Jacobson 1993). These studies have disclosed two salient features.

Trans-*acting Factors Can Affect the Accelerated Degradation of mRNAs That Contain Premature Terminators*

Gene products that regulate the degradation of mRNAs that prematurely terminate were found serendipitously. Culbertson and his co-workers (Leeds et al. 1991, 1992) attempted to isolate genes that increased the suppression of a frameshift mutation in the *his4* gene in the context of a low-level frameshift suppressor tRNA. (This strain has normal growth at 30°C but is auxotrophic for histidine at 37°C.) These authors found that some mutations (called *upf*, for up-frameshift) that allowed for the efficient expression of the His^+ phenotype did not alter the frequency of read-through at the frameshift mutation, but rather increased the level of the *his4* mRNA. It was subsequently shown that one of these genes,

UPF1, is required for the rapid degradation of yeast mRNAs that contain premature translation terminators (Leeds et al. 1991). At least one other gene, *UPF3*, has a similar phenotype (Leeds et al. 1992). The action of the *UPF1* gene appears to be specific for prematurely terminated mRNAs; the stability of a wide variety of normal mRNAs was not affected in the context of a *upf1* mutation (Leeds et al. 1991). It is not clear how the *UPF1* gene might function; it is not essential for growth, and it does not appear to affect the translation process. Sequence inspection of the *UPF1* gene demonstrates the presence of a zinc finger motif, suggesting that it could bind to RNA.

Nonsense-mediated mRNA Decay Has a Strong Position and/or Sequence-context Dependence

Peltz et al. (1993) used a series of mutations in the yeast *PGK1* gene to examine the *cis*-acting sequences that mediate the degradation of prematurely terminated mRNA. As is the case for many yeast mRNAs, the *PGK1* mRNA demonstrates a strong position dependence for nonsense-mediated mRNA decay. Whereas premature terminators near the 3′end of the coding region had no effect on its stability, terminators near the 5′end of the coding region reduced the stability of the mRNA by greater than tenfold. A careful examination of a series of mutations in which a terminator was moved progressively toward the 5′end of this 417-codon-translated region showed that most of the destabilizing effect occurred when the terminator was moved from codon 317 to codon 282, with another approximately 2.5-fold effect when it was further moved to codon 229. Furthermore, deletion mutants in which *PGK1* mRNA terminating at codon 23 (which is normally unstable) were rendered stable when most of the sequence downstream from the terminator was deleted. Similar results were also obtained with premature terminators in the *HIS4* gene (Hagan et al. 1995). These results suggest at least two possible interpretations. One is that a minimum length of the mRNA 3′to the terminator must be present in order for it to be a substrate for nonsense-mediated decay. Alternatively, a second interpretation is that the degradation of *PGK1* mRNA that has premature terminators requires specific sequences downstream from the terminator. Two results support the latter interpretation: (1) The addition of an extra untranslated region to the end of *PGK1* mRNAs that have premature terminators has no effect on their stability and (2) the region of the *PGK1* mRNA between codons 282 and 317 can confer nonsense-mediated mRNA degradation on an otherwise stable *PGK1* mRNA that has a terminator at codon 23 but with most of the rest of the mRNA deleted. A careful examination of this region showed that the presence of several AUG codons in this region are required for this effect. Recently, these researchers identified a sequence motif (TGYYYGATGYYYYY) that may be involved in nonsense-mediated mRNA decay; this sequence is thought to be able to base pair with a region of the 18S ribosomal mRNA (Zhang et al. 1995).

A Model for Nonsense-mediated mRNA Decay in Yeast

Studies on the *cis-* and *trans*-acting factors that affect the decay of prematurely terminated mRNAs in yeast have suggested a possible solution for the major dilemma of nonsense-mediated mRNA decay: How does the cell differentiate between a normal mRNA and one that has a premature terminator? In this model, once a ribosome terminates translation, it (or a subunit thereof) continues scanning the rest of the mRNA. If the ribosome encounters a particular sequence, it (aided perhaps by the action of the product of the *UPF1* gene) signals the cell to degrade the mRNA. This signal may include the presence of initiator codons, perhaps indicating the requirement for translational reinitiation or pausing for the degradation of the mRNA. Support for this model comes from the observation that short open reading frames in the 5′UTR of a yeast mRNA can cause destabilization of the mRNA (Oliviera and McCarthy 1995). Some of the features of this model are reminiscent of the translational regulation of the yeast *GCN4* gene (Hinnebusch 1990). In both cases, the scanning of a ribosome downstream from a terminator is required. It would be interesting to determine if the *upf* mutations would also affect the stability or translational regulation of the *GCN4* mRNA. Regardless of this outcome, it is apparent that the degradation of prematurely terminated mRNAs in yeast provides the most convincing examples of the link between translation and mRNA stability.

Premature Termination in Mammalian Cells

Many mutations that cause human disease are caused by nonsense mutations or frameshift mutations that cause premature termination. Frequently, but not always, mRNAs encoded by these mutated genes are found in lower abundance than their wild-type counterparts (Lim et al. 1992; Longo et al. 1992; Gotoda et al. 1993). Moreover, the effect of the mutations on mRNA levels are position-dependent; i.e., premature termination near the 3′end of the mRNA does not generally lower mRNA levels (Urlaub et al. 1989; Cheng et al. 1990). These observations lead one to suspect that there might be a pathway of nonsense-mediated mRNA decay functioning in a manner similar to the one in yeast. However, unlike the case in yeast, it could not be shown in mammalian cells that lower mRNA levels associated with premature termination were in fact mediated by cytoplasmic stability (Urlaub et al. 1989; Baserga and Benz 1992; Cheng and Maquat 1993). A closer look at two examples of lower mRNA levels caused by premature termination reveals some interesting differences between yeast and mammalian cells.

Dihydrofolate Reductase

Urlaub et al. (1989) examined a series of UV-induced mutations in the dihydrofolate reductase (DHFR) gene in Chinese hamster ovary cells; most of these mutations result in premature termination of translation. Moreover, premature termination in codons derived from exons 1 through 5 also reduced the abundance of the mutant DHFR mRNA, whereas premature termination in codons derived from the last exon did not reduce the level of the mRNA. Transcription rates of the DHFR gene were the same between wild-type and mutant genes, suggesting that the differences in the mRNA levels occurred posttranscriptionally. However, an examination of the decay rate of the mutant and wild-type DHFR mRNAs in the presence of actinomycin D did not reveal any differences in the stability of the cytoplasmic mRNA, implicating a nuclear event, such as splicing. To account for this effect, these authors suggested that cytoplasmic translation was coupled to splicing and transport of the mRNA from the nucleus. Alternatively, they proposed that there was a nuclear mechanism for examination of reading frames in internal exons. However, since this effect could not be recapitulated in transfected genes, it proved impossible to examine these models further.

Triosephosphate Isomerase

Mutations in the TPI gene that result in premature termination also reduce the levels of its mRNA (Daar and Maquat 1988). That the reductions in mRNA levels are caused by translational termination and not by altered mRNA structure is revealed by the fact that translation is coupled to lower RNA levels. If a translation-inhibiting hairpin is inserted into the 5′UTR of TPI, then the effect of the premature termination codon is abrogated. Moreover, expression of a suppressor tRNA that allows full-length translation of TPI-containing nonsense mutations raises the level of the mutant mRNA (Belgrader and Maquat 1993).

A careful examination of a series of mutations showed that there was a position dependence to this effect; if the premature terminator was found 3′ to a site near the end of the penultimate exon, then there was no effect on TPI RNA levels (Cheng et al. 1990). The position dependence of premature terminators on TPI mRNA levels is apparently due to *cis*-acting sequences in intron 6 (Cheng et al. 1994). Premature termination does not affect the transcription of the TPI gene, nor does it affect the level of unspliced pre-mRNA. Nevertheless, two results showed that the decrease in mRNA levels could not be accounted for by a decrease in the stability of cytoplasmic mRNA: (1) The decay of both mutant and wild-type cytoplasmic TPI mRNAs in the presence of actinomycin D showed that both mRNAs were equally stable (Cheng et al. 1990). (2) Using the c-*fos* promoter coupled to the wild-type or mutant genes, which allows a transcriptional pulse of TPI mRNA synthesis by serum addition, without the global decrease of transcription that actinomycin induces, these researchers showed that the cytoplasmic stability of TPI mRNA was not affected by premature termination (Cheng and Maquat 1993). However, examination of the levels and stability of nuclear RNA showed that the decrease in the cytoplasmic levels of prematurely terminated TPI mRNA could be accounted for by a decrease in the stability of fully spliced nuclear mRNA (Belgrader et al. 1994). Therefore, it appears that the translationally coupled degradation of prematurely terminated TPI mRNAs is the result of degradation of the nuclear mRNA before, or concomitant with, its transport to the cytoplasm. The degradation of this fully spliced mRNA is dependent on signals that were present in one or more of the introns before they were removed by splicing.

How Could Translation be Coupled to the Levels of Nuclear mRNA?

One possibility is that there might be a nuclear mechanism for scanning the reading frame of an mRNA before it is exported into the cytoplasm. This mechanism would also ensure that incompletely spliced mRNAs are not exported, since they would also most likely have premature terminators. However, so far there is not any evidence that such a nuclear scanning mechanism exists.

A second possibility is that the process of nuclear transport and cytoplasmic translation might be coupled. In this model, mRNAs would exit the nucleus 5′ end first, with ribosomes associating with the mRNA before transport is completed. If the ribosome terminates before transport is complete, then the mRNA could be degraded in the nucleus or at the nuclear pore. Interestingly, the very large Balbiani ring ribonucleoproteins of the salivary gland of *Chironomus tentans* larvae, which can be morphologically identified by electron microscopy, have been observed to exit through the nuclear pore 5′ end first (Mehlin et al. 1995). Therefore, directional transport of most mRNAs from the nucleus is a reasonable possibility.

Neither of these models has any definitive experimental evidence to back them up. Nevertheless, it does seem clear that there is likely to be communication between the nucleus and the cytoplasm.

Trans-*acting Factors in the Degradation of Prematurely Terminated mRNAs in Metazoans*

The UPF gene products that mediate the degradation of prematurely terminated mRNAs in yeast may have their counterparts in more complex organisms. In the nematode *Caenorhabditis elegans*, *unc-54* myosin heavy chain genes that contain premature terminators produce only low levels of mRNA, presumably due to increased turnover of the mutant mRNAs. However, in the presence of *smg*(–) mutations, the levels of the prematurely terminated mRNAs are increased to normal, whereas wild-type mRNA levels are not affected (Pulak and Anderson 1993). Therefore, the SMG gene products appear to be *trans*-acting regulators of prematurely terminated mRNAs, analogous to the UPF gene products of yeast. It will be interesting to determine if any of the *SMG* genes are structurally related to the yeast *UPF* genes as well.

Summary

Two questions are raised by the process of nonsense-mediated mRNA decay: First, why does the cell degrade prematurely terminated mRNA? One possibility is that it is probably worse for the cell to make no protein from a mutant mRNA than it is to make a truncated protein. If a mutant mRNA made no protein, that mutant would likely be a recessive one, and therefore not necessarily deleterious. However, a truncated protein could conceivably be a dominantly acting one if that protein was part of a multisubunit complex. Additionally, truncated proteins could be more likely to become misfolded, resulting in insoluble aggregates that might be detrimental to the cell. In support of this, it is interesting to note that one of the β-globin mutants in which a premature terminator is near the 3′ end, and thereby escapes degradation, produces a truncated protein and is a dominant thalassemia (Hall and Thein 1994). Similarly, *C. elegans* myosin heavy chain mutations resulting from premature terminators are recessive in a wild-type organism but are dominantly acting disruptors of thick filaments in the presence of *smg* mutations (Pulak and Anderson 1993). Another reason for the cell to degrade prematurely terminated mRNA is to degrade unspliced or partially spiced mRNA precursors that are exported prematurely. These RNAs would also most likely contain premature terminators. In support of this, one of the phenotypes of the *upf-1* mutants described earlier is that they accumulate unspliced precursor mRNAs (He et al. 1993).

Second, how does the cell know that a premature terminator is indeed "early"? Conceptually, an mRNA with a premature terminator superficially resembles an mRNA with a long 3′UTR. However, the presence of a long 3′UTR does not mark an mRNA for degradation. There are several possibilities that could distinguish between these two. One possibility is that there might be a sequence context for translational termination. For example, it has been shown that the sequence context around authentic termination codons are statistically nonrandom (Cavener and Ray 1991; Buckingham 1994). In this model, a nonsense mutation would likely not be in the right sequence context and thus mark the mRNA for degradation. However, this model could not account for the position dependence of nonsense-mediated mRNA decay.

A second possibility is that mRNAs might have in their translated regions sequences that mark the mRNA for degradation, unless those sequences are translated (Zhang et al. 1995). Premature termination would then expose those sequences to machinery that would cause the mRNA to be degraded.

A third possibility is that the difference between a premature terminator and an authentic one is that the premature terminator is followed by a length of open reading frame. If the ribosome continued scanning the mRNA following termination, and could somehow sense that another reading frame exists (perhaps by reinitiation at a downstream AUG), that process might mark the mRNA for degradation. In this model, the position effect of nonsense-mediated mRNA decay is accounted for by a presence of AUG codons downstream from the terminator. AUG codons in the 3′UTR would not act as markers for mRNA degradation because it is likely that they would not be followed by a long open reading frame.

These models are not necessarily mutually exclusive and may act in concert on certain mRNAs. At least some of the features of these models are subject to experimental verification. Nevertheless, regardless of the mechanism, it is clear that the degradation of prematurely terminated mRNAs provides a clear example of the coupling between translation and the degradation process.

MECHANISMS OF mRNA DEGRADATION

Are there any common features in the mechanisms by which mRNAs are degraded? In yeast, it appears that the process by which unstable mRNAs are degraded is the same as that which degrades most cellular mRNAs: The poly(A) tail is shortened, followed by removal of the 5′cap structure, after which a 5′exonuclease degrades the remainder of the mRNA (Decker and Parker 1993; Muhlrad et al. 1994, 1995; for review, see Decker and Parker 1994). Similarly, in mammalian cells, the removal of the poly(A) tail precedes the degradation of several unstable mRNAs (see above). However, whereas in mammalian cells the removal of the poly(A) tail requires that the mRNA be translated (Wilson and Treisman 1988; Laird-Offringa et al. 1990), in yeast, the mRNA need not be translatable for it to be a substrate for poly(A) removal and degradation (Beelman and Parker 1994; Muhlrad et al. 1994). Consequently, it remains to be determined whether there is a unifying mechanism for the degradation of mRNAs in yeast and metazoans, although it seems likely.

The mechanism of nonsense-mediated mRNA decay is probably distinct from that of the degradation of other unstable mRNAs. For prematurely terminated mRNAs, the removal of the poly(A) tail is not obligatory (Muhlrad and Parker 1994). Nevertheless, there is a clear role for translation in the degradation of prematurely terminated mRNAs, both in yeast and in mammals. Despite the surprising observations that in mammals, premature termination affects nuclear RNA (see above), it will be interesting to see if there are common mechanisms that all cells use to degrade these mRNAs.

REFERENCES

Aharon, T. and R.J. Schneider. 1993. Selective destabilization of short-lived mRNAs with the granulocyte-macrophage colony-stimulating factor AU-rich 3′noncoding region is mediated by a cotranslational mechanism. *Mol. Cell. Biol.* **13:** 1971–1980.

Bachurski, C.J., N.G. Theodorakis, R.M.R. Coulson, and D.W. Cleveland. 1994. An amino-terminal tetrapeptide specifies cotranslational degradation of β-tubulin but not α-tubulin mRNAs. *Mol. Cell. Biol.* **14:** 4076–4086.

Barker, G.F. and K. Beemon. 1991. Nonsense codons within the Rous sarcoma virus *gag* gene decrease the stability of unspliced viral mRNA. *Mol. Cell. Biol.* **11:** 2760–2768.

Baserga, S.J. and E.J. Benz, Jr. 1992. β-Globin nonsense mutation: Deficient accumulation of mRNA occurs despite normal cytoplasmic stability. *Proc. Natl. Acad. Sci.* **89:** 2935–2939.

Beelman C.A. and R. Parker. 1994. Differential effects of translational inhibition in *cis* and in *trans* on the decay of the unstable yeast MFA2 mRNA. *J. Biol. Chem.* **269:** 9687–9692.

Belgrader, P. and L.E. Maquat. 1993. Evidence to implicate translation by ribosomes in the mechanism by which nonsense codons reduce the nuclear level of human triosephosphate isomerase mRNA. *Proc. Natl. Acad. Sci.* **90:** 482–486.

Belgrader, P., J. Cheng, X. Zhou, L.S. Stephenson, and L.E. Maquat. 1994. Mammalian nonsense codons can be *cis* effectors of nuclear mRNA half-life. *Mol. Cell. Biol.* **14:** 8219–8228.

Ben Ze'ev, A., S.R. Farmer, and S. Penman. 1979. Mechanisms of regulating tubulin synthesis in cultured mammalian cells. *Cell* **17:** 319–325.

Bonnieu, A., P. Roux, L. Marty, P. Jeanteur, and M. Piechaczyk. 1990. AUUUA motifs are dispensable for rapid degradation of the mouse c-*myc* RNA. *Oncogene* **5:** 1585–1588.

Buckingham, R.H. 1994. Codon context and protein synthesis: Enhancements of the genetic code. *Biochimie* **76:** 351–354.

Caput, D., B. Beutler, K. Hartog, S. Brown-Shimer, and A. Cerami. 1986. Identification of common nucleotide sequence in the 3′untranslated region of mRNA molecules specifying inflammatory mediators. *Proc. Natl. Acad. Sci.* **83:** 1670–1674.

Caron, J.M., A.L. Jones, L.B. Ball, and M.W. Kirschner. 1985. Autoregulation of tubulin synthesis in enucleated cells. *Nature* **317:** 648–651.

Cavener, D.R. and S.C. Ray. 1991. Eukaryotic start and stop translation sites. *Nucleic Acids Res.* **19:** 3158–3192.

Chen, C.-Y. and A.-B. Shyu. 1994. Selective degradation of early-response gene mRNAs: Functional analyses of sequence features of the AU-rich elements. *Mol. Cell Biol.* **14:** 8471–8482.

Cheng, J. and L.E. Maquat. 1993. Nonsense codons can reduce the abundance of nuclear mRNA without affecting the abundance of pre-mRNA or the half-life of cytoplasmic mRNA. *Mol. Cell. Biol.* **13:** 1892–1902.

Cheng, J., M. Fogel-Petrovic, and L.E. Maquat. 1990. Translation to near the distal end of the penultimate exon is required for normal levels of spliced triosephosphate isomerase mRNA. *Mol. Cell. Biol.* **10:** 5215–5225.

Cheng, J., P. Belgrader, X. Zhou, and L.E. Maquat. 1994. Introns are *cis* effectors of the nonsense-codon-mediated reduction in nuclear mRNA abundance. *Mol. Cell. Biol.* **14:** 6317–6325.

Cleveland, D.W. and J.C. Havercroft. 1983. Is apparent autoregulatory control of tubulin synthesis nontranscriptionally controlled? *J. Cell. Biol.* **97:** 919–924.

Cleveland, D.W. and N.G. Theodorakis. 1994. Regulation of tubulin synthesis. In *Microtubules* (ed. J.S. Hyams and C.W. Lloyd), pp. 47–58. Wiley-Liss, New York.

Cleveland, D.W., M.F. Pittenger, and J.R. Feramisco. 1983. Elevation of tubulin levels by microinjection suppresses new tubulin synthesis. *Nature* **305:** 738–740.

Cleveland, D.W., M.A. Lopata, P. Sherline, and M.W. Kirschner. 1981. Unpolymerized tubulin modulates the level of tubulin mRNAs. *Cell* **25:** 537–546.

Cole, M.D. and S. Mango. 1990. *cis*-Acting determinants of c-*myc* mRNA stability. *Enzyme* **44:** 167–180.

Daar, I.O. and L.E. Maquat. 1988. Premature translation termination mediates triosephosphate isomerase mRNA degradation. *Mol. Cell. Biol.* **8:** 802–813.

Dani, C., J.M. Blanchard, M. Piechaczyk, S. El Sabouty, L. Marty, and P. Jeanteur. 1984. Extreme instability of *myc* mRNA in normal and transformed human cells. *Proc. Natl. Acad. Sci.* **81:** 7046–7050.

Decker, C.J. and R. Parker. 1993. A turnover pathway for both stable and unstable mRNAs in yeast: Evidence for a requirement for deadenylation. *Genes Dev.* **7:**

1632–1643.

———. 1994. Mechanisms of mRNA degradation in eukaryotes. *Trends Biochem. Sci.* **29:** 336–340.

Gay, D.A., S.S. Sisodia, and D.W. Cleveland. 1989. Autoregulatory control of β-tubulin mRNA stability is linked to translational elongation. *Proc. Natl. Acad. Sci.* **86:** 5763–5767.

Gay, D.A., T.J. Yen, J.T.Y Lau, and D.W. Cleveland. 1987. Sequences that confer β-tubulin autoregulation through modulated mRNA stability reside within exon 1 of a β-tubulin mRNA. *Cell* **50:** 671–679.

Gong, Z. and B.P. Brandhorst. 1988. Stabilization of tubulin mRNA by inhibitors of protein synthesis in sea urchin embryos. *Mol. Cell. Biol.* **8:** 3518–3525.

Gotoda, T., M. Kinoshita, H. Shimano, K. Harada, M. Shimada, J. Ohsuga, T. Teramoto, Y. Yazaki, and N. Yamada. 1993. Cholesteryl ester transfer protein deficiency caused by a nonsense mutation detected in the patient's macrophage mRNA. *Biochem. Biophys. Res. Commun.* **194:** 519–524.

Graves, R.A., N.B. Pandey, N. Chodchoy, and W.F. Marzluff. 1987. Translation is required for regulation of histone mRNA degradation. *Cell* **48:** 615–626.

Hagan, K.W., M.J. Ruiz-Echevarria, Y. Quan, and S.W. Peltz. 1995. Characterization of *cis*-acting sequences and decay intermediates involved in nonsense-mediated mRNA turnover. *Mol. Cell. Biol.* **15:** 809–823.

Hall, G.W. and S. Thein. 1994. Nonsense mutations in the terminal exon of the β-globin gene are not associated with a reduction in β-globin mRNA accumulation: A mechanism for the phenotype of dominant β-thalassemia. *Blood* **83:** 2031–2037.

He, F., S.W. Peltz, J.L. Donahue, M. Rosbash, and A. Jacobson. 1993. Stabilization and ribosome association of unspliced pre-mRNAs in a yeast upf1- mutant. *Proc. Natl. Acad. Sci.* **90:** 7034–7038.

Herrick, D.J. and J. Ross. 1994. The half-life of c-*myc* mRNA in growing and serum-stimulated cells: Influence of the coding and 3′untranslated regions and the role of ribosome termination. *Mol. Cell. Biol.* **14:** 2119–2128.

Hinnebusch, A.G. 1990. Involvement of an initiation factor and protein phosphorylation in translational control of GCN4 mRNA. *Trends Biochem. Sci.* **15:** 148–152.

Jones, T. and M.D. Cole. 1987. Rapid cytoplasmic turnover of c-*myc* mRNA: Requirement of the 3′-untranslated sequences. *Mol. Cell. Biol.* **7:** 4513–4521.

Koeller, D.M., J.A. Horowitz, J.L. Casey, R.D. Klausner, and J.B. Harford. 1991. Translation and stability of mRNAs encoding the transferrin receptor and c-*fos*. *Proc. Natl. Acad. Sci.* **88:** 7778–7782.

Kowalski, J. and D.T. Denhardt. 1989. Regulation of the mRNA for monocyte-derived neutrophil-activating peptide in differentiating HL60 promyelocytes. *Mol. Cell. Biol.* **9:** 1946–1957.

Lagnado, C.A., C.Y. Brown, and G.L. Goodall. 1994. AUUUA is not sufficient to promote poly(A) shortening and degradation of an mRNA: The functional sequence within AU-rich elements may be UUAUUUA(U/A)(U/A). *Mol. Cell. Biol.* **14:** 7984–7995.

Laird-Offringa, I.A. 1992. What determines the instability of c-*myc* proto-oncogene mRNA? *BioEssays* **14:** 119–124.

Laird-Offringa, I.A., P. Elfferich, and A.J. van der Eb. 1991. Rapid c-*myc* degradation does not require (A+U)-rich sequences or complete translation of the mRNA. *Nucleic Acids. Res.* **19:** 2387–2394.

Laird-Offringa, I.A., C.L. de Wit, P. Elfferich, and A.J. van der Eb. 1990. Poly(A) tail shortening is the translation-dependent step in c-*myc* mRNA degradation. *Mol. Cell. Biol.* **10:** 6132–6140.

Lau, L.F. and D. Nathans. 1987. Expression of a set of growth-related immediate-early genes in BALB/c 3T3 cells: Coordinate expression with c-*fos* or c-*myc*. *Proc. Natl. Acad. Sci.* **84:** 1182–1186.

Leeds, P., S.W. Peltz, A. Jacobson, and M.R. Culbertson. 1991. The product of the yeast *UPF1* gene is required for rapid turnover of mRNAs containing a premature translational termination codon. *Genes Dev.* **5:** 2303–2314.

Leeds, P., J.M. Wood, B.-S. Lee, and M.R. Culbertson. 1992. Gene products that promote mRNA turnover in *Saccharomyces cerevisiae*. *Mol. Cell. Biol.* **12:** 2165–2177.

Lim, S.-K. and L.E. Maquat. 1992. Human β-globin mRNAs that harbor a nonsense codon are degraded in murine erythroid tissues to intermediates lacking regions of exon I or exons I or II that have a cap-like structure at the 5′termini. *EMBO J.* **11:** 3271–3278.

Lim, S.-K., C.D. Sigmund, K.W. Gross, and L.E. Maquat. 1992. Nonsense codons in human β-globin mRNA result in the production of mRNA degradation products. *Mol. Cell. Biol.* **12:** 1149–1161.

Linial, M., N. Gunderson, and M. Groudine. 1985. Enhanced transcription of c-*myc* in bursal lymphomal cells requires continuous protein synthesis. *Science* **230:** 1126–1132.

Longo, N., S.D. Langley, L.D. Griffin, and L.J. Elsas II. 1992. Reduced mRNA and a nonsense mutation in the insulin-receptor gene produce heritable severe insulin resistance. *Am. J. Hum. Genet.* **50:** 998–1007.

Losson, R. and F. Lacroute. 1979. Interference of nonsense mutations with eukaryotic mRNA stability. *Proc. Natl. Acad. Sci.* **76:** 5134–5137.

Marzluff, W.F. and R.J. Hanson. 1993. Degradation of a non-polyadenylated messenger: Histone mRNA decay. In *Control of mRNA Stability* (ed. J. Belasco and G. Brawerman), pp. 267–290. Academic Press, San Diego.

Mehlin, H., B. Danholt, and U. Skoglund. 1995. Structural interaction between the nuclear pore complex and a specific translocating RNP particle. *J. Cell Biol.* **129:** 1205–1216.

Muhlrad, D. and R. Parker. 1994. Premature translation termination leads to mRNA decapping. *Nature* **370:** 578–581.

Muhlrad, D., C.J. Decker, and R. Parker. 1994. Deadenylation of the unstable mRNA encoded by the yeast *MFA2* gene leads to decapping followed by 5′→3′digestion of the transcript. *Genes Dev.* **8:** 855–866.

———. 1995. Turnover mechanisms of the stable yeast PGK1 mRNA. *Mol. Cell. Biol.* **15:** 2145–2156.

Oliviera, C.C. and J.E.G. McCarthy. 1995. The relationship between eukaryotic translation and mRNA stability: A short upstream open reading frame strongly inhibits translational initiation and greatly accelerates mRNA degradation in the yeast *Saccharomyces cerevisiae*. *J. Biol. Chem.* **270:** 8936–8943.

Pachter, J.S., T.J. Yen, and D.W. Cleveland. 1987. Autoregulation of tubulin expression is achieved through specific degradation of polysomal tubulin mRNAs. *Cell* **51:** 283–292.

Pandey, N.B. and W.F. Marzluff. 1987. The stem-loop structure at the 3′end of histone mRNA is necessary and sufficient for regulation of histone mRNA stability. *Mol. Cell. Biol.* **7:** 4557–4559.

Pei, R. and K. Calame. 1988. Differential stability of c-*myc* mRNAs in a cell-free system.

Mol. Cell. Biol. **8:** 2860–2868.

Peltz, S.W. and A. Jacobson. 1993. mRNA turnover in *Saccharomyces cerevisiae*. In *Control of mRNA stability* (ed. J. Belasco and G. Brawerman.), pp. 291–328. Academic Press, San Diego.

Peltz, S.W., A.H. Brown, and A. Jacobson. 1993. mRNA destabilization triggered by premature translation termination depends on at least three *cis*-acting sequence elements and one *trans*-acting factor. *Genes Dev.* **7:** 1737–1754.

Peterson, C. 1993. Translation and mRNA stability in bacteria: A complex relationship. In *Control of mRNA stability* (ed. J. Belasco and G. Brawerman), pp. 117–145. Academic Press, San Diego.

Pittenger, M.F. and D.W. Cleveland. 1985. Retention of autoregulatory control of tubulin synthesis in cytoplasts: Demonstration of a cytoplasmic mechanism that regulates the level of tubulin expression. *J. Cell Biol.* **101:** 1941–1952.

Pulak, R. and P. Anderson. 1993. mRNA surveillance by the *Caenorhabditis elegans smg* genes. *Genes Dev.* **7:** 1885–1897.

Raj, N.B.K. and P.M. Pitha. 1981. Analysis of interferon mRNA in human fibroblast cells induced to produce interferon. *Proc. Natl. Acad. Sci.* **78:** 7426–7430.

Ross, J., S.W. Peltz, G. Kobbs, and G. Brewer. 1986. Histone mRNA degradation in vitro: The first detectable step occurs at or near the 3′ terminus. *Mol. Cell. Biol.* **6:** 4362–4371.

Savant-Bhonsale, S. and D.W. Cleveland. 1992. Evidence for instability of mRNAs containing AUUUA motifs mediated through translation-dependent assembly of a >20S degradation complex. *Genes Dev.* **6:** 1927–1939.

Schiavi, S.C., C.L. Wellington, A.-B. Shyu, C.Y. Chen, M.E. Greenberg, and J.G. Belasco. 1994. Multiple elements in the c-*fos* protein-coding region facilitate mRNA deadenylation and decay by a mechanism coupled to translation. *J. Biol. Chem.* **269:** 3441–3448.

Schuler, G.D. and M.D. Cole. 1988. GM-CSF and oncogene mRNA stabilities are independently regulated in *trans* in a mouse monocytic tumor. *Cell* **55:** 1115–1122.

Shaw, G. and R. Kamen. 1986. A conserved AU sequence from the 3′ untranslated region of GM-CSF mRNA mediates selective mRNA degradation. *Cell* **46:** 659–667.

Shyu, A.-B., J. Belasco, and M.E. Greenberg. 1991. Two distinct destabilizing elements in the c-*fos* message trigger deadenylation as a first step in rapid mRNA decay. *Genes Dev.* **5:** 221–231.

Shyu, A.-B., M.E. Greenberg, and J. Belasco. 1989. The c-*fos* transcript is targeted for rapid decay by two distinct mRNA degradation pathways. *Genes Dev.* **3:** 60–72.

Sittman, D.B., R.A. Graves, and W.F. Marzluff. 1983. Structure of a cluster of mouse histone genes. *Nucleic Acids Res.* **11:** 6679–6697.

Stimac, E., V.E. Groppi, Jr., and P. Coffino. 1984. Inhibition of protein synthesis stabilizes histone mRNA. *Mol. Cell. Biol.* **4:** 2082–2087.

Stoeckle, M.Y. 1991. Post-transcriptional regulation of groα, β, γ, and IL-8 mRNAs by IL-1β. *Nucleic Acids Res.* **19:** 917–920.

———. 1992. Removal of a 3′ non-coding sequence is an initial step in degradation of groα mRNA and is regulated by IL-1. *Nucleic Acids Res.* **20:** 1123–1127.

Stoeckle, M.Y. and H. Hanafusa. 1989. Processing of 9E3 mRNA and regulation of its stability in normal and Rous sarcoma virus-transformed cells. *Mol. Cell. Biol.* **9:** 4738–4745.

Theodorakis, N.G. and D.W. Cleveland. 1992. Physical evidence for cotranslational regulation of β-tubulin mRNA degradation. *Mol. Cell Biol.* **12:** 791–799.

———. 1993. Translationally coupled degradation of tubulin mRNA. In *Control of mRNA stability* (ed. J. Belasco and G. Brawerman), pp. 219–238. Academic Press, San Diego.

Urlaub, G., P.M. Mitchell, C.J. Ciudad, and L.A. Chasin. 1989. Nonsense mutations in the dihydrofolate reductase gene affect RNA processing. *Mol. Cell. Biol.* **9:** 2868–2880.

Wellington, C.L., M.E. Greenberg, and J.G. Belasco. 1993. The destabilizing elements in the coding region of c-*fos* are recognized as RNA. *Mol. Cell. Biol.* **13:** 5034–5042.

Wilson, T. and R. Treisman. 1988. Removal of poly(A) and consequent degradation of c-*fos* mRNA facilitated by 3′ AU-rich sequences. *Nature* **336:** 396–399.

Wisdom, R. and W. Lee. 1991. The protein-coding region of c-*myc* mRNA contains a sequence that specifies rapid mRNA turnover and induction by protein synthesis inhibitors. *Genes Dev.* **5:** 232–243.

Yen, T.J., P.S. Machlin, and D.W. Cleveland. 1988a. Autoregulated instability of β-tubulin mRNAs by recognition of the nascent amino terminus of β-tubulin. *Nature* **334:** 580–585.

Yen, T.J., D.A. Gay, J.S. Pachter, and D.W. Cleveland. 1988b. Autoregulated changes in stability of polyribosome-bound β-tubulin mRNAs are specified by the first thirteen translated nucleotides. *Mol. Cell. Biol.* **8:** 1224–1235.

Zhang, S., M.J. Ruiz-Echevarria, Y. Quan, and S.W. Peltz. 1995. Identification and characterization of a sequence motif involved in nonsense-mediated mRNA decay. *Mol. Cell. Biol.* **15:** 2231–2244.

Zubiaga, A.M., J.G. Belasco, and M.E. Greenberg. 1995. The nonamer UUAUUUAUU is the key AU-rich sequence motif that mediates mRNA degradation. *Mol. Cell. Biol.* **15:** 2210–2230.

23
Regulatory Recoding

John F. Atkins[1] **and Raymond F. Gesteland**[1,2]
[1]Department of Human Genetics and
[2]Howard Hughes Medical Institute,
University of Utah
Salt Lake City, Utah 84112

Some messenger RNAs encode not only an amino acid sequence, but also special signals that alter the mechanism of ribosomal readout. These programmed alterations of decoding, termed recoding, include ribosomal frameshifts at particular sites, reading of stop codons as sense, and ribosomal jumping where some nucleotides are skipped in the mRNA. Recoding by frameshifting and stop codon read-through (Gesteland et al. 1992) is now known to be quite widely used for gene expression. In most cases, the function is to provide a set ratio between two products that have a common amino-terminal sequence. However, there are some cases where regulation is operative. Two known cases of autoregulatory frameshifting are in decoding the genes for mammalian antizyme and *Escherichia coli* release factor 2 (RF2). In other cases, frameshifting is part of a regulatory pathway. The level of frameshifting in yeast transposable elements and some bacterial insertion sequences is responsive to the physiological state of the cell and in turn governs the level of element mobility. Regulatory roles are only suspected in some of the other cases of frameshifting and stop codon read-through.

In programmed frameshifting, ribosomes initiate translation in the zero frame and translate conventionally to the shift site where some fraction of them are directed to one of the other two frames (+1 or –1) where they continue to the next stop codon in the new frame. Typically, only a minority of the ribosomes shift frame, whereas the majority continue on in the zero frame. The result is two protein products from the same mRNA that differ in their carboxy-terminal sequence. Usually, the lengths are quite different, and in some cases, the zero frame product is quite short and has no known function. In these latter cases, the act of synthesis, however, may be important to get ribosomes to the site where some can shift and go on to make the crucial elongated transframe product.

In the known cases of programmed +1 frameshifting, where the ribosome-tRNA complex slips forward along the mRNA by one nucleotide, a single tRNA is involved in the shift to the new frame. The base 3′ to the codon where the shift occurs is either involved in re-pairing with the shifty tRNA or occluded by the tRNA in the original frame. In either case, there is competition for use of this codon base between the shift event and standard decoding of the next codon in the zero frame. Often, this next codon has reduced ability to compete for decoding, being either a stop codon or a rare codon, providing an opportunity for regulation. A stop codon directly after a shift site makes that site especially slippery.

Programmed −1 frameshifts are different in that the great majority have tandem shift sites involving two tRNAs and no stop codon. In this case, A and P ribosomal sites are occupied by tRNAs that simultaneously shift −1 and re-pair in the new frame. Frequently, the codon-anticodon pairing between the tRNA in the A-site and the pre-slip codon is weak, perhaps contibuting to the tendency to slip and re-pair. This may be enhanced by correspondingly stronger re-pairing by this tRNA with the −1 codon. However, there is considerable latitude in the rules. For both +1 and −1 programmed frameshifting, additional mRNA structures or sequences are often necesssary to stimulate the level of frameshifting at the shift site to a useful efficiency.

AUTOREGULATORY FRAMESHIFTING

Mammalian Antizyme

A +1 ribosomal frameshift is required to decode mammalian antizyme, and the efficiency of this frameshifting is regulated by polyamines (Rom and Kahana 1994; Matsufuji et al. 1995). Antizyme tags ornithine decarboxylase (ODC), a key enzyme in polyamine synthesis, for proteolytic degradation by the 26S proteosome (Murakami et al. 1992b; Tokunaga et al. 1994; for review, see Hayashi and Canellakis 1989). High levels of polyamines result in high levels of frameshifting and hence antizyme, with resultant diminution of the amount of ODC and polyamines. The converse also holds such that with low polyamine levels, frameshifting decreases, resulting in less antizyme and longer-lived ODC (Fig. 1, top) (Rom and Kahana 1994; Matsufuji et al. 1995). Consequently, not only is ODC short-lived like other proteins that control important cellular processes (Goldberg and St. John 1976), but its turnover is also regulated. (A second less well characterized function of antizyme is that it represses polyamine uptake, thereby enabling sharper shut-down of polyamine accumulation in cells [Mitchell et al. 1994; Suzuki et al. 1994].)

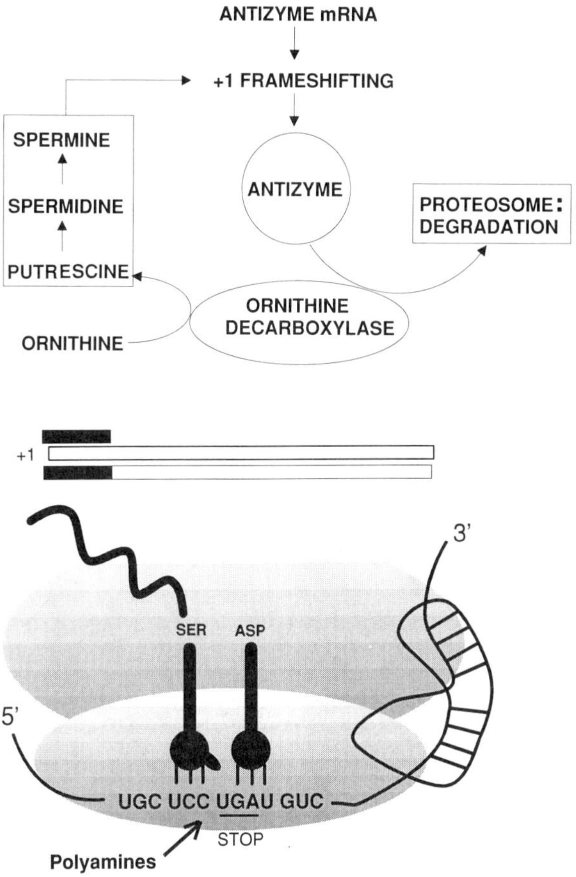

Figure 1 (*Top*) Autoregulatory circuit for polyamine synthesis. Polyamine concentration influences antizyme synthesis, via frameshifting, and antizyme in turn influences polyamine synthesis, via ornithine decarboxylase stability. (*Bottom*) Site of +1 frameshifting in decoding antizyme mRNA.

The first 35 amino acids of antizyme are encoded in the zero (initiating) frame, ORF1 (there may actually be 68 amino acids from ORF1 as there are two potential initiation codons for ORF1 and it is unclear which is used in vivo) (Miyazaki et al. 1992). The known antizyme functions are localized to the ORF2 product from the +1 frame, suggesting that the role of ORF1 is not for its protein product per se, but rather to provide recoding regulation by polyamines. The part of antizyme needed for interaction with ODC is encoded by ORF2. Cells stably transfected with ORF2 supplied with an in-frame initiator codon show antizyme activity that is not stimulated by exogenous polyamines (Murakami et al. 1992a). ORF1, through the act of its translation, provides a means to access

ORF2. However, it is not ruled out that ORF1 or its product may have some additional role.

The antizyme shift sequence, UCC-UGA-U, is unlike other known shift sites (Fig. 1, bottom). In vitro protein synthesis experiments in reticulocyte lysates established that serine (UCC) is the last amino acid encoded in the zero frame and aspartic acid (GAU) is the first amino acid from the +1 frame. Unlike other known cases of programmed frameshifting (except Ty3 as described below) peptidyl-tRNA^{Ser} does not re-pair in the new frame but instead either reads or occludes a fourth base, UCC U (Matsufuji et al. 1995). The stop codon in the initiating frame (UGA) is important for frameshifting, but any stop codon at this position will suffice (they all start with U). Shifting into the –1 frame is not detectable.

A pseudoknot 3′ of the antizyme shift sequence is also important. Although 3′ pseudoknots are well known for stimulating –1 frameshifting and stop codon read-through, this is the only known case where a downstream structure stimulates +1 frameshifting. However, the present evidence from in vitro experiments indicates that the antizyme pseudoknot is less important than its counterparts for –1 frameshifting, and it is also closer to its shift site (Matsufuji et al. 1995). Clearly, further work is required to characterize the antizyme pseudoknot and assess the significance of its differences from other pseudoknots. However, there are some provocative similarities. In the antizyme pseudoknot, there is a single base, A, between the two stems just as there is in the pseudoknot that stimulates mouse mammary tumor virus (MMTV) *gag pro* –1 frameshifting (Chen et al. 1995). In the MMTV pseudoknot, this A is wedged between the two stems, offsetting them from coaxial stacking producing a bent shape (Shen and Tinoco 1995). Shen and Tinoco (1995) suggest that the MMTV structure has some resemblance to a codon-anticodon complex. They raise the interesting possibility that this structure is recognized by a component of the translation apparatus responsible for stabilizing tRNA-codon interaction, and so influences frameshifting.

The original antizyme sequence was obtained from rat (Miyazaki et al. 1992), but both human (Tewari et al. 1994) and *Xenopus* (Ichiba et al. 1995) antizyme cDNA sequences also require frameshifting for their expression. The identity, or great similarity, of the shift sites and pseudoknots means that the basic mechanism is widely conserved. Curiously, when the antizyme frameshifting cassette is expressed in *Saccharomyces cerevisiae* cells, frameshifting occurs at the same site but is predominately –2 rather than +1 (Matsufuji et al. 1996) (–2 frameshifting yields an extra amino acid in the sequence). Although this finding raises interesting questions, it is discouraging for attempts to isolate meaningful mutants of ribosomal components that would be useful in the elucidation of the details of +1 antizyme frameshifting.

Perhaps the most striking facet of antizyme regulation is the stimulation of frameshifting by polyamines. The mechanism is unknown. The site of action of polyamines is not the pseudoknot since the lower level of frameshifting in the absence of the pseudoknot is still stimulated by polyamines. However, distortion of the decoding site is a possible mode of action (Matsufuji et al. 1995). Although it is very interesting that signals in mRNA can cause the ribosome to alter decoding in response to the concentration of small ubiquitous molecules, the mechanism will be difficult to dissect, however, because polyamines are an intimate part of the translation apparatus.

Bacterial Release Factor 2

RF2 causes polypeptide chain release at UGA and UAA codons, perhaps by direct interaction with stop codons in the decoding site on the 30S ribosomal subunit (Brown and Tate 1994). How this protein:RNA recognition is mediated is unknown. RF2 is the only release factor that acts at UGA (RF1 mediates release at UAA and UAG but not UGA). The number of RF2 molecules per *E. coli* cell increases from 5900 to 24,900 as the growth rate is increased from 0.3 to 2.4 doublings per hour (Adamski et al. 1994). The putative promoter for the RF2 gene, *prfB*, has a stringent discriminator (Kawakami et al. 1988), which presumably contributes to coordinate regulation in response to growth demands. There is also the added complexity that the specific *activity* of the expressed product is proportional to the level of expression, suggesting that a specific deactivation or activation mechanism may exist to control RF2 activity posttranslationally (Adamski 1992). However, much of the regulation of RF2 expression is via control of obligatory ribosomal frameshifting (Fig. 2). The efficiency of this shift can be as much as 40% or more in vivo with wild-type cells.

The first 25 amino acids of *E. coli* RF2 are encoded by the zero or initiating frame of the *prfB* gene, and the other 340 amino acids are encoded by its +1 frame (Fig. 2) (Craigen et al. 1985; Kawakami et al. 1988). The 25th and 26th zero frame codons are CUU UGA. To synthesize RF2, ribosomes shift +1 just before encountering the UGA stop codon. The cognate tRNA for CUU is $tRNA^{Leu}_{2}$ (anticodon 3′-GAG-5′). $tRNA^{Leu}$ (very likely $tRNA^{Leu}_{2}$) dissociates from the CUU leucine codon at position 25 and slips +1 to re-pair with the overlapping

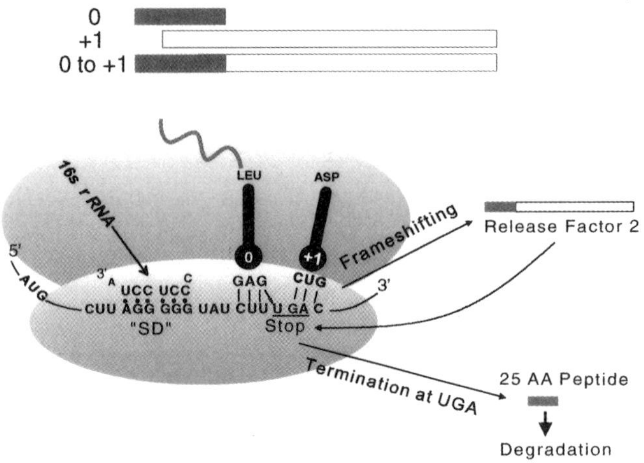

Figure 2 Elements important for the +1 frameshifting in decoding RF2 mRNA. The shift of $tRNA^{Leu}$ from pairing with CUU to pairing with UUU is influenced by the stop codon and the mRNA rRNA interaction of translocating ribosomes.

UUU (Weiss et al. 1987). Re-pairing requires first position U:G wobble pairing. $tRNA_2^{Leu}$ may be aided in re-pairing by lack of a bulky modification at the base 3′ to the anticodon (m_1G37) (see Curran 1993).

Direct evidence indicates that the obligatory frameshifting in RF2 decoding is autoregulatory. Exogenously added RF2 represses in vitro expression of the RF2 gene (Craigen and Caskey 1986) by decreasing frameshifting to just a few percent (Donly et al. 1990). In addition, diminution of RF2 levels in vivo by mutants in the RF2 gene leads to increased RF2 frameshifting to levels that can approach 100% (Kawakami and Nakamura 1990). The 3′ U of the UUU codon involved in re-pairing is the first base of the UGA codon (Weiss et al. 1987; Curran and Yarus 1989). The availability of the UGA codon and hence its first base is governed by the amount of RF2. When the RF2 level is low, UUU is available for re-pairing by peptidyl $tRNA^{Leu}$, and translation continues in the +1 frame to synthesize more RF2, thereby helping to correct the deficit. Conversely, in excess RF2, termination at the UGA codon is more likely to ensue, establishing competition between termination and frameshifting.

The small peptide originating from termination at codon 26, UGA, is rapidly degraded (Brown 1989; see Williams et al. 1989; Donly and Tate 1991). However, at the time the frameshifting occurs, the 25 mer is likely to be wholly contained within the exit channel of the ribosome. So far, there is no reason to think that this peptide affects the frameshifting since the 16-nucleotide frameshifting cassette containing only four of the codons for the 25 mer (see below) efficiently gives regulated frameshifting in heterologous contexts. However, a direct test has not been done, and in the case of the highly efficient 50-nucleotide translational bypass in decoding T4 gene *60*, part of the nascent peptide is important for recoding (Weiss et al. 1990a).

When removed from its RF2 gene environment, alterations of the CUU UGA shift site to CUU UXX, where UXX is not a stop codon, give only a very low level of frameshifting (Weiss et al. 1987). (In the RF2 gene context [see below], some sense codon substitutions can give up to 11% frameshifting [Curran and Yarus 1989], and this can be greatly elevated by a ribosomal mutant with increased accuracy [Sipley and Goldman 1993].) The stop codon elevates the level about tenfold and has been called a "shifty stop" (Weiss et al. 1987). The stop may act by causing a long pause in decoding. However, a mechanism that involves some early stage in the termination reaction needs to be seriously considered (Weiss et al. 1990b).

Considerable evidence exists that the recognition sites for release factors are effectively quadruplets (e.g., UGAN) (Brown et al. 1990). UGAC is the rarest context of the UGA and UAA terminators used in *E. coli* genes (Brown et al. 1993) and is the poorest termination signal (Poole et al. 1995). The nucleotide 3′ to the UGA terminator at codon 26 in the RF2 gene is C. This unfavorable context for termination contributes to high-level frameshifting (Poole et al. 1995). There is another case where a C 3′ to UGA is utilized to dampen the possibility of termination at that site (for review, see Tate and Brown 1992). In the *E. coli* formate dehydrogenase gene (*fdhF* and *fdhG*) where an internal UGA specifies selenocysteine and not termination, the nucleotide 3′ to the UGA is also C (Böck et al. 1991). UGA C is inefficient as a terminator in competition with suppressor tRNA in reporter systems (Miller and Albertini 1983; Stormo et al. 1986; Buckingham et al. 1990; Kopelowitz et al. 1992). Some of these suppressor studies revealed the importance of the 3′ codon (when it was a particular leucine codon which surprisingly is not decoded by a rare tRNA) rather than merely the 3′ base (Miller and Albertini 1983; Raftery et al. 1984), and results consistent with this have been found with some natural terminators (Atkins and Gesteland 1983; N.M. Wills et al., unpubl.). The codon immediately 3′ of the UGA stop codon at position 26 in the RF2 gene is CUA. $tRNA_3^{Leu}$, which decodes CUA, is a rare tRNA, but whether the 3′ codon rather than just the 3′ base effects release at codon 26 in the RF2 gene is unknown.

An additional and crucial signal in the RF2 gene stimulates the frameshifting fivefold above that which occurs at CUU UGA on its own. The additional element is a Shine-Dalgarno (SD) sequence located three bases 5′ of the CUU codon (Weiss et al. 1987, 1988b, 1990b; Curran and Yarus 1988). Its action is not dependent on the presence of the stop codon, i.e., it still acts with a CUU UXX shift site, where UXX is not a stop codon (Weiss et al. 1987). Base pairing between the anti-SD sequence at the 3′end of 16S rRNA in translocating ribosomes and the AGGGGG sequence 5′ of the shift site stimulates the shift (Weiss et al. 1988b). This implies that the 3′end of 16S rRNA of elongating ribosomes must scan mRNA for potential pairing during ribosome translocation. Pairing with an internal SD sequence presumably causes some conformational change at the decoding site that stimulates frameshifting. Although it has not yet been tested, the rRNA mRNA interaction may cause ribosome slowing or pausing and, if so, this would likely have a role, but spacing arguments (see below) suggest that the stimulatory effect is not solely due to pausing.

The distance of three nucleotides between the shift site and the SD sequence is critical: Moving it even one base in either direction greatly reduces frameshifting (Weiss et al. 1987). The spacing corresponds to the minimal distance between AUG start codons and their associated SD sequence (see Ringquist et al. 1992). In decoding the *dnaX* gene, SD interactions stimulate −1 frameshifting, and the ten-nucleotide distance between the shift site and the SD sequence is close to the maximal distance between AUG start codons and their associated SD sequences (Larsen et al. 1994). The reason for the spacing distinction between the RF2 gene and *dnaX* is unclear, but perhaps the rRNA:mRNA pairing results in tension in the −1 case or compression in the +1 case that is relieved by mRNA slippage in the appropriate direction (Larsen et al. 1994). Insertion of an additional SD sequence positioned 15 nucleotides upstream of the RF2 shift site strongly interferes with the stimulatory effect of the normal SD, as if sequestration of the rRNA anti-SD sequence by the upstream additional SD was maintained sufficiently long to preclude interaction with the correctly positioned SD (Weiss et al. 1990b).

The UGA C stop codon in the RF2 gene frameshifting cassette is two- to threefold more susceptible to in-frame read-through by suppressor tRNAs than the UGA G stop codon at the end of the RF2 gene (see Curran and Yarus 1988; Adamski et al. 1993). Previous suppression studies on these four-nucleotide terminator codons in "reporter" genes yielded the opposite result (see above), implying that the RF2 context is special. This increased suppression efficiency gives a measure of the extent to which RF2 is less effective at this site than at UGA codons in "normal" contexts. Recent experiments suggest that the SD interaction has a role in decreasing the efficiency of termination in addition to its direct stimulation of the frameshifting (B. Larsen et al., unpubl.).

Another feature to be considered is the nature of the penultimate amino acid encoded before the stop codon. In a test system, it has been shown that the acidic/basic property of the amino acid at that position affects termination efficiency at UGA stop codons (Mottagui-Tabar et al. 1994). This feature has not been tested in the RF2 context. Evidence has been found for a specific interaction between RF2 and the last peptidyl tRNA, serine/phenylalanine, in termination at UGA (Arkov et al. 1993), but this is not relevant to the required frameshifting in RF2 decoding as the last peptidyl tRNA is neither phenylalanine nor serine.

Release factor 3 (RF3) enhances release, in collaboration with RF1 and RF2, at the three stop codons but especially at UGA (Grentzmann et al. 1995). Consequently, the level of RF3 influences the level of termination at the codon-26 UGA terminator within the RF2 gene, and hence the amount of frameshifting and RF2 synthesis (Grentzmann et al. 1995). The amount of RF3 present during different growth conditions has not yet been determined, and thus its role in regulation of RF2 synthesis is unknown.

The genes for RF2 in both *Salmonella typhimurium* (Kawakami and Nakamura 1990) and *Bacillus subtilis* (Pel et al. 1992) show striking conservation of the frameshift sequence with their *E. coli* counterpart. The SD requirement for initiation in *B. subtilis* seems to be stricter than in *E. coli*, with less tolerance for spacing shorter than optimal (Vellanoweth and Rabinowitz 1992). The *Bacillus* RF2 gene shift sequence is AGG GGG ucu CUU UGA C compared to the *E. coli* sequence AGG GGG uau CUU UGA C. (The SD and shift/stop nucleotides are in capitals; note that the penultimate codon before the stop encodes a neutral amino acid in both cases.) This conservation is much greater than the 20.7% in rRNA between *E. coli* and *B. subtilis*, which are thought to have diverged 1500 million years ago (Ochman and Wilson 1987).

There is 37.2% identity between the amino acid sequences of *E. coli* RF1 and RF2, and many of the substitutions are conservative. This is likely to reflect a common ancestor for these two genes, rather than convergent evolution for partially overlapping binding sites and some common functions. Despite the apparently strong selection for the utilization of frameshifting in RF2 gene decoding, there is no programmed frameshifting in the synthesis of RF1 or RF3 (Craigen et al. 1985; Grentzmann et al. 1994; Mikuni et al. 1994). There may be some other mechanism for the control of RF1 synthesis. The gene for RF1, in both *E. coli* and *S. typhimurium*, is close to the 3′ end of the *hemA* gene which is terminated by UAGC, a putative "poor" stop. Elliott (1989) proposed that the concentration of RF1 is sensed by its action at this UAGC stop, with low levels of RF1 permitting read-through to the next stop codon. The second stop codon is UGAC (RF2-specific), and it is very close to the poor ribosome-binding site for the RF1 gene. Elliott proposed that termination at this UGA codon, which would be dependent on RF2, would "feed" ribosomes for initiation of RF1 synthesis (see Elliott and Wang 1991). Whether this model is right or not, the only evidence available hints that RF1 levels are governed by RF2 levels (for review, see Tate et al. 1993; Adamski et al. 1994).

Why RF2 and not RF1 should apparently be the primary regulatory focus is unknown. Although genes with UGA stop codons are used equally to those with UAA at low growth rates, genes with UAA are predominant at high growth rates. There is evidence for differential context effects for the release factors at UAA codons (Martin et al. 1988).

If the genes for both molecules derived from a common ancestor, as seems likely, the question arises as to whether the programmed frameshifting arose before the split and was lost from the ancestral RF1 gene or was acquired by the ancestral RF2 gene after the split. Either way, the degree of conservation of the signals between *E. coli* and *B. subtilis* suggests an ancient origin.

A Possible Case: Tryptophanase

Tryptophanase allows bacteria to use tryptophan as a sole carbon source. Expression of the tryptophanase operon of *E. coli* is induced by tryptophan in a growth medium lacking a catabolite-repressing carbon source. Induction requires the translation of a 24-codon gene located in the 319-nucleotide transcribed leader region preceding the structural gene for tryptophanase. This leader-coding region contains a single tryptophan codon, UGG, at position 12. Mutagenic and suppressor studies have shown that translation of the codon at position 12 by $tRNA^{Trp}$ is essential for induction (Gollnick and Yanofsky 1990 and references therein). Consistent with the importance of the tryptophan codon, it and its flanking sequence are highly conserved in *Proteus vulgaris* (Kamath and Yanofsky 1992). Despite much work (see Gish and Yanofsky 1993), the mechanism of tryptophan-mediated induction is unknown. Current studies implicate the leader peptide in induction and the participation of an out-of-frame stop codon, UGA, in the ρ-mediated termination that is relieved upon induction (K. Gish and C. Yanofsky, pers. comm.). How the out-of-frame stop codon, which occurs after codon 12, is sensed by ribosomes is not clear, but further work may reveal novel features of importance to our understanding of recoding.

Tryptophan Repressor

A low-level shift to the +1 frame has been reported in decoding the *E. coli* tryptophan repressor gene, *trpR* (Benhar and Engelberg-Kulka 1993). In a *trpR-lacZ* fusion construct, an intriguing translational bypass has been found (Benhar and Engelberg-Kulka 1993). The bypass uses a set of rules different from those of the T4 gene-*60* bypass (see below). Benhar et al. (1993) have shown that the level involved is dependent on the efficiency of translation initiation. However, it remains to be seen whether these findings are significant for *trpR* expression (see Engelberg-Kulka and Schoulaker-Schwarz 1994).

MOBILE GENETIC ELEMENTS AND REGULATORY FRAMESHIFTING

Transposition and the Generation of Variability

Transposition of resident mobile genetic elements is a major cause of mutation. In bacteria, this has been documented both in growing cells (Rodriguez et al. 1992) and in resting cells maintained in stabs (Naas et al. 1994). Interestingly, starvation conditions seem to increase the transposition of some mobile elements (Hall 1988; Shapiro and Higgins 1989; Mittler and Lenski 1990). Some of the diversity of genetic combinations created by insertion sequence (IS) movement is advantageous (see Arber 1991; Naas et al. 1994). In several bacterial IS elements, there is a direct correlation between frameshifting, which generates a fusion polypeptide, and transposition (Chandler and Fayet 1993). Similar considerations are likely to apply in yeast (Voytas and Boeke 1993; see below). Stationary-phase *E. coli* show higher frameshifting levels with certain test sequences (J. Gallant, unpubl.); the effect of starvation conditions is described below.

Yeast Ty Elements

Yeast transposable elements, Ty, replicate by a retrovirus-like mechanism involving an RNA intermediate and are widely dispersed throughout the genome of *S. cerevisiae*. Ty elements contain two genes *TYA* and *TYB* that are analogs of retroviral *gag* and *pol* genes (in Ty3, the genes

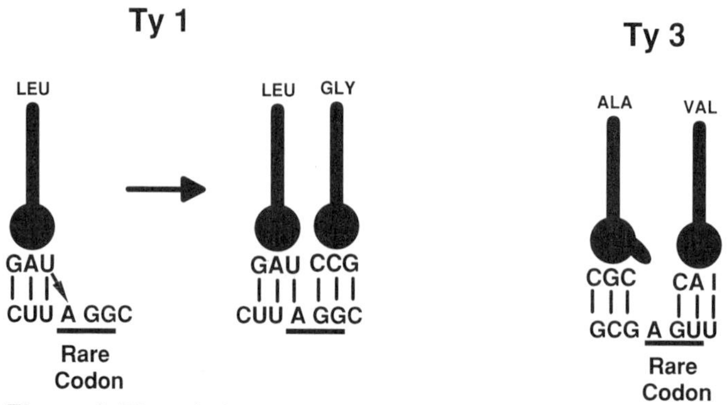

Figure 3 The +1 frameshifting in Ty1 decoding involves tRNA mRNA re-pairing, whereas that involved in Ty3 decoding utilizes mRNA base occlusion instead of re-pairing.

are called *GAG3* and *POL3* instead of *TYA* and *TYB*). In marked contrast to mammalian retroviruses, except spumaretroviruses (foamy viruses), *TYB* is in the +1 frame relative to *TYA*. Ribosome entry to the *TYB* gene requires +1 frameshifting near the end of the *TYA* gene. For Ty1, the frameshift signal is CUU AGG C, and the CUU decoding special leucine tRNA shifts +1 to re-pair with the underlined UUA triplet (Fig. 3). Subsequent decoding continues in this new frame (Belcourt and Farabaugh 1990). An important ingredient of the mechanism is that the zero frame AGG arginine codon is a rare codon with a correspondingly sparse tRNA (Belcourt and Farabaugh 1990). Overexpression of the single gene for this arginine tRNA results in reduced frameshifting (Belcourt and Farabaugh 1990; Xu and Boeke 1990) and consequently reduced Ty transposition since the TYA TYB fusion product is essential and rate limiting for transposition (Xu and Boeke 1990). The gene for the sparse arginine tRNA, *HSX1*, is involved in the heat shock response (Kawakami et al. 1992, 1993), and thus this system qualifies as being bona fide regulatory.

Another yeast transposon, Ty3, is distinctly different from Ty1 and has similarities to the *Drosophila* retrovirus *gypsy* (Hansen et al. 1988; Kim et al. 1994; Song et al. 1994). Transposition of Ty3 is dependent on the cell cycle (Menees et al. 1994). Frameshifting at the end of the Ty3 *gag* analog has also been investigated. A "hungry" rare codon is also involved in the corresponding Ty3 frameshifting, but it is an AGU serine codon. The shift site is GCG AGU U, but in this case, the tRNA for the initial zero frame codon GCG does not re-pair with CGA: Instead, the base A is skipped and GUU is read as the next codon (Farabaugh et al. 1993). Replacing the zero frame codon GCG with all other sense codons showed that only eight tRNAs can stimulate the +1 shift, consistent with there being special features of active peptidyl tRNAs (Vimaladithan and Farabaugh 1994). Interestingly, overproducing the tRNA for the first +1 frame codon stimulated the frameshifting (Pande et al. 1994) and provides evidence that the first codon has a role in the new frame, "fixing" the frameshifting event. However, further work is necessary to understand the mechanism for base skipping. The sequence of 14 nucleotides (CU AAC CGA UCU UGA, shown in codons of the initiating frame) 3′ to the shift site is important for the frameshifting (Farabaugh et al. 1993). It is not thought to be structured (P.J. Farabaugh et al., pers. comm.), but how it acts is unknown.

Initiation Factor 3 and Bacterial Insertion Sequence *911*

IS*911* encodes an *orfA* gene product, an *orfB* gene product (*orfB* begins 52 nucleotides 5′ of the *orfA* terminator), and an *orfAB* gene product. In the overlap region between the *orfA* and *orfB* ORFs is a "slippery" double-shift site A AAA AAG followed by a complicated stem-loop sequence. The structure stimulates −1 frameshifting to yield the *orfAB* gene fusion product important for transposition (for review, see Chandler and Fayet 1993). Immediately 5′ of the A AAA AAG shift site is an in-frame AUU codon preceded by an appropriately positioned SD sequence (Prère et al. 1990; Polard et al. 1991). This codon serves for initiation of synthesis of the OrfB protein whose function in transposition is unknown. AUU initiators are very uncommon and are only known elsewhere in the gene for initiation factor 3, *infC*, and in phage T4 gene *26** (Sacerdot et al. 1982; Nivinskas et al. 1992). The *infC* AUU initiator has been shown to be a site for autoregulation by initiation factor 3 (IF3) (Butler et al. 1987; Hartz et al. 1990; La Teana et al. 1993 and references therein). In the case of IS*911*, O. Fayet and colleagues (M.-F. Prère et al., unpubl.) have found that the level of *orfB* synthesis is influenced in the same manner by IF3 concentration. The obvious implication is that altered IF3 levels are an indicator of some adverse circumstances and that IS*911* uses this monitor to modulate the synthesis of at least one of the three products potentially important for transposition. However, there is at present no direct evidence for this model. The previous findings with *infC* mRNA could be interpreted as revealing a device for the maintenance of homeostasis. The IS*911* findings to date hint that IF3 levels may also have a sensory role.

Interestingly, IF3 is known to bind the 3′end of 16S rRNA of 30S ribosomal subunits (Wickstrom 1983). The SD sequence 5′of the IS*911* AUU initiator is essential for obtaining the maximum level of frameshifting (M.-F. Prère et al., unpubl.). The initial interpretation—that this was wholly a reflection of IF3 involvement—is very likely to be only partially true. The SD sequence is 11 bases 5′of the A AAA AAG shift sequence. This arrangement is virtually the same as that in *E. coli dnaX* –1 frameshifting discussed above where pairing between rRNA and mRNA is important for high-level frameshifting (Larsen et al. 1994). In IS*911* decoding, in translocating ribosomes near the end of the *orfA* gene, the 3′end of 16S rRNA presumably pairs with the mRNA SD sequence. Whether there is continued in-frame translation with nearby termination, initiation at the AUU initiator, or frameshifting at the A AAA AAG depends, in part, on IF3 levels (M.-F. Prère et al., unpubl.).

OTHER STUDIES ON LIMITATION OF AMINOACYL tRNA

Limitation of an aminoacyl tRNA reduces the potential for pairing by a cognate tRNA, and the waiting ribosomes adopt a much more open structure than normal (Öfverstedt et al. 1994). Such limitation is natural, commonly encountered, and potentially a serious problem. It is not surprising that this is at the center of several control systems. As considered above, limitation of particular aminoacyl tRNAs is important for Ty frameshifting. It, or its corollary, has also been found to mediate frameshifting in several other cases, as well as being involved in quite a number of unrelated control systems. The degree of aminoacylation of tRNAs is a sensitive indicator of cellular activity, and this measure in bacteria is used to operate several specific controls. Specific aminoacylation levels are detected by leader peptide gene translation and consequent attenuation in the expression of gram-negative amino acid biosynthetic operons. Uncharged tRNA directly pairs with leader RNA in gram-positive aminoacyl tRNA genes and effects attenuation by a different means (Grundy and Henkin 1994). tRNA aminoacylation levels also control certain translation factor genes, as well as the highly pleiotropic stringent control system (Gallant 1979; Cashel and Rudd 1987) and the expression of genes with rare codons early in the gene (Chen and Inouye 1994). Aminoacylation levels in turn are governed by several factors including the supply of amino acids and the expression levels of tRNAs in relation to synthetase levels.

Limitation of aminoacylated tRNA is also found in several human mitochondrial disease states, and the protein synthesis defect has been partially characterized in one case of myoclonus epilepsy with ragged red fiber encephalomyopathy (Enriquez et al. 1995). However, products arising from frameshifting have not yet been found.

Frameshifting as a Sensor

In vitro protein synthesis studies of MS2 RNA (Atkins et al. 1979; Dayhuff et al. 1986), in vivo studies of phage and bacterial frameshift mutants (Gallant and Foley 1980; Weiss et al. 1988a), and in vivo studies of overexpressed genes with tandem rare codons (Spanjaard et al. 1990) showed that alteration of the balance of aminoacylated tRNAs could have dramatic effects on ribosomal frameshifting (Gallant and Lindsley 1993). Even high-level expression of a mammalian gene in *E. coli* leads to enhanced ribosomal frameshifting as detected by increased β-galactosidase from *lacZ* frameshift mutants (Bogosian et al. 1990). It is not known if any of the signals identified in these studies are utilized for regulatory purposes, but it would be surprising if at least some of them are not. In fact, although the signals are different, as described above, it is clear that the +1 frameshifting utilized by mobile yeast Ty elements for generating their transposase is responsive to the level of certain charged tRNAs. In contrast, the sensing of aminoacyl tRNA limitation for programmed –1 frameshifting is much less clear. Model systems using the signals from programmed –1 shift sites show that it can occur, but there is as yet no clear evidence for its utilization.

Most programmed –1 frameshifting sites are of the general type X XXY YYZ, where X and Y, and sometimes even Z, can be the same base. These "double-shifty" sites permit two tRNAs to shift –1 in tandem and are used by viruses, IS elements, and at least one cellular gene, *E. coli dnaX*. Is the level of frameshifting at this type of signal also responsive to the relative amounts of aminoacylated tRNAs? In *E. coli*, a U-rich double-shifty sequence, U UUU UUA (the *gag pol* shift site in HIV-1) has been tested. When the UUA codon is "hungry" due to leucine starvation, frameshifting is elevated sixfold when starvation reduced growth rate sevenfold. On leucine starvation, two thirds of the frameshifting results in phenylalanine insertion rather than leucine insertion at the shift site. This is presumably due to –1 shifting before tRNA binding and incoming $tRNA^{Phe}$ pairing with the U UU codon (Yelverton et al. 1994), although unstable initial pairing of $tRNA^{Phe}$ to the UUA leucine codon cannot be ruled out. This ratio of phenylalanine to leucine is about double the ratio seen in unstarved *E. coli* and three times the ratio in reticulocyte lysates (Yelverton et al. 1994). In unstarved situations with different shift sites, the ratio of the two amino acids that correspond to the phenylalanine and leucine in the U UUA case just described may be quite different. For instance, with the HTLV-I shift cassette, the last four nucleotides of the shift site are A AAC, and in reticulocyte lysates, the amount of lysine compared to asparagine inserted is vanishingly small (Nam et al. 1993; AAA encodes lysine and AAC encodes asparagine). Starvation studies in *E. coli* done with a shift sequence with the same four last bases A AAA AAC from IS*1* show a substantial increase in frameshifting, over a broad range of values, in response to asparagine starvation, which may serve as an indicator of nitrogen starvation (E. Yelverton, pers. comm.). Starvation for asparagine could allow either the ribosome to slip back prior to tRNA attachment or unstable pairing by lysine tRNA at the AAC codon, which by analogy with other studies would be expected to facilitate frameshifting by permitting easier detachment (see Tsuchihashi and Brown 1992). The subsequent re-pairing step by a $tRNA^{Lys}$ would be to a cognate codon (AAA), and although the relationship between good re-pairing possibilities and frameshifting is complex, this should contribute to elevated frameshifting (for review, see Atkins and Gesteland 1995). With the tandem lysine codon shift sequence, A AAA AAG, a reduced supply of aminoacyl tRNA for the 3′ codon AAG by lysine starvation may not have the same effect on frameshifting since the −1 frame codon for re-pairing is also a lysine codon (cf. Fu and Parker 1994) and may be less useful for regulation by this means. In *E. coli*, A AAA AAG is much more shifty than A AAA AAC (which is used in IS*1*) (Weiss et al. 1989), but of course the nature of the stimulators greatly influences the level of programmed frameshifting. Furthermore, one study has shown that a stimulator influences the character of a shift, i.e., whether the amino acid encoded by the second slip codon is primarily that specified by the zero or −1 frame codon (Weiss et al. 1989). Clearly, the role of aminoacyl tRNA limitation in regulated −1 frameshifting is far from being resolved.

POLYPROTEINS AND RATIO SETTING

In the great majority of known programmed frameshifting cases, alteration of the ratio of products has not been reported. The ratio of shift and nonshift products appears set and not subject to regulation. However, the particular setting of the ratio varies widely. When frameshifting is used to generate polyproteins, the nonshift product is often used for viral structural purposes and is produced in large amounts, whereas the shift product, commonly a polymerase, is used for catalytic purposes and is produced in small amounts. The *gag-pol* –1 frameshifting in *S. cerevisiae* double-stranded RNA virus, L-A, occurs at 1.9%; the λG-T shift occurs at 4%; the MMTV *gag-pro* shift occurs at 23% (but a second shift, at the *pro-pol* junction, is required to give the Gag-Pro-Pol fusion and the component shifts have to be efficient to give a 5% ratio of Gag-Pro-Pol fusion to Gag product). In *E. coli dnaX*, decoding the –1 shift occurs at 50% and the two products are present in a 1:1 ratio in DNA polymerase III. Here, frameshifting is used to generate a truncated protein rather than a polyprotein.

The precise set ratio of Gag product to Gag-Pol product seems to be important for retroviruses and retrotransposons (see, e.g., Dinman and Wickner 1992), with increases or decreases in frameshifting levels likely to be deleterious. Viral intolerance of altered retrovirus ratios has provoked thoughts of the possible utility of potential drugs that would change frameshifting (or stop codon read-through) efficiency without upsetting standard decoding. A key consideration is the degree of flexibility permitted with host programmed frameshifts. However, since only one case of mammalian cellular ribosomal frameshifting, in decoding antizyme, has been identified to date, the potential detrimental effect is difficult to evaluate. Close relatives of retroviruses, pararetroviruses, use various strategies other than frameshifting or read-through to generate their Gag-Pol polyprotein, so even with a similar gene organization, different strategies for generating Pol products are possible. There is no evidence for regulation of the set ratio in these viruses either (see Rothnie et al. 1994).

The ratio of shift and nonshift products is set by the "slippery" nature of the shift sequence and the strength of the stimulatory signals. Efficient programmed –1 frameshifting sites have double-shift codons, X XXY YYZ, preceding a potential pseudoknot or stem-loop at a defined distance, 7 ± 2 (Jacks et al. 1988; Brierley et al. 1989; Weiss et al. 1989; ten Dam et al. 1990). Coronaviruses and MMTV *gag-pro* shifts use pseudoknots, whereas astrovirus (Marczinke et al. 1994), HTLV-II (Falk et al. 1993; Kollmus et al. 1994), and *dnaX* (see Tsuchihashi and Brown 1992) shifts use a stem-loop. However, not all –1 programmed frameshifting involves double slippery codons. Potato virus M utilizes a single slip-shift site (Gramstat et al. 1994), and the same is likely to be true for several bacterial IS elements (see Chandler and Fayet 1993). Furthermore, in phage T7 gene-*10* frameshifting, the shift site is at best a degenerate double-shift site (Condron et al. 1991). Putative +1 frameshifting sites are more difficult to identify. The known cases occur directly before stop or rare codons, but this on its own is not an easy identifier. When re-pairing is involved as in RF2 and Ty1, only one 3′ nucleotide is utilized for re-pairing, and in other cases, Ty3 and probably antizyme re-pairing is not utilized.

Viruses

Viral genomes are a rich source for recoding. This is especially true for plant viruses even though the efficiency of individual events in plant virus decoding tends to be low (for review, see Rohde et al. 1994; Miller et al. 1995). One issue raised by the frequent occurrence of recoding in viral expression is whether infection causes a change in host-protein synthesis to facilitate its particular kind of recoding event. This has been addressed for some retroviruses that utilize frameshifting at the end of their *gag* genes and also for murine leukemia virus (MLV) where ribosomes enter its *pol* gene by reading through the *gag* stop codon. The tRNAs that decode the shift site in several retroviruses are specifically undermodified on infection with the retrovirus in question (Hatfield et al. 1989). However, no alteration in the efficiency of frameshifting has been detected with the HIV-1 *gag-pol* shift sequence in cell lines from different tissues or species (Vickers and Ecker 1992; Reil et al. 1993; Cassan et al. 1994). Furthermore, the efficiency of frameshifting from a plasmid-borne copy of the HIV-1 shift cassette is unaffected by HIV-1 infection (Cassan et al. 1994; Reil et al. 1994). Earlier, it was reported that the minor glutamine tRNA that mediates read-through of the MLV *gag* UAG terminator was elevated in infected mouse cells (Kuchino et al. 1987). However, other investigators have found no alteration of glutamine tRNA or level of this programmed read-through on infection (Panganiban 1988; Feng et al. 1989; Berteaux et al. 1991).

Drosophila Gypsy

There is now good evidence that Gypsy is an invertebrate retrovirus (Kim et al. 1994; Song et al. 1994), and it is likely that –1 frameshifting (probably at a U-rich sequence similar to that in HIV-1) is required for ribosomes to enter its *pol* analog. Transposition of Gypsy is under the tissue-specific control of the *flamenco* gene, which, however, is likely to act on expression of the *env* analog (Pélisson et al. 1994) rather than at the presumptive frameshifting.

Nonproduct Roles

On occasion, the reason for recoding may not be to produce a "new" protein product per se, but rather to ensure ribosome movement into "special" locations (see Atkins et al. 1990). For instance, one speculative possibility is that the passage of some ribosomes into the 3′-untranslated region may influence mRNA half-life by affecting mRNA structure. Only a small number of specific suggestions have been made, among them are frameshifting near the end of the RNA phage MS2 synthetase gene affecting replicase progression (Dayhuff et al. 1986) and frameshifting at the end of the *lacZ* gene to deliver ribosomes to the *lacY* SD (see Weiss et al. 1987), but these have not been proven. Perhaps the ongoing investigations of the tryptophanase leader region decoding (see above) will reveal a nonproduct role.

SHUNTS AND HOPS

The leader sequences of cauliflower mosaic virus and other caulimoviruses are long (>600 nucleotides) and contain several ORFs. Detailed analysis has led to the proposal that ribosomes enter at the cap site and begin scanning. Subsequently, a "shunt" can occur such that the ribosomes transfer to the 3′end of the leader to initiate synthesis at the main ORF without scanning through the central portion of the leader (Fütterer et al. 1993). A related mechanism may be used for synthesis of the X protein of the parainfluenza virus, Sendai (Curran and Kolakofsky 1988). In *E. coli*, although direct entry of ribosomes to at least the first gene on a polycistronic mRNA is the norm, terminated but not released ribosomes can, with moderate efficiency, scan mRNA for restart codons at least 46 nucleotides distant (Adhin and van Duin 1990). Shunting of scanning ribosomes has not been detected in *E. coli*, but a bypass of 50 nucleotides by translating ribosomes is being studied in detail (Huang et al. 1988; Weiss et al. 1990a; Herbst et al. 1994). The hop, which occurs in decoding phage T4 gene *60*, is highly efficient, but no regulatory function is known.

STOP CODON READ-THROUGH

Although there are many known cases of functional read-through of stop codons, none are known to be regulated. Before the *E. coli* gene for RF2 was sequenced, the possibility existed that its expression would be regulated by read-through of an in-frame UGA stop codon early in the gene. As described above, sequencing revealed that it does have UGA as codon

26, but expression and regulation require frameshifting rather than stop codon read-through. Why? Perhaps stop codon read-through cannot be made to operate at the required 30–50% efficiency without elaborate mechanisms. None of the many known cases of stop codon read-through with insertion of one of the 20 standard amino acids occur with more than 12% frequency, even though one category of read-through, that found in MLV and related retroviruses, uses an elaborate structure, a pseudoknot, as stimulator. The decoding of certain UGA codons as the 21st amino acid, selenocysteine, occurs at 100% efficiency, but a special tRNA is required. In *E. coli*, elegant work has revealed the crucial nature of a stem-loop 3′ of the UGA selenocysteine codon and a special elongation factor (Böck et al. 1991). (In mammalian cells, a critical mRNA element in the 3′-untranslated region is required instead of the 3′-flanking stem-loop [Berry et al. 1993].) One component of the optimization of selenocysteine insertion is presumably the minimization of termination and release factor access. Could the mRNA structure component of the *E. coli* mechanism be modified to permit 40% read-through of UGA codons with insertion of a standard amino acid and be regulated by RF2? Quite a number of cases of the utilization of read-through of UAG or UGA stop are known but none are known for UAA, which is the most commonly used terminator. In general, a very small, special 3′ context is required to make UGA an inefficient terminator and thus prone to read-through. In addition, in at least some cases of UAG read-through, the identity of the six 3′ nucleotides (Skuzeski et al. 1991; Zerfass and Beier 1992a) or even a pseudoknot is important (Wills et al. 1991, 1994; Feng et al. 1992) to stimulate read-through. Starting with the initial finding of UGA stop codon read-through in the RNA phage Qβ (Weiner and Weber 1973) and later in numerous plant viruses (see Pelham 1978; Schmitt et al. 1992; Zerfass and Beier 1992b; Brault et al. 1995; for review, see Miller et al. 1995) and a variety of animal viruses, such as retroviruses (Philipson et al. 1978; Yoshinaka et al. 1985), and some alphaviruses (Li and Rice 1993; Shirako and Strauss 1994), stop codon read-through is quite frequently used in viral gene expression. Interestingly, in the latter case, there is a hint that the level of UGA read-through, which in reticulocyte lysates is temperature-dependent, may affect the timing of the regulation switch and be different in insect and mammalian hosts (see Wang et al. 1994; Shirako and Strauss 1994).

In contrast to viruses, programmed read-through of stop codons is hardly known in chromosomal gene expression. There is an in-frame UGA stop codon within the *kelch* gene of *Drosophila* species that is known to be read-through (Xue and Cooley 1993), However, it is not known whether the levels of termination and read-through product vary in different tissues or developmental stages.

Independently of read-through, the translation termination process can be involved in the regulated expression of downstream genes. This is best exemplified in the translational control of the *GCN4* gene of *S. cerevisiae* where the ten bases downstream from a stop codon are important for a strong translational barrier (Grant and Hinnebusch 1994; see Hinnebusch, this volume). Ribosome release is also known to modulate the basal level of *E. coli trp* operon expression (Roesser and Yanofsky 1988).

FRAMESHIFTING AND READ-THROUGH AFFECT HUMAN DISEASE

No diseases are yet known to be due to defects in recoding, and no drugs have been identified that affect recoding and not standard decoding. Although the ratio of Gag to Gag-Pol product of retroviruses is known to be critical, it is unclear how useful recoding specific drugs would be for their effects on retroviruses. However, the leakiness of premature UGA stop codons and of certain frameshift mutations in certain contexts is important to disease progression. Mutants of the human cystic fibrosis transmembrane conductance regulator with UGA instead of the Gly-542 codon or Arg-553, which are about one third of the way through the gene, cause less severe pulmonary problems than some missense mutants (Cuppens et al. 1990; Cutting et al. 1990; Kerem et al. 1990), presumably because the UGA is decoded as sense at some efficiency. A study with UGA premature stop codons at the equivalent positions of a member of the same protein family in yeast revealed that one was read-through with an efficiency of 10% (Fearon et al. 1994). Similarily, frameshift mutant leakiness has been described for patients from several families with defects in their carbonic anhydrase II, whose loss results in osteopetrosis. Patients in each of five families have the same single-base deletion. Compensatory ribosomal frameshifting 12 codons 3′ of the mutation site yields a low level of active product with less severe consequences than expected. Even though the lesion in each of the families is the same, they manifest different levels of symptoms (Hu et al. 1994, 1995). The reason for this has not yet been determined.

It will be interesting to see whether frameshift mutant leakiness influences the course of dominant inherited diseases as well, although this will be more difficult to investigate than in recessive inherited diseases. One class of late-onset familial polyposis patients has fewer polyps and later-onset cancer than classical cases (Leppert et al. 1990) despite having either frameshift mutations or premature stops relatively early in an important region of the adenomatous polyposis coli (*APC*) gene (Spirio et al. 1993). Ribosomal frameshifting or read-through has not been shown to be responsible for the better prognosis in this instance, but sensitive tests are required in this and other cases. Not only can frameshift mutant leakiness alleviate a disease, but, on occasion, it can also help an infectious agent evade therapy. The drug acyclovir is frequently used to treat herpes infections. It is activated by herpes thymidine kinase. Viruses can become resistant to acyclovir by inactivation of their thymidine kinase (tk), but null mutants fail to reactivate from latent infections of sensory ganglia. Several acyclovir-resistant frameshift mutants of the viral thymidine kinase gene yield a low level of active thymidine kinase, which permits reactivation from latency and other pathogenic phenotypes (Hwang et al. 1994). Despite its benign appearance when first discovered in the early 1970s, the ribosomal frameshifting responsible for frameshift mutant leakiness is a two-edged sword that would be nice to control.

LOOKOUT

How best to spot further cases of frameshifting? Since two out of the three known cases of cellular frameshifting are regulatory, the subject is of considerable interest (retrotransposons and IS elements, although integrated in nuclear DNA, are not included in this count). Searches of the databases for adjacent or overlapping ORFs in different frames, especially where each has homology with different domains of a single protein, are under way. Although it is clearly desirable to keep sequencing errors in the databases to a low level, care needs to be taken not to sanitize the entries of programmed frameshifting. It seems likely that recoding is less frequent in decoding cellular mRNAs than in viruses, retrotransposons, and IS elements with their compact genomes. However, detailed knowledge is often necessary to spot such events especially if they are below the 10% level. Direct inspection of sequences for shifty sites can be rewarding. Cases of –1 frameshifting are often flagged by double slippery sequences, X XXY YYZ, but especially with +1 frameshifting, predictions are currently difficult. For instance, it is suspected that +1 frameshifting may be required for ribosomes to enter the *pol* gene of foamy viruses (spumaretroviruses), but sequence inspection of the appropriate region has not been productive. Another approach is to search for pseudoknots with tightly defined lengths of the stems and loop 1 and a maximum size for loop 2. Following the work of Shen and Tinoco

(1995) and Chen et al. (1995), this is being done with or without the important wedged A between the two stems. This analysis is generating numerous candidates which require analysis. However, if many people keep a lookout, the outlook is good!

ACKNOWLEDGMENTS

We thank Fran Adamski, Bente Larsen, Senya Matsufuji, and Norma Wills for their comments on the manuscript and C. Yanofsky for communicating unpublished information. R.F.G. is an Investigator of the Howard Hughes Medical Institute, and J.F.A. is supported by a grant from the National Institutes of Health (RO1-GM-48152).

REFERENCES

Adamski, F.M. 1992. "Regulation of prokaryote release factors 1 and 2." Ph.D. thesis, University of Otago, Dunedin, New Zealand.

Adamski, F.M., B.C. Donly, and W.P. Tate. 1993. Competition between frameshifting, termination and suppression at the frameshift site in the *Escherichia coli* release factor 2 mRNA. *Nucleic Acids Res.* **21:** 5074–5078.

Adamski, F.M., K.K. McCaughan, F. Jørgensen, C.G. Kurland, and W.P. Tate. 1994. The concentration of polypeptide chain release factors 1 and 2 at different growth rates of *Escherichia coli*. *J. Mol. Biol.* **238:** 302–308.

Adhin, M.R. and J. van Duin. 1990. Scanning model for translational reinitiation in eubacteria. *J. Mol. Biol.* **213:** 811–818.

Arber, W. 1991. Elements in microbial evolution. *J. Mol. Evol.* **33:** 4–12.

Arkov, A.L., S.V. Korolev, and L.L. Kisselev. 1993. Termination of translation in bacteria may be modulated via specific interaction between peptide chain release factor 2 and the last peptidyl-tRNA (Ser/Phe). *Nucleic Acids Res.* **21:** 2891–2897.

Atkins, J.F. and R.F. Gesteland. 1983. Resolution of the discrepancy between a gene translation-termination codon and the deduced sequences for release of the encoded polypeptide. *Eur. J. Biochem.* **137:** 509–516.

———. 1995. Discontinuous triplet decoding with or without re-pairing by peptidyl tRNA. In *tRNA: Structure, biosynthesis and function* (ed. D. Söll and U.L. RajBhandary), pp. 471–490. American Society for Microbiology, Washington, D.C.

Atkins, J.F., R.B. Weiss, and R.F. Gesteland. 1990. Ribosome gymnastics—Degree of difficulty 9.5, style 10.0. *Cell* **62:** 413–423.

Atkins, J.F., R.F. Gesteland, B.R. Reid, and C.W. Anderson. 1979. Normal tRNAs promote ribosomal frameshifting. *Cell* **18:** 1119–1131.

Belcourt, M.F. and P.J. Farabaugh. 1990. Ribosomal frameshifting in the yeast retrotransposon Ty:tRNAs induce slippage on a 7 nucleotide minimal site. *Cell* **62:** 339–352.

Benhar, I. and H. Engelberg-Kulka. 1993. Frameshifting in the expression of the *E. coli trpR* gene occurs by the bypassing of a segment of its coding sequence. *Cell* **72:** 121–130.

Benhar, I., C. Miller, and H. Engelberg-Kulka. 1993. Frameshifting in the expression of the *Escherichia coli trpR* gene is modulated by translation initiation. *J. Bacteriol.* **175:** 3204–3207.

Berry, M.J., L. Banu, J.W. Harney, and P.R. Larsen. 1993. Functional characterization of the eukaryotic SECIS elements which direct selenocysteine insertion at UGA codons. *EMBO J.* **12:** 3315–3322.

Berteaux, V., J.P. Rousset, and M. Cassan. 1991. UAG readthrough is not increased *in vivo* by Moloney murine leukemia virus infection. *Biochimie* **73:** 1291–1293.

Böck, A., K. Forchhammer, J. Heider, and C. Baron. 1991. Selenoprotein synthesis: An expansion of the genetic code. *Trends Biochem. Sci.* **16:** 463–467.

Bogosian, G., B.N. Violand, P.E. Jung, and J.F. Kane. 1990. Effect of protein overexpression on mistranslation in *Escherichia coli.* In *The ribosome: Structure, function and evolution* (ed. W.E. Hill et al.), pp. 546–558. American Society for Microbiology, Washington, D.C.

Brault, V., J.F.J.M. van den Heuvel, M. Verbeek, V. Ziegler-Graff, A. Reutenauer, E. Herrbach, J.-C. Garaud, H. Guilley, K. Richards, and G. Jonard. 1995. Aphid transmission of beet western yellows luteovirus requires the minor capsid read-through protein P74. *EMBO J.* **14:** 650–659.

Brierley, I., P. Digard, and S.C. Inglis. 1989. Characterization of an efficient coronavirus frameshifting signal: Requirement for an RNA pseudoknot. *Cell* **57:** 537–547.

Brown, C.M. 1989. "The decoding of termination codons." M.Sc. thesis, University of Otago, New Zealand.

Brown, C.M. and W.P. Tate. 1994. Direct recognition of mRNA stop signals by *Escherichia coli* polypeptide chain release factor two. *J. Biol. Chem.* **26:** 33164–33170.

Brown, C.M., M.E. Dalphin, P.A. Stockwell, and W.P. Tate. 1993. The translational termination data base. *Nucleic Acids Res.* **21:** 3119–3123.

Brown, C.M., P.A. Stockwell, C.N.A. Trotman, and W.P. Tate. 1990. The signal for the termination of protein synthesis in procaryotes. *Nucleic Acids Res.* **18:** 6339–6345.

Buckingham, R.H., P. Sörensen, F.T. Pagel, K.A. Hijazi, B.H. Mims, D. Brechemeir-Baey, and E.J. Murgola. 1990. Third position base changes in codons 5′ and 3′ adjacent UGA codons affect UGA suppression *in vivo. Biochim. Biophys. Acta* **1050:** 259–262.

Butler, J.S., M. Springer, and M. Grunberg-Manago. 1987. AUU-to-AUG mutation in the initiator codon of the translation initiation factor IF3 abolishes autocontrol of its own gene (*infC*) in vivo. *Proc. Natl. Acad. Sci.* **84:** 4022–4025.

Cashel, M. and K.E. Rudd. 1987. The stringent response. In Escherichia coli *and* Salmonella typhimurium: *Cellular and molecular biology* (ed. F. Neidhardt et al.), vol. 2, pp. 1410–1438. American Society for Microbiology, Washington, D.C.

Cassan, M., N. Delaunay, C. Vaquero, and J.-P. Rousset. 1994. Translational frameshifting at the *gag-pol* junction of human immunodeficiency virus type 1 is not increased in infected T-lymphoid cells. *J. Virol.* **68:** 1501–1508.

Chandler, M. and O. Fayet. 1993. Translational frameshifting in the control of transposition in bacteria. *Mol. Microbiol.* **7:** 497–503.

Chen, G.-F.T. and M. Inouye. 1994. Role of the AGA/AGG codons, the rarest codons in global gene expression in *Escherichia coli. Genes Dev.* **8:** 2641–2652.

Chen, X., M. Chamorro, S.I. Lee, L.X. Shen, J.V. Hines, I. Tinoco, and H.E. Varmus. 1995. Structural and functional studies of retroviral RNA pseudoknots involved in ribosomal frameshifting: Nucleotides at the junction of the two stems are important for efficient ribosomal frameshifting. *EMBO J.* **14:** 842–852.

Condron, B.G., R.F. Gesteland, and J.F. Atkins. 1991. An analysis of sequences stimulating frameshifting in the decoding of gene 10 of bacteriophage T7. *Nucleic Acids Res.* **19:** 5607–5612.

Craigen, W.J. and C.T. Caskey. 1986. Expression of peptide chain release factor 2 requires high efficiency frameshift. *Nature* **322:** 273–275.

Craigen, W.J., R.G. Cook, W.P. Tate, and C.T. Caskey. 1985. Bacterial peptide chain release factors: Conserved primary structure and possible frameshift regulation of release factor 2. *Proc. Natl. Acad. Sci.* **82:** 3616–3620.

Cuppens, H., P. Marynen, C. de Boeck, F. de Baets, E. Eggermont, H. van den Berge, and J.J. Cassiman. 1990. A child, homozygous for a stop codon in exon 11, shows milder cystic fibrosis symptoms than her heterozygous nephew. *J. Med. Genet.* **27:** 717–719.

Curran, J.F. 1993. Analysis of effects of tRNA:message stability on frameshift frequency at the *Escherichia coli* RF2 programmed frameshift site. *Nucleic Acids Res.* **21:** 1837–1843.

Curran, J. and D. Kolakofsky. 1988. Scanning independent ribosomal initiation of the Sendai virus X protein. *EMBO J.* **7:** 2869–2874.

Curran, J.F. and M. Yarus. 1988. Use of tRNA suppressors to probe regulation of *Escherichia coli* release factor 2. *J. Mol. Biol.* **203:** 75–83.

———. 1989. Rates of aminoacyl-tRNA selection at 29 sense codons *in vivo. J. Mol. Biol.* **209:** 65–77.

Cutting, G.R., B.S. Kasch, B.J. Rosenstein, L.-C. Tsui, H.H. Kazazian, and S.E. Antonarakis. 1990. Two patients with cystic fibrosis, nonsense mutations in each cystic fibrosis gene, and mild pulmonary disease. *N. Engl. J. Med.* **323:** 1685–1689.

Dayhuff, T.J., J.F. Atkins, and R.F. Gesteland. 1986. Characterization of ribosomal frameshift events by protein sequence analysis. *J. Biol. Chem.* **261:** 7491–7500.

Dinman, J.D. and R.B. Wickner. 1992. Ribosomal frameshifting efficiency and *gag*/*gag-pol* are critical for yeast M_1 double-stranded RNA virus propagation. *J. Virol.* **66:** 3669–3676.

Donly, B.C. and W.P. Tate. 1991. Frameshifting by eukaryotic ribosomes during expression of *Escherichia coli* release factor 2. *Proc. R. Soc. Lond. Biol. Sci.* **244:** 207–210.

Donly, B.C., C.D. Edgar, F.M. Adamski, and W.P. Tate. 1990. Frameshift autoregulation in the gene for *Escherichia coli* release factor 2: Partly functional mutants in frameshift enhancement. *Nucleic Acids Res.* **18:** 6517–6522.

Elliott, T.C. 1989. Cloning, genetic characterization, and nucleotide sequence of the *hemA-prfA* operon of *Salmonella typhimurium. J. Bacteriol.* **171:** 3948–3960.

Elliott, T.C. and X. Wang. 1991. *Salmonella typhimurium prfA* mutants defective in release factor 1. *J. Bacteriol.* **173:** 4144–4154.

Engelberg-Kulka, H. and R. Schoulaker-Schwarz. 1994. Regulatory implications of translational frameshifting in cellular gene expression. *Mol. Microbiol.* **11:** 3–8.

Enriquez, J.A., A. Chomyn, and G. Attardi. 1995. MtDNA mutation in MERRF syndrome causes defective aminoacylation of $tRNA^{Lys}$ and premature translation termination. *Nat. Genet.* **10:** 47–55.

Falk, H., N. Mador, R. Udi, A. Panet, and A. Honigman. 1993. Two *cis*-acting elements control ribosomal frameshift between human T-cell leukemia virus type II *gag* and *pro* genes. *J. Virol.* **67:** 6273–6277.

Farabaugh, P.J., H. Zhao, and A. Vimaladithan. 1993. A novel programmed frameshift expresses the POL3 gene of retrotransposon Ty3 of yeast: Frameshifting without tRNA slippage. *Cell* **74:** 93–103.

Fearon, K., V. McClendon, B. Bonetti, and D.M. Bedwell. 1994. Premature translation termination mutations are efficiently suppressed in a highly conserved region of yeast Ste6p, a member of the ATP-binding cassette (ABC) transporter family. *J. Biol. Chem.* **269:** 17802–17808.

Feng, Y.-X., D.L. Hatfield, A. Rein, and J.G. Levin. 1989. Translational readthrough of the murine leukemia virus *gag* gene amber codon does not require virus induced alteration of tRNA. *J. Virol.* **63:** 2405–2410.

Feng, Y.-X., H. Yuan, A. Rein, and J.G. Levin. 1992. Bipartite signal for read-through suppression in murine leukemia virus mRNA: An eight-nucleotide purine-rich sequence immediately downstream of the *gag* termination codon followed by an RNA pseudoknot. *J. Virol.* **66:** 5127–5132.

Fu, C. and J. Parker. 1994. A ribosomal frameshifting error during translation of the *argI* mRNA of *Escherichia coli. Mol. Gen. Genet.* **243:** 434–441.

Fütterer, J., Z. Kiss-László, and T. Hohn. 1993. Nonlinear ribosome migration on cauliflower mosaic virus 35S RNA. *Cell* **73:** 789–802.

Gallant, J. 1979. Stringent control in *E. coli. Annu. Rev. Genet.* **13:** 393–415.

Gallant, J. and D. Foley. 1980. On the cause and prevention of mistranslation. In *Ribosomes, structure, function and genetics* (ed. G. Chambliss et al.), pp. 615–638, University Park Press, Baltimore, Maryland.

Gallant, J. and D. Lindsley. 1993. Ribosome frameshifting at hungry codons: Sequence rules, directional specificity and possible relationship to mobile element behaviour. *Biochem. Soc. Trans.* **21:** 817–821.

Gesteland, R.F., R.B. Weiss, and J.F. Atkins. 1992. Recoding: Reprogrammed genetic decoding. *Science* **257:** 1640–1641.

Gish, K. and C. Yanofsky. 1993. Inhibition of expression of the tryptophanase operon in *Escherichia coli* by extrachromosomal copies of the *tna* leader region. *J. Bacteriol.* **175:** 3380–3387.

Goldberg, A.L. and A.C. St. John. 1976. Intracellular protein degradation in mammalian and bacterial cells (part 2). *Annu. Rev. Biochem.* **45:** 747–803.

Gollnick, P. and C. Yanofsky. 1990. $tRNA^{Trp}$ translation of leader peptide codon 12 and other factors that regulate expression of the tryptophan operon. *J. Bacteriol.* **172:** 3100–3107.

Gramstat, A., D. Prüfer, and W. Rohde. 1994. The nucleic acid-binding zinc finger protein of potato virus M is translated by internal initiations well as by ribosomal frameshifting involving a shifty stop codon and a novel mechanism of P-site slippage. *Nucleic Acids Res.* **22:** 3911–3917.

Grant, C.M. and A.G. Hinnebusch. 1994. Effect of sequence context at stop codons on efficiency of reinitiation in *GCN4* translational control. *Mol. Cell. Biol.* **14:** 606–618.

Grentzmann, G., D. Brechemier-Baey, V. Heurgue-Hamard, and R.H. Buckingham. 1995. Function of polypeptide chain release factor RF-3 in *Escherichia coli. J. Biol. Chem.* **270:** 10595–10600.

Grentzmann, G., D. Brechemier-Baey, V. Heurgue, L. Mora, and R.H. Buckingham. 1994. Localization and characterization of the gene encoding release factor RF3 in *Escherichia coli. Proc. Natl. Acad. Sci.* **91:** 5848–5852.

Grundy, F.J. and T.M. Henkin. 1994. Conservation of a transcription antitermination mechanism in aminoacyl-tRNA synthetase and amino acid biosynthesis genes in gram-positive bacteria. *J. Mol. Biol.* **235:** 798–804.

Hall, B.G. 1988. Adaptive evolution that requires multiple spontaneous mutations. 1. Mutations involving an insertion sequence. *Genetics* **120:** 887–897.

Hansen, L.J., D.L. Chalker, and S.B. Sandmeyer. 1988. Ty3, a yeast retrotransposon associated with tRNA genes, has homology to animal retroviruses. *Mol. Cell. Biol.* **8:** 5245–5256.

Hartz, D., J. Binkley, T. Hollingsworth, and L. Gold. 1990. Domains of initiator tRNA and initiation codon crucial for initiator tRNA selection by *Escherichia coli* IF3. *Genes Dev.* **4:** 1790–1800.

Hatfield, D., Y.-X. Feng, B.J. Lee, A. Rein, J.G. Levin, and S. Oroszlan. 1989. Chromatographic analysis of the aminoacyl tRNAs which are required for translation of codons at and around the ribosomal frameshift sites of HIV, HTLV-1 and BLV. *Virology* **173:** 736–742.

Hayashi, S. and E.S. Canellakis. 1989. Ornithine decarboxylase antizymes. In *Ornithine decarboxylase: Biology, enzymology, and molecular genetics* (ed. S. Hayashi), pp. 47–58. Pergamon Press, New York.

Herbst, K.L., L.M. Nichols, R.F. Gesteland, and R.B. Weiss. 1994. A mutation in ribosomal protein L9 affects ribosomal hopping during translation of gene 60 from bacteriophage T4. *Proc. Natl. Acad. Sci.* **91:** 12525–12529.

Hu, P.Y., A. Waheed, and W.S. Sly. 1995. Partial rescue of human anhydrase II frameshift mutation by ribosomal frameshift. *Proc. Natl. Acad. Sci.* **92:** 2136–2140.

Hu, P.Y., A.R. Ernst, W.S. Sly, P.J. Venta, L.A. Skaggs, and R.E. Tashian. 1994. Carbonic anhydrase II deficiency: Single-base deletion in exon 7 is the predominant mutation in Caribbean hispanic patients. *Am. J. Hum. Genet.* **54:** 602–608.

Huang, W.M., S. Ao, S. Casjens, R. Orlandi, R. Zeikus, R. Weiss, D. Winge, and M. Fang. 1988. A persistent untranslated sequence within bacteriophage T4 DNA topoisomerase gene 60. *Science* **239:** 1005–1012.

Hwang, C.B.C., B. Horsburgh, E. Pelosi, S. Roberts, P. Digard, and D.M. Coen. 1994. A net +1 frameshift permits synthesis of thymidine kinase from a drug-resistant herpes simplex mutant. *Proc. Natl. Acad. Sci.* **91:** 5461–5465.

Ichiba, T., S. Matsufuji, and S. Hayashi. 1995. Nucleotide sequence of *Xenopus laevis* ornithine decarboxylase antizyme. *Biochim. Biophys. Acta* **1262:** 83–86.

Jacks, T., H.D. Madhani, F.R. Masiarz, and H.E. Varmus. 1988. Signals for ribosomal frameshifting in the Rous sarcoma virus *gag-pol* region. *Cell* **55:** 447–458.

Kamath, A.V. and C. Yanofsky. 1992. Characterization of the tryptophanase operon of *Proteus vulgaris. J. Biol. Chem.* **267:** 19978–19985.

Kawakami, K. and Y. Nakamura. 1990. Autogenous suppression of an opal mutation in the gene encoding peptide chain release factor 2. *Proc. Natl. Acad. Sci.* **87:** 8432–8436.

Kawakami, K., Y.H. Jönsson, G.R. Björk, H. Ikeda, and Y. Nakamura. 1988. Chromosomal location and structure of the operon encoding peptide-chain-release factor 2 of *Escherichia coli. Proc. Natl. Acad. Sci.* **85:** 5620–5624.

Kawakami, K., B.K. Shafer, D.J. Garfinkel, J.N. Strathern, and Y. Nakamura. 1992. Ty element-induced temperature-sensetive mutations of *Saccharomyces cerevisiae. Genetics* **131:** 821–832.

Kawakami, K., S. Pande, B. Faiola, D.P. Moore, J.D. Boeke, P.J. Farabaugh, J.N. Strathern, Y. Nakamura, and D.J. Garfinkel. 1993. A rare tRNA-Arg(CCU) that regulates Ty1 element ribosomal frameshifting is essential for Ty1 retrotransposition in *Saccharomyces cerevisiae. Genetics* **135:** 309–320.

Kerem, B.-S., J. Zielenski, D. Markiewicz, D. Bozon, E. Gazit, J. Yahav, D. Kennedy,

J.R. Riordan, F.S. Collins, J.M. Rommens, and L.-C. Tsui. 1990. Identification of mutations in regions corresponding to the two putative nucleotide (ATP)-binding folds of the cystic fibrosis gene. *Proc. Natl. Acad. Sci.* **87:** 8447–8451.

Kim, A., C. Terzian, P. Santamaria, A. Pélisson, N. Prud'homme, and A. Bucheton. 1994. Retroviruses in invertebrates: The gypsy retrotransposon is apparently an infectious retrovirus of *Drosophila melanogaster*. *Proc. Natl. Acad. Sci.* **91:** 1285–1289.

Kollmus, H., A. Honigman, A. Panet, and H. Hauser. 1994. The sequences of and distance between two *cis*-acting signals determine the efficiency of ribosomal frameshifting in human immunodeficiency virus type I and human T-cell leukemia viris type II in vivo. *J. Virol.* **68:** 6087–6091.

Kopelowitz, J., C. Hampe, R. Goldman, M. Reches, and H. Engelberg-Kulka. 1992. Influence of codon context on UGA suppression and readthrough. *J. Mol. Biol.* **225:** 261–269.

Kuchino, Y., H. Beier, N. Akita, and S. Nishimura. 1987. Natural UAG suppressor glutamine tRNA is elevated in mouse cells infected with Moloney murine leukemia virus. *Proc. Natl. Acad. Sci.* **84:** 2668–2672.

Larsen, B., N.M. Wills, R.F. Gesteland, and J.F. Atkins. 1994. rRNA-mRNA base pairing stimulates a programmed -1 ribosomal frameshift. *J. Bacteriol.* **176:** 6842–6851.

La Teana, A., C.L. Pon, and C.O. Gualerzi. 1993. Translation of mRNAs with degenerate initiation triplet AUU displays high initiation factor 2 dependence and is subject to initiation factor 3 repression. *Proc. Natl. Acad. Sci.* **90:** 4161–4165.

Leppert, M., R. Burt, J.P. Hughes, W. Samowitz, Y. Nakamura, S. Woodward, E. Gardiner, J.-M. Lalouel, and R. White. 1990. Genetic analysis of an inhereted predisposition to colon cancer with a variable number of adenomatous polyps. *N. Engl. J. Med.* **322:** 904–908.

Li, G. and C.M. Rice. 1993. The signal for translational read-through of a UGA codon in Sindbis virus RNA involves a single cytidine residue immediately downstream of the termination codon. *J. Virol.* **67:** 5062–5067.

Marczinke, B., A.J. Bloys, T.D.K. Brown, M.M. Willcocks, M.J. Carter, and I. Brierley. 1994. The human astrovirus RNA-dependent RNA coding region is expressed by ribosomal frameshifting. *J. Virol.* **68:** 5588–5595.

Martin, R., M. Weiner, and J. Gallant. 1988. Effects of release factor context at UAA codons in *Escherichia coli*. *J. Bacteriol.* **170:** 4714–4717.

Matsufuji, S., T. Matsufuji, N.M. Wills, R.F. Gesteland, and J.F. Atkins. 1996. Reading two bases twice: Mammalian antizyme frameshifting in yeast. *EMBO J.* **15:** (in press).

Matsufuji, S., T. Matsufuji, Y. Miyazaki, Y. Murakami, J.F. Atkins, R.F. Gesteland, and S. Hayashi. 1995. Autoregulatory frameshifting in decoding mammalian ornithine decarboxylase antizyme. *Cell* **80:** 51–60.

Menees, T.M. and S.B. Sandmeyer. 1994. Transposition of the yeast retroviruslike element Ty3 is dependent on the cell cycle. *Mol. Cell. Biol.* **14:** 8229–8240.

Mikuni, O., K. Ito, J. Moffat, K. Matsumura, K. McCaughan, T. Nobukuni, W. Tate, and Y. Nakamura. 1994. Identification of the *prfC* gene, which encodes peptide-chain-release factor 3 of *Escherichia coli*. *Proc. Natl. Acad. Sci.* **91:** 5798–5802.

Miller, J.H. and A.M. Albertini. 1983. Effects of surrounding sequence on the suppression of nonsense codons. *J. Mol. Biol.* **164:** 59–71.

Miller, W.A., S.P. Dinesh-Kumar, and C.P. Paul. 1995. Luteovirus gene expression. *Crit. Rev. Plant. Sci.* **14:** 179–211.

Mitchell, J.L.A., G.G. Judd, A. Bareyal-Leyser, and S.Y. Ling. 1994. Feedback repression of polyamine transport is mediated by antizyme in tissue-culture cells. *Biochem. J.* **299:** 19–22.

Mittler, J.E. and R.E. Lenski. 1990. Excisions of Mu from *E. coli* MCS2 are not directed mutations. *Nature* **344:** 173–175.

Miyazaki, Y., S. Matsufuji, and S. Hayashi. 1992. Cloning and characterization of a rat gene encoding ornithine decarboxylase antizyme. *Gene* **113:** 191–197.

Mottagui-Tabar, S., A. Björnsson, and L.A. Isaksson. 1994. The second to last amino acid in the nascent peptide as a codon context determinant. *EMBO J.* **13:** 249–257.

Murakami, Y., S. Matsufuji, Y. Miyazaki, and S. Hayashi. 1992a. Destabilization of ornithine decarboxylase by transfected antizyme gene expression in hepatoma tissue culture cells. *J. Biol. Chem.* **267:** 13138–13141.

Murakami, Y., S. Matsufuji, T. Kameji, S. Hayashi, K. Igarashi, T. Tamura, K. Tanaka, and A. Ichihara. 1992b. Ornithine decarboxylase is degraded by the 26S proteosome without ubiquitination. *Nature* **360:** 597–599.

Naas, T., M. Blot, W.M. Fitch, and W. Arber. 1994. Insertion sequence-related genetic variation in resting *Escherichia coli* K-12. *Genetics* **136:** 721–730.

Nam, S.H., T.D. Copeland, M. Hatanaka, and S. Oroszlan. 1993. Characterization of ribosomal frameshifting for expression of *pol* gene products of human T-cell leukemia virus type 1. *J. Virol.* **67:** 196–203.

Nivinskas, R., R. Vaiskunaite, and A. Raudonikiene. 1992. An internal AUU codon initiates a smaller peptide encoded by bacteriophage T4 baseplate gene 26. *Mol. Gen. Genet.* **232:** 257–261.

Ochman, H. and A.C. Wilson. 1987. Evolution in bacteria: Evidence for a universal substitution rate in cellular genomes. *J. Mol. Evol.* **26:** 74–86.

Öfverstedt, L.-G., K. Zhang, S. Tapio, U. Skoglund, and L.A. Isaksson. 1994. Starvation *in vivo* for aminoacyl-tRNA increases the spatial separation between the two ribosomal subunits. *Cell* **79:** 629–638.

Pande, S., H. Vimaladithan, H. Zhao, and P.J. Farabaugh. 1994. Pulling the ribosome out of frame +1 at a programmed frameshift site by cognate binding of aminoacyl-tRNA. *Mol. Cell. Biol.* **15:** 298–304.

Panganiban, A.T. 1988. Retroviral *gag* gene amber codon suppression is caused by an intrinsic *cis*-acting component of the viral mRNA. *J. Virol.* **62:** 3574–3580.

Pel, H.J., M. Rep, and L.A. Grivell. 1992. Sequence comparison of new prokaryotic and mitochondrial members of the polypeptide chain release factor family predicts a five-domain model for release factor structure. *Nucleic Acids Res.* **20:** 4423–4428.

Pelham, H.R.B. 1978. Leaky UAG termination codon in tobacco mosaic virus RNA. *Nature* **272:** 469–471.

Pélisson, A., S.U. Song, N. Prud'homme, P.A. Smith, A. Bucheton, and V.G. Corces. 1994. Gypsy transposition correlates with the production of a retroviral envelope-like protein under the tissue-specific control of the *Drosophila flamenco* gene. *EMBO J.* **13:** 4401–4411.

Philipson, L., P. Andersson, U. Olshevsky, R. Weinberg, D. Baltimore, and R. Gesteland. 1978. Translation of MuLV and MSV RNAs in nuclease-treated reticulocyte extracts: Enhancement of the gag-pol polypeptide with yeast suppressor tRNA. *Cell* **13:** 189–199.

Polard, P., M.F. Prère, M. Chandler, and O. Fayet. 1991. Programmed translational frameshifting and initiation at an AUU codon in gene expression of bacterial insertion sequence IS911. *J. Mol. Biol.* **222:** 465–477.

Poole, E.S., C.M. Brown, and W.P. Tate. 1995. The identity of the base following the stop codon determines the efficiency of *in vivo* translational termination in *Escherichia coli*. *EMBO J.* **14:** 151–158.

Prère, M.-F., M. Chandler, and O. Fayet. 1990. Transposition in *Shigella dysenteriae*: Isolation and analysis of IS911, a new member of the IS3 group of insertion sequences. *J. Bacteriol.* **172:** 4090–4099.

Raftery, L.A., J.B. Egan, S.W. Cline, and M. Yarus. 1984. Defined set of cloned termination suppressors: *In vivo* activity of isogenetic UAG, UAA, and UGA suppressor tRNAs. *J. Bacteriol.* **158:** 849–859.

Reil, H., H. Kollmus, U.H. Weidle, and H. Hauser. 1993. A heptanucleotide sequence mediates ribosomal frameshifting in mammalian cells. *J. Virol.* **67:** 5579–5584.

Reil, H., M. Höxter, D. Moosmayer, G. Pauli, and H. Hauser. 1994. CD4 expressing human 293 cells as a tool for studies in HIV-1 replication: The efficiency of translational frameshifting is not altered by HIV-1 infection. *Virology* **205:** 371–375.

Ringquist, S., S. Shinedling, D. Barrick, L. Green, J. Binkley, G.D. Stormo, and L. Gold. 1992. Translation initiation sites in *Escherichia coli*: Sequences within the ribosome binding site. *Mol. Microbiol.* **6:** 1219–1229.

Rodriguez, H., E.T. Snow, U. Bhat, and E.L. Loechler. 1992. An *Escherichia coli* plasmid-based mutational system in which *supF* mutants are selectable—Insertion elements dominate the spontaneous spectra. *Mutat. Res.* **270:** 219–231.

Roesser, J.R. and C. Yanofsky. 1988. Ribosome release modulates basal level expression of the *trp* operon of *Escherichia coli*. *J. Biol. Chem.* **263:** 14251–14255.

Rohde, W., A. Gramstat, J. Schmitz, E. Tacke, and D. Prüfer. 1994. Plant viruses as model systems for the study of non-canonical translation mechanisms in higher plants. *J. Gen. Virol.* **75:** 2141–2149.

Rom, E. and C. Kahana. 1994. Polyamines regulate the expression of ornithine decarboxylase antizyme *in vitro* by inducing ribosomal frame-shifting. *Proc. Natl. Acad. Sci.* **91:** 3959–3963. (Correction **91:** 9195).

Rothnie, H.M., Y. Chapdelaine, and T. Hohn. 1994. Pararetroviruses and retroviruses: A comparative review of viral structure and gene expression strategies. *Adv. Virus Res.* **44:** 1–67.

Sacerdot, C., G. Fayat, P. Dessen, M. Springer, J.A. Plumbridge, M. Grunberg-Manago, and S. Blanquet. 1982. Sequence of a 1.26-kb DNA fragment containing the structural gene for *E. coli* initiation factor3: Presence of an AUU initiator codon. *EMBO J.* **1:** 311–315.

Schmitt, C., E. Balmori, G. Jonard, K.E. Richards, and H. Guilley. 1992. *In vitro* mutagenesis of biologically active transcripts of beet necrotic yellow vein virus RNA2: Evidence that a domain of the 75-kDa readthrough protein is important for virus assembly. *Proc. Natl. Acad. Sci.* **89:** 5715–5719.

Shapiro, J.A. and N.P. Higgins. 1989. Differential activity of a transposable element in *Escherichia coli* colonies. *J. Bacteriol.* **171:** 5975–5986.

Shen, L.X. and I. Tinoco. 1995. The structure of an RNA pseudoknot that causes efficient frameshifting in mouse mammary tumor virus. *J. Mol. Biol.* **247:** 963–978.

Shirako, Y. and J.H. Strauss. 1994. Regulation of Sindbis virus RNA replication: Uncleaved P123 and nsP4 function in minus-strand RNA synthesis, whereas cleaved products from P123 are required for efficient plus-strand RNA synthesis. *J. Virol.* **68:** 1874–1885.

Sipley, J. and E. Goldman. 1993. Increased ribosomal accuracy increases a programmed translational frameshift in *Escherichia coli*. *Proc. Natl. Acad. Sci.* **90:** 2315–2319.

Skuzeski, J.M., L.M. Nichols, R.F. Gesteland, and J.F. Atkins. 1991. The signal for a leaky UAG stop codon in several plant viruses includes the two downstream codons. *J. Mol. Biol.* **218:** 365–373.

Song, S.U., T. Gerasimova, M. Kurkulos, J.D. Boeke, and V.G. Corces. 1994. An env-like protein encoded by a *Drosophila* retroelement: Evidence that *gypsy* is an infectious retrovirus. *Genes Dev.* **8:** 2046–2057.

Spanjaard, R.A., K. Chen, J.R. Walker, and J. van Duin. 1990. Frameshift suppression at tandem AGA and AGG codons by cloned tRNA genes: Assigning a codon to *argU* tRNA and T4 $tRNA^{Arg}$. *Nucleic Acids Res.* **18:** 5031–5036.

Spirio, L., S. Olschwang, J. Groden, M. Robertson, W. Samowitz, G. Joslyn, L. Gelbert, A. Thliveris, M. Carlson, B. Otterud, H. Lynch, P. Watson, P. Lynch, P. Laurent-Puig, R. Burt, J.P. Hughes, G. Thomas, M. Leppert, and R. White. 1993. Alleles of the APC gene: An attenuated form of familial polyposis. *Cell* **75:** 951–957.

Stormo, G.D., T.D. Schneider, and L. Gold. 1986. Quantitative analysis of the relationship between nucleotide sequence and functional activity. *Nucleic Acids Res.* **14:** 6661–6679.

Suzuki, T., Y. He, K. Kashiwagi, Y. Murakami, S. Hayashi, and K. Igarashi. 1994. Antizyme protects against abnormal accumulation and toxicity of polyamines in ornithine decarboxylase-overproducing cells. *Proc. Natl. Acad. Sci.* **91:** 8930–8934.

Tate, W.P. and C.M. Brown. 1992. Translational termination: "Stop" for protein synthesis or "Pause" for regulation of gene expression. *Biochemistry* **31:** 2443–2450.

Tate, W.P., F.M. Adamski, C.M. Brown, M.E. Dalphin, J.P. Gray, J.A. Horsfield, K.K. McCaughan, J.G. Moffat, R.J. Powell, K.M. Timms, and C.N.A. Trotman. 1993. Translational stop signals: Evolution, decoding for protein synthesis and recoding for alternative events. In *The translational apparatus: Structure, function, regulation, evolution* (ed. K.H. Nierhaus et al.), pp. 253–262. Plenum Press, New York.

ten Dam, E.B., C.W.A. Pleij, and L. Bosch. 1990. RNA pseudoknots: Translational frameshifting and readthrough on viral RNAs. *Virus Genes* **4:** 121–136.

Tewari, D.S., Y. Qian, R.D. Thornton, J. Pieringer, R. Taub, E. Mochan, and M. Tewari. 1994. Molecular cloning and sequencing of a human cDNA encoding ornithine decarboxylase antizyme. *Biochim. Biophys. Acta* **1209:** 293–295.

Tokunaga, F., T. Goto, T. Koide, Y. Murakami, S. Hayashi, T. Tamura, K. Tanaka, and A. Ichihara. 1994. ATP-and antizyme-dependent endoproteolysis of ornithine decarboxylase to oligopeptides by the 26S proteosome. *J. Biol. Chem.* **269:** 17382–17385.

Tsuchihashi, Z. and P.O. Brown. 1992. Sequence requirements for efficient translational frameshifting in the *Escherichia coli dnaX* gene and the role of an unstable interaction between $tRNA^{Lys}$ and an AAG lysine codon. *Genes Dev.* **6:** 511–519.

Vellanoweth, R.L. and J.C. Rabinowitz. 1992. The influence of ribosome-binding-site elements on translational efficiency in *Bacillus subtilis* and *Escherichia coli* in vivo. *Mol. Microbiol.* **6:** 1105–1114.

Vickers, T.A. and D.J. Ecker. 1992. Enhancement of ribosomal frameshifting by oligonucleotides targeted to the HIV *gag-pol* region. *Nucleic Acids Res.* **20:** 3945–3953.

Vimaladithan, A. and P.J. Farabaugh. 1994. Special peptidyl-tRNA molecules promote translational frameshifting without slippage. *Mol. Cell. Biol.* **14:** 8107–8116.

Voytas, D.F. and J.D. Boeke. 1993. Yeast retrotransposons and tRNAs. *Trends Genet.* **9:** 421–427.

Wang, Y.-F.M., S.G. Sawicki, and D.L. Sawicki. 1994. Alphavirus nsP3 functions to form replication complexes transcribing negative-strand RNA. *J. Virol.* **68:** 6466–6475.

Weiner, A.M. and K. Weber. 1973. A single UGA codon functions as a natural termination signal in the coliphage Qβ coat protein cistron. *J. Mol. Biol.* **80:** 837–855.

Weiss, R.B., W.M. Huang, and D.M. Dunn. 1990a. A nascent peptide is required for ribosomal bypass of the coding gap in bacteriophage T4 gene 60. *Cell* **62:** 117–126.

Weiss, R.B., D.M. Dunn, J.F. Atkins, and R.F. Gesteland. 1987. Slippery runs, shifty stops, backward steps, and forward hops: –2, –1, +2, +5, and +6 ribosomal frameshifting. *Cold Spring Harbor Symp. Quant. Biol.* **52:** 687–693.

———. 1990b. Ribosomal frameshifting from –2 to +50 nucleotides. *Prog. Nucleic Acid Res. Mol. Biol.* **39:** 159–183.

Weiss, R.B., D. Lindsley, B. Falahee, and J. Gallant. 1988a. On the mechanism of ribosomal frameshifting at hungry codons. *J. Mol. Biol.* **203:** 403–410.

Weiss, R.B., D.M. Dunn, A.E. Dahlberg, J.F. Atkins, and R.F. Gesteland. 1988b. Reading frame switch caused by base-pair formation between the 3′ end of 16S rRNA and the mRNA during elongation of protein synthesis in *Escherichia coli. EMBO J.* **7:** 1503–1507.

Weiss, R.B., D.M. Dunn, M. Shuh, J.F. Atkins, and R.F. Gesteland. 1989. *E. coli* ribosomes re-phase on retroviral frameshift signals at rates ranging from 2 to 50 percent. *New Biol.* **1:** 159–169.

Wickstrom, E. 1983. Nuclease mapping of the secondary structure of the 49-nucleotide 3′ terminal cloacin fragment of *Escherichia coli* 16S rRNA and its interactions with initiation factor 3. *Nucleic Acids Res.* **11:** 2035–2052.

Williams, J.M., B.C. Donly, C.M. Brown, F.M. Adamski, and W.P. Tate. 1989. Frameshifting in the synthesis of *Escherichia coli* polypeptide chain release factor 2 on eukaryotic ribosomes. *Eur. J. Biochem.* **186:** 515–521.

Wills, N.M., R.F. Gesteland, and J.F. Atkins. 1991. Evidence that a downstream pseudoknot is required for translational read-through of the Moloney murine leukemia virus *gag* stop codon. *Proc. Natl. Acad. Sci.* **88:** 6991–6995.

———. 1994. Pseudoknot-dependent read-through of retroviral *gag* termination codons: Importance of sequences in the spacer and loop 2. *EMBO J.* **13:** 4137–4144.

Xu, H. and J.D. Boeke. 1990. Host genes that influence transposition in yeast: The abundance of a rare tRNA regulates Ty1 transposition frequency. *Proc. Natl. Acad. Sci.* **87:** 8360–8364.

Xue, F. and L. Cooley. 1993. *kelch* encodes a component of intercellular bridges in *Drosophila* egg chambers. *Cell* **72:** 681–693.

Yelverton, E., D. Lindsley, P. Yamauchi, and J.A. Gallant. 1994. The function of a ribosomal frameshifting signal from human immunodeficiency virus-1 in *Escherichia coli. Mol. Microbiol.* **11:** 303–313.

Yoshinaka, Y., I. Katoh, T.D. Copeland, and S. Oroszlan. 1985. Murine leukemia virus protease is encoded by the *gag-pol* gene and is synthesized through suppression of an amber termination codon. *Proc. Natl. Acad. Sci.* **82:** 1618–1622.

Zerfass, K. and H. Beier. 1992a. Pseudouridine in the anticodon G psi A of plant cytoplasmic tRNA (Tyr) is required for UAG and UAA suppression in the TMV-specific context. *Nucleic Acids Res.* **20:** 5911–5918.

———. 1992b. The leaky UGA termination codon of tobacco rattle virus RNA is suppressed by tobacco chloroplast and cytoplasmic tRNAs(Trp) with CmCA anticodon. *EMBO J.* **11:** 4167–4173.

24

Mammalian Ribosomes: The Structure and the Evolution of the Proteins

Ira G. Wool, Yuen-Ling Chan, and Anton Glück
Department of Biochemistry and
Molecular Biology, The University of Chicago
Chicago, Illinois 60637

Ribosomes are universal, essential, and complicated; they mediate protein synthesis in all organisms in our biosphere and thereby link genotype to phenotype. The imperative in research on ribosomes is to determine their structure and thus be able to account for their function. This led to the isolation, the characterization, and the determination of the structures of mammalian (rat) ribosomal proteins and nucleic acids (Wool 1979, 1986; Wool et al. 1990). Mammalian ribosomes are composed of two subunits that are designated by their sedimentation coefficients: The smaller is 40S and the larger is 60S. The subunits associate—they are held together by noncovalent bonds, perhaps by magnesium salt bridges—and form the functional 80S ribosome. The 40S subunit has a single molecule of RNA, designated 18S rRNA, and 33 proteins, whereas the 60S subunit has three molecules of RNA, designated 5S, 5.8S, and 28S rRNA, and 47 proteins. The covalent structures of the four species of rat rRNAs (5S, 5.8S, 18S, and 28S) have been established (Nazar et al. 1975; Aoyama et al. 1982; Chan et al. 1983, 1984; Torczynski et al. 1983; Hadjiolov et al. 1984), and there are rational proposals for their secondary structures (Chan et al. 1984; Hadjiolov et al. 1984; Wool 1986; Raué et al. 1988). Eighty-two proteins have been isolated from rat ribosomes, and the complete amino acid sequences in 75 of these proteins have been either determined directly from the protein or deduced from the sequences of nucleotides in cDNAs (for references, see Table A1 in the Appendix at the end of this chapter). An understanding of the chemistry is essential for a solution of the structure, and the structure in turn is needed for a coherent molecular account of the function of the organelle in protein synthesis. Moreover, a compilation of data (the primary structure of the approximately 80 proteins) is a general resource that has been useful to many, including scientists whose research does not touch directly on ribosomes nor on protein synthesis. It was anticipated from the start that the amino acid sequences would help in understanding the evolution of the ribosomal proteins, in unraveling their functions, in defining the rules that govern their interaction with the rRNAs, and in uncovering those amino acid sequences that direct the proteins to the nucleolus for assembly on nascent rRNA. There have also been the following unanticipated findings: (1) the recognition of DNA-binding motifs, which has led to speculations on the origin of the ribosomal proteins, and (2) an awareness of the bifunctional nature of some ribosomal proteins, i.e., that some have a function apart from the ribosome and from protein synthesis (Wool 1993).

This chapter takes stock of what has been learned of the characteristics of the 75 rat ribosomal proteins and of the mRNAs that encode them and considers the relationship of mammalian ribosomal proteins to those from other species.

RAT RIBOSOMAL PROTEINS: THE AMINO ACID SEQUENCES

The amino acid sequences in 75 rat ribosomal proteins have been determined (for references, see Table A1 in the Appendix). Several sequences were initially determined by Edman degradation of peptides using micro-manual techniques, but all 75 have now been obtained with recombinant methods. It is likely that at least five additional proteins exist: S1, for which there is an unpublished amino acid sequence for the homologous human protein; L6, L10, and L14, for which there are partial amino acid sequences; and L10a, for which there is as yet no amino acid sequence. In addition, a number of proteins once thought to be unique probably are not (e.g., L20 most likely does not exist) and the following pairs are probably identical: L25 is L30, S22 is L32, L16 is L12, and L33 is S24. If our bookkeeping is correct and no new proteins are uncovered, mammalian ribosomes have 80 proteins.

An average mammalian ribosomal protein has a molecular weight of 18,500 (the range is 47,280 for L4 to 3,454 for L41) and contains 164 amino acids (the range is 421 to 25). The protein is very basic; it has a pI of 11.05 (the range is 4.07 for P1 to 13.46 for L41) and contains 22 moles% arginine and lysine and only 9 moles% aspartic and glutamic acids. The protein is likely to contain a number of clusters of basic residues and to have several short amino acid repeats.

STRUCTURAL MOTIFS IN RIBOSOMAL PROTEINS

Clusters of Basic Residues

The basic amino acids in ribosomal proteins tend to occur in clusters. Indeed, this is such a commonplace occurrence that it would be tedious to present extensive documentation; instead, we give a single illustration. Rat L5 associates with 5S rRNA; 5S rRNA•L5 RNP complexes actually appear to be discrete subparticles of the ribosome (Blobel 1971; Terao et al. 1975; Nazar et al. 1979). These complexes can be reconstituted independently of other ribosomal components, and they can be removed from, and added back to, ribosomes as a unit. L5 has 64 basic residues (26 arginyl, 33 lysyl, and 5 histidyl), i.e., 22 moles%. Most of these occur in groups of three or four consecutive residues and are concentrated in two regions.

Clusters of Acidic Residues

Only a relatively small number of rat ribosomal proteins have clusters of acidic residues. Those of four amino acids or more include L5, ^{213}EEDE; L22, ^{120}DEEEEEDED; L31, ^{94}EDED; P0, ^{301}EESEESDED; P1, ^{99}EESEESEDD; P2, ^{100}EESEESDDD; S8, ^{132}EEEE; and S9, ^{189}DDEEED (the superscript designates the position of the amino acid in the sequence). These occur, in general, at or near the carboxyl terminus. We do not know the function of these clusters, but they are reminiscent of similar amino acid sequences in the high mobility group of nucleic-acid-binding proteins and may have a similar function in ribosomes (Walker et al. 1978; Watson and Dixon 1981).

Amino Acid Sequence Repeats

A common structural feature of ribosomal proteins is the repetition of short amino acid sequences, generally of three to eight residues. Because these repeats are common, and because little is known of their origins or of their contribution to the structure or the function of the protein, only a few prominent examples are discussed here.

A number of short amino acid sequences are repeated in L13a: The tetrapeptides VLDG and LLGR each occur twice, and the pentapeptide LGRLA, which overlaps with the previous repeat, occurs twice as well. One other possible duplication occurs in L13a: ^{25}KQVLLGRK and ^{172}KQLLRLRK.

Protein S2 has an exceptionally large number of glycines, 44 of 293 residues or 15 moles%; 26 are among the amino-terminal 53 amino acids. This region is also rich in arginine, and all 11 of the latter are followed by glycine. Most strikingly, the region has three kinds of repetitive sequences: three consecutive GGPs, two successive RGGFs, and eight tandem RG repeats. In all, the dipeptide RG occurs 15 times in S2. RG dipeptides and the RGGF motif are also found in a number of nucleolar proteins: nucleolin (Lapeyre et al. 1986, 1987; Caizergues-Ferrer et al. 1989; Bourbon and Amalric 1990), fibrillarin (Lischwe et al. 1985; Christensen and Fuxa 1988; Schimmang et al. 1989; Henriquez et al. 1990; Aris and Blobel 1991), and the single-strand nucleic-acid-binding protein SSB1, which is the major yeast nucleolar protein (Jong et al. 1987). What all of these proteins share is nucleolar localization and binding to RNA. Thus, the RG tandem repeats and the RGGF boxes in S2 may well direct the protein to the nucleolus and mediate binding there to nascent rRNA.

Rat ribosomal protein L7 has at its amino terminus five tandem repeats of 12 residues each; these repeats are all close variants of the first one, ^{7}KKKKVAAALGTL. It is worth noting that the number of amino-terminal tandem repeats varies in homologous mammalian L7 ribosomal proteins: six in the mouse proteins and four in the human proteins. The repeats are not found in the *Dictyostelium discoideum* or *Saccharomyces cerevisiae* homologs of mammalian L7 and hence may be characteristic of higher eukaryotes.

Although amino acid sequence repeats are a striking feature of the structure of a number of mammalian ribosomal proteins, their function is a matter of surmise. This is at least in part because so little is known of the function of any individual ribosomal protein.

Shared Amino Acid Sequences in the P-type Proteins

All large ribosomal subunits have a set of proteins that are distinguished by their chemical, structural, and functional properties (Terhorst et al. 1973; Matheson et al. 1980). The prototypes are *Escherichia coli* L12 and L10. L12 is acidic, which is in itself a distinction, and it also has an unusually large number of alanyl residues and no aromatic amino acids (Wittmann-Liebold et al. 1977; Luer and Wong 1979). There are four copies (two dimers) of L12 in ribosomes (Subramanian 1975); the stoichiometry of all of the other ribosomal proteins is one. The amino-terminal serine of about half of the L12 molecules is acetylated (Brot and Weissbach 1981); this form has been designated L7 from its coordinates on two-dimensional gels and the two together have been designated L7/L12, although they are the products of a single gene. No phenotype has been attached to this posttranslational modification. Two dimers of L12 form an assembly with a single copy of L10; the complex forms in solution and is a discrete substructure that is located in, and at the base of, the stalk in the 50S ribosomal subunit (Pettersson et al. 1976; Brot and Weissbach 1981; Möller and Maassen 1986; Traut et al. 1986). The acidic proteins are involved in the binding of the elongation factors EF1A (formerly EF-Tu) and EF2 (formerly EF-G), the initiation factor IF2, and the release factor RF and are necessary for the hydrolysis of GTP (Stöffler 1974; Weissbach 1980; Brot and Weissbach 1981).

Eukaryotic ribosomes contain proteins that are related to *E. coli* L12 and L10. These proteins have been designated P0, P1, and P2—P because they are phosphorylated (Tsurugi et al. 1978; Towbin et al. 1982). The P-proteins have chemical and physical properties similar to those of the *E. coli* homologs, but the P-proteins also have a conserved, indeed, almost identical, sequence of 17 amino acids near their carboxyl termini, KEESEESDDDMGFGLFD (Rich and Steitz 1987; Wool et al. 1991). This conserved sequence is an epitope recognized by autoantibodies in an appreciable number of patients with systemic lupus erythematosus (Elkon et al. 1985, 1986); a rise in the titer of these autoantibodies sometimes correlates with the onset of episodes of psychosis (Bonfa and Elkon 1986).

Homodimers of P1 and of P2 are integrated with a single copy of P0 into a coherent structure; the complex has the composition $P1_2 \cdot P2_2 \cdot P0$ (Elkon et al. 1986; Uchiumi et al. 1987). P1 and P2 can be stripped selectively from 60S ribosomal subunits with ethanol and high concentrations of monovalent salts (Stöffler et al. 1974; Wool and Stöffler 1974; MacConnell and Kaplan 1980, 1982). The core particles deficient in P1 and P2 cannot catalyze protein synthesis; they do not associate with eEF1A nor eEF2 and hence do not bind aminoacyl-tRNA nor hydrolyze GTP (MacConnell and Kaplan 1980, 1982). Monoclonal antibodies against an epitope in the conserved carboxy-terminal amino acid sequence also inhibit the binding of eEF1A and eEF2 to ribosomes and the hydrolysis of GTP (Uchiumi et al. 1990).

Zinc Finger Motifs

Rat ribosomal proteins S27 and S29 have zinc finger domains of a type that frequently mediates binding to DNA. The zinc finger domains in S27 and S29 have the form $-\underline{C}-X_2-\underline{C}-X_{14,15}-\underline{C}-X_2-\underline{C}-$; i.e., they are of the C_2–C_2 variety rather than the more common C_2–H_2 kind. The secondary and tertiary structures of zinc fingers of these two types are

Table 1 Ribosomal proteins with zinc finger motifs

RS27	37[a]	C	PG	C	YKITTVFSHAQTVVL	C	VG	C
RS29	21	C	RV	C	SNRHGLIRKYGLNM	C	RQ	C
BsuS14	23	C	ER	C	GRPHSVIRKFKL	C	RI	C
HmS14	26	C	QR	C	GREQGLVGKYDIWL	C	RQ	C
McS14	24	C	NH	C	GRPHAVLKKFGI	C	RL	C
MvS14	17	C	KR	C	GRKGPGIIRKYGLDL	C	RQ	C
ScMRP2	78	C	VDS		GHARFVLSDFRL	C	RYQFR	
EcS14	63	C	RQT		GRPHGFLRKFGL		SRIKV	
RL37a	39	C	SF	C	GKTKMKRRAVGIWH	C	GS	C
RL37	18	C	RR	C	GSKAYHLQKST	C	GK	C
RL40	20	C	RK	C	YARLHPRAVN	C	RKKK	C
RS27a	45	C	PSDE	C	GAGVFMASHFDRHY	C	GK	C
BstL32	29	C	PN	C	GEWKLAHRV	C	KA	C
BsuL36	11	C	EK	C	KVIRRKGKVMVI	C	ENPK	H
EcL36	11	C	RN	C	KIVKRDGVIRVI	C	SAEPK	H

(R) Rat; (Bsu) *Bacillus subtilis*; (Hm) *Halobacterium marismortui*; (Mc) *Mycoplasma capricolum*; (Mv) *Methanococcus vannielii*; (ScM) *Saccharomyces cerevisiae* mitochondria; (Ec) *Escherichia coli*; (Bst) *Bacillus stearothermophilus*.

[a]The number designates the position of the first amino acid in the sequence.

different; indeed, they may not be related (Schwabe and Rhodes 1991). However, proteins of both types share the potential to coordinate a zinc ion, and almost all bind to nucleic acids, most to DNA and a few to RNA (Berg 1986, 1990; Klug and Rhodes 1987). For example, transcription factor IIIA (TFIII), in which the zinc finger domain occurs in tandem nine times (Miller et al. 1985), binds both to the internal control region of 5S rDNA genes and to 5S rRNA transcripts (Brown 1984). In S27 and S29, the amino acid sequence between the internal cysteinyls, which we refer to as the linker sequence, is dominated by basic and hydrophobic residues and contains at least one aromatic amino acid.

A search of our library of more than 1600 ribosomal protein amino acid sequences revealed that rat L37 and L37a have similar zinc finger domains, and variants are to be found in S27a and L40 (Table 1). Analysis of rat liver ribosomes and ribosomal subunits by atomic absorption spectroscopy has indicated the presence of appreciable amounts of zinc; for 80S ribosomes, the stoichiometry (moles of zinc per mole of ribosomal particle) ranges from 3 to 6; for 60S subunits, from 3 to 6; and for 40S subunits, from 3 to 5 (Chan et al. 1993).

A number of eubacterial and archaebacterial ribosomal proteins have C_2–C_2 zinc-binding motifs (Table 1), and members of the S14 family, which are related to rat S29, are of special interest. Homologous S14 proteins from several species have the entire element, whereas *E. coli* S14 has a degenerate form of the domain (Table 1). In the alignment with the other proteins, a cysteine in *E. coli* S14 occurs at what would be the initial position; the other three are absent. The hydrophilic and hydrophobic character of the linker sequence, however, is preserved in *E. coli* S14, and 8 of the 12 residues share identity with amino acids at the same position in *Bacillus subtilis* S14. The yeast nucleus-encoded mitochondrial ribosomal protein MRP2 is also related to the S14 family; however, it has only the two cysteinyls that correspond to the first and third sites in the full motif. One interpretation of these findings is that *E. coli* S14 and yeast mitochondrial MRP2 once had full C_2–C_2 motifs and that parts were lost during divergent evolution; there is, of course, no direct evidence for this bias.

It is not known whether ribosomal proteins with the C_2–C_2 motifs bind zinc and if they do whether the structure formed by the coordination of the metal participates in binding to rRNA. It is possible that the zinc finger domains are the vestiges of a former function, for example, the binding to DNA, preserved in ribosomal proteins that have evolved other means for associating specifically with rRNA. This is consistent with the persistence in *E. coli* S14 and in yeast MRP2 of only a part of the motif, recognizable but presumably no longer capable of coordinating a metal ion. We would argue that it is likely that the same amino acid sequences and/or structures in the *B. subtilis* and *E. coli* S14 proteins are used to bind to 16S rRNA since the binding site is conserved (Stern et al. 1989). If this assumption is correct, then the conclusion is that the proteins do not use the canonical secondary and tertiary structures that are formed by association with zinc in binding to rRNA because *E. coli* S14 lacks the capacity to bind the metal and hence to form the structures. One reconciliation is that the binding to rRNA employs the side chains of the conserved basic, hydrophobic, and aromatic amino acids in the linker region rather than the finger structure per se.

Basic Region-Leucine Zipper Motifs

Rat ribosomal protein L13a has at its amino terminus six heptad repeats of a leucine-zipper-like motif. A typical leucine zipper is a protein dimerization domain containing four or five heptad repeats of hydrophobic and nonpolar residues that pack together in a parallel α-

helical coiled-coil (O'Shea et al. 1989; Ellenberger et al. 1992). These repeats were first recognized because every seventh residue in the heptad repeat was leucine (Landschulz et al. 1988), but leucine can be replaced by other hydrophobic residues (Hemmerich et al. 1993). In L13a, only two of the hydrophobic residues in the seventh position in the heptad repeats are leucine; the others are isoleucine and valine (twice each). Leucine zippers occur primarily in transcription factors (Harrison 1991; Ellenberger et al. 1992); however, they are also found, albeit less frequently, in other proteins (Busch and Sassone-Corsi 1990).

Leucine zippers frequently have an amino-adjacent domain containing clusters of basic amino acids that mediate binding to DNA (Vinson et al. 1989). The amino-terminal leucine zipper of L13a is not preceded by a basic region. The carboxy-terminal region of L13a, however, does have a bZIP motif—a leucine zipper preceded by a basic domain—that is related to one in the amino-terminal region of rat ribosomal protein L7. The bZIP motif in L7 has been shown to mediate protein-protein dimerization and stable binding to DNA and to RNA; the binding of L7 to nucleic acids has some specificity since association with 28S rRNA is favored over that with 18S rRNA (Hemmerich et al. 1993). We searched our collection of rat ribosomal protein amino acid sequences for potential bZIP elements and found a number of candidates in addition to L7 and L13a: L9, L12, L35, L37a, S2, and S9. The acidic ribosomal proteins of the P-type (P0, P1, and P2) have also been reported to have hydrophobic zipper domains (Tsurugi and Mitsui 1991). All have some of the features of the element, but none conforms entirely to the canonical structure; nevertheless, they retain a sufficient number of the characteristics to convince us that they are related to bZIP elements and at least have the potential to mediate dimerization and nucleic acid binding.

Finally, it is relevant that at least one other DNA-binding motif is to be found in a ribosomal protein, albeit a eubacterial protein. The carboxyl terminus of *E. coli* L12 has a helix-turn-helix motif (Rice and Steitz 1989) very much like that found in proteins that bind to DNA and regulate transcription.

PHOSPHORYLATED RIBOSOMAL PROTEINS

The small ribosomal subunit protein S6 is the major, but not the sole, phosphoprotein in mammalian ribosomes (Gressner and Wool 1974a,b; Wool 1979; Leader 1980); the other phosphoproteins P0, P1, and P2 reside in the large subparticle (Tsurugi et al. 1978; Towbin et al. 1982). Five seryl residues are clustered at the carboxyl terminus of S6 (Gressner and Wool 1974a; Chan and Wool 1988), and subsets of these residues appear to be phosphorylated by different protein kinases in response to different stimuli (DuVernay and Traugh 1978; Del Grande and Traugh 1982; Freedman and Jamieson 1982; Wettenhall and Cohen 1982; Wettenhall et al. 1982; Jefferies and Thomas, this volume). What is remarkable is the extraordinary number and variety of the agents, and of the alterations in the cellular milieu, that affect the phosphorylation of S6 (for references, see Wool 1979; Chan and Wool 1988). These include hormones, growth factors, toxic substances, viral infection, cellular transformation, serum treatment, fertilization, phorbol esters, and heat shock. Despite this large catalog of stimuli that affect the phosphorylation state of S6, little is known of the function of the protein and hence how its function is altered by this posttranslational modification. A general correlation exists between an increase in the phosphorylation of S6 and an increase in protein synthesis; there are, however, important exceptions. For example, cycloheximide and puromycin, antibiotics that inhibit the synthesis of protein, are very effective stimuli for the phosphorylation of S6 (Gressner and Wool 1974b).

Two proposals concerning the function of ribosomal protein S6 and of the effect of phosphorylation have been given serious consideration. The first is that S6 forms a part of the mRNA-binding domain in the 40S ribosomal subunit and that phosphorylation of the protein favors the attachment of certain classes of mRNAs or messenger ribonucleoproteins (Terao and Ogata 1979; Wool 1979; Takahashi and Ogata 1981; Duncan and McConkey 1982; Thomas et al. 1982; see Jefferies and Thomas, this volume). It must be emphasized that no substantial, convincing evidence exists that phosphorylation of S6 affects the activity of the ribosome either in overall protein synthesis or in any of the partial reactions and much evidence to the contrary (Leader et al. 1981). The second possibility is that phosphorylation of S6 serves no physiological purpose; i.e, that S6 is an adventitious substrate for protein kinases that have other specific, physiological substrates (Chan and Wool 1988). Although this negative hypothesis is difficult to prove, indirect evidence supports it.

The yeast homolog of mammalian S6 is YS6 (also referred to as S10); the yeast protein has only two phosphorylation sites, rather than the five sites of the mammalian protein. Nevertheless, the phosphorylation of YS6 is affected by stimulation of growth, germination, sporulation, and heat shock, as well as other conditions (cf. Johnson and Warner 1987). To test the function of phosphorylation and dephosphorylation of YS6, the two chromosomal genes encoding the protein were inactivated, and cells were transfected with a plasmid containing a mutant YS6 cDNA in which the two serines that are phosphorylated were replaced with alanines (Johnson and Warner 1987). The mutant YS6 was incorporated into ribosomes but of course was not phosphorylated. Nevertheless, the mutant strain grew normally, i.e., like wild-type cells, even during changes in growth conditions, for example, during sporulation. Moreover, the cells had a normal response to heat shock (YS6 is usually dephosphorylated in heat shock). In a competitive growth experiment (Johnson and Warner 1987), equal numbers of mutant and wild-type cells were mixed together and diluted into fresh medium—a strong stimulus for growth and for YS6 phosphorylation—whenever growth became stationary. During each cycle, the YS6 of wild-type cells was phosphorylated and dephosphorylated. After 70 generations, the fraction of mutant and wild-type cells was unchanged, i.e., there was no selection against cells with the mutant YS6 that could not be phosphorylated. It appears then that the phosphorylation of the two serines in YS6 serves no discernible function. This is consistent with the lack of a demonstrable function for the acetylation of the amino-terminal serine of *E. coli* L12 referred to earlier. Although two of the four L12 molecules in *E. coli* ribosomes are modified by acetylation, a null mutation in the L12 acetylase gene has no phenotype.

The phosphate groups of the serines in P0, P1, and P2 are stable, i.e., they are unaffected by changes in the physiological milieu. Phosphorylation of these proteins is necessary, however, for their assembly into 60S ribosomal subunits (MacConnell and Kaplan 1980, 1982). Core particles depleted of P-proteins are unable to catalyze eEF1- and eEF2-dependent reactions; these functions are restored by adding the proteins to depleted cores. However, if the P-proteins are dephosphorylated, they lose their ability to bind to 60S cores; binding is restored by phosphorylation of the proteins prior to reconstitution.

RIBOSOMAL PROTEINS THAT ARE CARBOXYL EXTENSIONS OF UBIQUITIN OR UBIQUITIN-LIKE PROTEINS

Unlike the majority of ribosomal proteins, which are the unprocessed primary products of the translation of their mRNAs, three are formed by cleavage from a larger hybrid protein: S27a and L40 are the carboxyl extensions of ubiquitin (Finley et al. 1989; Redman and Rechsteiner 1989; Baker and Board 1991) and S30 is the carboxyl extension of a ubiquitin-like protein (Olvera and Wool 1993). Ubiquitin has no intrinsic enzymatic activity, rather it marks proteins for degradation (Hershko 1991). Covalent conjugation to the targeted protein is by formation of an isopeptide bond between the carboxy-terminal glycine of ubiquitin and the ε-amino group of a lysine residue in the acceptor protein. The attachment of ubiquitin to the substrate is critical for nonlysosomal degradation of the protein. In contrast to ubiquitin, whose sequence of amino acids is conserved, there is a family of ubiquitin-like proteins that deviate from each other, but these proteins have sufficient identity with ubiquitin to substantiate a relationship. Rat ribosomal protein S30 is formed by cleavage from one of these ubiquitin-like proteins (Olvera and Wool 1993).

The suggestion is that the ubiquitin moieties, and perhaps the ubiquitin-like protein, serve as chaperones in ribosome biogenesis (Finley et al. 1989). The evidence adduced for this proposal is neither convincing nor conceptually satisfying. It offers no justification for the necessity for a covalently attached chaperone for the assembly into subunits of these three ribosomal proteins (S27a, S30, and L40) when the 70 or more others appear to manage without assistance. Moreover, the proposal implies that cleavage of the fusion protein occurs on the ribosome after assembly. The site where the cleavage occurs is not known, but the ubiquitin-like S30 polyprotein can be cleaved in a reticulocyte lysate in the absence of ribosome assembly. Moreover, no protease activity has been found among the ribosomal proteins, although the proteolytic enzyme might only associate transiently with the particle while it is being assembled in the nucleolus. Although the proposal that the organization into a polyprotein favors, or is essential for, the assembly of the ribosomal protein into particles has not been established, there are no reasonable alternatives.

RAT RIBOSOMAL PROTEIN mRNAs

General Features

Rat ribosomal protein mRNAs are relatively short, on the average about 700 nucleotides in length, since they encode small proteins. A typical mRNA has in its 5′-untranslated region (5′UTR) an initial sequence of 4–20 pyrimidines followed by a GC-rich stretch of approximately 40 nucleotides and ending in an AUG codon for the initiation of translation. The 3′UTR begins with a termination codon, has an AU-rich region of approximately 50 nucleotides, and ends with a long poly(A) tail.

5′-Untranslated Region

Pyrimidine sequences are found at the immediate 5′end of most, if not all, eukaryotic ribosomal protein mRNAs; the presence of a polypyrimidine stretch can be taken to confirm that the mRNA includes the region where transcription started. Of the 75 rat ribosomal protein cDNAs we have analyzed, 40 have this polypyrimidine stretch; they range in length from 4 to 20 nucleotides. A consensus sequence cannot be derived, but 10 of the 40 have the sequence CTTTCC, and a variant ($CT_{2\text{ or }3}C_{1\text{ or }2}$) is in most of the others. This motif may be a promoter that binds a *trans*-acting factor that accounts for the regulation (and perhaps the coordination) of the translation of ribosomal protein mRNAs (Perry and Meyuhas 1990; Wool et al. 1990; Levy et al. 1991; Meyuhas et al., this volume).

Downstream from the polypyrimidine stretch is a GC-rich 5′UTR (60–80% G and C). Thus, the region is likely to be structured, and GC helices can, of course, be formidable barriers to translation. How these are overcome is not known, but encumbered leader sequences strongly imply that the synthesis of ribosomal proteins is regulated at the initiation of translation.

The initiation codon occurs most frequently in the context (A/G)(A/C)CAUGG, a close approximation to the consensus for all vertebrates (Cavener and Ray 1991; Kozak 1991) and to the experimentally derived optimum (A/G)CCAUGG (Kozak 1991). In 74 of 75 rat ribosomal protein mRNAs, initiation is at the first AUG; the exception is the mRNA for rat L5.

The 5′UTR of rat ribosomal protein L5 has an open reading frame that begins with an AUG codon at position –90 and ends with an UAG codon at position –60; hence, it encodes ten amino acids (Tamura et al. 1987). It is not known if this short peptide is synthesized and if it is synthesized, what the significance might be. No other open reading frames have been found in the 5′UTR of rat ribosomal protein mRNAs. Codon usage in rat ribosomal protein mRNAs is not unusual (cf. Wada et al. 1991).

3′-Untranslated Region

The frequency of usage of the three termination codons in rat ribosomal protein mRNAs is 56%, UAA; 28%, UGA; and 16%, UAG. This is distinctive since in vertebrates in general (Cavener and Ray 1991), and in other rat proteins in particular (Wada et al. 1991), UGA is the most common termination codon. The more frequent use of UAA for termination may reflect the general character of the 3′UTR of rat ribosomal protein mRNAs; i.e., they are more AU-rich than most vertebrate mRNAs. The generalization for rat ribosomal protein mRNAs is that the 5′UTR is GC-rich, and the 3′UTR is AU-rich.

The 3′UTR in 54 of the 75 rat ribosomal protein mRNAs has the hexamer AAUAAA that directs posttranscriptional cleavage-polyadenylation of the 3′end of the precursor mRNA (Proudfoot 1991). Three have the sequence AUUAAA, whereas CAUAAA and AAUAUA occur once each. The other 16 presumably lack a signal sequence because their cDNAs lack a complete 3′UTR. The poly(A) tract usually begins 14 nucleotides (the range is 4 to 27) from the signal hexamer.

Number of Copies of Rat Ribosomal Protein Genes

We have determined the number of genes encoding the protein for 59 of the rat ribosomal proteins; the average is 12 copies. The prominent exception is rat S5, which is encoded by a single gene. Although there are multiple copies of most mammalian ribosomal protein genes, in no instance (with the exception of the separate alleles encoding HS4X and HS4Y; Fisher et al. 1990) has it been shown that more than one of the genes is functional; the presumption is that the others are retroposon pseudogenes.

The intron-containing functional human ribosomal protein genes are being mapped, and the chromosomal locations of 66 have been determined (N. Kenmochi and D.C. Page, pers. comm.).

EVOLUTION OF MAMMALIAN RIBOSOMAL PROTEINS

A subsidiary purpose in the determination of the amino acid sequences of the rat ribosomal proteins was from the start to learn something of their evolution. The assumption that the ribosome arose on a single occasion is supported by compelling evidence for a relationship of the rRNAs of all species, albeit the relationship being more apparent in the comparisons of secondary structure than of nucleotide sequences. Whether the ribosomal proteins are also related has theoretical consequences since it bears on whether the ribosome diverged before or after its conversion from an RNA to an RNP machine. Early evidence (Wool and Stöffler 1974) suggested that the ribosomal proteins from distant species were related, although divergent evolution has blurred the relationship. It is now clear that there are related proteins in the ribosomes of eubacteria, archaebacteria, and eukaryotes.

Mammalian Ribosomal Proteins

The amino acid sequences of the ribosomal proteins of different mammalian species are particularly close. This is exemplified by a comparison of the sequences from rat and humans, the two mammalian species for which the data sets are largest: 78 rat sequences (75 complete, 3 partial) and 77 human sequences (75 complete, 2 partial). Most of the human ribosomal protein amino acid sequences are a by-product of the Human Genome Project and most were recognized by their near identity with the rat sequences. Several of the sequences are identified for the first time here (see Table A2 in the Appendix). Of the 78 rat sequences, 75 can be correlated with a human sequence; the amino acid sequences of the human L29, L34, and L36 proteins have not yet been determined. Of the 77 human sequences, 76 can be correlated with a rat sequence; the additional human sequence is S1 (completed, but not yet published).

The amino acid sequences of related rat and human ribosomal proteins are near identical; the average for all 72 comparisons of complete sequences is 99% and for 32 comparisons, it is 100%. For one comparison, S15a, the apparent identity is only 91%, but the alignment improves if one assumes three frameshifts in the human DNA sequence. For the L4 and S4Y proteins, it is only 93%; however, S4Y is a special case. The identity of rat and human ribosomal proteins is so close that a difference of more than a few residues is cause to be suspicious of an error in the determination (cf. HS15a and HL15). We note that there are two isoforms of HS24 derived from alternate splicing of the mRNA; the first is identical to RS24 (Xu et al. 1994) and the second lacks the carboxy-terminal three amino acids.

There are two human transcribed genes for the ribosomal protein S4, one on the X chromosome and a second on the Y chromosome (Fisher et al. 1990). What is extraordinary is that the amino acid sequences encoded in the S4X and S4Y alleles differ at 19 of 263 positions; extraordinary since, as we have pointed out, there are seldom more than a few amino acid differences in homologous mammalian ribosomal proteins. This great a deviation is more than one would expect to find in a comparison with *Xenopus laevis* proteins (Wool et al. 1990). The S4Y gene is located in the sex-determining region of the Y chromosome and S4X escapes X-inactivation (Fisher et al. 1990). Both genes are transcribed in human cells; mRNAs specific for S4X and S4Y can be demonstrated by Northern hybridization, and ribosomes from male humans have 90% S4X and 10% S4Y (Zinn et al. 1994). Thus, there are male- and female-type ribosomes.

Yeast Ribosomal Proteins

There no longer can be any doubt that the ribosomal proteins from all eukaryotic species are derived from a common set of ancestral genes. This conviction is substantiated by comparison of the sequences of amino acids in rat and yeast ribosomal proteins, two evolutionarily distant eukaryotic species for which there are large data sets (see Table A3 in the Appendix). Of the 78 rat ribosomal proteins, 65 can be related to a yeast protein (58 to complete sequences and 7 to partial sequences). Two rat proteins are related to two separate yeast proteins: RP1 to YP1α and YP1β and RP2 to YP2α and YP2β. The complete amino acid sequences of the rat homologs of YL6, YL10, and YL14 have not yet been determined, but we know from unpublished partial sequences that they are related to RL6, RL10, and RL14. The yeast counterparts of 13 rat ribosomal proteins have not yet been determined.

The average identity in the alignment of the complete amino acid sequences of related rat and yeast ribosomal proteins is 60%; the range for the complete sequences is from 40% for RP1 and YP1β to 88% for RL41 and YL41. The data are sufficient to provide confidence that most if not all of the ribosomal proteins from the two species are homologous and that it will be possible to establish a protein to protein correlation. This has had at least one practical consequence: The correlation has been used to establish a uniform nomenclature for the two species (Wool et al. 1991) that we follow here. This substitutes a measure of order for the chaos that has confounded the designation of individual yeast ribosomal proteins (cf. the large number of synonyms for individual yeast proteins).

Archaebacterial Ribosomal Proteins

The ribosomes of archaebacteria are thought to be transition particles in the evolution of the organelle. Thus, it has been frequently remarked that the amino acid sequences of archaebacterial ribosomal proteins are closer to their eukaryotic homologs than to their eubacterial homologs, whereas the organization of archaebacterial ribosomal protein genes mimics that of the eubacteria.

For a correlation of the amino acid sequences of rat and archaebacterial ribosomal proteins, *Halobacterium marismortui* offers the largest set of completed sequences. If there is no *H. marismortui* sequence, the correlation is made to some other halophile, or, as a last resort, to whatever archaebacterial sequence is available. Of the 78 rat ribosomal proteins, 49 can be related to an archaebacterial ribosomal protein (see Table A4 in the Appendix). Rat P1 and P2 are related to a single *H. marismortui* protein L12 and the percent identity is about the same for the separate comparisons. There are eight archaebacterial ribosomal protein amino acid sequences (four complete and four partial) for which no rat counterpart is known. For the 49 related sequences (all complete), the average amino acid identity is 34%; the range is from 17% (RL18 and HmL29) to 55% (RS23 and HhS12). The reality of the relationship of RL18 and HmL29 is reinforced by pairwise comparisons using yeast YL18: 55% identity for RL18 and YL18, and 29% identity for YL18 and HmL29.

Eubacterial Ribosomal Proteins

The challenge at the beginning was to determine if an evolutionary relationship exists between the proteins of mammalian and eubacterial ribosomes. There had to be a strong presumption for the relationship because it seemed likely that the ribosome was invented once and it was clear that the structures of the rRNAs are homologous. The question was whether it would be possible to trace the chemical spoor. Although we had argued from the start for common ancestors for eukaryotic and eubacterial ribosomal proteins, it has only been possible to accrue convincing evidence in the past few years. What has made the difference more than any other factor is the accumulation of near complete sets of amino acid sequences of mammalian (rat) and eubacterial (*E. coli*) ribosomal proteins.

Twenty-four rat and *E. coli* ribosomal proteins can be correlated directly; for eight additional pairs (RS13-EcS15, RP0-EcL10, RP1-EcL12, RL12-EcL11, RL27a-EcL15, RL5-EcL18, RL26-EcL24, and RL7-EcL30), an initial statistically weak correlation is reinforced by pairwise comparisons using the amino acid sequences of either the yeast or the archaebacterial counterpart or both. An example is the correlation of rat P0 and *E. coli* L10 using the related *H. marismortui* protein. The alignment of the amino acid sequences of RP0 and HmL10 yields 82 identities in 313 possible matches (26% identity), whereas the RP0 and EcL10 alignment yields only 19 identities in 110 possible matches (17% identity). An alignment of EcL10 and HmL10 sequences shows 40 identities in 159 possible matches (25% identity). Thus, the sequence of amino acids in HmL10 is about as close to rat P0 as to EcL10, and because of this three-way comparison, the rat P0-EcL10 linkage is reinforced. Another example is the correlation of rat L26 and *E. coli* L24. The alignment of the amino acid sequences of RL26 and HmL24 yields 37 identities in 119 possible matches (31% identity), whereas the RL26 and EcL24 alignment yields only 15 identities in 74 possible matches (20% identity). An alignment of EcL24 and HmL24 sequences shows 17 identities in 80 possible matches (21% identity). Thus, the sequence of amino acids in HmL24 is closer to rat L26 than to EcL24; this follows a pattern we have noted before. However, the operon organization of the HmL24 gene resembles that of the EcL24 gene (Arndt 1990), providing indirect reinforcement for the validity of the rat L26-EcL24 relationship. The amino acid sequences of the archaebacterial ribosomal proteins serve in this way as a sort of molecular Rosetta stone to relate eukaryotic and eubacterial proteins.

The correlation of rat P2 and *E. coli* L12 depends on the assumption that the amino and carboxyl termini were transposed during evolution (Lin et al. 1982); there is substantial evidence that these two proteins are functionally equivalent (see earlier). One additional pair of proteins is related: EcS1 is related to a human homolog whose sequence has been determined but not yet published (the identity is 37%). The average amino acid identity for the related rat and *E. coli* ribosomal proteins is 27%; the range is 17% (RP0-EcL10) to 42% (RS14-EcS11).

Of the 33 mammalian (32 rat and 1 human) and 21 *E. coli* small ribosomal subunit proteins whose amino acid sequences have been determined, 16 are homologous (see Table A5 in the Appendix). Among these 16 are 7 *E. coli* proteins that are crucial for ribosome assembly (Held et al. 1974), including 6 of the 7 proteins (EcS4, EcS7, EcS8, EcS13, EcS15, EcS17, and EcS20) that bind independently and stoichiometrically to 16S rRNA. In addition, of the 17 *E. coli* ribosomal proteins necessary for the assembly of the large subunit (Nierhaus 1991), 11 have rat homologs: EcL2, EcL3, EcL10, EcL11, EcL12, EcL15, EcL18, EcL22, EcL23, EcL24, and EcL29. Thus, most of the proteins critical for the assembly of *E. coli* 30S and 50S ribosomal subunits are conserved. Although this reinforces the conclusion that mammalian and eubacterial ribosomal proteins share a common ancestor, it can be argued that all or most of the proteins are conserved but that this subset is easier to identify, perhaps because their function in assembly constrains variations in the amino acid sequence. It should be noted that the argument has been made, by analogy with the rRNAs, that it will be easier to see the identity of mammalian and eubacterial ribosomal proteins when their three-dimensional structures are known.

What does homology of a significant number of eukaryotic and eubacterial ribosomal proteins say about the evolution of the particle? It supports the argument that the ancestor of contemporary ribosomes was an RNP particle substantially similar to the organelle we know today—that the ribosomal proteins were added to the rRNAs before the divergence into eubacteria, archaebacteria, and eukaryotes. Any other scenario requires an unacceptable number of a priori assumptions.

Table 2 Phylogenetic distribution of homologs of rat ribosomal proteins

40S Subunit			60S Subunit		
Group[a] I	II	III	Group[a] I	II	III
Sa	S4	S1[b]	P0	L4	L6[b]
S2	S19	S3a	P1	L18	L7a
S3	S24	S6	P2	L18a	L10[b]
S5	S27a	S7	L3	L19	L10a[b]
S9		S8	L5	L21	L13
S11		S10	L7	L24	L14[b]
S13		S12	L8	L30	L15
S14		S17	L9	L31	L22
S15		S21	L11	L32	L27
S15a		S25	L12	L35a	L28
S16		S27	L13a	L36a	L29
S18		S26	L17	L37	L34
S20		S28	L23	L39	L36
S23		S30	L23a		L37a
S29			L26		L38
			L27a		L40
			L35		L41

[a]Rat ribosomal proteins in Group I have homologs in the eubacterial and archaebacterial kingdoms, those in Group II have homologs only in the archaebacterial kingdom, and those in Group III are unique to eukaryotes.

[b]The complete amino acid sequence has not been determined; thus, relegation to Group III is provisional.

Ribosomal Proteins Unique to Eukaryotes

Since eubacterial ribosomes have 54 proteins and eukaryotes have approximately 80, it follows that some must be unique to the latter. It is now possible to make a preliminary identification of those proteins (Table 2). Of the 80 rat ribosomal proteins, 32 have a homolog in the eubacterial and archaebacterial kingdoms (group I proteins in Table 2), 17 have homologs in the archaebacterial but not the eubacterial kingdoms (group II), and 31 are unique to eukaryotes (group III). The amino acid sequences for 5 proteins (S1, L6, L10, L10a, and L14) have not been completed, and thus their assignment to group III is provisional. It has been obvious for a long time that proteins were added to ribosomes after the divergence of eubacteria and archaebacteria, and of archaebacteria and eukaryotes. What is new is that it is now possible to make a provisional identification of these proteins. The question that all of this occasions is what was the reason for the addition of the proteins? One answer, albeit one that begs the question, is that more proteins were needed because of the expansion of the rRNA. But then why more rRNA, to which question, of course, there is no answer. A second possibility is that more proteins were needed because of the addition of functions. The assembly of eukaryotic ribosomes is a more complicated process at least in part because of the extensive intracellular traffic it requires. The proteins are synthesized in the cytoplasm and transported to the nucleus and ultimately to the nucleolus where they are assembled on nascent rRNA. Some of the ribosomal proteins may be involved only in the assembly process and some may be needed only for the processing of the 45S rRNA precursor to mature 18S, 28S, and 5.8S rRNAs; 5S rRNA, perhaps in a complex with L5, must be recruited to the nucleolus from another site in the nucleus where it is transcribed. Finally, the 40S and 60S ribosomal subunits are relocated to the cytoplasm. It is also possible that the additional proteins participate in aspects of the biochemistry of protein synthesis—perhaps in the initiation of translation, which appears to be more complicated in eukaryotes—or in the regulation of translation, which is more frequent in eukaryotes.

ON THE ORIGIN OF THE RIBOSOMAL PROTEINS

A question that has nagged at the minds of many who work on ribosomes is where did the proteins come from—what were their origins? It is a tenet of our faith that the *Ur*-ribosome had only RNA. This is a belief that is supported by the mounting evidence that the rRNAs are responsible for the basic biochemistry of protein synthesis: for the binding of mRNA, aminoacyl-tRNA and the initiation, elongation, and termination factors; for peptide bond formation; and for translocation. The ribosomal proteins are now viewed as a later evolutionary embellishment and are deemed to facilitate the folding of the rRNA and the maintenance of an optimal configuration (Stern et al. 1989), perhaps in this way endowing protein synthesis with speed and accuracy (Wool et al. 1990). This may be too severe a restriction on the role of the proteins in ribosome function; nevertheless, it is likely that RNA preceded the proteins and hence it is pertinent to ask from whence the latter came.

The occasion for the transition during evolution from a ribosome that had only RNA to an RNP machine may have coincided with, or been a response to, the appearance of nucleases that would have put an RNA ribosome at risk. Two possible story lines for the origin of the ribosomal proteins are (1) that they were designed specifically for the ribosome or (2) that they were coopted from among a set of preexistent proteins that already had defined functions. The two possibilities are by no means exclusive, nor is it likely that all the proteins were added at one time. If the latter conjecture has substance, the proteins most likely to have been recruited would have been those that already had the capacity to bind to nucleic acids. Considered from this perspective, the zinc finger, bZIP, and helix-turn-helix motifs in ribosomal proteins may have been used earlier to bind to DNA or to other RNAs. A possible archetype is TFIIIA, which may have evolved to bind to the internal control region of the 5S rRNA gene and only later came to associate with 5S rRNA.

A paradigm for the process by which rRNA could have coopted preexisting nucleic-acid-binding proteins, and indeed the likelihood of its having happened, is exemplified by the *E. coli* basic proteins NS1 and NS2 (Mende et al. 1978). These proteins, which have been likened to eukaryotic histones, bind to single- and double-stranded DNAs and to RNA (Berthold and Geider 1976). What is significant is that NS1 and NS2 are found with native *E. coli* 30S ribosomal subunits in near stoichiometric amounts but not with native 50S subunits or 70S couples (Suryanarayana and Subramanian 1978). The binding to 30S subunits is specific and the affinity is strong. There is no evidence, however, that either NS1 or NS2 has a function in protein synthesis: They are not in 70S ribosomes and they are not needed for translation of mRNA, and antibodies specific for NS1 and NS2 do not inhibit protein synthesis (Suryanarayana and Subramanian 1978). Nevertheless, the observations that the DNA-binding proteins NS1 and NS2 can associate specifically with 30S ribosomal subunits presumably by interaction with a particular site on 16S rRNA provide a measure of credence to our proposal concerning the origin of the ribosomal proteins. NS1 and NS2 can be viewed as DNA-binding proteins auditioning for a role in ribosome structure by association with native 30S subunits.

That NS1 and NS2 are not required for protein synthesis does not militate against the proposal: Indeed, a considerable number of authentic *E. coli* ribosomal proteins are absent from ribosomes in various mutants, and the loss is compatible with residual protein synthesis and growth (Dabbs 1985). Proteins may have been added to ribosomes initially only to coat and protect the RNA. DNA-binding proteins would be eminently suitable for the purpose and many bind to RNA. The proteins, either from the beginning or perhaps only later, tuned the rRNA and thereby optimized the higher-order structure and pari passu conferred velocity and fidelity on protein synthesis. Still later the proteins may have assumed other functions. NS1 and NS2 may be in early stages of this evolutionary process.

This raises a related matter: Received wisdom holds that the ribosome has a single function, to catalyze the synthesis of protein. The corollary of this axiom is that any protein not directly or indirectly involved in, or necessary for, protein synthesis is not a ribosomal protein. This reasoning dominated the strategy underlying the identification, purification, and characterization of the ribosomal proteins and was responsible for designating as contaminants many proteins more or less tenaciously, and more or less specifically, associated with the particle. The reasoning was heuristic and served those who applied it well. However, the bias inherent in the reasoning may have a defect—it may be wrong. Indeed, there is a prominent exception: *E. coli* ribosomes participate with stringent factor in the synthesis of guanosine tetraphosphate and guanosine pentaphosphate (Block and Haseltine 1974; Cashel and Gallant 1974). Although the synthesis of the guanosine polyphosphates is a by-product of ribosomes idling during protein synthesis, it is not necessary for the synthesis of protein, and stringent factor is not considered a ribosomal protein. The lesson is that one should be inclined to an open mind concerning the possibility that ribosomes and ribosomal proteins participate in other cellular functions.

THE BIFUNCTIONAL NATURE OF RIBOSOMAL PROTEINS

If rRNA coopted preexisting proteins, and if they retained their original function, then ribosomal proteins would be bifunctional or even multifunctional. There is evidence for bifunctionality for a number of ribosomal proteins. The ribosomal protein that comes to mind first is *E. coli* S1, which is a component of the replicase of some RNA phages (Kamen 1975). In Qβ, for example, replication is catalyzed by a complex containing the product of the viral replicase gene and three host factors: elongation factors EF1A and EF1B (formerly EF-Ts), and ribosomal protein S1. The latter is required for the initiation of replication on the plus strand of the viral RNA.

A second *E. coli* ribosomal protein, S10, participates with bacteriophage λ N protein and NusB in antitermination of transcription (Friedman et al. 1981). Apparently, NusB binds to N-modified transcription complexes by interacting with *E. coli* ribosomal protein S10, and this interaction is essential for efficient antitermination (Mason et al. 1992).

Three bacteriophage T4 structural genes have self-splicing group I introns that are processed in the usual manner by two transesterification reactions initiated by nucleophilic attack of a guanosine at the 5′ splice site (for references, see Coetzee et al. 1994). Although the T4 introns are spliced in vitro in the absence of proteins, evidence exists for the involvement of accessory factors during in vivo splicing. For one thing, splicing of group I introns is appreciably more efficient in vivo than in vitro. In accord with this, a number of proteins enhance in vitro *cis*- and *trans*-splicing of T4 introns; several are *E. coli* ribosomal proteins, and the most effective is ribosomal protein S12 (Coetzee et al. 1994). The suggestion is that S12 modulates splicing by promoting association of distant pairing elements and by preventing formation of catalytically inert structures.

There is a most unexpected instance of a separate function for a eukaryotic ribosomal protein. Damage to DNA in mammalian cells occurs all the time and the lesions are ordinarily promptly and efficiently repaired by an ensemble of enzymes (Imlay and Linn 1988). Among these enzymes is one that has been designated apurinic/apyrimidinic endonuclease III (APIII), which catalyzes incision on the 3′side of an apurinic or an apyrimidinic site (Kim and Linn 1988). Xeroderma pigmentosum is a rare autosomal recessive disease that is characterized by hypersensitivity to UV irradiation. Because of this sensitivity, patients with the disease are at great risk of developing sunlight-induced skin cancers. Seven complementation groups for the disease have been defined; group D lacks APIII activity. Linn and colleagues purified APIII and determined the sequences of amino acids in a number of proteolytic fragments (S. Linn, pers. comm.); the sequences were the same as those established before for regions of rat ribosomal protein S3 (Chan et al. 1990). This extraordinary counterintuitive finding raised the possibility that the purified protein contained predominantly S3 but was contaminated with small amounts of APIII. To test the possibility, a rat ribosomal protein S3 cDNA clone was expressed in *E. coli*. The expressed protein had APIII activity. Moreover, both the protein that was originally isolated from mammalian cells and the protein expressed in *E. coli* reacted with an antiserum raised against rat ribosomal protein S3. Thus, S3 appears to be both a ribosomal component and a DNA-repair endonuclease. In keeping with these findings, a cDNA encoding *Drosophila melanogaster* ribosomal protein S3 (the homolog of the rat protein) when expressed in *E. coli* also has an activity that cleaves DNA at apurinic/apyrimidinic sites; this enzyme is an AP lyase (Wilson et al. 1994).

S3 is not the only ribosomal protein involved in DNA repair. A *Drosophila* gene that encodes a separate apurinic/apyrimidinic endonuclease (AP3) was identified using antibody to a related human enzyme

(Kelley et al. 1989); the *Drosophila* gene was used in turn to probe a HeLa cell cDNA library. The single human cDNA that was identified encodes a protein that has 66% identity with the *Drosophila* AP3 gene and is identical to human ribosomal protein P0 (Rich and Steitz 1987), suggesting that this ribosomal protein is also an AP endonuclease. That ribosomal protein P0 is induced in human cells by bifunctional alkylating agents that are used in the treatment of cancer and that cause DNA damage supports this deduction (Grabowski et al. 1992).

Finally, damage to DNA by UV irradiation in cells deficient in repair induces the ribosomal protein L7a (Ben-Ishai et al. 1990). A function for L7a in DNA repair has not been established; nevertheless, the observation that it is induced by DNA damage and the precedent of S3 and P0 raise a suspicion that it may have a role in the process.

There are findings that can be construed to indicate that ribosomal proteins serve as specific transcription factors during development. In humans, there are two transcribed genes for the ribosomal protein S4, one on the X chromosome and a second on the Y chromosome (Fisher et al. 1990). S4Y is in a region of the Y chromosome that has been linked to certain forms of Turner Syndrome (for references and discussion, see Fisher et al. 1990). Turner females have gonadal insufficiency, and a variety of anatomic abnormalities including short stature, webbing of the neck, lymphadema, and aortic coarctation. The Turner phenotype obviously is a manifestation of a serious anomaly of development, and it is not the only one associated with a mutation, usually a deficiency, of a ribosomal protein. A collection of mutations at about 50 separate loci in *Drosophila* have been designated *Minute* (Lindsley and Grell 1967). Flies with the *Minute* phenotype have delayed larval development, diminished viability, reduced body size, decreased fertility, thin bristles, and etching of the abdomen. The *Minute* phenotype is also a manifestation of development gone awry. The *Minute* genes that have been identified encode ribosomal proteins (Kongsuwan et al. 1985; Andersson and Lambertsson 1990).

The explanation given to account for the Turner and *Minute* phenotypes is haploinsufficiency for a ribosomal protein (Fisher et al. 1990). The presumption is that a ribosomal protein insufficiency leads to an insufficiency of competent ribosomes at one or more critical stages of development. One cannot gainsay that this is a logical and economical explanation; it is a simple means of accounting for complex phenotypes consistent with the available evidence. However, this explanation implies a lack of the regulation of the synthesis of ribosomal proteins, something that is barely credible given the large body of evidence to the contrary

(Warner et al. 1985), i.e., that their synthesis is precisely adjusted so that equimolar amounts of each are available for packaging with stoichiometric amounts of rRNA. The expectation would be that if the S4Y gene were inactivated, there would be an increase in the transcription of the S4X allele or an increase in the translation of the S4X mRNA. Indeed, if the S4X gene accounts for 90% of the protein in ribosomes, the adjustment necessary would be small.

An alternate explanation is that during development, certain ribosomal proteins regulate the expression of genes in a positive or negative manner by binding to DNA or that they participate with the RNA polymerases in transcription of selected genes. S4Y, for example, might serve as a transcription factor for genes that endow the individual with characteristics that we associate with maleness. With respect to this proposal, it may be significant that the degree of repression of the translation of individual ribosomal protein mRNAs in early *Drosophila* embryos varies during development (Al-Atia et al. 1985; Kay and Jacobs-Lorena 1985). A similar observation—selective repression of the translation of ribosomal protein mRNAs during development—has been made for *Xenopus* as well (Pierandrei-Amaldi et al. 1982), which may reflect a need for specific ribosomal proteins (i.e., the mRNAs whose translation is not repressed) as factors for the regulation of development. This interpretation, of course, may be biased. There most certainly are observations not entirely consistent with this explanation. The effects of combinations of two or three mutations at *Minute* loci do not increase the severity of the phenotype, suggesting that the loci code for gene products of similar function and favoring the interpretation that the phenotype arises from their contribution to ribosome function. It is, however, also possible that the genes encode transcription factors with similar and coordinated functions in development.

The proposal that ribosomal proteins participate in development finds support in the observation that P-element-induced lethal mutations in the 5′ regulatory region of the gene encoding *D. melanogaster* ribosomal protein S6, and which cause a reduction in the abundance of transcripts, affect hematocytes that mediate the insect immune response (Watson et al. 1992; Stewart and Denell 1993). The mutants display melanotic tumors characteristic of mutations affecting the immune system, and the moribund insects develop grossly hypertrophied hematopoietic organs because of increased cell proliferation and extra rounds of endoreduplication in hematopoietic cells. Thus, S6 appears to be required for tumor suppression in the hematopoietic system. Whether this is an extraribosomal effect of S6 during development or whether tumorigenesis is a manifestation of an insufficient number of ribosomes is not known. However, the phenotype of the S6-deficient mutants is distinctly different from that of *Minute*, and this favors the former explanation that S6 has an extraribosomal effect during larval development.

The membranes of the nucleus, the endoplasmic reticulum, and the Golgi apparatus are dispersed during mitosis and reassembled later (for references, see Bowman et al. 1992). There is in *Xenopus* oocytes a peripheral membrane protein, p27, associated with vesicles formed from the nuclear envelope and from the endoplasmic reticulum (K. Sullivan and K. Wilson, unpubl.). During mitosis, p27 is phosphorylated, whereas in interphase it is not. The sequences of amino acids in two tryptic peptides from *Xenopus* p27 were determined (K. Sullivan and K. Wilson, unpubl.): All 12 residues of one peptide and 12 of 13 residues of the other peptide were identical to amino acid sequences in rat ribosomal protein S8 (Chan et al. 1987). Thus, the *Xenopus* homolog of rat ribosomal protein S8 is also a membrane-associated protein whose phosphorylation and dephosphorylation are correlated temporally with nuclear envelope disassembly at the beginning of mitosis and reassembly at its conclusion. Phosphorylation of p27/S8 also provides a plausible means for the partition of the protein between the ribosome and intracellular membranes.

Bone has the potential, especially under pathological conditions, to develop almost anywhere in the body. Moreover, bone formation can be induced by transplantation to ectopic sites of various tissues or tissue extracts (for references, see Ito 1992). Demineralized bone matrix is particularly effective, and attempts have been made to identify the responsible inducing factor. One highly purified fraction with bone-inducing activity was prepared from a murine osteosarcoma (Ito 1992); the protein was isolated and identified from the amino-terminal amino acid sequence as ribosomal protein L32 (Rajchel et al. 1988). Thus, L32 would appear to be a morphogen capable of inducing cartilage formation, and ultimately endochondrial bone, from mesenchymal cells.

Iron in the circulation is bound to transferrin and is taken up by cells through a transferrin-receptor-mediated mechanism, but how iron, which is insoluble at neutral pH and potentially toxic, is distributed within cells is not known. In a search for intracellular ligands, an iron-binding protein was isolated from rat liver (Furukawa et al. 1992); the amino acid sequence of tryptic peptides established that the iron carrier is ribosomal protein P2 (Lin et al. 1982).

These examples suggest that second functions of ribosomal proteins are the rule rather than the exception and lead us to predict that more will be discovered. In addition, other RNP assemblies, the spliceosome is a prime example, may contain bifunctional proteins of which some may be ribosomal proteins. Indeed, we know that the prediction has substance. Human B lymphocytes infected with Epstein-Barr virus synthesize two small nuclear RNAs called EBER 1 and 2 (Epstein-Barr-encoded RNAs). The EBERs bind proteins to form nuclear RNPs; both bind the La antigen and EBER 1 is also associated with a second protein designated EAP (EBER-associated protein) (Toczyski and Steitz 1991). EAP is ribosomal protein L22 (Toczyski et al. 1994).

Is there a precedent for the proposal we make for the origins of the ribosomal proteins and for their bifunctionality? Perhaps there is a paradigm. The crystallins are structural proteins of the lens that contribute to the refraction of light. Some crystallins have no other known function; others, especially species-specific crystallins, are identical to enzymes found in lesser amounts in other tissues (for references and discussion, see Piatigorsky and Wistow 1991). Both of the α-crystallins, αA and αB, are related to heat shock proteins; the β- and γ-crystallins of vertebrates are distantly related to the dormancy proteins of microorganisms, which are also induced by stress. The most abundant mammalian-specific crystallin is identical to aldehyde dehydrogenase. In various species, the crystallins include lactate dehydrogenase-B, argininosuccinate lyase, and α-enolase (birds and reptiles); NADPH-dependent reductases (frogs); and glutathione *S*-transferases (cephalopods) (Piatigorsky and Wistow 1991). The enzyme crystallins are bifunctional, just as some ribosomal proteins appear to be; bifunctionality seems to have been acquired by modification of gene expression, a phenomenon called gene sharing (Piatigorsky and Wistow 1991). The presumption is that a transcriptional or posttranscriptional modification has allowed recruitment of the enzyme to the lens. Later, there may have been evolutionary pressure for changes in the structure of the enzyme that improve its function as a crystallin but are neutral with regard to its catalytic activity, or there may have been gene duplication with separation of the functions. There are examples consistent with both modes of evolution (Piatigorsky and Wistow 1991). The crystallin enzymes are present in lens in amounts that exceed reasonable catalytic need. Indeed, in some cases, the enzymes no longer have catalytic activity because of posttranslational modification or because of alteration of the amino acid sequence in the case of gene duplication and separate evolution. Thus, lens crystallins share features with ribosomal proteins, including one or more separate functions and the presumption of a means (most likely posttranslational modification) to effect partition of the protein to separate sites.

Finally, it must be recognized that the primordial sequence of events may have been the other way around from what we have postulated, that, for example, the *E. coli* ribosomal protein S1 was adopted by small RNA phages to increase the efficiency of their replicase and that the mammalian ribosomal proteins S3 and P0 were adopted as endonucleases for DNA repair.

ACKNOWLEDGMENTS

We acknowledge with gratitude the contribution of many colleagues over the past 20 years to the research summarized in this chapter. We are especially grateful to Veronica Paz and Joe Olvera for their assistance in the enterprise and to Katsuyuki Suzuki for his valuable role in the compilation and the analysis of the data. Arlene Timosciek provided help in the preparation of the manuscript.

REFERENCES

Al-Atia, G.R., P. Fruscolini, and M. Jacobs-Lorena. 1985. Translational regulation of mRNAs for ribosomal proteins during early *Drosophila* development. *Biochemistry* **24:** 5798–5803.

Andersson, S. and A. Lambertsson. 1990. Characterization of a novel *Minute*-locus in *Drosophila melanogaster*: A putative ribosomal protein gene. *Heredity* **65:** 51–57.

Aoyama, K., S. Hidaka, T. Tanaka, and K. Ishikawa. 1982. The nucleotide sequence of 5S RNA from rat liver ribosomes. *J. Biochem.* **91:** 363–367.

Aris, J.P. and G. Blobel. 1991. cDNA cloning and sequencing of human fibrillarin, a conserved nucleolar protein recognized by autoimmune antisera. *Proc. Natl. Acad. Sci.* **88:** 931–935.

Arndt, E. 1990. Nucleotide sequence of four genes encoding ribosomal proteins from the 'S10 and spectinomycin' operon equivalent region in the archaebacterium *Halobacterium marismortui. FEBS Lett.* **267:** 193–198.

Baker, R.T. and P.G. Board. 1991. The human ubiquitin-52 amino acid fusion protein gene shares several structural features with mammalian ribosomal protein genes. *Nucleic Acids Res.* **19:** 1035–1040.

Ben-Ishai, R., R. Scharf, R. Sharon, and I. Kapten. 1990. A human cellular sequence implicated in *trk* oncogene activation is DNA damage inducible. *Proc. Natl. Acad. Sci.* **87:** 6039–6032.

Berg, J.M. 1986. Potential metal-binding domains in nucleic acid binding proteins. *Science* **232:** 485–487.

———. 1990. Zinc finger domains: Hypotheses and current knowledge. *Annu. Rev. Biophys. Biophys. Chem.* **19:** 405–421.

Berthold, V. and K. Geider. 1976. Interaction of DNA with DNA-binding proteins. The characterization of protein HD from *Escherichia coli* and its nucleic acid complexes. *Eur. J. Biochem.* **71:** 443–449.

Blobel, G. 1971. Isolation of a 5S RNA-protein complex from mammalian ribosomes.

Proc. Natl. Acad. Sci. **68:** 1881–1885.

Block, R. and W.A. Haseltine. 1974. *In vitro* synthesis of ppGpp and pppGpp. In *Ribosomes* (ed. M. Nomura et al.), pp. 747–762. Cold Spring Harbor Laboratory, Cold Spring Harbor, New York.

Boman, A.L., M.R. Delannoy, and K.L. Wilson. 1992. GTP hydrolysis is required for vesicle fusion during nuclear envelope assembly *in vitro. J. Cell Biol.* **116:** 281–294.

Bonfa, E. and K.B. Elkon. 1986. Clinical and serological associations of the anti-ribosomal P protein antibody. *Arthritis Rheum.* **29:** 981–985.

Bourbon, H.M. and F. Amalric. 1990. Nucleolin gene organization in rodents: Highly conserved sequences within three of the 13 introns. *Gene* **88:** 187–196.

Brot, N. and H. Weissbach. 1981. Chemistry and biology of *E. coli* ribosomal protein L12. *Mol. Cell. Biochem.* **36:** 47–63.

Brown, D.D. 1984. The role of stable complexes that repress and activate eucaryotic genes. *Cell* **37:** 359–365.

Busch, S.J. and P. Sassone-Corsi. 1990. Dimers, leucine zippers and DNA-binding domains. *Trends Genet.* **6:** 36–40.

Caizergues-Ferrer, M., P. Mariottini, C. Curie, B. Lapeyre, N. Gas, F. Amalric, and F. Amaldi. 1989. Nucleolin from *Xenopus laevis*: cDNA cloning and expression during development. *Genes Dev.* **3:** 324–333.

Cashel, M. and J. Gallant. 1974. Cellular regulation of guanosine tetraphosphate and guanosine pentaphosphate. In *Ribosomes* (ed. M. Nomura et al.), pp. 733–746. Cold Spring Harbor Laboratory, Cold Spring Harbor, New York.

Cavener, D.R. and S.C. Ray. 1991. Eukaryotic start and stop translation sites. *Nucleic Acids Res.* **19:** 3185–3192.

Chan, Y.L. and I.G. Wool. 1988. The primary structure of rat ribosomal protein S6. *J. Biol. Chem.* **263:** 2891–2896.

Chan, Y.L., J. Olvera, and I.G. Wool. 1983. The structure of rat 28S ribosomal ribonucleic acid inferred from the sequence of nucleotides in a gene. *Nucleic Acids Res.* **11:** 7819–7831.

Chan, Y.L., K.R.G. Devi, J. Olvera, and I.G. Wool. 1990. The primary structure of rat ribosomal protein S3. *Arch. Biochem. Biophys.* **283:** 546–550.

Chan, Y.L., R. Gutell, H.F. Noller, and I.G. Wool. 1984. The nucleotide sequence of a rat 18S ribosomal ribonucleic acid gene and a proposal for the secondary structure of 18S ribosomal ribonucleic acid. *J. Biol. Chem.* **259:** 224–230.

Chan, Y.L., A. Lin, V. Paz, and I.G. Wool. 1987. The primary structure of rat ribosomal protein S8. *Nucleic Acids Res.* **15:** 9451–9459.

Chan, Y.L., K. Suzuki, J. Olvera, and I.G. Wool. 1993. Zinc finger-like motifs in rat ribosomal proteins S27 and S29. *Nucleic Acids Res.* **21:** 649–655.

Christensen, M.E. and K.P. Fuxa. 1988. The nucleolar protein, B-36, contains a glycine and dimethylarginine-rich sequence conserved in several other nuclear RNA-binding proteins. *Biochem. Biophys. Res. Commun.* **155:** 1278–1283.

Coetzee, T., D. Herschlag, and M. Belfort. 1994. *Escherichia coli* proteins, including ribosomal protein S12, facilitate *in vitro* splicing of phage T4 introns by acting as RNA chaperones. *Genes Dev.* **8:** 11575–1588.

Dabbs, E.R. 1985. Mutant studies on the prokaryotic ribosome. In *Structure, function, and genetics of ribosomes* (ed. B. Hardesty and G. Kramer), pp. 733–748. Springer-Verlag, New York.

Del Grande, R.W. and J.A. Traugh. 1982. Phosphorylation of 40-S ribosomal subunits by cAMP-dependent, cGMP-dependent and protease-activated protein kinases. *Eur. J. Biochem.* **123:** 421–428.

Duncan, R. and E.H. McConkey. 1982. Preferential utilization of phosphorylated 40S ribosomal subunits during initiation complex formation. *Eur. J. Biochem.* **123:** 535–538.

DuVernay, V.H. and J.A. Traugh. 1978. Two-step purification of the major phosphorylated protein in reticulocyte 40S ribosomal subunits. *Biochemistry* **17:** 2045–2049.

Elkon, K.B., A.P. Parnassa, and C.L. Foster. 1985. Lupus antibodies target ribosomal P proteins. *J. Exp. Med.* **162:** 459–471.

Elkon, K., S. Skelly, A. Parnassa, W. Möller, W. Danho, H. Weissbach, and N. Brot. 1986. Identification and chemical synthesis of a ribosomal protein antigenic determinant in systemic lupus erythematosus. *Proc. Natl. Acad. Sci.* **83:** 7419–7423.

Ellenberger, T.E., C.J. Brandl, K. Struhl, and S.C. Harrison. 1992. The GCN4 basic region leucine zipper binds DNA as a dimer of uninterrupted alpha helices: Crystal structure of the protein-DNA complex. *Cell* **71:** 1223–1237.

Finley, D., B. Bartel, and A. Varshavsky. 1989. The tails of ubiquitin precursors are ribosomal proteins whose fusion to ubiquitin facilitates ribosome biogenesis. *Nature* **338:** 394–401.

Fisher, E.M.C., P. Beer-Romero, L.G. Brown, A. Ridley, J.A. McNeil, L.G. Lawrence, H.F. Willard, F.R. Bieber, and D.C. Page. 1990. Homologous ribosomal protein genes on the human X and Y chromosomes: Escape from X inactivation and possible implications for Turner syndrome. *Cell* **63:** 1205–1218.

Freedman, S.D. and J.D. Jamieson. 1982. Hormone-induced protein phosphorylation. III. regulation of the phosphorylation of the secretagogue-responsive 29,000-dalton protein by both Ca^{2+} and cAMP *in vitro*. *J. Cell Biol.* **95:** 918–923.

Friedman, D.I., A.T. Schauer, M.R. Baumann, L.S. Baron, and S.L. Adhya. 1981. Evidence that ribosomal protein S10 participates in control of transcription termination. *Proc. Natl. Acad. Sci.* **78:** 1115–1118.

Furukawa, T., T. Uchiumi, R. Tokunaga, and S. Taketani. 1992. Ribosomal protein P2, a novel iron-binding protein. *Arch. Biochem. Biophys.* **298:** 182–186.

Grabowski, D.T., R.O. Pieper, B.W. Futscher, W.A. Deutsch, L.C. Erickson, and M.R. Kelley. 1992. Expression of ribosomal phosphoprotein P0 is induced by antitumor agents and increased in Mer^- human tumor cell lines. *Carcinogenesis* **13:** 259–263.

Gressner, A.M. and I.G. Wool. 1974a. The phosphorylation of liver ribosomal proteins *in vivo*. Evidence that only a single small subunit protein (S6) is phosphorylated. *J. Biol. Chem.* **249:** 6917–6925.

———. 1974b. The stimulation of the phosphorylation of ribosomal protein S6 by cycloheximide and puromycin. *Biochem. Biophys. Res. Commun.* **60:** 1482–1490.

Hadjiolov, A.A., O.I. Georgiev, V.V. Nosikov, and L.P. Yavachev. 1984. Primary and secondary structure of rat 28S ribosomal RNA. *Nucleic Acids Res.* **12:** 3677–3693.

Harrison, S. 1991. A structural taxonomy of DNA-binding domains. *Nature* **353:** 715–719.

Held, W.A., B. Ballou, S. Mizushima, and M. Nomura. 1974. Assembly mapping of 30S ribosomal proteins from *E. coli:* Further studies. *J. Biol. Chem.* **249:** 3103–3111.

Hemmerich, P., A. von Mikecz, F. Neumann, O. Sözeri, G. Wolff-Vorbeck, R. Zoebelein, and U. Krawinkel. 1993. Structural and functional properties of ribosomal protein L7 from human and rodents. *Nucleic Acids Res.* **21:** 223–231.

Henriquez, R., G. Blobel, and J.P. Aris. 1990. Isolation and sequencing of NOP1. A yeast gene encoding a nucleolar protein homologous to a human autoimmune antigen. *J. Biol. Chem.* **265:** 2209–2215.

Hershko, A. 1991. The ubiquitin pathway for protein degradation. *Trends Biochem. Sci.* **16:** 265–268.

Imlay, J.A. and S. Linn. 1988. DNA damage and oxygen radical toxicity. *Science* **240:** 1302–1309.

Ito, Y. 1992. Purification and partial identification of bone-inducing protein from a murine osteosarcoma. *Biochem. J.* **284:** 847–854.

Johnson, S.P. and J.R. Warner. 1987. Phosphorylation of the *Saccharomyces cerevisiae* equivalent of ribosomal protein S6 has no detectable effect on growth. *Mol. Cell. Biol.* **7:** 1338–1345.

Jong, A.Y.S., M.W. Clark, M. Gilbert, A. Oehm, and J.L. Campbell. 1987. *Saccharomyces cerevisiae* SSB1 protein and its relationship to nucleolar RNA-binding proteins. *Mol. Cell. Biol.* **7:** 2947–2955.

Kamen, R.I. 1975. Structure and function of the Qβ RNA replicase. In *RNA phages* (ed. N.D. Zinder), pp. 203–234. Cold Spring Harbor Laboratory, Cold Spring Harbor, New York.

Kay, M.A. and M. Jacobs-Lorena. 1985. Selective translational regulation of ribosomal protein gene expression during early development of *Drosophila melanogaster. Mol. Cell. Biol.* **5:** 3583–3592.

Kelley, M.R., S. Venugopal, J. Harless, and W.A. Deutsch. 1989. Antibody to a human DNA repair protein allows for cloning of a *Drosophila* cDNA that encodes an apurinic endonuclease. *Mol. Cell. Biol.* **9:** 965–973.

Kim, J. and S. Linn. 1988. The mechanisms of action of *E. coli* endonuclease III and T4 UV endonuclease (endonuclease V) at AP sites. *Nucleic Acid Res.* **16:** 1135–1141.

Klug, A. and D. Rhodes. 1987. 'Zinc fingers': A novel protein motif for nucleic acid recognition. *Trends Biochem. Sci.* **12:** 464–469.

Kongsuwan, K., Y. Quiang, A. Vincent, M.C. Frisardi, M. Rosbash, J.A. Lengyel, and J. Merriam. 1985. A *Drosophila Minute* gene encodes a ribosomal protein. *Nature* **317:** 555–558.

Kozak, M. 1991. An analysis of vertebrate mRNA sequences: Intimations of translational control. *J. Cell Biol.* **115:** 887–903.

Landschulz, W.H., P.F. Johnson, and S.L. McKnight. 1988. The leucine zipper: A hypothetical structure common to a new class of DNA binding proteins. *Science* **240:** 1759–1764.

Lapeyre, B., H. Bourbon, and F. Amalric. 1987. Nucleolin, the major nucleolar protein of growing eukaryotic cells: An unusual protein structure revealed by the nucleotide sequence. *Proc. Natl. Acad. Sci.* **84:** 1472–1476.

Lapeyre, B., F. Amalric, S.H. Ghaffari, S.V.V. Rao, T.S. Dumbar, and M.O. Olson. 1986. Protein and cDNA sequence of a glycine-rich, dimethylarginine-containing region located near the carboxyl-terminal end of nucleolin (C23 and 100 kDa). *J. Biol. Chem.* **261:** 9167–9173.

Leader, D.P. 1980. The control of phosphorylation of ribosomal proteins. In *Molecular aspects of cellular regulation* (ed. P. Cohen), vol. 1, pp. 203–233. Elsevier Biomedical, Amsterdam.

Leader, D.P., A. Thomas, and H.O. Vorrma. 1981. The protein synthetic activity *in vitro* of ribosomes differing in the extent of phosphorylation of their ribosomal proteins. *Biochim. Biophys. Acta* **656:** 69–75.

Levy, S., D. Avni, N. Hariharan, R.P. Perry, and O. Meyuhas. 1991. Oligopyrimidine tract at the 5′ end of mammalian ribosomal protein mRNAs is required for their translational control. *Proc. Natl. Acad. Sci.* **88:** 3319–3323.

Lin, A., B. Wittmann-Liebold, J. McNally, and I.G. Wool. 1982. The primary structure of the acidic phosphoprotein P2 from rat liver 60 S ribosomal subunits. *J. Biol. Chem.* **257:** 9189–9197.

Lindsley, D.L., and E.H. Grell. 1967. Genetic variations of *Drosophila melanogaster*. Carnegie Inst. Washington Publ. **627:** 1–471.

Lischwe, M.A., R.L. Ochs, R. Reddy, R.G. Cook, L.C. Yeoman, E.M. Tan, M. Reichlin, and H. Busch. 1985. Purification and partial characterization of a nucleolar scleroderma antigen (M_r = 34,000; pI, 8.5) rich in N^G,N^G-dimethylarginine. *J. Biol. Chem.* **260:** 14304–14310.

Luer, C.A. and K.P. Wong. 1979. Conformation of *Escherichia coli* ribosomal protein L7/L12 in solution: Hydrodynamic spectroscopic, and conformation prediction studies. *Biochemistry* **18:** 2019–2027.

MacConnell, W.P. and N.O. Kaplan. 1980. The role of ethanol extractable proteins from the 80S rat liver ribosome. *Biochem. Biophys. Res. Commun.* **92:** 46–52.

———. 1982. The activity of the acidic phosphoproteins from the 80S rat liver ribosome. *J. Biol. Chem.* **257:** 5359–5366.

Mason, S.W., J. Li, and J. Greenblatt. 1992. Direct interaction between two *Escherichia coli* transcription antitermination factors, NusB and ribosomal protein S10. *J. Mol. Biol.* **223:** 55–66.

Matheson, A.T., W. Möller, R. Amons, and M. Yaguchi. 1980. Comparative studies of the structure of ribosomal proteins with emphasis on the alanine-rich, acidic ribosomal, ‘A’ protein. In *Ribosomes—Structure, function and genetics* (ed. G. Chambliss et al.), pp. 297–332. University Park Press, Baltimore.

Mende, L., B. Timm, and A.R. Subramanian. 1978. Primary structures of two homologous ribosome-associated DNA-binding proteins of *Escherichia coli*. *FEBS Lett.* **96:** 395–398.

Miller, J., A.D. McLachlan, and A. Klug. 1985. Repetitive zinc-binding domains in the protein transcription factor IIIA from *Xenopus* oocytes. *EMBO J.* **4:** 1609–1614.

Möller, W. and J.A. Maassen. 1986. On the structure, function and dynamics of L7/L12 from *Escherichia coli* ribosomes. In *Structure, function, and genetics of ribosomes* (ed. B. Hardesty and G. Kramer), pp. 309–325. Springer-Verlag, New York.

Nazar, R.N., T.O. Sitz, and H. Busch. 1975. Structural analyses of mammalian ribosomal ribonucleic acid and its precursors. Nucleotide sequence of 5.8S ribonucleic acid. *J. Biol. Chem.* **50:** 8591–8597.

Nazar, R.N., M. Yaguchi, G.E. Willick, C.F. Rollin, and C. Roy. 1979. The 5S RNA binding protein from yeast (*Saccharomyces cerevisiae*) ribosomes. Evolution of the eukaryotic 5S RNA binding protein. *Eur. J. Biochem.* **102:** 573–582.

Nierhaus, K.H. 1991. The assembly of prokaryotic ribosomes. *Biochimie* **73:** 739–755.

Olvera, J. and I.G. Wool. 1993. The carboxyl extension of a ubiquitin-like protein is rat ribosomal protein S30. *J. Biol. Chem.* **268:** 17967–17974.

O’Shea, E.K., R. Rutkowski, and P.S. Kim. 1989. Evidence that the leucine zipper is a coiled coil. *Science* **243:** 538–542.

Perry, R.P. and O. Meyuhas. 1990. Translational control of ribosomal protein production in mammalian cells. *Enzyme* **44:** 83–92.

Pettersson, I., S.J.S. Hardy, and A. Liljas. 1976. The ribosomal protein L8 is a complex of L7/L12 and L10. *FEBS Lett.* **64:** 135–138.

Piatigorsky, J. and G. Wistow. 1991. The recruitment of crystallins: New functions precede gene duplication. *Science.* **252:** 1078–1079.

Pierandrei-Amaldi, P., N. Campioni, E. Beccari, I. Bozzoni, and F. Amaldi. 1982. Expression of ribosomal-protein genes in *Xenopus laevis* development. *Cell* **30:** 163–171.

Proudfoot, N. 1991. Poly(A) signals. *Cell* **64:** 671–674.

Raué, H.A., J. Klootwijk, and W. Musters. 1988. Evolutionary conservation of structure and function of high molecular weight ribosomal RNA. *Prog. Biophys. Mol. Biol.* **51:** 77–129.

Rajchel, A., Y.L. Chan, and I.G. Wool. 1988. The primary structure of rat ribosomal protein L32. *Nucleic Acids Res.* **16:** 2347.

Redman, K.L. and M. Rechsteiner. 1989. Identification of the long ubiquitin extension as ribosomal protein S27a. *Nature* **338:** 438–440.

Rice, P.A. and T.A Steitz. 1989. Ribosomal protein L7/L12 has a helix-turn-helix motif similar to that found in DNA-binding regulatory proteins. *Nucleic Acid Res.* **17:** 3757–3762.

Rich, B.E. and J.A. Steitz. 1987. Human acidic ribosomal phosphoproteins P0, P1, and P2: Analysis of cDNA clones, *in vitro* synthesis, and assembly. *Mol. Cell. Biol.* **7:** 4065–4074.

Schimmang, T., D. Tollervey, H. Kern, R. Frank, and E.C. Hurt. 1989. A yeast nucleolar protein related to mammalian fibrillarin is associated with small nucleolar RNA and is essential for viability. *EMBO J.* **8:** 4015–4024.

Schwabe, J.W.R. and D. Rhodes. 1991. Beyond zinc fingers: Steroid hormone receptors have a novel structural motif for DNA recognition. *Trends Biochem. Sci.* **16:** 291–296.

Stern, S., T. Powers, L.M. Changchien, and H.F. Noller. 1989. RNA-protein interactions in 30S ribosomal subunits: Folding and function of 16S rRNA. *Science* **244:** 783–790.

Stewart, M.J. and R. Denell. 1993. Mutations in the *Drosophila* gene encoding ribosomal protein S6 cause tissue overgrowth. *Mol. Cell. Biol.* **13:** 2524–2535.

Stöffler, G. 1974. Structure and function of the *Escherichia coli* ribosome: Immunochemical analysis. In *Ribosomes* (ed. M. Nomura et al.), pp. 615–667. Cold Spring Harbor Laboratory, Cold Spring Harbor, New York.

Stöffler, G., I.G. Wool, A. Lin, and K.H. Rak. 1974. The identification of the eukaryotic ribosomal proteins homologous with *Escherichia coli* proteins L7 and L12. *Proc. Natl. Acad. Sci.* **71:** 4723–4726.

Subramanian, A.R. 1975. Copies of protein L7 and L12 and heterogeneity of the large subunit of *Escherichia coli* ribosome. *J. Mol. Biol.* **95:** 1–8.

Suryanarayana, T. and A.R. Subramanian. 1978. Specific association of two homologous DNA-binding proteins to the native 30S ribosomal subunits of *Escherichia coli. Biochim. Biophys. Acta* **520:** 342–357.

Takahashi, Y. and K. Ogata. 1981. Ribosomal proteins cross-linked to natural mRNA by UV irradiation of rat liver polymerase. *J. Biochem.* **90:** 1549–1552.

Tamura, S., Y. Kuwano, T. Nakayama, S. Tanaka, T. Tanaka, and K. Ogata. 1987. Molecular cloning and nucleotide sequence of cDNA specific for ribosomal protein L5. *Eur. J. Biochem.* **168:** 83–87.

Terao, K. and K. Ogata. 1979. Proteins of small subunits of rat liver ribosomes that interact with poly(U). II. Crosslinks between poly(U) and ribosomal proteins in 40S subunits induced by UV irradiation. *J. Biochem.* **86:** 605–617.

Terao, K., Y. Takahashi, and K. Ogata. 1975. Differences between the protein moieties of active subunits and EDTA-treated subunits of rat liver ribosomes with specific references to a 5S rRNA protein complex. *Biochim. Biophys. Acta* **402:** 230–237.

Terhorst, C., W. Möller, R. Laursen, and B. Wittmann-Liebold. 1973. The primary structure of an acidic protein from 50S ribosomes of *Escherichia coli* which is involved in GTP hydrolysis dependent on elongation factor G and T. *Eur. J. Biochem.* **34:** 138–152.

Thomas, G., J. Martin-Perez, M. Sigmann, and A.M. Otto. 1982. The effect of serum, EGF, $PGF_{2\alpha}$ and insulin on S6 phosphorylation and the initiation of protein and DNA synthesis. *Cell* **30:** 235–242.

Toczyski, D.P.W. and J.A. Steitz. 1991. EAP, a highly conserved cellular protein associated with Epstein-Barr virus small RNAs (EBERs). *EMBO J.* **10:** 459–466.

Toczyski, D.P., A.G. Matera, D.C. Ward, and J.A. Steitz. 1994. The Epstein-Barr virus (EBV) small RNA EBER1 binds and relocalizes ribosomal protein L22 in EBV-infected human B. lymphocytes. *Proc. Natl. Acad. Sci.* **91:** 3463–3467.

Torczynski, R., A.P. Bollon, and M. Fuke. 1983. The complete nucleotide sequence of the rat 18S ribosomal RNA gene and comparison with the respective yeast and frog genes. *Nucleic Acids Res.* **11:** 4879–4890.

Towbin, H., H. Ramjoue, H. Kuster, D. Liverani, and J. Gordon. 1982. Monoclonal antibodies against eukaryotic ribosomes: Use to characterize a ribosomal protein not previously identified and antigenically related to the acid phosphoproteins P1/P2. *J. Biol. Chem.* **257:** 12709–12715.

Traut, R.R., D.S. Tewari, A. Sommer, G.R. Gavino, H.M. Olsen, and D.G. Glitz. 1986. Protein topography of ribosomal functional domains: Effects of monoclonal antibodies to different epitopes in *Escherichia coli* protein L7/L12 on ribosome function and structure. In *Structure, function, and genetics of ribosomes* (ed. B. Hardesty and G. Kramer), pp. 286–308. Springer-Verlag, New York.

Tsurugi, K. and K. Mitsui. 1991. Bilateral hydrophobic zipper as a hypothetical structure which binds acidic ribosomal protein family together on ribosomes in yeast *Saccharomyces cerevisiae*. *Biochem. Biophys. Res. Comm.* **174:** 1318–1323.

Tsurugi, K., E. Collatz, K. Todokoro, N. Ulbrich, H.M. Lightfoot, and I.G. Wool. 1978. Isolation of eukaryotic ribosomal proteins: Purification and characterization of the 60S ribosomal subunit proteins La, Lb, Lf, P1, P2, L13′, L14, L18′, L20, and L38. *J. Biol. Chem.* **253:** 946–955.

Uchiumi, T., R.R. Traut, and R. Kominami. 1990. Monoclonal antibodies against acidic phosphoproteins P0, P1, and P2 of eukaryotic ribosomes as functional probes. *J. Biol. Chem.* **265:** 89–95.

Uchiumi, T., A.J. Wahba, and R.R. Traut. 1987. Topography and stoichiometry of acidic proteins in large ribosomal subunits from *Artemia salina* as determined by crosslinking. *Proc. Natl. Acad. Sci.* **84:** 5580–5584.

Vinson, C.R., P.B. Sigler, and S.L. McKnight. 1989. Scissors-grip model for DNA recognition by a family of leucine zipper proteins. *Science* **246:** 911–916.

Wada, K., Y. Wada, H. Doi, F. Ishibashi, T. Gojobori, and T. Ikemura. 1991. Codon usage tabulated from the GenBank genetic sequence data. *Nucleic Acids Res.* **19:** 1981–1986.

Walker, J.M., J.R.B. Hastings, and E.W. Johns. 1978. A novel continuous sequence of 41 aspartic and glutamic residues in a non-histone chromosomal protein. *Nature* **271:** 281–282.

Warner, J.R., E.A. Elion, M.D. Dabeva, and W.F. Schwindeger. 1985. The ribosomal genes of yeast and their regulation. In *Structure, function, and genetics of ribosomes* (ed. B. Hardesty and G. Kramer), pp. 719–732. Springer-Verlag, New York.

Watson, D.C. and G.H. Dixon. 1981. Amino acid sequence homologies between the high-mobility-group proteins, HMG-T from trout testis and HMG-1 and -2 from calf thymus: Is the poly-aspartic-glutamic acid polypeptide within the main chain? *Biosci. Rep.* **1:** 167–175.

Watson, K.L., K.D. Konrad, D.F. Woods, and P.J. Bryant. 1992. *Drosophila* homolog of the human S6 ribosomal protein is required for tumor suppression in the hematopoietic system. *Proc. Natl. Acad. Sci.* **89:** 11302–11306.

Weissbach, H. 1980. Soluble factors in protein synthesis. In *Ribosomes—Structure, function, and genetics* (ed. G. Chambliss et al.), pp. 477–411. University Park Press, Baltimore.

Wettenhall, R.E.H. and P. Cohen. 1982. Isolation and characterization of cyclic AMP-dependent phosphorylation sites from rat liver ribosomal protein S6. *FEBS Lett.* **140:** 263–269.

Wettenhall, R.E.H., P. Cohen, B. Caudwell, and R. Holland. 1982. Differential phosphorylation of ribosomal protein S6 in isolated rat hepatocytes after incubation with insulin and glucagon. *FEBS Lett.* **148:** 207–213.

Wilson, D.M. III, W.A. Deutsch, and M.R. Kelley. 1994. *Drosophila* ribosomal protein S3 contains an activity that cleaves DNA at apurinic/apyrimidinic sites. *J. Biol. Chem.* **269:** 25359–25364.

Wittmann-Liebold, B., S.M.L. Robinson, and M. Dzionara. 1977. Prediction for secondary structure of six proteins from the 50S subunit of the *Escherichia coli* ribosome. *FEBS Lett.* **81:** 204–213.

Wool, I.G. 1979. The structure and function of eukaryotic ribosomes. *Annu. Rev. Biochem.* **48:** 719–754.

———. 1986. Studies of the structure of eukaryotic (mammalian) ribosomes. In *Structure, function, and genetics of ribosomes* (ed. B. Hardesty and G. Kramer), pp. 391–411. Springer Verlag, New York.

———. 1993. The bifunctional nature of ribosomal proteins and speculations on their origins. In *The translational apparatus* (ed. K.H. Nierhaus et al.), pp. 727–737. Plenum Press, New York.

Wool, I.G. and G. Stöffler. 1974. Structure and function of eukaryotic ribosomes. In *Ribosomes* (ed. M. Nomora et al.), pp. 417–461. Cold Spring Harbor Laboratory, Cold Spring Harbor, New York.

Wool, I.G., Y.L. Chan, A. Glück, and K. Suzuki. 1991. The primary structure of rat ribosomal proteins P0, P1, and P2 and a proposal for a uniform nomenclature for mammalian and yeast ribosomal proteins. *Biochimie* **73:** 861–870.

Wool, I.G., Y. Endo, Y.L. Chan, and A. Glück. 1990. Structure, function, and evolution of mammalian ribosomes. In *The ribosome: Structure, function, and evolution* (ed. W.E. Hill et al.), pp. 203–214. American Society for Microbiology, Washington, D.C.

Xu, L., G.P. He, A. Li, and H-S Ro. 1994. Molecular characterization of the mouse ribosomal protein S24 multigene family: A uniquely expressed intron-containing gene with cell-specific expression of three alternatively spliced mRNAs. *Nucleic Acids Res.* **22:** 646–655.

Zinn, A.R., R.K. Alagappan, L.G. Brown, I.G. Wool, and D.C. Page. 1994. Structure and function of ribosomal protein S4 genes on the human and mouse sex chromosomes. *Mol. Cell. Biol.* **14:** 2485–2492.

Appendix

Table A1 Rat ribosomal proteins

Protein	No. of residues	m.w.	pI	GenBank Access. no.	Protein	No. of residues	m.w.	pI	GenBank Access. no.
				40S subunit					
Sa	295	32823	4.65	D25224	S15a	129	14698	10.60	X77953
S2	293	31211	10.66	X57432	S16	145	16304	10.59	X17665
S3	243	26643	10.00	X51536	S17	134	15368	10.36	K02933
S3a	263	29794	10.18	X75161	S18	152	17707	11.47	X57529
S4	263	29596	10.60	X14210	S19	144	15944	10.85	X51707
S5	204	22863	10.04	X58465	S20	119	13364	10.43	X51537
S6	249	28683	11.32	J03538	S21	83	9121	8.83	X79059
S7	194	22113	10.58	X53377	S23	142	15666	10.97	X77398
S8	207	23928	10.73	X06423	S24	133	15413	11.26	X51538
S9	193	22360	10.93	X66370	S25	125	13733	10.58	X62482
S10	165	18917	10.53	X13549	S26	115	13007	11.47	X02414
S11	157	18287	10.72	K03250	S27	83	9339	9.82	X59375
S12	132	14515	7.00	J02824	S27a	80	9397	10.25	X81839
S13	150	17080	10.96	X53378	S28	69	7836	11.19	X59277
S14	151	16248	10.53	X15040	S29	55	6541	10.47	X59051
S15	145	17040	10.81	D11388	S30	59	6643	12.65	X62671
				60S subunit					
P0	316	34178	5.86	X15096	L23	140	14856	10.95	X58200
P1	114	11490	4.07	X15097	L23a	156	17684	10.90	X65228

(*Continued on following pages.*)

P2	115	11684	4.24	X15098	L24	157	17767	11.75	X78443
L3	403	46106	10.63	X62166	L26	145	17266	11.08	X14671
L4	421	47280	11.39	X82180	L27	135	15666	10.99	X07424
L5	296	34298	10.07	M17419	L27a	147	16476	11.59	X52733
L7	260	30310	11.32	M17422	L28	136	15707	12.51	X52619
L7a	265	29863	11.07	X15013	L29	155	17183	12.28	X68283
L8	257	28007	11.51	X62145	L30	114	12652	9.98	K02932
L9	192	21879	10.38	X51706	L31	124	14331	10.99	X04809
L11	178	20239	9.97	X62146	L32	135	15730	11.81	X06483
L12	165	17834	9.88	X53504	L34	117	13498	12.16	X14401
L13	210	24094	11.99	X78327	L35	122	14412	11.53	X51705
L13a	202	23330	11.48	X68282	L35a	109	12422	11.32	X03475
L15	203	24000	12.11	X78167	L36	104	12128	12.09	X68284
L17	184	21383	10.63	X58389	L36a	105	12311	11.04	M19635
L18	187	21530	12.28	M20156	L37	96	10939	12.24	X66369
L18a	176	20718	11.14	X14181	L37a	91	10143	10.86	X14069
L19	196	26971	11.97	J02650	L38	69	8081	10.56	X57007
L21	159	18322	10.83	X15216	L39	50	6271	13.05	X82551
L22	128	14779	9.45	X78444	L40	52	6177	10.75	X82636
					L41	25	3454	13.46	X82550

References:

Sa Tohgo, A., S. Takasawa, H. Munakata, H. Yonekura, N. Hayashi, and H. Okamoto. 1994. Structural determination and characterization of a 40 kDa protein isolated from rat 40S ribosomal subunit. *FEBS Lett.* **340:** 133–138.

S2 Suzuki, K., J. Olvera, and I.G. Wool. 1991. Primary structure of rat ribosomal protein S2. A ribosomal protein with arginine-glycine tandem repeats and RGGF motifs that are associated with nucleolar localization and binding to ribonucleic acids. *J. Biol. Chem.* **266:** 20007–20010.

S3 Chan, Y.L., K.R.G. Devi, J. Olvera, and I.G. Wool. 1990. The primary structure of rat ribosomal protein S3. *Arch. Biochem. Biophys.* **283:** 546–550.

S3a Chan, Y.L., J. Olvera, V. Paz, and I.G. Wool. The primary structure of rat ribosomal proteins S3a and S3b. S3a is the V-fos transformation effector and the homologue of the yeast mitochondria import-protein *MFT1*. In preparation.

S4 Devi, K.R.G., Y.L. Chan, and I.G. Wool. 1989. The primary structure of rat ribosomal protein S4. *Biochim. Biophys. Acta* **1008:** 258–262.

S5 Kuwano, Y., J. Olvera, and I.G. Wool. 1992. The primary structure of rat ribosomal protein S5. A ribosomal protein present in the rat genome in a single copy. *J. Biol. Chem.* **267:** 25304–25308.
S6 Chan, Y.L. and I.G. Wool. 1988. The primary structure of rat ribosomal protein S6. *J. Biol. Chem.* **263:** 2891–2896.
S7 Suzuki, K., J. Olvera, and I.G. Wool. 1990. The primary structure of rat ribosomal protein S7. *FEBS Lett.* **271:** 51–53.
S8 Chan, Y.L., A. Lin, V. Paz, and I.G. Wool. 1987. The primary structure of rat ribosomal protein S8. *Nucleic Acids Res.* **15:** 9451-9459.
S9 Chan, Y.L., V. Paz, J. Olvera, and I.G. Wool. 1993. The primary structure of rat ribosomal protein S9. *Biochem. Biophys. Res. Commun.* **193:** 106–112.
S10 Glück, A., Y.L. Chan, A. Lin, and I.G. Wool. 1989. The primary structure of rat ribosomal protein S10. *Eur. J. Biochem.* **182:** 105–109.
S11 Tanaka, T., Y. Kuwano, K. Ishikawa, and K. Ogata. 1985. Nucleotide sequence of cloned cDNA specific for rat ribosomal protein S11. *J. Biol. Chem.* **260:** 6329–6333.
S12 Lin, A., Y.L. Chan, R. Jones, and I.G. Wool. 1987. The primary structure of rat ribosomal protein S12. The relationship of rat S12 to other ribosomal proteins and a correlation of the amino acid sequences of rat and yeast ribosomal proteins. *J. Biol. Chem.* **262:** 14343–14351.
S13 Suzuki, K., J. Olvera, and I.G. Wool. 1990. The primary structure of rat ribosomal protein S13. *Biochem. Biophys. Res. Commun.* **171:** 519–524.
S14 Paz, V., Y.L. Chan, A. Glück, and I.G. Wool. 1989. The primary structure of rat ribosomal protein S14. *Nucleic Acids Res.* **17:** 9484.
S15 Kitagawa, M., S. Takasawa, N. Kikuchi, T. Itoh, H. Teraoka, H. Yamamoto, and H. Okamoto. 1991. *rig* encodes ribosomal protein S15. The primary structure of mammalian ribosomal protein S15. *FEBS Lett.* **283:** 210–214.
S15a Chan, Y.L., J. Olvera, V. Paz, and I.G. Wool. 1994. The primary structure of rat ribosomal protein S15a. *Biochem. Biophys. Res. Commun.* **200:** 1498–1504.
S16 Chan, Y.L., V. Paz, J. Olvera, and I.G. Wool. 1990. The primary structure of rat ribosomal protein S16. *FEBS Lett.* **263:** 85–88.
S17 Nakanishi, O., M. Oyanagi, Y. Kuwano, T. Tanaka, T. Nakayama, H. Mitsui, Y. Nabeshima, and K. Ogata. 1985. Molecular cloning and nucleotide sequences of cDNAs specific for rat ribosomal proteins S17 and L30. *Gene* **35:** 289–296.
S18 Chan, Y.L., V. Paz, and I.G. Wool. 1991. The primary structure of rat ribosomal protein S18. *Biochem. Biophys. Res. Commun.* **178:** 1212–1218.
S19 Suzuki, K., J. Olvera, and I.G. Wool. 1990. The primary structure of rat ribosomal protein S19. *Biochimie* **72:** 299–302.
S20 Chan, Y.L. and I.G. Wool. 1990. The primary structure of rat ribosomal protein S20. *Biochim. Biophys. Acta* **1049:** 93–95.
S21 Itoh, T., E. Otaka, and K. Matsui. 1985. Primary structures of ribosomal protein YS25 from *Saccharomyces cerevisiae* and its counterparts from *Schizosaccharomyces pombe* and rat liver. *Biochemistry* **24:** 7418–7423.
S23 Kitaoka, Y., J. Olvera, and I.G. Wool. 1994. The primary structure of rat ribosomal protein S23. *Biochem. Biophys. Res. Commun.* **202:** 314–320.
S24 Chan, Y.L., V. Paz, J. Olvera, and I.G. Wool. 1990. The primary structure of rat ribosomal protein S24. *FEBS Lett.* **262:** 253–255.
S25 Chan, Y.L. and I.G. Wool. 1992. The primary structure of rat ribosomal protein S25. *Biochem. Biophys. Res. Commun.* **186:** 1688–1693.
S26 Kuwano, Y., O. Nakanishi, Y. Nabeshima, T. Tanaka, and K. Ogata. 1985. Molecular cloning and nucleotide sequence of DNA complementary to rat ribosomal protein S26 messenger RNA. *J. Biochem.* **97:** 983–992.
S27 Chan, Y.L., K. Suzuki, J. Olvera, and I.G. Wool. 1993. Zinc finger-like motifs in rat ribosomal proteins S27 and S29. *Nucleic Acids Res.* **21:** 649–655.
S27a Chan, Y.L., K. Suzuki, and I.G. Wool. 1995. Carboxyl extensions of two rat ubiquitin fusion proteins are ribosomal proteins S27a and L40. *Biochem. Biophys. Res. Commun.* **215:** 682–690.
S28 Chan, Y.L., J. Olvera, and I.G. Wool. 1991. The primary structure of rat ribosomal protein S28. *Biochem. Biophys. Res. Commun.* **179:** 314–318.
S29 Chan, Y.L., K. Suzuki, J. Olvera, and I.G. Wool. 1993. Zinc finger-like motifs in rat ribosomal proteins S27 and S29. *Nucleic Acids Res.* **21:** 649–655.
S30 Olvera, J. and I.G. Wool. 1993. The carboxyl extension of a ubiquitin-like protein is rat ribosomal protein S30. *J. Biol. Chem.* **268:** 17967–17974.

P0	Wool, I.G., Y.L. Chan, A. Glück, and K. Suzuki. 1991. The primary structure of rat ribosomal proteins P0, P1, and P2 and a proposal for a uniform nomenclature for mammalian and yeast ribosomal proteins. *Biochimie* **73:** 861–870.
P1	Wool, I.G., Y.L. Chan, A. Glück, and K. Suzuki. 1991. The primary structure of rat ribosomal proteins P0, P1, and P2 and a proposal for a uniform nomenclature for mammalian and yeast ribosomal proteins. *Biochimie* **73:** 861–870.
P2	Wool, I.G., Y.L. Chan, A. Glück, and K. Suzuki. 1991. The primary structure of rat ribosomal proteins P0, P1, and P2 and a proposal for a uniform nomenclature for mammalian and yeast ribosomal proteins. *Biochimie* **73:** 861–870.
L3	Kuwano, Y. and I.G. Wool. 1992. The primary structure of rat ribosomal protein L3. *Biochem. Biophys. Res. Commun.* **187:** 58–64.
L4	Chan, Y.L., J. Olvera, and I.G. Wool. 1995. The primary structures of rat ribosomal proteins L4 and L41. *Biochem. Biophys. Res. Commun.* **214:** 810–818.
L5	Chan, Y.L., A. Lin, J. McNally, and I.G. Wool. 1987. The primary structure of rat ribosomal protein L5. A comparison of the sequence of amino acids in the proteins that interact with 5 S rRNA. *J. Biol. Chem.* **262:** 12879–12886.
L7	Lin, A., Y.L. Chan, J. McNally, D. Peleg, O. Meyuhas, and I.G. Wool. 1987. The primary structure of rat ribosomal protein L7. The presence near the amino terminus of L7 of five tandem repeats of a sequence of 12 amino acids. *J. Biol. Chem.* **262:** 12665–12671.
L7a	Nakamura, H., T. Tanaka, and K. Ishikawa. 1989. Nucleotide sequence of cloned cDNA specific for rat ribosomal protein L7a. *Nucleic Acids Res.* **17:** 4875.
L8	Chan, Y.L. and I.G. Wool. 1992. The primary structure of rat ribosomal protein L8. *Biochem. Biophys. Res. Commun.* **185:** 539–547.
L9	Suzuki, K., J. Olvera, and I.G. Wool. 1990. The primary structure of rat ribosomal protein L9. *Gene* **93:** 297–300.
L11	Chan, Y.L., J. Olvera, V. Paz, and I.G. Wool. 1992. The primary structure of rat ribosomal protein L11. *Biochem. Biophys. Res. Commun.* **185:** 356–362.
L12	Suzuki, K., J. Olvera, and I.G. Wool. 1990. The primary structure of rat ribosomal protein L12. *Biochem. Biophys. Res. Commun.* **172:** 35–41.
L13	Olvera, J. and I.G. Wool. 1994. The primary structure of rat ribosomal protein L13. *Biochem. Biophys. Res. Commun.* **201:** 102–107.
L13a	Chan, Y.L., J. Olvera, A. Glück, and I.G. Wool. 1994. A leucine zipper-like motif and a basic region-leucine zipper-like element in rat ribosomal protein L13a. Identification of the tum$^-$ transplantation antigen P198. *J. Biol. Chem.* **269:** 5589–5594.
L15	Chan, Y.L., J. Olvera, and I.G. Wool. 1994. The primary structure of rat ribosomal protein L15. *Biochem. Biophys. Res. Commun.* **201:** 108-114.
L17	Suzuki, K. and I.G. Wool. 1991. The primary structure of rat ribosomal protein L17. *Biochem. Biophys. Res. Commun.* **178:** 322–328.
L18	Devi, K.R.G., Y.L. Chan, and I.G. Wool. 1988. The primary structure of rat ribosomal protein L18. *DNA* **7:** 157–162.
L18a	Aoyama, Y., Y.L. Chan, O. Meyuhas, and I.G. Wool. 1989. The primary structure of rat ribosomal protein L18a. *FEBS Lett.* **247:** 242–246.
L19	Chan, Y.L., A. Lin, J. McNally, D. Peleg, O. Meyuhas, and I.G. Wool. 1987. The primary structure of rat ribosomal protein L19. A determination from the sequence of nucleotides in a cDNA and from the sequence of amino acids in the protein. *J. Biol. Chem.* **262:** 1111–1115.
L21	Devi, K.R.G., Y.L. Chan, and I.G. Wool. 1989. The primary structure of rat ribosomal protein L21. *Biochem. Biophys. Res. Commun.* **162:** 364–370.
L22	Chan, Y.L. and I.G. Wool. 1995. The primary structure of rat ribosomal protein L22. *Biochim. Biophys. Acta* **1260:** 113–115.
L23	Chan, Y.L., V. Paz, and I.G. Wool. 1991. The primary structure of rat ribosomal protein L23. *Biochem. Biophys. Res. Commun.* **178:** 1153–1159.
L23a	Suzuki, K. and I.G. Wool. 1993. The primary structure of rat ribosomal protein L23a. The application of homology search to the identification of genes for mammalian and yeast ribosomal proteins and a correlation of rat and yeast ribosomal proteins. *J. Biol. Chem.* **268:** 2755–2761.
L24	Chan, Y.L., J. Olvera, and I.G. Wool. 1994. The primary structure of rat ribosomal protein L24. *Biochem. Biophys. Res. Commun.* **202:** 1176–1180.
L26	Paz, V., J. Olvera, Y.L. Chan, and I.G. Wool. 1989. The primary structure of rat ribosomal protein L26. *FEBS Lett.* **251:** 89–93.
L27	Tanaka, T., Y. Kuwano, K. Ishikawa, and K. Ogata. 1988. Nucleotide sequence of cloned cDNA specific for rat ribosomal protein L27. *Eur. J. Biochem.* **173:** 53–56.

L27a Wool, I.G., Y.L. Chan, V. Paz, and J. Olvera. 1990. The primary structure of rat ribosomal proteins: The amino acid sequences of L27a and L28 and corrections in the sequences of S4 and S12. *Biochim. Biophys. Acta* **1050:** 69–73.

L28 Wool, I.G., Y.L. Chan, V. Paz, and J. Olvera. 1990. The primary structure of rat ribosomal proteins: The amino acid sequences of L27a and L28 and corrections in the sequences of S4 and S12. *Biochim. Biophys. Acta* **1050:** 69–73.

L29 Chan, Y.L., J. Olvera, V. Paz, and I.G. Wool. 1993. The primary structure of rat ribosomal protein L29. *Biochem. Biophys. Res. Commun.* **192:** 583–589.

L30 Nakanishi, O., M. Oyanagi, Y. Kuwano, T. Tanaka, T. Nakayama, H. Mitsui, Y. Nabeshima, and K. Ogata. 1985. Molecular cloning and nucleotide sequences of cDNAs specific for rat ribosomal proteins S17 and L30. *Gene* **35:** 289–296.

L31 Tanaka, T., Y. Kuwano, T. Kuzumaki, K. Ishikawa, and K. Ogata. 1987. Nucleotide sequence of cloned cDNA specific for rat ribosomal protein L31. *Eur. J. Biochem.* **162:** 45–48.

L32 Rajchel, A., Y.L. Chan, and I.G. Wool. 1988. The primary structure of rat ribosomal protein L32. *Nucleic Acids Res.* **16:** 2347.

L34 Aoyama, Y., Y.L. Chan, and I.G. Wool. 1989. The primary structure of rat ribosomal protein L34. *FEBS Lett.* **249:** 119–122.

L35 Suzuki, K., J. Olvera, and I.G. Wool. 1990. The primary structure of rat ribosomal protein L35. *Biochem. Biophys. Res. Commun.* **167:** 1377–1382.

L35a Tanaka, T., K. Wakasugi, Y. Kuwano, K. Ishikawa, and K. Ogata. 1986. Nucleotide sequence of cloned cDNA specific for rat ribosomal protein L35a. *Eur. J. Biochem.* **154:** 523–527.

L36 Chan, Y.L., V. Paz, J. Olvera, and I.G. Wool. 1993. The primary structure of rat ribosomal protein L36. *Biochem. Biophys. Res. Commun.* **192:** 849–853.

L36a Gallagher, M.J., Y.L. Chan, A. Lin, and I.G. Wool. 1988. Primary structure of rat ribosomal protein L36a. *DNA* **7:** 269–273.

L37 Chan, Y.L., V. Paz, J. Olvera, and I.G. Wool. 1993. The primary structure of L37—A rat ribosomal protein with a zinc finger-like motif. *Biochem. Biophys. Res. Commun.* **192:** 590–596.

L37a Tanaka, T., Y. Aoyama, Y.L. Chan, and I.G. Wool. 1989. The primary structure of rat ribosomal protein L37a. *Eur. J. Biochem.* **183:** 15–18.

L38 Kuwano, Y., J. Olvera, and I.G. Wool. 1991. The primary structure of rat ribosomal protein L38. *Biochem. Biophys. Res. Commun.* **175:** 551–555.

L39 Lin, A., J. McNally, and I.G. Wool. 1984. The primary structure of rat liver ribosomal protein L39. *J. Biol. Chem.* **259:** 487–490.

L40 Chan, Y.L., K. Suzuki, and I.G. Wool. 1995. Carboxyl extensions of two rat ubiquitin fusion proteins are ribosomal proteins S27a and L40. *Biochem. Biophys. Res. Commun.* **215:** 682–690.

L41 Chan, Y.L., J. Olvera, and I.G. Wool. 1995. The primary structures of rat ribosomal proteins L4 and L41. *Biochem. Biophys. Res. Commun.* **214:** 810–818.

Table A2 Comparison of the amino acid sequences of rat and human ribosomal proteins

Rat	Human	GenBank Access. no.[a]	Identities residues	%
RSa	HSa	J03799	292/295	99
RS2	HS2	X17206	221/221	100
RS3	HS3	S42658	239/243	98
RS3a	HS3a	M84711	262/264	99
RS4	HS4X	M58458	263/263	100
	HS4Y	M58459	244/263	93
RS5	HS5	U14970	201/204	99
RS6	HS6	M77232	249/249	100
RS7	HS7	Z25749	194/194	100
RS8	HS8	X67247	208/208	100
RS9	HS9	U14971	189/194	97
RS10	HS10	U14972	163/165	99
RS11	HS11	X06617	158/158	100
RS12	HS12	X53505	130/132	98
RS13	HS13	L01124	151/151	100
RS14	HS14	M13934	150/151	99
RS15	HS15	J02984	145/145	100
RS15a	HS15a	X62691	117/129	91
RS16	HS16	M60854	146/146	100
RS17	HS17	M13932	132/135	98
RS18	HS18	X69150	152/152	100
RS19	HS19	M81757	144/145	99

Rat	Human	GenBank Access. no.[a]	Identities residues	%
RP0	HP0	M17885	308/316	97
RP1	HP1	M17886	112/114	98
RP2	HP2	M17887	112/115	97
RL3	HL3	X73460	398/403	99
RL4	HL4	L20868	387/417	93
RL5	HL5	U14966	291/296	98
RL6[b]	HL6	X69391	38/49	78
RL7	HL7	X52967	238/248	96
RL7a	HL7a	M36072	266/266	100
RL8	HL8	Z28407	257/257	100
RL9	HL9	D14531	190/192	99
RL10[b]	HL10	M73791	24/28	86
RL11	HL11	L05092	178/178	100
RL12	HL12	L06505	164/165	99
RL13	HL13	X64707	203/211	96
RL13a	HL13a	X56932	195/203	96
RL14[b]	HL14[b]	Z20414	34/37	92
RL15	HL15	L25899	197/201	98
RL17	HL17	X53777	183/184	99
RL18	HL18	L11566	182/188	97
RL18a	HL18a[b]	L05093 M77998	175/176	99

RS20	HS20	L06498	119/119	100	RL19	HL19	X63527	196/196	100
RS21	HS21	L04483	79/83	95	RL21	HL21	U14967	157/160	98
RS23	HS23	D14530	143/143	100	RL22	HL22	X59357	126/128	98
RS24	HS24	U12202	133/133	100	RL23	HL23	X52839	140/140	100
RS25	HS25	M64716	125/125	100	RL23a	HL23a[b]	U02032	154/156	99
RS26	HS26	X69654	114/115	99	RL24	HL24	M94314	157/157	100
RS27	HS27	L19739	81/84	95	RL26	HL26	L07287	143/145	99
RS27a	HS27a	X63237	79/80	99	RL27	HL27	L19527	136/136	100
RS28	HS28	L05091	69/69	100	RL27a	HL27a	U14968	143/148	97
RS29	HS29	U14973	56/56	100	RL28	HL28	U14969	131/137	96
RS30	HS30	X65921	59/59	100	RL30	HL30	X79238	115/115	100
					RL31	HL31	X15940	125/125	100
					RL32	HL32	X03342	135/135	100
					RL35	HL35	U12465	121/123	98
					RL35a	HL35a	X52966	108/110	98
					RL36a	HL36a	M15661	104/106	98
					RL37	HL37	L11567	97/97	100
					RL37a	HL37a	X66699	92/92	100
					RL38	HL38	Z26876	70/70	100
					RL39	HL39	L05096	47/51	92
					RL40	HL40	X56998	52/52	100
					RL41	HL41	Z12962	25/25	100

[a]The accession numbers are for the human ribosomal proteins.
[b]An incomplete sequence.

Table A3 Comparison of the amino acid sequences of rat and yeast ribosomal proteins

	Saccharomyces cerevisiae			Identities	
Rat	Std. nomen.	GenBank Access. no.[a]	Synonyms	residues	%
			40S Subunit		
RSa	YSa	M88277	*NAB1*	145/251	58
RS2	YS2	M38029	S4; YS5; YP9; RP12; *SUP44*	147/254	58
RS3	YS3	L31405	S3; YS3; RP13	155/240	65
RS3a	YS3a	X55360	RP10; *MFT1*	148/254	58
RS4	YS4	M64293	S7; YS6; RP5	187/261	72
RS6	YS6	Z36050	S10; YS4; RP9	149/236	63
RS8	YS8	Z26879	S14; YS9; RP19	126/200	63
RS9	YS9	D00724	S13; YS11; RP21; *SUP46*	129/193	67
RS11	YS11	L15408	S18; YS12; RP41	108/152	71
RS13	YS13[b]	P05756[c]	S27a; YS15	22/40	55
RS14	YS14	M16126	RP59; *CRY1*	108/137	79
RS15	YS15	D11386	S21; RP52	86/142	61
RS15a	YS15a	X01962	S24; YS22; RP50	100/130	77
RS16	YS16[b]	P26787[c]	RP61R	10/25	40
RS17	YS17	K02480	RP51	76/132	58
RS18	YS18	Z46659	no identification	102/144	71
RS19	YS19	X02635	S16a; YS16; RP55	72/142	51
RS20	YS20	U11582	YHL015w	66/119	55
RS21	YS21	X07811	S26; YS25	45/83	54
RS23	YS23	M96570	S28; YS14; RP37; LA26	111/143	78
RS24	YS24	Z38060	RP50	75/134	56
RS25	YS25	X03013	S31; YS23; RP45	50/108	46
RS26	YS26	U10563	*RPS26*	74/115	64
RS27a	YS27a	X05730	S37; YS24; *UBI3*	51/76	67
RS28	YS28	X00128	S33; YS27	47/67	70
RS29	YS29	D14676	S36; YS29	36/56	64
			60S Subunit		
RP0	YP0	X06959	A0; L10e	170/312	54
RP1	YP1α	M26503	A1; L12eIIA	48/106	45
	YP1β	M26507	L44′; L12eIIB	42/106	40
RP2	YP2α	M26504	A2; L44; L12eIB	52/106	49
	YP2β	M26505	L45; YL44c; YPA1; L12eIA	57/109	52
RL3	YL3	J01351	L3; YL1; RP1; *TCM1*	260/387	67
RL4	YL4	J03195	L2; YL2; RP2	203/351	58

Table A3 (*continued.*)

	Saccharomyces cerevisiae			Identities	
Rat	Std. nomen.	GenBank Access. no.[a]	Synonyms	residues	%
RL5	YL5	M94864	L1a; YL3; YL4	143/290	51
RL6[b]	YL6	D10225	L17; YL16; RP18?	7/16	44
RL7	YL7	X62627	L6; YL8; RP11	116/240	48
RL7a	YL7a	X17204	L4; YL5; RP6	143/255	56
RL8	YL8[b]	P05736[c]	L5; YL6; RP8	26/40	65
RL9	YL9	X60190	L8; YL11; RP25	95/189	50
RL10[b]	YL10	X78887	*GRC5*	21/28	75
RL11	YL11	X01029	L16; YL22; RP39	120/174	69
RL12	YL12	X51519	L15; YL23	115/165	70
RL13a	YL13a	Z38059	L21; YL15; RP22; RP23	111/198	56
RL14[b]	YL14	Z28006	YKL006w	21/38	55
RL15	YL15	D14675	L13; YL10; RP15?	146/203	72
RL17	YL17	Z28180	L20; YL17	98/183	54
RL18	YL18	X02635	RP28	102/185	55
RL19	YL19	Z36751	L23; YL14; RP33?	107/189	57
RL21	YL21	M86408	*URP1*	93/160	58
RL22	YL22[b]	P05749[c]	L1c; YL31; RP4?	16/51	31
RL23	YL23	X01694	L17a; YL32	106/137	77
RL23a	YL23a	X01014	L25; YL25; RP61L	88/142	62
RL24	YL24	K02650	L30; YL21; RP29	73/155	47
RL26	YL26[b]	P05743[c]	L33; YL33	20/40	50
RL27a	YL27a	X01573	L29; YL24; *CYH2*; RP62?	93/148	63
RL29	YL29[b]	P05747[c]	YL43	30/39	77
RL30	YL30	J03457	L32; YL38; RP73	61/105	58
RL31	YL31	X01441	L34; YL28	65/112	58
RL32	YL32	Z35853	YBL092w	81/110	74
RL35	YL35	L02328	*SOS1*	70/120	58
RL35a	YL35a	X57969	L37; YL37; RP47	55/105	52
RL36	YL36[b]	P05745[c]	L39; YL39	18/40	45
RL36a	YL36a	M62391	L41; YL27; YP44	79/104	76
RL37	YL37[d]		L43; YL35; YP55	43/84	51
RL39	YL39	X01963	L46; YL40	32/51	63
RL40	YL40	X05728	*UBI1*	42/52	81
RL41	YL41	X16065	L47; YL41	22/25	88

[a]The accession numbers are for the yeast ribosomal proteins.

[b]An incomplete sequence.

[c]A SWISS-PROT accession number.

[d]Itoh, T., K. Higo, E. Otaka, and S. Osawa. 1980. Studies on the primary structures of yeast ribosomal proteins. In *Genetics and evolution of RNA polymerase, tRNA and ribosomes* (ed. S. Osawa et al.), pp. 609–624. University of Tokyo Press, Tokyo. This sequence is not present in the databases.

Table A4 Comparison of the amino acid sequences of rat and archaebacterial ribosomal proteins

		GenBank	Identities				GenBank	Identities	
Rat	Archaebacteria[a]	Access. no.[b]	residues	%	Rat	Archaebacteria	Access. no.[a]	residues	%
RSa	HmOrfMSG	M76567	76/140	54	RP0	HmL10	X51430	82/313	26
RS2	HmS5	X58395	88/210	42	RP1	HmL12	X51430	26/108	24
RS3	HmS3[c]	J05222	74/228	32	RP2	HmL12	X51430	26/112	23
RS4	HmS3[c]	X55311	64/231	28	RL3	HmL3	J05222	119/334	36
RS5	HmS7	P32552[d]	83/199	42	RL4	HmL6[c]	J05222	80/246	33
RS9	HmS4	X87833	59/171	35	RL5	HmL18	X58395	60/187	32
RS11	HmS17	X55311	41/111	37	RL7	HmL30[c]	X58395	28/132	21
RS13	HmS15[c]	J04062	52/149	35	RL8	HmL2	J05222	121/232	52
RS14	HmS11	M87833	59/128	46	RL9	HmL6[c]	X58395	59/178	33
RS15	HmS19	J05222	63/135	47	RL11	HmL5[c]	X58395	68/174	39
RS15a	HmS8	X58395	49/129	38	RL12	HmL11	X51430	29/157	18
RS16	HmS9	M76567	56/131	43	RL13a	HmL13	M76567	49/138	36
RS18	HmS13	X87833	53/151	35	RL17	HmL22	J05222	51/154	33
RS19	HmS12	P19952[d]	45/136	33	RL18	HmL29[c]	M76567	20/115	17
RS20	HmS10	X16677	35/100	35	RL18a	HmL32	P14125[d]	11/55	20

RS23	HhS12	X57144	77/141	55	RL19	HmL24[c]	X58395	52/148	35
RS24	HmS15[c]	X70117	25/102	25	RL21	HmL31	P12734[d]	23/95	24
RS27a	HmSH	X70117	14/44	32	RL23	HmL14	X55311	60/132	45
RS29	HmS14	X58395	17/42	40	RL23a	HmL23	J05222	38/84	45
					RL24	HmL21/22	P14116[d]	18/66	27
					RL26	HmL24[c]	X55311	37/119	31
					RL27a	HmL15	X63127	27/91	30
					RL30	MvORF1	X15970	39/105	37
					RL31	HmL30[c]	X55007	25/81	31
					RL32	HmL5[c]	X58395	36/132	27
					RL35	HmL29[c]	J05222	20/70	29
					RL35a	PwORFY	M83987	25/87	29
					RL36a	HmLA	P32411[d]	21/92	23
					RL37	HmL35e	P32410[d]	25/55	45
					RL39	HmL46e	X55007	19/49	39

[a]The species are Hh, *Halobacterium halobium*; Hm, *Halobacterium marismortui*; Mv, *Methanococcus vannielii*; Pw, *Pyrococcus woesei*.
[b]The accession numbers are for the archaebacterial ribosomal proteins.
[c]Because of confusion in the nomenclature, there are pairs of *H. marismortui* proteins with different amino acid sequences related to separate rat proteins.
[d]A SWISS-PROT accession number.

Table A5 Comparison of the amino acid sequences of rat and *E. coli* ribosomal proteins

Rat	*E. coli*[a]	Identities	
		residues	%
RSa	EcS2	48/235	20
RS2	EcS5	40/154	26
RS3	EcS3	48/204	24
RS5	EcS7	32/132	24
RS9	EcS4	38/131	29
RS11	EcS17	28/83	34
RS13	EcS15[b]	14/82	17
RS14	EcS11	53/127	42
RS15	EcS19	27/91	30
RS15a	EcS8	18/119	24
RS16	EcS9	44/129	34
RS18	EcS13	37/117	32
RS20	EcS10	28/102	27
RS23	EcS12	33/112	29
RS29	EcS14	13/56	23
RP0	EcL10[b]	19/110	17
RP1	EcL12[b,c]	21/112	28
RP2	EcL12[c]	28/113	25
RL3	EcL3	42/156	27
RL5	EcL18[b]	32/117	27
RL8	EcL2	70/233	30
RL7	EcL30[b]	13/44	30
RL9	EcL6	34/161	21
RL11	EcL5	44/143	31
RL12	EcL11[b]	28/140	20
RL13a	EcL13	30/121	25
RL17	EcL22	23/110	21
RL23	EcL14	38/106	36
RL23a	EcL23	27/90	30
RL26	EcL24[b]	15/74	20
RL27a	EcL15[b]	6/24	25
RL35	EcL29	21/63	33

[a]The amino acid sequences of all of the *E. coli* ribosomal proteins are in Wittmann-Liebold, B. 1984. *Adv. Protein Chem.* **36:** 56–78.

[b]A statistically weak direct correlation was reinforced by pairwise comparisons using the yeast and archaebacterial homologs.

[c]The amino and carboxyl termini were transposed as in Lin, A. et al. 1982. *J. Biol. Chem.* **257:** 9189–9197.

25

Genetics of Mitochondrial Translation

Thomas D. Fox
Section of Genetics and Development
Cornell University
Ithaca, New York 14853-2703

CELLULAR ORGANIZATION

Cellular genetic systems of eukaryotic microorganisms, and most tissues of multicellular organisms, must produce thousands of different proteins with widely different fates. This contrasts sharply with the role of mitochondrial genetic systems, which carry out the specialized and limited task of supplying only a few proteins that are subunits of energy-transducing complexes imbedded in the inner membrane. Although this unequal division of labor has been observed in all eukaryotic species examined, considerable species to species variation exists in the spectrum of proteins coded in mitochondrial DNA (mtDNA), the arrangement of their genes in mtDNA, the structure of their messenger RNAs, and even the genetic code used to specify their amino acid sequences. This chapter cannot attempt to cover mitochondrial gene expression broadly (for this, see Attardi and Schatz 1988; Gray 1989; Costanzo and Fox 1990; Hanson and Folkerts 1992; Dieckmann and Staples 1994). Instead, the focus will be on the budding yeast *Saccharomyces cerevisiae* as a model system in which it has been possible to bring genetic tools to bear on the study of mitochondrial translation in vivo, making occasional comparative comments on other species.

With the exception of a single mitochondrially coded ribosomal protein, all of the known proteins comprising the yeast mitochondrial translation system are coded by nuclear genes, synthesized in the cytoplasm, and imported into the organelles (see Fig. 1) (for review, see Costanzo and Fox 1990; Dieckmann and Staples 1994; Pel and Grivell 1994). Thus, more than 100 nuclear genes are required to allow the translation of eight major mitochondrially coded mRNAs. Among the proteins coded by these nuclear genes are homologs of prokaryotic initiation factor 2 (IF2), elongation factors EF1A (also called EF-Tu) and EF2 (also called EF-G), and release factor 1 (RF1) (Nagata et al. 1983; Vambutas

 0-87969-458-0/96 $5 + .00

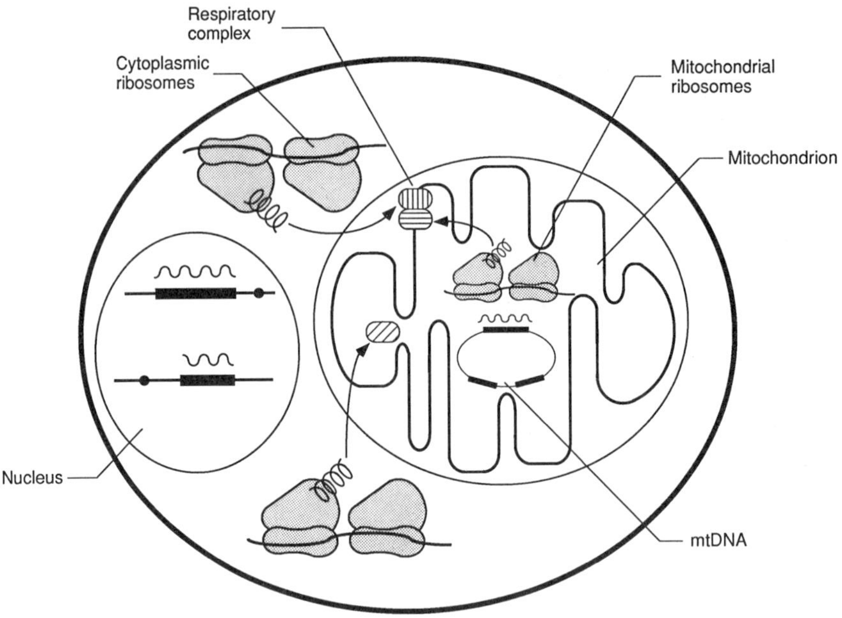

Figure 1 Translation in two compartments of eukaryotic cells. Translation of mitochondrially coded mRNAs in the matrix yields some subunits of respiratory complexes (*horizontally striped*). Cytoplasmic translation of nuclearly coded mRNAs (upper polysome) also yields some respiratory complex subunits (*vertically striped*). Cytoplasmic translation (lower polysome) yields most protein components of the mitochondrial genetic system (*diagonally striped*).

et al. 1991; Pel et al. 1992b). In addition, nuclear genes encoding 28 of the approximately 80 yeast mitochondrial ribosomal proteins have been identified and sequenced. However, only 13 of these code for proteins that have recognizable homology with known ribosomal proteins of prokaryotes (Grohmann et al. 1994), demonstrating the considerable divergence of mitochondria from their prokaryotic ancestors. Although most of the known nuclearly encoded components of the mitochondrial system function only in mitochondria, there are several examples of enzymes involved in tRNA biosynthesis and charging that are shared with the nucleocytoplasmic system (Martin and Hopper 1982; Ellis et al. 1986; Natsoulis et al. 1986; Dihanich et al. 1987; Chatton et al. 1988; Li et al. 1989; Chen et al. 1992; Chiu et al. 1992).

The known RNA components of the yeast mitochondrial translation system are all coded in mtDNA. However, a cytoplasmic $tRNA^{Lys}$ appears to be present in yeast mitochondria (Martin et al. 1979), and tRNA import apparently has an important role in mitochondrial gene expression in plants and trypanosomes (for review, see Schneider 1994).

The mitochondrial translation system is probably highly specialized for the synthesis of hydrophobic membrane proteins. Consistent with this notion, about half of the ribosomes are recovered associated with the inner membrane after disruption of yeast mitochondria (Spithill et al. 1978; Marzuki and Hibbs 1986; McMullin and Fox 1993). Available evidence indicates that at least some mitochondrial translation products are inserted into membranes cotranslationally (for review, see Poyton et al. 1992). To date, no mitochondrial complex homologous to the signal recognition particle (SRP)-related system of *Escherichia coli* (Poritz et al. 1990) has been reported. However, circumstantial evidence (discussed below) suggests that yeast mitochondrial translation may be coupled to the inner membrane by mRNA-specific translational activators, a feature apparently unique to organellar systems.

GENETIC ANALYSIS

The study of translation in mitochondria has been hampered by the fact that no mRNA-dependent in vitro system has ever been developed from mitochondria of any species. Although isolated mitochondria are capable of protein synthesis (McKee and Poyton 1984), such "in organello" systems offer the researcher few advantages over whole cells. For this reason, experimental data bearing on the mechanistic details of mitochondrial translation are largely lacking.

In contrast to the biochemical difficulties, the well-developed tools of genetic analysis in yeast offer a unique opportunity to study mitochondrial translation in vivo (Bolotin-Fukuhara and Grivell 1992). Mitochondrial translation is not required for cell viability in the presence of fermentable sugars, allowing the isolation and analysis of both mitochondrial and nuclear mutants incapable of growing on glycerol or ethanol due to defective organellar translation (Dujon 1981; Tzagoloff and Dieckmann 1990). Furthermore, both nuclear and mitochondrial genes can be manipulated in their respective genomes virtually at will (Stearns et al. 1990b; Fox et al. 1991; Rothstein 1991), allowing specific alteration of the participants in in vivo reactions. For example, by modifying promoters in mtDNA and using a plasmid-borne gene in the nucleus to supply T7 RNA polymerase targeted to mitochondria, specific mitochondrial transcripts can be overproduced (Pinkham et al. 1994). The only major factor complicating genetics in the yeast mitochondrial system is that a *complete* loss of mitochondrial translation causes instability of the wild-type mitochondrial chromosome, for unknown reasons (Myers et al. 1985).

Genetic analysis of mitochondrial translation in other organisms is generally very difficult. However, a number of interesting mutations in mitochondrial genes encoding tRNAs have been identified associated with maternally inherited human diseases that affect the muscular and nervous systems (Shoffner et al. 1990; Moraes et al. 1993; Wallace 1993). The cells of affected individuals contain both wild-type and mutant mtDNAs in varying ratios, and the severity of their symptoms generally increases with the proportion of mutant mtDNA (Wallace 1992). The quantitative and qualitative effects of these human mutations on mitochondrial translation can be analyzed in cultured cells (Chomyn et al. 1991). Interestingly, a *Drosophila* mutation termed *tko* (for technical knockout) affects a nuclear gene coding for a homolog of *E. coli* ribosomal protein S12, suggesting that it may have a defect in mitochondrial translation (Royden et al. 1987). This mutation causes a "bang sensitive" phenotype (the flies become briefly paralyzed after the bottle they are in is banged on the bench), suggesting a defect in the function of the nervous system caused by deficient mitochondrial translation, and thus a possible parallel with human mitochondrial diseases. In plants, very few characterized mitochondrial mutations have been described. However, deletion of maize mitochondrial genes encoding ribosomal proteins, by the mutation *NCS3*, leads to stunted growth and white stripes in the leaves (Hunt and Newton 1991).

Not only is yeast an excellent model for the study of mitochondrial translation per se, but its mitochondrial system also has advantages for the genetic study of translation generally. Because the rRNA genes are present in single copy in mtDNA, it has been possible to select mutations, many in highly conserved positions, leading to temperature sensitivity (Daignan-Fornier and Bolotin-Fukuhara 1988; Cui and Mason 1989), drug resistance (Dujon 1981; Li et al. 1982), and nonsense suppression (Shen and Fox 1989). Recently, the fact that mitochondrial translation is not essential for viability has allowed the demonstration of a functional role for a highly conserved ribose methylation in the peptidyl transferase center of the large rRNA, and the identification of the modifying enzyme (Sirum-Connolly and Mason 1993).

INFERENCES FROM mRNA STRUCTURE AND RNA-BINDING PROTEINS

All of the reading frames encoding the eight major yeast mitochondrial translation products begin with an AUG codon. However, the anatomy of these eight mRNAs (for review, see Grivell 1989; Dieckmann and Staples 1994) does not immediately suggest a mechanism for translation initiation. Although the 5′-untranslated regions (5′UTRs) contain sequences complementary to the 3′end of the 15S rRNA (Li et al. 1982), these sequences cannot function strictly as Shine-Dalgarno sites (Shine and Dalgarno 1974) since their positions vary between –8 and –107 relative to initiator AUG codons. Indeed, a chimeric mRNA lacking any of these sequences was translated normally (Costanzo and Fox 1988), and mutation of the putative Shine-Dalgarno sequence in the *COB* (*c*yt*o*chrome *b*) mRNA had no detectable effect on translation (Mittelmeier and Dieckmann 1995). A possible consensus surrounding the eight AUG initiation codons is suggested by the occurrence of A at positions –25, –13, –6, –3, +6, +15, and +18 in seven out of the eight mRNAs, whereas at position +12, A is present in all eight messages (Folley and Fox 1991). However, the biological significance of this observation remains to be determined.

Simple scanning of the 5′UTRs by ribosomes (Sherman and Stewart 1982; Kozak 1989) is ruled out by mRNA structure. Seven of the eight major mRNAs have long 5′UTRs, ranging approximately from 300 to 950 bases, that contain AUG triplets upstream of their initiation codons (Grivell 1989; Dieckmann and Staples 1994). Although the possible significance of these upstream AUG triplets has not been investigated carefully, mutation of the single upstream AUG triplet in the *COX3* (*c*yt*o*chrome o*x*idase subunit III) mRNA 5′UTR had no detectable effect on gene expression (Wiesenberger et al. 1995). The 5′ends of yeast mitochondrial mRNAs are said to be uncapped (Pel and Grivell 1994), and four of them are derived by processing of larger precursor transcripts (Dieckmann and Staples 1994). No proteins homologous to cap-binding proteins have been reported in this system, although, as discussed below, translation of most if not all yeast mitochondrial mRNAs depends on activator proteins that functionally interact with the 5′UTRs.

It is interesting in this context to compare the yeast and mammalian mitochondrial systems. In various mammalian species, initiation occurs not only at AUG codons, but occasionally also at the vertebrate mitochondrial methionine codon AUA, as well as isoleucine codons AUU and AUC (Fox 1987). Mammalian mitochondrial mRNAs lack significant 5′UTRs (Attardi and Schatz 1988). Thus, translational activator proteins, if they exist in mammals, must operate through other parts of the mRNAs, and a classical Shine-Dalgarno pairing mechanism can be ruled out. However, sequences downstream from the initiation codon could be involved in ribosome binding, as in the case of some leaderless mRNAs in *E. coli* (Shean and Gottesman 1992). In vitro, the small subunit of mammalian mitochondrial ribosomes will bind to synthetic mitochondrial mRNAs (Denslow et al. 1989; Liao and Spremulli 1989). The reaction is not competed by tRNA or rRNA. However, the binding appears to be nonspecific insofar as it can be competed by poly(U) (Liao and Spremulli 1990). Furthermore, although the bound ribosomal subunits partially protected the mRNA from degradation, the protected fragments did not arise from a specific region of the mRNA (Liao and Spremulli 1989). Nevertheless, addition of a mitochondrial extract (but not *E. coli* translation factors and tRNAs) stimulated protection of mRNA 5′ ends by the ribosomes (Denslow et al. 1989).

Two protein complexes that might have a role in translation have been identified by biochemical studies of mRNA-binding proteins from yeast mitochondria. One complex, consisting of at least three different polypeptides, binds specifically to the consensus sequence 5′-AAUAA-(U/C)AUUCUU-3′ which occurs at the 3′ ends of all yeast mitochondrial mRNAs (Hofmann et al. 1993; Min and Zassenhaus 1993). Whether this sequence and its associated complex have a direct role in translation is unknown. The other protein, which has recently been identified as the NAD^+-dependent isocitrate dehydrogenase, is an abundant species that binds to the 5′ UTRs of all mitochondrial mRNAs in vitro (Papadopoulou et al. 1990; Dekker et al. 1991, 1992; Elzinga et al. 1993). It has been proposed that this enzyme, which joins a growing list of enzymes with RNA-binding activities (Hentze 1994), might function as a translational repressor in mitochondria (Dekker et al. 1992).

INITIATION

mRNA-specific Translational Activation

Among the first nuclear mutants of yeast shown to have defects in mitochondrial gene expression were strains that failed to produce individual mitochondrially translated subunits of cytochrome *c* oxidase (Ebner et al. 1973; Cabral and Schatz 1978). Subsequent work in several laboratories has led to the identification of recessive nuclear mutations that cause the specific loss of five different mitochondrially coded proteins (coxI, coxII, coxIII, cytochrome *b*, and ATPase9) and to the demonstration that many of these mutations lie in genes coding mRNA-specific translational activators (for review, see Tzagoloff and Dieckmann 1990; Pel and Grivell 1994; also see below). It seems likely that similar activators for the other three mitochondrially coded proteins exist, but they have not yet been identified in mutant collections. Positive control of translation is relatively unusual (Gold 1988), although similar phenomena have been observed in translation of the bacteriophage λ *c*III mRNA (Altuvia et al. 1987) and the phage Mu *mom* mRNA (Wulczyn and Kahmann 1991). In addition, translation initiation at internal ribosome entry sites on some viral and cellular mRNAs in animal cell cytoplasms (Pelletier and Sonenberg 1988; Jackson et al. 1990; Macejak and Sarnow 1991; Witherell and Wimmer 1994) may resemble the positive control system of yeast mitochondria.

A note on genetic nomenclature: The first translational activator genes identified were designated *PET* in keeping with the nonrespiratory, and consequently small (*petite*), colony-size phenotype caused by the mutations. Mutations in other activator genes cause the same growth phenotype, but the genes have been designated with letters symbolizing their specific mitochondrial functions, as indicated below.

COX3 *mRNA*

The best studied of the mRNA-specific activator systems works on the *COX3* mRNA. Translation of this mRNA requires the mitochondrially located protein products of three unlinked nuclear genes, *PET54, PET122,* and *PET494* (Müller et al. 1984; Costanzo and Fox 1986; Costanzo et al. 1986, 1989; Fox et al. 1988; Kloeckener-Gruissem et al. 1988; McMullin and Fox 1993). Submitochondrial fractionation experiments have revealed that the PET54 protein is present both as a soluble species in the matrix and as a peripherally bound inner membrane protein, in approximately equal amounts (McMullin and Fox 1993). Although the PET122 and PET494 proteins have only been detected in strains that artificially overproduce them, their behavior during alkaline carbonate extraction indicates that they are probably integral membrane proteins (McMullin and Fox 1993).

Several lines of evidence have demonstrated that PET54, PET122, and PET494 interact with each other to form a complex. First, a missense mutation affecting PET54 was allele-specifically suppressed by a missense mutation affecting PET122 (Brown et al. 1994). The second set of experiments relied on the two-hybrid system, in which the ability of two proteins to bind to each other is revealed by their ability to bring together the separated domains of the transcriptional activator GAL4 (Fields and Song 1989). In this system, PET54 interacted with both PET122 and with PET494 (Brown et al. 1994). Although PET122 and PET494 failed to interact with each other in a pairwise combination, overproduction of unmodified PET54 protein bridged the interaction between PET122 and PET494 fusion proteins (N.G. Brown et al., unpubl.), arguing strongly for the existence of a trimeric complex. Finally, antiserum directed against PET494 coimmunoprecipitates both PET54 and PET122 from a solubilized extract of mitochondria (C.A. Butler et al., unpubl.). Thus, there appears to be a membrane-bound complex that activates translation of the highly hydrophobic coxIII protein, which is in turn inserted into the membrane. In addition to its *COX3* translational activation function, PET54 has a role in the splicing of a mitochondrial intron present in the *COX1* pre-mRNA of some, but not all, strains of yeast (Valencik and McEwen 1991). It seems likely that the membrane-bound PET54 participates in translational activation, whereas the soluble PET54 participates in RNA splicing and possibly recruitment of the soluble *COX3* mRNA to the membrane.

The *COX3* mRNA-specific translational activator complex acts on a target that maps genetically to the 613-base mRNA 5′UTR. This was demonstrated by examining the in vivo translation of chimeric mRNAs bearing the *COX3* 5′UTR and either the cytochrome *b* structural gene (Costanzo and Fox 1988) or the *COX2* structural gene (Mulero and Fox 1993b). In both cases, translation of the chimeric mRNAs depended on the *COX3*-specific activators. Analysis of deletion mutations affecting the 5′UTR, and their revertants, indicates that the translational activation target lies in the upstream half of the 5′UTR, in a region of 151 bases (between –480 and –330 relative to the start codon) (Costanzo and Fox 1988, 1993; Wiesenberger et al. 1995). In the prokaryotic cases of positive translational control, the activator proteins appear to work by antagonizing intrinsically negative elements in the mRNAs (Altuvia et al. 1987; Wulczyn and Kahmann 1991). No such negative element could be identified genetically in the *COX3* mRNA, suggesting that the mRNA is simply inert translationally in the absence of activator proteins (Wiesenberger et al. 1995).

It is highly likely that direct physical interactions occur between the translational activator proteins and the *COX3* mRNA 5′UTR in vivo. The strongest evidence for this comes from analysis of nuclear mutations that suppress mitochondrial deletions within the mRNA 5′UTR. In one case, a 5′UTR mutation that causes cold-sensitive translation was suppressed by a glutamine to leucine substitution at position 195 of the PET122 protein (Costanzo and Fox 1993). In another case, a nonconditional 5′UTR mutation was also suppressed by mutations tightly linked to *PET122* (Wiesenberger et al. 1995). These interactions were highly allele-specific, as evidenced by the fact that the suppressor of the cold-sensitive 5′UTR mutation failed to suppress the nonconditional 5′UTR mutation, and vice versa. In addition, both of these 5′UTR mutations were weakly suppressed by overproduction of PET494, suppressed more strongly by co-overproduction of PET494 plus PET122, and, in the case of the nonconditional mutation, suppressed most strongly by overproduction of all three activator proteins (Costanzo and Fox 1993; Wiesenberger et al. 1995). These results suggest that a protein-RNA interaction weakened by the 5′UTR mutations can be partially restored by increasing activator concentration in vivo. Finally, the defective phenotype of the cold-sensitive 5′UTR mutation was strongly enhanced by a *pet54* missense mutation (phenylalanine to alanine at position 244) that had little effect on respiratory growth in an otherwise wild-type strain. This synthetic defective phenotype (Huffaker et al. 1987; Stearns et al. 1990a) suggests a functional interaction between PET54 and the 5′UTR (Brown et al. 1994). (The phenylalanine at PET54 position 244 lies in a short region of sequence similarity to the phage T4 translational repressor protein regA, and appears to correspond to regA Phe-106, which can be cross-linked to nucleic acid [Webster et al. 1992].) These genetic data are supported by the recent observation that all three activator proteins can be cross-linked to RNA in vitro by UV radiation, albeit nonspecifically (C.A. Butler et al., unpubl.).

The *COX3* mRNA-specific translational activator subunit PET122 also interacts functionally with the mitochondrial ribosome. Carboxy-terminal truncations of the PET122 protein, caused either by a short deletion or by a nonsense mutation, lead to loss of function (Haffter et al. 1990). However, these mutations can be suppressed by unlinked mutations in nuclear genes encoding three proteins of the mitochondrial ribosomal small subunit that are required generally for mitochondrial translation: PET123 (McMullin et al. 1990), MRP1 (Myers et al. 1987; Dang et al. 1990; Haffter et al. 1991), and MRP17 (Haffter and Fox 1992). These suppressors do not act on other *pet122* mutations, indicating that they compensate for a specific defect caused by PET122 truncation. The three ribosomal proteins may interact with each other, since several double-mutant combinations exhibit synthetic defective phenotypes (Haffter et al. 1991; Haffter and Fox 1992). It seems likely that the PET122 protein contacts the ribosomal small subunit, but direct evidence for such a physical interaction has not been obtained to date. The same *pet122* truncation mutations are also weakly suppressed by inactivation of *PET127*, a gene of unknown function whose product is apparently too large to be a ribosomal protein (Haffter and Fox 1992).

COX2 *mRNA*

Translation of the *COX2* mRNA is specifically activated by the product of the *PET111* nuclear gene (Poutre and Fox 1987). The PET111 protein is targeted to mitochondria (Strick and Fox 1987) and is found tightly associated with the inner membrane in cells overproducing the protein (Strick 1988). The target of *PET111* action maps genetically to the 54-base *COX2* mRNA 5′UTR: Translation of a chimeric mRNA, comprising the *COX2* 5′UTR and the *COX3* structural gene, is dependent on *PET111* and independent of *COX3*-specific activation (Mulero and Fox 1993b). Indeed, the PET111 protein probably interacts directly with the *COX2* mRNA 5′UTR, based on suppression studies similar to those described above. Deletion of the G residue at position –24 of the *COX2* mRNA 5′UTR results in greatly reduced translation and a leaky nonrespiratory growth phenotype. However, this single-base deletion is strongly suppressed by a chromosomal *PET111* missense mutation (alanine to threonine at position 652) and weakly suppressed by overproduction of the wild-type PET111 protein (Mulero and Fox 1993a).

A phylogenetic comparison of known or presumed *COX2* mRNA 5′UTR sequences from the budding yeasts *S. cerevisiae, Torulopsis glabrata, Kluyveromyces lactis, K. thermotolerans,* and *Hansenula saturnus* revealed that they all contain the sequence UCUAA between 18 and 37 bases upstream of the translation initiation codon but are not otherwise highly homologous (Hardy and Clark-Walker 1990; Clark-Walker and Weiller 1994). This conserved sequence is immediately downstream from the G residue deleted in the suppressible 5′UTR mutation, suggesting that it could be part of a PET111 recognition site. Consistent with this idea, the chromosomal *PET111* suppressor worked very weakly on a 5′UTR mutation in which three new bases were inserted in place of the deleted G just upstream of UCUAA, but it failed to suppress a 29-base deletion that removed the conserved sequence.

Cytochrome b (COB) *mRNA*

Two nuclear genes, *CBS1* and *CBS2* (cytochrome *b* synthesis), specifically activate translation of the *COB* mRNA through a site that maps genetically to its 954-base 5′UTR (Rödel 1986; Rödel and Fox 1987). The *COB* 5′UTR sequences necessary for activator-dependent translation lie in the regions –954 to –898 and/or –170 to –1 (Mittelmeier and Dieckmann 1995). Both the CBS1 and CBS2 proteins are specifically located in mitochondria of wild-type cells and are membrane-bound: In detergent and salt solubilization studies, the CBS1 protein behaved like an integral membrane protein, whereas CBS2 behaved like a peripheral membrane protein (Michaelis and Rödel 1990; Michaelis et al. 1991). An antiserum directed against CBS2 reacted with a protein present in purified mitochondrial ribosomal small subunits, suggesting that CBS2 could be a ribosomal protein (Michaelis et al. 1991). Although no evidence for functional or physical interactions among CBS1, CBS2, and the *COB* mRNA 5′UTRs has been reported, the parallels with *COX3* translational activation are strong. Mutations in a third gene, *CBP6* (*c*ytochrome *b* *p*rocessing), apparently also block translation of the *COB* mRNA (Dieckmann and Tzagoloff 1985). However, *CBP6* must either work differently from *CBS1* and *CBS2* or have additional functions, since the loss of *CBP6* activity cannot be bypassed by attaching the 5′UTR of the mitochondrial *ATP9* (*ATP*ase subunit *9*) mRNA to the *COB*-coding sequence (Dieckmann and Staples 1994).

COX1 *and* ATP9 *mRNAs*

Translation of the *COX1* mRNA is blocked by mutations in two genes, *MSS51* (*m*itochondrial *s*plicing *s*ystem, a functional designation based on an indirect effect) (Decoster et al. 1990) and *PET309* (Manthey and McEwen 1995). Neither of these gene products has been located within the cell, and their mechanism of action has not been explored in detail. *PET309* appears to work through the *COX1* mRNA 5′UTR, since a *pet309* null mutation is bypassed by mtDNA rearrangements that place a new 5′UTR on the *COX1*-coding sequence (Manthey and McEwen 1995). No such bypass suppressors of the *mss51* mutation have been isolated (Decoster et al. 1990).

Translation of the *ATP9* mRNA is apparently blocked by mutations in two genes, *ATP13* (Ackerman et al. 1991) and *AEP1* (*a*tp *e*xpression *p*rotein) (Payne et al. 1991, 1993). The target of these likely translational activators has not been established. However, the *ATP9* mRNA 5′UTR is capable of activating translation of downstream structural genes in chimeric mRNAs (Rödel et al. 1985; Costanzo and Fox 1986), and an *ATP9* 5′UTR mutation causes temperature-sensitive translation (Ooi et al. 1987).

Do Activators Tether Translated mRNAs to the Inner Membrane?

A model emerges from the studies of *COX3* translation in which the inner-membrane-bound translational activator complex recognizes both the mRNA 5′UTR and the mitochondrial ribosome. Translation of the mRNA depends on this mediation of the mRNA-ribosome interaction

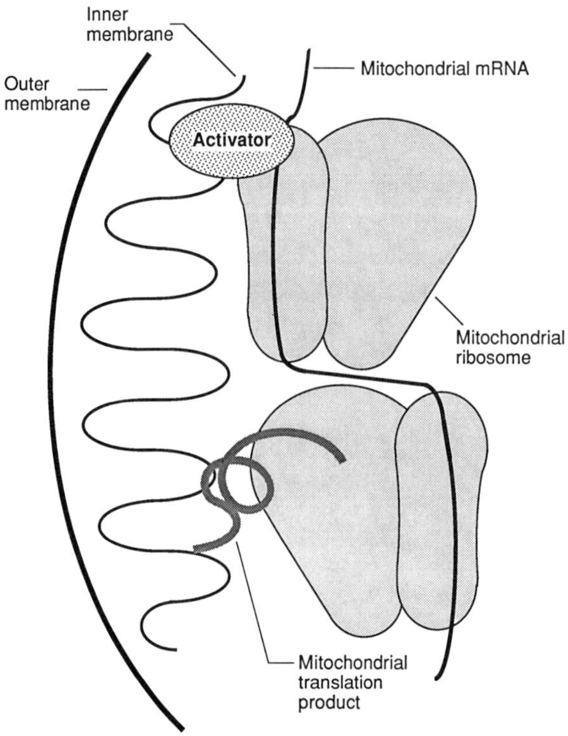

Figure 2 mRNA-specific translation activators may tether initiating mRNA-ribosome complexes to the inner membrane, the destination of most mitochondrial gene products (shown here as a gray helix being inserted cotranslationally).

and can therefore only occur at the membrane surface. This arrangement is appealing in view of the highly hydrophobic nature of coxIII and most other mitochondrial gene products. The data available for translation of other mitochondrial mRNAs, although less complete, are consistent with this picture (Fig. 2). This model for the initiation of cotranslational insertion into the mitochondrial membrane is somewhat analogous to the mechanism by which translation of some mRNAs in the cytoplasm is targeted to the endoplasmic reticulum. In that case, the SRP complex interacts with the signal sequence in the amino terminus of the nascent polypeptide and slows translation elongation until the arrested complex binds to a receptor located in the membrane (Walter and Lingappa 1986; Ng and Walter 1994).

It is puzzling that each mitochondrial mRNA appears to require its own specific activators for translation. Perhaps this complexity allows for topological distinctions between the sites of synthesis for each mitochondrially coded protein. Such organization could promote efficient assembly of the energy-transducing complexes in the inner membrane. Although this complexity could be advantageous, it cannot be absolutely required, since at the level of respiratory growth, the 5′UTRs of different mRNAs are functionally interchangeable: For example, coxIII translated from a chimeric mRNA with the 5′UTR of the *COB* mRNA finds its way into active cytochrome *c* oxidase, allowing cells to respire.

Initiation Codon Selection

Translation of yeast mitochondrial mRNAs apparently initiates with the incorporation of fMet at initiator AUG codons (the amino termini of unprocessed proteins are blocked) (Poyton et al. 1992). The importance of AUG for translation initiation has been established in vivo by site-directed mutagenesis of both the *COX2* (Mulero and Fox 1994) and *COX3* (Folley and Fox 1991) initiation codons to AUA (an alternative methionine codon in yeast mitochondria [Hudspeth et al. 1982]). In both cases, these mutations had no effect on mRNA levels but reduced translation by about sevenfold and produced leaky nonrespiratory phenotypes.

Interestingly, the residual translation in both mutants appeared to initiate at the altered initiation codon, not at the next downstream AUG. This conclusion, based on SDS-gel mobility of the mutant products, was especially robust in the case of *COX2*, since amino-terminal processing of the coxII precursor protein could be blocked genetically, producing accurate size standards for both possible initiation events (Mulero and Fox 1994). This conclusion is important because it indicates that sequence or structural information in addition to the AUG codon must be used to select the initiation site. Thus, when the AUG is mutated to AUA, translation initiation is severely impaired, but there is sufficient information remaining to allow ribosomes to select the proper initiation site at low frequency. The mutant AUA codons themselves almost certainly cannot specify initiation sites, since there are three AUA triplets in the 54-base 5′UTR of the *COX2* mRNA and 98 AUA triplets in the 613-base 5′UTR of the *COX3* mRNA.

What, in addition to AUG, specifies the initiation site? It could be the weak consensus, noted above, that surrounds the AUG initiation codons (Folley and Fox 1991) or some as yet unidentified sequence or structure near the initiation site. Analysis of the revertants of a deletion mutation that removed bases in the 5′UTR immediately upstream of the *COX3* initiation codon (–329 to –1) suggests that the precise sequence in that region is not critical for start site selection (Costanzo and Fox 1993;

Wiesenberger et al. 1995). (This deletion leaves the putative consensus relatively intact.) However, deletion of *COB* mRNA 5′UTR bases −32 to −5, and their replacement by an *Eco*RI site, prevented translation of the cytochrome *b* protein but caused the production of a novel polypeptide that migrated slightly faster on SDS gels (Mittelmeier and Dieckmann 1995), suggesting that the deletion might have shifted initiation to a downstream AUG codon. Thus, the role of sequences immediately flanking the initiation codon remains unresolved.

An alternative possibility for choosing initiation codons in this system is that the mRNA-specific translational activator complexes, bound to sites in the 5′UTRs, might position ribosomes over the correct AUGs. At first glance, the flexibility in positioning of the 5′UTR activation sites with respect to initiation codons in various deleted and chimeric mRNAs appears to argue against this hypothesis (Rödel et al. 1985; Costanzo and Fox 1986, 1988, 1993; Wiesenberger et al. 1995). However, the distance between these sites and initiation codons measured in nucleotides tells us little about their three-dimensional spatial relationships. Indeed, these altered, but translatable, mRNAs were *selected* by virtue of their ability to function and thus could mimic the wild-type structure.

Several genetic phenomena could be interpreted as suggesting direct interactions between mRNA-specific activation sites in 5′UTRs and initiation codons. The most striking is a single base insertion in the *COX3* 5′UTR (U inserted between bases −428 and −427) that exhibits no phenotype in an mRNA with the wild-type initiation codon but abolishes translation from the mutant (AUG to AUA) initiation codon (Costanzo and Fox 1995). This insertion may subtly alter the activation site since it lies in the region of the 5′UTR encoding the activation site (Wiesenberger et al. 1995) and also enhances the defective phenotype of a *pet54* missense mutation (Costanzo and Fox 1995). The other genetic phenomena consistent with this hypothesis are that alterations in the dosage of wild-type translational activator genes (Folley and Fox 1991; Mulero and Fox 1994) and activator gene mutations that suppress other 5′UTR mutations (see above) both affect translation of *COX2* and *COX3* mRNAs containing AUA initiation codons (Costanzo and Fox 1993; Mulero and Fox 1993a). However, these results must be interpreted with caution since it is clear that translation of the AUA initiation codon in the *COX3* mRNA is sensitive to indirect effects: The mutation can be suppressed by alterations in the cytoplasmic translation system (Folley and Fox 1994) and by a mutation affecting the *COX1* mRNA-specific activator, *PET309* (D.F. Steele et al., unpubl.).

However initiation codons are selected in yeast mitochondria, it has not been possible to reconstitute this activity in a soluble in vitro system. Such a system was recently developed in which translation of poly(U) was relatively active. However, although the triplet ApUpG stimulated the synthesis of formylmethionyl-puromycin by mitochondrial ribosomes in this system, the *COX2* mRNA did not (Dekker et al. 1993).

ELONGATION

The yeast mitochondrial system is highly amenable to the study of translational elongation through the isolation and analysis of informational suppressors. As in *E. coli*, mutations affecting the ribosome can reduce accuracy, leading to suppression. In one case, a suppressor of mitochondrial UAA mutations was mapped to a nuclear gene, termed *NAM9* (*n*uclear *a*ccommodation of *m*itochondria), which encodes the yeast mitochondrial homolog of *E. coli* ribosomal protein S4, the product of the well-known *ramA* gene (Boguta et al. 1992). Interestingly, this suppressor is specific for UAA mutations and thus does not cause generalized ribosomal ambiguity. In another case, a mitochondrial UAA suppressor that failed to suppress amber, frameshift, or missense mutations (Fox and Staempfli 1982; Shen and Fox 1989) was found to be a mutation in the mitochondrial gene for the small rRNA that substituted A for the highly conserved G corresponding to G517 of the *E. coli* 16S rRNA at the base of the "530-loop" (Shen and Fox 1989). When mutations at this position were tested in *E. coli*, they caused a general ribosomal ambiguity phenotype (O'Connor et al. 1992).

Frameshift mutations at a specific run of U residues in the *COX2* gene are naturally leaky (Fox and Weiss-Brummer 1980). Interestingly, ribosomal frameshifting at this site is greatly reduced by a paromomycin resistance mutation located near the 3′ end of the small rRNA (Weiss-Brummer and Hüttenhofer 1989). A suppressor mutation in the anticodon stem of a mitochondrially coded $tRNA^{Ser}$ was thought to increase frameshifting at this site in *COX2* (Hüttenhofer et al. 1990), although it now appears that the mechanism by which this suppressor acts may have more to do with effects on protein sequence than on frameshifting per se (Sakai et al. 1991).

TERMINATION

As noted above, null mutations that completely block yeast mitochondrial translation destabilize the wild-type mtDNA, resulting in the production of cells bearing large deletions of mtDNA (*rho*⁻ mutants) (Myers et al. 1985). However, less severe mutations in the same genes sometimes produce nonrespiratory growth without completely destabilizing mtDNA (Myers et al. 1985). Analysis of a nuclear gene defined by two such mutations has led to the identification of the mitochondrial release factor, mRF1. This protein is homologous to the eubacterial proteins RF1 (specific for UAA and UAG) and RF2 (specific for UAA and UGA) (Pel et al. 1992b). Since UGA is not a termination codon in yeast mitochondria (Fox 1987), mRF1 is probably the functional homolog of RF1. This protein functions as a release factor in vivo, since overproduction of mRF1 strongly reduces efficiency of UAA suppression by ribosomal ambiguity mutations in both the small rRNA and the ribosomal protein homologous to S4, discussed above (Pel et al. 1992b, 1993). The missense mutations that defined this gene altered residues separated by a single amino acid position. Since both mutant proteins behave in vivo as though they were defective in binding to mitochondrial ribosomes, these mutations appear to define a ribosome-binding domain of the protein (Pel et al. 1993). As would be expected for a protein generally required for translation in this system, a deletion of the nuclear gene encoding mRF1 destabilized wild-type mtDNA (Pel et al. 1992b). Interestingly, however, both original missense mutations cause relatively specific defects in the translation of the *COX2* mRNA and in the splicing and/or translation of the *COX1* mRNA precursor (Pel et al. 1992a,b). The basis for these phenomena remains unknown.

TRANSLATIONAL MODULATION OF MITOCHONDRIAL GENE EXPRESSION

The levels of respiratory complexes in *S. cerevisiae* mitochondria are regulated by the availability of oxygen (Woodrow and Schatz 1979) and glucose (Polakis et al. 1965). The limited data on mitochondrial gene expression in wild-type cells (for review, see Costanzo and Fox 1990; Fox and Shen 1992) suggest that translational modulation may have a role in regulating mitochondrial genes in response to glucose repression. The case is particularly clear for *COX3*: The mRNA level is unaffected by glucose (Baldacci and Zennaro 1982; Zennaro et al. 1985), whereas the rate of coxIII synthesis rises substantially after release from glucose repression (Falcone et al. 1983). Similar comparisons for other mitochondrial genes are less clear since mRNA levels are typically three- to sixfold higher in derepressed cells than in glucose-repressed cells (Baldacci and Zennaro 1982; Zennaro et al. 1985; Mueller and Getz 1986). However, a recent study employing T7 RNA polymerase to overproduce the *COX2* mRNA in mitochondria suggests that its expression is limited posttranscriptionally: An artificial fivefold increase in mRNA levels actually decreased the level of coxII protein by approximately 50% (Pinkham et al. 1994).

Due to the lack of a reporter gene system for mitochondria, studies of gene expression within the organelle have had to rely either on in vivo labeling of cells poisoned with cycloheximide or on steady-state measurements of protein levels, neither of which is ideal. The former is clearly not a normal physiological condition and the latter may be a very poor approximation of translation rates, especially for components of multisubunit complexes. Standard reporter genes could not be used in yeast mitochondria because the genetic code alterations (Fox 1987) prevented their functional expression. To overcome this problem, a gene specifying the soluble matrix enzyme acetylornithine aminotransferase has recently been synthesized in the yeast mitochondrial genetic code (D.F. Steele and T.D. Fox, unpubl.). This enzyme, normally encoded by the nuclear gene *ARG8*, can be assayed quantitatively in crude cellular extracts. Insertion of this synthetic gene, *ARG8*^{m}, into the mitochondrial genome in place of the *COX3*-coding sequence leads to the production of assayable enzyme and complements the Arg^{-} phenotype of a nuclear *arg8* deletion. Importantly, its expression is controlled by the *COX3* mRNA-specific translational activator complex. This reporter gene for yeast mitochondria will greatly increase the scope and quantitative accuracy of studies on mitochondrial gene expression and should allow the in vivo study of membrane targeting within the organelle.

The mRNA-specific translational activators are possible candidates for modulators of mitochondrial gene expression. Indeed, expression of the nuclear genes *PET494* and *PET54* is regulated in response to glucose and oxygen approximately in parallel with cytochrome oxidase levels, although *PET54* expression is about tenfold higher than *PET494* expression (Marykwas and Fox 1989; Fox and Shen 1992). Furthermore, the *COX3* initiation codon mutation described above, which reduces expression to the point that it becomes rate-limiting for respiratory growth, makes growth sensitive to reduced gene dosage of *PET122* and *PET494* (but not *PET54*), suggesting that these activator proteins may be present at near rate-limiting levels (Folley and Fox 1991). Translation of the *COX2* mRNA with a mutant initiation codon was similarly sensitive to *PET111* gene dosage (Mulero and Fox 1994).

It is difficult to rationalize why translation of different mRNAs encoding subunits of a single enzyme should be regulated by distinct activators. Nevertheless, the phenomenon may be quite general. mRNA-specific translational activation, mediated through mRNA 5′UTRs, also appears to modulate gene expression in the chloroplast, the other energy-transducing organelle with a genetic system (Rochaix 1992; Sakamoto et al. 1993; Staub and Maliga 1994; Zerges and Rochaix 1994). Translational activation in vertebrate mitochondria, if it occurs, must work slightly differently since, as noted above, vertebrate mitochondrial mRNAs lack 5′UTRs. In any event, there is evidence for translational regulation of mammalian mitochondrial gene expression (Whitfield and Jefferson 1990; Polosa and Attardi 1991).

ACKNOWLEDGMENTS

I thank M.C. Costanzo for carefully reading the manuscript and J.E. McEwen for communicating results prior to publication. Work in my laboratory is supported by National Institutes of Health grant GM-29362.

REFERENCES

Ackerman, S.H., D.L. Gatti, P. Gellefors, M.G. Douglas, and A. Tzagoloff. 1991. *ATP13*, a nuclear gene of *Saccharomyces cerevisiae* essential for the expression of subunit 9 of the mitochondrial ATPase. *FEBS Lett.* **278:** 234–238.

Altuvia, S., H. Locker-Giladi, S. Koby, O. Ben-Nun, and A.G. Oppenheim. 1987. RNase III stimulates the translation of the *c*III gene of bacteriophage λ. *Proc. Natl. Acad. Sci.* **84:** 6511–6515.

Attardi, G. and G. Schatz. 1988. Biogenesis of mitochondria. *Annu. Rev. Cell Biol.* **4:** 289–333.

Baldacci, G. and E. Zennaro. 1982. Mitochondrial transcripts in glucose-repressed cells of *Saccharomyces cerevisiae*. *Eur. J. Biochem.* **127:** 411–416.

Boguta, M., A. Dmochowska, P. Borsuk, K. Wrobel, A. Gargouri, J. Lazowska, P. Slonimski, B. Szczesniak, and A. Kruszewska. 1992. *NAM9* nuclear suppressor of mitochondrial ochre mutations in *Saccharomyces cerevisiae* codes for a protein homologous to S4 ribosomal proteins from chloroplasts, bacteria, and eukaryotes. *Mol. Cell. Biol.* **12:** 402–412.

Bolotin-Fukuhara, M. and L.A. Grivell. 1992. Genetic approaches to the study of mitochondrial biogenesis in yeast. *Antonie Leeuwenhoek* **62:** 131–153.

Brown, N.G., M.C. Costanzo, and T.D. Fox. 1994. Interactions among three proteins that specifically activate translation of the mitochondrial *COX3* mRNA in *Saccharomyces cerevisiae*. *Mol. Cell. Biol.* **14:** 1045–1053.

Cabral, F. and G. Schatz. 1978. Identification of cytochrome *c* oxidase subunits in nuclear yeast mutants lacking the functional enzyme. *J. Biol. Chem.* **253:** 4396–4401.

Chatton, B., P. Walter, J.-P. Ebel, F. Lacroute, and F. Fasiolo. 1988. The yeast *VAS1* gene encodes both mitochondrial and cytoplasmic Valyl-tRNA synthetases. *J. Biol. Chem.* **263:** 52–57.

Chen, J.Y., P.B.M. Joyce, C.L. Wolfe, M.C. Steffen, and N.C. Martin. 1992. Cytoplasmic and mitochondrial transfer RNA nucleotidyltransferase activities are derived from the same gene in the yeast *Saccharomyces cerevisiae*. *J. Biol. Chem.* **267:** 14879–14883.

Chiu, M.I., T.L. Mason, and G.R. Fink. 1992. *HTS1* encodes both the cytoplasmic and mitochondrial histidyl-tRNA synthetase of *Saccharomyces cerevisiae*: Mutations alter the specificity of compartmentation. *Genetics* **132:** 987–1001.

Chomyn, A., G. Meola, N. Bresolin, S.T. Lai, G. Scarlato, and G. Attardi. 1991. In vitro genetic transfer of protein synthesis and respiration defects to mitochondrial DNA-less cells with myopathy-patient mitochondria. *Mol. Cell. Biol.* **11:** 2236–2244.

Clark-Walker, G.D. and G.F. Weiller. 1994. The structure of the small mitochondrial DNA of *Kluyveromyces thermotolerans* is likely to reflect the ancestral gene order in fungi. *J. Mol. Evol.* **38:** 593–601.

Costanzo, M.C. and T.D. Fox. 1986. Product of *Saccharomyces cerevisiae* nuclear gene *PET494* activates translation of a specific mitochondrial mRNA. *Mol. Cell. Biol.* **6:** 3694–3703.

———. 1988. Specific translational activation by nuclear gene products occurs in the 5′ untranslated leader of a yeast mitochondrial mRNA. *Proc. Natl. Acad. Sci.* **85:** 2677–2681.

———. 1990. Control of mitochondrial gene expression in *Saccharomyces cerevisiae*. *Annu. Rev. Genet.* **24:** 91–113.

———. 1993. Suppression of a defect in the 5′-untranslated leader of the mitochondrial *COX3* mRNA by a mutation affecting an mRNA-specific translational activator protein. *Mol. Cell. Biol.* **13:** 4806–4813.

———. 1995. A point mutation in the 5′-untranslated leader that affects translation of the mitochondrial *COX3* mRNA. *Curr. Genet.* **28:** 60–66.

Costanzo, M.C., E.C. Seaver, and T.D. Fox. 1986. At least two nuclear gene products are specifically required for translation of a single yeast mitochondrial mRNA. *EMBO J.* **5:** 3637–3641.

———. 1989. The *PET54* gene of *Saccharomyces cerevisiae:* Characterization of a nuclear gene encoding a mitochondrial translational activator and subcellular localization of its product. *Genetics* **122:** 297–305.

Cui, Z. and T.L. Mason. 1989. A single nucleotide substitution at the *rib2* locus of the yeast mitochondrial gene for 21S rRNA confers resistance to erythromycin and cold-sensitive ribosome assembly. *Curr. Genet.* **16:** 273–279.

Daignan-Fornier, B. and M. Bolotin-Fukuhara. 1988. Mutational study of the rRNA in yeast mitochondria: Functional importance of T_{1696} in the large rRNA gene. *Nucleic Acids Res.* **16:** 9299–9306.

Dang, H., G. Franklin, K. Darlak, A. Spatola, and S.R. Ellis. 1990. Discoordinate expression of the yeast mitochondrial ribosomal protein MRP1. *J. Biol. Chem.* **265:** 7449–7454.

Decoster, E., M. Simon, D. Hatat, and G. Faye. 1990. The *MSS51* gene product is required for the translation of the *COX1* mRNA in yeast mitochondria. *Mol. Gen. Genet.* **224:** 111–118.

Dekker, P.J.T., B. Papadopoulou, and L.A. Grivell. 1991. Properties of an abundant RNA-binding protein in yeast mitochondria. *Biochimie* **73:** 1487–1492.

———. 1993. In-vitro translation of mitochondrial mRNAs by yeast mitochondrial ribosomes is hampered by the lack of start-codon recognition. *Curr. Genet.* **23:** 22–27.

Dekker, P.J.T., J. Stuurman, K. van Oosterum, and L.A. Grivell. 1992. Determinants for binding of a 40 kDa protein to the leaders of yeast mitochondrial mRNAs. *Nucleic Acids Res.* **20:** 2647–2655.

Denslow, N.D., G.S. Michaels, J. Montoya, G. Attardi, and T.W. O'Brien. 1989. Mechanism of mRNA binding to bovine mitochondrial ribosomes. *J. Biol. Chem.* **264:** 8328–8338.

Dieckmann, C.L. and R.R. Staples. 1994. Regulation of mitochondrial gene expression in *Saccharomyces cerevisiae. Int. Rev. Cytol.* **152:** 145–181.

Dieckmann, C.L. and A. Tzagoloff. 1985. Assembly of the mitochondrial membrane system: *CBP6,* a yeast nuclear gene necessary for the synthesis of cytochrome *b. J. Biol. Chem.* **260:** 1513–1520.

Dihanich, M.E., D. Najarian, R. Clark, E.C. Gillman, N.C. Martin, and A.K. Hopper. 1987. Isolation and characterization of *MOD5,* a gene required for isopentenylation of cytoplasmic and mitochondrial tRNAs of *Saccharomyces cerevisiae. Mol. Cell. Biol.* **7:** 177–184.

Dujon, B. 1981. Mitochondrial genetics and functions. In *The molecular biology of the yeast* Saccharomyces: *Life cycle and inheritance* (ed. J.N. Strathern et al.), p. 605–635. Cold Spring Harbor Laboratory, Cold Spring Harbor, New York.

Ebner, E., T.L. Mason, and G. Schatz. 1973. Mitochondrial assembly in respiration-deficient mutants of *Saccharomyces cerevisiae.* II. Effect of nuclear and extrachromosomal mutations on the formation of cytochrome *c* oxidase. *J. Biol. Chem.* **248:** 5369–5378.

Ellis, S.R., M.J. Morales, J.-M. Li, A.K. Hopper, and N.C. Martin. 1986. Isolation and characterization of the *TRM1* locus, a gene essential for the $m^{2}_{2}G$ modification of both mitochondrial and cytoplasmic tRNA in *Saccharomyces cerevisiae. J. Biol. Chem.* **261:** 9703–9709.

Elzinga, S.D.J., A.L. Bednarz, K. van Oosterum, P.J.T. Dekker, and L.A. Grivell. 1993. Yeast mitochondrial NAD^+-dependent isocitrate dehydrogenase is an RNA-binding protein. *Nucleic Acids Res.* **21:** 5328–5331.

Falcone, C., M. Agostinelli, and L. Frontali. 1983. Mitochondrial translation products during release from glucose repression in *Saccharomyces cerevisiae. J. Bacteriol.* **153:** 1125–1132.

Fields, S. and O.-K. Song. 1989. A novel genetic system to detect protein-protein interactions. *Nature* **340:** 245–246.

Folley, L.S. and T.D. Fox. 1991. Site-directed mutagenesis of a *Saccharomyces cerevisiae* mitochondrial translation initiation codon. *Genetics* **129:** 659–668.

———. 1994. Reduced dosage of genes encoding ribosomal protein S18 suppresses a mitochondrial initiation codon mutation in *Saccharomyces cerevisiae. Genetics* **137:** 369–379.

Fox, T.D. 1987. Natural variation in the genetic code. *Annu. Rev. Genet.* **21:** 67–91.

Fox, T.D. and Z. Shen. 1992. Positive control of translation in organellar genetic systems. In *Protein synthesis and targeting in yeast* (ed. J.E.G. McCarthy and M.F. Tuite), p. 157–166. Springer-Verlag, Berlin.

Fox, T.D. and S. Staempfli. 1982. Suppressor of yeast mitochondrial ochre mutations that maps in or near the 15S ribosomal RNA gene of mtDNA. *Proc. Natl. Acad. Sci.* **79:** 1583–1587.

Fox, T.D. and B. Weiss-Brummer. 1980. Leaky +1 and -1 frameshift mutations at the same site in a yeast mitochondrial gene. *Nature* **288:** 60–63.

Fox, T.D., M.C. Costanzo, C.A. Strick, D.L. Marykwas, E.C. Seaver, and J.K. Rosenthal. 1988. Translational regulation of mitochondrial gene expression by nuclear genes of *Saccharomyces cerevisiae. Philos. Trans. R. Soc. Lond. B Biol. Sci.* **319:** 97–105.

Fox, T.D., L.S. Folley, J.J. Mulero, T.W. McMullin, P.E. Thorsness, L.O. Hedin, and M.C. Costanzo. 1991. Analysis and manipulation of yeast mitochondrial genes. *Methods Enzymol.* **194:** 149–165.

Gold, L. 1988. Posttranscriptional regulatory mechanisms in *Escherichia coli*. *Annu. Rev. Biochem.* **57:** 199–233.

Gray, M.W. 1989. Origin and evolution of mitochondrial DNA. *Annu. Rev. Cell Biol.* **5:** 25–50.

Grivell, L.A. 1989. Nucleo-mitochondrial interactions in yeast mitochondrial biogenesis. *Eur. J. Biochem.* **182:** 477–493.

Grohmann, L., M. Kitakawa, K. Isono, S. Goldschmidt-Reisin, and H.-R. Graack. 1994. The yeast nuclear gene *MRP-L13* codes for a protein of the large subunit of the mitochondrial ribosome. *Curr. Genet.* **26:** 8–14.

Haffter, P. and T.D. Fox. 1992. Suppression of carboxy-terminal truncations of the yeast mitochondrial mRNA-specific translational activator PET122 by mutations in two new genes, *MRP17* and *PET127*. *Mol. Gen. Genet.* **235:** 64–73.

Haffter, P., T.W. McMullin, and T.D. Fox. 1990. A genetic link between an mRNA-specific translational activator and the translation system in yeast mitochondria. *Genetics* **125:** 495–503.

———. 1991. Functional interactions among two yeast mitochondrial ribosomal proteins and an mRNA-specific translational activator. *Genetics* **127:** 319–326.

Hanson, M.R. and O. Folkerts. 1992. Structure and function of the higher plant mitochondrial genome. *Int. Rev. Cytol.* **141:** 129–172.

Hardy, C.M. and G.D. Clark-Walker. 1990. Nucleotide sequence of the cytochrome oxidase subunit 2 and *val*-tRNA genes and surrounding sequences from *Kluyveromyces lactis* K8 mitochondrial DNA. *Yeast* **6:** 403–410.

Hentze, M.W. 1994. Enzymes as RNA-binding proteins: A role for (di)nucleotide-binding domains? *Trends Biochem. Sci.* **19:** 101–103.

Hofmann, T.J., J.J. Min, and H.P. Zassenhaus. 1993. Formation of the 3′ end of yeast mitochondrial mRNAs occurs by site-specific cleavage two bases downstream of a conserved dodecamer sequence. *Yeast* **9:** 1319–1330.

Hudspeth, M.E.S., W.M. Ainley, D.S. Shumard, R.A. Butow, and L.I. Grossman. 1982. Location and structure of the *var1* gene on yeast mitochondrial DNA: Nucleotide sequence of the 40.0 allele. *Cell* **30:** 617–626.

Huffaker, T.C., M.A. Hoyt, and D. Botstein. 1987. Genetic analysis of the yeast cytoskeleton. *Annu. Rev. Genet.* **21:** 259–284.

Hunt, M.D. and K.J. Newton. 1991. The NCS3 mutation: Genetic evidence for the expression of ribosomal protein genes in *Zea mays* mitochondria. *EMBO J.* **10:** 1045–1052.

Hüttenhofer, A., B. Weiss-Brummer, G. Dirheimer, and R.P. Martin. 1990. A novel type of +1 frameshift suppressor: A base substitution in the anticodon stem of a yeast mitochondrial serine-tRNA causes frameshift suppression. *EMBO J.* **9:** 551–558.

Jackson, R.J., M.T. Howell, and A. Kaminski. 1990. The novel mechanism of initiation of picornavirus RNA translation. *Trends Biochem. Sci.* **15:** 477–483.

Kloeckener-Gruissem, B., J.E. McEwen, and R.O. Poyton. 1988. Identification of a third nuclear protein-coding gene required specifically for posttranscriptional expression of the mitochondrial *COX3* gene in *Saccharomyces cerevisiae*. *J. Bacteriol.* **170:** 1399–1402.

Kozak, M. 1989. The scanning model for translation: An update. *J. Cell Biol.* **108:**

229–241.
Li, J.-M., A.K. Hopper, and N.C. Martin. 1989. N^2,N^2-dimethylguanosine-specific tRNA methyltransferase contains both nuclear and mitochondrial targeting signals in *Saccharomyces cerevisiae*. *J. Cell Biol.* **109:** 1411–1419.
Li, M., A. Tzagoloff, K. Underbrink-Lyon, and N.C. Martin. 1982. Identification of the paromomycin-resistance mutation in the 15S rRNA gene of yeast mitochondria. *J. Biol. Chem.* **257:** 5921–5928.
Liao, H.-X. and L.L. Spremulli. 1989. Interaction of bovine mitochondrial ribosomes with messenger RNA. *J. Biol. Chem.* **264:** 7518–7522.
———. 1990. Effects of length and mRNA secondary structure on the interaction of bovine mitochondrial ribosomes with messenger RNA. *J. Biol. Chem.* **265:** 11761–11765.
Macejak, D.G. and P. Sarnow. 1991. Internal initiation of translation mediated by the 5′ leader of a cellular mRNA. *Nature* **353:** 90–94.
Manthey, G.M. and J.E. McEwen. 1995. The product of the nuclear gene *PET309* is required for translation of mature mRNA and stability or production of intron-containing RNAs derived from the mitochondrial *COX1* locus of *Saccharomyces cerevisiae*. *EMBO J.* **14:** 4031–4043.
Martin, N. and A.K. Hopper. 1982. Isopentenylation of both cytoplasmic and mitochondrial tRNA is affected by a single nuclear mutation. *J. Biol. Chem.* **257:** 10562–10565.
Martin, R.P., J.-M. Schneller, A.J.C. Stahl, and G. Dirheimer. 1979. Import of nuclear deoxyribonucleic acid coded lysine-accepting transfer ribonucleic acid (anticodon C-U-U) into yeast mitochondria. *Biochemistry* **18:** 4600–4605.
Marykwas, D.L. and T.D. Fox. 1989. Control of the *Saccharomyces cerevisiae* regulatory gene *PET494:* Transcriptional repression by glucose and translational induction by oxygen. *Mol. Cell. Biol.* **9:** 484–491.
Marzuki, S. and A.R. Hibbs. 1986. Are all mitochondrial translation products synthesized on membrane-bound ribosomes? *Biochim. Biophys. Acta* **866:** 120–124.
McKee, E.E. and R.O. Poyton. 1984. Mitochondrial gene expression in *Saccharomyces cerevisiae*: Optimal conditions for protein synthesis in isolated mitochondria. *J. Biol. Chem.* **259:** 9320–9331.
McMullin, T.W. and T.D. Fox. 1993. *COX3* mRNA-specific translational activator proteins are associated with the inner mitochondrial membrane in *Saccharomyces cerevisiae*. *J. Biol. Chem.* **268:** 11737–11741.
McMullin, T.W., P. Haffter, and T.D. Fox. 1990. A novel small subunit ribosomal protein of yeast mitochondria that interacts functionally with an mRNA-specific translational activator. *Mol. Cell. Biol.* **10:** 4590–4595.
Michaelis, U. and G. Rödel. 1990. Identification of CBS2 as a mitochondrial protein in *Saccharomyces cerevisiae*. *Mol. Gen. Genet.* **223:** 394–400.
Michaelis, U., A. Körte, and G. Rödel. 1991. Association of cytochrome *b* translational activator proteins with the mitochondrial membrane: Implications for cytochrome *b* expression in yeast. *Mol. Gen. Genet.* **230:** 177–185.
Min, J. and H.P. Zassenhaus. 1993. Identification of a protein complex that binds to a dodecamer sequence found at the 3′ ends of yeast mitochondrial mRNAs. *Mol. Cell. Biol.* **13:** 4167–4173.
Mittelmeier, T.M. and C.L. Dieckmann. 1995. In vivo analysis of sequences required for translation of cytochrome *b* transcripts in yeast mitochondria. *Mol. Cell. Biol.* **15:**

780–789.

Moraes, C.T., F. Ciacci, E. Bonilla, V. Ionasescu, E.A. Schon, and S. Dimauro. 1993. A mitochondrial tRNA anticodon swap associated with a muscle disease. *Nat. Genet.* **4:** 284–288.

Mueller, D.M. and G.S. Getz. 1986. Steady state analysis of mitochondrial RNA after growth of yeast *Saccharomyces cerevisiae* under catabolite repression and derepression. *J. Biol. Chem.* **261:** 11816–11822.

Mulero, J.J. and T.D. Fox. 1993a. Alteration of the *Saccharomyces cerevisiae COX2* 5′-untranslated leader by mitochondrial gene replacement and functional interaction with the translational activator protein PET111. *Mol. Biol. Cell* **4:** 1327–1335.

———. 1993b. *PET111* acts in the 5′-leader of the *Saccharomyces cerevisiae* mitochondrial *COX2* mRNA to promote its translation. *Genetics* **133:** 509–516.

———. 1994. Reduced but accurate translation from a mutant AUA initiation codon in the mitochondrial *COX2* mRNA of *Saccharomyces cerevisiae. Mol. Gen. Genet.* **242:** 383–390.

Müller, P.P., M.K. Reif, S. Zonghou, C. Sengstag, T.L. Mason, and T.D. Fox. 1984. A nuclear mutation that post-transcriptionally blocks accumulation of a yeast mitochondrial gene product can be suppressed by a mitochondrial gene rearrangement. *J. Mol. Biol.* **175:** 431–452.

Myers, A.M., M.D. Crivellone, and A. Tzagoloff. 1987. Assembly of the mitochondrial membrane system: *MRP1* and *MRP2,* two yeast nuclear genes coding for mitochondrial ribosomal proteins. *J. Biol. Chem.* **262:** 3388–3397.

Myers, A.M., L.K. Pape, and A. Tzagoloff. 1985. Mitochondrial protein synthesis is required for maintenance of intact mitochondrial genomes in *Saccharomyces cerevisiae. EMBO J.* **4:** 2087–2092.

Nagata, S., Y. Tsunetsugu-Yokota, A. Naito, and Y. Kaziro. 1983. Molecular cloning and sequence determination of the nuclear gene coding for mitochondrial elongation factor Tu of *S. cerevisiae. Proc. Natl. Acad. Sci.* **80:** 6192–6196.

Natsoulis, G., F. Hilger, and G.R. Fink. 1986. The *HTS1* gene encodes both the cytoplasmic and mitochondrial histidine tRNA synthetases of *S. cerevisiae. Cell* **46:** 235–243.

Ng, D.T.W. and P. Walter. 1994. Protein translocation across the endoplasmic reticulum. *Curr. Opin. Cell Biol.* **6:** 510–516.

O'Connor, M., H.U. Goeringer, and A.E. Dahlberg. 1992. A ribosomal ambiguity mutation in the 530 loop of *E. coli* 16S rRNA. *Nucleic Acids Res.* **20:** 4221–4227.

Ooi, B.G., H.B. Lukins, A.W. Linnane, and P. Nagley. 1987. Biogenesis of mitochondria: A mutation in the 5′-untranslated region of yeast mitochondrial *oli1* mRNA leading to impairment in translation of subunit 9 of the mitochondrial ATPase complex. *Nucleic Acids Res.* **15:** 1965–1977.

Papadopoulou, B., P. Dekker, J. Blom, and L.A. Grivell. 1990. A 40 kd protein binds specifically to the 5′-untranslated regions of yeast mitochondrial mRNAs. *EMBO J.* **9:** 4135–4143.

Payne, M.J., E. Schweizer, and H.B. Lukins. 1991. Properties of two nuclear *pet* mutants affecting expression of the mitochondrial *oli1* gene of *Saccharomyces cerevisiae. Curr. Genet.* **19:** 343–351.

Payne, M.J., P.M. Finnegan, P.M. Smooker, and H.B. Lukins. 1993. Characterization of a second nuclear gene, *AEP1*, required for expression of the mitochondrial *OLI1* gene in *Saccharomyces cerevisiae. Curr. Genet.* **24:** 126–135.

Pel, H.J. and L.A. Grivell. 1994. Protein synthesis in mitochondria. *Mol. Biol. Rep.* **19:**

183–194.
Pel, H.J., A. Tzagoloff, and L.A. Grivell. 1992a. The identification of 18 nuclear genes required for the expression of the yeast mitochondrial gene encoding cytochrome *c* oxidase subunit I. *Curr. Genet.* **21:** 139–146.
Pel, H.J., C. Maat, M. Rep, and L.A. Grivell. 1992b. The yeast nuclear gene *MRF1* encodes a mitochondrial peptide chain release factor and cures several mitochondrial RNA splicing defects. *Nucleic Acids Res.* **20:** 6339–6346.
Pel, H.J., M. Rep, H.J. Dubbink, and L.A. Grivell. 1993. Single point mutations in domain-II of the yeast mitochondrial release factor mRF1 affect ribosome binding. *Nucleic Acids Res.* **21:** 5308–5315.
Pelletier, J. and N. Sonenberg. 1988. Internal initiation of translation of eukaryotic mRNA directed by a sequence derived from poliovirus RNA. *Nature* **334:** 320–325.
Pinkham, J.L., A.M. Dudley, and T.L. Mason. 1994. T7 RNA polymerase-dependent expression of COXII in yeast mitochondria. *Mol. Cell. Biol.* **14:** 4643–4652.
Polakis, E.S., W. Bartley, and G.A. Meek. 1965. Changes in the activities of respiratory enzymes during the aerobic growth of yeast on different carbon sources. *Biochem. J.* **97:** 298–302.
Polosa, P.L. and G. Attardi. 1991. Distinctive pattern and translation control of mitochondrial protein synthesis in rat brain synaptic endings. *J. Biol. Chem.* **266:** 10011–10017.
Poritz, M.A., H.D. Bernstein, K. Strub, D. Zopf, H. Wilhelm, and P. Walter. 1990. An *E. coli* ribonucleoprotein containing 4.5S RNA resembles mammalian signal recognition particle. *Science* **250:** 1111–1117.
Poutre, C.G. and T.D. Fox. 1987. *PET111*, a *Saccharomyces cerevisiae* nuclear gene required for translation of the mitochondrial mRNA encoding cytochrome *c* oxidase subunit II. *Genetics* **115:** 637–647.
Poyton, R.O., D.M.J. Duhl, and G.H.D. Clarkson. 1992. Protein export from the mitochondrial matrix. *Trends Cell Biol.* **2:** 369–375.
Rochaix, J.D. 1992. Post-transcriptional steps in the expression of chloroplast genes. *Annu. Rev. Cell Biol.* **8:** 1–28.
Rödel, G. 1986. Two yeast nuclear genes, *CBS1* and *CBS2*, are required for translation of mitochondrial transcripts bearing the 5′-untranslated *COB* leader. *Curr. Genet.* **11:** 41–45.
Rödel, G. and T.D. Fox. 1987. The yeast nuclear gene *CBS1* is required for translation of mitochondrial mRNAs bearing the *cob* 5′-untranslated leader. *Mol. Gen. Genet.* **206:** 45–50.
Rödel, G., A. Körte, and F. Kaudewitz. 1985. Mitochondrial suppression of a yeast nuclear mutation which affects the translation of the mitochondrial apocytochrome *b* transcript. *Curr. Genet.* **9:** 641–648.
Rothstein, R. 1991. Targeting, disruption, replacement and allele rescue: Integrative DNA transformation in yeast. *Methods Enzymol.* **194:** 281–301.
Royden, C.S., V. Pirrotta, and L.Y. Jan. 1987. The *tko* locus, site of a behavioral mutation in *D. melanogaster*, codes for a protein homologous to prokaryotic ribosomal protein S12. *Cell* **51:** 165–173.
Sakai, H., R. Stiess, and B. Weiss-Brummer. 1991. Mitochondrial mutations restricting spontaneous translational frameshift suppression in the yeast *Saccharomyces cerevisiae*. *Mol. Gen. Genet.* **227:** 306–317.
Sakamoto, W., K.L. Kindle, and D.B. Stern. 1993. *In vivo* analysis of *Chlamydomonas*

chloroplast *petD* gene expression using stable transformation of β-glucuronidase translational fusions. *Proc. Natl. Acad. Sci.* **90:** 497–501.

Schneider, A. 1994. Import of RNA into mitochondria. *Trends Cell Biol.* **4:** 282–286.

Shoffner, J.M., M.T. Lott, A.M.S. Lezza, P. Seibel, S.W. Ballinger, and D.C. Wallace. 1990. Myoclonic epilepsy and ragged-red fiber disease (MERRF) is associated with a mitochondrial DNA $tRNA^{Lys}$ mutation. *Cell* **61:** 931–937.

Shean, C.S. and M.E. Gottesman. 1992. Translation of the prophage λ *cI* transcript. *Cell* **70:** 513–522.

Shen, Z. and T.D. Fox. 1989. Substitution of an invariant nucleotide at the base of the highly conserved "530-loop" of 15S rRNA causes suppression of mitochondrial ochre mutations. *Nucleic Acids Res.* **17:** 4535–4539.

Sherman, F. and J.W. Stewart. 1982. Mutations altering initiation of translation of yeast iso-1-cytochrome *c*; contrasts between the eukaryotic and prokaryotic initiation process. In *The molecular biology of the yeast* Saccharomyces: *Metabolism and gene expression* (ed. J.N. Strathern et al.), p. 301–333. Cold Spring Harbor Laboratory, Cold Spring Harbor, New York.

Shine, J., and L. Dalgarno. 1974. The 3′-terminal sequence of *Escherichia coli* 16S ribosomal RNA: Complementarity to nonsense triplets and ribosome binding sites. *Proc. Natl. Acad. Sci.* **71:** 1342–1346.

Sirum-Connolly, K. and T.L. Mason. 1993. Functional requirement of a site-specific ribose methylation in ribosomal RNA. *Science* **262:** 1886–1889.

Spithill, T.W., M.K. Trembath, H.B. Lukins, and A.W. Linnane. 1978. Mutations of the mitochondrial DNA of *Saccharomyces cerevisiae* which affect the interaction between mitochondrial ribosomes and the inner mitochondrial membrane. *Mol. Gen. Genet.* **164:** 155–162.

Staub, J.M. and P. Maliga. 1994. Translation of *psbA* mRNA is regulated by light via the 5′-untranslated region in tobacco plastids. *Plant J.* **6:** 547–553.

Stearns, T., M.A. Hoyt, and D. Botstein. 1990a. Yeast mutants sensitive to antimicrotubule drugs define three genes that affect microtubule function. *Genetics* **124:** 251–262.

Stearns, T., H. Ma, and D. Botstein. 1990b. Manipulating the yeast genome using plasmid vectors. *Methods Enzymol.* **185:** 280–297.

Strick, C.A. 1988. "A study of the expression and protein product of the yeast positive regulatory gene *PET111*." Ph.D. thesis, Cornell University, Ithaca, New York.

Strick, C.A. and T.D. Fox. 1987. *Saccharomyces cerevisiae* positive regulatory gene *PET111* encodes a mitochondrial protein that is translated from an mRNA with a long 5′ leader. *Mol. Cell. Biol.* **7:** 2728–2734.

Tzagoloff, A. and C.L. Dieckmann. 1990. *PET* genes of *Saccharomyces cerevisiae*. *Microbiol. Rev.* **54:** 211–225.

Valencik, M.L. and J.E. McEwen. 1991. Genetic evidence that different functional domains of the *PET54* gene product facilitate expression of the mitochondrial genes *COX1* and *COX3* in *Saccharomyces cerevisiae*. *Mol. Cell. Biol.* **11:** 2399–2405.

Vambutas, A., S.J. Ackerman, and A. Tzagoloff. 1991. Mitochondrial translational-initiation and elongation factors in *Saccharomyces cerevisiae*. *Eur. J. Biochem.* **201:** 643–652.

Wallace, D.C. 1992. Diseases of the mitochondrial DNA. *Annu. Rev. Biochem.* **61:** 1175–1212.

———. 1993. Mitochondrial diseases: Genotype versus phenotype. *Trends Genet.* **9:**

128–133.

Walter, P. and V.R. Lingappa. 1986. Mechanism of protein translocation across the endoplasmic reticulum membrane. *Annu. Rev. Cell Biol.* **2:** 499–516.

Webster, K.R., S. Keill, W. Konigsberg, K.R. Williams, and E.K. Spicer. 1992. Identification of amino acid residues at the interface of a bacteriophage T4 regA protein-nucleic acid complex. *J. Biol. Chem.* **267:** 26097–26103.

Weiss-Brummer, B. and A. Hüttenhofer. 1989. The paromomycin resistance mutation (*par*^{r}-454) in the 15S rRNA gene of the yeast *Saccharomyces cerevisiae* is involved in ribosomal frameshifting. *Mol. Gen. Genet.* **217:** 362–369.

Whitfield, C.D. and L.M. Jefferson. 1990. Elevated mitochondrial RNA in a chinese hamster mutant deficient in the mitochondrially encoded subunits of NADH dehydrogenase and cytochrome *c* oxidase. *J. Biol. Chem.* **265:** 18852–18859.

Wiesenberger, G., M.C. Costanzo, and T.D. Fox. 1995. Analysis of the *Saccharomyces cerevisiae* mitochondrial *COX3* mRNA 5′-untranslated leader: Translational activation and mRNA processing. *Mol. Cell. Biol.* **15:** 3291–3300.

Witherell, G.W. and E. Wimmer. 1994. Encephalomyocarditis virus internal ribosomal entry site RNA-protein interactions. *J. Virol.* **68:** 3183–3192.

Woodrow, G. and G. Schatz. 1979. The role of oxygen in the biosynthesis of cytochrome *c* oxidase of yeast mitochondria. *J. Biol. Chem.* **254:** 6088–6093.

Wulczyn, F.G. and R. Kahmann. 1991. Translational stimulation: RNA sequence and structure requirements for binding of com protein. *Cell* **65:** 259–269.

Zennaro, E., L. Grimaldi, G. Baldacci, and L. Frontali. 1985. Mitochondrial transcription and processing of transcripts during release from glucose repression in "resting cells" of *Saccharomyces cerevisiae*. *Eur. J. Biochem.* **147:** 191–196.

Zerges, W. and J.-D. Rochaix. 1994. The 5′ leader of a chloroplast mRNA mediates the translational requirements for two nucleus-encoded functions in *Chlamydomonas reinhardtii*. *Mol. Cell. Biol.* **14:** 5268–5277.

26

Control of Translation Initiation in Prokaryotes

Harry O. Voorma
Department of Molecular Cell Biology
Utrecht University
Hugo R. Kruytbuilding
3584 CH Utrecht, The Netherlands

The initiation phase of translation involves the binding of the initiator tRNA, formyl-methionyl-tRNA$_f$ (fMet-tRNA$_f$), and messenger RNA to the 70S ribosome, recognition of the initiator codon, and the precise phasing of the reading frame of the mRNA. The process not only results in the selection of a specific mRNA for translation, but also determines the rate at which the encoded protein is synthesized since the initiation phase is thought to be rate-limiting in most circumstances. The pathway of initiation (Fig. 1) involves the dissociation of 70S ribosomes into 30S and 50S subunits, the binding of fMet-tRNA$_f$ and mRNA to the 30S subunit, the junction of the 50S subunit with the 30S preinitiation complex, and the ejection of initiation factors.

The pool of free 30S particles is controlled by two initiation factors, IF1 and IF3, which shift the equilibrium of the 70S particle and its subunits toward dissociation. The 30S subunit, carrying IF3, IF1, and probably IF2 as well, interacts with mRNA, fMet-tRNA$_f$, and GTP through a series of intermediates giving rise to the 30S initiation complex. IF2 is implicated in recognizing the acylated aminoacyl end of fMet-tRNA$_f$, thereby assuring that fMet-tRNA$_f$, not some other aminoacyl-tRNA, is bound to the 30S subunit. IF3 recognizes the anticodon stem of fMet-tRNA$_f$ and monitors the anticodon interaction with the initiation codon. The resulting complex places the fMet-tRNA$_f$ in the P-site of the 30S subunit (de Smit and van Duin 1990a; McCarthy and Brimacombe 1994). The junction with the 50S ribosomal subunit then takes place, triggering GTP hydrolysis and release of the initiation factors and GDP. The 70S initiation complex is then competent to enter the elongation cycle of protein synthesis.

During the cyclic elongation phase, elongation factor EF1A (formerly EF-Tu) forms a ternary complex with an aminoacyl-tRNA and GTP, the ternary complex binds at the ribosomal A-site, GTP hydrolysis occurs,

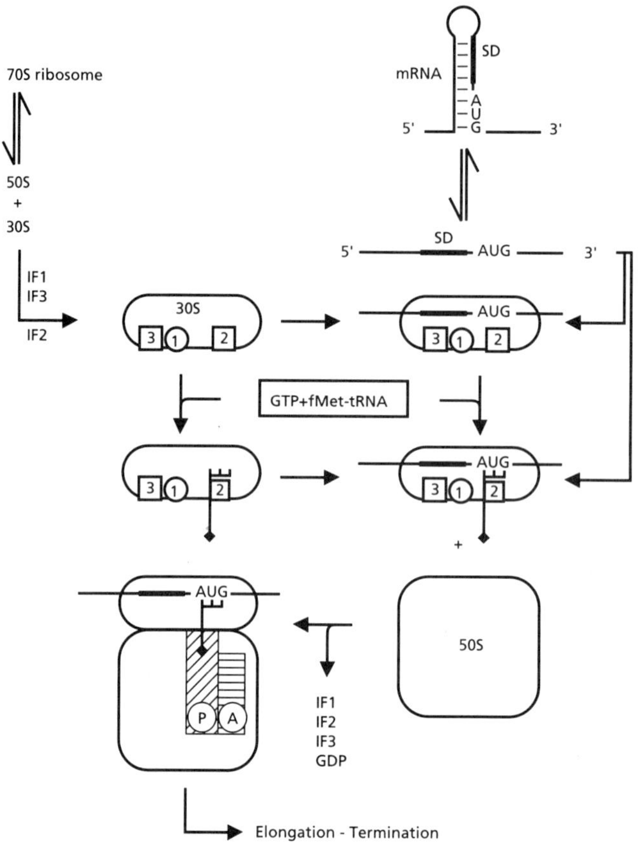

Figure 1 General scheme for translation initiation in prokaryotes. The mRNA is shown in a folded and unfolded configuration. The interaction with 30S ribosomal subunits occurs with the unfolded mRNA.

and the binary complex (EF1A•GDP) is ejected. EF1A is recycled by means of a GDP-GTP exchange reaction catalyzed by EF1B (formerly EF-Ts), thereby allowing EF1A to form another ternary complex. Following ejection of EF1A•GDP, peptide bond formation occurs in a reaction catalyzed by the 50S subunit. Then, translocation of the mRNA and the newly formed peptidyl-tRNA from the A-site to the P-site occurs in a reaction involving EF2 (formerly EF-G) and the hydrolysis of GTP. The resulting ribosomal complex contains a new codon in the A-site ready for a subsequent round of aminoacyl-tRNA binding. When the ribosome reaches any one of the termination codons UAA, UAG, or UGA, termination occurs. Release factor RF1 recognizes UAA and UAG, whereas RF2 recognizes UAA and UGA. Release factor activities are enhanced by a third termination factor, RF3, that binds and hydrolyzes GTP, and the completed polypeptide chain is hydrolyzed from its tRNA and is released from the ribosome. A ribosome release factor, RF4 (formerly RRF), then promotes the dissociation of mRNA and tRNAs from the 70S ribosome.

How are the translational initiation regions of an mRNA recognized by the 30S ribosomal subunit? The recognition is largely governed by two RNA-RNA interactions and by the extent of secondary structure in the mRNA (for review, see Gualerzi and Pon 1990). One of the RNA-RNA interactions is the codon-anticodon interaction between fMet-$tRNA_f$ and the (usually) AUG initiation codon. The second involves base pairing between the Shine-Dalgarno sequence, a purine-rich sequence in the mRNA located 5–13 nucleotides upstream of the initiation codon, and the anti-Shine-Dalgarno sequence, at the 3′end of the 16S rRNA. These two RNA-RNA interactions stabilize the 30S preinitiation complex in precisely the proper reading frame. The secondary structure content of the ribosome docking site on the mRNA also is of crucial importance; binding of the 30S ribosome requires that the mRNA be unfolded throughout the approximately 35-nucleotide binding region. There are many regions within mRNAs that contain a potential Shine-Dalgarno sequence and properly placed AUG but do not serve as initiation sites because they are masked by secondary structure. In addition, RNA-protein interactions, especially those involving ribosomal protein S1, contribute to mRNA binding as well (Boni et al. 1991).

This chapter focuses on the Shine-Dalgarno interaction, the AUG codon, the spacing between these two elements, the role of S1, the role of initiation factors, and the role of mRNA secondary structure in translational control. A number of examples of translational control in prokaryotes are then examined. Finally, a comparison is drawn between prokaryotic and eukaryotic initiation, because it seems unexpected that such a well-tuned recognition mechanism as that operative in prokaryotes should be essentially ignored in the eukaryotic initiation process.

THE SHINE-DALGARNO INTERACTION

Shine and Dalgarno (1974) were the first to recognize that ribosome-binding sites of bacteriophage mRNAs contain all or a substantial part of the sequence 5′-GGAGGU-3′. On the basis of their sequence determination of the 3′terminus of 16S rRNA, the unique sequence 5′-ACCUCC-3′ was identified adjacent to the terminal UUA_{OH}. They proposed that the purine-rich sequence in the mRNA, subsequently called

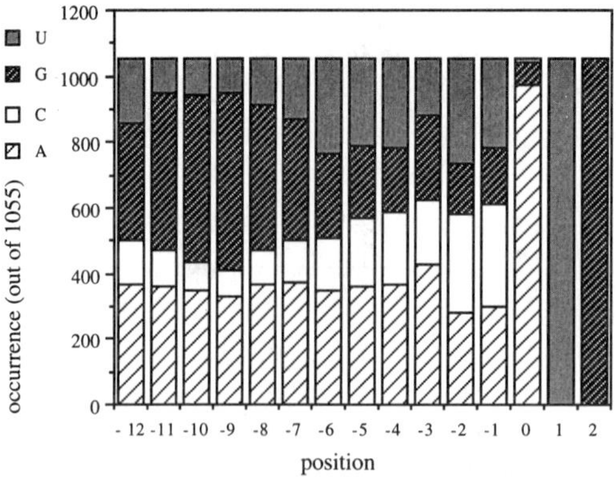

Figure 2 Occurrence of U, G, C, and A in ribosome-binding sites. (Adapted from Barrick et al. [1994] and Rudd and Schneider [1992].)

the Shine-Dalgarno (SD) region, interacts with the complementary sequence in the 16S rRNA, now called the anti-Shine-Dalgarno (ASD) sequence, thereby promoting the binding of mRNA to the 30S ribosomal subunit. It was found that ribosome-binding sites, determined by initiation complex protection of a fragment of mRNA from nuclease attack (Steitz 1975), contained such SD sequences. The hypothesis was confirmed by the isolation from initiation complexes of a nuclease-resistant dimer between the ribosome-binding site in the mRNA and the ASD region at the 3′terminus of 16S rRNA (Steitz and Jakes 1975). A compilation of 1055 ribosome-binding sites from *Escherichia coli* genes (Rudd and Schneider 1992) showed a striking frequency of purines at positions –7 to –12 (Fig. 2) (Barrick et al. 1994). The SD region upstream from the initiation codon has the consensus sequence 5′-UA**AGGAG**GUGA-3′, in which the bases in boldface are most frequently encountered. Specific mRNAs vary considerably in the number of nucleotides that complement the ASD sequence, ranging from as few as two to nine or more.

Insight into the fundamental properties of the interaction between 16S rRNA and mRNA resulted from measurements of the formation of complexes in vitro with mRNA, purified 30S ribosomal subunits, and uncharged tRNA (or simply the anticodon stem and loop) by a process called "toeprinting" (Hartz et al. 1988). The toeprint assay involves the annealing of a cDNA primer downstream from the putative initiation region of an mRNA, followed by extension of the primer by reverse transcriptase. Extension is inhibited when the reverse transcriptase meets a stably bound 30S subunit, thereby generating a specific length of extended DNA which can be sized on a sequencing gel. A toeprint signal was obtained 15 nucleotides downstream from the first letter of the codon recognized by the tRNA in the complex. The surprising result was that any tRNA can serve to promote complex formation, provided the cognate codon is within the correct range of the SD sequence. The strength of the toeprints was proportional to the translational efficiency in vivo of the mRNAs being tested, indicating that physicochemical parameters dominate complex formation. Thus, the information required for 30S binding to mRNA is simply the SD-ASD interaction and a codon-anticodon interaction placed a proper distance downstream. Initiation factors are not required to form such complexes. However, when IF2 and IF3 are added, stable complexes form only with fMet-tRNA$_f$ and other aminoacyl-tRNAs are excluded (Hartz et al. 1989). The presence of initiation factors therefore serves to assure that fMet-tRNA$_f$ and not some other aminoacyl-tRNA binds to the 30S subunit, and thus to assure that translation begins with AUG (or a closely related codon). A recent report that toeprints can be obtained even in the absence of tRNA suggests that the SD interaction may be dominant in the binding of mRNAs to 30S subunits (Ringquist et al. 1993).

Recent studies of ribosome structure support the view that the ASD region of 16S rRNA lies close to bound mRNA. Cross-linking studies employed mRNAs in which thioU residues were placed in the spacer region between the SD sequence and the initiation codon (Rinke-Appel et al. 1991; Dontsova et al. 1992). Three cross-links at nucleotides 665, 1360, and 1530 were detected; these are in regions of RNA secondary structure universally conserved in all ribosomes, despite the absence of the ASD sequence in the eukaryotic ribosome. The location of the mRNA on the ribosomal surface is proposed to run along the head-body junction of the 30S subunit. The region upstream of the AUG, the SD sequence included, runs on the side of the large lateral protuberance, whereas the A- and P-site anticodons are located in the middle, facing the 50S subunit (Montesano-Roditis and Glitz 1994). In the absence of tRNA, the mRNA is held on the ribosome primarily by the ASD-SD interaction, forming the so-called standby complex. Upon binding of the fMet-tRNA$_f$, the mRNA becomes firmly fixed in the downstream region, especially at the A-site (+4 to +6). In a recent model of the 30S subunit, nucleotides 532, 1052, 1395, and 1402 of rRNA, all of which cross-link to mRNA, are close together in the three-dimensional structure (Brimacombe 1995).

Nuclease treatment of initiation complexes protects about 35 nucleotides of the mRNA (Steitz 1975), from about –20 to +15, suggesting that multiple contacts of the mRNA with the 30S subunit are possible. An even larger area (–35 to +20) is protected from hydroxyl radical attack (Hüttenhofer and Noller 1994), although protection occurs only to +5 in the absence of tRNA in the A-site. Analyses of ribosome-binding sites have identified preferred residues at various positions within the protected region (Ringquist et al. 1992), but how these bases are sensed by the ribosome is not known in detail. There is some evidence that the second codon, especially in combination with a sub-optimal initiation codon, may facilitate translation initiation (Gold 1988).

THE INITIATION CODON

AUG is preferred as the initiation codon over other codons, although GUG and UUG are used with a lower frequency of 8% and 1%, respectively (Gold 1988; McCarthy and Gualerzi 1990). An exceptional case is AUU which functions as the initiation codon in the *infC* gene encoding initiation factor IF3. A model for autogenous regulation of *infC* (Gold et al. 1984) predicts that when the IF3 level is high, translation of mRNAs with non-AUG initiation codons such as that for *infC* is inhibited, whereas when the level is low, IF3-independent translation of *infC* mRNA is favored over other cellular mRNAs containing AUG initiators. The model is consistent with the view that IF3 functions to destabilize the interaction of fMet-$tRNA_f$ with non-AUG codons on the 30S subunit (Gualerzi and Pon 1990). An AUU to AUG mutation increases about 40-fold the in vivo expression of *infC* genes in both *E. coli* and *Bacillus stearothermophilus,* due to loss of autorepression (Butler et al. 1987; Gold 1988; La Teana et al. 1993). When IF3 binds to the 30S subunit, it may shield nucleotides present in 16S rRNA that otherwise could interact with the translation initiation region of *infC* mRNA.

THE SPACING BETWEEN THE SD SEQUENCE AND THE AUG

A further important feature of the ribosome-binding site is the spacing between the initiation codon and the SD sequence. The allowed distance separating these two elements has been extensively studied by analysis of naturally existing ribosome-binding sites as well as by analysis of a great number of mutated ribosome-binding sites. Although the efficiency of translational initiation differs between natural and synthetic mRNAs, the spacing is similar and amounts to six to eight nucleotides between the SD sequence and the AUG (Ringquist et al. 1992; Barrick et al. 1994). Hartz et al. (1991b) determined the influence of spacing on the formation of

30S preinitiation complexes with mRNAs having either AAGGA or UAAGGAGG as the SD sequence, by inserting increasing numbers of A residues. For mRNAs with AAGGA, the minimum spacing requirement appears to be five bases, with an optimum of five to seven, and for mRNAs with the longer UAAGGAGG sequence, three to four bases with an optimum of four to eight bases were found. The longer SD sequence UAAGGAGG gives three to six times more expression than the shorter AAGGA sequence for every spacing between two and eight residues. This suggests that for each SD sequence, there is a minimal spacing for translation and that a rather precise physical distance is required between the 3′ end of 16S rRNA and the anticodon of the fMet-$tRNA_f$ located in the P-site. Chen et al. (1994) also undertook a systematic study of ribosomal-binding sites by inserting 0–15 A/T residues between either TAAGG or GAGGT and the ATG upstream of the CAT reporter gene and found a preferred spacing of the SD sequence and the AUG of five nucleotides. Thus, for each SD sequence, an optimal although relatively broad spacing exists.

THE ROLE OF RIBOSOMAL PROTEIN S1

It is possible that specific protein-RNA interactions take place during initiation; the best candidate to have binding targets on mRNA is undoubtedly ribosomal protein S1. Its amino-terminal domain is associated with the 30S subunit by protein-protein interactions, whereas the elongated and flexible carboxy-terminal domain, containing RNA-binding centers with high affinity for pyrimidine-rich sequences, provides the ribosome with an "mRNA catching arm" (Gribskov 1992). Oligo(U) sequences are located upstream of the coat protein gene on phage Qβ and f2 RNAs, and a great number of *E. coli* mRNAs possess pyrimidine-rich sequences upstream of the SD sequence that may represent S1-binding sites (Boni et al. 1991). A recent report provides evidence that *E. coli* 30S ribosomal subunits recognize the translational start sites of mRNAs lacking the SD sequence, i.e., alfalfa mosaic virus RNA4 and tobacco mosaic virus RNA, and form stable ternary complexes only in the presence of S1 and IF3, in which the latter is crucial in facilitating the codon-anticodon interaction (Tzareva et al. 1994). By using the extension inhibition technique, 70S ribosomes and initiator tRNA yielded a toeprint at +15 from the first base of the initiation codon of a transcript derived from T4 gene *32,* whereas ribosomes lacking S1 yielded a drastically diminished toeprint (Hartz et al. 1991a). Neither 70S nor 70S(–S1) ribosomes yielded a toeprint in the absence of initiator tRNA; furthermore, antibodies raised against S1 strongly inhibited ternary complex formation provided they were added to the 70S ribosomes prior to mRNA (Hartz et al. 1991a). Site-directed mutagenesis of a U8 sequence upstream of the SD region of *rnd*, an *E. coli* gene encoding RNase D, showed that removal of only two U residues has a profound negative effect on translation (Zhang and Deutscher 1992). Taken together, these findings suggest that the affinity of protein S1 for U-rich sequences has an important role in the interaction of *rnd*-mRNA with 30S ribosomes.

A sequence of events has been proposed for initiation complex formation (see Fig. 1), starting with S1-dependent formation of a 30S-mRNA binary complex, followed by the SD-ASD interaction. The initiator tRNA then binds to form a 30S preinitiation complex, in which the tRNA at the P-site is not yet base paired with the initiation codon. Finally, the complex rearranges to form a 30S initiation complex, accomplishing base pairing with the initiation codon (Gualerzi and Pon 1990; Hartz et al. 1991b).

THE ROLE OF INITIATION FACTORS IF1, IF2, AND IF3

The prokaryotic initiation factors stimulate the on-rate of 30S initiation complex formation. IF2 binds fMet-tRNA$_f$ and GTP and positions fMet-tRNA$_f$ into the P-site of the 30S subunit. It favors the binding of aminoacyl-tRNAs with blocked α-amino groups and thereby helps to exclude noninitiator tRNAs from binding to the 30S subunit. Perhaps most importantly, it is a kinetic effector of 30S and 70S initiation complex formation. IF2 also functions as a ribosome-dependent GTPase which serves to more rapidly eject the factor following 70S initiation complex formation. IF3 binds to 30S subunits, acts as an anti-association factor, and increases the affinity of 30S subunits for IF1 and IF2. It inspects the correctness of the fMet-tRNA$_f$ anticodon stem and the P-site codon-anticodon interaction, leading to ejection of noninitiator tRNAs or interactions with noninitiation codons. The function of IF1 is less clear, but it appears to stimulate both IF2 and IF3 activities.

Taking into account the intracellular concentration of 30S subunits and the initiation factors and association constants of their interactions, it may be assumed that all native 30S subunits contain bound initiation factors (Gualerzi and Pon 1990). It appears that none of the initiation factors influences the SD interaction, because no effect on the association constants of the mRNA•30S complexes is detected. It appears that in the absence of initiation factors, the mRNA preferentially occupies a ribosomal "standby site," corresponding to the SD interaction. In the presence of factors, the mRNA is shifted toward a site closer to the P-site, where the factors exert their kinetic influence (Gualerzi and Pon 1990).

THE CONTROL OF TRANSLATION BY mRNA SECONDARY STRUCTURE

There are numerous examples of translational control of prokaryotic gene expression (reviewed extensively by McCarthy and Gualerzi [1990] and by Gold [1988]). The regulatory mechanisms involve two major types: repression by *trans*-acting proteins that prevent the formation of competent initiation complexes, and modulation of mRNA secondary structure within the ribosome-binding site that occludes ribosome binding. To provide the reader with insight into such mechanisms, a number of recently studied systems have been selected for description in this section.

Bacteriophage MS2

A major part in the control of gene expression in RNA bacteriophages is played by RNA folding, which can deny ribosomes access to translational initiation regions. The A (maturation) protein, lysis, and replicase cistrons in MS2 RNA are not accessible to ribosomes in the folded state of the RNA. The translational blocks for the replicase and lysis proteins are released by translation of the open coat protein cistron, during which RNA structure is melted, allowing initiation of their translation. Thus, expression of both genes is translationally coupled to that of the coat gene. To avoid the situation where phage RNA serves at the same time as template for ribosomes and for replication, the concentration of replicase complex increases during early infection and eventually reaches a concentration that enables it to bind at the initiation region of the coat gene and thereby to inhibit coat synthesis. As replication proceeds, viral RNA begins to exceed the number of replicase complexes, releasing the inhibition at the coat cistron. Later in the infection cycle, the coat protein reaches a sufficiently high concentration to act as a repressor of replicase synthesis by binding to the initiation start site of the replicase. The synthesis of the A protein is independently regulated by a process that is coupled to positive RNA strand transcription. The repressive secondary structure forms relatively slowly in the nascent transcript, allowing a few ribosomes to bind and initiate A-protein synthesis prior to formation of the inhibitory secondary structure (see below). Thus, bacteriophage MS2 provides us with four well-defined regulatory systems: the classic SD interaction at the open coat protein initiation site; translational coupling of downstream cistrons; a *trans*-acting protein repressor-mRNA system impairing ribosome entry; and coupling translation and replication (for review, see van Duin 1988).

Studies of initiation complex formation at the MS2 coat protein gene have provided insight into the role of secondary structure in regulating the efficiency of initiation (de Smit and van Duin 1990a,b). The coat cistron AUG and SD region are found in a stem-loop structure of intermediate stability. A large number of mutations were created by site-directed mutagenesis that affected the stability of this structure and the efficiency of coat protein synthesis. The overall conclusion from these studies is that translation initiation occurs only on unstructured ribosome-binding sites that are in equilibrium with their structured conformations. Therefore, the efficiency of initiation correlates with the fraction of time that the ribosome-binding site is in the unfolded conformation. There is a direct competition between intramolecular base pairing within the ribosome-binding site of the mRNA and intermolecular base pairing between the SD sequence and the rRNA. The SD sequence emerges as an element that affects the affinity of the mRNA for the 30S subunit, increasing the ability of the ribosome to compete with secondary structure. In other examples, the efficiency of λ repressor and antitermination factor N synthesis can be understood as the result of direct competition between formation of secondary structure and binding of ribosomes to the initiation region of these cistrons (de Smit and van Duin 1990b, 1994a). Whenever translation is unambiguously controlled by the stability of the local secondary structure, the relationship is quantitatively the same as for the MS2 coat protein gene, regardless of the position of the initiation codon in the structure (de Smit and van Duin 1994b).

In the expression of the A protein, the kinetics of RNA folding has a regulatory role. It is suggested that the 5′ terminus of the positive strand of MS2 RNA folds into a cloverleaf, i.e., a three stem-loop structure closed by a long-distance interaction with a region located 80 nucleotides downstream (Groeneveld et al. 1995). The downstream region contains the SD element of the A-protein cistron. Mutational analysis suggests that a ribosome can initiate at the A-protein cistron during the short interval between the synthesis of the downstream region (and ribosome-binding site) and the formation of the long-range secondary structure between the upstream and downstream sequences. As soon as the folding is completed, accessibility of ribosomes to the A-protein initiation site is lost.

Antisense RNA

Regulation by antisense RNA involves the synthesis of a short transcript that possesses a high degree of complementarity with a second RNA. Hybridization of the antisense RNA with the second RNA represses the latter's function, e.g., protein synthesis when an mRNA is involved. The antisense locus may overlap the targeted gene or the loci may not be linked at all (Pines and Inouye 1986; Delihas 1995).

An example of overlapping loci occurs in the control of gene expression in the transposon IS*10* (Simons 1988). IS*10* encodes a transposase (*tnp*) whose expression is rate-limiting for transposition. Two promoters are involved: one that generates the mRNA encoding the transposase, and another, called pOUT, that is located within the coding region of *tnp* but is in the opposite orientation. The 5′ ends of pOUT and *tnp* mRNAs are complementary for 35 base pairs, including the ribosome-binding site. The resulting duplex structure (termed pot for *p*rotection from *o*utside *t*ranscription) thereby prevents ribosome binding. The analysis of point mutations in the pot structure reveals that the effectiveness of the inhibitory RNA structure correlates well with the thermodynamic stability, but a number of mutations appear to cause a decrease in the rate at which the antisense RNA forms the inhibitory structure (Ma et al. 1994).

Antisense RNA generated from an unlinked locus has been implicated in the relative expression of two outer membrane proteins, OmpF and OmpC, which function as passive diffusion pores for small hydrophilic molecules (Mizuno et al. 1984; Pines and Inouye 1986). An increase in osmolarity causes a decrease in OmpF production and an increase in OmpC. The *micF* gene, coding for a 93-nucleotide *m*RNA-*i*nterfering *c*omplementary *micF* RNA (4.5S) and a rarer 174-nucleotide *micF* RNA form (6S), lies just upstream of *ompC* but is expressed in the opposite orientation. Both the *micF* and *ompC* genes are activated by high osmolarity. The *micF* RNA product forms a hybrid with *ompF* mRNA and inhibits its translation by blocking the SD sequence and the initiation codon; it also may result in destabilization of the mRNA (see Fig. 3a). Thus, this mechanism assures a constant amount of OmpC and OmpF in the cell.

Pseudoknots and *Trans*-acting Proteins

Regulation of translation may proceed through competition between a repressor protein and the ribosome for an overlapping binding site in the translation initiation region. Binding of the repressor protein prevents

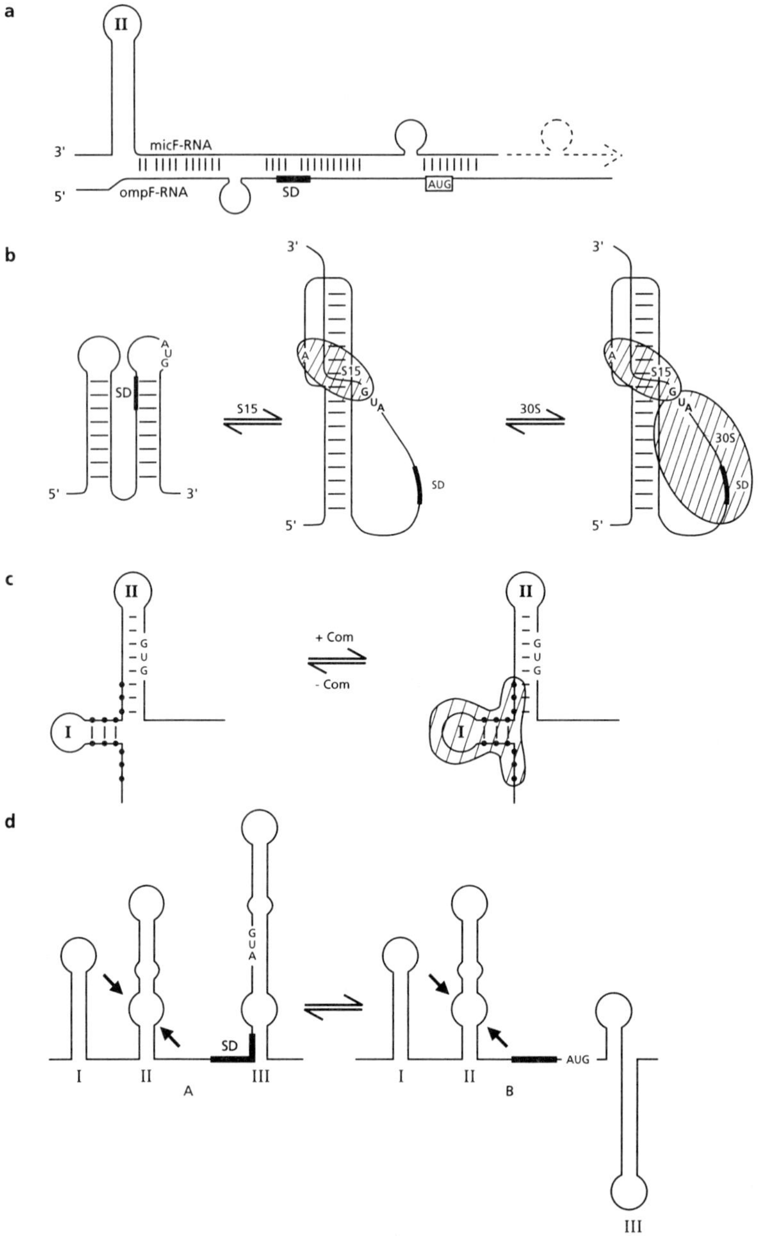

Figure 3 (*See facing page for legend.*)

binding of the ribosome (see above for an example of the coat protein of bacteriophage MS2). However, an alternative repression mechanism involves the repressor protein trapping the ribosome on its initiation site and preventing further assembly of the 30S initiation complex (Portier and Grunberg-Manago 1993; Spedding and Draper 1993; Spedding et al. 1993). An example of such translational regulation is found in the expression of ribosomal protein S15. This protein is able to control its own translation by binding to a region of the S15 mRNA that can fold in two mutually exclusive conformations, either a structure with two hairpins or one with a pseudoknot that is recognized and stabilized by S15. The bound S15 prevents the conversion of an inactive ribosome-mRNA complex to a productive 30S initiation complex by trapping the ribosome in an inactive transitory stage, as illustrated in Figure 3b. Both in vivo mutational analysis and in vitro toeprinting experiments confirm this model (Philippe et al. 1993, 1994). Thus, the biologically relevant control element, in this case the pseudoknot, exists in a variety of topologically equivalent structures and is not only recognized by S15, but shaped by it as well (Philippe et al. 1995).

Extensive toeprinting and physicochemical studies with ribosomal protein S4 mRNA and the α operon mRNA, which encodes four ribosomal proteins as well as the α-subunit of RNA polymerase, suggest a mechanism similar to that of S15 regulation. The model proposes two mRNA conformations, "active" and "inactive," which are in rapid equilibrium. The inactive conformation binds S4 to a complex pseudoknot structure in which the GUG initiation codon appears to be inaccessible to the initiator tRNA (Spedding and Draper 1993; Spedding et al. 1993). Many pseudoknot structures have been detected in the past few years and their roles in processes such as translational autoregulation or ribosomal frameshifting have been studied in great detail (ten Dam et al. 1992; Pleij 1994).

Figure 3 (*a*) Antisense *micF* RNA hybridizes with *ompF* mRNA and blocks the SD sequence and initiation codon. (Adapted from Pines and Inouye 1986.) (*b*) Interaction of ribosomal protein S15 and pseudoknot on S15 mRNA traps the 30S ribosome in an inactive transitory stage. (Adapted from Philippe et al. 1994, 1995.) (*c*) Translational activation of bound Com protein is caused by partial unfolding of stem-loop II. (Adapted from Wulczyn and Kahmann 1991.) (*d*) Translational activation in *CIII* mRNA occurs by RNase III, which catalyzes the transition from the inactive A into the active B structure, without cleaving the mRNA at the cleavage sites (indicated by arrows). (Adapted from Altuvia et al. 1991.)

Translational Activation by Com

Expression of the *mom* gene of bacteriophage Mu is regulated at both the transcriptional and translational levels (Wulczyn and Kahmann 1991). The Mom protein plays a part in DNA modification that protects Mu DNA from host restriction endonucleases. The phage protein Com is required for efficient translation of *mom* mRNA. *com* is the first gene of the *mom* operon; *com* and *mom* partially overlap and are cotranscribed. A stretch of 53 nucleotides around the GUG initiation codon of *mom* confers the Com dependency on *mom* mRNA translation. This region forms a stable hairpin structure in which the *mom* GUG is sequestered. The structure is responsible for inefficient translation, since mutations that destabilize the structure lead to efficient Com-independent translation. Translation of *com* in *cis* is not essential for regulation because expression of the Com protein in *trans* stimulates *mom* translation, even when the 5′end of *com* is deleted (Wulczyn et al. 1989). In vitro footprinting provided evidence that the binding site for Com extends from nucleotide −32 to −18, and chemical and enzymatic probing of the mRNA suggest that the translational activating sequence, TAS (loop I in Fig. 3c), is folded into a stem-loop structure. Results from both in vivo footprinting and oligonucleotide probing show that binding of Com protein leads to partial unfolding of stem-loop II. This results in exposing the *mom* SD sequence and the GUG initiation codon, thereby allowing ribosomes to bind and initiate translation.

RNase III-mediated Control

RNase III often plays a part in the onset of mRNA degradation, but it also has been implicated in the translational regulation of several bacteriophage and *E. coli* genes. In the λ p_L transcript, several RNase III processing sites are present, one of which is found upstream of the *CIII* coding sequence. As shown in Figure 3d, *CIII* mRNA is found in equilibrium between two alternative structures, A and B. In the A conformation, the SD sequence and the AUG are inaccessible to the 30S subunit, whereas in the B conformation, the AUG and SD sequence are accessible and translation is efficient (Altuvia et al. 1987). Furthermore, translation of *CIII* mRNA is greatly reduced in a host defective in RNase III and thus depends on its presence. Both alternative structures are recognized by RNase III, but the A structure is processed much more efficiently. Nevertheless, processing per se seems not to be the mechanism for translational stimulation. It is proposed that RNase III catalyzes the transition from the inactive A structure to the active B structure, rather than cleaving the *CIII* mRNA (Altuvia et al. 1991).

Attenuation of *ermC* mRNA Translation

The basis of erythromycin resistance is the posttranscriptional N^6, N^6-dimethylation of adenine nucleotides in 23S rRNA, which markedly reduces the affinity of ribosomes for erythromycin. The *ermC* mRNA encoding the N^6, N^6-dimethyladenine methylase contains a leader RNA with a strong SD sequence followed by an open reading frame of 19 amino acids. This uORF is out of frame with the *ermC*-coding sequence. In addition, because of the presence of six complementary segments, several hairpin loop structures can be formed in the leader. In the most stable structure, the methylase AUG and part of the SD sequence are sequestered, preventing the translation of *ermC* mRNA. The induction model proposes that in the presence of erythromycin, ribosomes translating the leader peptide stall during synthesis of the hydrophilic carboxy-terminal portion, since erythromycin appears to block more severely the transpeptidation reaction involving hydrophilic amino acids (Dubnau 1984). As a consequence of ribosome stalling, the downstream *ermC* mRNA attains an open configuration that permits initiation at the methylase initiation codon. The apparent paradox is that erythromycin as an inhibitor of peptide bond formation acts as an inducer of protein synthesis. An explanation may be that a minor fraction of ribosomes is methylated and provides a pool of resistant ribosomes for the translation of the methylase.

A COMPARISON WITH EUKARYOTIC TRANSLATION INITIATION

Recognition of the ribosome-binding site on prokaryotic mRNA depends on its secondary structure and its nucleotide sequence, which comprises the SD sequence, the initiation codon, and the spacing between them. The 30S ribosomal subunit, in possession of IF1, IF2, and IF3, scans the mRNA for the ribosome-binding site and initially recognizes the SD sequence. The three initiation factors stimulate the on-rate of 30S initiation complex formation. IF2 positions the fMet-tRNA in the P-site of the ribosome and IF3 locks the mRNA into the initiation position, inspects the codon-anticodon interaction, and clears the ribosome of uncharged tRNAs or improper aminoacyl-tRNAs. Productive initiation occurs only at sites that are devoid of secondary structure, as prokaryotes do not appear to possess RNA helicase activity.

One can imagine that the prokaryotic 30S preinitiation complex, consisting of a 30S subunit, IF1, IF2, IF3, fMet-tRNA$_f$, and GTP, is in essence not much different from the 43S preinitiation complex found in eukaryotes. The difference lies in the mechanism of mRNA binding. Prokaryotic mRNAs do not have to be prepared for ribosome binding, other than to possess an unstructured region. In contrast, eukaryotic mRNAs require prior interaction with a set of initiation factors before the 40S ribosome binds, and they utilize a 5′cap-dependent mechanism for selection of the AUG initiation codon. Initiation factor eIF4F first recognizes the 5′cap structure and together with eIF4B melts any secondary structure in the 5′-untranslated region (see Merrick and Hershey, this volume). The helicase-requiring step likely was introduced because eukaryotic mRNAs are fully synthesized prior to translation and often contain a large amount of secondary structure. Since the helicase activity begins by recognition of the 5′cap structure, ribosome binding occurs at the 5′terminus of mRNAs. The subsequent mechanism of scanning down the mRNA results in a biased recognition of the 5′-proximal AUG as the initiator codon. When unwinding of the 5′-untranslated region is accomplished, the well-equipped ribosome docking mechanism as designed for the prokaryotic system could be applied. Both eIF2 and eIF3 are present on 40S subunits together with Met-$tRNA_i$ and GTP and contribute to the search for the AUG on a stretch of nucleotides only somewhat larger than the size of the ribosome-binding site in prokaryotes (Thomas et al. 1992). Thus, recognition of the eukaryotic initiation site involves the same parameters as found in prokaryotes, possibly including mRNA and rRNA interactions, except for the SD interaction which may be replaced by interactions involving the much more complex factors, eIF3 and eIF2.

The situation is quite different for picornavirus and a limited number of cellular mRNAs because they bypass the cap-dependent scanning mechanism by a process called internal initiation (see Jackson, this volume). Ribosomes bind to an internal sequence on these messengers called the internal ribosome entry site (IRES). Base pairing between eukaryotic 18S rRNA and picornavirus RNA is possible, which is reminiscent of the base pairing between the SD sequence and prokaryotic 16S rRNA (Scheper et al. 1994). Thus, it is anticipated that picornaviruses may be capable of using the well-designed system of RNA recognition employed by the small prokaryotic ribosomal subunit.

ACKNOWLEDGMENT

I thank Dr. Adri Thomas for helpful comments on the manuscript.

REFERENCES

Altuvia, S., D. Kornitzer, S. Kobi, and A.B. Oppenheim. 1991. Functional and structural elements of the mRNA of the *cIII* gene of bacteriophage lambda. *J. Mol. Biol.* **218:** 723–733.

Altuvia, S., H. Locker-Giladi, S. Koby, O. Ben-Nun, and A.B. Oppenheim. 1987. RNase III stimulates the translation of the *cIII* gene of bacteriophage λ. *Proc. Natl. Acad. Sci.* **84:** 6511–6515.

Barrick, D., K. Villanueba, J. Childs, R. Kalil, T.D. Schneider, C.E. Lawrence, L. Gold, and G.D. Stormo. 1994. Quantitative analysis of ribosome binding sites in *E. coli. Nucleic Acids Res.* **22:** 1287–1295.

Boni, I.V., D.M. Isaeva, M.L. Musychenko, and N.V. Tzareva. 1991. Ribosome-messenger recognition: mRNA target sites for ribosomal protein S1. *Nucleic Acids Res.* **19:** 155–162.

Brimacombe, R. 1995. The structure of ribosomal RNA: A three-dimensional jigsaw puzzle. *Eur. J. Biochem.* **230:** 365–383.

Butler, J.S., M. Springer, and M. Grunberg-Manago. 1987. AUU-to-AUG mutation in the initiator codon of the translation initiation factor IF3 abolishes translational auto-control of its own gene (*infC*) *in vivo. Proc. Natl. Acad. Sci.* **84:** 4022–4025.

Chen, H., M. Bjerknes, R. Kumar, and E. Jay. 1994. Determination of the optimal aligned spacing between the Shine-Dalgarno sequence and the translation initiation codon of *Escherichia coli* mRNAs. *Nucleic Acids Res.* **22:** 4953–4956.

Delihas, N. 1995. Regulation of gene expression by *trans*-encoded antisense RNAs. *Mol. Microbiol.* **15:** 411–414.

de Smit, M.H. and J. van Duin. 1990a. Secondary structure of the ribosome binding site determines translational efficiency: A quantitative analysis. *Proc. Natl. Acad. Sci.* **87:** 7668–7672.

———. 1990b. Control of prokaryotic translational initiation by mRNA secondary structure. *Prog. Nucleic Acid Res. Mol. Biol.* **38:** 1–35.

———. 1994a. Translational initiation on structured messengers. Another role for the Shine–Dalgarno interaction. *J. Mol. Biol.* **235:** 173–184.

———. 1994b. Control of translation by mRNA secondary structure in *Escherichia coli.* A quantitative analysis of literature data. *J. Mol. Biol.* **244:** 144–150.

Dontsova, O., S. Dokudovskaya, A. Kopylov, A. Bogdanov, J. Rinke-Appel, N. Jünke, and R. Brimacombe. 1992. Three widely separated positions in the 16S RNA lie in or close to the ribosomal decoding region; a site-directed cross-linking study with mRNA analogues. *EMBO J.* **11:** 3105–3116.

Dubnau, D. 1984. Translational attenuation: The regulation of bacterial resistance to the macrolide-lincosamide-streptogramin B antibiotics. *CRC Crit. Rev. Biochem.* **16:** 103–132.

Gold, L. 1988. Posttranscriptional regulatory mechanisms in *Escherichia coli. Annu. Rev. Biochem.* **57:** 199–233.

Gold, L., G. Stormo, and R. Saunders. 1984. *Escherichia coli* translational initiation factor IF3: A unique case of translational regulation. *Proc. Natl. Acad. Sci.* **81:** 7061–7065.

Gribskov, M. 1992. Translation initiation factors IF-1 and eIF-2α share an RNA-binding motif with prokaryotic ribosomal protein S1 and polynucleotide phosphorylase. *Gene* **119:** 107–111.

Groeneveld, H., K. Thimon, and J. van Duin. 1995. Translational control of maturation-protein synthesis in phage MS2: A role for the kinetics of RNA folding? *RNA* **1:** 79–88.

Gualerzi, C.O. and C.L. Pon. 1990. Initiation of mRNA translation in prokaryotes. *Biochemistry* **29:** 5881–5889.

Hartz, D., D.S. McPheeters, and L. Gold. 1989. Selection of the initiator tRNA by *Escherichia coli* initiation factors. *Genes Dev.* **3:** 1899–1912.

———. 1991a. Influence of mRNA determinants on translation initiation in *Escherichia coli*. *J. Mol. Biol.* **218:** 83–97.

Hartz, D., D.S. McPheeters, L. Green, and L. Gold. 1991b. Detection of *Escherichia coli.* ribosome binding at translation initiation sites in the absence of tRNA. *J. Mol. Biol.* **218:** 99–105.

Hartz, D., D.S. McPheeters, R. Traut, and L. Gold. 1988. Extension inhibition analysis of translation initiation complexes. *Methods Enzymol.* **164:** 419–425.

Hüttenhofer, A. and H.F. Noller. 1994. Footprinting mRNA-ribosome complexes with chemical probes. *EMBO J.* **13:** 3892–3901.

La Teana, A., C.L. Pon, and C.O. Gualerzi. 1993. Translation of mRNAs with degenerate initiation triplet AUU displays high initiation factor 2 dependence and is subject to initiation factor 3 repression. *Proc. Natl. Acad. Sci.* **90:** 4161–4165.

Ma, C.K., T. Kolesnikow, J.C. Rayner, E.L. Simons, H. Yim, and R.W. Simons. 1994. Control of translation by mRNA secondary structure: The importance of the kinetics of structure formation. *Mol. Microbiol.* **14:** 1033–1047

McCarthy, J.E.G. and R. Brimacombe. 1994. Prokaryotic translation: The interactive pathway leading to initiation. *Trends Genet.* **10:** 402–407.

McCarthy, J.E.G. and C. Gualerzi. 1990. Translational control of prokaryotic gene expression. *Trends Genet.* **6:** 78–85.

Mizuno, T., M.-Y. Chou, and M. Inouye. 1984. A unique mechanism regulating gene expression: Translational inhibition by a complementary RNA transcript (micRNA). *Proc. Natl. Acad. Sci.* **81:** 1966–1970.

Montesano-Roditis, L. and D.G. Glitz. 1994. Tracing the path of messenger RNA on the *Escherichia coli* small ribosomal subunit. Immune electron microscopy using defined oligodeoxynucleotide analogs of mRNA. *J. Biol. Chem.* **269:** 6458–6470.

Philippe, C., L. Bénard, C. Portier, E. Westhof, B. Ehresmann, and C. Ehresmann. 1995. Molecular dissection of the pseudoknot governing the translational regulation of *Escherichia coli* ribosomal protein S15. *Nucleic Acids Res.* **23:** 18–28.

Philippe, C., F. Eyermann, L. Bénard, C. Portier, B. Ehresmann, and C. Ehresmann. 1993. Ribosomal protein S15 from *Escherichia coli* modulates its own translation by trapping the ribosome on the mRNA initiation loading site. *Proc. Natl. Acad. Sci.* **90:** 4394–4398.

Philippe, C., L. Bénard, F. Eyermann, C. Cachia, S.V. Kirillov, C. Portier, B. Ehresmann, and C. Ehresmann. 1994. Structural elements of *rpsO* mRNA involved in the modulation of translational initiation and regulation of *E. coli* ribosomal protein S15. *Nucleic Acids Res.* **22:** 2538–2546.

Pines, O. and M. Inouye. 1986. Antisense RNA regulation in prokaryotes. *Trends Genet.* **2:** 284–287.

Pleij, C.W.A. 1994. RNA pseudoknots. *Curr. Opin. Struct. Biol.* **4:** 337–344.

Portier, C. and M. Grunberg-Manago. 1993. Regulation of ribosomal protein mRNA translation in bacteria—The case of S15. In *Translational regulation of gene expression* (ed. J. Ilan), vol. 2, pp. 23–47. Plenum Press, New York.

Ringquist, S., M. MacDonald, T. Gibson, and L. Gold. 1993. Nature of the ribosomal mRNA track: Analysis of the ribosome-binding sites containing different sequences and secondary structures. *Biochemistry* **32:** 10254–10262.

Ringquist, S., S. Shinedling, D. Barrick, L. Green, J. Binkley, G.D. Storma, and L. Gold. 1992. Translation initiation in *Escherichia coli*: Sequences within the ribosome-binding site. *Mol. Microbiol.* **6:** 1219–1229.

Rinke-Appel, J., N. Jünke, K. Stade, and R. Brimacombe. 1991. The path of mRNA through the *Escherichia coli* ribosome; site-directed cross-linking of mRNA analogues carrying a photo-reactive label at various points 3′ to the decoding site. *EMBO J.* **10:** 2195–2202.

Rudd, K.E. and T.D. Schneider. 1992. Compilation of *E. coli* ribosome binding sites. In *A short course in bacterial genetics: A laboratory manual and handbook for* Escherichia coli *and related bacteria* (ed. J.H. Miller), pp. 17.19–17.45. Cold Spring Harbor Laboratory Press, Cold Spring Harbor, New York.

Scheper, G.C., H.O. Voorma, and A.A.M. Thomas. 1994. Basepairing with 18S ribosomal RNA in internal initiation of translaticn. *FEBS Lett.* **352:** 271–275.

Shine, J. and L. Dalgarno. 1974. The 3′-terminal sequence of *Escherichia coli* 16S ribosomal RNA: Complementarity to nonsense triplets and ribosome binding sites. *Proc. Natl. Acad. Sci.* **71:** 1342–1346.

Simons, R.W. 1988. Naturally occurring antisense RNA control—A brief review. *Gene* **72:** 35–44.

Spedding, G. and D.E. Draper. 1993. Allosteric mechanism for translational repression in the *Escherichia coli* α operon. *Proc. Natl. Acad. Sci.* **90:** 4399–4403.

Spedding, G., T.C. Gluick, and D.E. Draper. 1993. Ribosome initiation complex formation with the pseudoknotted α operon messenger RNA. *J. Mol. Biol.* **229:** 609–622.

Steitz, J.A. 1975. Ribosome recognition of initiator regions in the RNA bacteriophage genome. In *RNA phages* (ed. N.D. Zinder), pp. 319–352. Cold Spring Harbor Laboratory, Cold Spring Harbor, New York.

Steitz, J.A. and K. Jakes. 1975. How ribosomes select initiator regions in mRNA: Basepair formation between the 3′ terminus of 16S rRNA and the mRNA during initiation of protein synthesis in *Escherichia coli*. *Proc. Natl. Acad. Sci.* **72:** 4734–4738.

ten Dam, E., K. Pleij, and D. Draper. 1992. Structural and functional aspects of RNA pseudoknots. *Biochemistry* **31:** 11665–11676.

Thomas, A.A.M., G.C. Scheper, and H.O. Voorma. 1992. Hypothesis: Is eukaryotic initiation factor 2 the scanning factor? *New Biol.* **4:** 404–407.

Tzareva, N.V., V.I. Makhno, and I.V. Boni. 1994. Ribosome-messenger recognition in the absence of the Shine-Dalgarno interactions. *FEBS Lett.* **337:** 189–194.

van Duin, J. 1988. The single-stranded RNA bacteriophages. In *The bacteriophages* (ed. R. Calendar), vol. 1, pp. 117–167. Plenum Press, New York.

Wulczyn, F.G. and R. Kahmann. 1991. Translational stimulation: RNA sequence and structural requirements for binding of Com protein. *Cell* **65:** 259–269.

Wulczyn, F.G., M. Bölker, and R. Kahmann. 1989. Translation of the bacteriophage Mu *mom* gene is positively regulated by the phage *com* gene product. *Cell* **57:** 1201–1210.

Zhang, J. and M.P. Deutscher. 1992. A uridine-rich sequence required for translation of prokaryotic mRNA. *Proc. Natl. Acad. Sci.* **89:** 2605–2609.

Index

Aconitase
iron cofactor, 337, 341, 344
IRP1 homology, 340–342, 346–347, 354–355
regulation by oxidative stress, 352
structure, 341–342, 347
Actinomycin D, inhibition of protein translation, 5–6, 9
Activin, mesoderm induction in frogs, 421
Acyclovir, resistant frameshift mutants, 674
Adenomatous polyposis coli (*APC*) gene, recoding and disease progression, 673–674
S-Adenosylmethionine decarboxylase (AdoMetDC), upstream open reading frame and translation efficiency, 181–186
Adenovirus
gene expression in infection, 576
genome, 576
infectious cycle, 576–577, 584–585
inhibition of host-cell antiviral response, 580–582
ribosome shunting, 587
suppression of host-cell translation, mechanisms
dephosphorylation of eIF4E, 586–587, 590
mRNA binding by 100K protein, 590–591
RNA transport inhibition, 584–585
VA RNA_I role, 591–592
translation, repression and activation, 526
tripartite leader in mRNA, 577, 586
VA RNAs, 580–582, 591–592
AdoMetDC. *See* *S*-Adenosylmethionine decarboxylase
Angiotensin II, induction of eIF4E phosphorylation, 250, 255
Antizyme, mammalian
frameshifting
induction by polyamines, 654–655, 657
open reading frames, 655
shift sequence, 656
tagging of ornithine decarboxylase for degradation, 654
APIII. *See* Apurinic/apyrimidinic endonuclease III
APC. *See* Adenomatous polyposis coli gene
Apurinic/apyrimidinic endonuclease III (APIII)
activity in ribosomal proteins
P0, 707, 711
S3, 706, 711
deficiency in xeroderma pigmentosa, 706
ATP9 mRNA, translational activation, 743
AUG. *See* Initiator codon; Upstream initiation codons
Avian sarcoma-leukosis retrovirus, upstream open reading frames and RNA packaging, 190–191

Bacteriophage f2, translational control, 8
Bacteriophage MS2

Bacteriophage MS2 (*continued*)
kinetics of A protein mRNA folding, 768
translational control, 767–768
Bacteriophage Mu, translational control, 17
Bacteriophage T4
prr exclusion system, 526–527
self-splicing introns, 705–706
translational repressors, 526
tRNA cleavage, 527
Basic fibroblast growth factor, initiator codons, 188

Calcium
cell cycle regulation, 301–303
effects on activity
eEF2 and phosphorylation, 295–307
eIF2, 154, 307–310
PKR, 154
translational control during heat shock, 274, 277
CaM kinase III
autophosphorylation, 304
calcium dependence, 304
cell cycle regulation, 301–303
expression
regulation, 305–306
tissue levels, 304–305
kinases, 304
phosphorylation of eEF2, 296–297
structure, 303
substrate specificity, 297–298, 304
cAMP-dependent protein kinase (PKA)
CaM kinase III phosphorylation, 304
eIFβ phosphorylation, 147
induction of GCN4 translation, 233–234
Cap. *See* m^7G cap
Casein kinase II
eIF2B phosphorylation, 148
eIF2β phosphorylation, 147–148
messenger ribonucleoprotein particle phosphorylation, 328–329
Cauliflower mosaic virus (CaMV)
ribosome shunting, 519, 671
stimulating upstream open reading frames, 188–190
Cell cycle
eEF2 role, 301–303
eIF4E role, 246, 249–251, 256–260
signal transduction pathways, 245, 389
translation control and cell growth, 245–246
c-*fos*, mRNA degradation, 644
specifying domains, 636
translational coupling, 636–638
c-*mos* mRNA
polyadenylation role
maturation, 414–416
meiosis, 491–492
translation during oocyte maturation, 413–415
c-*myc*
initiator codons, 187–188
mRNA degradation
deadenylation, 637
half-life, 637
specifying sequences, 637–638
mRNA targeting in eIF4E-induced transformation, 259–260
Cordycepin, inhibition of polyadenylation, 482, 496
COX1 mRNA, translational activation, 743
COX2 mRNA
frameshift mutations, 747
translational activation, 742
COX3 mRNA
PET protein complex, role in translation, 739–741
tethering to inner membrane, 743–744
translational activation, 739–741
5′-untranslated region, 740–741
Crystallins
bifunctionality, 710
homology with other proteins, 710
Cyclin-dependent kinase, mRNA deadenylation in embryogenesis, 493, 495
Cyclins
cyclin D1 mRNA targeting in eIF4E-induced transformation, 259
oocyte maturation and translation, 412–413
regulation of maternal mRNAs, 413
Cycloheximide
effects
gene expression, 311
polysome packing, 14–15
rpmRNA translation, 376
elongation inhibition, 14, 376
Cytochrome *b*, yeast mRNA
CBS protein role in translation, 742–743
translational activation, 742–743

Cytoskeleton
destruction by viruses, 596–597
role in translational control of mRNAs, 23, 438

Dihydrofolate reductase (DHFR), nonsense-mediated mRNA degradation, 643
Double-stranded RNA-activated kinase. *See* PKR

EBV. *See* Epstein-Barr virus
eEF1A
coregulation with rpmRNA, 373
cytoskeleton maintenance, 23
genes, 56
GTPase, 52
guanine nucleotide exchange, 52, 54–55
ligands, 52, 54, 56
posttranslational modification, 56
structure, 55–56
eEF1B
elongation role, 57
structure, 57
eEF2
ADP-ribosylation by diphtheria toxin, 19, 57, 296
coregulation with rpmRNA, 373
GTP binding, 298–299
phosphorylation, 19, 296, 310–311
calcium role, 300–301
cell cycle regulation, 301–303
extent, 300
kinase. *See* CaM kinase III
mechanism of inhibition, 298–299
NMDA effects, 301
protein phosphatase-2A, 306
sites, 296–297, 300
posttranslational modification, 57–58
regulation of synthesis, 307
structure, 57–58, 297
eEF3
elongation role, 58
structure, 58
EF1A. *See* Elongation factor 1A
EGF. *See* Epidermal growth factor
eIF1, initiator codon recognition, 130
eIF1A
ligands, 38, 120
structure, 38
translational control, 393
eIF2
accessory proteins, 19
calcium effects on activity, 154, 307–310
cellular stress and activation, 153–155
guanine nucleotide exchange, 41–42, 116–118, 119, 125–126, 128–129, 140, 200
heat shock effect on activity, 153–155, 279–281
initiator codon recognition, 123
kinases, 18, 126, 129, 139, 141–153, 307, 527–528
methionyl-tRNA$_i$ binding, 116
nutritional deprivation and phosphorylation, 157, 199
phosphatases, 147, 160–161, 235
phosphorylation, 7, 18, 43, 116, 126–127, 129–131, 140, 153–155, 157–160, 200, 236, 280, 307–310, 527–528, 613
recycling, 125–128, 130
regulation of expression, 117
RNA binding, 119–120, 123
site-directed mutagenesis of phosphorylation sites, 159–160
structure, 41–42
subunits, 41–42, 116, 140
eIF2B
allosteric regulation, 127–128
enhancers of activity, 127
homology with GCG/GCN3 complex, 220–221
inhibition by phosphorylated eIF2, mechanism, 224–226
interaction with eIF2, 43, 126–127
kinases, 128, 140–146, 148
nucleotide binding, 43
phosphorylation, 128, 140–141
subunits, 42–43, 125, 220
eIF2C
size, 43
stimulation of ternary complex formation, 120
eIF3
genes in yeast, 121
ligands, 39–40, 120–121
structure, 38, 40
eIF4A
ATPase, 46, 247
genes, 45
ligands, 45
phosphorylation, 260–261

eIF4A (*continued*)
structure, 46
eIF4B
initiator codon recognition, 123–124
ligands, 46
phosphorylation, 261
RNA helicase, 247–248
structure, 46–47
eIF4E
binding proteins and phosphorylation, 47, 253–256, 402–403, 529, 565
expression levels, 47, 249
initiation role, 49
malignant transformation role, 246, 256–260
mesoderm induction in frogs, 421
mRNA targets in transformation, 258–260
phosphorylation, 48, 249–253
dephosphorylation by viruses, 528–529, 586–587, 590, 612
effect on translation rates, 252–253
heat shock effects, 281
kinases, 251–252
phosphatases, 252
regulation of cell growth, 249–251
sites, 250, 257
stimuli, 250–251
recognition of m^7G cap, 44, 47, 248–249, 528
regulation of cell growth, 249–253
translational control of rpmRNA, 376–377
eIF4F
efficiency and mRNA type, 91–92
heat shock effect on activity, 279–282, 284–285
modification by viruses, 556–557
mRNA cap structure binding, 248–249, 284–285, 556
RNA helicase, 248
subunits, 47–48, 247
eIF4G
cleavage in picornavirus infection, 8, 19, 48, 97–98, 248, 511
2A protease
discovery, 562–563
purification, 563–564
trans-activation, 559
eIF3 role, 564
inhibition of host translation, 560–562, 565
L-protease, 564–565
products, 561–562, 565
genes, 48
ligands and binding sites, 247–248, 562
phosphorylation, 48, 261
eIF5
formation of 80S complex, 49, 73, 124
GTPase, 124–126
structure, 50
eIF6, ligands, 40, 120
Elongation
aminoacyl-tRNA binding, 52
determination of rate, 13
mitochondria, 747
peptide bond formation, 54
regulation, 15
translocation, 54–55
Elongation factor 1A (EF1A)
recycling, 760
translation in prokaryotes, 759–760
Elongation factors. *See also individual factors*
mammalian factors, table of properties, 35
yeast factors, table of properties, 39
Encephalomyocarditis virus. *See* Picornavirus
Epidermal growth factor (EGF), induction of eIF4E phosphorylation, 250
Epstein-Barr virus (EBV)
exploitation of host translational apparatus, 516–517
L22 binding of RNA, 710
*erbA*α, initiator codons, 187
eRF1
structure, 59
termination role, 59
Erythromycin, prokaryotic resistance and translational control, 773

fem-3
point mutation element and translational control, 418–419
sex determination in *Caenorhabditis elegans*, 416, 418–420, 434
Ferritin
iron sequestration, 337–338
structure, 338
translational control. *See also* Iron-responsive element
discovery, 338–339
interleukin-1β response, 353–354
oxidative stress, 352–353

Fibroblast growth factor (FGF), receptor mRNA and mesoderm induction in frogs, 420
Frameshifting
amino acid starvation sensor role, 667–668
autoregulation, 653, 658–660
detection from sequence database search, 674–675
human disease role, 673–674
limitation of aminoacyl tRNA, 666–667
mammalian antizyme, 654–657
nonproduct roles, 671
programmed frameshifting, 653–654, 656, 662, 667–668
3′ pseudoknot role, 656
ratio setting, 668–670
reading frames, 655, 657–658
RF2, 657–662
RF3 role, 661
shift sequence, 656, 659, 661
Shine-Dalgarno sequence role, 660–661, 666
transposable elements and regulatory frameshifting, 663–666
tryptophanase, evidence, 662–663
tryptophan repressor, 663
viruses, 522–523

GCN1 role, 232–233
GCN2
eIF2α phosphorylation, 142, 159–160, 204, 217
homology with other kinases, 217
kinase activation
starvation mechanism, 146, 204, 217, 226–233, 238
uncharged tRNA role, 227–230
oligomerization, 231–232
protein kinase domain, 142, 146, 217
regulation of GCN4 translation, mechanism, 217–224
ribosome association, 230–231
GCN3
GCD/GCN3 complex
eIF2 association, 221
eIF2B homology, 219–220
regulation of GCN4 translation, 219–220
GCN4
application as translation regulation model, 237–238, 642
eIF2 phosphorylation and translational control, 199–201, 224
GCD role in translation, 204–205, 217–219
induction by starvation, 199–200, 217, 233–235
mutation phenotypes, 204–205, 217–218
reinitiation following translation of uORF1, 211–213
suppression of reinitiation at upstream open reading frames, 213–217, 237
trans-acting regulators of translation, 204–207
upstream open reading frames and translational control, 199–201, 204–205, 208–217, 236
GLC7
dephosphorylation of eIF2, 235–236
essentiality in yeast, 236
Glycogen synthase kinase-3 (GSK-3)
eIF2B phosphorylation, 148
regulation, 148
GM-CSF. *See* Granulocyte macrophage–colony-stimulating factor
Granulocyte macrophage–colony-stimulating factor (GM-CSF), mRNA degradation, 635–636, 638
GSK3. *See* Glycogen synthase kinase-3
GYPSY, frameshifting, 670

HCR. *See* Hemin-controlled repressor
Heat shock
adaptation in mammalian cells, 276
effect on initiating factor activity
eIF2, 153–155, 279–281
eIF4E, 281
eIF4F, 279–282, 284
translational control, 9–10
amino acid analog effects, 274
calcium effects, 274, 277
preferential translation of heat shock proteins, 271, 282–286
preinitiation complex inhibition, 279
repression of normal translational activity, 271–275
restoration of normal translation, 275–277
ribosome disaggregation, 278

Heat shock (*continued*)
translational thermotolerance, 271, 274–277
Hemin-controlled repressor (HCR)
activation, 148–151
disulfide bonds and regulation, 149
eIF2α phosphorylation, 142, 159–160
heat shock protein association, 149–151, 285–286
inhibition by p67, 159
phosphatase, 161
phosphorylation, 149
protein kinase domain, 142–143
tissue specificity of expression, 143
Hemoglobin, synthesis, 6–7, 149
Hepatitis B virus, upstream open reading frames and RNA packaging, 190
Hepatitis C virus, IRES characteristics, 96–97
HisRS, homology with GCN2, 227–230
Histone
mRNA degradation
3′ structure role, 634–635
translational coupling, 634–635
synthesis coupling to DNA synthesis, 634
HIV-1. *See* Human immunodeficiency virus type 1
HS4, nonsense-mediated mRNA degradation, 640–641
HSP70
association with hemin-controlled repressor, 285–286
autoregulation, 279, 285–286
role in translational recovery following heat shock, 275–277
trans-acting factors, 286
5′-untranslated region and preferential translation of mRNA, 283–287
HSP90, association with hemin-controlled repressor, 149–151, 285–286
Human immunodeficiency virus type 1 (HIV-1)
frameshifting, 670
infectious phase switches, 510
inhibition of PKR, 535–536

IF1. *See* Initiation factor 1
IF2. *See* Initiation factor 2
IF3. *See* Initiation factor 3
Influenza virus
double infection with adenovirus, 610
inhibition of PKR
discovery, 613–614
P58 inhibition, 536, 614–616, 625
mortality of infection, 607
reverse genetics technology, 624
suppression of host cell translation, mechanisms
eIF4E dephosphorylation, 612
nuclear RNA degradation, 608–609
table, 608
5′-untranslated region of viral mRNA, role, 611–612
viral mRNA structure and translation efficiency, 610–611
temporal regulation of translation, 612
transfection/infection assay, 611
Initiation. *See also* Methionyl–tRNA$_i$
definition, 71
determination of rate, 14
inhibition, 14
internal initiation
assays, 90
eukaryotes, 93, 95–97, 102
trans-acting factors, 97–101, 103
mechanism, 32, 36–52, 200
mitochondria, 738–747
poly(A) tail role, 458–459
prokaryotes. *See also* Shine–Dalgarno sequence
autogenous regulation, 764
comparison with eukaryotic initiation, 773
initiation codon, 764
initiation factors, 759, 766–767
pathway, 759, 766, 773–774
recognition of initiation site by 30S ribosomal subunit, 761
ribosome-binding sites on mRNA, 762–764
S1 role, 765–766
toeprint assay, 762–763
translational control, 767–774
rate-limiting steps, 51, 139, 389
reinitiation following termination, 59, 74–76, 177, 521–522
site availability and viral autoregulation, 523–526
site selection
eukaryotes, 81–83
prokaryotes, 76–78

Initiation factor 1 (IF1), initiation of translation in prokaryotes, 759, 766–767
Initiation factor 2 (IF2), initiation of translation in prokaryotes, 759, 766–767
Initiation factor 3 (IF3)
  and frameshifting, 665–666
  initiation of translation in prokaryotes, 759, 766–767
  insertion sequence, 911
Initiation factors. *See also individual factors*
  comparison between prokaryotes and eukaryotes, 72–74
  mammalian factors, table of properties, 34–35
  phosphorylation, 7, 18
  yeast factors, table of properties, 39
Initiator codon. *See also* Upstream initiation codons
  alternative in-frame initiators in eukaryotes, 187–189, 191
  effect of upstream AUG, 37
  leaky scanning, 45
  methionyl-tRNA$_i$ binding, 32, 71, 121–122
  recognition, 43–45, 122–124
  selection in mitochondria
    sequence, 736, 745
    translational activators, 746
    5′-untranslated region role, 745–746
  sequences, 36, 174, 187
Insertion sequence (IS), induction of frameshifting, 664–666
Insulin, induction of eIF4E phosphorylation, 250
Interferon
  types, 579
  virus infection protection, 578–580
Internal ribosome entry site (IRES)
  biological advantages in viral infection, 519
  evolution, 555–556
  hepatitis C virus, 96–97
  mutation studies, 553–554
  neurovirulence determinants in poliovirus, 559–560
  protein interactions
    cell extracts, 557
    initiation factors, 556–557
    La, 557
    polypyrimidine-tract-binding protein, 558
  stimulation of translation by binding proteins, 17
  structure in picornaviruses, 93, 95–97, 517–518, 551–552
    encephalomyocarditis virus, 554–555
    poliovirus, 553–554
    secondary structure, 553–556
  translation activation after viral infection, 558–559
IRE. *See* Iron-responsive element
IRES. *See* Internal ribosome entry site
Iron
  binding by P2, 709
  effect on IRP2 half-life, 350
  iron-sulfur enzymes, 337, 352
  sequestration in ferritin, 337–338
  toxicity, 337
Iron-responsive element (IRE)
  associated genes, 335, 351–352, 355
  identification in ferritin mRNA, 339
  initiation factor binding, 353
  IRP-binding proteins
    expression levels in tissues, 350
    interplay, 350–352
    IRP1
      aconitase homology and activity, 341–342, 346–347, 354–355
      iron-sulfur cluster, 343–346, 353
      mediation of iron sensing, 343–345, 355
      mutational analysis, 345–347
      phosphorylation, 354
      RNA binding, 347–348
      sequence, 341
      steric inhibition of translation, 343, 354
    IRP2
      discovery, 349
      homology with IRP1, 349
      regulation by iron, 349–350
    purification, 340–341
    RNA-binding sites, 339–340, 347–348, 351, 355
    types, 335
  positioning, 342
  structure, 335–336, 339–340
  translational repression, 16–17, 342–343
IRP1. *See* Iron–responsive element
IRP2. *See* Iron–responsive element
IS. *See* Insertion sequence

La
  interaction with IRES, 557
  internal initiation role, 99–100
L-A virus, SKI system, 538
Leucine zipper, ribosomal proteins, 691–692
lin-4, translational regulation and *C. elegans* life cycle, 427–429
lin-14, translational regulation and *C. elegans* life cycle, 427–428, 434
15-Lipoxygenase (LOX), translational repression and red blood cell differentiation, 430–431
LOX. *See* 15-Lipoxygenase

MAPK. *See* Mitogen-activated protein kinases
Maternal pattern formation
  anterior-posterior axis establishment, 423, 425
  coordinate activation of mRNA, 422–423
  mRNA translational control
    bicoid, 422–423, 425, 492
    caudal, 424–425
    Drosophila, 421–422
    glp-1, 425–426, 441
    hunchback, 423–426
    nanos, 423, 425, 441
  translational repression and embryonic asymmetry, 421, 423–425
Messenger ribonucleoprotein (mRNP) particle
  abundance in cultured cells, 12
  core protein
    functions, 324–325
    p50
      effect on RNA structure, 322
      structure, 322
    phosphorylation, 328–329
    Y-box proteins
      homology with cold-shock protein, 324
      phosphorylation, 436
      sequence specificity, 436
      types, 323–324
  discovery, 320
  free cytoplasmic particles, 321
  masked particles, 321, 328, 432
  polysome-bound particles, 321
  protein content, 321
  RNA-binding proteins and translational control, 24
  structure, 329–330
  3′-untranslated region of mRNA and masking, 326–328
Messenger RNA (mRNA). *See also individual mRNAs*
  cap. *See* m^7G cap
  cis-acting elements, 16–17
  cytoskeleton association, 23, 438
  decay. *See* Messenger RNA degradation
  discovery, 3, 505
  efficiency of utilization, 1–2, 11–12
  initiation at 3′ end, 24
  localization, 23
  masking
    developmental masking, 443
    proteins, 319, 325–330
  nuclear transport, 644–645
  poly(A) tail. *See* Poly(A) tail
  ribosomal protein mRNA. *See* Ribosomal protein
  ribosome binding, 43–45, 246–249
  secondary structure and translation, 16, 37
  sequestration in eggs, 5–6
  18S rRNA pairing and initiation site selection, 81–83
  stored mRNA, 320, 325
  translational enhancers in prokaryotes, 78–81
  3′-untranslated region
    AU-rich elements, 327, 454, 635–637
    derepression role, 438–440
    developmental regulation of translation, 411–412
    masking role, 326–328, 432–435
    mechanisms of translational repression
      5′cap interference, 437
      cell localization, 438
      nucleating assembly of a repressive structure, 436–437
      poly(A) tail shortening, 437–438
    negative control elements, 434, 439–440
  5′-untranslated region
    binding proteins, 320
    heat shock proteins, 283–287
    length, 36, 173
Messenger RNA degradation. *See also individual mRNAs*

AU-rich element mRNAs, 635–637
c-*myc*, 637–638
deadenylation and degradation, 452–453, 467–468, 471, 637, 647
half-life and cellular role, 631
histone, 634–635
inhibition by protein synthesis inhibition, 631, 638, 640
mechanisms, 452–453, 467–471
nonsense-mediated degradation
biological rationale, 645–646
mammals, 642–645, 647
metazoans, 645
recognition of premature terminator, 646
yeast, 639–642
stability, 17–18, 311
translational coupling, 632, 634–639
β-tubulin, 632–634
Methionyl-tRNA
eEF1A binding, 115
structural distinction from initiator tRNA, 113–114
Methionyl-$tRNA_i$
binding
ATP, 42, 118–119
eIF2, 115–118, 200
initiator codon, 32, 71, 121–122
purification of complexes, 113
40S ribosomal subunit, 32, 40–41, 116, 120–122
deacylation, 127, 130
GDP inhibition, 41
initiation complex in prokaryotes, 72, 759, 766–767
structural distinction from elongation tRNA, 113–114
synthetase reaction, 114–115
yeast
genes, 114
sequences, 114
m^7G cap
accessibility, 36
analogs, 377
cap-independent initiation, 89–93
decapping mechanisms, 468, 470
discovery, 505
effect on translation, 497
interaction with 3′-untranslated region, 437, 468, 470
methyltransferase association with poly(A) polymerase, 497–498
recognition by eIF4E, 44
uncapped mRNA translation in viruses, 421, 549
Minute, mutation and *Drosophila* development, 707–708
Mitochondrial translation
elongation, 747
genetic analysis in yeast, 735–736
informational suppressors, 747
initiation codon selection, 745–747
in organello systems, 735
membrane proteins, 735, 743–744
mRNA
binding proteins, 738
codon usage, 733, 736, 745
structure, 736–738
tethering to inner membrane, 743–745
translational activation, 738–745
5′-untranslated region, 737, 740–741, 750
nuclear encoding of translation components, 733–734
reporter genes, 749
subcellular localization, 733–734
termination, 747–748
translational control in yeast, 739
glucose repression, 748–749
oxygen stress, 748
PET protein role, 749
tRNA
import, 734–735
mutation and disease, 736
Mitogen-activated protein (MAP) kinases
kinase, 255
phosphorylation of eIF4E-binding proteins, 253–256, 402–403
signal transduction role, 254, 400
stimulation by insulin, 255
Myogenesis, tropomyosin feedback loop, 431–432, 442

Nerve growth factor (NGF)
induction of eIF4E phosphorylation, 250
regulation of CaM kinase III expression, 305–306
NGF. *See* Nerve growth factor
Nitric oxide (NO), effect on iron-sulfur cluster enzymes, 352–353
NO. *See* Nitric oxide
Nonsense codon, recognition in nucleus, 18

ODC. *See* Ornithine decarboxylase
2′,5′-Oligoadenylate synthetase
  activation by interferon, 579
  antiviral activity, 530–533
  double-stranded RNA activation, 531–532
  inhibition by vaccinia virus, 583
Olins box, translation enhancement, 79, 81–82
Omega sequence, translation enhancement, 79–80
Omp. *See* Outer membrane protein
Ornithine decarboxylase (ODC)
  mRNA targeting in eIF4E-induced transformation, 259
  tagging for degradation, 654
Outer membrane protein (Omp), antisense RNA and translational control, 769

P58
  cloning, 614
  homology with other proteins, 614
  inhibitor of P58 (I-P58), 615–616
  malignant transformation role, 616
  PKR inhibition in influenza virus infection, 614–616, 625
PABP. *See* Poly(A)-binding protein
Pactamycin, initiation inhibition, 14
α-Pal, transcription factor for eIF2α, 117
PAN. *See* Poly(A) nuclease
Pattern formation. *See* Maternal pattern formation
PDGF. *See* Platelet-derived growth factor
PGK1, nonsense-mediated mRNA degradation, 641–642
Picornavirus
  cap-independent initiation, 90–91
  cap recognition system, inactivation, 560
  eIF4G cleavage in infection, 8, 19, 48, 97–98, 248, 511
    2A protease
      discovery, 562–563
      purification, 563–564
      trans-activation, 559
    eIF3 role, 564
    inhibition of host translation, 560–562, 565
    L-protease, 564–565
    products, 561–562, 565
  IRES
    evolution, 555–556
    mutation studies, 553–554
    neurovirulence determinants in poliovirus, 559–560
    protein interactions
      cell extracts, 557
      initiation factors, 556–557
      La, 557
      polypyrimidine-tract-binding protein, 558
    structure, 93, 95–97, 551–552
      encephalomyocarditis virus, 554–555
      poliovirus, 553–554
      secondary structure, 553–556
    translation activation after viral infection, 558–559
  shutoff of host-cell translation, 7–8
PKA. *See* cAMP-dependent protein kinase
PKC. *See* Protein kinase C
PKR
  activation, 145, 152–154, 531–532, 613, 615, 620
  antiviral activity, 530–533, 579
  apoptosis role, 144, 156
  autophosphorylation, 144, 152–153, 310, 531, 581
  calcium activation, 154
  cellular roles, 155–157, 161
  double-stranded RNA binding, 145–146, 152–153, 310, 531–532, 613, 615, 620
  eIF2α phosphorylation, 142, 159–160
  I-κB substrate, 161
  inhibition
    discovery, 613–614
    P58, 158, 536, 614–616, 625
    p67, 158–159
  interferon-induced antiviral response, 155, 158
  isoelectric point, 144
  phosphatase, 161
  protein kinase domain, 142–144
  proteolysis, 145–146
  subcellular localization, 144–145
  tumor suppressor activity, 156–158, 246, 616
  viral countermeasures
    double-stranded RNA analogs, 534–535
    double-stranded RNA sequestration,

534, 580–583
down-regulation of levels, 536
inhibition of activation, 535–536
Plant viruses, exploitation of host translational apparatus, 515–516
Platelet-derived growth factor (PDGF)
induction of eIF4E phosphorylation, 250, 255
signal transduction
PKR, 156
S6 kinase, 400–401
Poliovirus. *See* Picornavirus; Sabin vaccine
Poly(A)-binding protein (PABP)
closed-loop model of translation initiation, 464–467
essentiality in yeast, 462
negative regulation of decapping, 468
oligomerization, 463
properties, 456
sequestration by viral mRNA, 595–596
translation role, 462–464
Poly(A) nuclease (PAN)
poly(A)-binding protein dependence, 456–457
properties, 456–457
Poly(A) polymerase
cap methyltransferase association, 497–498
phosphorylation and activation, 439
subcellular localization, 489–491
Poly(A) tail
cleavage-polyadenylation specificity factor, 489
closed-loop model of translation initiation, 464–467
c-*mos* mRNA maturation role, 414–416
cordycepin inhibition of polyadenylation, 482, 496
cytoplasmic polyadenylation elements
binding proteins
assay, 487–488
isolation, 488
Orb homology, 488
role in polyadenylation, 488–489
developmental role, 486–487
sequences, 485–486
developmental regulation
cis-acting elements
embryogenesis, 486
invertebrate development, 486–487
deadenylation
embryogenesis, 493, 495
oocyte maturation, 492–493
translational repression, 495–496
mouse oocyte maturation, 484, 495–496
sea urchin eggs, 482
trans-acting factors, 487–491
translational control mechanisms, 496–498
Xenopus oocyte, 483–486, 491–493
discovery, 481
effect on mRNA translational efficiency
in vitro, 458–459
in vivo, 459–462
exogenous poly(A) and translation inhibition, 463
interference with 5′ cap, 437
length
assays, 453–454
homogeneity in newly synthesized tails, 452
kinetics of shortening, 454–456, 467
mechanisms of shortening, 456–457
mRNA aging effects, 452–455
regulation, 455
shortening effects on mRNA
decapping, 468
decay, 452–453, 467–468, 471, 637, 647
repression, 437–439, 461, 498
polyadenylation
cleavage site in processing, 451
signals, 434, 439–441, 485–486
structure, 17, 320, 451
translational derepression pathways, 439–440
Polypyrimidine-tract-binding protein (PTB)
interaction with IRES, 557
internal initiation role, 100–101, 103
translational control of rpmRNA, 381
Polysome
determination of amount, 15
packing, 14–15
Protein kinase C (PKC)
eIF2β phosphorylation, 147
eIF4E phosphorylation, 251–252
Protein kinase, RNA-dependent. *See* PKR
Protein phosphatase-2A, dephosphorylation of eEF2, 306
Protein synthesis. *See* Translation

PTB. *See* Polypyrimidine-tract-binding protein
Puromycin, inhibition of initiation, 120–121, 129

Rapamycin
S6 kinase, blocking of activation, 380, 394–396
sensitivity of eIF4E-binding protein phosphorylation, 255
Ras
eIF4E
ras dependence of phosphorylation, 400–401
substitution in transformation assay, 257
inhibition by GAP, 257
Red blood cell, 15-lipoxygenase role in differentiation, 430
Release factor 1 (RF1)
mitochondria, 748
stop codon recognition, 760
Release factor 2 (RF2)
abundance and cell growth rate, 657
homology with RF1, 661–662
stop codon specificity, 657, 662, 760
Release factor 3 (RF3), enhancement of termination, 761
Reovirus
genera, 616–617
genome, 617
interferon sensitivity, 620–622
mRNA translational efficiency, regulation, 622–623
porcine rotavirus, 624
suppression of host-cell translation, mechanisms
NSP3 role, 624
PKR inhibition, 620–622, 624
σ3 role, 617–622, 625
table, 619
viral mRNAs
abundance, 617
structure, 618
Reticulocyte, translational control, 4, 6–7
Retrovirus, exploitation of host translational apparatus, 516–517
RF1. *See* Release factor 1
RF2. *See* Release factor 2
RF3. *See* Release factor 3
Ribonuclease III (RNase III), translational control in prokaryotes, 772
Ribonuclease L (RNase L)
antiviral activity, 530–533
double-stranded RNA activation, 531–532
Ribonucleotide reductase, masking of mRNA, 432–433
Ribosomal protein. *See also* S6
APIII activity
P0, 707, 711
S3, 706, 711
developmental roles, 707–708
evolution
additional proteins in eukaryotes, 702–703
archaebacteria, 799–700
eubacteria, 700–701
mammals, 697–698
NS1 and NS2 in bacteria, 704
origin possibilities, 703–704
phylogenetic distribution of homologs in rat, 702
yeast, 698–699
gene copy number in rat, 697
guanosine polyphosphate synthesis in bacteria, 705
iron binding by P2, 709
isoelectric points, 686, 721–722
L22 binding of Epstein-Barr virus RNA, 710
L32 induction of bone formation, 709
messenger RNA (rpmRNA)
bimodal distribution with respect to activity, 365–366
coregulation with elongation factor mRNA, 371, 373, 375
length in rat, 695
sequences around the cap site, 367–370, 372, 374
storage and recruitment, 366–367, 380
5′-terminal oligopyrimidine tract
binding proteins, 380–383
conservation, 370–371
termination codon usage in rat, 696
trans-acting factors, 375–376, 380–383
translational *cis*-regulatory element, 370–371
translational efficiency, 377
3′-untranslated region, 696–697
5′-untranslated region, 695–696
phosphorylation, 19, 692–694, 709

rat proteins, table of properties, 721–722
S1 roles
prokaryotic initiation, 765–766
RNA phage replication, 705, 711, 765
S4 genes and sex differences, 698, 707–708
S10 participation in phage antitermination of transcription, 705
S12 modulation of T4 intron splicing, 706
sequence homology between species
archaebacteria and rat, 699–700, 702, 730–731
Escherichia coli and rat, 700–702, 732
human and rat, 697–698, 726–727
yeast and rat, 698–699, 728–729
sequencing, 686
stoichiometry, 363
structure
acidic residue clusters, 687
amino acid sequence repeats, 687–688
basic residue clusters, 687
leucine zipper, 691–692
shared sequences in P-type proteins, 688–689
zinc finger motifs, 689–691
translational control
autogenous feedback in prokaryotes, 377–378
cell cycle, 366
developmental regulation, 365
eIF4E role, 376–377
free ribosomes and feedback inhibition, 378–379
growth dependence, 364, 367
hormonal regulation, 365
S6 phosphorylation role, 379–380
translational repression, 364
ubiquitin-like proteins, 694–695
Ribosomal RNA (rRNA)
initiation site selection, 81–83, 124
interaction with NS1 and NS2, 704
secondary structure, 685
zinc role in protein binding, 691
Ribosome
abundance and translation efficiency, 12–13
aminoacyl-tRNA binding, 52
assay of active ribosomes, 13
dissociation of, 80S ribosomes, 37–38
frameshifting. *See* Frameshifting
heat-induced disaggregation, 278
hopping, 16, 44, 523, 671, 671
initiation complexes
40S complex, 32, 40–41, 73, 121–122
80S complex, 49–50, 73
mRNA binding, 43–45, 246–249
nucleotide occupancy, 11
pausing, 11, 623
peptide bond formation, 54
proteins. *See* Ribosomal protein
rate of elongation, 11
recycling, 466–467
translocation, 54–55
zinc binding stoichiometry, 690
Ribosome scanning
backward scanning, 186–187, 521–522
cap-independent initiation, 89–93, 103
frameshifting in viruses, 522–523, 669–670, 673
GCN4 translation, 199–201, 204–205, 208–217
initiator codon recognition, 43–45, 122–124, 173, 175–176
leaky scanning, 45, 176, 191, 520–521
linear search versus random diffusion, 86–88
mechanism, 83–93
off-rate, 87–88
ribosome shunting, 88–89, 190, 213, 519–520, 587, 671
RNA helicase migration, 84–85
40S subunit migration, 84–86
RNA. *See* Messenger RNA; Ribosomal RNA; Transfer RNA
RNA phage
exploitation of host translational apparatus, 515
S1 role in replication, 705, 711, 765
translational coupling, 524–525
translation, repression and activation, 525–526
Rous sarcoma virus, gag nonsense-mediated mRNA degradation, 639–640

σ3
double-stranded RNA binding
binding site, 618–620
specificity, 618

σ3 (*continued*)
mRNA translation efficiency, 622–623, 625
PKR inhibition, 620–622
vaccinia virus insertion mutation, 621–622
S6
mitogenic stimulation, 392–393
$p70^{s6k}/p85^{s6k}$ kinase
cloning, 396–397
discovery, 396
domains, 398–399
immunofluorescence studies, 398
kinase, 399
phosphorylation sites, 391
purpose of isoforms, 397–398
rapamycin inhibition, 380, 394–396, 399
signaling pathway, 400–401
phosphorylation, 19, 255, 307, 379–380, 382, 391–396, 401–403, 692–694
regulation of rpmRNA translation, 379–380
rRNA contact, 392
subcellular localization, 391–392
translation initiation role, 393–394
tumor suppression in hematopoietic system, 708–709
yeast homolog, 693–694
Sabin vaccine
development, 559
neurovirulence determinance, 559–560
Scanning. *See* Ribosome scanning
Sea urchin egg, translational control, 5–6
Shine-Dalgarno sequence
base pairing with 16S rRNA, 761–763
consensus sequence, 761–762
initiation site selection, 76–78
role in RF2 frameshifting, 660–661, 666
S1 affinity, 765–766
secondary structure, 761
spacing between sequence and initiation codon, 764–765
strength, 16, 74, 78–79
Signal peptide, discovery, 9
Signal recognition particle (SRP), discovery, 10
Sparsomycin, elongation inhibition, 14
Spermatogenesis
Drosophila, 430
mouse protamine-1 translational control, 430
mouse protamine-1 translational control, 430
SRP. *See* Signal recognition particle
Stop codon. *See* Termination
SUI1, initiator codon recognition, 123, 129

Termination
codons, 55
mechanism, 55
mitochondria, 747–748
reinitiation following termination, 59, 74–76, 177, 521–522
stop codon read-through, 16, 523, 653, 671–674
Termination factors. *See also individual factors*
mammalian factors, table of properties, 35
yeast factors, table of properties, 39
Tissue plasminogen activator
mRNA masking, 327
poly(A) tail in oocyte maturation, 484, 496
TPI. *See* Triosephosphate isomerase
tra-2
direct repeat elements and translational repression, 417–418, 434
sex determination in *C. elegans*, 416–420
Transferrin receptor
control of biosynthesis, 337
iron-responsive element, 335, 351
Transfer RNA (tRNA). *See also* Methionyl-tRNA
aminoacylation level and frameshifting, 666–668
cleavage by bacteria as phage defense, 536–537
cleavage by viruses, 527
mitochondria import, 734–735
mutation and disease, 736
viral synthesis, 505
Translation. *See also* Elongation; Initiation; Mitochondrial translation; Termination
assays, 13
chaperones, 23
history of research, 2–4, 31–32
limiting factors

activity of translation machinery, 11
efficiency of mRNA, 12
rate of elongation, 11
ribosome abundance, 12–13
rate-limiting steps, 13–15, 51, 295
Translational control
biological rationale
fine control, 21
flexibility, 22
immediacy of control, 20
large gene regulation, 21
oogenesis, 442–443
reversibility, 20–21
spatial control, 21–22
systems lacking transcriptional control, 21
cascades, 441–442
definition, 10
feedback loops, 442
global control, 4–5
initiation, 4, 10
mechanisms, 15–19
networks, 441
physiological stimuli, 8–9
prokaryotes
antisense RNA, 769
attenuation control, 181, 773
bacteriophage MS2, 767–768
Com translational activation, 772
mRNA secondary structure modulation, 767–768
ribonuclease III mediation, 772
S15 pseudoknot, 771
trans-acting protein repression, 767, 769, 771
reticulocytes, 4, 6–7
sea urchin eggs, 5–6
secretory proteins, 9–10
selective control, 4–5
targets, 15–19
virus-infected cells, 7–8
Triosephosphate isomerase (TPI), nonsense-mediated mRNA degradation, 643–644
Tropomyosin, translational control in myogenesis, 431–432, 442
trp operon, attenuation control, 181
Tryptophanase, evidence for frameshifting, 662–663
Tryptophan repressor, frameshifting, 663
β-Tubulin, mRNA degradation
cellular factors, 633
elongation role, 634
MREI sequence autoregulation, 632–633
poly(A) tail in development, 483
stimulation by cycloheximide, 632
5′-terminal oligopyrimidine tract, 371
Tumor necrosis factor-α
induction of eIF4E phosphorylation, 250
mRNA masking, 327
Turner Syndrome, S4Y gene mutation, 707–708
Ty transposable elements
genes, 663–664
induction of frameshifting, 664–665

Ubiquitin
attachment to receptor proteins, 694–695
coregulation with rpmRNA, 375
processing into ribosomal proteins, 694
5′-terminal oligopyrimidine tract, 375
Upstream initiation codons (uAUG)
effects on translation
basic transcription element-binding protein, 179
cell-free translation assay, 179
cytomegalovirus pp150, 179, 181
nonoverlapping upstream open reading frames, 178–181
overlapping upstream open reading frames, 186–187
platelet-derived growth factor, 178
retinoic acid receptor, 179
sequence-dependent upstream open reading frames
attenuation control in prokaryotes, 181
eukaryotic examples, 181–183, 186
peptide products and translation inhibition, 185
translation inhibition mechanisms, 185–186
stimulating upstream open reading frames, 188–190, 192
theory, 175–178
in-frame initiators, 187–188
initiation frequency, 183–184
organization of open reading frames in transcripts, 174–175
prevalence
eukaryotic genes, 174

Upstream initiation codons (*continued*)
herpesvirus thymidylate synthase, 175
proto-oncogenes, 174
termination codons in upstream open reading frames, 184
translational control of GCN4, 199–201, 204–205, 208–217
viral RNA packaging role, 190–191

Vaccinia virus
E3L expression, 582–584
gene expression in infection, 577–578
genome, 576, 578
infectious cycle, 577–578
inhibition of host-cell antiviral response, 582–584
inhibition of PKR, 534–536
K3L expression, 582–584
translational control in infected cells
evidence, 592–593
initiation factor modification, 594
5′ poly(A) mRNA role, 594–596
RNA polymerase inhibition, 592
viral mRNA abundance, 593
uncoating process, 577
Virus. *See also individual viruses*; Internal ribosome entry site
contributions to translation research, 505–506, 538–539
enzymes and infection, 508, 511, 515–517
exploitation of host translational apparatus, 514–517, 527–530, 539
frameshifting , 522–523, 669–670, 673
genome structure, 507
host-cell shutoff, 512–514, 560–562, 565–566, 590–597, 608–625
host defense, 506, 530–533
infectious phase switches, 510
initiation site availability and autoregulation, 523–526
life cycles, 508–510
particle sizes, 507
pathogenicity, 511–512
PKR inhibition. *See* PKR
ribosome hopping, 671
ribosome shunting, 671
sites of replication, 511
stop codon read-through, 672
unorthodox translational tactics, 517–523

Y-box protein. *See* Messenger ribonucleoprotein particle

Zinc
role in rRNA-binding proteins, 691
stoichiometry in ribosomes, 690
Zinc finger, ribosomal proteins, 689–691